FOURTH EDITION

HARTMANN'S PLANT SCIENCE

Growth, Development, and Utilization of Cultivated Plants

Margaret J. McMahon
Department of Horticulture and Crop Science
The Ohio State University

Anton M. Kofranek
Emeritus
Department of Environmental Horticulture
University of California at Davis

Vincent E. Rubatzky
Emeritus
Department of Vegetable Crops
University of California at Davis

PEARSON

Prentice
Hall

Upper Saddle River, New Jersey
Columbus, Ohio

Library of Congress Cataloging-in-Publication Data
McMahon, Margaret.
   Hartmann's plant science : growth, development, and utilization of cultivated plants /
Margaret J. McMahon, Anton M. Kofranek, Vincent E. Rubatzky.—4th ed.
      p. cm.
   Rev. ed. of: Plant science / Hudson T. Hartmann, William J. Flocker, Anton M. Kofranek.
   ISBN 0-13-114075-2
   1. Plants, Cultivated. 2. Botany, Economic. I. Title: Plant science. II. Kofranek, Anton
M. III. Rubatzky, Vincent E. IV. Hartmann, Hudson Thomas. Plant science. V.
Title.
SB91.P56 2007
631—dc22

                                                                    2005051456

Editor in Chief: Vernon Anthony
Editor: Jill Jones-Renger
Production Editor: Alexandrina Benedicto Wolf
Production Coordination: Carlisle Publishing Services
Design Coordinator: Diane Ernsberger
Cover Designer: Ali Mohrman
Cover Images: Index Stock
Production Manager: Matt Ottenweller
Marketing Manager: Jimmy Stephens

This book was set in Times by Carlisle Publishing Services. It was printed and bound by Courier Kendallville, Inc. The cover was printed by Coral Graphics Services, Inc.

Pearson Education Ltd.   Pearson Education Australia Pty. Limited
Pearson Education Singapore Pte. Ltd.   Pearson Education North Asia Ltd.
Pearson Education Canada, Ltd.   Pearson Educación de Mexico, S.A. de C.V.
Pearson Education—Japan   Pearson Education Malaysia Pte. Ltd.

10 9 8 7 6 5 4 3 2 1
ISBN 0-13-114075-2

This fourth edition of *Hartmann's Plant Science: Growth, Development, and Utilization of Cultivated Plants* is dedicated to Dr. Hudson T. Hartmann. Dr. Hartmann died in 1994 and is remembered as a committed scientist and dedicated teacher. Dr. Hartmann's name has been added as a part of the new title of this book in recognition of his long-lasting contributions to the field of plant science. We gratefully acknowledge those contributions.

About the Authors

Margaret (Peg) J. McMahon
Associate Professor of Horticulture and Crop Science,
The Ohio State University

Peg is a fourth generation horticulturist and grew up working on the family farm and in the ornamental and vegetable greenhouses owned by her family. She majored in Horticulture at The Ohio State University, earning her B.S. in Agriculture in 1970. After graduation she worked for fifteen years as a grower and propagator for Yoder Brothers, Inc., a multinational greenhouse company, starting at their Barberton, Ohio, location, transferring to operations in Salinas, CA, and then Pendleton, SC. She was responsible at one time or another for the production or propagation of nearly all of the company's crops.

While in South Carolina she enrolled in the graduate school at Clemson University. She earned an M.S. in Horticulture and a Ph.D. in Plant Physiology in 1988 and 1992, respectively. Peg's masters project focused on the early detection of chilling injury in tropical and subtropical foliage plants. Her Ph.D. research was in photomorphogenesis, specifically the development and use of far-red absorbing filters as a technology to reduce unwanted stem elongation in greenhouse crop production. She continues her research in far-red absorbing filters. Far-red absorbing filters are now coming into the product lines of several agricultural plastics manufacturers partly as a result of her research.

In 1994 she accepted a faculty position in the Department of Horticulture and Crop Science (H&CS) at The Ohio State University. Her responsibilities are in teaching and research. She teaches several H&CS courses, including the introductory and senior capstone crop science classes as well as greenhouse production courses. The introductory and capstone classes are required for all students with majors and minors in Horticulture and Crop Science.

Dr. McMahon has published over 20 peer-reviewed articles on photomorphogenesis, teaching methods, and other subjects.

Anton M. Kofranek
Retired as Professor Emeritus from University of California at Davis in 1987

Dr. Kofranek obtained his B.S. in Horticulture at the University of Minnesota in 1947, an M.S. in Plant Physiology at Cornell University in 1949, and his Ph.D. in Plant Physiology at Cornell University in 1950. His first academic appointment was as Instructor in the Department of Horticulture at the University of California at Los Angeles in September 1950. In 1968 when the department was transferred to the University of California at Davis and renamed as the Department of Environmental Science, he also relocated and continued his floricultural research and teaching at Davis until retirement in 1987.

His research specializations dealt with photoperiod research for numerous floricultural crops. Other interests included postharvest physiology and nutritional investigations of flower crops. Dr. Kofranek produced over 200 publications. He co-authored *Hartmann's Plant Science: Growth, Development, and Utilization of Cultivated Plants* and *The Azalea Manual.*

Throughout his career, Dr. Kofranek was highly respected and appreciated by persons in all segments of the floriculture trade, and he received numerous awards and recognition from producers, wholesalers, landscape architects, and other allied industry segments. He has been repeatedly recognized for his efforts by the California Flower Growers. He also received considerable academic recognition, and in 1979 was installed as a Fellow of the American Society for Horticultural Sciences. Dr. Kofranek carried out sabbatical study programs in The Netherlands, Israel, England, South Africa, and at Cornell University. He was a consultant with USAID programs with Egypt and India. His extensive travels led to collaboration with scientists in several countries.

Vincent E. Rubatzky
Retired as University of California Extension Vegetable Specialist, Emeritus in 1995

Dr. Rubatzky obtained his B.S. in Vegetable Crops at Cornell University in 1956, his M.S. in Plant Physiology at Virginia Polytechnic Institute in 1958, and his Ph.D. in Plant Physiology and Horticulture at Rutgers University in 1964. His first academic position was as Extension Vegetable Specialist at the University of California at Davis in 1964, where he conducted his extension outreach and research program until retirement.

His research dealt with variety development and evaluation, harvest mechanization, crop physiological disorders, and crop scheduling and cultural practices. He also served as a liaison with several California vegetable industry organizations and groups. For several years Dr. Rubatzky taught a class on Evolution, Biology, and Systematics of Vegetables, as well as a Field Study of the California Vegetable Industry course. Sabbatical studies and consultations within Poland, France, United Kingdom, Italy, The Netherlands, Peru, and Ecuador provided many travel experiences. In 1995, he was elected as an American Society for Horticultural Sciences Fellow, Man of the Year by the Pacific Seedman's Association in 1987, and received additional recognition from other organizations. Following retirement Dr. Rubatzky continued to serve on a voluntary basis as Specialist, Emeritus, until mid-1999. Presently residing in New Jersey, he collaborates with colleagues in vegetable production research in California and New Jersey.

Dr. Rubatzky has produced many applied research and extension publications. He co-authored *Hartmann's Plant Science: Growth, Development, and Utilization of Cultivated Plants; World Vegetables: Principles, Production, and Nutritive Values;* and *Carrots and Related Vegetable Umbelliferae,* and was an editor for the *Third International Symposium for Diversification of Vegetable Crops,* the *Atlas of the Traditional Vegetables in China,* and several other publications on vegetables.

Preface

Enter the fascinating world of plant science by studying the fourth edition of *Hartmann's Plant Science*. Discover why we depend on plants and the people who understand them for our survival. Find out how plants provide sustenance for our bodies and add enjoyment to our lives. Learn how to grow, maintain, and utilize plants to benefit people and the environment. Whether your interests range from running the family farm to managing a tournament golf course, or directing an international business, rewarding, challenging, and fulfilling careers are open to anyone skilled in plant science.

Human survival absolutely depends on the ability of plants to capture solar energy and convert that energy to a form that can be used as food. The captured energy stored in plant tissues also provides fiber and oil for fuel, clothing, and shelter. The production of plants that meet our needs for survival is an important application of the knowledge of plant science; however, the essentials for nutrition and shelter can be provided by relatively few plant species. Life would be boring, though, if those few were the only species produced for our needs. Fortunately for those who dislike boredom, thousands of plant species can add enjoyment to life by providing a variety of flavors and textures in food and fiber. Other plants brighten our lives when used in landscaping and interior decoration. The importance of turfgrass in athletic and outdoor recreation sites is evident around the world. Animal feeds are another critical use of plants. Animals provide nutrition and variation in our diets, along with materials for clothing and shelter. They also reduce labor in many parts of the world and add pleasure to our lives through recreation or as pets.

Plants have tremendous economic impact in developed and developing nations. The career opportunities created by the need for people with an understanding of plant growth are unlimited. *Hartmann's Plant Science* is written for anyone with an interest in how plants are grown and utilized for maintaining and adding enjoyment to human life. The beginning chapters of the text provide the fundamentals of botany, plant physiology, and environmental factors affecting plant growth so that students unfamiliar with plant science can comprehend them easily. The later chapters integrate the aforementioned topics into strategies of producing plants for human use as food, fiber, and recreation.

The fourth edition of *Hartmann's Plant Science* has been updated to include the most recent statistics, production methods, and issues concerning the production and utilization of plants. This edition has been revised and reorganized to reflect the changing information needs of those who grow plants for a living. New information has been added, and some of the information found in earlier editions has been moved to a different chapter or section. In some cases, content has been reduced or eliminated, most notably in the detailed crop production information previously included in Unit III.

Unit I focuses on ecology and addresses the human and environmental factors and issues that influence how, why, and where plants are grown. The unit includes more information on how natural ecosystems influence plant cultivation than was provided in previous editions. A new section describes the scientific method and research in plant science. An ecological paradigm that includes plant production and utilization efficiency, environmental compatibility, economic viability, and social responsibility is the basis for Unit I, and the concept of that paradigm continues throughout the book.

Unit II addresses the biological basis of plant science. Plant physiology and biochemistry, genetics, propagation, biodiversity, germplasm preservation, water relations, and mineral nutrition are the major topics of Unit II. Because soil/plant water relations along with mineral nutrition are so important, those topics now have their own chapters.

Unit III presents the general principles of growing plants and the application of those principles to the production and/or use of major commodity groups. The book's editor and the chapter authors feel that teaching how to apply general concepts and methods to specific situations is the key to having a useful combination introductory textbook and reference book.

The editors thank all those who contributed to earlier editions of this book. Those contributors are Curtis Alley, A. H. Allison, Victor Ball, Robert F. Becker, Brian L. Benson, Alison Berry, Itzhak Biran, Arnold J. Bloom, James W. Boodley, Robert A. Brendler, Royce Bringhurst, Thomas G. Byrne, Jack Canny, Will Carlson, Kenneth Cockshull, Charles A. Conover, Beecher Crampton, Began Degan, Frank Dennis, Francis DeVos, Dominick Durkin, Roy Ellerbrock, James E. Ellis, Clyde Elmore, Thomas W. Embledon, Harley English, Elmer E. Ewing, Michael Farhoomand, Gene Galleta, Melvin R. George, Marvin Gerdts, Ernest M. Gifford, Victor L. Guzman, L. L. Haardman, Abraham H. Halevy, Richard W. Harris, R. J. Henning, Charles Heuser, James E. Hill, Jackson F. Hills, Karl H. Ingebretsen, Lee F. Jackson, Subodh Jain, Merle H. Jensen, Robert F. Kasmire, Thomas A. Kerby,

Dale Kester, Paul F. Knowles, C. Koehler, Harry C. Kohl, Dale W. Kretchman, Harry Lagerstedt, Pierre Lamattre, Robert W. Langhans, Roy A. Larson, Robert A. H. Legro, Andrew T. Leiser, Gil Linson, Warner L. Lipton, Oscar A. Lorenz, James Lyons, Harry J. Mack, John H. Madison, Vern L. Marble, George Martin, Shimon Mayak, Keith S. Mayberry, Richard Mayer, Donald N. Maynard, Arthur McCain, Charles A. McClurg, Harry A. Mills, Franklin D. Moore III, Yoram Mor, Julia Morton, James L. Ozbun, Jack L. Paul, William S. Peavy, Nathan H. Peck, Jack Rabin, Allan R. Rees, Michael S. Reid, Charles Rick, Frank E. Robinson, J. Michael Robinson, Norman Ross, Edward J. Ryder, Kay Ryugo, Roy M. Sachs, Robert W. Scheuerman, Art Schroeder, John G. Seeley, William Sims, Donald Smith, L. Arthur Spomer, George L. Staby, W. B. Storey, Vernon T. Stoutemyer, Walter Stracke, Mervin L. Swearingin, Herman Tiessen, Edward C. Tigchelaar, Ronald Tukey, Benigno Villalon, Stephen Weinbaum, Ortho S. Wells, Bernard H. Zandstra, Naftaly Zieslin, and Frank W. Zink. The authors who wrote or revised chapters in the current edition are listed on the opening page of his or her respective chapter.

Margaret J. McMahon
Anton M. Kofranek
Vincent E. Rubatzky

Additional Resources

ONLINE SUPPLEMENTS ACCOMPANY THE TEXT

An online Instructor's Manual is also available to instructors through the McMahon catalog page at **www.prenhall.com**. Instructors can search for a text by author, title, ISBN, or by selecting the appropriate discipline from the pull down menu at the top of the catalog home page. To access supplementary materials online, instructors need to request an instructor access code. Go to **www.prenhall.com**, click the **Instructor Resource Center** link, and then click **Register Today** for an instructor access code. Within 48 hours after registering you will receive a confirming e-mail including an instructor access code. Once you have received your code, go to the site and log on for full instructions on downloading the materials you wish to use.

A companion website is available to both instructors and students at **www.prenhall.com/mcmahon**.

AGRICULTURE SUPERSITE

This site is a free on-line resource center for both students and instructors in the Agricultural field. Located at **http://www.prenhall.com/agsite**, students will find additional study questions, job search links, photo galleries, Power-Points, *The New York Times* eThemes archive, and other agricultural-related links.

Instructors will find a complete listing of Prentice Hall's agriculture titles, as well as instructor supplements supporting Prentice Hall Agriculture textbooks available for immediate download. Please contact your Prentice Hall sales representative for password information.

THE NEW YORK TIMES THEMES OF THE TIMES FOR AGRICULTURE

Taken directly from the pages of *The New York Times,* these carefully edited collections of articles offer students insight into the hottest issues facing the industry today. These supplements can be packaged along with the text.

AGRIBOOKS: A CUSTOM PUBLISHING PROGRAM FOR AGRICULTURE

Just can't find the textbook that fits *your* class? Here is your chance to create your own ideal book by mixing and matching chapters from Prentice Hall's agriculture textbooks. Up to 20% of your custom book can be your own writing or come from outside sources. Visit us at **http://www.prenhall.com/agribooks**.

Brief Contents

Contents

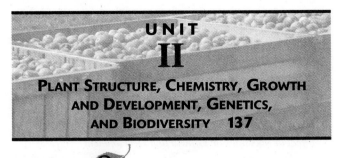

UNIT II

PLANT STRUCTURE, CHEMISTRY, GROWTH AND DEVELOPMENT, GENETICS, AND BIODIVERSITY   137

U N I T
III
CROPS: THEIR PRODUCTION
AND UTILIZATION SYSTEMS 305

CHAPTER 16

Field Crops Grown for Food, Fiber, Fuel, and Other Industrial Uses 307

CHAPTER 17

Forage Crops and Rangelands 335

CHAPTER 18

Vegetable Production 348

CHAPTER 19

Temperate Fruit and Nut Crops 368

CHAPTER 20

Tropical and Subtropical Crops and Crop Production Systems   411

CHAPTER 21

Nursery Production   441

CHAPTER 22

Landscape Plants: Evergreen Trees and Shrubs, Deciduous Trees and Shrubs, and Herbaceous Plants   450

CHAPTER 23

Floriculture   484

CHAPTER 24

Turfgrasses   501

CHAPTER 25

Residential and Public Landscapes   524

CHAPTER 26

Harvest, Post-Harvest Handling, Storage, Distribution, and Marketing   537

UNIT
I

ENVIRONMENTAL, CULTURAL, AND SOCIAL FACTORS THAT INFLUENCE THE CULTIVATION AND UTILIZATION OF PLANTS

CHAPTER

1

History, Trends, Issues, and Challenges in Plant Science

Michael Knee and Margaret McMahon

KEY LEARNING CONCEPTS

After reading this chapter, you should be able to:

- ◆ Understand why plant science is a dynamic and changing discipline.
- ◆ Appreciate the role that plant science has played and continues to play in the world economy and culture.
- ◆ Understand why modern plant scientists take into consideration production efficiency, economic viability, environmental compatibility, and social responsibility when researching the solution to a problem.
- ◆ Explain the principles of research in plant science and begin to evaluate scientific reports for use in making decisions for the successful cultivation of plants.

HISTORY

As citizens of the twenty-first century, we tend to pride ourselves on how we have used agriculture to shape the modern world to serve and please us. We have reason to feel that way—our agricultural practices have changed the world. But if we were to use H. G. Wells's time machine to transport us back 150 million years, we would see many plants very similar to those common in our century. We would see many of the same trees that grow in our world, along with other members of the **angiosperms,** the group of plants to which grasses, flowers, vegetables, and shrubs belong. Dinosaurs would be feeding on these plants or on other dinosaurs and smaller creatures. As time progresses, the dinosaurs would disappear and other animals would appear and evolve, and some would die out. Plant life would change, some as a result of changes in climate, and become the plants we know today. During this time, humans had no influence on changes in life forms that occurred.

Humans as a race appeared around 3 million years ago and modern man, *Homo sapiens,* appeared about 28,000 years ago. For thousands of years, *H. sapiens* existed without doing much to change how plants grew. As hunters and gatherers, the nomadic tribes followed herds of animals and gathered plant materials along the way. The plants they probably gathered would have been some of the same nuts, grains, and fruits we eat today. Other plants known today would have also provided shelter.

Then something happened around 12,000 to 10,000 years ago (perhaps earlier according to recent archaeological finds) that had a dramatic impact not only on human lifestyle but the entire global ecosystem. Humans began the purposeful growing or **cultivation** of plants to improve the supply of materials obtained from these plants. The science of understanding the cultivation of plants, plant science, was born. Plant cultivation is believed to have started in areas now making headline news for other reasons: Iraq and other tropical and subtropical regions in the Middle East and Africa. By cultivating plants, humans reduced the need to travel to follow the food supply. Those who did the traveling became traders more than gatherers. Commerce began when goods from distant places were transported and sold or traded to a local population. Many of these products were plants or plant products. Along with trading the goods, ideas about cultivating plants also spread. By 6,000 to 5,000 years ago, crops were being grown in Europe and Mexico. However, the types of crops grown in the various areas differed for reasons that will be explained in later chapters.

At first, the developing lifestyle had little effect on global ecosystems; as trading increased, however, commercial or urban centers developed. The population of the urban centers increased along with the need to bring in more food and other plant products from rural areas to support that population. However, the citizens of these urban areas became less and less aware of how the plants they used were produced. The **rural/urban interface** developed. In many areas, this interface was a very distinct line of demarcation with a defense wall separating urban from rural areas. As populations grew, the urban areas spread out away from the central urban area. The rural areas were forced to move outward too, quite often into areas less favorable for growing plants.

As demand for plant products increased, cultivation methods were developed to help production keep up with the demand. Some of the earliest products of the industrial revolution were farm implements for crop cultivation, such as plows and planting and harvesting equipment. The closely related disciplines of agronomy and horticulture were born and developed. **Agronomy** is the study of field-grown crops such as wheat, soybeans, corn, forages, and those used for industrial purposes that require relatively low input during the growing part of their life cycle. **Horticulture** is the study of crops that require more intense and constant care, from planting through delivery to the consumer. Examples of horticulture crops include most fruits and vegetables that can be eaten raw or with little processing and ornamentals. There are gray areas where the disciplines overlap. Tomatoes grown for processing and turfgrass are examples of crops that can be designated agronomic or horticultural.

TRENDS AND ISSUES AFFECTING PLANT SCIENCE

Background

Traditionally crops are commodities that are exchanged for money in the marketplace. In the case of recreation areas such as golf courses, the use of that field is exchanged for money. According to classical economics, the price of commodities (or user fees) results from the balance between the supply of commodities (recreational areas) and the demand for them. The supply should be influenced by the availability of the resources and raw materials required for production, and the demand should be influenced by the value that consumers perceive in the commodity.

Many people have drawn attention to the mismatch between ecological and money values. If excess fertilizer from a crop pollutes a river, this cost of production will not be reflected in the price of the commodity. On the other hand, many people may enjoy seeing an ornamental tree planted by someone else, and it can also help to absorb the carbon dioxide (CO_2) produced by cars. Economists call costs and benefits that are not included in the price externalities. Some argue that environmental degradation occurs when no one owns a resource. If land is under common ownership, it will be overcropped or overgrazed because it is in no one's interest to conserve the resource. Thus, they refer to the general problem of the tragedy of the commons. According to this view, if everything is privately owned, individuals have an interest in its preservation, those who use resources have to pay the owners, and the value of the resources is reflected in the price of the commodity.

Classical economics assumes that buyers and sellers are fully informed of the value of commodities in the market and make decisions that maximize their welfare. However, many people make unwise choices on both sides of the market relationship. Farmers in the American Southwest in the 1930s contributed to soil erosion on their own farms through their crop management practices. The market for tobacco products remains strong after fifty years of health warnings. Individuals may not know all of the consequences of their decisions, and it may take the input from many people to arrive at a full accounting. On the other hand, people are capable of making communal decisions to conserve resources that do not belong to any individual. The soil conservation service was born out of the dust bowl experience. Agricultural research and education grew out of a perception of a wider public good to be obtained by applying scientific principles to food production. Now it seems we need to go beyond production to look at ecological consequences and the consequences for the individuals who consume the produce.

Dependence on the market to regulate crop production practices assumes that ecological values can be translated into cash value and that ecology is a subcomponent of the market rather than the other way around. The problem is partly one of time scale. Financial analysis discounts future values relative to current value. Low yields in future years because of soil degradation appear to be less important than high yields this year. The idea of **sustainability** as a guiding principle in crop production is an attempt to transcend this mind-set. Energy transactions are central to ecological relationships. Crop production requires energy inputs at several points and provides us with energy in food or embodied in nonfood commodities.

Ecologist Howard Odum has suggested that we use energy accounting to try to correct some of the distortions introduced by cash accounting in agriculture and other industries. Energy comes in many forms, and every energy transaction is accompanied by losses. To arrive at a common basis for accounting, Odum proposed that all forms of energy should be related back to the primary source, which is usually the sun. Odum's concept of embodied energy, or **emergy,** is the amount of solar energy represented in the resource or commodity (Fig. 1–1). Emergy accounting can be related back to financial accounting by averaging costs across all forms of energy according to their solar energy equivalent. If energy converts raw materials into valuable commodities, then there is no reason why one form of emergy should be regarded as any more valuable than another. Emdollar accounting results in some interesting shifts

| TABLE 1–1 | Emergy Content of Commodities and Prices Calculated on a Uniform Cost of $0.86 for 10^{12} Solar Emjoules (Calculated Average for the Whole US Economy) |

Commodity	Emergy content (10^9 Solar Emjoules/g)	Calculated Price (Emdollars/Unit)
Corn grain	14.5	5.68/pound
Cotton	23.1	9.04/pound
Soybeans	9.9	3.87/pound
Sweetcorn	2.5	0.54/ear
Tomatoes	16.0	6.26/pound
Beef (range-fed)	48.5	19.00/pound
Eggs	107	62.00/dozen
Milk	33.7	13.20/pint
Gasoline	3.0	7.50/gallon

Source: Brandt-Williams, S. 2002. University of Florida.

of value (Table 1–1). The dollar prices of agricultural commodities appear to be much too cheap and natural ecosystems with little or no cash value turn out to have high emdollar value in terms of services such as flood and pollution control.

Domestic Trends and Issues

In the United States, the prices of many agricultural commodities continued to fall in real terms in the last quarter of the twentieth century. Even though yields continued to increase, income per acre declined (Fig. 1–2). This is one reason that it has been impossible to break away from the system of price support for selected commodities. In 2000, the subsidies amounted to half of farm income. The payments are supposed to help farms stay in business but end up as one more factor encouraging consolidation in the industry: large farms get more government assistance than small farms do. The number of farms decreased as individual holdings got larger in the second half of the century.

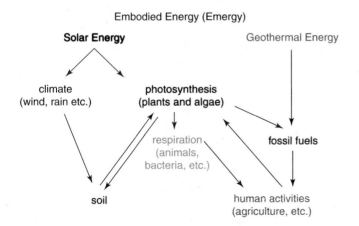

Figure 1–1 Derivation of embodied energy, or emergy, from primary energy sources. *Source:* Michael Knee.

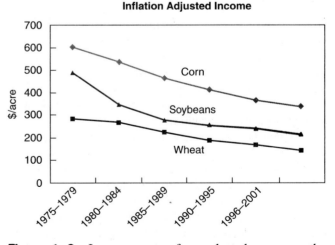

Figure 1–2 Income per acre from selected crops over the past twenty-five years, corrected for inflation by estimating on the basis of 2001 dollar value. *Source:* Michael Knee.

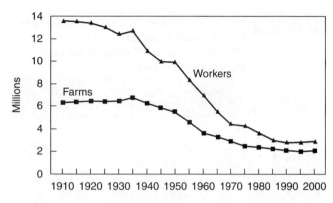

Figure 1–3 Number of farms and farm workers during the last century (1910–2000). *Source:* USDA NASS,

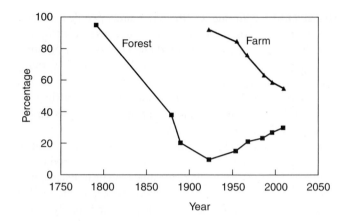

Figure 1–4 Percentage of forested land and farm area in Ohio since European settlement. *Source:* Michael Knee.

The number of farm workers also continued its long-term decline so that there is now a little more than one full-time worker for each farm. The numbers of farms and farm workers seem to have stabilized toward the end of the century, and these numbers may be minimum sustainable values (Fig. 1–3). Most farms are run as part-time businesses, and about 10 percent of the farms remaining in the United States account for 70 percent of production. The profitability of farming has been helped to some extent by diversifying the uses of staple crops (e.g., ethanol and syrup from corn and tofu from soybeans) and by adopting alternative crops. The area of farmland has also declined from its high point at the beginning of the twentieth century. In contrast to many other countries, the United States has seen an overall increase in the area of woodland (Fig. 1–4). Some of the increase comes from the conversion and/or reversion of farmland to woodland.

The increase in US population during the last century was mainly in urban areas so that rural population decreased in relative terms from 60 percent in 1900 to 25 percent in 2000. In the upper Midwest, the rural population fell in absolute terms from the middle of the century. This decline was associated with a loss of economic and cultural vitality in rural communities. Such communities were more likely to survive if there were large towns that provided an economic stimulus to the surrounding areas. This may be the reason why rural populations persisted in Ohio and other states with several large towns and cities.

Urban populations have increased in every state of the Union. The increases were most marked in the coastal states, the East and West, and along the shores of the Great Lakes. Although this is classed as urban development, it is more accurate to call it suburban. Average lot sizes for new homes are about 0.15 hectare, or 0.4 acre. The spacious lifestyle of the suburbs depends on personal transportation for access to work, shops, and leisure and this accounts for much of our energy demand. Many people have criticized this and other aspects of suburban sprawl, but it leads to new opportunities.

The new homes are an expanding market for landscape supplies and services. Surviving farms can market directly to the surrounding population. Families can enjoy a visit to the local farm to buy or pick their own produce. This interaction may help maintain contact and understanding between the mass of the population and the few remaining farmers.

Another related development is the growth of golf as an industry. Golf is now equivalent to about two-thirds of major crop sales and involves 12 percent of the US population. The number of golf courses has increased threefold in the past fifty years.

Global Trends and Issues

Two major trends that will affect crop production and the global environment are the increases in human population and energy use. After two centuries of exponential growth, world population shows signs of stabilizing at about 9 billion in 2050 (Fig. 1–5). Energy use is projected to rise about twice as fast as population because of economic development (Fig. 1–6). People have long argued about the potential for the future growth of population in relation to resource use and about the related question of the earth's carrying capacity. Estimates of carrying capacity have varied from the low billions to the trillions.

As you will learn in Chapter 3, it is possible to feed an adult on a plant diet from about 0.2 hectares of land, and this is what will be available on average when the world population reaches 8 billion. However, the yields we achieve in industrial countries need to be achieved worldwide. These yields in turn require that inputs, at least of fertilizer and probably pesticides, have to rise in other countries to match those in North America and Europe. If other countries mechanize crop production, it will likely entail further energy consumption because mass migration from rural areas to the cities is underway on all continents. While agriculture accounts for only about 2 percent of energy consumption in North America and Europe, it would amount to about 10 percent of energy demand in the rest of the world.

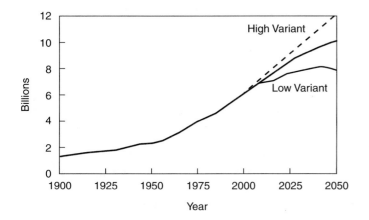

Figure 1–5 Projections of world population growth. *Source: http://www.usgcrp.gov/usgcrp/seminars/.*

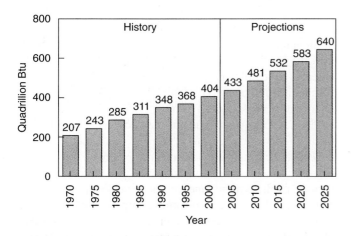

Figure 1–6 History and projection of world energy consumption, 1970–2025. *Sources:* **History:** Energy Information Administration (EIA), *International Energy Annual 2001,* DOE/EIA-0219(2001) (Washington, DC, February 2003), web site *www.eia.doe.gov/iea/.* **Projections:** EIA, System for the Analysis of Global Energy Markets (2003).

A balance among agricultural energy consumption, food supply, and population is achievable, but it depends on certain assumptions. The main assumption is that everyone in the world has equal access to the food supply. At the moment, the food supply and the ability to buy it are unequally spread among the countries of the world. Many countries in Africa, South America, and Asia would need to become much richer to develop their own agriculture or purchase the food that they need.

China alone represents a major challenge to the world's food supply. It has about 20 percent of the world's population but less than 7 percent of the world's cropland. It had about 0.2 hectares per person in 1950 and now has about 0.08. It has managed to feed itself and actually improve peo-

ple's diet over the past decade. However, with its population set to increase by an additional 25 percent before stabilizing at 1.6 billion, it is doubtful that this trend can continue. Fertilizer use is high and further increases will not bring much additional yield. There are already problems of nutrient pollution and soil degradation, and the area of agricultural land is shrinking because of urban development.

The United States has about twice as much cropland as China but one-fifth of its population. It is one of a handful of countries with a relative excess of cropland in relation to population. Others include Canada, Australia, and the Russian Federation. China is undergoing rapid industrial development and now supplies other industrial (or postindustrial) nations with a wide range of the manufactured articles that they used to make. In the future it is likely to be a major customer for US agricultural products. Other countries may try to follow China's road to economic development. Even if they do not, it is likely that China's demand for food will raise prices of agricultural commodities, which could be good news for US producers. Of course, US consumers are likely to be unhappy as food becomes more expensive after many years of falling prices.

Grain-fed meat production is increasing in China, which increases the land required to meet its dietary needs and cannot be sustained on a global scale. To provide an adequate diet for everyone, cropland needs to be devoted to feeding people. Animal production needs to be based on more marginal land that will support grazing but not crops. But overgrazing and desertification are a constant threat in arid lands.

A 50 percent increase in global energy use is expected over the next fifty years, and it seems likely that most of this energy will come, as it does now, from combustion of fossil fuels. The rate of discovery of new reserves has slowed, but for some reason the oil industry is confident that supplies will extend beyond the forty years of known reserves at current rates of consumption. Gas reserves are a little higher: at about sixty years (but little of this is in the United States). Coal reserves are considerably higher: at over 200 years. Coal is more evenly distributed, and its consumption is rising faster than consumption for gas and oil outside the United States and Europe. So atmospheric carbon dioxide will continue increasing for the next fifty years, and other gaseous pollutants will increase as coal consumption rises.

The argument about the connections between rising atmospheric CO_2 and rising temperatures and other weather changes is almost over. Evidence overwhelmingly supports the connections. Productivity of some crops could increase with rising CO_2 levels, as long as rainfall patterns are not disrupted. High temperatures could lead to decreases in yields of some crops, particularly rice. Climatic changes impose additional stress on natural plant communities, and many doubt whether forest ecosystems can adapt fast enough to survive the changes. For example, American beech may become extinct over a large part of its present range, and it may not be able to spread north fast enough to take advantage of new habitat. The other atmospheric pollutants that accompany CO_2 emissions from fossil fuels will cause more damage to crops and wild plants.

Global energy demand is only a fraction of the energy captured in photosynthesis and an even smaller fraction of solar radiation reaching the earth. About 40 percent of the world's population relies on biomass as their primary energy source, but these same people consume very little energy by our standards. Gathering fuel for firewood consumes more and more of these people's time as it progressively degrades or eradicates native vegetation. Sustainable production of biofuels seems like the answer to problems of energy supply. However, this approach would require massive changes in land use and a large investment in equipment for conversion of biomass to usable fuels. The major growth in fuel requirements is predicted for use in transportation, but biomass-to-fuel conversion processes (such as corn to ethanol) may not generate enough of an energy profit to support this growth.

The demand for food and fuel is leading to deforestation in many tropical countries. The loss of biodiversity and environmental quality is unfortunate for the resident population, but it also has implications for us. The tropical forests are a major **sink,** or depository, for the carbon dioxide that we generate through our fossil fuel consumption. Losing that depository contributes to increasing atmospheric CO_2 levels and global warming.

International trade is being progressively liberalized, removing tariffs and duties that restricted imports of food and other commodities. Industrial nations are also under pressure to eliminate the subsidies that favor their own producers. This change will allow poor countries to export commodities such as sugar and grains to the rich countries. Classical economics predicts that production will become more efficient and profitable in the long run when the market is no longer distorted by the subsidies. This prediction does not seem to account for the fact that poor countries often cannot get good prices even for commodities such as bananas or coffee that are not produced by the United States and Europe. A side effect of the globalization of trade is that pests and diseases are spread around the world more easily. A recent example is the introduction of the Emerald Ash Borer (*Agrilus planipennis*) into the Great Lakes region from wooden pallets carrying goods from China to the Port of Detroit. The borer is threatening the survival of all ash species, much as Dutch Elm disease decimated the elm population in the United States a century ago.

Bioterrorism is a threat to agriculture that has become a major concern recently. Terrorists may be able to introduce pests or chemicals that will destroy a good portion of crop production in certain areas. Because of this fear, quarantines and restrictions to free trade may increase. In 2003 and again in 2004, some geranium growers in the United States were ordered by the United States Department of Agriculture (USDA) to destroy much of their geranium crop because it came in from abroad infested with *Ralstonia solanacearum*

race 3 biovar 2, which is listed in the Agriculture Bioterrorism Act of 2002. *Ralstonia* is a bacteria that is endemic in some parts of the world but is not found in the United States. However, it was not fear of what *Ralstonia* would do to geraniums that prompted the USDA's action. *Ralstonia* is a serious pathogen of several of our food crops. The potential of the fungus to devastate those crops is the reason it has been listed as a bioterror organism. Although the disease was imported unintentionally, its designation as a bioterror organism prompted the need to destroy the geranium crops. The negative financial impact on foreign and domestic geranium growers was significant.

Research

Many of the changes in crops and their role in the world have prompted plant scientists to change the focus of their research. For several centuries, plant scientists studied ways to improve crop productivity in a cost-effective way. They studied light, soil, water, and temperature and developed ways of managing or monitoring those factors to influence or predict plant growth. Improved understanding of plant genetics lead to breeders developing plants that would produce more reliably. A scientific approach to pests and their management reduced losses to those factors. Traditional economic analysis was used to see if the new production methods were cost-effective. Great gains were made, but as you have read, increasing evidence showed that some agricultural practices were having negative effects on the environment. Agriculture became the focus of public scrutiny, and a negative public opinion of agriculture developed. This opinion was exacerbated by the fact that most urban dwellers today have very little understanding of their dependence on agriculture. Plant scientists have to find ways to increase crop productivity and keep crops safe from bioterrorist activity without negatively affecting the environment. When calculating the cost-effectiveness of a production procedure, costs no longer include just material and labor costs. The cost to repair the resulting environmental damage must also be factored into cost analysis.

Although it may seem like a relatively simple calculation to factor in environmental costs, it is not. Assessing environmental impact and the cost to repair negative impact can be a challenge. It can be difficult to predict what will happen when a new production practice moves from the lab or experimental field to the real world. For example, **genetically modified organisms (GMOs)** or plants have been created that dramatically reduce the need for pesticides and field tillage. Pesticide runoff and erosion are reduced, which is beneficial for the environment. But there is concern that heritable traits from the GMOs will escape and become part of the wild or native plant populations. There is fear that wild plants with these traits may have serious negative impacts on the ecosystem in which they grow. Although there is little scientific evidence showing that these traits would transfer to wild plants and then become established in the population, it cannot be proven either way. As a result of the uncertainty, many parts of the world forbid the use of GMOs, while in other areas their use is fairly common. Heated debates rage worldwide over the pros and cons of GMOs, and they show little sign of relenting.

Organic farming has been proposed as a solution to many problems related to crop production and environmental impact. Organic farming does not allow the use of GMOs and certain types of chemicals for pest control and fertilization. Some organic farms have proved to be successful, thus demonstrating that the process works. However, it is not known if organic farming has the capability to produce the quantities of crop products needed in today's world. Also there is great debate regarding what constitutes organic farming.

On the other hand, the public often wants to see plants growing "perfectly"—a very unnatural state for plants. Generally this occurs when plants are used in leisure and recreational settings. Immaculately groomed and weed-free landscapes, flawless floral arrangements, and impeccable golf courses and athletic fields are sources of pride to those who own or use them. Ask any golf course superintendent what his greens committee would say if the fairways and greens had even a few weeds or insect and/or disease problems in them. Maintaining perfect plants almost certainly has some degree of negative environmental impact.

Plant scientists and growers can find all the current uncertainty and controversy about their fields discouraging and wonder why we should even bother to try to solve seemingly insurmountable problems. We must remember that, although most plants grow very well without human intervention, the cultivated grains and grasses, fruits, vegetables, and ornamentals have become dependent on human intervention to survive. Without cultivation, these plants would die out after several generations and be replaced by hardier species such as wild grasses and thistles (even if those wild plants have picked up "escaped" GMO traits). But the dependence is mutual. The estimated 6.5 billion people now living in the world depend on the survival of cultivated plants for nourishment and to provide quality to their lives. The global population cannot survive as hunters and gathers. The need for plant scientists to increase knowledge about crop plants and their place in the ecosystem and the need for professionals who know how to use that knowledge to grow plants in environmentally sound ways will not disappear. In fact, it should increase, perhaps dramatically if the world population increases as predicted.

Setting debate and uncertainty aside, what plants do we depend on? We know that photosynthesis is the underlying process that makes all life on earth dependent on all plants, not just the cultivated ones. Photosynthesis and its role in human survival and the global ecosystem will be explained in greater detail in later chapters. When we look at cultivated plants, however, we see that only a small percentage of all the plant species in existence feed the world's

people either directly or indirectly (through animals). These plants are:

1. Cereal crops—wheat, maize (corn), rice, barley, oats, sorghum, rye, and millet. (Over half the world's food supply comes from the photosynthetic activity of these crops.)
2. Roots and tubers—potatoes, sweet potatoes, and cassavas.
3. Oil crops—soybeans, corn, peanuts, palm, coconuts, sunflowers, olive, and safflower.
4. Sugar—sugar cane and sugar beets.
5. Fruit crops—bananas, oranges, apples, pears, and many others.
6. Vegetable crops—tomatoes, lettuce, carrots, melons, asparagus, and so forth. (Fruits and vegetables add to the variety and palatability of our daily meals and supply much-needed vitamins and minerals.)

Table 1–2 shows some of the common crops ranked in relation to the calories and proteins produced per unit of land area. Not all of the total production of food materials becomes available for human consumption. Much is lost during harvesting, transportation, and marketing, primarily from attacks by insects, diseases, birds, and rodents. Also, some of the production is saved to be used as seed for future plantings.

A different kind of energy loss occurs when plants are used to produce human food in the form of animal products. For example, it takes about 10 kilograms (22 pounds) of grain (which could be consumed by humans directly) to produce 1 kilogram (2.2 pounds) of beef. But some of the kinds of feed consumed by beef cattle is ordinarily not eaten by humans. A bushel of corn consumed as whole corn meal would meet the daily energy and protein requirements of 23 people, but when this same bushel is fed to chickens and consumed as eggs, it meets the energy requirements of only two persons and the protein requirements of eight. Nevertheless, meat,

TABLE 1–2	Some Important Food Crops Ranked According to Calorie and Protein Production per Unit of Land Area	
Rank	**Calories Produced per Unit Area**	**Protein Produced per Unit Area**
1	Sugar cane	Soybeans
2	Potato	Potato
3	Sugar beets	Corn
4	Corn	Peanuts
5	Rice	Sorghum
6	Sorghum	Peas
7	Sweet potato	Beans
8	Barley	Rice
9	Peanuts	Barley
10	Winter wheat	Winter wheat

Source: USDA, IR-1 Potato Introduction Station, Sturgeon Bay, Wisconsin.

milk, and eggs are important in the human diet because they contain proteins of the proper quality (balanced quantities of essential amino acids), as well as some of the necessary minerals and vitamins.

Assessing the world's food situation involves other factors besides the utilitarian one of meeting minimal food needs. When the people of a country become more affluent, they want and can afford to purchase a greater proportion of their protein requirements in the form of the more palatable animal products—steaks, chops, eggs, processed meats, and dairy products. This shift in food consumption patterns, part and parcel of the modern world's "rising expectations," coupled with the tremendous increase in world population, especially in developing regions (see Fig. 1–7), requires continuing increases in the world's food-producing capability. Much of the world's best agricultural land is already under cultivation, although there is still unused productive land awaiting development in Argentina, Brazil, Canada, Sudan, and Australia. However, in most if not all developed countries such as the United States (see Fig. 1–3), Japan, and those in Europe, farmland is being lost forever to industrial, residential, or recreational development. In addition, in the United States, even though the number of farms is decreasing, the size of farms is increasing. The loss of the small family farm and the increase in corporate megafarms have become a serious social and political issue. Plant scientists are being asked not only to increase productivity and profitability of existing crops, but also to find and develop new crops or new uses for old crops.

In the mid-1970s various projections implied that the world was on the brink of famine or ecological disasters due to desperate food needs. But this assessment changed in the 1980s, especially in the less developed countries, by improved production technology and greater incentives to use

it. Agricultural research made new **cultivars** of high-yielding wheat, rice, corn, and other crops available to highly populated developing countries. (Cultivar = *culti*vated *vari*ety.) Much of this improved technology can be attributed to assistance from agricultural researchers in the United States and other developed countries working with less developed countries.

But there are still areas where people starve or are badly malnourished. These underfed populations are mostly in the sub-Saharan regions of Africa, where drought, wars, and political instability are major problems, along with high human fertility rates, low per capita income, and insufficient monetary investment in agricultural production. In addition to the above-mentioned factors, poor distribution of available food contributes to starvation in many areas, including areas of developed countries. The United States Department of Agriculture estimates that by 2009, food consumption in thirty of sixty-seven developing countries is likely to lag behind minimal nutritional standards. While food distribution is expected to improve in many developing regions, poor distribution of food to those thirty countries is expected to intensify by 2009. Improving the nutritional value of food, so that the food distributed provides better nourishment, is a challenge to current and future plant scientists.

BENEFITS FROM CULTIVATED PLANTS

Imagine for a moment torrential rains falling on a landscape of rolling hills completely denuded of plants. It is easy to visualize the increasing worthlessness of such a landscape as it slowly erodes into the rivers, lakes, and oceans. The fertile topsoil disappears, uncovering a rocky base of little value to humans or animals. However, plants help prevent such erosion because their roots act as webs to hold the soil in place. In addition, their leaves and branches slow the force of the falling water, permitting it to trickle down and become stored in the depths of the soil, gravel, and rocks rather than running in torrents over the surface. Also, as the grasses, shrubs, and trees age and die, their decomposing roots, stems, and leaves add to the soil's mass, forming a humus material that constitutes a favorable environment for the germination of seeds and the growth of seedlings. For eons, the presence of plants on the earth's surface has protected the fragile topsoil layer, which is the natural habitat for the roots of both naturally occurring and cultivated plants, from the wearing effects of rain and winds. Erosion-controlling techniques are a necessary part of any cultivation system.

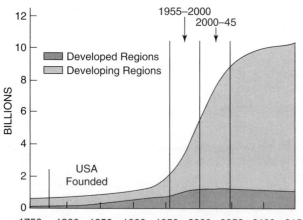

Figure 1–7 Trends and Projections (UN Medium Variant) in World Population Growth, 1750–2150. *Source:* World Resources Institute (WRI) in collaboration with United Nations Environment Programme (UNEP), United Nations Development Programme (UNDP), and World Bank. 1996. World Resources 1996–1997: The Urban Environment. Washington, DC: WRI.

Wood and Wood Products from Timber Trees

The world's forests have provided wood for human shelter since the dawn of civilization. For centuries wood has also served as a source of fuel for warmth and for cooking. Fortunate indeed are those countries endowed by nature with luxuriant stands of tall timber trees. Unlike coal, gas, and oil, wood is a renewable natural resource. With proper management, forests can be enlarged and made more productive by proper tree selection and care. It is essential that the importance of wood for a nation's future be realized and that forests not be exploited for the immediate profits of the day.

Wood and wood products provide material for construction and other utilitarian uses, such as home furnishings. The ornamental value of trees used for landscape and other aesthetic purposes such as carving is nearly inestimable. The biochemicals found in some wood species provide products for industry and medicine. Industrial products include latex, pitch, and resin. Wood, of course, is used for purposes other than building homes and other structures. For example, in the United States, billions of cubic feet of wood per year are used in the manufacture of paper.

Textiles from Fiber-Producing Crops

Fiber crops such as cotton, flax, and hemp supply much of the fabric that clothe us and shelter us. Cotton, hemp, and also jute are used to make string, twine, rope, and burlap. Hemp (*Cannabis sativa*) is a crop that has a deep root system, which can aid in reconditioning a compacted soil. Unfortunately, hemp grown for fiber is the same species as marijuana. Marijuana plants produce the drug tetrahydrocannabinol (THC). As a result, the production of hemp for fiber is illegal in many areas of the world. Although closely related, the two types of *Cannabis* are very different in morphological characteristics. Fiber hemp has the long, unbranched, and sturdy stems required for fiber production. Hemp leaves are relatively small with low THC content. In contrast, the marijuana-type plant is highly branched with larger leaves that have high THC levels. Morphological identification of one plant from the other is not difficult. Because the hemp plant has environmental benefit as a field crop and because the fiber from hemp produces an excellent clothing fabric, many countries, such as Canada, are once again allowing it to be grown. Currently, hemp production in the United States is illegal but increasing pressure on government agencies is forcing a reevaluation of that policy.

Drugs and Medicines

Plants are the source of many drugs and medicines. From the earliest history of modern humans, the shaman and wise elders of a tribe used herbs to treat diseases and disorders among tribal members. Much of this wealth of plant lore is in danger of being lost as indigenous cultures are absorbed into modern civilizations. Ethanobotany, the study of plant usage by indigenous cultures and the preservation of that knowledge about plants, has become an important branch of plant science. One example of a modern-day medicine that can trace its roots to usage by Native Americans is aspirin. Salicylic acid, the pain-relieving component of aspirin, is found in the bark of willow trees (*Salix*). Native Americans chewed willow bark to relieve pain.

Many other modern medicines are directly derived from plant materials. Examples include quinine, which is used to treat malaria, and the heart medication, digitalis. Morphine, opium, codeine, and heroin are all pain-relieving extracts of the opium poppy (*Papaver somniferum*). Cocaine, another painkiller, is derived from the coca plant (*Erythroxylum coca*). However, each of these pain-relieving drugs also has addictive properties. Addiction to these chemicals has become a major problem worldwide, as is finding alternative crops for farmers who grow illegal crops. Tobacco (*Nicotiana tabacum*) is a crop grown for recreational use. However, because of the connection of the drug nicotine to cancer and other health problems, the demand for the crop has dropped. Tobacco farmers are now looking for alternative crops to grow.

One of the newest plants to be recognized for its medicinal benefits is a yew (*Taxus*) that produces taxol, a chemical that shows promise in the treatment of cancer. The taxol-producing species grows wild in the Pacific Northwest and is now being evaluated as a potential crop plant for areas where it would grow well.

Plants also provide latex, pitch, waxes, essential oils, perfumes, species, and other products. Although relatively small crops, each one can provide a significant portion of a region's or a farmer's income.

Plants for Ornamental and Recreational Purposes

It would be hard to find a part of the world, even the poorest, where the aesthetic and recreational value of plants is not important. You only have to watch the summer Olympics to see athletes from almost every country, no matter how poor, running, jumping, or throwing on fields of well-groomed turfgrass. The appearance and playability of fields is vitally important to many sports. Residential and commercial landscaping is a business that generates billions of dollars annually in the United States alone. Even the most humble of homes anywhere in the world are likely to have flowers for decoration at times throughout the year. It is not realistic to give a representative list here of plants used for ornamental and recreational purposes because species grown for these uses outnumber by far the species used in other commodity areas, and the number increases every year.

MEETING THE CHALLENGES IN PLANT SCIENCE

As you read earlier in this chapter, plant scientists face new challenges, many of them more social than production-improvement driven. In fact, plant scientists are sometimes challenged about why they continue to study crop production improvement when production efficiency has increased yield per acre dramatically, especially in developed countries. Currently, enough food is produced to feed the world's population. However, malnutrition and starvation exist in both developed and undeveloped countries mostly because social and political issues prevent the distribution of food to those who need it. If the social issues cannot be resolved, it will likely become the responsibility of plant scientists and growers to find creative ways to produce food crops locally in starvation-prone areas.

In the farming communities of developed countries, increased costs, low profit margins, and surpluses have decreased farm income. As a result, many small family farms have gone out of business, and the land has been sold for development or to become part of a large megafarm, especially in the United States (see Fig. 1–3). The loss of family farms and the farming community is itself a serious social problem. The solutions will no longer lie primarily in increasing productivity of traditional crops. New uses for old crops and production strategies for new crops will have to be developed to allow the family farm to remain viable.

You have already learned that environmental issues pose a challenge to plant scientists and growers. Recently, many of the improvements in production were developed not only to reduce labor and increase productivity and profit, but also to allow farmers to be better stewards of the environment. High-oil corn has been bred to yield a product that can be used to replace petrochemicals in some industrial uses. No- or low-till farming reduces labor costs and is less detrimental to the soil than traditional cultivation practices are. However, no- or low-till farming requires the use of herbicides to control weeds. To make herbicide use more effective and to reduce the amount of herbicide used, genetically modified organisms (GMOs) (or crops) were developed that are resistant to a very effective herbicide, Round-Up®. Weeds are susceptible to the chemical, but Round-Up Ready® crops are resistant. As a result, only one or two applications of a single herbicide is required in a season to get the same or better weed control compared to multiple applications of several herbicides in nonresistant fields.

A genetically modified corn (Bt®) synthesizes a naturally occurring protein that is lethal to the larva of many species of Lepidoptera, such as corn borer. The production of a natural larval toxin in corn has reduced the need to spray insecticides for control of a very destructive pest. Rice has been genetically modified to produce beta-carotene in the grain (golden rice). Beta-carotene (vitamin A) provides a critical nutritional element that is predicted to save the eyesight of millions of children in areas where vitamin A deficiency causes blindness.

Genetically modified crops such as those mentioned above could not have been developed without the technology of gene transfer from one organism to another. Traditional breeding could not impart the traits because the trait does not exist in the gene pool of those crops or species with which the crops are breeding compatible. However, not all GMOs are the result of gene transfer. Sometimes an **endogenous** gene is altered. In the case of Flavr Savr® tomatoes, the gene that produces the enzyme that softens the ripening fruit is scrambled so that the enzyme produced is dysfunctional. The fruit ripens but softens much more slowly than unmodified fruit. The shelf life of the harvested fruit is increased while the traditional tomato flavor is maintained.

As you read earlier, however, there is considerable public concern with GMOs. The public has demonstrated a negative response to each of the aforementioned uses of GMOs. There is fear the Round-Up Ready® gene will escape into the native weeds adjacent to the fields where Round-Up Ready® crops are being grown, thus creating superweeds. The Monarch butterfly (a member of the Lepidoptera species) was thought to be threatened by Bt® corn. Golden rice cannot be grown in many of the areas where it is needed most because of concern about the beta-carotene gene escaping into the traditional rice crop. Flavr Savr® tomatoes have disappeared from the market because chefs and diners refused to buy or eat them out of fear of what the altered gene would do to human health if consumed. Although most of these concerns are likely to be unfounded, it will be up to those who study and work with crops to find ways to show that using these and other production methods does not create undue risk to the global ecosystem, which is easy to say but much harder to do when it comes to public opinion. However, one approach could work.

Perhaps the best way for plant scientists to meet today's challenges is to include the ecological paradigm of agriculture in all scientific studies. A model for this concept was developed in the College of Food, Agricultural, and Environmental Sciences at The Ohio State University (OSU). As described by the college:

> [t]his model is built on four areas of focus: production efficiency, economic viability, environmental compatibility, and social responsibility. A pyramid [Fig. 1–8] has been chosen to provide a visual representation of this model. Each wall of the pyramid represents one of the four dimensions. Four equal walls provide support and strength to each other and emphasize the critical need for balance and integration of the four areas. Disregarding one of the dimensions would weaken the structure significantly; operating without considering two dimensions would cause the system to collapse.

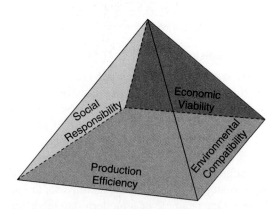

Figure 1–8 Pyramid representing the four sides of the ecological paradigm of scientific study in the agricultural sciences. *Source:* Adapted from College of Food, Agricultural, and Environmental Sciences, Ohio State University, Columbus, OH.

This approach is not unique to OSU. Most if not all institutions where plant science is studied have seen the need to take the same or a similar approach.

CHANGE THROUGH SCIENTIFIC INQUIRY

Hypothesis

Scientific inquiry or research begins with an objective and some idea of how the objective can be met. The objective might be to solve a problem or improve a procedure. The idea can be stated as a hypothesis, such as, "This is the source of the problem" or "This is how the process can be improved." A simple problem might be determining if a common landscape species will grow well in compacted soil. You may have observed what appears to be suitable growth in compacted soils, but maybe you would like more scientific evidence before you start using that species in compacted landscape soil.

From your observations or other information (such as claims from seed companies), you can create the hypothesis that the species will have the ability to grow acceptably well in compacted soils. From this hypothesis, a simple experiment can be developed to test the validity of the hypothesis. The two species can be grown in compacted and porous soil, and the differences in plant growth between the two soils (treatments) can be measured. From our hypothesis, we can predict that compaction would not affect growth when compared to the species growing in porous soil. If this prediction turns out to be correct, then we have confirmed our hypothesis.

If our supposed clay-tolerant species did not grow as well in compacted soil as in porous soil, perhaps our hypothesis was false or perhaps we might have achieved better results if the experiment had been designed differently. By doing more experiments, we can gain confidence that our hypothesis is correct. Our confidence can be increased if we can explain why the growth was better in porous soil. We can set up more experiments that look at sub-hypotheses; for example, maybe the root system's growth was different in the two treatments.

The example above is a relatively simple problem. At the other end of the scale are complex ecosystem problems such as determining if agricultural practices have a role in creating the dead zone of the Gulf of Mexico. The dead zone is an area off the coast of Louisiana where an unusually large number of marine life dies each summer. It is generally not wise to focus on a single hypothesis to solve a complex problem such as an ecosystem problem. It is quite likely that there is no single cause of the problem. For example, the dead zone may also be caused by industrial activity, natural causes, and/or something else. It is best to develop several hypotheses that represent as many explanations as we can imagine. We can then look for experimental evidence that will support or refute the different hypotheses. After extensive research, we can build a plausible description that is consistent with the data our experiments have generated. From this research, we may be able to suggest ways to correct the problem.

Experiments

Controlled experiments are the most certain way of obtaining reliable information on which to base decisions. In large and complex problems, a full-scale experiment is most likely impossible for testing our hypotheses. However, we can

---

UNITS OF MEASURE FOR VARIOUS SYSTEMS

All scientific studies present data in scientific units. Although they may be less familiar than US units, they are often much easier to work with and they are used in scientific literature (even in agronomy and horticulture). There are different "flavors" of metric system. Older systems such as cgs (centimeter, gram, second) are being replaced by Système Internationale (SI). This table lists some corresponding units and common conversions that might help as you move through the course.

Quantity	Old Metric	SI	US	Conversion
Length	Centimeter (cm)	meter (m)	foot (ft)	1 m = 3′ 3″
Area	cm^2, hectare (ha)	m^2, km^2	ft^2, acre	1 ha = 2.47 acre
Volume	cm^3, liter (L)	m^3	gallon	1 L = 0.265 gal
Mass	g, ton (1,000 kg)	kg	pound (lb)	1 kg = 2.2 lb
Force	kg	Newton (N)	pound	1 kg = 9.81 N
Pressure	$kg\ cm^{-2}$	Pascal (Pa)	psi	1 kPa = 0.15 psi
Energy	calorie (cal)	Joule (J)	btu	1 kJ = 0.95 btu
Power	dyne	Watt (W)	hp	1 kW = 1.34 hp
Temperature	°C	°K	°F	°C = (F − 32)* ⅝
Light	lux	$mol\ m^{-2}\ s^{-1}$	foot candle	

Common prefixes: micro (µ)(÷1,000,000), milli (m)(÷1,000), kilo (k)(×1,000), mega (M)(×1,000,000)

isolate parts of the problem and study those parts, eventually putting it all together to address the complex situation. Experiments are artificial situations where we try to exclude everything except our immediate area of interest, such as plant response to compacted soil or the effect of erosion sediment on marine life.

The key to a meaningful experiment is to be sure that the only factor affecting the outcome is the factor that we set out to study (controlled experiment). If we are testing whether one species of plant is less affected by soil compaction than another, we need to be sure it is only compaction that is affecting one species more than another. For example, we would select plants of the same age to evaluate for response to compaction. If we had plants of different ages, we cannot be sure that age was not a factor in response to soil compaction.

In a typical experiment, plants of uniform characteristics are planted in compacted soil and others with the same characteristics are grown in uncompacted soil. The type of soil is called the independent variable. Plant growth is the dependent variable because its value depends on changes in the soil. Compacted soil is called the treatment. Uncompacted soil is called the control, or check. Controls allow for the comparison needed to draw conclusions about treatment effects. In practice, it is difficult or even impossible to make everything completely uniform. Replication and randomization are used to avoid unintentional bias resulting from lack of uniformity. Replication means that we base our conclusions on more than one observation on a single experimental unit (Fig. 1–9). The average of the response of experimental units to a treatment is what we base our conclusions on so there is less chance that an aberrant individual is the basis for a conclusion. Working with more than one individual is the first step to ruling out chance variation as a factor that can affect our conclusions.

Figure 1–10   Plants showing natural variation in height from genetic and/or environmental conditions. *Source:* Michael Knee.

Randomization takes into account that no two plants are exactly alike. Even cuttings from the same plant may have come from different size branches or have other differences, such as mutations (Fig. 1–10). Randomly selecting the plants to be treated helps to reduce the chance that all the plants for one treatment are inherently different from those for another treatment. For example, if from a group of plants chosen for an experiment only the short ones were used for the compact soil treatment while the tall plants grew in noncompact soil, then the difference in height might carry all the way through the experiment and the conclusion that the plants grew best in noncompact soil would not be accurate. By mixing up the size of the plants within the treatments, changes in plant growth would more likely be the result of the treatment (Fig. 1–11).

Blocking is another way to reduce the effect of variation. An example of blocking is to arrange plants in groups by height, then subdividing those groups so that each height group receives all the treatments, not just one. Similarly, in a field trial, an experimental field can be divided into treatment plots where the treatments are replicated. These plots are randomly assigned treatments (Figs. 1–12 and 1–13). With this method, you reduce the chance that another factor,

Figure 1–9   Randomized arrangement of pots containing compacted (dark-colored) and noncompacted (control) soil (light-colored). *Source:* Michael Knee.

Figure 1–11   Randomized placement of plants with different heights at time of planting into pots with compacted (dark-colored) and noncompacted (light-colored) soil. *Source:* Michael Knee.

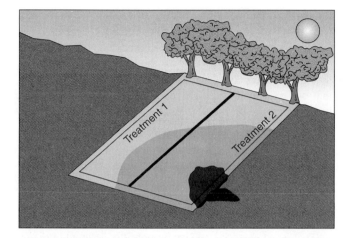

Figure 1–12 Example of a field experiment where the treatments (1 and 2) are represented only in the experiment. The darker portions of the field are wet areas that could distort or confound the data and increase the chance of making a Type I or Type II error. *Source:* Michael Knee.

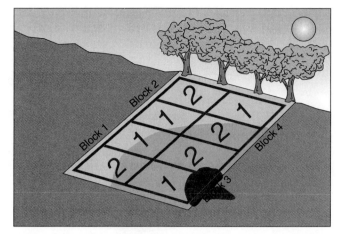

Figure 1–13 Example of a field experiment with multiple plots or blocks in which each of the two treatments (1 and 2) are randomly placed. Blocking helps to separate the effects of the wet areas from the effects of the treatment, reducing the chance of making a Type I or Type II error. *Source:* Michael Knee.

such as a wet area of the field, influences only one treatment. The more you know about the area you are using for an experiment, the more effective the blocking design can be. In greenhouse experiments, pots with compacted and noncompacted soil can be arranged randomly on a bench to eliminate the effect of drafts and drying out on those pots on the perimeter of the bench. Controlled experiments are ideal for studying one or a few variables; however, one of their disadvantages is that treatment effects on factors other than the defined dependent variable are often not considered. Other parts of the ecosystem may be detrimentally affected.

Although current scientific inquiry into agricultural and horticultural problems sometimes involves simple situations such as determining how insect-resistant one new cultivar is compared to another, or what fertilizer rate is the best for flower development of a newly introduced ornamental, most research now does involve some aspect of ecosystem study. We have learned the hard way that plants do not grow in isolation and what we do to help in one aspect of plant growth may have detrimental consequences elsewhere. These oversights created the suspicion that the public attaches to scientific study and the reassurances from the scientific community that a procedure is safe. For example, organochlorine insecticides were widely adopted after they were shown to be effective against a wide range of agricultural pests. After many years, however, the chemicals accumulated in animals at the end of the food chain. They were particularly harmful to birds of prey because they caused the birds' eggshells to be thin and brittle. Failure to hatch young led to a collapse in the populations of many hawks. Pesticides are now tested for effects on nontarget organisms, but no amount of research can ever prove that any chemical will be safe under all conditions of use and for all time.

The complexity of ecological research requires the use of different methods than experimental research that has one or only a few variables. Ecological research is often descriptive of a situation within an area that represents the larger ecosystem. Here, correlation analysis generates a descriptive model of the relationship between or among variables. Determining the scale of the area to be described is analogous to deciding what factor is to be the subject of an experiment. The scale has to be sufficiently large to represent the ecosystem but small enough to be effectively described. Many more variables are often recorded than in an experiment in a laboratory or greenhouse. In an experiment, variables that we do not want to influence the treatment must be controlled or held constant. In the real world, this condition is usually not possible, so the extra variables are recorded in case they are found to have an influence in the system. Dependent and independent variables are less clearly defined in descriptive research than in experimental research. The appropriateness of the research method chosen depends on the factor or factors to be studied.

Examples of ecological research may be the study of the effects of farm runoff on the aquatic life in a single creek that feeds a larger river (Fig. 1–14). By studying the creek, inferences about the larger river's watershed can be made. Likewise, it might take fifty years to see the consequences of clear-cutting a rainforest to create farmland. However, by studying other rainforests that have been cut at different times during the past, we can look for patterns in the data that might show what happens to deforested land over time. In the case of Bt® corn, a more comprehensive study of the ecosystem of the Monarch butterfly, not just of how the butterfly larva was affected when fed only Bt® corn pollen, revealed that Bt® corn pollen was one of the

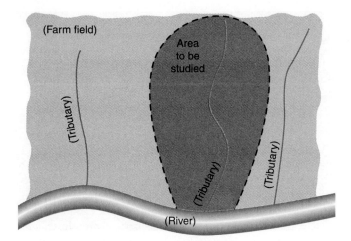

Figure 1–14 Correlative studies study a small area and make correlations to a larger area. For example, a section of a field that has a small tributary stream running through it can be examined closely for effects from a farming practice. Inferences can be made regarding how other tributaries in other parts of the field are also affected by that practice. *Source:* Margaret McMahon, The Ohio State University.

least of the butterfly's threats. The conclusions from that study showed that the original public outcry stating that Bt® corn threatened the Monarch population was unfounded. Many other factors, including natural predators and pesticides (many of which no longer had to be used on Bt® corn), were much more harmful to the butterfly than the Bt® corn pollen.

Data Interpretation

So how does a scientist arrive at conclusions from a set of experimental data? The data must be analyzed statistically to see if the original hypothesis is supported or not. Generally we say that the null hypothesis (meaning the treatment had no effect) was accepted or rejected. If rejected, then our treatment most likely had an effect. But we can never be sure that it was our treatment and only our treatment that caused a difference in the dependent variable. We usually assign a percentage probability that something else was the cause. In most cases, if we can say that there was a 95 percent or greater chance that the treatment caused the effect, then we reject the null hypothesis and claim our treatment had an effect. If the probability is less than 95 percent, we accept the null hypothesis and say that our results were nonsignificant.

To determine the probability, the natural variability in the population being tested has to be compared to the variation caused by the experimental treatments. In our compacted soil experiment, suppose we look at the number of roots of the intolerant plant in the pots of compacted soil and the number of roots in the pots of loose soil. We may see a wide range of numbers in both situations. If the variation in the pots is great enough, then the compacted pots may be no different from the control, even though the two treatments may have different average root sizes. We would have to say that our treatment had no effect. For example, suppose we had four pots of porous soil and four pots of compacted soil. The average number of roots in the loose soil was fifteen, and the average number of roots in the compacted soil was ten. We would be tempted to say that the plant roots grew better in the porous soil. However, suppose we looked at the individual numbers and saw that the porous pots had twenty-two, eight, twenty-one, and nine roots and the compacted pots had fourteen, twelve, eight, and six. There may be so much natural variation (error) that the differences in the two averages could have occurred without any treatment. One reason for having as much replication of experimental units as possible is to get a better estimation of natural variation.

Statistical procedures test relationships between variables. In an experiment, we test the model that an independent variable or treatment causes an effect in a dependent variable. In complex systems where an experimental model cannot be constructed, correlation analysis is used to generate a descriptive model of relationships between models. We can use the relationship model to determine whether it would be worthwhile to change a variable under our control, such as fertilizer use in a watershed, or whether changing the variable would be unlikely to have any effect.

No statistical method is completely reliable. There is always a risk of saying that there was an effect when there really was not, an error called a Type 1 error. Conversely, a Type 2 error occurs when we say that results were nonsignificant when they were. That is why it is important when reporting experiments to describe in detail how they were done. Thus, future researchers can compare their results and methods to yours. It is not failure to have the results of a well-conducted experiment later contradicted by other researchers. The difference in results from different experimental situations quite often lead to a better understanding of how an ecosystem works.

As you can see, understanding the principles of scientific study is important to plant scientists, who must be trained in these principles. However, it is equally important that others understand the principles. Anyone who works with growing plants for a living has to be able to read and interpret the scientific literature of his or her discipline. Without this ability, it is likely that decisions will not be based on sound scientific findings. We hope that students using this book will become knowledgeable in the basic principles of plant science so they can grow plants productively and cost-effectively, and communicate confidently and effectively to the public that the methods they are using are environmentally sound.

SUMMARY AND REVIEW

Humans and cultivated plants are mutually dependent on one another. Cultivated plants require attention from humans to survive. Humans need the plants to fill basic nutritional needs and to add quality and enjoyment to their lives. Early plant scientists studied ways to improve crop productivity to meet the demands of a growing population. Plant scientists were successful at increasing production and today we are capable of feeding everyone on earth; however, problems of starvation and uneven supply and distribution of crop commodities continue to plague the world. In addition, the population continues to grow, placing higher and higher demands on our farms, which are shrinking in number.

Complicating the situation is the fact that many improvements in production were shown in time to be detrimental to the global ecosystem. Plant scientists and growers had to find ways to grow crops without creating undue risk to the global ecosystem. Because of past mistakes, however, new production practices, such as the use of GMOs to reduce pesticide use or improved shipping quality of produce, are met with suspicion because of the fear that the new methods will also have negative effects. Public opinion is proving to be a powerful force in plant science. The difficulty is exacerbated by the fact that much of the population in many countries resides in urban areas and has little understanding of how food and other plants are produced.

Today's plant scientist and grower must have a firm understanding of the relationship among production efficiency, economic viability, environmental compatibility, and social responsibility when studying ways to improve productivity. Integrating that understanding with a sound scientific approach to research in plant science will help restore public confidence by reducing the chance that a procedure will have an undesirable side effect. A sound approach begins with the formation of a hypothesis, designing the proper experiments to test that hypothesis, and finally analyzing and interpreting the data gathered correctly. Randomization and replication in experiments are important to reduce the chance of making a Type I or Type II error.

FOOD FOR THOUGHT

1. Crop varieties were once owned and intimately known by individuals in primitive societies. What has happened and is likely to happen in the future to crop germplasm and our knowledge of it?

2. Government agencies such as the Environmental Protection Agency and the Natural Resources Conservation Service are often attacked for interfering with people's lives and businesses. Do local people or businesses always know best? Look for specific examples in the news.

3. Karl Marx thought that human labor creates value in commodities that can be sold in the marketplace. Howard Odum seems to suggest that value comes from the energy embodied in the commodity. Where do you think the value of a particular crop comes from?

4. Many countries support their crop production industries with direct subsidies or price supports. Do you see any reason to continue these payments? Why has it been so hard to stop them?

5. In many poor countries, you are likely to see a lot more cars on the roads than tractors in the fields. Does this situation present a problem?

6. Are you comfortable with a world in which China is the main supplier of manufactured goods and the United States is one of the main suppliers of agricultural products?

7. Supermarkets now often feature fair-trade and organic products. Are these products examples of niche markets or could they become the norm?

8. A scientific report states that a new herbicide had no effect on turfgrasses used on golf course greens. After publication of that report, further investigation revealed that a Type II error had been made and that the herbicide slowly weakened bentgrass, a turfgrass commonly used on greens. (If you want, substitute corn, flowering annuals, or any other crop for the turfgrass.) What do you think would be the consequences of this situation?

SUPPLEMENTARY READING

ANONYMOUS. 2002. *The state of food and agriculture.* Rome: Food and Agriculture Organization of the United Nations (FAO). Available online at **http://www. fao.org/DOCREP/004/y6000e/y6000e00.htm**.

ODUM , H. T. 1996. *Environmental accounting: Emergy and environmental decision making.* New York: John Wiley & Sons, Inc.

POWERS, L., and R. MCSORLEY. 2000. *Ecological principles of agriculture.* Albany, N.Y.L: Thomson Learning.

SMITH, B. D. 1995. *The emergence of agriculture.* New York: Freeman.

UNITED STATES DEPARTMENT OF AGRICULTURE. National Agriculture Statistics Service Publications. Available online at **http://www.usda.gov/nass/**.

CHAPTER 2

Terrestrial Ecosystems

Michael Knee

KEY LEARNING CONCEPTS

After reading this chapter, you should be able to:

♦ Appreciate the fundamental importance of plants in all terrestrial ecosystems.
♦ Understand the role of other organisms in ecosystems.
♦ Explain how climate determines the kind of vegetation that occurs naturally in different parts of the world.
♦ Appreciate the potential and actual productivity of natural and crop ecosystems in different areas.
♦ Understand the dynamics of change in soils and vegetation over time.

Almost all terrestrial ecosystems depend on the ability of plants to capture energy from the sun and store it in complex organic molecules. Climate determines the kinds of plants that grow naturally in an area and the rate at which they grow. All of the other organisms in the ecosystem depend on and are influenced by the nature of the vegetation and its productivity. Natural ecosystems are usually more complex than crop ecosystems; they illustrate the gamut of ecological relationships and ecosystem processes and provide a reference point from which we can evaluate crop ecosystems. Generally, the more a crop ecosystem differs from the ecosystem that would occur naturally in an area, the more difficult it is to sustain and the more inputs it requires.

ECOSYSTEM COMPONENTS

An ecosystem consists of a community of organisms in a physical environment (Fig. 2–1). The community consists of populations of individual species; different kinds of plants, animals, fungi, bacteria, and so on. A population can be defined as all of the individuals of a species that inhabit a particular environ-

Figure 2–1 An ecosystem consists of a community of organisms in a physical environment (consisting of soil and atmosphere in the case of a terrestrial ecosystem). *Source: Michael Knee.*

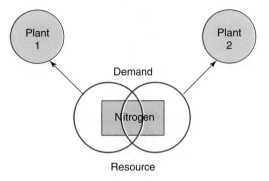

Figure 2–2 Competition occurs when plants of the same or different species attempt to use a resource that is in limited supply. *Source: Michael Knee.*

ment. The way that we define the environment determines the boundaries of the population. Often we are thinking of a limited area in which the individuals share the available resources. In an isolated area of woodland, a field, or a greenhouse, the plants may be competing for light, water, or nutrients. However, ecological relationships may extend over larger areas, particularly for animals that can roam from one place to another. Even for plants, pollination may occur between individuals in physically separate environments many miles apart. It is fairly obvious that plants found naturally on different continents belong to different populations, but if there is any possibility of an ecological interaction between individuals of a species, then they can be viewed as belonging to the same population.

We tend to think of ecosystems in terms of the plants and animals that we might see on a visit to a "natural" area in a state park or wilderness location. However, anywhere organisms exist is, in a strict sense, an ecosystem. Many of the most important organisms and ecological processes in all ecosystems are invisible to us because the organisms are small or because they and the processes occur in the soil. A cornfield or an area of turfgrass may look too simple to qualify as an ecosystem, but a whole community of organisms interact with the single species of plant, both above and below the ground. Ecology does not disappear when we grow plants in a home or a greenhouse. The physical environment, including the growing medium, may be artificial, but it will be colonized by other organisms, some harmful and some beneficial. Although books and television programs about nature may focus on the animals and birds that enliven natural ecosystems, many of the functions and ecological interactions that sustain the ecosystems are carried out by small invertebrate animals, fungi, and bacteria, and many of these organisms are in the soil. The microflora and -fauna are, relatively speaking, even more important in crop ecosystems, where we generally try to exclude macrofauna (birds and mammals) and noncrop macroflora (in other words, weeds). What distinguishes these crop ecosystems is that some of the

organisms, processes, and interactions that would sustain a natural ecosystem are absent or heavily modified by human intervention. For example, inputs of nutrients are necessary to promote crop growth, or chemicals are needed to suppress disease. We can describe crop production in terms of these inputs and ignore the ecological processes and interactions. However, the goal of sustainable production requires that we minimize the inputs and maximize the contribution of ecosystem processes. By observing the full range of interactions and processes in natural ecosystems, we can learn how to manage crop ecosystems to achieve the goal of sustainability.

The organisms in an ecosystem interact according to the nature of the species and their role in the ecosystem. Each individual organism draws on the resources of the ecosystem, and competition occurs when more than one organism draws on a resource that is in short supply. The resource might be required for growth or the reproduction of the organism. Plants require light, water, and nutrients for growth and may require pollinators and dispersal agents for reproduction (Fig. 2–2).

In ecosystems with a complete cover of vegetation, the most intense competition among plants is likely to be for light. A plant with a greater leaf area is likely to capture more light, particularly if the leaves are positioned above those of another plant. Trees can capture most of the light and are the dominant vegetation in many parts of the world. However, they have to invest considerable resources in the trunk and branches that support their leaves, which is one factor that limits the size of trees. Vines such as Virginia creeper, English ivy, and poison ivy can compete for light with less structural investment by using the support provided by the trees.

Usually a single species cannot utilize all of the resources or the whole of any one resource in the environment. For example, some light does penetrate the canopy in most forests, and herbaceous plants, such as ferns, are adapted to growth in the low-light environment of the forest floor. Many of the trees in temperate forests are deciduous, and for a period in the spring, light is available for herbaceous flowering

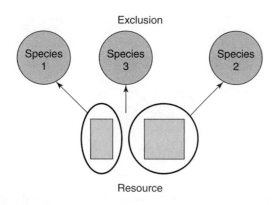

Figure 2–3 Species 1 is able to occupy a niche because species 2 does not use all of the resources. However, species 3 is subject to exclusion because it also needs the resources, but none remain. *Source:* Michael Knee.

plants such as bloodroot, *Trillium,* and *Hepatica* to grow, flower, and set seed before the trees develop their full canopy. The shade-adapted and spring-flowering plants occupy niches. Niches exist when a resource is partitioned so that different portions of it are accessible to only certain species (Fig. 2–3).

Although a critical environmental resource may be required for a species to find its niche in an ecosystem, each species has more or less strict requirements for a long list of environmental resources. In the fullest sense, the niche includes everything that is required for an organism to flourish in the environment. For a plant, this list can include light intensity, availability of water, nutrient concentrations, and the presence of the right pollinators. If two plants have identical environmental requirements, they are trying to occupy the same niche and competition will be more severe than it is for two plants with different requirements. Competition is thus more intense between plants of the same species than it is between plants of different species. For species to coexist in an ecosystem, they must occupy different niches. It can be difficult to see how species requirements differ, for example, when looking at different species of deciduous trees that make up the forest canopy. However, species are separated by their physiological requirements as much as by their more obvious morphological differences.

In healthy, natural ecosystems, every available niche tends to be filled, and consequently resources are fully exploited. It is difficult for new species to colonize the environment because it lacks open niches or unexploited resources, and the result is called exclusion (see Fig. 2–3). Thus, annual herbaceous plants cannot colonize mature woodland. If seeds blow in and germinate, the development of the foliage in the canopy during the spring will close off the light before the annual plants can flower and set seed. On the other hand, if some trees come down in a storm, annual plants may be able to colonize the open area or gap until new trees grow and the canopy closes.

Crop ecosystems have one or a few species that do not exploit all of the resources of the environment. Thus, niches tend to be available for other plants to colonize. We call the other plants weeds, and their presence is a problem to the extent that their resource requirements overlap those of the crop species.

After resources have been captured, they may be lost to predation or parasitism. Plants are consumed by a wide range of animals, which we call herbivores. (We use the term predator for animals that consume other animals). Herbivores include mollusks (slugs and snails), arthropods (insects and mites), mammals, and birds. The diversity and numbers of herbivores underline the importance of plants as a food source in terrestrial ecosystems. Many believe that herbivores do not compete with each other or come close to consuming all of the plant food resources of natural ecosystems because their numbers are kept in check by predatory animals in the same or other groups. There is a wide range of small predators among the arthropods, and there are many insect-eating birds and mammals. Large predators that feed on birds and mammals are less common because ecosystems cannot support many of them. As the number of herbivores increases, they become easy prey for the predators. Crop ecosystems are usually more vulnerable to herbivory because the number of predators is low or they are nonexistent. Thus, it may be necessary to use chemicals or other means to control slugs, caterpillars, rodents, birds, or deer.

Parasites are organisms that derive their nutrition by living in or on the tissues of another organism, often producing symptoms of disease such as swellings or discolored tissue. Plant parasites include many kinds of viruses, bacteria, and fungi. Some parasites can grow only in the host and are called obligate. Other parasites can grow outside the plant and are called facultative. All viruses are obligate parasites, whereas fungi and bacteria may be obligate parasites or facultative (when they commonly exist outside the plant in the soil). A few parasitic plants, such as dodders and mistletoes, can infect plants in natural and crop ecosystems. Other classes of plant parasites include nematode worms and some insects and mites, particularly those that form galls on leaves and stems.

Parasitic animals differ from herbivores in the same animal groups because they cause a disease rather than simply eating their way through the tissue. In a disease, the physiology and often the morphology of the host is altered as its resources are redirected toward the pathogen or disease-causing organism. Parasitism is a long-term relationship between a host and a pathogen, in which the host does not usually die because this would also entail the death of the pathogen. Parasitoids are animals, often insects, that spend the juvenile phase of their life cycle in the tissues of another insect. When they emerge as adults, the host is killed. Parasitoids can be helpful in crop ecosystems when they infect herbivorous insects such as caterpillars.

Some associations between organisms are mutually beneficial rather than antagonistic. In natural ecosystems, the

Figure 2–4 Many plant roots form a symbiosis with fungi, a relationship in which the plant supplies sugar to the fungus and the fungus draws mineral nutrients from the soil, which it passes on to the plant. *Source:* Michael Knee.

roots of most plants are colonized by fungi that draw nutrients from the soil and pass some of them on to the plant. In return, the plant passes sugar and other organic molecules to the fungus. The fungi are known as mycorrhizae and can either live on the surface of the root (ectomycorrhizae) or invade the root tissue (endomycorrhizae) (Fig. 2–4). This relationship is an example of symbiosis, which involves a permanent and close association between two organisms that are known as symbionts. Mycorrhizae are less common in crop ecosystems, particularly on herbaceous plants, than in the wild. The nutritional requirements of crops are often supplied by fertilizers and fungi are not needed to scavenge for them. Lichens that are found on tree trunks, rocks, and the soil surface are formed by the symbiosis of a fungus with an algae species. Another kind of plant symbiosis involves a bacterial symbiont that forms the nitrogen-fixing nodules on the roots of plants in the *Fabaceae* and a few other families. Nitrogen-fixing plants are important for maintaining the nitrogen supply in natural ecosystems. The legume family, or *Fabaceae,* includes many important crops such as beans, peas, and alfalfa. A more general term for a mutually beneficial association between organisms is commensalism. In addition to symbiosis, commensalism includes looser and less permanent associations such as that between a plant and its pollinating insect.

Although parasitic organisms may be the most conspicuous fungi and bacteria in crop ecosystems, most microorganisms are saprophytes that digest dead plant and animal material at or below the soil surface (Fig. 2–5). Saprophytes are essential to the survival of the ecosystem because they recycle nutrients that would otherwise be tied up in dead organisms. In natural ecosystems, a few plant species are saprophytic. Saprophytes are aided in their activity by detritovores that break up large pieces of organic matter as they consume it. Earthworms are an important part of this group, which also includes other kinds of worms, many insects, and small mammals. Detritovores and saprophytes may be less important in crop ecosystems than in nature because crops

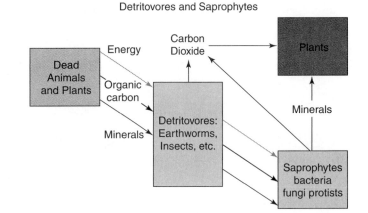

Figure 2–5 Recycling is essential for maintaining the supply of nutrients in natural ecosystems, and soil microorganisms are the primary agents in this process. *Source:* Michael Knee.

and crop residues are removed and the nutrients are replaced by synthetic fertilizers. They become more important, however, as we try to recycle crop waste and minimize fertilizer inputs to develop more sustainable production systems.

Primary producers (mainly plants), herbivores, parasites, commensals, detritovores, and saprophytes exist in all terrestrial ecosystems. In a natural ecosystem, each of these roles is filled by many species of organism that exist in a complex web of relationships (Fig. 2–6). Nutrients are passed continuously from one organism to another along with the energy and carbon compounds that sustain the whole ecosystem and that originally came from plants. The system tends to be highly conservative. Many organisms compete for the resources available at any level in the food chain. If there is any gap in the utilization of resources, species are recruited to fill it. Crop ecosystems are based on one or a few plant species, which are usually intended to be harvested rather than allowing resources to be consumed by other species. While the crop is growing, however, surplus resources exist that could be exploited by competitors, herbivores, and parasites. We have attempted to manage these problems by mechanical and chemical means. By eliminating species, we attempt to simplify the ecosystem and minimize ecological interactions. More recently, we have begun to think about introducing species that could help control some of the problems, companion plants, predators, parasitoids, and so on.

THE BIOMES

We often like to go to different climatic zones for vacations, but if you get the chance to travel to a similar zone, say, from eastern North America to northern Europe, you may be struck by the fact that similar but not identical trees are present in the forests. You might see oak, maple, beech, and ash but not the same species that you would see in the United States. The explorer Alexander von Humboldt noticed that

Simplified Food Web for the Deciduous Forest

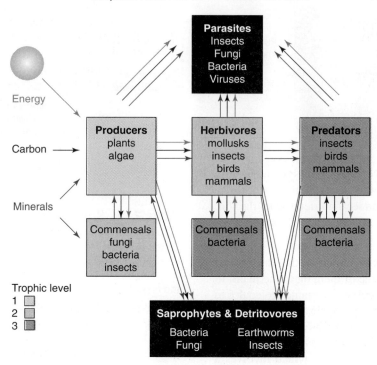

Figure 2–6 Natural ecosystems operate with inputs of carbon, energy, and water and are highly conservative of nutrients because of a complex web of ecological relationships. *Source:* Michael Knee.

there was a similar progression of vegetation types as he moved north or south from the equator or ascended in altitude in different parts of the world. He realized that the same kind of vegetation tended to occur in similar climatic conditions around the globe. The vegetation appears similar because the plants are functionally similar but not necessarily the same species. The functional groups include deciduous broad-leaved trees, coniferous evergreens, succulent plants, tall grasses, spring-flowering bulbs, and so on. Today we call the recurrent vegetation types biomes; a **biome** is a collection of ecosystems with similar climate, soil, and plant composition.

Climate is the main influence on the type of vegetation that develops. The most important climatic variables are temperature, rainfall (or more strictly precipitation), and any seasonal variation in both. A key component in the relationship between climate and vegetation is soil. As we will see in Chapter 6, soils are produced from parent material, such as rock, by the interaction of climate and organisms. Plants are key components of the soil ecosystem and influence all of the other organisms present. Although a world map may show the natural distribution of the different biomes across all of the continents, many of these areas have now been converted to agriculture or urban development. The soils that developed in the biomes differ greatly in their suitability for crop production. In general, the greater the difference between a crop production system and the preexisting biome, the more

difficult it is to sustain that system. Sometimes it may even prove to be impossible.

Temperature is primarily influenced by latitude, and within latitudes, it is influenced by the height of the land above sea level. The other factor, moisture availability, is affected by patterns of air circulation around the globe. Air moves in one direction in the winds that we experience at ground level and then rises to the upper atmosphere to flow in the opposite direction, before returning to ground level. This circulation is rather like a donut-shaped mass of air, with the wind moving around it from the hole in the middle to the outside and back. A series of three such systems operates between the equator and each pole (Fig. 2–7).

At the equator, temperatures are high throughout the year. The trade winds from the north and south carry moisture from the oceans and converge in this region. The warm, moisture-laden air rises and cools down so that the moisture condenses as rain. The warmth and readily available water are favorable for plant growth throughout the year, and all kinds of other organisms can benefit from the high productivity of the plants. The rainforests of South America, Africa, India, and southern Asia developed under these conditions and make up the most productive biome, with the greatest range of biological diversity on earth.

The forest is dominated by broad-leaved and evergreen trees whose dense canopy captures most of the light. Rain-

Figure 2–7 Three main belts of surface winds exist in each hemisphere. Each one determines patterns of precipitation and the vegetation that develops. *Source:* Michael Knee.

forest soils are shallow, and the tree roots spread outward rather than down. The trees gain stability from their shallow root systems by forming buttress roots. Because of the low-light levels, only a few shade-tolerant plants grow under the canopy; however, the trees provide support to many **lianes.** Also many herbaceous plants grow on the trunks and branches of the trees. The roots of these epiphytes never reach the soil; they catch their water as it runs down the surface of the tree or from the air itself with specialized aerial roots.

Much of the life of the forest occurs in the canopy, unseen by humans and other groundlings. Most of the nutrients in the ecosystem are also in the canopy and are cycled from one organism to another without passing through the soil. Because of high rates of growth, nutrients are rapidly taken up by plants and rapidly recycled from dead organisms, so rainforest soils contain little in the way of free nutrients. After the forest is cleared, good crops can be raised for a year or two, but the nutrients are soon exhausted. To make matters worse, nutrients are washed out by the rains, and the soils themselves are unstable in the absence of vegetation. So production of annual crops is not easily sustainable in this region.

About 30° north and south of the equator are bands of divergence in the global air circulation. Air moves away from these latitudes in the trade winds toward the equator and in the westerlies toward the temperate latitudes. This movement is powered by a downdraft of air that carries moisture away, and it is most pronounced over large land masses. It gives rise to the deserts of North and South America, Africa, Asia, and Australia, which make up the largest biome on the planet. By day, temperatures are warm, but at night the absence of moisture in the air permits rapid loss of heat so that it can be surprisingly cold. The plants of this environment are highly specialized and grow slowly or intermittently. They can be slow-growing perennials with succulent stems or leaves that help conserve water. They can also be fast-growing annuals that take advantage of occasional rains to complete their life cycles while the moisture is available. Because of slow growth, there is little accumulation of organic matter that can be recycled to build up soil fertility in these ecosystems. Be-

cause temperatures can be favorable to plant growth and deserts occupy about two-fifths of the land surface, many people have had the dream of "making the desert bloom." This requires massive amounts of water that can be economically transported or diverted to the desert region. Fertilizers must be used to compensate for the low fertility of the soils. All water sources contain small amounts of salts, even if they are so-called fresh water. When the desert is irrigated, these salts are left behind after the water evaporates from the soil or through plant transpiration. There is usually not enough water to leach the salts down through the soil and into groundwater or streams. Desert soils are difficult to manage in the long term because the salts tend to accumulate to levels that inhibit the growth of or kill plant roots.

Moving north or south away from the deserts, the climate becomes moister but temperatures are not continuously favorable for plant growth. In latitudes between 40° and 50° and when there is more than about 75 cm of rainfall, temperate deciduous forest develops. This development occurs on all of the land masses in the northern hemisphere, but there is little land area in these latitudes and temperate deciduous forest is absent south of the equator. During the summer, the trees in these forests can be almost as productive as those in the tropical rainforests. Growth nearly ceases in the winter, when most of the trees have lost their leaves. Although lianes are present in old-growth forests of this type, the canopy is not as favorable a habitat for epiphytes as in the rainforest. However, understory shrubs and herbaceous plants can exploit the light that comes through the canopy in the spring before the new leaves are fully formed. (The **understory** is composed of the plants that grow under the canopy of other taller plants.) The botanical diversity of the Appalachian forests of North America is a distant second to the rainforest, but it is more readily visible to a wanderer in the sunlit carpets of spring ephemerals than it is to a ground-based visitor to the comparative (and permanent) gloom of the rainforest.

The annual leaf fall adds a certain amount of nutrients to the litter layer of the temperate deciduous forest. However, the leaves contain only a fraction of the nutrients locked up in the permanent structure of the tree, and trees manage to withdraw much of the nitrogen and phosphorus from their leaves before they are shed. So forest soils develop slowly and tend to have only a thin organic-rich layer. Their fertility is soon exhausted when crops are grown after the forest is cleared.

The area between the deserts and the forests consist of grasslands and savannas. Tree growth in grassland and savanna is limited to some extent by lack of water, but fire and grazing animals are also important in maintaining these ecosystems. Grassland vegetation survives and may even benefit from a certain level of grazing that woody plants cannot tolerate. Also dead grass is readily ignited by lightning strikes. When fires sweep through grassland, most trees are killed but grasses and herbaceous perennials, which have their growing points at or below the soil surface, typically survive. Grasslands and savannas were favored by early people and our first efforts at land management may have been to set fires to provide habitat for animals that we were trying to encourage into or deliberately keep in herds. We still find prairies, meadows, and forest glades attractive and may not fully realize our own role in their ecology.

Grassland plants accumulate nutrients in the roots and stems that are renewed from year to year. In temperate climates, the old stems and roots are not fully recycled each year and organic matter builds up over time. This organic matter contains reserves of nutrients and makes up a large part of the black earths that are characteristic of moist grasslands around the world. The nutrients in such soils can sustain crop production for many years, and the former grasslands of North America, Europe, and Asia have proved to be the world's most productive agricultural land. Agriculture itself originated in such a region, the so-called fertile crescent that occupied river valleys from present-day Egypt to Turkey and Iran. Grassland becomes less productive as it merges with drier regions on its margin. In the United States, for example, the tall grass prairie of the central states gives way to short grass prairie to the west. Attempts to raise productivity with irrigation can meet the same problems of salinization as in the desert. Arid lands can all too easily be turned to desert by overgrazing, and soil erosion can be a devastating consequence of attempted cultivation. The extent to which human beings are responsible for the spread of deserts throughout the world is debatable. Climatic changes, such as unusual and persistent droughts, are certainly involved in the process, but questionable agricultural practices can increase the risk of degradation. This interaction contributed to the US dust bowl of the 1930s, and recovery from it required massive changes in land management. Arid lands are generally sparsely populated, but the quarter billion or so people who live in such areas around the world are among the most vulnerable to climatic change.

At about 60°N, the convergence of arctic winds and the westerlies from lower latitudes creates an upcurrent leading to precipitation, just as at the equator. Of course, it is much colder at this latitude than at the equator, and the long winter allows for only a short season of plant growth each year. Coniferous trees make up the dominant vegetation in this boreal forest, or taiga, and plant diversity is much lower than in the temperate and tropical forests. Because the conifers are mostly slow-growing evergreens, recycling and the availability of nutrients is lower than in the deciduous forest. Nutrients are also depleted as water drains through soils that are wet for most of the year. This type of vegetation is almost absent from the southern hemisphere because there is little land around 60°S. However, taiga occurs in a band from North America through northern Europe and Asia, making it the second largest biome after the desert. Of course, coniferous trees are a valuable crop for lumber and paper making, but there is little prospect for large-scale production of other kinds of crops in this area.

In the tundra, north of the taiga, the ground is permanently frozen except for a surface layer throughout the year. The short melt period in the summer allows only a low vegetation of dwarf shrubs, sedges, and grasses to develop. Similar vegetation occurs in high mountains. Some of the plants have showy flowers that make the tundra briefly more colorful than the taiga. Alpine plants are attractive for specialist collectors, but there is almost no prospect for crop production in the tundra itself.

The biomes are described in terms of the vegetation that existed before human interference, which began several thousand years ago to change the vegetation to provide food directly for humans or for animals that could be eaten by humans. At this stage, almost all the biomes have been influenced to some extent by human activity and some have been almost destroyed. Grasslands have been the most extensively modified because of their value for agriculture. In the United States, only 2 percent of the tallgrass prairie remains. Temperate forest has nearly all been harvested at some time and much was converted to farmland, although small areas have returned to woodland in many parts of the United States and Europe over the past fifty years. These historic changes have been followed by clearance of the tropical forests in the twentieth century. This development is widely regarded as an ecological catastrophe, but it is not yet clear whether we can prevent it (or have any right to expect it to stop).

Even when small areas of ecosystems are allowed to remain in or return to their natural state, they do not function in the original way. One of the most pervasive causes of change has been the widespread practice of land drainage. All of the kinds of ecosystems that made up the biomes contained wet areas, which were often the most productive areas because water was permanently available. Wetlands also acted as ecological buffer zones, providing refuge for animals and sources of moisture for plant life during dry seasons. Drainage proved necessary for most kinds of crop production in moist temperate areas, but this practice left remaining natural areas more vulnerable to periods of drought.

Crop production has displaced many preexisting ecosystems, and it is more realistic to describe global ecology in terms of new biomes such as cereal fields, rice paddies, or pasture. Just about all of the land that can be cultivated easily is now used for the production of crops. With the pressure of world population, we need to produce more food. It is also desirable to increase the diversity of food available to many people who currently depend on one or two staple crops. Is it realistic to increase the land area available for crops? Substantial gains could be made only from biomes that have not been exploited so far. Although tomatoes can be grown in greenhouses in Alaska, the temperature limitations on crop production in the taiga and tundra cannot be overcome on a large scale. Lack of water can be overcome more easily and irrigation will continue to be important for crops in many parts of the world. Fresh water is a limited resource in all parts of the world where it could be most useful, however, and it is not realistic to contemplate massive conversion of desert areas to any kind of crop production. In addition, marginal areas are often the most vulnerable to environmental degradation. When plant growth is slow, recovery from disturbance is also slow, and other organisms that depend on plant life can easily lose their place in the ecosystem. The low productivity of the ecosystems limits the earnings of people that depend on them and can encourage overexploitation. Many argue that we should focus on increasing the productivity of existing crop areas rather than expand into currently unproductive land.

PRODUCTIVITY OF TERRESTRIAL ECOSYSTEMS

The relative productivity of different biomes has already been discussed. As we modify more of these areas for human use, it is important to know what effect this modification has on global productivity. The productivity of ecosystems is an indication of how much we can expect to harvest from them without causing loss of species or collapse of the whole ecosystem. Some of the most obvious examples of the failure to consider sustainable harvesting practices are the exploitation of fish stocks and the wood supply in many parts of the world. On a larger scale, productivity is related to the carbon balance of the world. Plant productivity can be defined in terms of the amount of carbon taken from the atmosphere by photosynthesis. When we change ecosystems for our own use, we may also change the rate at which carbon dioxide is removed from the air, which will have an effect on global climate. We are thinking increasingly in terms of changes that we can implement to maximize atmospheric carbon removal, one way of accomplishing this objective is to maximize plant productivity.

Photosynthesis fixes carbon into organic compounds, but this process is not quite equivalent to plant productivity. The plant uses a portion of the organic compounds for its own maintenance. At the end of a growing season, less carbon is retained than was originally fixed. Net primary productivity (NPP) is carbon fixed minus the amount consumed by the plant and returned to the atmosphere through respiration.

It is quite easy to measure the carbon accumulation by a single plant, but it is difficult to measure for complex communities consisting of many species under field conditions. Two-thirds of our planet is covered by water, and it is at least as difficult to estimate the productivity of the oceans as it is for the land. Estimates of productivity have varied greatly over the years, but it is now thought that although terrestrial ecosystems occupy about 30 percent of the surface of the planet, they account for about 55 percent of global photosynthetic productivity. The world's terrestrial ecosystems can be divided into three broad groups of high, intermediate, and low productivity (Fig. 2–8). Tropical forest and savanna are the most productive biomes and account for about half of terrestrial productivity, although they make up less than one-quarter of the land area. Temperate forest is almost as productive but contributes about 10 percent of total terrestrial productivity because it occupies only about 7 percent of the

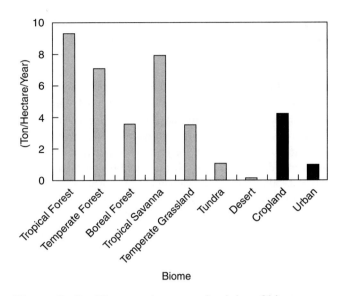

Figure 2–8 The net primary productivity of biomes represents the amount of carbon fixed in photosynthesis for a given area in a year. *Source:* Michael Knee.

land surface. Temperate grassland, boreal forest, and cropland are about half as productive as temperate and tropical forest and tropical savanna. The contributions to global productivity from these biomes with intermediate productivity tend to match their share of land area (8 percent for temperate grassland, 6 percent for boreal forest, and 10 percent for cropland). One-third of the land surface is occupied by tundra, permanent ice, desert, and semidesert, which contribute about 4 percent of global productivity. Urban areas have about the same productivity as tundra and occupy about 1.5 percent of land surface. In all parts of the world, wetlands tend to be the most productive areas; on average their contribution to productivity is three times their share of land area but because this is only about 1.8 percent, their total contribution is small.

In terms of carbon fixation, cropland is usually less productive than the biomes that it replaced. This is particularly true for annual crops because they do not fully occupy the land for most of the year. Of course, crop productivity is not usually the same as net primary productivity of an ecosystem. The two might be similar for a biomass crop that is harvested for energy production, and in a sense the productivity of ornamental plants is often the same as total biomass production. However, ornamentals are required to satisfy other demands than simply size and mass. Sometimes their growth may be deliberately restricted, as with dwarf varieties or the use of chemical growth inhibitors.

Usually crop productivity is measured in terms of a specific part of the plant that is harvested, rather than the whole plant. Part of the challenge of production is to maximize this part in relation to total biomass, or in other words to maximize the **harvest index**. In principle, annual crops are wasteful of primary productivity because most of the fixed carbon is discarded at harvest and must be fixed again in the following year to make a new corn or tomato plant. However, perennial crops must allocate part of the fixed carbon to their survival structures and if this is not the part that is harvested, less fixed carbon may be gained than from an annual crop. An old and cruel joke for apple growers is that their major crop is firewood. Some of the most productive crops are perennial plants from which the survival structures are harvested for consumption. Such crops have historically sustained people with limited land and resources, such as the Irish who grew potatoes in the nineteenth century, and the 10 percent of the world's population that today depends on cassava for its nutrition.

We have not yet exploited the inherent productivity of wetlands, with the exception of rice and a few minor crops, probably because it is more difficult to work in flooded fields than on dry land.

SUCCESSION

Descriptions of biomes can leave the impression that they consist of communities of plants and other organisms that would always be there in the absence of human interference. However, we know that organisms have evolved, landforms have changed, and climate has varied over time. The change in plant communities over time is known as **succession.** It is often described in terms of the development of vegetation from a blank-slate situation such as a rock surface, a pool of water, or bare soil (Fig. 2–9). These situations may have arisen because of natural causes (an earthquake, volcanic eruption, or the passage of a glacier) or because of human interference.

If there is no soil, succession must begin with soil formation. A rock surface can be colonized by lichens, which trap windblown dust particles and eventually decompose to form a thin layer of soil that can be colonized by mosses. Further soil accumulation allows herbaceous plants to move in. Water might be colonized by floating plants, which decompose and sink to the bottom of a pond that may also be receiving sediment from streams. As the pond becomes

shallower, bottom-rooting plants with emergent leaves can colonize, followed by wetland plants. If some kind of soil is already present, fast growing herbaceous plants will exploit the space but will give way to slower-growing plants that can use resources more efficiently. The composition of the ecosystem is always evolving toward the group of organisms that can best exploit the available resources at that time. Succession begins with pioneer species that can survive under special conditions and tend to get displaced as the ecosystem develops. As niches become available in the ecosystem, they are filled by the best available candidates.

The theoretical end point of succession is known as the **climax community.** At one time, this point was thought of as a permanent assembly of species that were linked together and would always return after any kind of disturbance. Now it is clear that there is no end point to succession and that species just happen to occur together, not because of necessary associations. The climax community can be seen as a group of species within the broad structure of the biome that is appropriate to the climatic conditions of the area. Over much of the eastern United States, this would be temperate deciduous forest with a mix of forest trees, shrubs, and herbaceous plants. Although we can characterize typical associations such as oak-hickory forest, the makeup of the ecosystem varies with location and has changed over time.

In spite of the difficulties inherent in the idea of a climax community, the broad idea of succession is sound and has implications for crop production. Many crop ecosystems are maintained at an early successional stage where they invite colonization by pioneer species, which we also know as weeds. It is easy to see that without cultivation, weeds would move in, followed by a thicket of woody plants, and the process would culminate in a forest of the longer-lived tree species. The further from the climax community we try to stay, the more work and inputs are required to maintain the ecosystem. The success of the corn belt can be explained by the fact that a mix of tallgrass species in the original prairie was replaced by another tallgrass that happens to be physiologically similar, that being a C_4 plant. In the temperate zone, we have been lucky often to be able to make drastic changes in the vegetation without suffering excessive losses of soil fertility and structure. In more difficult climates, the consequences have not been so positive, and it can be a good idea to model crop ecosystems on the broad structure of the climax vegetation.

Once a climax community is established, no other species may be able to invade because all the niches in the ecosystem are filled and no surplus resources are available to exploit. This outcome is related to the idea of the climax community as a well-defined group of mutually dependent species. However, alien species do invade natural ecosystems, often with quite disastrous consequences, and this outcome is part of the evidence that no necessary association exists between the species in a climax community. Long-standing species may be well-adapted to secure the resources that they need from the environment, but they are also vulnerable to pathogens and

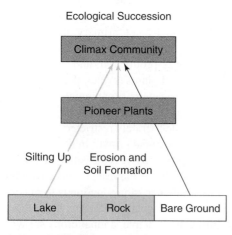

Ecological Succession

Figure 2–9   All locations in a geographic area tend to develop the same kind of vegetation over time. *Source:* Michael Knee.

herbivores, which are equally well-established in the environment. An alien species may be able to exploit the resources, but it may be immune to the diseases and unpalatable to the herbivores. These characteristics give it a competitive advantage until the pathogens and herbivores catch up with it.

Human beings have a long history of moving plants and other organisms around the world, deliberately or by accident. Many ecosystems have been irreversibly altered by introduced species. In recent years, the movement of organisms across national borders has become much more highly regulated and actively policed. Travel and trade have also increased greatly, however, and there is no sign that the flow of potentially damaging organisms has decreased.

SUMMARY AND REVIEW

Plants provide the energy that drives almost all terrestrial ecosystems. An ecosystem consists of populations of different species of organisms in a physical environment. Crop ecosystems require inputs because they usually lack some of the organisms and processes that sustain natural ecosystems. The organisms can interact in several ways to exploit the resources of the environment. Competition is likely to occur between individuals of the same species or closely related species. A single species cannot fully exploit all the resources in an environment. Resources are partitioned to provide niches that are occupied by different species. A natural ecosystem may not have any unoccupied niches, but crop ecosystems often provide niches that can be occupied by other plants. After their capture by plants, resources may be lost to herbivores and parasites. Crop ecosystems may lack organisms that naturally regulate these losses. Small animals, fungi, and bacteria help to recycle nutrients in natural ecosystems and make them available to plants. Nutrients are often withdrawn from crop ecosystems when the crop is harvested, and they cannot generally be sustained by natural recycling.

The natural ecosystems that occur in different parts of the world can be grouped together in biomes. Each biome is a collection of ecosystems with similar climate, soil type, and vegetation, and these three elements are interrelated. Temperature and precipitation are the main climatic factors. Temperature is determined mainly by latitude and height above sea level. Precipitation patterns are related to the predominant wind patterns that occur in three bands in the northern and southern hemispheres. Temperature and precipitation are consistently high in the equatorial region, leading to continuously high rates of growth and low nutrient availability in the soil for the tropical rainforest. The divergence of wind systems north and south of the equatorial zone leads to dry conditions and the low productivity of the desert biome. In the northern hemisphere, the moist temperate zone is occupied by deciduous hardwood forest, which is highly productive in summer but dormant in winter. On the major continental land masses, the region between forest and desert was naturally occupied by grasslands. While these areas were not as productive as forests, highly fertile soils developed, and they became some of the world's most productive agricultural areas. To the north of the temperate deciduous forest, coniferous trees predominate in the taiga, or boreal forest; further north still, the subsoil is permanently frozen and trees cannot grow.

Many of the ecoregions that were most suitable for crop production have now been destroyed. The remaining natural areas of the planet are too dry or too cold for agricultural exploitation. Terrestrial ecosystems account for about half of global carbon fixation, but this proportion has been reduced by conversion to agriculture and urban development. Soils and ecosystems develop over time in a process known as succession to a stable climax community. Most crop ecosystems are maintained at an early, unstable successional stage that requires significant energy input to maintain.

FOOD FOR THOUGHT

1. For a crop ecosystem of your choice, list the ecological interactions between and among organisms that could occur. (Do not forget the soil or growing medium!)

2. What happens to unfilled niches or underutilized resources in crop ecosystems? What problems do they cause?

3. It may seem strange that some of the most productive parts of the earth have not proved suitable for crop production and that many people live in unproductive areas. What can be done to address these problems?

4. Can you think of ways that crop ecosystems could be made as productive as natural ecosystems or their potential productivity could be more fully exploited?

SUPPLEMENTARY READING

CHRISTOPHERSON, R. W. 2004. *Geosystems: An Introduction to Physical Geography,* Fifth Edition. Upper Saddle River, N.J.: Prentice Hall.

CHAPTER

Human Impact on Ecosystems

Michael Knee

KEY LEARNING CONCEPTS

After reading this chapter, you should be able to:

♦ Understand the ways in which we modify ecosystems through our demand for plant products and services.
♦ Appreciate the scale of human impact on terrestrial ecosystems.
♦ Think of ways to balance population demand with available resources.

INTRODUCTION

In Chapter 2, we looked at the natural forces shaping the terrestrial ecosystems of the world and the interactions that occur within them. We considered how resources are allocated and cycled within ecosystems and began to see how human beings have modified the ecosystems. In many parts of the world, human beings are now the dominant organisms shaping the environment and utilizing its primary productivity. Ten thousand years ago, at the beginning of the agricultural revolution, early farmers could afford to burn vegetation to clear areas for crops without worrying about the ecological consequences (which were largely unknown, of course). At that time, perhaps 5 million people lived on earth. Today, with a global population of more than 6 billion people, the impact on the environment is impossible to ignore. As we try to fit even more people onto the planet (Fig. 3–1), we need to be more aware of our use of its resources. In this chapter, we will consider most of the ways in which we make use of plants and the ecological impact of that use. For institutional rather than scien-

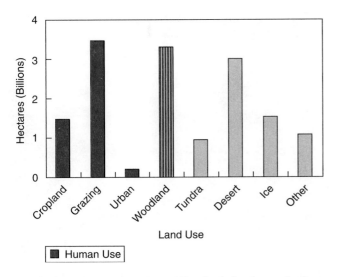

Figure 3–1 Global land use. The shaded columns indicate land used by people. Woodland is striped because it includes areas that are natural and unexploited, natural but managed or harvested, and planted purely for economic use. In a sense, all areas are used for carbon dioxide absorption, if nothing else. *Source: http//www.usgcrp.gov*

tific reasons, forest crops will not be included, although we will look to forests for the absorption of much of the carbon dioxide that we generate in our lives.

HUMAN FOOTPRINT

Our impact on natural ecosystems is both physical and biological. Worldwide we now deliberately choose or influence the organisms that inhabit about 35 percent of the land surface of the earth. As we saw in the previous chapter, much of the remaining area is desert, with low productivity and biological diversity (Fig. 3–1). Thus, we exclude or even deliberately exterminate organisms in many of the most diverse and productive regions of the earth. Some ecologists allege that we have started an era of mass extinction of species like those that have occurred in geological history from meteoric impacts or other physical causes. This proposition is somewhat controversial, and people often argue about the value of preserving individual species or populations of species. However, we cannot escape the fact that we are increasingly presented with the choice about whether to try to preserve plants and animals from extinction or to allow them to disappear. Having admitted the existence of this debate, we will spend the rest of the chapter focusing on the physical cost of providing the crop plants that we utilize or consume. The physical cost comprises the land and energy inputs required in crop production.

Land and energy requirements are the basis for two ways of evaluating the environmental impact of human activity. Accounting for energy inputs enables us to evaluate the efficiency of technological systems such as crop production, and we can look at energy demand in relation to availability on a national or global level. The land area required for the same activities is a measure of their contribution to the human footprint. The total **human footprint** is an estimate of the land area required by an individual, a geographic area, a sociopolitical group, or the world population taken as a whole. It accounts for the space required to provide all of the items we use in our lives and to absorb all of the wastes that we generate. Thus, crop production contributes to our footprint not just through the area required to grow crops, but also the areas required to provide inputs of fuel and chemicals, to transport the produce, and to absorb the wastes produced.

A footprint analysis often takes account of the area of land without regard for the intensity of use. Thus, a square meter of land occupied by a home is not differentiated from a square meter of forest that provides lumber or pasture that provides meat or dairy products. But the home uses the area almost entirely for human purposes, whereas the forest or pasture can support many other organisms. A more sophisticated equation is:

$$\text{Footprint} = \text{area used} \times \text{intensity of use}$$

This equation implies that we can reduce our footprint by reducing the area or the intensity of our activities. This calculation is particularly relevant to agriculture and horticulture, including landscape installation and maintenance. For many industries, the intensity of land use is 1, implying that they use 100 percent of the resource area. However, crop production need not completely displace various plants and animals from the space that is used. The global human footprint is said to exceed the area available by 20 percent. While it is desirable that some wilderness areas should remain comparatively free from human activity, it is unrealistic to expect that people withdraw from many of the areas of the globe that they now occupy. Although governments have generally supported the maximum intensity of crop production, they can begin to support greater biodiversity in crop production systems. This kind of policy would be consistent with attempts to rely on ecological processes rather than manufactured inputs to achieve production goals, such as maintenance of soil fertility or reduction in pests.

Much of our footprint on the planet is related to energy use. Most of our energy consumption is based on fossil fuels, and a footprint is associated with extraction, processing, and distribution of these fuels. However, a far greater footprint arises from the carbon dioxide produced as the fuel is burned to generate energy. This requires 8 hectares of forest for each person in the United States. Cropland makes a smaller contribution to carbon absorption because most of the carbon is returned to the atmosphere as the crop is consumed. The maintenance of forest for carbon fixation is a low-intensity

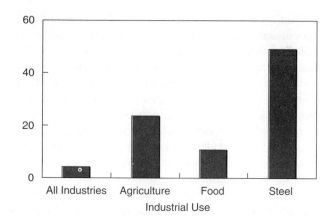

Figure 3–2 Energy use by US industries in 2000, in relation to contribution to gross domestic product. Agriculture appears to be energy intensive partly because profit margins are low. Food manufacture adds value to its raw materials and appears more energy efficient. *Source: http://www.eosdis.gov*

Figure 3–3 Energy in the food system includes production, transportation, processing, and storage. *Source:* Margaret McMahon, The Ohio State University.

component of our total footprint because we can allow natural ecosystems to perform this function. It is important that the most productive forests in equatorial Asia, Africa, and South America should not be converted to farmland. Indeed we should probably pay for them to be maintained because we use their services.

Agriculture accounts for only about 2 percent of US energy use, so it can make only a minor contribution to energy conservation. However, because it accounts for less than 1 percent of gross domestic product, it is using above average amounts of energy and has to be regarded as an energy intensive industry (Fig. 3–2). Production of fertilizer accounts for nearly half of agricultural energy use in the United States. If the whole world adopted a similar mechanized and high-input agriculture, it would require about 10 percent of current global energy consumption. Most of the energy used for our food supply is consumed after produce leaves the farm. It is difficult to account for all of the energy inputs between the farm and the dinner table, but the food sector has been estimated to account for up to 10 percent of total energy consumption. Surprisingly, less than 10 percent of this energy, or 1 percent of total energy consumption, is used in transport. Much of the energy is consumed in food marketing and preparation in the home (Fig. 3–3). If the rest of the world adopted this aspect of our lifestyle, it would use up about half of the world's energy supply.

Footprint and energy accounting are complementary and related to each other, although no simple equivalence exists between the two. As we saw, use of fossil fuels entails land use both on the supply and consumption side. In terms of supply, these fuels minimize our present-day footprint by using the past productivity of the earth. A decision to use renewable energy often enlarges our footprint, whether it is land devoted to wind farms, solar panels, or biomass production.

PLANTS FOR HUMAN USE
Nutrition

In addition to simple minerals and water, a healthy diet requires an energy source such as carbohydrate or fat, protein, linoleic acid and various vitamins, minerals, and water. Apart from the minerals and water, these requirements are complex organic molecules that animals do not produce. Plants are essentially the only terrestrial organisms that convert inorganic carbon, oxygen, hydrogen nitrogen, and sulfur to organic forms through photosynthesis. This characteristic is why they are called **autotrophs** (meaning "self-feeder"), whereas humans and just about everything else that live on land are **heterotrophs** (or "other feeders") because they feed off autotrophs or other heterotrophs.

Adults need 2,000 to 3,000 kcalories of energy per day, depending on their size and level of activity. This energy can be provided by carbohydrates, which are typically found in plant foods. So twenty-four to thirty-six slices of bread is one way to get our daily energy requirement. Lipids provide energy in more condensed form, and 9 to 14 oz of oil can provide our energy for the day. Fats are more characteristic of animal foods, but many health problems are associated with consumption of animal fats partly because they tend to be saturated fats, which lack double bonds in their fatty acid side chains. Unlike animals, plants can make polyunsaturated fatty acids (PUFAs) such as linoleic and linolenic acids. Apart from the health benefits of PUFAs, we need linoleic acid to make hormones such as prostaglandin, and plant oils are a more direct source than animal fats. We can also derive energy from protein-rich foods such as meat. This approach is wasteful and probably unhealthy because of the need to excrete the excess nitrogenous material. Most nutritionists believe that reliance on plant foods as an energy source is healthy because the energy providing carbohydrates is associated with indigestible fiber, which protects us from colon disorders and other so-called diseases of affluence. Of course, these benefits come only with unrefined carbohydrates, such as whole wheat or brown, unpolished rice. Purified carbohydrates such as sugar or starch are not beneficial and carry their own health risks.

An adult needs about 70 grams of protein in a day. Plant foods, especially cereal grains and pulses (peas and beans), often provide enough protein along with the energy that they supply. However a plant diet may not satisfy our protein requirement because it may be deficient in one or more of the amino acids that we require. The essential amino acids are leucine, isoleucine, valine, threonine, methionine, phenylalanine, tryptophan, and lysine; children also require histidine. Plants make these and the rest of the twenty amino acids that make up proteins, but plant proteins often have only a low proportion of some of the essential amino acids, particularly the sulfur-containing acid, methionine, and the basic amino acid, lysine. No common plant food provides the right balance of amino acids, so we would need to overeat to meet our dietary requirements. An exception is the so-called miracle grain quinoa (*Chenopodium quinoa*), which provides enough of the essential amino acids along with its energy content. Animals, being more closely related to us, have a similar amino acid composition and generally provide better-quality protein. Eggs are generally regarded as providing perfect protein and are given a protein score of 100. Individual plant foods tend to have protein scores of 30 or 40. A balanced diet can be achieved on a plant diet supplemented with a small amount of meat or other animal food. Alternatively, the deficiencies of individual plant foods can be remedied to a large extent by combining them. Grains tend to be low in methionine, whereas pulses are low in lysine. A mixture of a grain and a pulse provides a more balanced protein, perhaps with a score of 60 or 70. Many cultures in different parts of the world have based their diet on such mixtures: rice and soybeans in China, wheat and chick peas in the Middle East, corn and beans in South America.

Carbohydrates, lipids (fats), and proteins are the bulk constituents of our diet. We also need small amounts of other organic molecules, called vitamins. We have some ability to make our own vitamin D, depending on the exposure of our skin to sunlight. All of the other vitamins are plant or microbial products. Most vitamins can be obtained from fruits, vegetables, or grains. Grains and grain flours are a poor source of some vitamins, particularly if they are polished or refined. Vitamin C is found only in fresh fruits and vegetables. Vitamin A is derived from carotene, which is found in green, yellow, or orange fruits, vegetables, and grains. It is perhaps ironic that great publicity was given to the development of golden rice as a cure for vitamin A deficiency in the Third World, whereas we consume white sweet corn and asparagus in richer nations. Plant foods do not provide vitamin

B_{12} (cyanocobalamin), which is manufactured by bacteria, especially in the guts of ruminant organisms, so it is available to us in dairy produce and meat.

In addition to the organic components, we require six major inorganic nutrients (calcium, phosphorus, magnesium, sodium, potassium, and chloride) and seven micronutrients. All can be obtained from plant foods, although they are often more abundant in animal foods. We can assume most of the time that we are getting enough of the nutrients in the food that we normally eat, but calcium, phosphorus, iron, and iodine are the most likely to be deficient and are called critical nutrients.

If an adult needs 3,000 calories of energy per day and if average US yields can be achieved, it would take about 0.15 hectares of wheat or 0.06 hectares of corn to satisfy this dietary requirement for a year. To improve the protein quality of the diet, beans can be substituted for one-quarter of the grain intake. Again assuming average US yields of soybeans, this diet would take about 0.05 hectares. Yields of soybeans are about the same as those for wheat, so the total area required (0.16 hectares) would not change much on a diet of beans and wheat. However, corn yields are much higher, and a diet of corn and beans would take about 0.09 hectares. This diet would be boring, and it would lack important vitamins (particularly A and C) and minerals. United States Department of Agriculture (USDA) recommended intakes could be supplied by about 400 grams of fruit and vegetables per day, which would require another 0.02 hectares of land (once more assuming typical US yields). So the wheat-based diet would require a total of 0.16 hectares, and the corn-based diet 0.11 hectares (Fig. 3–4). These figures are optimistic because cereal yields in most countries are lower than in the United States, and soybeans are the highest yielding of the pulse

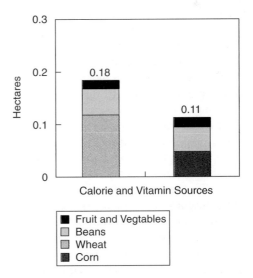

Figure 3–4 Amount of land required to provide an adult with vitamins from 400 g fruits and vegetables, 750 kcal from soybeans, and 2,250 kcal from either wheat or corn. *Source:* Michael Knee.

TABLE 3–1	Energy Costs of Supplying Fruits and Vegetables out of Season*			
Commodity	**Provided By**	**Energy Input (kJ/kg)**	**Food Energy (kJ/kg)**	**Relative Energy Cost***
Strawberries	Road from Florida to Ohio	1,400	1,250	1.12
Broccoli	Road from California to Ohio	2,800	1,110	2.53
Lettuce	Greenhouse production	230,000	500	460

* The relative energy cost is estimated by dividing energy input by food energy value. A positive value indicates that the energy put into producing and shipping a crop is greater than its calorie value.
Source: Michael Knee.

crops. The peas and beans that are mostly consumed by people have much lower yields. It is questionable whether the inputs that are used to attain US yields are physically available or environmentally desirable on a worldwide scale. The United States is exceptionally fortunate in having stable and productive soils and an excess of cropland relative to its population. Worldwide there are about 0.25 hectares (0.6 acres) of cropland for each person. With realistic inputs and yields, this land is about enough to provide everyone with a vegetarian diet.

In addition to the land required to grow crops, we should count the area needed to provide the inputs of energy and chemicals that are used in modern production systems. However, chemical inputs of fertilizer and pesticides actually reduce the footprint because they increase the yield from a given area. Area is also required to absorb the wastes generated by food production. In the past, this factor tended to be ignored, with the result that groundwater and surface water became contaminated. Now we might try to apply chemicals that minimize leaching and to provide some kind of buffer vegetation to prevent soil particles and fertilizer from running off from cropland into streams and lakes. Plant waste can often be recycled by composting or incorporation into soil or by feeding it to animals. Animal wastes are often spread on the land and can cause major problems if they cannot be recycled in the production of additional feed crops.

Crop production has become concentrated and specialized in regions and countries around the world. For example, California and Florida produce more than 60 percent of the fruits and vegetables grown in the United States, which means that crops are often transported thousands of miles between production and consumption points. Although some argue from an ecological perspective that we should consume local produce, the energy costs of transport are much lower than for out-of-season production in a greenhouse (Table 3–1). On the other hand, the nutrient flows arising from food transport do present some problems. The sewerage system transporting human wastes can also be viewed as part of the agricultural footprint. Although some municipalities have found ways of composting this waste, it cannot be recycled in food crop production because it is contaminated with heavy metals and other chemicals. Several countries in Africa and South America produce high-value horticultural crops for export to Europe and North America. These crops tend to have a high potassium content, so an essential plant nutrient is being exported and may not be easy to replace.

In moist temperate parts of the world, it is easy to forget how much water is needed to produce crops. It can take a ton of water to produce a kilogram of grain, and the water needed to produce our food is at least 500 times our direct intake. In dry areas, a large land surface may be needed to collect water to be used for irrigation to grow crops. This necessity causes political problems in places like California and outright hostility in the Middle East.

Each person in an industrialized country requires about 1 hectare of cropland, which is about 12 percent of their total footprint (Table 3–2). While it seems that food consumption is not a major part of the footprint, the area is far more than once thought necessary. Furthermore, if everyone lived in an industrial economy, agriculture would occupy all of the world's land. Because most of the presently unused land is too cold or too dry to support crop production, agricultural use is clearly impossible. The problem arises because people are not eating beans and corn, these crops are being fed to animals, and only 10 percent of the energy is being recovered in the meat or other animal products. Because of specialization, the crops are often produced on one farm and then fed to animals on another farm that may be hundreds of miles away. The crop grower needs fertilizer to maintain high yields, whereas the stock farm has the problem of disposing of animal wastes, which contain the same nutrients as the fertilizer. This method is apparently the most profitable way to raise animals, and it is cheaper to buy inorganic fertilizer than to take animal waste back to the corn fields.

TABLE 3–2	Ecological Footprint for an Inhabitant of Guernsey*
Resource	**Area Required (Hectares)**
Fossil energy	5.55
Crops	0.99
Pasture	0.59
Forest	0.84
Sea	0.31
Total	8.28

* Guernsey is a small island with an affluent population in the English Channel. It is easier to keep track of their consumption and waste generation than it would be for the whole of the United States.
Source: Reprinted from Ecological Economics, Volume 32, No. 1. C. Simmons et al., pp. 375–380, 2000 with permission from Elsevier.

Forage, Fiber, and Fuel

Ruminant animals can extract energy from plant materials that are indigestible for us, so if they are fed on forage crops or allowed to graze on pasture, the energy recovery can be as high as 50 percent of what we might have obtained from crops grown for human consumption. Soil problems and climatic limitations make it unlikely that we can increase the area of cropland substantially. However, much land that is unsuitable for crop production can provide grazing for animals. Thus, it should be possible to provide a varied diet for a world population advancing toward 10 billion if we reserve the best land for the production of crops for human consumption and confine animals to grazing on marginal land or limit their consumption to crop wastes. Grazing needs to be managed so that it does not lead to degradation of the soils and vegetation that supports it. On every continent, large land areas have been converted to unproductive scrub or near-desert through overgrazing. The difficulty of managing sustainable grazing is greatest in tropical countries, where the demand for food is often most acute. In principle, pasture can support higher levels of biodiversity than can fields from which crops are harvested. European countries are increasingly seeing the advantage of maintaining countryside that is ecologically diverse and also attractive for visitors.

Historically, farming provided many useful products in addition to food. Most of our clothing and textiles in general were made of plant and animal fibers. Animal skins were used to make most of our footware and some of our clothing. Agriculture made some contribution to structural materials used for houses and other buildings. Straw was used as a roofing material and combined with clay to make bricks. Forestry was (and still is) a much more important source of construction materials and fuel. Agriculture provided plant and animal fats that were burned for light but was of minor importance as a source of heat. Forestry and farming compete in providing paper and card; the cheaper products are usually made from wood pulp and the higher grades are made from herbaceous plant fibers such as cotton. Plants also provided dyes, perfumes, medicines, and other specialized chemical ingredients. Today most of these nonfood materials have been more or less replaced by synthetic materials, often produced from petrochemicals.

Only about 2.5 percent of cropland is devoted to fiber crops, both in the United States and worldwide. Cotton (*Gossypium spp.*) is the most important and best-known plant fiber crop. Jute (*Corchorus spp.*) is probably the next most important; it is grown in tropical countries such as Bangladesh and is used for coarse string and fabrics, such as burlap. The market for natural fibers is stagnant or declining. Although cotton is and is likely to remain a major crop, it is unlikely that production and the area of fiber crops will increase in the near future.

In the last few years, corn has become a major fuel crop in the United States. The grain is milled and its starch content is converted with enzymes into glucose, which in turn is converted to ethanol by fermentation. The ethanol must be concentrated by distillation, and the fermentation residue can be used as animal feed. The ethanol is added to gasoline to raise its octane value and prevent engine knock. Energy is used in the process, and the net gain in fuel energy may be as low as 50 percent. However, the electrical energy input could be from coal- or gas-burning power stations, so the process represents one way of converting energy to a form that can be readily used for transportation and reduces the need for imported petroleum. About 8 percent of the US corn crop is being used to generate less than 1 percent of the gasoline consumed. Even increasing its contribution to 10 percent would mean a major restructuring of US agriculture. Corn can also be imported for ethanol production, but this approach would increase dependence on imports and would come at the expense of food production in other countries.

Although corn is one of the most productive crops, it is not the ideal fuel crop. It requires high-quality land and inputs of fertilizer for efficient production, and the gasohol process uses the most nutritious part of the plant, which is less than 50 percent of the total dry weight. It would be more advantageous if the fuel crop could be grown on low-quality land and if the fuel could be derived from the whole plant or waste after high-value crops have been harvested. Various tropical grasses are being investigated as biofuel crops in warmer countries, whereas fast-growing woody plants, such as willows, are being used in temperate countries. Energy generation is usually by direct combustion of the dry crop to produce electricity. No economically viable process converts cellulose to ethanol at present, which is unfortunate because the critical need in the United States and other western countries is for fuel that can be used for transportation. Electricity demand can be met easily from coal and gas reserves for the foreseeable future.

Apart from fiber, the major nonfood use of crops is for oil-based products. Tropical oil crops, such as coconut (*Cocos nucifera*) and oil palm (*Elaeis guineensis*), are the main sources of oils used in soaps and detergents. Temperate oil crops, such as canola (*Brassica rapa*) and soybean (*Glycine max*), are not suitable for these uses but they are used in the manufacture of lubricants and plastics. Until the last century, most medicines were derived from plants. Today, about 25 percent of pharmaceuticals are based on plant products, and the rest are produced synthetically. Concerns about the future availability of petroleum and the pollution produced by the petrochemical industry have led to suggestions that we should return to using plants as chemical feedstock for detergents, plastics, pharmaceuticals, and other industrial chemicals. Biotechnology could be used to increase the production of existing chemicals or introduce new chemicals into crop plants. For this approach to be economically feasible, the chemical must be sufficiently valuable and the market must be large enough to recover development and production costs.

The current annual value of the US plastics industry, at $110 billion, exceeds that of major agricultural commodities, at $90 to $100 billion. Genetically manipulated canola and soybeans could be used to produce adipic acid, which is used in the manufacture of nylon. For agriculture to capture this $2 billion market, about 5 percent of the total US crop area would be required for canola and 15 percent for soybean. Small molecules like adipic acid must be processed to produce plastic polymers such as nylon. The direct production of plastics in plants is a more attractive possibility. *Arabidopsis* plants have been engineered to produce small amounts of polyhydroxybutyrate, which suggests that production of plastics in plants is feasible. Much development is required, however, and a major commitment of cropland would be necessary to make a significant contribution to the total consumption of plastics. The most rewarding applications of biotechnology may be in the production of high-value therapeutic and diagnostic proteins in plants. These proteins are impossible to produce synthetically and are required only in small quantities. So production of transgenic plants expressing proteins such as vaccines against malaria or rabies could be a profitable small-scale enterprise.

Before the industrial revolution, most of the fuels and raw materials (except for metals) used to manufacture everyday goods were of biological origin. These sources were gradually supplanted by coal and oil, but even in 1930 about 30 percent of raw materials came from plants. At that time, the dominant source was coal, whereas today over 80 percent of chemicals used in industry are petroleum products. Some suggest that this trend will be reversed so that 50 percent of chemical feedstock will be biological by 2050. However, this reversal will require a massive restructuring of the chemical industry and diversion of land from food production. The present industrial uses of crops help to raise the market price of agricultural commodities. While this trend is welcome for an industry that has seen a continuous fall in real prices for at least 100 years, market forces may limit the proportion of agricultural produce that can be diverted in this way. Some people argue that the remaining reserves of petroleum should be kept for chemical feedstock because it will be particularly difficult to find substitutes for all the diverse ways in which oil is converted to consumer goods.

Ornamental and Recreational Uses

After looking at the challenges of obtaining food and industrial resources from plants, it may seem frivolous to consider the ways in which plants give pleasure. However, strong reasons for consuming or using plants in these ways lead to some of the most profitable plant-based industries. In rich countries, food consumption is driven as much by hedonic (pleasure-giving) factors as nutritional requirements. We consume particular foods because of the sensory stimuli of appearance, texture, taste, and aroma that they provide. Some of the plant products that we consume provide plea-

sure but no nutritional benefit; they may even be harmful to our health. Tobacco and various forms of alcohol clearly fall in this category, although they also represent some of the most profitable ways of using land. Similarly, tropical countries can find it much more profitable to grow coffee or tea for export than to produce staple food crops. One of the most pervasive and insidious pleasures underlies our demand for sweet foods. Sugar provides energy but we do not really need it in our diet. Sugarcane in the tropics and sugar beet in temperate countries tend to be more profitable than staple food crops. In the United States, the bulk of the market for sweetener has been taken over by corn syrup. About 8 percent of the corn crop is used in this way, which (like ethanol production) helps raise corn prices.

Since neolithic times, human beings have coevolved with agricultural plants and animals. People have shaped plants and animals through selection and breeding, and these organisms have influenced our physiology, psychology, and sociology. As cities developed, people were less intimately involved with plants and animals in their daily lives. It is even possible to forget that bread, meat, or milk were once part of living organisms. However, people retain an affinity for plants and animals: our desire for pets and ornamental plants may be an aspect of a deep-seated human tendency that E. O. Wilson has called "biophilia." One of the most persistent, if minimal, expressions of biophilia is the desire for cut flowers or pot plants in the home or workplace. In cool-temperate areas, these plants are mostly produced in greenhouses. Because of the high market value of floral crops, greenhouse production can be more profitable than vegetables production. As we saw with vegetables, it is often cheaper in energy and economic terms to produce cut flowers in warmer countries and then transport them to the United States or Europe. Flowers are generally more perishable than vegetables, however, and it is usually necessary to ship them by air, so the energy advantage may not be so great.

Lack of space and limited transportation encouraged intensive development of cities in Europe in the nineteenth century. There was little room for plants where people lived. In the twentieth century, the United States developed a new kind of city with extensive open space between houses and other buildings. While 75 percent of Americans are said to live in urban areas, it is more accurate to say that most of them live in suburbs. While built-up land is said to account for 10 percent of the total land area, much of this area supports vegetation of one kind or another (Fig. 3–5). The suburban landscape has become, in effect, a major biome. The plants in urban areas tend to be those that we like to look at rather than those that we need for food or fiber. In areas that receive little attention, relics of native vegetation, with a proportion of weedy exotics, will persist. However, most of the area that is not under buildings or blacktop is covered by turf. Some of this area is devoted to organized sports, but mostly it is maintained for informal use or merely appearance. In ad-

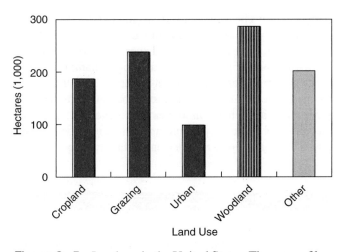

Figure 3–5 Land use in the United States. The areas of human use are shaded. Woodland is highlighted because some of it is unused (compare with Fig 3–1). *Source:* Michael Knee.

dition to turf, there are often shade trees and, in residential or office areas, shrubs and herbaceous ornamentals.

Nurseries, greenhouses, and sod farms provide plants for the urban environment. These crops with purely aesthetic value are produced in about 6,800 hectares of greenhouses and 280,000 hectares of open area, which corresponds to about 0.15 percent of the US crop area or about 1 percent of the area used for corn. Yet the combined value of the ornamental crops is about 50 percent of the value of corn. Ornamental crops are much more labor intensive than are other crops, accounting for nearly 20 percent of US agricultural labor costs. They may also require more intensive use of fertilizers and pesticides than does traditional agriculture. Cities and suburbs themselves occupy an area equivalent to about half the area of cropland in the United States. Built-up areas are expanding at a rate equivalent to 0.5 percent of crop area per year. Some argue that this expansion occurs at the expense of agriculture. However, urbanization will increase farm prices by taking land out of production and reducing the supply of agricultural products.

The scale of urban areas and the cultivated landscapes within them means that agriculture can no longer be re-

garded as the only significant source of pollution from fertilizers and pesticides. About the same rate of fertilizer application is recommended for turf as for corn or other cereals. Environmental Protection Agency (EPA) data imply that more pesticides, mostly herbicides, are applied to lawns than to agronomic crops. The productivity promoted by these chemical applications is mostly unwanted. Fruits, seeds, and even flowers are regarded as a nuisance when they drop from ornamental plants onto sidewalks. Almost every homeowner is encumbered by a surplus of grass clippings throughout the summer and leaves in the fall. In the last century, farming became increasingly specialized, so single crops are grown over large areas. Turfgrasses reproduce this monoculture situation in the landscape. Species and cultivars have been selected for climatic zones, but still a large part of the economic and environmental cost of landscapes results from the attempt to maintain green, weed-free grass in unfavorable climates. Ornamental horticulture has developed in the opposite direction: with a tremendous variety of trees, shrubs, and herbaceous plants being grown. A single landscape may include plants from different climatic zones: spruce from the boreal forest, cannas from the tropics, yucca from the desert, and hostas from the temperate forest. The downside to this humanmade diversity is that, after introduction to the United States, some ornamental plants have escaped to become agricultural weeds or have invaded natural areas.

Early European settlers struggled to grow food and make a congenial home out of the North American forest and prairie. Their energies went into removing native vegetation, draining soils, and eliminating animals that were regarded as pests. Most of the native flora and fauna did not seem attractive or appeared to be a threat to people, crops, or domestic animals. New animals and plants were brought deliberately or by accident to repopulate the land. Some people have argued recently that we should use native plants to try to create landscapes with a more distinctively American style and heritage. However, strong social and cultural reasons lead most people to continue to plant landscapes using the introduced species.

SUMMARY AND REVIEW

Agricultural and other human uses now dominate the ecology of nearly all of the most productive areas of the earth. Crop plants and farm animals have replaced the organisms that inhabited these areas, even to the extent that we are causing extinction of other species. We can summarize our impact on global ecology in terms of the land or the energy that is used for human purposes, including crop production. Land use is summarized in the concept of the human footprint. While crop production represents a small fraction of total footprint and energy use, the intensity of land and energy use

is higher than for other industries. The food supply system uses up to four times more energy after it leaves the farm than was used in production. The human nutritional requirements for carbohydrate, fat, protein, vitamins, and minerals can be met from plant sources, with the important exception of vitamin B_{12}. Assuming average US yields, adequate nutrition can be obtained from the less than 0.25 hectares of cropland that is available for each of the 6 billion people on the planet. Energy and nutrients are lost when crops are fed to animals, and less people can be fed in this situation. However, meat and dairy products can be produced by animals that graze on land unsuitable for crop production. The wastes generated by the food production system and by human and animal consumption of the food are part of the human footprint. Historically, farming provided raw materials for the manufacture of textiles, fuels, and various chemicals, including dyes, perfumes, and medicines. Today, most of these materials are produced from synthetic (usually petrochemical) sources. As oil supplies diminish, however, there is renewed interest in plant sources. The production of sugar, alcohol, and tobacco can be highly profitable; they provide enjoyment, but they are unnecessary or even harmful. Other crops are appreciated for their visual appeal. For the urban population, flowers, turfgrass, shrubs, and trees are an important part of the environment. These crops are at least as demanding as agriculture for energy and chemical inputs and generate significant quantities of waste. Depending on aesthetic taste, landscape plants can maintain or displace native biodiversity.

FOOD FOR THOUGHT

1. What kinds of information can influence a decision to pursue a new development in crop production that might endanger populations of naturally occurring organisms? Consider some specific examples, such as plowing up or improving grassland (effects on native plants and ground-nesting birds), use of a new pesticide (effects on nontarget organisms), or landscaping a long-abandoned industrial site (refuge for many wild plants and animals).

2. Ohio State Dean of Agriculture Bobby Moser has said, "[We] are learning to tread lightly on the land." Can you think of examples of how we are accomplishing this objective right now and how we can go even further?

3. Eighty percent of our food energy comes directly or indirectly from six major crops (wheat, corn, rice, sweet potato, white potato, and cassava or manioc). Many people in the world eat little other than one of these crops, whereas others eat a lot of meat from animals that have been fed on these crops. Do you consider this situation to be a problem?

4. Do you believe the predictions about returning to biological sources for fuel and chemical raw materials over the next fifty years? What adjustments in crop production and in society would be necessary?

5. What factors influence the selection of ornamental plants for our landscapes? Are any of these factors related to ecological criteria?

SUPPLEMENTARY READING

Chambers, N., C. Simmons, and M. Wackernagel. 2001. *Sharing Nature's Interest: Ecological Footprints as an Indicator of Sustainability.* London: EarthScan Publications.

Wackernagel, M., and W. Rees. 1996. *Our Ecological Footprint: Reducing Human Impact on the Earth.* Gabriola Island, British Columbia: New Society Publishers.

Wilson, E. O. 1996. *Biophilia.* Cambridge, Mass.: Harvard University Press.

on geological conditions, water can remain in these aquifers for many years, or it may gradually flow back to the surface through springs or into surface bodies of water.

The **hydrologic cycle** is a global cycle in which evaporation and precipitation are exactly equal over time and on a global scale (Fig. 4–5). Everyday observation tells us that the balance between the two processes varies from day to day in any one relatively small place. Persistent imbalances also occur so that one process exceeds the other over long periods of time. The relationship between precipitation and evaporation is a major factor defining the climate of a region and its suitability for plant or crop growth. The Köppen system of classification of climate combines temperature and moisture conditions: A is tropical and rainy; B is dry; C is warm, temperate, and rainy; D is snow forest; and E is polar. People live and crops are grown primarily in zone C. Although a lot of water occurs in zones D and E, the water is mostly unavailable because it is frozen; these water zones are in effect ice deserts. Most of the earth's land surface falls in zones B, D, and E.

If we want to find out the suitability of an area for crop production or determine the day-to-day water needs of crops, we need to know the relationship between the supply of water or precipitation and the demand in terms of evapotranspiration. Precipitation is fairly easy to measure with a rain gauge, but we may need to account for losses through runoff or drainage. Evapotranspiration is more difficult to estimate because it is influenced by several factors. In a closed system, evaporation from a water surface is determined by the water content of the air above the surface. The amount of water in the air can be expressed as a percentage by volume, percent relative humidity, or partial pressure. **Relative humidity (RH)** expresses the actual water content relative to the maximum amount that the air can hold at a given temperature, which is known as the **saturated vapor pressure (SVP).** Partial pressure indicates the contribution of a gas to the total pressure of the atmosphere, which is around 102 kPa (15

TABLE 4–1	Effect of Temperature Change on Vapor Pressure Deficit at a Relative Humidity of 70 Percent	
	10°C	**30°C**
Vapor pressure (kPa)	0.86	2.97
Vapor pressure deficit (kPa)	0.37	1.27
Saturated vapor pressure (kPa)	1.23	4.24

Source: Michael Knee.

lb/in.^{2}) at sea level. At a given temperature, air comes to equilibrium with water and RH is 100 percent at all temperatures, but the percentage volume that the air can hold, or SVP, increases with temperature. It is also possible to identify a temperature for any actual vapor pressure or percentage volume at which RH would be 100 percent; this point is called the dew point. RH tends to be 100 percent when it has just rained or when the temperature drops and the dew point is reached because the partial pressure becomes equal to the SVP. At other times, RH is less than 100 percent and the difference between the actual vapor pressure and the SVP is called the **vapor pressure deficit (VPD).** Evaporation and plant transpiration are influenced by VPD rather than humidity. Much more water evaporates at 70 percent humidity at 30°C than at 10°C because the VPD is much higher (Table 4–1).

The rate of evapotranspiration at a given VPD is increased with the input of solar radiation, temperature, and air movement. Researchers such as Penman and Monteith have developed complex equations that relate evapotranspiration to these factors. Both expected and actual values for evapotranspiration have important implications for crop selection and management. Potential evapotranspiration (Et$_0$) specifies how much water would be lost if it was available. This loss would occur from an ideal open surface of pure water in equilibrium with the environment. Standard evaporation pans located in a crop production area can be used to estimate Et$_0$, but it is more usually calculated from data provided by a weather station. Actual Et may be much lower than Et$_0$. The water use of a crop can be expressed as a crop coefficient, which is the fraction of Et$_0$ that the crop would transpire when well-supplied with water. This coefficient is typically 0.6 to 0.8 for plants in the temperate zone. Water evaporates from the soil, and for a bare surface of saturated soil, the rate may be similar to ET$_0$. However, evaporation decreases sharply as the soil dries out, and the moisture is drawn from lower and lower levels. When soil is covered by vegetation, air movement and input of solar radiation tends to be low at the soil surface. The humidity under these conditions is higher than above the crop, and evaporation from a wet soil surface may be 10 percent of ET$_0$ or less.

The balance between evapotranspiration and precipitation defines whether the climate is moist or dry and whether irrigation is likely to be necessary to produce crops. In the United States, east of about 95°W, there is a surplus of

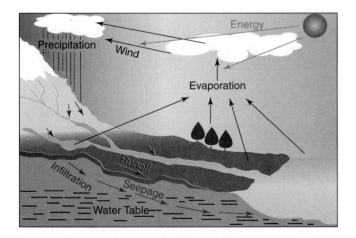

Figure 4–5   The hydrologic cycle. *Source:* Michael Knee.

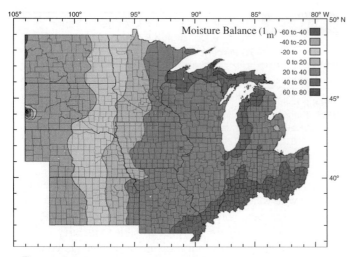

Figure 4–6   The balance between potential evapotranspiration and precipitation in the central United States. East of 95°, precipitation exceeds evapotranspiration. Precipitation is spread nearly uniformly only throughout the year. However, high evapotranspiration rates in the summer often exceed precipitation during that season, causing drought conditions at times. *Source:* USDA.

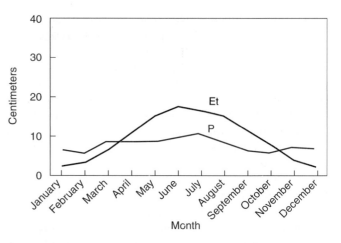

Figure 4–7   Seasonal variation in precipitation and evapotranspiration in Indianapolis, Indiana. *Source:* U.S. Department of Commerce.

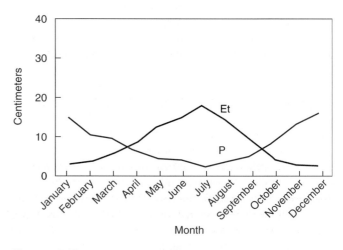

Figure 4–8   Seasonal variation in precipitation and evapotranspiration in Seattle, Washington, which has a Mediterranean climate. *Source:* U.S. Department of Commerce.

precipitation over Et_0, but rainfall is spread throughout the year and Et_0 regularly exceeds precipitation in the summer (Fig. 4–6 and 4–7). The situation is more extreme in areas such as the Pacific Northwest, where there is a Mediterranean climate and most of the rain falls in the winter (Fig. 4–8). (A **Mediterranean climate** is characterized by cool, wet winters and warm, dry summers.) Soil typically holds about 100 cm of water in the plant root zone. After the soil is saturated, the excess water runs off or drains away. It is important for spring planting that the soil should be fully charged with water, but waterlogged conditions are harmful for most crops. In many parts of the United States, it is necessary to install drainage systems to get rid of a seasonal excess of water. Successful water management involves regulating the availability of water to meet evaporative demand. Of course, if water is not present, evaporation does not occur and actual Et is much lower than Et_0. For most crops, this situation means that in the short term, they stop growing; if the drought persists, they will die.

Although most of the world's crops are grown in areas with adequate rainfall for plant growth, water availability is the one aspect of climate that can be and is modified on a large scale. Irrigation has a long history. Water was supplied to crops through open irrigation channels in the Middle East in prehistoric times. The water flowed under gravity to furrows in the fields much as it does in many parts of the world today. Now, large areas are often irrigated with water piped under pressure to some form of overhead irrigation. This method can result in losses because the water may fall outside the crop area and a proportion evaporates before it reaches the root zone. Smaller-scale irrigation of horticultural crops, including ornamentals, often aims to supply the water more directly to the plant roots through trickle-feed pipes or spray emitters close to the soil surface. In a moist climate, water can be held in local ponds or lakes to supply irrigation during periods of water deficit. In a dry climate, irrigation may supply the entire water needs of the crop. The water may be drawn from a distant source or from underground aquifers.

Pests and diseases may be less troublesome for irrigated crops grown in dry climates than in moist areas where the crop itself might grow more readily. On the other hand, the world's fresh-water supplies are limited, and it is getting increasingly difficult and expensive to maintain water supplies in the western United States and many countries. It is important to supply water efficiently so that enough reaches the plant roots and as little as possible is lost through evaporation, runoff, or drainage. Irrigation managers increasingly use estimates of evapotranspiration to decide how much water to apply. However, conservative use of water itself leads to problems in dry climates. Irrigation water adds small amounts of various salts to the soil. If these salts are not washed away by natural rainfall, they can build up to concentrations that inhibit root growth. Large areas of land have been damaged by salinization in many parts of the world.

An alternative to irrigation is to select a crop that has a low water demand. This approach is particularly relevant for landscape horticulture in the southwestern states, where water supply is at a premium. Researchers are trying to develop xeriscapes using locally adapted plants to replace those that are more appropriate for the eastern United States. Conversely, wetland plants are making a comeback in the midwestern states. The heavy postglacial soils of this region were drained over the past two centuries when agriculture was introduced, leading to the elimination of wetlands. This development led to its own problems of erosion and flooding

as water ran directly off fields, carrying soil particles into drainage channels that could not cope with the sudden flow. Wetlands act as a buffer zone, temporarily holding water to spread the flow and catch sediments. Furthermore, wetland vegetation removes nutrients from excess fertilizer washing into the drainage water and helps to prevent eutrophication of lakes, ponds, and waterways.

As with other climatic variables, plants adapt to the water availability of the region where they originated. Sometimes plants have developed drought-avoidance strategies. For example, many plants in the Mediterranean area grow and flower during the late winter when it is wet and go dormant over the summer when it is dry. Many of our ornamental spring flowering bulbs originate from this region. Other plants tolerate drought through various physiological mechanisms. Many desert succulents minimize transpiration by absorbing carbon dioxide at night and keeping their stomata closed during the day. Some grasses, called warm-season grasses, have a mechanism that lets them reduce transpiration and water loss but still maintain high photosynthetic rates during prolonged hot and dry spells. Corn, sugarcane, and several turfgrasses and forage grasses fall into this group.

SUMMARY AND REVIEW

Climate is not the same as weather. It is defined by the prevailing weather conditions of an area. The input of solar radiation influences all other aspects of climate. Vegetation and human interference modify climate, which in turn influences natural vegetation and the possibilities for crop production in an area. Most of the solar radiation reaching the earth's surface is infrared, and about 40 percent is photosynthetically active radiation between 400 and 700 nm. Away from the equator, the intensity of solar radiation decreases and the duration of daylight varies throughout the year. Cloud cover further decreases light intensity. Plants are adapted to the light conditions of their natural habitat. Day length influences processes such as flowering, leaf-fall, and bud-break. The spectral quality of light reflecting from or passing through foliage changes, which affects the growth of neighboring plants. It is uneconomic to supply high-intensity artificial light on a large scale. Supplementary lighting may be used to control development and flowering of greenhouse crops.

Only a small proportion of the world's water is available to plants, but this amount is continuously replenished through the hydrologic cycle. Most crops are grown in the small proportion of the earth's surface with a moist temperate climate. The availability of water is determined by the supply (rainfall and irrigation) relative to the demand, which can be summarized as evapotranspiration (Et). Potential evapotranspiration (Et_0) is influenced by solar energy input, temperature, air movement, and vapor pressure deficit (VPD). VPD is one way of expressing the dryness of the air and is the difference between its actual water content and the

water holding capacity, both expressed as partial pressures. Actual Et is usually less than ET_0. It is influenced by water availability and the nature of the vegetation. Seasonal estimates of rainfall and Et can be used to determine what crops to grow in a region and whether irrigation will be necessary. Daily rainfall variation can be used to manage irrigation supply. Water availability is the only aspect of climate that can be modified on a large scale for crop production. However, increasing problems of water availability and salinization of soils result from irrigation in many parts of the world. Drainage is sometimes necessary to manage seasonal excess of water, but maintenance of wetlands can help to control flooding and the pollution of lakes and waterways with nutrients and sediments. Plants differ in their requirements for water, and water management can involve selection of plants with low water needs.

As with other aspects of global ecology, we have to think of crop plants as part of the hydrologic cycle. Thus, we must consider how the demand for water for crops can be met without disrupting the cycle. We also need to think about the effects of crop production on water quality. Fertilizers and pesticides applied to crops can be transported through groundwater and runoff into streams and rivers. The timing of the application of a chemical in relation to rainfall affects how much reaches its target in the plant and how much enters the hydrologic cycle.

Although it is necessary for plant growth, water can cause problems. We already mentioned the need for drainage in many growing areas. Excessive rains can still cause flooding, and plants vary in their tolerance of partial or complete submergence. Gases diffuse much more slowly in water than in air, and flooded plants tend to be injured by oxygen deficiency and excess carbon dioxide. Hail storms can devastate almost any crop, which is why growers may specifically insure against hail damage. As you will learn in Chapter 6, water is an important agent in the weathering processes that lead to soil formation. In the eastern half of the United States, water is also the main agent of soil erosion. On slopes and difficult soils, erosion can be a major constraint on crop production.

FOOD FOR THOUGHT

1. How would sloping ground affect solar radiation available to crops at different latitudes?

2. To what extent can slope compensate for the decrease of available light with increasing latitude?

3. Both temperature and day length vary with season. Why might it be safer for a plant to adjust its growth and development according to day length rather than temperature?

4. Crops are increasingly being grown under no-till conditions where ground is not plowed between harvest of the last crop and sowing the next. How do you expect this approach to affect the local hydrologic cycle?

5. A plant in the sun loses more water than does one in the shade, even if the atmosphere is at the same temperature and humidity in the sun and in the shade. What are some possible reasons for this difference?

6. Can you think of ways to avoid some of the problems of crop production in arid lands mentioned in the chapter?

SUPPLEMENTARY READING

CHRISTOPHERSON, R. W. 2004. *Geosystems: An Introduction to Physical Geography,* Fifth Edition. Upper Saddle River, N.J.: Prentice Hall.

CHAPTER

5

Climate— Temperature, Air Movement, and Composition

Michael Knee and Margaret McMahon

KEY LEARNING CONCEPTS

After reading this chapter, you should be able to:

- ◆ Understand the interaction between climatic variables.
- ◆ Understand how temperature, air movement, and air composition vary from place to place.
- ◆ Appreciate what can be done to modify these factors to improve crop growth.

As we have seen, climate and weather are largely determined by the input of solar radiation, which drives all the other climatic variables. Although sunlight and water are vital for plant life, temperature may be the most important factor in the decision about whether a crop can be grown or when to plant a crop in a given location. We generally think of sunshine as providing warmth, but wind and water move huge amounts of thermal energy around the planet. The air that we breathe is maintained by plants, and plants have a key role in the cycling of nitrogen and carbon from the air through the global ecosystem, which includes our food supply.

TEMPERATURE

In Chapter 4, we saw how the amount of solar radiation reaching the earth's surface decreases with increasing latitude. The energy balance of the earth is maintained by radiating energy at longer wavelengths back into space. Although the total amounts received and radiated are equal, the balance varies across the earth's surface. Radiant heat loss does not vary with latitude as much as solar radiation. Thus, a net energy gain occurs at the equator and a net loss occurs at higher and lower latitudes. Gain and loss are equal at about 35° to 40°. These imbalances are the underlying reason for decreasing temperatures as we move north or south from the equator. Temperature variations are moderated by convection currents in the oceans and atmosphere that carry huge amounts of heat from the equator to higher and lower latitudes. The amount of energy received in this way varies greatly according to geographic position relative to ocean currents and prevailing winds. Thus, western Europe enjoys mild winters because of the northward drift of water in the Atlantic, whereas the central United States suffers from blasts of arctic air coming down from Canada.

Temperature is affected by altitude. As with other gases, the temperature of air drops as pressure decreases through adiabatic cooling (which is the basis of mechanical refrigeration). The pressure of air at a given location is generated by the mass of air above that location. As air rises, its pressure decreases and its temperature drops. In dry air, the decrease is about 10°C for 1,000 m (or 5.5°F for 1,000 ft); in moist air, it is 6°C for 1,000 m (or 3°F for 1,000 ft). This feature means that temperate fruits and vegetables can be produced at high altitudes in tropical countries such as Kenya.

The earth's surface absorbs solar energy and radiates it back at longer wavelengths. Water vapor can absorb this infrared radiation. Thus, cloud cover limits radiative heat loss, which can be particularly important at night. In areas with little cloud cover, temperatures can swing violently between day and night, a feature often noticed by desert travelers. Conversely, severe freezes almost never occur in the wet winters that are characteristic of Mediterranean climates. The variation in winter weather in temperate climates is partly a matter of the degree of cloud cover. However, convective heat transfer is an important factor in both weather and climate. Water and air provide the two media for convective heat transfer, which can occur at any scale between a field or city block and a whole country or continent. Thus, northern Europe is warmed by the Gulf Stream from the south Atlantic, but the west coast of the United States is cooled by the California current from the north Pacific. Alberta clippers bring sudden freezing temperatures to the central United States, but chinook winds can cause sudden snowmelt on the east face of the Rockies.

It takes more heat input to raise the temperature of water than that of the ground, and water cools down more slowly. Thus, large bodies of water can moderate temperature changes in adjacent land during the day and across longer time periods. The Salinas Valley in California is more favorable for temperate vegetable production than the San Joaquin Valley because it is up to 10°C cooler in summer. Yet Salinas is only slightly closer to the Pacific Ocean (Fig. 5–1). On the other hand, grape production is possible around the eastern shore of Lake Erie because winter temperatures are up to 5°C warmer than a few kilometers inland. Although temperature generally falls with altitude, this pattern can undergo a temporary inversion, which can have local effects on crops growing on slopes or on undulating ground. This type of change tends to happen on cloudless nights when heat radiates from the ground to the sky. Cold air then flows down the slope or into hollows, leaving warm air above (Fig. 5–2). Crops at the bottom of the slope or in the hollows may be damaged or killed by the sudden cold, whereas those growing a few feet up the slope may be undamaged.

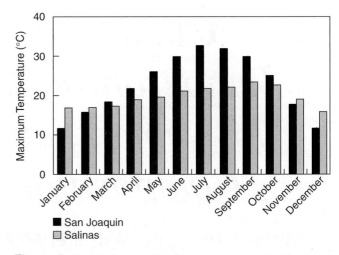

Figure 5–1   Salinas is slightly nearer to the Pacific Ocean than San Joaquin and enjoys much cooler summers. *Source:* Michael Knee.

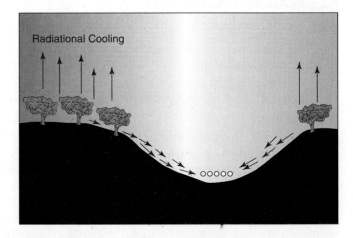

Figure 5–2   On a clear night, heat lost from the ground leads to cooling of the air, which rolls down slopes and leaves warmer air above it. *Source:* Michael Knee.

Urban development has a major effect on temperature at all times of the year. In winter, buildings radiate heat to the surrounding environment. In summer, buildings and paved surfaces absorb solar radiation and release it as so-called sensible heat over time. In the suburbs and countryside, much of the radiation is absorbed by vegetation and is used to drive transpiration. Most of the rainfall is drained away from urban areas as soon as it falls. Because of the absence of standing water and vegetation, the city does not enjoy the evaporative cooling that lowers the temperature in rural areas (Fig. 5–3). At all times of the year, buildings provide a barrier to air movement, and carbon dioxide emissions form a blanket that reflects long-wave radiation back to earth. The cumulative result of these factors is that cities can be as much as 5°C warmer than the surrounding countryside. A dome of warm air exists over cities, forming the so-called urban heat island. Warm air flows out toward the suburbs, which can be 2° to 3°C warmer than the countryside.

A defining feature of the climate of the temperate zone is that temperatures are not conducive to plant growth throughout the year. The seasonal variation in temperature has different implications for annual and perennial plants. Winter temperatures are unimportant for annuals, which survive the winter season as seeds. Perennial plants have two strategies for winter survival. Many herbaceous perennials die down to ground level; they have underground storage structures with buds, from which new growth begins each year. These plants avoid the coldest weather because soil temperatures do not fall nearly as low as air temperatures. Woody perennials and some herbaceous plants survive above ground. Their stems become dormant and acquire a degree of cold hardiness that varies with plant species. Thus, a plant's survival depends on temperatures never falling below the minimum that it can tolerate. US plant hardiness zones are defined according to the minimum temperatures that can be expected. Over the years, growers have gathered expe-

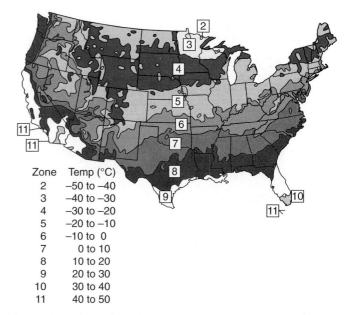

Zone	Temp (°C)
2	−50 to −40
3	−40 to −30
4	−30 to −20
5	−20 to −10
6	−10 to 0
7	0 to 10
8	10 to 20
9	20 to 30
10	30 to 40
11	40 to 50

Figure 5–4   United States Department of Agriculture (USDA) plant hardiness zones are defined by usually expected minimum temperatures, not the average winter temperature or the extreme lows recorded.

rience about the minimum temperatures that particular plants can survive (Fig. 5–4). We can match plants to zones and generally be confident of their survival. However, the plant hardiness zone map shows normal minimum temperatures; occasional extreme temperatures can still result in plant death.[1]

Springtime temperatures are critical for both annuals and perennials. Many seeds can be planted only after the danger of frost has passed, and spring-flowering perennials, including many fruit crops, can be injured by late frosts. Maps showing expected dates of last frost can be used to decide when to sow seeds and to select areas suitable for fruit production (Fig. 5–5). Other maps can show the expected dates of first frost in the autumn. For many crops, this point marks the end of the growing season. If a mature crop cannot be harvested by this time, there is little point in growing it (Fig. 5–6). A cotton plant grows well in the summer in Ohio, but the growing season is just not long enough to allow mature cotton bolls to be harvested.

Plants generally show a parabolic relationship between growth and temperature. A species has a minimum temperature, below which growth does not occur; an optimum temperature for growth; and a maximum temperature, above which injury occurs. The optimum varies according to the geographic origin and season of growth of the plant. For most plants in the temperate zone, the minimum is between 0° and 5°C, and the maximum is around 35°C. Many tropical plants are injured below 10°C and show optimal growth around 35°C (Fig. 5–7). Spring-flowering plants and winter

Figure 5–3   It feels hotter in the city than in the country because heat is absorbed and radiated by buildings and paved surfaces rather than being used to drive evapotranspiration.

_____
[1]The United States Department of Agriculture (USDA) plant hardiness zone map is being revised. The revision is expected to be published in late 2005.

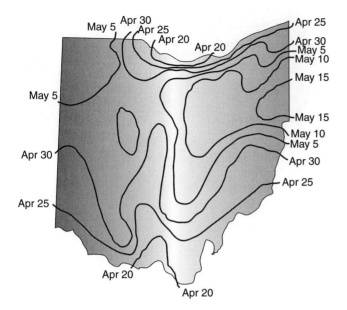

Figure 5–5   Variation in date when there is a 50 percent chance that no more frosts will occur in Ohio. Similar data is available for other states at *http://www.ncdc.noaa.gov/oa/climate/stateclimatologists.html.*

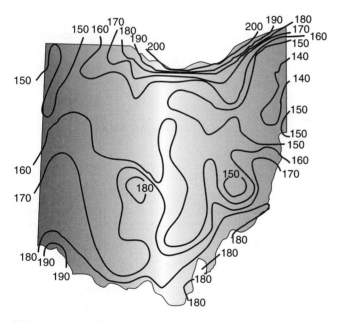

Figure 5–6   Map showing the growing seasons for the state of Ohio. The growing season is defined as the time between the average dates for the last frost in the spring and the first frost in the autumn. Similar data is available for other states at *http://www.ncdc.noaa.gov/oa/climate/stateclimatologists.html.*

annual weeds may grow well at low temperatures. (The America Horticulture Society [AHS] has produced a heat zone map but it has not been as widely used as the plant hardiness zone map mainly because we know less about the heat tolerance of plants than about their cold tolerance.)

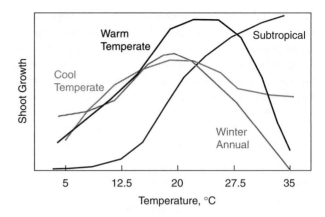

Figure 5–7   Growth of different kinds of plants in relation to temperature.

Many methods have been used to manipulate the temperature around growing crops throughout their growth or at critical times, and temperature manipulation is the major function of greenhouses. Methods can range from unheated, plastic-covered structures to oil- or gas-heated glass houses. The simplest greenhouses reduce heat loss by convection and radiation. In the past, various kinds of cover and wind protection have been used to promote plant growth early in the spring. Now, the most commonly used methods involve polyethylene covers, which can be applied to the soil to raise its temperature for seed germination or used in the form of a tunnel covering the whole plant, at least for part of its growth cycle. This method can be enough to prevent injury in marginally hardy crops during the winter or to prolong the growing season for fruits, vegetables, or flowers. More sophisticated greenhouses with full temperature control may allow for year-round production of certain crops. By their very nature, however, greenhouses are poorly insulated structures, and even floral crops are not sufficiently valuable to offset the energy costs of providing heat. Furthermore, the available light may be a limitation on production in the winter. Much of the heat requirement is at night. On a sunny day in winter, the heat gain of even the most primitive greenhouse may require ventilation to lower the temperature.

Local temperatures can be altered by the presence of a body with a high thermal mass that absorbs solar energy during the day and radiates it over time. In natural situations, this body could be a cliff face, but horticulturalists down the ages have used walls to provide shelter and stored heat to grow tender crops. Sometimes the input of solar radiation was augmented by fireplaces and flues built into the structure of the wall.

The vulnerability of fruit blossom to spring frosts has led to several methods of short-term temperature manipulation. The danger arises mainly from radiational cooling of the ground on clear nights, leading to an inversion of cold air close to the ground with warmer air above. One way of counteracting this phenomenon is to mix the air layers with a wind machine. Various fuels have been used in heaters to provide frost protection in orchards. Water can also be sprayed over trees that are in danger of freezing. The trees become coated with ice, but as the water freezes, it gives up latent heat that may be sufficient to prevent injury because the buds freeze at a slightly lower temperature than water.

AIR MOVEMENT AND COMPOSITION

Winds are driven by differences in air pressure resulting from corresponding differences in the input of solar energy. Air movement around the earth is dominated by three belts of wind in both the northern and southern hemispheres. Although the winds move air masses between different latitudes, the rotation of the earth gives them an easterly or westerly shift via the Coriolis effect. The trade winds blow toward the equator from a northeast direction in the upper atmosphere and a southeast direction in the southern. Between about 30° and 50° latitude, winds blow in opposite directions: southwest in the northern hemisphere and northwest in the southern. Beyond 50°, polar winds blow predominantly from the northeast in the northern hemisphere and southeast in the southern. These surface-level winds are accompanied by upper-level winds blowing in the opposite direction. Each belt of wind is part of a toroidal (doughnut-shaped) air mass about 8 km (5 m) deep that encircles the earth. The convergence of the trade winds at the equator leads to an updraft of air, from which rain falls as it cools. The divergence of prevailing winds around 30° latitude is associated with a downdraft that carries moisture away and results in low rainfall (Fig. 5–8).

Air circulation tends to be more orderly in the southern hemisphere than in the northern, where large land masses capture the sun's heat and give rise to high pressure areas that break up the circulation patterns. The high pressure areas shift, leading to changes in the weather. In the central United States, summer weather is dominated by southeast winds bringing moisture-laden air from the Gulf of Mexico,

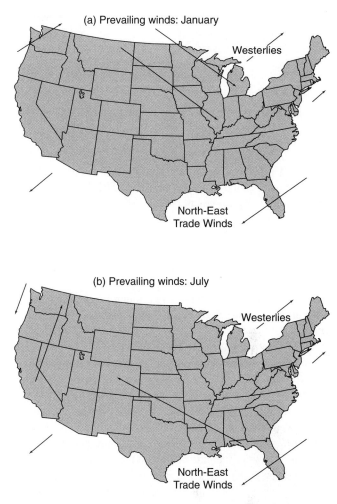

Figure 5–9 Prevailing winds in the United States in summer and winter.

whereas in the winter, northwest winds bring polar air (Fig. 5–9). Sudden shifts can occur in these patterns, particularly in the spring and autumn, leading to unstable weather. When a cold front meets a mass of warm, moist air, powerful air currents can lead to thunderstorms. When crosswinds blow through the rising clouds, conditions are set for tornado development.

On a more local level, bodies of water affect air circulation because they warm up more slowly than the land does in the daytime and cool more slowly at night. This characteristic leads to a sea breeze toward the land in the morning and an offshore wind in the evening. In mountainous areas, breezes tend to blow up valleys and hillsides as they warm up during the day, but cool air flows in the opposite direction at night. Buildings tend to slow down air circulation in urban areas; however, depending on the height and distribution of buildings, areas of turbulence or convergence of winds can produce strong gusts. Air movement tends to be slowed down by vegetation. This effect is most noticeable under a continuous canopy of herbaceous or woody vegetation. Air tends to move across the top of the canopy, which forms an

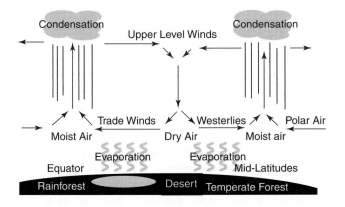

Figure 5–8 Major global air and water circulation patterns.
Source: Michael Knee.

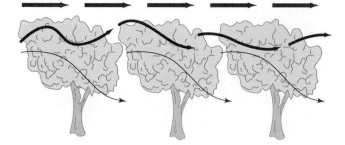

Figure 5–10   Air movement within plant canopies is generally slower than outside the canopies. The thicker the line, the faster the air movement. *Source:* Michael Knee.

albedo layer (Fig. 5–10). Humidity tends to be higher within the canopy than above it because of limited air exchange. Because transpiration is related to vapor pressure deficit and wind speed, water loss tends to be much lower for plants in continuous stands than for isolated plants of the same species.

Wind can be a major cause of damage to crops, either because of direct physical damage or excessive transpiration. In exposed areas, it may be necessary to construct windbreaks or grow shelter belts of tall vegetation. However, plants are mostly adapted to grow with a certain amount of physical disturbance by wind. When this adaptation is absent, they tend to become tall and weak-stemmed, which is a problem for woody plants that are staked in nursery production and for plants grown in a greenhouse. For greenhouse plants, it can be beneficial to provide physical stimulation by brushing the foliage on a regular basis.

In some areas, wind has been an agent of soil formation. Loess soils are formed by the accumulation of wind-blown particles. As with water, however, wind is an agent of soil erosion, which is a problem particularly in the drier parts of the western United States and other parts of the world.

ATMOSPHERIC COMPOSITION

Dry air is comprised of 78 percent nitrogen (N), 20.9 percent oxygen (O), and about 1.1 percent other gases. (As we saw in Chapter 4, water vapor is normally a significant proportion of the atmosphere, but we will ignore that fact in this section.) Nearly all of the earth's nitrogen is in the atmosphere (77.5 percent) or in the lithosphere (rock and soil, 22.4 percent). The rest is mainly in the hydrosphere, and a tiny amount is in living organisms. Life would cease if this tiny fraction could not be replenished continuously by converting atmospheric nitrogen into forms that can be absorbed by plants and passed up the food chain. Nitrogen oxides are formed from free nitrogen and oxygen by lightning during thunderstorms, and they fall to earth as nitrates in rainwater. Soil bacteria, both free-living and in association with plant roots, convert nitrogen to ammonia. In the last century, the

same reaction was developed on an industrial scale in the Haber-Bosch process. The ammonia fertilizer from this process makes an essential contribution to the levels of crop productivity we depend on for survival. On the other hand, oxides of nitrogen are generated as by-products of gasoline combustion in vehicles and have become pollutants in the atmosphere. The oxides dissolve in rainwater and contribute to the soil and water nitrogen pool wherever they fall. This effect is mostly undesirable in natural ecosystems and contributes to eutrophication of bodies of water, the disappearance of nitrogen-fixing plant species (which are no longer competitive), and apparently forest decline in the United States and Europe.

The oxygen in the atmosphere that we depend on for our survival is produced by plants and algae during photosynthesis. Land plants account for about 60 percent of global photosynthesis, which keeps the balance between oxygen and carbon dioxide (CO_2) at about 600 to 1. By competing so effectively for carbon dioxide and keeping it at low levels, plants and algae modify the climate of the earth. As is now well known, carbon dioxide absorbs infrared radiation from the earth so that heat is retained in the atmosphere. This phenomenon is loosely referred to as the greenhouse effect (although this is *not* how greenhouses work). The level of oxygen in the atmosphere is so high (209,000 $\mu l \, l^{-1}$) that it is not much affected by changes in the balance of carbon dioxide production and consumption. This balance has a much greater impact on carbon dioxide levels that are currently at 370 $\mu l \, l^{-1}$ (or parts per million) and are rising at about 1.3 $\mu l \, l^{-1}$ per year. Carbon dioxide levels undergo local fluctuation depending on the amount of photosynthesis occurring. Thus, concentrations can fluctuate between day and night and between winter and summer by up to 20 $\mu l \, l^{-1}$ (Fig. 5–11). The increasing combustion of fossil fuels and the clearance of natural vegetation on all continents have caused a 30 percent increase in atmospheric carbon dioxide over the past two centuries. There is now little doubt that this rise in carbon dioxide is contributing to an upward trend in temperature, particularly in higher latitudes. Whereas this trend has many negative effects, crop yields could actually increase, although crops are not generally as effective in removing carbon dioxide from the atmosphere as is natural vegetation.

Human activity has added other gases to the atmosphere that are more obviously pollutants than is CO_2. Sulfur dioxide is another product of combustion, particularly from coal. It is directly injurious to plants. Along with nitrogen oxides, it contributes to acid rain, which has caused problems in terrestrial and aquatic habitats in areas around coal-burning power stations. Ozone (O_3) is formed from molecular oxygen by photochemical reactions with other pollutants, including oxides of nitrogen (Fig. 5–12). Plants vary in their sensitivity to ozone, which is also injurious to people. Ozone is beneficial in the upper atmosphere, however, where it is formed directly from oxygen by ultraviolet radiation and is itself a strong absorber of ultraviolet (UV) light, which

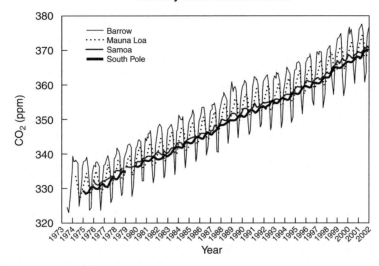

Figure 5–11   Generally, rising carbon dioxide levels show seasonal variation, depending on the photosynthetic activity of vegetation. Data shown here through May 2002. *Source: http://www.cmdl.noaa.gov.*

Figure 5–12   The UV component of solar radiation promotes ozone formation when the pollutant nitrogen dioxide is present in the lower atmosphere. *Source:* Michael Knee.

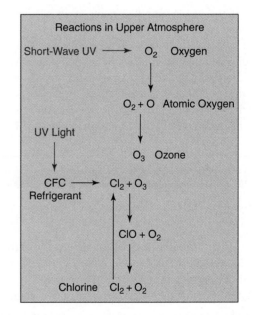

Figure 5–13   The UV component of solar radiation promotes ozone destruction when chlorofluorocarbons are present in the upper atmosphere. *Source:* Michael Knee.

would be damaging for all life if it reached the earth's surface. Chlorfluorocarbons (CFCs) from aerosol spray cans and leaking refrigeration equipment convert ozone in the upper atmosphere back to oxygen (Fig. 5–13). The danger of increasing UV exposure led to an international ban on the use of CFCs, although it will be many years before they disappear from the atmosphere. Several hydrocarbons are present in the atmosphere in trace quantities. The most abundant is methane, which arises from sewage treatment and intensive animal production as well as natural sources, such as wetlands. It is not very toxic, but like CO_2, it is a greenhouse gas. Ethylene is produced by plants and several microbes; it is one of the plant hormones and is active at about $0.1\ \mu l\,l^{-1}$. Its

effects can be benign or even beneficial, such as its role in the promotion of fruit ripening, but it also causes premature senescence (aging) of plant tissues and abnormal growth. It is a minor product of the combustion of fossil fuels and can be troublesome for plants in enclosed spaces, especially greenhouses. Although enhanced levels tend to occur in urban areas, it is not clear that this level has had any adverse effects on plant growth.

In addition to the gases, all air contains particulate matter, which can also be increased by human activity. Soil cultivation and major construction projects cause soil particles to be released into the atmosphere. In addition to causing loss of soil, the dust cloud can block sunlight, clog the stomatal pores in plant leaves, and create a general nuisance for people. The problem is greatest in dry seasons and arid areas. Burning of forests and other natural vegetation in preparation for agriculture or other development also contributes to airborne particulates. Volcanic eruptions are natural sources of dust. Dust from local sources such as the eruption of Mount St. Helens, the burning of the Indonesian rainforests, or agricultural operations in China can have a worldwide impact, reducing light levels and temperatures. (The greenhouse effect might be worse but for all the dust we have kicked up!) A fraction of the number of the particulates consists of living organisms, plant pollen, microbial spores, and insect eggs. These "particles" also move around the world, causing pest and disease outbreaks on plants as well as hay fever in humans.

Attempts to modify the atmosphere focus on preventing the emission of pollutants that can be harmful to plant, animal, and human life. Carbon dioxide is sometimes added to the atmosphere in greenhouses to promote crop growth, and optimistic people suggest that rising levels will be beneficial for field crops. General interest has developed in agricultural practices that will lead to sequestration of CO_2, which means that carbon will somehow be taken out of circulation, or its reentry into the carbon cycle will be delayed. Of course, when most crops are consumed, their carbon content is returned to the atmosphere as CO_2, so the benefit of each crop may be temporary. In natural ecosystems, carbon can be sequestered in organic debris that gets incorporated into the soil or in the stems of long-lived woody plants. There is interest in cultivation practices, such as low- or no-till cultivation, that could increase carbon sequestration in agricultural soils.

SUMMARY AND REVIEW

As long as the water supply remains at an adequate level, temperature is the main factor that determines whether a crop can be grown in an area. Temperature is influenced by solar radiation and heat transfer by air and water. Many aspects of crop growth and its management are influenced by temperature conditions. Local features modify these conditions and can be manipulated to improve crop growth. Air movements are driven by heat input from the sun. They can directly affect crop growth; modify temperature; and carry water vapor, soil particles, pests and diseases, and other atmospheric components to or from a crop area. Plants have an important role in maintaining atmospheric composition. Plants require nitrogen for growth, and most of it is in the atmosphere. Plant growth is affected by particulates and gaseous pollutants in the atmosphere.

FOOD FOR THOUGHT

1. Why don't the lines defining the edges of plant hardiness zones run parallel to lines of latitude?

2. It is often tempting to push hardiness by introducing tender plants to cooler zones. What are some factors that might encourage this practice? How would you decide whether it is worth the risk for a particular plant?

3. Some people like turf and some people like trees, and they can get into arguments about which needs more water or which absorbs more carbon dioxide. It is difficult to decide one way or the other, but what are some factors to consider on both sides of the debate?

4. You have probably heard of the Gaia hypothesis, developed by James Lovelock, that the earth is a huge self-regulating system, almost like a superorganism. What features discussed in this chapter tend to support or negate this view?

5. It seems that olives can now be grown and harvested in London. Considering the ways in which the climate of urban and rural areas differ, which do you think is the better environment for plant growth?

SUPPLEMENTARY READING

CHRISTOPHERSON, R. W. 2004. *Geosystems: An Introduction to Physical Geography,* Fifth Edition. Upper Saddle River, N.J.: Prentice Hall.

CHAPTER 6

Soil and Managing Soil, Soil Water, and Fertility

Don Eckert

KEY LEARNING CONCEPTS

After reading this chapter, you should be able to:

♦ Understand that soil ecology is a complex system.
♦ Explain the factors that influence soil formation and give soil its physical and chemical characteristics.
♦ Explain how soil characteristics influence plant growth.
♦ Understand why soil conservation is an important practice in the cultivation of plants.
♦ Know the different types of irrigation systems and when each type is used.
♦ Understand the basic principles of managing soil fertility.

People often refer to soil dismissively as dirt and probably regard it as dead stuff. However, it is an important part of the living world; along with climate, it determines the composition and function of the ecosystems that can exist in different areas. Plants live in the soil as much as they live in the air. Many if not most of the ecological interactions between plants and other organisms occur below ground. Most of the individual organisms in an ecosystem and a large part of the biomass exist in the soil. Over time, the composition and properties of the soil are influenced by soil organisms. As you will learn later in the chapter, soil is the product of the interaction of parent material, climate, topography, and soil organisms over time. Soil is a key element in the cultivation of plants and, to some degree, it can be manipulated to influence plant growth.

Plants require oxygen, carbon dioxide, water, and certain mineral elements in order to grow and develop. Some of the required oxygen and carbon dioxide is exchanged through the leaves and roots; oxygen and mineral elements are also absorbed by the roots from the soil solution. This chapter discusses the role of soil, soil water, and soil air in providing plants with anchorage, essential mineral nutrients, and water.

DEFINITION OF SOIL

Soils differ around the world, but they are basically natural bodies composed of weathered rock, air, water, decomposed organic material, and various organisms, all working together in a complex ecosystem capable of supporting the growth of land plants. Soil is a very complex and dynamic system of many interacting factors and components that affect and are affected by plants. Soil consists of:

1. A solid fraction; that is, rock fragments and minerals.
2. An organic fraction; the decayed and decaying residues of plants, microbes, and soil animals.
3. A liquid fraction, including water and dissolved minerals.
4. A soil atmosphere or soil air.

The voids found between the solids are called **pore spaces.** Pore spaces vary in size and continuity and are an important physical property of soils.

    The kind of soil and its ability to support plant growth are influenced by the relative amounts of each of these four components. Typically, a natural soil is approximately half solids and half pore space. The solid phase of the soil might average 40 to 45 percent mineral and 5 to 10 percent organic material by volume, while the pore space might be filled with a mixture of water and air, the ratio depending on the wetness or dryness of the soil. All of these relationships vary from one soil type to the next, and can be affected significantly by the way a soil is managed.

Kinds of Rocks

Parent rocks contain the nutrients that will be found in parent material, and later in the soil. Rocks are made up of consolidated material, unconsolidated material, or both (Fig. 6–1).

1. **Igneous rocks** (i.e., lava, magma) are formed from the hardening of various kinds of molten rock material and are composed of minerals such as quartz and feldspar.
2. **Sedimentary rocks** are generally unconsolidated and composed of rock fragments that have been transported and deposited by wind, water, or glaciers. Limestone, sandstone, and shale are examples.
3. **Metamorphic rocks** form from igneous or sedimentary rocks that have been subjected to sufficiently high pressures and temperatures to change their structure and composition. Slate, gneiss, schist, and marble are examples of metamorphic rocks.

Figure 6–1 Parent rocks from which soil can be formed. (A) Igneous. (B) Sedimentary. (C) Metamorphic. *Source: Rosa Maria Marquez.*

FACTORS INVOLVED IN SOIL FORMATION

Soil is derived from rocks, minerals, and decaying organic matter. The two major processes in soil formation are (1) accumulation and (2) transformation of the parent material. **Parent material** accumulates from the breakdown of rocks by weathering. This process must occur before soil can begin to form. The parent material accumulates as an unconsolidated mass that later differentiates into characteristic layers called horizons. A **horizon** is a distinct layer of soil having physical and/or chemical differences resulting from soil-forming processes as seen in a vertical cross section. Differentiation occurs by mechanical separation and/or transformation of the parent material. As the process continues, the horizons generally become more distinguishable and finally develop into a soil profile. A **soil profile** is a vertical section of soil extending through all its horizons from the surface to the parent material.

    The factors responsible for soil formation are: (1) parent material, (2) climate, (3) biology, (4) topography, and (5) time.

Parent Material

The formation and accumulation of material by chemical and physical weathering of parent rocks is the first step in the development of soil.

Chemical Weathering

Chemical weathering entails four distinct processes.

Dissolution is the process by which the constituents of parent material dissolve in water or weak, naturally occurring acids and are leached away. Chloride, nitrate, and sulfate salts are generally very soluble in water and can be leached by rainfall alone. Other materials, such as carbonates, dissolve very slowly in water, but they dissolve much more quickly when acted upon by organic acids produced when organic matter decomposes or the carbonic acid that forms when atmospheric carbon dioxide combines with rainwater.

Hydration adds molecular water to another compound to form a hydrated material more vulnerable to pulverization. An example is calcium sulfate ($CaSO_4$) absorbing water to form gypsum ($CaSO_4 \cdot 2\,H_2O$), a hydrated calcium sulfate:

$$CaSO_4 + 2\,H_2O \rightarrow CaSO_4 \cdot 2\,H_2O$$

Hydrolysis is the reaction between a compound and water to form a more soluble product. In the following hydrolysis reaction, potassium ions (K^+) are made more available to plants by the reaction of the slowly soluble feldspar mineral ($KAlSi_3O_8$) with water (H_2O) to form soluble potassium hydroxide (KOH):

$$KAlSi_3O_8 + H_2O \rightarrow HAlSi_3O_8 + KOH$$

Oxidation reactions form oxides of parent material by reaction with oxygen. For example, ferrous oxide (FeO) reacts with oxygen (O_2) to yield ferric oxide (Fe_2O_3), a product more oxidized than the reactant:

$$4FeO + O_2 \rightarrow 2Fe_2O_3$$

Physical Weathering

Changes in temperature greatly affect the rate of physical weathering. Differential rates of contraction and expansion caused by temperature changes bring about cracking and peeling of the outer layers of rocks by a process called **exfoliation.** A second process occurs due to the presence of different materials within a rock, each with its own characteristic coefficient of expansion. Because of these differences, sudden large temperature changes cause uneven expansion or contraction, cracking the rocks. A third process is the cracking of rocks caused by the expansion of water as it freezes in rock fissures.

The mechanical action of glaciers causes rocks embedded in the ice to scrape against other rocks as the glacier moves. This action grinds the rocks into increasingly smaller rock fragments. This physical process is powerful: it reached a tremendous magnitude during the Ice Ages (Pleistocene epoch). The product of glacial weathering is called **glacial till,** and it comprises rock particles ranging in size from clay to boulders. This material is deposited beneath, beside, and at the terminus of the melting glacier.

Physical weathering is also caused by moving water, as in stream erosion, sheet erosion, rill erosion, or wave action (Fig. 6–2). The action is similar to that of glaciers. Water from rains and melting snow moves rapidly down streambeds, carrying parent rock fragments of varying sizes. As these fragments move along, they are gradually worn down to create smaller and smaller particles, eventually forming parent material. In arid regions, wind acts similarly to water. Coarse sand particles (parent material) are swept along the ground with sandblasting action wearing away other larger parent rocks.

The action of plant roots can sometimes physically break down parent rocks. For example, a tree root growing into a crack in a rock can ultimately fracture the rock. While this example is not considered weathering, it is a physical soil-forming process. In this case, some chemical weathering must occur first to provide nutrients for the plants.

Figure 6–2 Massive gullies formed by severe stream erosion during periods of heavy rainfall. These gullies are beyond reclamation by practicable methods. *Source:* USDA Soil Conservation Service.

Climate

The climate affects soil formation. In areas of high rainfall, soils are often highly leached and acidic in reaction. Chemical weathering proceeds at a rapid rate, especially if high rainfall is coupled with high temperature. The fertility level of soils formed under high rainfall is generally low because many of the plant nutrients are leached from the root zone. Many of these soils are red or yellow in color, indicating a relatively high percentage of iron oxide, which remains after other elements have been removed.

On the other hand, soils developed in arid climates are not highly leached. Calcium and magnesium carbonates tend to accumulate, and chemical weathering proceeds at a much slower rate. Soils formed under arid conditions often contain excessive quantities of salts other than carbonates, and are not productive until the amounts of salts are reduced. Land can be desalinated by flooding with water and leaching the salts downward through the soil profile. Fields treated in this manner must possess excellent subsurface drainage, either natural or improved by human intervention. If heavily salted water accumulates in the soil, it can cause the soil to disperse and become waterlogged. Such degradation can render soils useless for future production unless expensive remediation practices are employed.

Biology

Just as in above-ground ecology, plants are the base of the food chain in the soil. Between 10 and 50 percent of the plant's dry matter is in its root system. Roots have three functions: they support and anchor the plant stem, they absorb water and nutrients, and they provide storage for food reserves. To keep the plant supplied with water and nutrients, roots must grow. While the root hairs in the zone behind the root tip are absorbing material from the soil, the tip is moving ahead, exploring the gaps between soil particles for fresh supplies that will be needed for future growth.

Plant roots secrete chemicals into the soil that help in the absorption of nutrients. They also secrete mucilage, which helps lubricate the passage of the tip between soil particles. The root tip is covered with a protective cap from which cells are sacrificed continually (about 20,000 a day from a single corn root). A plant's fine, absorbing roots do not grow indefinitely; old roots die off and are replaced by new ones so that more than half of the root system will be turned over in a year. So plant roots are continually adding organic matter to the soil quite apart from any that drops from above ground or is taken by root parasites or herbivores.

The soil is home to a great diversity of organisms from all of the five kingdoms: Plantae, Animalia, Fungi, Protista, and Monera. These organisms are involved in all possible types of ecological interaction, and many of them have profound effects on soil chemistry, structure, and function. Many soil organisms have not yet been classified or characterized.

Bacteria are the most abundant soil organisms; some of them are parasites of plants and other soil organisms, but many are involved in decomposition and recycling of nutrients in soil organic matter. Some have the ability to fix atmospheric nitrogen.

Fungi may also be parasites, but most of the fungi in the soil are saprophytes, breaking down soil organic matter. The roots of nearly all plants in the wild are associated with mycorrhizal fungi, which assist in plant uptake of nutrients and water.

Soil Protista include algae that can carry out photosynthesis at the soil surface and protozoa that are predators of soil bacteria. Soil animals include nematodes, molluscs (slugs and snails), annelids (including earthworms), and arthropods (mites, insects, millipedes, spiders, etc.). These groups can be subdivided into herbivores, microbivores, predators, and detritovores.

The nonliving organic components of the soil are residues of plants and animals. The amount and kind of these organisms are influenced by the climate. For example, the climate of an area determines the quantity of biomass produced on a site, which in turn influences the soil-forming processes.

Vegetation aids soil formation by supplying organic matter in the form of dying and decomposing plants. Grasses decompose into a different kind of organic residue than do trees. Also, the amount of organic matter varies according to the amount and type of vegetation (Fig. 6–3). Peat soils form where reeds, sphagnum moss, and grasses grow and where decomposition of the organic matter is minimized. Soils in arid regions normally contain low amounts of organic matter because of the limited growth of desert grasses, shrubs, and cacti.

The amount of organic matter in soil is influenced by the difference between the rates of accumulation and decomposition of organic material. In cases where decomposition rates are very high, the organic fraction accumulates very slowly. Thus, the amount of organic matter remains low. Consider the case where temperatures are high and rainfall or irrigation keeps the soil moist. Under these conditions, the rate of decomposition nearly equals the rate of accumulation, and the organic matter content changes little over time. On the other hand, in cold areas or sometimes under anaerobic conditions, decomposition is inhibited and organic matter accumulates. Finally, when a soil is tilled, introducing additional oxygen into the soil profile, the decomposition rate can exceed the accumulation rate and organic matter is lost from the soil. Many factors affect the rates of accumulation or decomposition, all of which play a role in determining the amount of organic matter present in a given soil at a particular time and location.

The type of root system produced by the different plant species also influences soil formation. Dense fibrous root systems of grasses often lead to permeable soils, such as some of the Mollisols (prairie soils) formed in the central United States and in some areas of eastern Europe.

Figure 6–3 The organic matter produced from the decomposition of forest litter is different from that produced from prairie grasses. In this example, the forest is not so dense that it excludes grasses, and both types of vegetation are present and produce organic matter. *Source:* USDA Soil Conservation Service.

Topography

Topography influences drainage and runoff. Steep slopes are subject to erosion because water flows downhill quickly, with little percolating into the soil. Resulting erosion leads to the formation of shallower soils because the soil surface is removed almost as quickly as it is formed. In depressional areas, however, deposition of eroded material leads to formation of deeper soils. Gentle slopes that are heavily covered with vegetation slow the water flow and allow more time for water to percolate into the soil, permitting the development of a well-defined profile. The same is true of flatter areas. Rapid surface runoff causes more erosion, and if vegetation is removed or absent, deep gullies can be cut into the sloping land (Fig. 6–4).

Topography has a marked effect on climate. High altitudes mean lower soil and air temperatures, which influence the amount and type of vegetation. On average, air temperature decreases 1°C for each 300 meters (2°F/1,000 ft) of elevation for the first 10 to 15 km (6 to 9 mi) of altitude. Thus, an increase in elevation changes the kind and type of vegetation. Grasses and deciduous trees grow at lower elevations,

with coniferous evergreens at higher elevations. Above the timber line, little or no vegetation exists because of the lower temperature and humidity.

Time

Hard-to-decompose rocks such as granite require millions of years to form parent material; softer rocks such as limestone require less time. Interactions between biological and chemical agents reacting with parent material over long periods of time differentiate the soil into horizons. Soils without well-developed horizons are classified as young soils, even though the parent material may have been present for a great many centuries. For example, some desert soils, constantly shifted and transported by wind, are young soils because they lack profile development. Also, silt is often moved along in rivers, deposited, and then moved again, leaving little chance for profile development. Such soils remain pedologically young no matter how many years they may have been in existence.

As they develop, mature soils differentiate into well-defined profiles consisting of three principal horizons (Fig. 6–5). The surface layer, the **A horizon,** varies in depth and contains most of the plant roots. This leached zone often lacks some of the important mineral nutrients, but it does contain the largest amount of organic matter. The organic matter makes the A horizon permeable and dark-colored, and the A horizon is normally the zone of greatest biological activity in the soil profile.

Below the A horizon is the **B horizon,** the **zone of accumulation.** Plant nutrients, silts, clays, and other materials from the upper layer are leached into and accumulate in this horizon. The color is generally lighter than that of the A horizon, and less organic matter is present (although the roots of deep-rooted plants do reach into the B horizon).

The **C horizon** consists of unweathered to slightly weathered material from which the A and B horizons are formed. It can also include accumulated calcium carbonates or other salts.

The factors affecting soil formation are obviously interrelated. Parent material affects, along with other factors, the capacity of the soil to support plant life, which in turn influences the kind of vegetation. Topography and temperature also influence vegetative growth. Temperature interacts with organic matter. Topography, temperature, and time influence parent rock conversion into parent material, each interacting to yield different soil formation and soil characteristics. Thus, each factor affects and is affected by all the others. It would be difficult to say that any one is more important than another in soil formation.

PHYSICAL PROPERTIES OF SOIL

Soil Texture

An important physical property of soil is its texture. **Soil texture** is defined as the percentage of sand, silt, and clay particles in a soil. Soil particles vary in size from coarse rock

Figure 6–4 Sloping topography accelerates rill erosion, which removes topsoil. Constant removal of topsoil exposes the parent material to weathering and more rapid soil formation. *Source:* USDA Soil Conservation Service.

A
Horizon
Zone of Leaching,
Accumulations, and
Intensive Weathering.
Zone of Organic Matter
Accumulation.
0 to 65 cm (0 to 25 in.)

B
Horizon
Zone of Accumulation of
Products from A Horizon.
Moderate Weathering.
65 to 90 cm (25 to 36 in.)

C
Horizon
Zone of Parent Material.
Slight Weathering.
90+ cm (36+ in.)

Bedrock

Figure 6–5 An idealized soil profile.

TABLE 6–1	Dimensions of Soil Particle Size Classes		
Diameter of Particles			
Millimeters 2.0	0.05		0.002 or less
Inches 0.080	0.002		0.00008 or less
Gravel, stones	Sand	Silt	Clay
Particles visible with the naked eye		Particles visible under microscope	Particles visible under electron microscope

fragments (>2 mm) to those so small (<0.002 mm) that an electron microscope is needed to observe them (Table 6–1). To measure soil texture, the soil particles are separated into their respective sizes and the percentage in each size category is calculated. A textural classification is then made with the aid of a soil textural triangle (Fig. 6–6).

Soil texture influences many of the soil's properties related to crop production (Table 6–2). The distribution of different particle sizes determines the ability of soils to hold and transmit water. Soil with a high percentage of sand loses water quickly, retaining little for plant use. Plants growing in these soils will experience water deficits sooner after wetting than those grown in loam or clay soils. Texture also influences soil aeration. In soils largely composed of very fine clay particles, movement of both air and water can be lim-

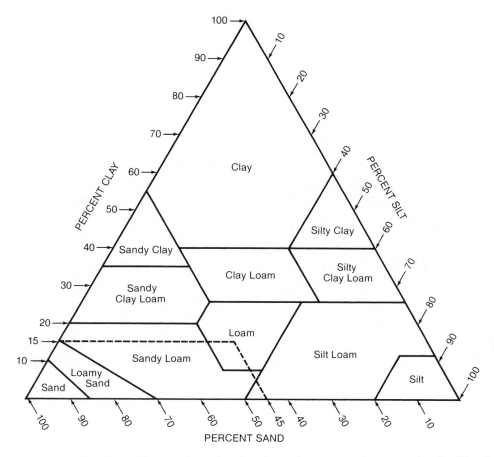

Figure 6–6 A soil textural triangle. To illustrate how the triangle works, assume that a sample of soil has been analyzed and found to contain 45 percent sand, 15 percent clay, and 40 percent silt. On the triangle, locate the 45 percent value on the sand (bottom) axis and draw a line parallel to the lines in the direction indicated by the arrow. Next, locate either the 15 percent value for clay on the clay axis (left side of triangle) or the 40 percent silt on the silt axis (right side of triangle), and draw another parallel line along either of the two axes in the direction indicated by the respective arrows. The two lines intersect in the loam area of the triangle; thus, this soil is classified as a loam. *Source:* USDA.

TABLE 6–2	Some Soil Properties Influenced by Soil Texture		
	Textural Class		
Soil Property	**Sand**	**Silt Loam**	**Clay**
Aeration	Excellent	Good	Poor
Cation exchange	Low	Medium	High
Drainage	Excellent	Good	Poor
Erodibility*	Easy	Moderate	Difficult
Permeability*	Fast	Moderate	Slow
Temperature (spring)	Warms fast	Warms moderately	Warms slowly
Tillage	Easy	Moderate	Difficult
Water-holding capacity	Low	Moderate	High

* By water.

ited. Plant roots need oxygen for respiration, and soils with low rates of gaseous diffusion restrict respiration and plant growth. Also, many beneficial soil microorganisms require well-aerated soils.

A soil consisting of a mixture of 40 percent sand, 40 percent silt, and 20 percent clay produces a loam soil that retains sufficient water for good plant growth and permits its movement without restricting aeration. Due to their excellent water-holding properties, loams are usually excellent soils for crop production.

The ease of tillage (plowing, disking, cultivating) is influenced by soil texture. Sandy or loam soils at the proper moisture content are easier to till than clay soils. Root penetration is sometimes restricted in soils of high clay content. Other factors being equal, most crops grow better in loam soils than in either sandy or clay soils.

Soil Structure

Soil structure is defined as the arrangement of primary soil particles into secondary units, that is, the manner in which individual primary particles clump and hold together. The secondary unit (aggregate) is a clump of soil particles that acts as an individual larger particle with specific characteristics. The kind of soil structure is determined not only by the relative amounts of each primary particle but also by the manner in which these particles are arranged into aggregates. The size and form of aggregation is known as the structure of soil.

Descriptive words are used to classify soil structure, for example, prismatic, subangular blocky, blocky, columnar, platy, and granular. These words describe the shape, character, and appearance of the aggregates (Fig. 6–7). Aggregates may vary from a fraction of a centimeter to several centimeters in diameter and may be held together strongly or weakly.

Structure can be changed by mechanical operations. Tilling the soil when its moisture level is higher than field capacity can destroy good structure. (**Field capacity** is that amount of water retained in the soil after gravitational drainage of excess water.) Tillage at a moisture level slightly below field capacity is less likely to damage soil structure. Soil with crusts and compacted layers can have their structure improved, though often only temporarily, by tillage.

A soil with good structure has soil pores that are large enough to transmit water and air without restriction, yet small enough to retain some water against the pull of gravity. To form pores of this size requires soil aggregates about 1.5 to 6 mm (0.06 to 0.24 in.) in diameter. Smaller aggregates form pores that are too small to permit adequate drainage and soil aeration, whereas larger aggregates form pores that are too large to hold much water, even though some is held within the aggregate itself. Good soil structure provides a favorable air and water relationship required for plant growth and microbial activity.

Aeration

Soil pore spaces serve as passageways for the transmission of oxygen needed for respiration by plant roots and microorganisms, and for the escape of waste carbon dioxide. Soil aeration can become limiting when soil pores are filled with water, when the soil is compacted, or when a crust forms at the surface. The relative amounts of oxygen, nitrogen, and carbon dioxide vary in the soil. Generally, soil air contains less oxygen but more carbon dioxide than the above-ground atmosphere. Concentrations of 5 percent carbon dioxide have been found in soil air, whereas the above-ground atmosphere contains about 0.03 percent. Under flooded conditions, oxygen content in the soil is often low. Many plants suffer when oxygen concentrations in the soil atmosphere drop below 10 to 12 percent or if carbon dioxide concentrations rise above 5 percent. Studies have shown that oxygen deficiency in soils causes poor plant growth more frequently than does carbon dioxide excess.

Figure 6–7 Several kinds of soil structure. (A) Platy. (B) Prismatic. (C) Columnar. (D) Angular block. (E) Subangular blocky. (F) Granular.

Golf course greens and athletic fields often have sand as the primary component supporting the turf. The use of sand allows golf course and athletic field superintendents to manage fertilization and irrigation programs with less difficulty than when finer-textured soils are used. However, without the finer particles of clay and silt, the soil matrix is too coarse to provide stability. One of the challenges facing turfgrass scientists is to develop materials that hold the sand in place, providing stability and allowing a vigorous root system to develop.

Soil Compaction

The use and size of farm equipment have increased considerably in recent years, and it is common to see heavy equipment make many trips across the fields. Trucks heavily loaded with harvested crops travel over the land. Any of these can seriously compact soil (Fig. 6–8). Grazing animals can also compact soils when pasture soils are wet.

Soil compaction is a serious problem, directly related to land preparation, tillage, and particularly to harvesting operations (Fig. 6–9). The term **soil compaction** is difficult to define precisely and often means different conditions to different people. To growers, a compacted soil is best described as one with abnormally high bulk density with very small pore spaces. Soil is not compacted all at once. The problem develops slowly, worsening each time heavy equipment is used. Often years pass before the problem reveals itself in declining crop yield or quality. Research has found no easy or simple solution to the problem. The problem can be recognized and alleviated somewhat by proper soil management and organic matter additions. Soil compaction is especially serious for the fresh market vegetable farmer because vegetables are often harvested when the soil is wet. Farmers in irrigated regions sometimes irrigate up to the time of harvest to keep plants fresh and turgid, then quickly harvest the crop, often using

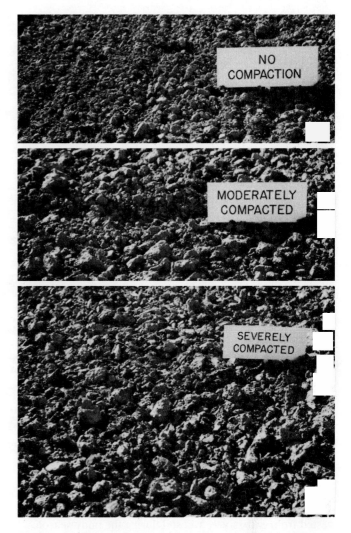

Figure 6–9 Seedbed prepared on soils having different degrees of compaction, using normal tillage operations such as plowing, double disking, and harrowing with spike-toothed harrow. Note excessive cloddiness on the moderately and severely compacted soils.

heavy equipment. Also, wholesale market prices for fresh vegetables frequently change dramatically from day to day. Thus, growers may harvest the crop even though the soil moisture is high and the danger of soil compaction great (Fig. 6–10).

Whenever a soil becomes more compacted, the process is accompanied, among other changes, by an increase in **bulk density.** In simple terms, an increase in bulk density means that the soil has become more dense through a loss of pore space.

Compacting the soil causes the particles to be pressed together, reducing pore space. This reduction in pore space, however, occurs to a greater extent in larger soil pores and to a lesser extent in smaller pores. Consequently, the selective reduction of the larger-sized pores leads to reduced water infiltration and gas exchange with the atmosphere (Figs. 6–11 through 6–13).

Figure 6–8 Heavy earth-moving equipment such as this or other heavy machinery can contribute to the serious problem of soil compaction and especially so when the soil is moist. This equipment, which is used to level land, can easily carry 20 MT (19.5 t). *Source: The Daily Democrat, Woodland-Davis, California.*

Figure 6–10   The upper portion of a soil profile showing a compacted layer about 25 to 30 cm (10 to 12 in.) below the soil surface. This layer was caused by repeated traffic over the field when soil was wet. The photo was taken soon after a crop of tomatoes was harvested from the field. The previous fall and early winter, the field had been planted to spinach, which was harvested from excessively wet soil during the rainy season.

Figure 6–11   Water infiltration rates were measured on each of three plots mechanically compacted to three densities. Tests were made in the fall after compaction and then repeated the following spring on the lightly, moderately, and severely compacted plots. Obviously, severe compaction decreases the water infiltration rate.

Figure 6–12 Soil samples taken from the top 10 cm (4 in.) of adjacent test plots of a noncompacted (*left*) and compacted (*right*) soil. The photographs are magnified 40 times. Note the greater porosity in noncompacted soil samples.

Soil compaction impedes root penetration and thereby limits the volume of soil from which plant roots can extract nutrients and water. This can result in nutrient deficiencies, particularly of phosphorus. The reduction in porosity can result in a greater accumulation of carbon dioxide in the soil.

Soil crusting, similar to soil compaction, appears as a dense, hard layer of varying thickness on the soil surface. Soil crusts result from the action of raindrops or irrigation, which disperse surface aggregates. Crusting is particularly severe when the amount of water applied exceeds the infiltration rate. Soil crusts restrict seedling emergence, gas exchange, and water infiltration (Fig. 6–14).

In some instances a small degree of soil compaction is beneficial. Some compaction after seeding assures intimate contact between soil and seeds, enhancing imbibition of soil water by the planted seeds and thus increasing chances of germination. Also, some soil fumigation procedures entail purposeful soil crusting. The crust slows the leaking of the injected toxic gases into the atmosphere and thus makes the fumigation more efficient.

Soil compaction is easier to prevent than to correct. The risk of compaction can be reduced by good soil management. This includes using tillage implements of the lightest possible weight at the proper soil moisture content, avoiding unnecessary and expensive tillage, keeping equipment off the land when the soil is wet, and leaving green manure crops or crop residues on the soil surface.

CHEMICAL PROPERTIES OF SOIL

Effect of Climate

The type of parent material predominately influences the chemical characteristics of young soils. As weathering proceeds, soils tend to show the effects of climate, and the resemblance to parent material lessens or disappears. The chemical properties of the soil are determined largely by the colloid-sized (not visible with an ordinary microscope) aluminosilicate clay minerals.

In the temperate zones, chemical weathering is less intense in arid regions than it is in humid regions. Soluble salts released by weathering are not lost by leaching from soils in arid regions.

Figure 6–13 Soil samples from noncompacted and compacted fields were air-dried and screened to simulate tillage. Water was allowed to infiltrate into each for four hours. Note that much more water infiltrated into the noncompacted soil, showing that the ill effects of previous compaction still remained.

In tropical zones, with higher temperature and more rainfall than temperate regions, weathering and leaching are greater. The silicate and aluminosilicate minerals are more weathered, resulting in soils known as **Oxisols.** These soils contain high concentrations of iron and aluminum oxides, are generally red to reddish brown in color, and are low in fertility and organic matter. Even though large amounts of vegetation are produced, organic matter is low because dead plant matter is rapidly decomposed by high microbial activity.

In the cold humid regions, forest vegetation, mainly conifer trees, combine with climatic factors to produce **Spodosol** soil groups. The silica content of these soils is high in the surface layer in contrast to the iron and aluminum content in the Oxisols because the parent material of these soils is usually silica sand. These soils are highly leached and inherently low in plant nutrients. Organic compounds produced by decaying conifer needles form acid solutions that dissolve iron and aluminum oxides and basic compounds (calcium and magnesium salts). The ions are leached to lower depths in the soil profile and redeposited there together with some aluminum and dissolved organic matter to produce a dense layer 60 to 90 cm (2 to 3 ft) below the surface. These soils can be identified by an ashy, bleached white layer immediately above the dense layer. This layer is known as the **E horizon,** and it forms between the A and B horizons.

Cation Exchange Capacity

An important property of clay and of the organic humus fraction of the soil is its ability to attract and hold cations—positively charged ions, some of which are essential plant

Figure 6–14 This 25 percent stand of potato plants (*left*) resulted from severe soil crusting that prevented and delayed shoot emergence. A loss of this magnitude necessitates redisking and replanting the field. The partly excavated cross section of a raised bed (*right*) exposes the strength and thickness of a soil crust. It is apparent why potato shoots would have difficulty penetrating this barrier. *Source:* Herman Timm.

nutrients. Clay colloids carry thousands of negative charges throughout the clay particle and at the broken edges of the clay's alumina and silica layers. Thus, a clay colloid acts as a large, highly negatively charged particle (anion) (Fig. 6–15). A soil's capacity to hold cations is called its **cation exchange capacity (CEC).** It is called exchange capacity because cations are held very loosely by the colloids, and can be replaced, or "exchanged" with cations in the soil water (often called the **soil solution**). Ions on the CEC are held tightly enough, however, to prevent their leaching in percolating water. The abbreviation CEC is also used to refer to the physical entity (cation exchange complex) that holds exchangeable cations in the soil's solid phase. Organic matter has a greater net negativity than clay and has a higher CEC. The CEC is measured in terms of centimoles of posi-

tive charge per kg of soil ($cmol_+ kg^{-1}$), which has a numerical value equivalent to the now obsolete measurement, milliequivalents per 100 g ($meq\ 100g^{-1}$). Not all cations are attracted to or held by the CEC with equal energy. The strength of attraction for some cations when present in equivalent amounts is:

$$calcium\ (Ca^{++}) > magnesium$$
$$(Mg^{++})\ potassium\ (K^+) > ammonium\ (NH_4^+)$$
$$> sodium\ (Na^+) > hydrogen\ (H^+)$$

Most plant nutrients are cations and can be held by soil particles. The most notable exception is nitrate (NO_3^-), which is not held and is readily leached out of the root zone. Cation exchange capacity is influenced somewhat by soil acidity: it is greater under alkaline than under acidic conditions. The difference is slight in mineral soils, but it can be substantial in soils containing large quantities of organic matter.

Soil Acidity and Alkalinity

The acidity or alkalinity of the soil is expressed as its pH. pH is defined as the negative logarithm of H_3O^+ activity (usually approximated by molar concentration) in the soil solution:

$$pH = -\log [H_3O^+]$$

It does not include any H^+ adsorbed to the CEC or otherwise not dissolved in solution. pH is not a fixed characteristic of soil and, depending on a number of conditions, varies over time. Values for pH vary considerably among soils, ranging from about 4.0 for an acid soil to 10.0 for an alkaline soil (Fig. 6–16).

Plants do not grow well in highly acid or highly alkaline soils, although within these extremes plants differ in their pH tolerance. Some grow well in acid soil, others grow better in alkaline soil, others in neutral soil (pH 7.0), and still others grow well over a wider pH range (Table 6–3).

Soils in climates with high rainfall and humidity generally tend to be acid, while those found in arid climates tend to be alkaline. In wet climates, the base elements (sodium, potassium, calcium, and magnesium) are removed from the

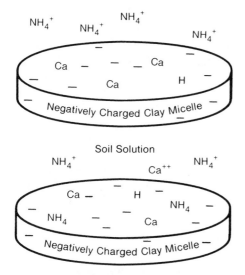

$$Ca\ clay + 4NH_4^+ \longrightarrow (NH_4)_2\ clay + Ca^{++} + 2NH_4^+$$

Figure 6–15 If a soil colloid made up of numerous negatively charged clay micelles is treated with fertilizer containing ammonium cations, the added ammonium cations will replace an equivalent amount of another cation, such as calcium, on the clay colloid and force Ca^{++} ions into the soil solution.

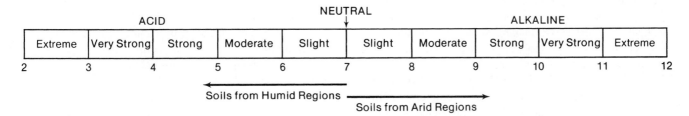

Figure 6–16 The pH for many mineral soils ranges from about 4 to 10. The pH range for most agricultural soils lies between 5 and 8.5.

TABLE 6–3	Soil pH Range for Optimum Growth of Selected Plants		
4.5 to 5.5	**5.5 to 6.5**	**6.5 to 7.5**	

4.5 to 5.5	5.5 to 6.5	6.5 to 7.5
Azalea	Barley	Alfalfa
Bent grass	Bean (snap, lima)	Apple
Blueberry	Carrot	Asparagus
Camellia	Chrysanthemum	Beets (sugar, table)
Chicory	Corn (field, sweet)	Broccoli
Cranberry	Cucumber	Cabbage
Dandelion	Eggplant	Cauliflower
Endive	Fescue	Celery
Fennel	Garlic	Chard
Fescue	Oats	Hydrangea
Gardenia	Peas	(sepals become
Hydrangea	Pepper	pink)
(sepals become	Poinsettia	Leek
blue)	Pumpkin	Lettuce (head, Cos)
Potato	Radish	Muskmelon
Poverty grass	Rye	Onion
Red top	Squash	Parsnip
Rhododendron	Strawberry	Soybean
Rhubarb	Timothy	Spinach
Shallot	Tobacco	Sweet clover
Sorrel	Tomato	
Sweet potato	Turnip	
Watermelon	Wheat	

TABLE 6–4	Characteristics of Saline and Sodic Soils*		
Soil	**Electrical Conductivity**	**Exchangeable Sodium**	**pH**
Saline	>4 mmhos/cm†	<15%	<8.5
Saline-sodic	>4 mmhos/cm	>15%	<8.5
Nonsaline-sodic	<4 mmhos/cm	>15%	>8.5

* Specific thresholds are approximate because many factors affect the measured characteristics.

† mmhos = millimhos = 1/1,000 mho. The mho is the reciprocal of ohm, a unit of resistance.

soil by leaching as well as by the harvested crops that have absorbed them. As the base elements are lost, the exchange sites on the clay colloids become occupied with hydrogen ions, making the soil acid.

Normally, sudden and large changes in soil pH do not occur. Change is generally gradual, especially with fine-textured soils. These and other soils resist such change due to **buffering.**

When necessary or desirable, soil pH can be altered by adding certain materials. The pH of an acid soil can be increased by adding basic amendments or fertilizers. Most often, calcium carbonate ($CaCO_3$), or agricultural lime, is used. This finely ground limestone is spread evenly over the surface and tilled into the soil. The CO_3^{-2} ions in the calcium carbonate neutralize acidity by combining with H^+ ions to produce carbon dioxide and water. Estimating the amount of lime required involves several factors that include the soil pH, texture, organic matter content, structure, crops to be grown (Table 6–3), and the fineness of grind (mesh).

In arid regions, the soil pH is usually neutral or alkaline. Calcium is abundant, making liming undesirable. More likely, the pH has to be lowered to make iron, manganese, and zinc more available to the plants.

Lowering the soil pH can be performed by adding acid-forming chemicals such as aluminum sulfate [$Al_2(SO_4)_3$], ferrous sulfate ($FeSO_4$), and elemental sulfur (S). These sulfur-containing compounds lead to the formation of sulfuric acid (H_2SO_4) and acidify the soil. Alternatively, though less effective, acidic organic materials—such

as peat and decomposed plant materials—can also be mixed with the soil to lower pH.

The prolonged use of chemical fertilizers that are residually acid such as ammonium sulfate [$(NH_4)_2SO_4$], ammonium nitrate (NH_4NO_3), and ferrous sulfate ($FeSO_4$) tend to make the soil acid.

Residually basic (alkaline) fertilizers make the soil reaction more alkaline. Examples include sodium nitrate ($NaNO_3$), potassium nitrate (KNO_3), calcium nitrate [$Ca(NO_3)_2$], and calcium carbonate ($CaCO_3$).

Saline and Sodic Soils

Saline soils contain unusually large quantities of soluble salts. The soluble salts are typically chlorides and sulfates of calcium (Ca^{++}), magnesium (Mg^{++}), and sodium (Na^+), although other soluble cationic salts may contribute as well. Sodic soils differ from saline soils because a large percentage (over 15 percent) of the total cation exchange sites of the soil are occupied specifically by sodium ions (Table 6–4). Displacement of the sodium (Na^+) ion is the main objective in reclamation of a sodic soil. Agriculturally, saline and sodic soils are problem soils that require special handling for successful farming. Excessive amounts of soluble salts are harmful to plants and, when cations are predominantly monovalent (with a single charge), they have adverse effects on soil structure. Soils can be classified on the basis of the kind and amount of salts present, as shown in Table 6–4.

SOIL ORGANISMS

A microscopic examination of a soil sample reveals a wide variety of animal and plant life, some beneficial and essential to human well-being, and some harmful, often causing problems for people, their livestock, or crop plants. The animals include earthworms, gophers, insects, mice, millipedes, mites, moles, nematodes, slugs, snails, sowbugs, and spiders. Plant and plantlike organisms in the soil include actinomycetes, algae, bacteria, and fungi. Certain beneficial fungi live in symbiotic associations with plant roots; the association is called **mycorrhiza.**

Soil organisms act both chemically and physically. They digest crop residues and other organic matter enzymatically, and may physically move the residues from one place to another, mixing it with the soil. Earthworms and burrowing animals mix large quantities of material with the soil mass. The kind and amount of soil organisms depend on several factors, including climate, vegetation, soil pH, fertility level, soil temperature, and soil moisture.

Consider the ways that soil organisms increase crop productivity. The roots of higher plants are a good source of organic matter. Their decomposition by soil organisms produces organic acids and gluelike materials that bind soil particles together to form the aggregates necessary for good soil structure. Following the decomposition of roots, open channels are left in the soil, improving drainage and aeration. Organisms also decompose stems, leaves, and other crop residues.

Nitrogen fixation, sulfur oxidation, and nitrification are processes carried on by soil bacteria essential to higher plants. Legume crops with roots and inoculated with *Rhizobium,* a nitrogen-fixing bacterium, are frequently grown as a nitrogen source for succeeding crops. As the legume crop grows, the bacteria convert unavailable atmospheric nitrogen (N_2) into nitrogenous compounds that the legumes use while the plant furnishes energy to the bacteria. Such a relationship between two dissimilar organisms living together for mutual benefit is called **symbiosis.** When the legume crop dies or is plowed under and decomposes, the nitrogen compounds become a source of nitrogen for a succeeding crop.

Elemental sulfur is not immediately available to higher plants; it must first be oxidized to the sulfate form. Autotrophic *Thiobacillus* bacteria bring about this transformation through a complicated series of reactions. Under certain conditions, autotrophic bacteria oxidize iron and manganese to compounds that are less soluble and thus less available to plants. The action of these bacteria helps prevent toxic amounts of iron and manganese from being taken up by the plants.

Not all soil organisms are beneficial. Some of the most injurious plant pests are soil borne. For example, nematodes attack and destroy plants in a wide range of species. *Phylloxera,* an aphid that attacks grape roots, devastated large vineyard areas until resistant rootstock cultivars were developed. Pathogenic soil-borne bacteria and fungi are also responsible for significant crop losses.

SOIL ORGANIC MATTER

The soil's organic matter content has a profound effect on its biological, chemical, and physical properties. Through the decomposition of organic matter, many nutrients become available to crop plants. Organic matter provides food and energy for soil organisms. Most organic matter, except for a small animal fraction, comes from plants. By weight, about 90 percent is made up of carbon, hydrogen, and oxygen. The remainder is usually nitrogen, sulfur, phosphorus, potassium, calcium, and magnesium plus a minute amount of microelements.

The speed of organic matter decomposition varies according to its chemical composition. It is rapid for simple carbohydrates and slow for fats and lignins. Essentially the decomposition reaction is the oxidation of carbon compounds to carbon dioxide, water, and energy:

$$(CH_2O)_n + O_2 \rightarrow CO_2 + H_2O + \text{energy}$$

Proteins, fats, and other complex compounds decompose in a multitude of reactions to form amino acids, ammonia, nitrates, phosphates, carbon dioxide, and others. After complete decomposition, a complex, amorphous, colloidal substance called **humus** remains that is resistant to further decomposition. This is the material that helps improve soil structure, imparts the dark color to the soil mineral fraction, and increases the soil's water-holding and cation exchange capabilities. For example, the cation exchange capacity of a mineral soil ranges from about 10 to 50 cmol$_+$ kg^{-1}, while the capacity of humus ranges from about 100 to 300 cmol$_+$ kg^{-1}. Being colloidal in size, humus acts similarly to clay colloids in cation exchange reactions, but it is composed chiefly of carbon, hydrogen, and oxygen with small amounts of other elements, while clay colloids largely consist of aluminum, silicon, and oxygen.

Carbon:Nitrogen Ratio

A widely used practice for improving the physical condition of the soil is to incorporate crop residues, green manure crops, or other organic matter. This is a beneficial practice, but one must be aware of the resulting effect it has on the carbon:nitrogen (C:N) ratio. Under natural conditions, there is a close relationship between the amount of carbon and the amount of nitrogen in the soil. This ratio is nearly constant at about 12 parts of carbon to one part nitrogen, worldwide. Variations when present seem to correlate with climate, especially temperature and rainfall. For instance, the C:N ratio tends to be smaller (less carbon, more nitrogen) in arid and warmer regions than in humid and cooler regions. Incorporation of the straw residue from a high-yielding grain crop having a high C:N ratio (50:1) will result in a change in the soil's C:N ratio. Soil organisms, primarily fungi and bacteria, rapidly multiply because of the large source of added carbon food (Fig. 6–17). These organisms also require large amounts of nitrogen for their own growth. Consequently, they tie up the soil nitrogen, causing a temporary nitrogen deficiency for the growing crop.

With time, the organic residues continue to decompose until the soil organisms exhaust their supply of food and begin to die. The nitrogen from the decomposing organisms is then returned to the soil and made available to crops again at about the original nitrogen level. Eventually, soil organisms convert the excess carbon in the straw into carbon dioxide, and the C:N ratio of the straw falls to a ratio that is indistinguishable from that of the other soil organic matter.

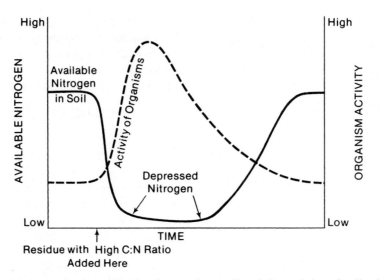

Figure 6–17   The relationship between the available nitrogen in a soil and the activity of soil microorganisms after a heavy application of organic matter with a high C:N ratio.

SOIL DEGRADATION

Natural soils are not static entities; their properties change constantly, but the quality of the soil is usually maintained when these changes occur in response to the slow changes that normally occur in the natural environment. When conditions change abruptly, however, resulting effects on the soil can lead to severe degradation and loss of productivity. Major forms of degradation related to mismanagement include accelerated erosion, loss of organic matter, compaction (already discussed), waterlogging, and salinization.

All upland soils experience small episodes of erosion yearly, but the severity of that erosion is usually minimized by the protection of permanent vegetative cover. When that cover is removed, the soil surface is exposed directly to the erosive forces of wind, rainfall, and runoff. Accelerated erosion removes soil faster than new soil can be formed, reducing the depth of the productive topsoil. If erosion continues unabated, eventually all of the rich A horizon can be lost, forcing crops to root wholly in the undesirable B horizon. Even if the entire A horizon is not lost, erosion preferentially removes smaller particles (including organic matter) first, degrading topsoil productivity long before it is lost completely. Erosion can be accelerated by practices that remove vegetative cover, including overgrazing, intensive and repeated tillage, and deforestation, all common worldwide practices in food production.

Loss of soil organic matter is another facet of soil degradation aggravated by these practices and others. The amount of organic matter in a soil profile is dictated by the balance between additions (mainly plant residues) and losses (mainly as carbon dioxide during respiration). Additions are regulated by the amount of residue the ecosystem supplies on a continuing basis, while losses are limited by the usually low concentrations of oxygen in the soil air. Practices such as overgrazing, crop residue removal, and forest clearing obviously reduce the quantity of organic material available for addition to the soil, but practices such as drainage and tillage have effects that can be even more damaging than these. Drainage removes excess water from the soil, increasing the air (and oxygen) concentration in the soil. Tillage stirs the soil, increasing pore space (though only temporarily) and increasing contact between the soil solids and the atmosphere. Both practices provide aerobic microorganisms with extra oxygen, which they use to respire more carbon than they would under undisturbed conditions. Loss of organic matter eventually leads to deterioration of soil structure and loss of water-holding capacity, among other effects. Unfortunately, most cultivated soils worldwide have suffered significant losses of organic matter since they were brought into production.

Waterlogging and salinization are degradations related to the introduction of cropping and irrigation practices into arid and semiarid regions. Many arid-region soils exist under conditions where only a few millimeters of water may percolate below the root zone each year. Cropping when growing plants are present only part of the year or adding excess water with irrigation can increase this quantity significantly. Excess water can dissolve, transport, and concentrate soluble salts that would have remained undisturbed under natural conditions. Saltwater may be released from the soil on the lower reaches of hillsides, forming saline seeps that eventually make the soil below the seep too salty to support plant life. Saltwater can also accumulate in the production field itself, eventually rendering it too salty for further use. In many instances, the salts are compounds of sodium, which cause the soil to disperse, destroying its structure. Dispersed soils do not transmit water to any degree and are prone to becoming almost permanently waterlogged and useless for further production. Both salinization and waterlogging are widespread, exceptionally severe problems that threaten the viability of agriculture in arid regions worldwide.

Soil Improvement

Often, the best strategy for improving a degraded soil is just to leave it alone for several years. Significant soil improvement can often be accomplished by establishing a permanent cover crop that includes a deeply rooting species, such as a mixed grass-alfalfa cover crop, which ensures sufficient fertility to get improvement started, and then simply watching the cover crop grow. After a few years, the crop will provide fresh organic material throughout the profile, provide an environment for the reestablishment of soil organisms, and aid in rebuilding soil structure while protecting the soil from erosion. One of the major benefits of using a living cover crop for remediation is that roots eventually penetrate deeply into the soil and dry it to deeper depths. This promotes repeated shrinking and swelling of the soil mass, which aids in the formation of pore space and structure deep in the profile.

If establishing a permanent cover over a long period of time is not feasible, adding large quantities of organic material can often help improve a soil, particularly if the degradation is not too deep or severe. Repeated additions of organic matter provide a constant supply of food and raw materials to soil organisms, allowing populations to flourish and create compounds needed to promote soil aggregation. It takes much more organic material than most people realize to improve a soil this way, but it is an effective method if it is practiced repeatedly.

On larger areas, such as corn fields, no-tillage crop production may provide a way to improve the soil. In such systems, crops are planted into the undisturbed remains of the previous year's crop, with no pre-plant soil disturbance. Crop residues protect the soil from erosion, and lack of tillage allows soil organisms to flourish. Repeated use of the practice improves the structure, organic matter content, and infiltration capacity of the surface soil after a few years. Improvement slowly moves downward through the profile.

Salt-affected soils (whether natural or degraded) require different approaches for improvement. Excess salts must be removed before other practices can be employed. Desalinization often involves installing drainage improvements deep in the profile and flushing the soil with fresh water. Salts are removed in the drainage tailwater. If the soil is sodic, additional measures must be taken to remove excess sodium from the CEC before the structure can be restored, usually by incorporating large quantities of a calcium salt such as gypsum into the soil before it is flushed. Calcium is bound preferentially over sodium to the CEC, so when the gypsum dissolves, its calcium displaces sodium from the CEC and prevents its readsorption, forcing it to leach and leave the soil in the tailwater. Reclaiming saline and particularly sodic soils can be extremely expensive and is generally recommended only when high-value crops are to be grown.

LAND PREPARATION

Major purposes of land preparation are to: (1) level the land where needed; (2) incorporate crop residues, green manure, and cover crops; (3) prepare and maintain a seedbed in good tilth (**tilth** is a subjective term for the physical condition of the soil with respect to its capability to provide a good environment—aeration and porosity—for optimizing crop production); (4) help control weeds, diseases, and insects; (5) improve the physical condition of the soil; and (6) help control erosion where needed.

In general, **tillage** is defined as the mechanical manipulation of soil to provide a favorable environment for crop growth. Soil moisture condition is a factor in the effectiveness of tillage. Unfavorable soil conditions—too dry, too wet—will result in ineffective tillage and will damage soil structure. Tillage is done with a wide variety of equipment and for various purposes. The seedbed should provide an environment conducive to rapid germination of seeds and growth. For most crop plants, such a seedbed is one in which the surface soil is loose and free of clods (Fig. 6–18). The subsoil is permeable

Figure 6–18 A well-prepared raised seedbed that might be used for small seeded vegetable crops such as lettuce or carrots (*above*). It should be free from unwanted crop residues or excessive cloddiness (*below*).

to air and water and has adequate drainage and aeration. It should not be water-logged or anaerobic (without oxygen).

Plowing

Often, the first step in seedbed preparation is to plow the land. When large amounts of crop residues are left on the field from a preceding crop, they are often chopped with a disk or a rotary stalk cutter before plowing. Plows invert the soil and cover the trash, but they often leave the soil in large linear lumps that must be reduced in size.

A farmer has the choice between two plow types, the moldboard or the disk plow, each adapted to certain soil characteristics. Moldboard plows, of which there are many variations, range in size from a single moldboard, or bottom, to a gang of plows that turns twelve furrows (twelve bottoms) simultaneously. Each moldboard shears and inverts a slice of soil commonly 15 to 30 cm (6 to 12 in.) deep and 30 to 60 cm (12 to 24 in.) wide as it moves along, leaving the top of the slice at the bottom of the furrow (Fig. 6–19). Moldboard plows are used when the soil is sufficiently moist to allow the plow to pass through easily but not so wet as to cause the furrow slice to stick to the face of the moldboard. If the soil is either too dry or too wet, excessive power is required and poor plowing results. The ideal moisture content for plowing loam soils is slightly less than field capacity.

Two-way reversible plows (flip-over plows) are used to eliminate "dead" furrows (unfilled furrows). They are also used in hilly areas for contour plowing and in irrigated areas where dead furrows hinder irrigation. These plows have right- and left-hand moldboards mounted so that one series plows in one direction, then at the end of the row, the moldboards are mechanically rotated into position so that the second series can plow the return trip (Fig. 6–20). These two-way plows always throw the furrow slice in the same direction regardless of the direction the plow travels.

Figure 6–20 Five moldboard plows, traveling from left to right, are in the soil turning over furrows. At the end of the field on the return trip from right to left, the plow is rotated so that the moldboards now in the air turn the furrows. This plow turns all furrows in the same direction, thereby eliminating "dead" furrows in the middle of the field. This type of moldboard plow is used in irrigated areas to help maintain leveled fields. *Source:* Allis-Chalmers.

Moldboard plows are used on bare fields, small grain stubble, corn stubble, sod pastures, and hay crop fields. The operation can be done in spring, summer, or fall if soil moisture content is satisfactory.

Chisel plows are primary tillage tools consisting of curved shanks spaced widely along a tool bar. They are normally operated at depths similar to those attained with moldboard plows, but they shatter rather than invert the soil in the plow layer. Chiseling can be done more quickly than moldboard plowing and leaves significant quantities of crop residue on the soil surface to provide erosion control. For these reasons, chisel plows are rapidly replacing moldboard plows as the primary tillage tool on many farms, particularly in the Midwest.

Many farmers prefer to plow in late summer or early fall, especially on small grain stubble fields. Fall plowing allows the clods to "slake" (crumble) because of alternate freezing and thawing in the winter. It also distributes the labor load by moving the plowing from the busier spring season to the fall.

The disk plow (not to be confused with the disk harrow) consists of a series of large concave disks 60 to 75 cm (24 to 30 in.) in diameter that are set at an angle to the forward movement and cut into the soil while rotating as the plow moves forward. There can be three to ten or more disks on a plow. Soil moisture conditions are less critical for disk plow operation. These plows are better adapted to dry, hard soils or soils too sticky for a moldboard plow (Fig. 6–21). In other respects, the two types of plow serve the same purpose and accomplish the same results, except that the disk plow generally does not invert the soil or cover crop residues as completely.

Figure 6–19 A four bottom moldboard plow turning under a green manure legume crop. Note how well the green manure crop is being buried. *Source:* Allis-Chalmers.

Figure 6–21   The disk plow accomplishes essentially the same objective as the moldboard plow. However, the disk plow can operate when soil conditions are either too wet or too dry for the moldboard plow. The disks rotate as the plow moves forward, rolling the furrow slice over.

Figure 6–23   A stubble disk is similar to a disk harrow except that the stubble disk has half-circle notches around the periphery of the disks to help cut through dry stubble residues.

Figure 6–22   A disk harrow cutting and covering crop residue near prior seedbed preparation. *Source:* Allis-Chalmers.

Disking

Disk harrows are used to reduce the size of larger soil clods by fracturing them with cleavage and pressure. Disking generally follows plowing, but under some conditions disking can substitute reasonably well for plowing. If the soil is in good tilth, a satisfactory seedbed can be prepared by disking alone.

Disk harrows are general-purpose tillage implements consisting of gangs of concave disks. Most disks have two gangs, one behind the other. In operation, the front gang breaks in the middle into a ∧-shaped configuration so that the sideway forces will balance each other when half of the gang throws the soil to the right and the other half throws it to the left. The rear gang breaks into a-shaped configuration and throws the soil in the opposite direction (Fig. 6–22). Many

larger disks break into operating position by allowing one side of the front gang to break forward while the same side of the rear gang breaks back. This forms a <-shaped configuration (Fig. 6–23).

The depth of penetration is regulated by adjusting the angle of the gangs. Farmers normally set disks to till the soil at shallower depths than do plows, and they may use disks to incorporate fertilizer and pesticides in the upper layers of the plowed soil. Disking to a depth of 10 to 15 cm (4 to 6 in.) is common. The size of the implement varies considerably. Some small tractor-mounted disks cut swaths of 180 cm (6 ft), whereas other units cut swaths up to 12 m (40 ft) wide. A special-purpose disk, called a stubble disk, has semicircular notches cut around the periphery of the disk blade (Fig. 6–23). The notches help cut crop residues more effectively, and in some situations stubble discs are used in lieu of plowing.

Harrowing

The function of the harrow is to reduce further the size of soil clods left after disking, to smooth the soil surface, and to do small-scale leveling. Harrowing also destroys small weeds. This operation generally follows disking. Frequently, farmers attach a harrow behind the disk and do both operations simultaneously. This is a final touch to seedbed preparation, unless beds are to be formed for irrigated row crops.

A wide variety of harrows are used. The principal types are: (1) spike-tooth; (2) spring-tooth (Fig. 6–24); (3) chain or drag; and (4) cultipackers, packers, mulchers, and corrugated rollers. The fourth group crushes clods by applying pressure and tends to break up hard, dry clods better than drag-type harrows. They also pack the soil slightly, reducing large air spaces. Harrowing is becoming less common and necessary, particularly for agronomic crops, as new planters become more capable of operating in uneven seedbeds.

Figure 6–24 A spring-tooth harrow is sometimes used in seedbed preparation because it is effective in breaking up soil crusts, reducing clod sizes, and destroying small weeds. *Source:* Allis-Chalmers.

Listing and Ridging

In some areas, row crops are planted on ridges formed by listers. A lister is a plow equipped with two moldboards that cuts a furrow slice two ways—half to the right and half to the left. This forms a ridge of soil commonly about 20 to 25 cm (8 to 10 in.) high and of variable width at the base. Listers can be equipped with attachments to list, plant, and fertilize in one operation. Some farmers flatten the tops of the ridges with a roller, drag, or bed shaper before planting. For production of some row crops (e.g., corn and soybeans), ridges are often built and maintained in the standing crop using large cultivators.

Cultivation

Cultivation is the tillage between seedling emergence and crop harvest. The main reason for cultivating is to control weeds, but other benefits are improved water infiltration and soil aeration on soils that crust, the conservation of soil moisture, loosening compacted soils (Fig. 6–25), and in some cases help with insect control. Some farmers claim that cultivation is neither necessary nor beneficial. Certainly, some row crops are cultivated much more frequently and deeper than necessary—a practice wasteful of time and energy. Cultivating equipment can be divided into four main classes: (1) row-crop cultivators, (2) field cultivators, (3) rotary hoes, and (4) rototillers.

Row-crop cultivators have various shaped steel shovels that manipulate the soil (Fig. 6–26). The shovels on most equipment are short, narrow, curved, or pointed. For shallow cultivation, wide, thin, horizontal, knifelike blades are used. In fields where vine weeds appear, disks replace the shovels.

Field cultivators are not designed for row crops. They penetrate deeper than row-crop cultivators and are used on fallow ground or stubble fields to control weeds. These cultivators have longer and stronger shovels or sweeps. Some have

Figure 6–25 The shank spacing depth makes considerable difference when loosening compacted soil. These shanks were spaced about 50 cm (20 in.) apart, and operate about 40 cm (15 in.) deep. In this case, however, the shanks were set too shallow. They did not get under the compacted soil layer to break it up, and may in fact have created more compaction. Generally, the closer the spacing and the more sufficient the depth, the more effective the soil loosening. *Source:* William E. Wildman.

Figure 6–26 A multiple-row crop cultivator equipped with cultivation shovels is loosening soil and destroying weeds in this soybean field. The enclosed driver cabs in modern tractors are often air conditioned. Many have computer systems and other types of electronic equipment. *Source:* International Harvester.

spring teeth (Fig. 6–27). Field cultivators are often used for secondary tillage following plowing, instead of disk harrows.

Rotary hoes can be used on crops that are drilled or broadcast planted as well as for row crops. They are especially useful on young crops that are too small for other types of cultivators. Also, they are more efficient when operated at higher speeds than other types. Rotary hoes are made up of gangs of rimless wheels whose spokes resemble slightly curved fingers mounted on a horizontal axle. Generally the equipment is made and used in tandem gangs.

Figure 6–27 A field cultivator (ripper) is used to break up plowsoles or hardpans formed by repeatedly plowing at the same depth. This cultivator is particularly useful in semiarid regions where, because of drought and high winds, it is best to leave the soil protected with a covering of plant residue to reduce wind erosion. *Source:* Deere and Company.

Rototillers or rotary plows are used on small plots. Such implements, powered by gasoline engines or by tractors, have an assembly of rotating knives or tines mounted on a horizontal axle. These machines cut a swath of variable width, depending on machine size. The knives rotate vertically and cut into the soil on the downward stroke. The depth may be relatively shallow or down to 15 cm (6 in.). Rototillers are used for seedbed preparation as well as for cultivation. They mulch and shape seedbeds after listing. Often herbicides are applied to the soil prior to powered incorporation. The chemicals are mixed into the soil to a depth about half the length of the tiller's teeth.

Deep Tillage

Some farmers, especially those in the western part of the United States where irrigation is prevalent, use deep tillage to improve problem soils. Extra heavy equipment is used for deep tillage when the soil is dry. One type of implement called a slip plow uses a V-shaped blade that slips along horizontally from 120 to 180 cm (4 to 6 ft) below the soil surface and lifts the soil mass about 15 cm (6 in.) as it passes through. This shatters the soil profile and breaks any deep, hard, cemented layers (Fig. 6–28).

Another deep-tillage tool is the deep moldboard plow used to turn a furrow slice 150 cm (5 ft) or more deep. The tool serves to bury surface salts.

Another deep-tillage tool is the ripper or deep chisel (Fig. 6–29). This implement consists of one or several shanks that penetrate the soil from 60 to 120 cm (2 to 4 ft). It shatters hardpans best when the soil is dry. Unfortunately, some soils become recompacted and the operation needs to

Figure 6–28 The V-shaped blade of this slip plow is pulled horizontally well below the soil surface to break up hard soil layers. *Source:* USDA Soil Conservation Service.

Figure 6–29 An example of a deep chisel (ripper) used to break deep compacted hardpans. This procedure is used only when justified because the high amount of energy required makes this an expensive operation. *Source:* William E. Wildman.

be repeated every three to seven years, depending on the nature of the compacted layer. In the alluvial desert of the southwest United States, deep ripping is used instead of deep plowing to keep the accumulated soil salts buried.

Deep tillage is expensive, and sometimes it does not materially increase crop yields. In established orchards, it can damage trees by severely cutting the roots.

Conservation Tillage Systems

As noted earlier, tillage is performed prior to and after planting for several reasons. One is to allow the planter to establish good seed–soil contact to promote successful crop establishment. Another, and perhaps the major, reason is to control unwanted, competitive vegetation. As technology has advanced, planters capable of planting into unpulverized soil (Fig. 6–30) have been developed and herbicides have appeared that control almost all weed species without a need

Figure 6–30 A multi-row planter planting corn directly into an alfalfa field with no previous land preparation (no till). *Source:* Allis-Chalmers.

Figure 6–31 A rice field ready to be planted. The levees follow contour lines to allow uniform water depth within each paddy. *Source:* USDA Soil Conservation Service.

for tillage. Farmers and advisers are also realizing that reducing the intensity of or eliminating tillage can protect the soil from accelerated erosion by leaving a blanket of crop residue on the surface, thus protecting it from wind and rain.

Conservation tillage systems often involve one pass with a chisel plow, disk harrow, or field cultivator, followed by planting, and spraying to control weeds. One leveling operation may be necessary following a chisel plowing. Yields are usually equivalent to those achieved with moldboard plowing, provided careful attention is paid to all phases of the production process. When conservation tillage produces lesser yields, it is usually because something else (drainage, fertility, crop rotation, disease or insect control) was neglected. Because crop rotation provides so many production benefits, it is usually considered a major factor influencing the success of crops produced using conservation tillage practices.

The ultimate form of conservation tillage is no-till, a practice in which no tillage is performed prior to planting and the maximum amount of residue is left on the soil surface prior to planting. Weed control is achieved entirely with herbicides. This practice can be extremely effective for controlling erosion and has been heavily promoted by conservation agencies over the past thirty years. It has the additional benefit of usually increasing crop yields over plowing on well drained soils on slopes because the residue cover promotes greater water infiltration and reduces the occurrence of moisture stress during the summer. On more poorly drained soils, however, drainage improvements are essential to prevent the soil from staying too wet in the spring. Excessive wetness can produce several conditions that restrict yield. A combination of improved drainage and crop rotation usually allows farmers to produce yields with no-till planting equivalent to those obtained following use of the moldboard plow.

Land Leveling

Irrigated land generally benefits from being level, especially if flood (Fig. 6–31) or furrow irrigation is used and row crops

are grown. Exceptions are many; some vineyards, orchards, and some high-value vegetables are grown very successfully on rolling land in contoured rows with sprinkler or drip irrigation. Contoured rows put each plant in the row at the same elevation as other plants in the same row.

Land is leveled to permit water to flow and spread evenly over the soil surface without causing erosion. In considering the land's suitability for leveling, the land's productive capacity and the method of irrigation to be used are evaluated. Features that render a site unsuitable include: (1) excessively permeable soil, (2) soil that is very shallow, and (3) rough topography (excessive grading will be needed). Land leveling can also be used to remove excess surface water from poorly drained fields.

If leveling the land is feasible, heavy equipment will be needed (Fig. 6–32). Timing is important. Land should not be leveled in the rainy season because leveling of wet soil subjects it to compaction.

Soil Fumigation

For some crops, the soil must be fumigated before seedbed preparation. These usually are high-value crops where the potential for pest damage is severe enough to justify treatment. Fumigation is normally too expensive to use extensively on most row or field crops. Certain chemicals are used to fumigate soil and destroy harmful bacteria, fungi, and nematodes as well as many weed seeds. A widely used soil fumigant is methyl bromide (CH_3Br). This toxic gas is colorless and odorless, and is usually mixed with chloropicrin (tear gas). Chloropicrin is used to indicate the presence of the more toxic CH_3Br gas because it is nontoxic in small amounts, but it causes discomfort to the eyes. Chloropicrin also has some effect as a fumigant. Methyl bromide will no longer be allowed for most, if not all, soil fumigation after 2005. At the time this text was revised, no suitable substitute was available.

Figure 6–32 Land-leveling equipment using laser technology has greatly contributed to the practice of land leveling. The procedure can now be performed more rapidly and more accurately. At the top of the raised mast in the photograph is a receiver that intercepts the horizontal laser beam emitted from a stationary source located in or near the field. The interception of the laser beam influences the position of the land plane blade. The received signal automatically causes the land plane blade to be lowered (to cut deeper into the soil) or raised as needed to level the field. *Source:* William E. Wildman.

Nursery employees often prefer to use steam instead of chemicals to partially sterilize, or pasteurize, their soil. A closed container with some means of admitting live steam is used. The temperature at the center of the soil mass is brought to above 71°C (160°F) for thirty minutes to destroy disease-causing organisms. Some nurseries use aerated steam at about 60°C (140°F) for thirty minutes to do the same job.

Artificial Soils

Several products are available for growers that are prepared and marketed under descriptions such as "artificial media," "potting soils," "soilless mixes," "container mixes," and so on. Most artificial medias are utilized for production of greenhouse and nursery crops in pots. They can be bought premixed or custom-blended by a grower. They are prepared for very specific uses, but most are designed to facilitate rapid seedling emergence, rooting, and early plant development, and allow for relatively easy management of water and fertilization. They are usually prepared to hold significant amounts of plant-available water after irrigation but also to drain readily and thus avoid waterlogging that can damage or kill roots.

The factors that influence soilless mix characteristics are the same as those that influence native soils. Particle size and structure determine porosity, CEC and water-holding characteristics. Handling influences compaction and particle

degradation. Pot-filling machines are specifically designed to avoid destroying particle size or compacting the mix as it is blended then put into pots. Degradation can also occur over time when the mix contains organic materials that compost with age.

The media can contain a blend of many different components, such as organic soil, sand, peat moss, and expanded vermiculite. Other commonly found ingredients include coir fiber from coconut husks; bark; sawdust; perlite (a special kind of sand that is heated to extreme temperatures until it pops like popcorn); and calcined clay, which contains small (kitty-litter size) particles of kiln-fired clay. Each component gives the mix a certain characteristic. For example, a typical mix may contain sand for large pore spaces and good drainage, vermiculite to hold water and some nutrients, and sphagnum peat moss for its nutrient-holding capacity and to give the mix an acidic pH. The addition of a wetting agent is often needed to allow the mix to take up water when the ingredients have been dry when first added. Peat moss is very good at holding water when it is already wet, but it repels water when it gets dry. (If you have ever tried to moisten a dry pot containing peat moss, you know what we mean.) Many mixes also contain a small amount of a complete fertilizer called a starter charge, which provides nutrients during the early stages of growth when the plant needs fertilizer but is too small to remove enough water to allow for application of a fertilizer solution.

SOIL CONSERVATION

Soil degradation began long before people started farming, but the process was accelerated by permanent agriculture and land tillage. In America, erosion and soil depletion became problems as soon as settlers migrated from Europe, mainly because of the clearing of land for crop production. As population increased, more land was cleared, cropped, and depleted of nutrients. Much of the westward movement in the United States was a search for new, more fertile lands; eastern soils had lost their productivity by continuous cropping. Under the Homestead Act, the US government encouraged the movement by offering free land to those who would move west and settle.

George Washington and Patrick Henry were among the earliest American land conservationists. In his final message to Congress in 1796, Washington urged the creation of a board of agriculture. More than a half-century later, in 1862, Lincoln established the Department of Agriculture. Little interest in soil conservation was felt for the next seventy-five years because of the availability of new, western lands. During the 1920s, a soil surveyor, H. H. Bennett, called attention to the waste and depletion of America's greatest natural resource—land. Finally, in 1929, Congress established ten soil conservation experiment stations, and assigned personnel to study and gather information on erosion control measures.

The Natural Resources Conservation Service (NRCS) was established in 1933 in the Department of Interior as one means of helping the United States recover from the Great Depression of 1929 to 1935. The need for immediate soil conservation became evident on Black Sunday, April 14, 1935, because that day the most severe dust storm in U.S. history completely blotted out the noonday sun (Fig. 6–33). During the summer and fall of 1935, the skies over Washington, D.C., and New York were darkened with topsoil blown from Texas, Oklahoma, and other prairie states. That same year, the US government created the first erosion control agency ever established by any nation. The first soil conservation act soon followed, charging the agency with the responsibility of cooperating with farmers to demonstrate effective land management and erosion control practices (Fig. 6–34).

Soil conservation is the preservation and extension of the life of soil by using land wisely, keeping it in its most productive state for the present and future generations. Lands best suited for grazing of animals are planted to sod crops.

Hilly or mountainous land is kept in trees, which are harvested as timber. Plowing up grassland prairie soils and planting them in row crops in semiarid regions without irrigation has proven to be disastrous. They should be left as grasslands. It does not take long for wind or water to erode and remove the fertile topsoil and form gullies (Figs. 6–35 and 6–36). One of the complicating factors in soil conservation is our dependence on the soil for food and fiber. The land must be used but at the same time saved for future use. To do both takes wise land management.

Extent of Erosion

Recent soil surveys in several countries show that vast areas of productive land have been damaged beyond recovery.

Figure 6–35 The beginning of a deep gully. Special care and treatment will be required to conserve this land. If it is left in grassland, however, further erosion can be prevented.

Figure 6–33 Two consecutive years of drought, followed by high winds, blew immeasurable amounts of fertile topsoil from areas of Texas and Oklahoma into the Atlantic Ocean. This catastrophe created what came to be known as the dust bowl on Black Sunday, April 14, 1935. The late afternoon sun, a circular ball above the auto, is barely visible through the dust cloud. *Source:* Library of Congress.

Figure 6–34 Catastrophic soil erosion caused by drought and high winds occurred in the Texas panhandle during the early 1930s. Vast areas of previously productive farmland were devastated. *Source:* Library of Congress.

Figure 6–36 Rill erosion has begun in this orchard, a forerunner of severe gully formation if not checked soon. *Source:* USDA Soil Conservation Service.

Erosion continues to be a critical concern in almost every agricultural region of the world. Erosion is particularly severe where intense torrential rains are frequent. Unfortunately, the need for food in some nations has overshadowed the danger of uncontrolled erosion. This is particularly true in East Africa, the Yellow River basin in China, Eastern Europe, Latin America, and parts of Australia, India, and the United States—all plagued with serious and widespread erosion. In the United States, erosion has damaged nearly 110 million hectares (272 million acres). The soil erosion problem in these nations is not over the entire nation, but it is usually severe in hilly high rainfall areas and/or in the semiarid regions and especially in those areas where irrigation is not used. At times when the demand for food crops (at home and abroad) is high, economic pressure is put on the farmer to seed land to grain or other row crops when the soil should remain as grassland. This practice contributes to erosion.

Factors Affecting Erosion

An important factor in erosion control is the amount of plant cover. Land covered with sod or trees loses little, if any, soil, while barren land can quickly lose considerable topsoil. The intensity, duration, and distribution of rainfall are also factors. A torrential rain of short duration on land with little plant cover causes severe soil losses while a gentle, evenly distributed rain causes less. Topography of the land is also a factor. Level land is less likely to erode than sloping land (Fig. 6–37). The soil's physical properties affect erosion. Deep permeable soils that absorb water are less likely to erode than shallow slowly permeable soils.

Figure 6–37 Water does not erode level land. Leveling, as shown here, is one way of preventing erosion by water where irrigation is necessary. *Source:* USDA Soil Conservation Service.

Gently sloping land can be cropped if proper erosion controls are used, but row crops that require tillage for weed control should never be planted in rows that run up and down steep hills. Row crops on gentle slopes require contoured rows. Sod crops or crops planted by broadcast methods should be used to reduce hillside erosion losses.

Methods of Conservation

The appropriate method of soil conservation depends on the topography, soil type, cropping and livestock system, and climate. To help with these decisions, the US farmer can call on the Natural Resources Conservation Service (NRCS). A professional conservationist will survey and classify the soil into one of eight broad land-capability classes according to its best use with least erosion (Table 6–5). The important consideration is that each parcel of land is managed according to its needs. This means that land not suitable for any type of agriculture, even though unaffected by erosion, should be left for wildlife and recreation; forest land should be used to produce trees, range and grassland should be used to produce forage, and land suitable for cultivation should be reserved for crop production.

Each kind of farming needs its own special conservation practices, but even with careful land management, additional measures are often necessary to improve land use. In addition to the conservation tillage methods described earlier in this chapter, the following subsections describe ways to conserve soil.

Grass Waterways

These are strips of land of varying width permanently seeded to a grass sod. They conduct water to drainage outlets and control runoff from sloping land with cultivated crops. Waterways are used with contours or terraces that drain into them.

Contour Tillage

One easy cropping practice that reduces losses of topsoil is to till the land on the contour (level elevation) instead of up and down the hill. The land is plowed and the crop rows planted and cultivated around the slope, always at the same elevation from end to end. The rows are curved and sometimes come together in points. The ridges left by the tillage tools form small dikes to catch water, allowing more time for it to percolate into the soil instead of running down the hill.

Contour Strip-Cropping

This effective practice is used to conserve both soil and water. Soil conservation is enhanced by alternating strips of solid-planted crops with row crops; for instance, strips of grain or hay crops can alternate with corn or sugar beets (Fig. 6–38). The strips always run on the contour. In some cases this practice reduces erosion by more than half of what it would be if either crop were planted alone.

TABLE 6–5	Land Capability Classes	
Class	**Use**	**Conservation Practices Needed**
I	Few limitations. Suitable for wide range of plants. Can be used for row crops, pasture, range, woodland, and wildlife. Not subject to overflow.	Needs ordinary management practices—fertilizer, lime, cover, or green-manure crops, conservation of crop residue, animal manures, and crop rotations.
II	Some limitations. Choice of crop plants reduced. With proper land management, land can be used for cultivated crops, pasture, range, woodland, or wildlife.	Limitations few and easy to apply. Problems may include gentle slopes, moderate susceptibility to wind or water erosion, less than ideal soil depth, slight salinity. May require special conservation practices, water-control devices, or tillage methods, terraces, strip cropping, contour tillage, special crop rotations, and cover crops.
III	Severe limitations reduce choice of plants and/or require special conservation practices. May be used for cultivated crops, pasture, range, woodland, or wildlife.	May require drainage and cropping systems that improve soil structure. Organic matter additions might be needed. In irrigated areas, soils may have high water table, high salinity, or sodic accumulations. Soils may be slowly permeable.
IV	Severe limitations that reduce choice of plants. Requires very special management. Limited use for cultivated crop but can be used for pasture, range, woodland, or wildlife.	Limited cultivated crops because of steep slopes, susceptibility to wind or water erosion, effects of past erosion, shallow soils, overflows, poor drainage, salinity, adverse climate. May be suitable for orchards and ornamental trees and shrubs. Special practices needed to prevent soil blowing and to conserve moisture.
V	Land limited in use—generally not suitable for cultivation. Little or no erosion hazard but has other limitations. Use limited to pasture, range, woodland, or wildlife.	May be nearly level but has excessive wetness, frequent overflow, rocks, or climate variations. Cultivation of common crop not feasible but pastures can be improved and benefits from proper management can be expected.
VI	Severe limitations make the land unsuitable for cultivation. Restricted to pasture, range, woodland, or wildlife.	Pastures can be improved by seeding, liming, fertilizing, water control with contour furrows, drainage ditches, etc. Have severe limitations that cannot be corrected, thus not suitable for cultivated crops. Some soils can be used for crops such as sodded orchards, berries, etc.
VII	Severe limitations make land unsuitable for cultivation. Use limited to grazing, woodland, or wildlife.	Physical condition of soils prevents range or pasture improvement practices. Restrictions are more severe than those of Class VI. Can be used for grazing. May be possible to seed some areas.
VIII	Limitations preclude use for commercial plant production. Use restricted to recreation, wildlife, or water supply.	Cannot be expected to yield any significant return from crops, grasses, trees, but benefits from wildlife use and watershed protection or recreation are possible. Class VIII includes badlands, rock outcrops, sand beaches, river wash, mine tailings, etc.

Source: USDA Handbook 210, 1973. A detailed description is available at *http://soils.usda.gov.*

Terraces

Terraces are used on long gentle slopes to decrease runoff and to increase water infiltration. On gently rolling land, terraces are low broad mounds that follow the contour and retain water that would otherwise run down the slope. The terraces are constructed with a slight grade so excess water flows slowly to an outlet, often a grass waterway. Terraces are also used in some places on steep slopes (Fig. 6–39). They have been used for centuries in the Andes, China, and many other parts of the world where insufficient flat arable land is available. In Thailand, rice is grown on steep, terraced hillsides.

Wind Erosion

Wind erodes land by removing topsoil just as water does. As with water erosion, the best protection against wind erosion is to provide vegetative cover for the land during periods of high winds. Tillage methods such as stubble mulching have been helpful. Leaving the soil surface rough or cloddy reduces wind velocity at the soil surface, and windbreaks are

Figure 6–38 Contour strip cropping is useful for conserving water as well as reducing soil losses by wind erosion. In this case a cereal grain crop has been alternately planted with a hay crop. Sometimes if the area is subject to sudden but infrequent torrential rains, permanent flat, broad terraces are formed on the contour. *Source:* R. L. Haaland.

Figure 6–39 Terrace cultivation of avocado trees in southern California. This method enables the use of land areas not normally suitable for cultivation. Usually, high-value crops and favorable climatic environments are important factors in the decision to use the terrace method.

helpful. These vary in size from tall trees to hedges planted close together perpendicular to the prevailing wind.

WATER MANAGEMENT

The relationship between soil and water in the soil as it relates to plant growth is an important concept of plant physi-

Figure 6–40 Excessive soluble salts in soils and from applied irrigation water may accumulate in lower areas of a field or where drainage is poor. High levels of soluble salts interfere with plant growth. The loss of plants in parts of this field was attributed to excessive salts but was also aggravated by extensive soil crusting. *Source:* William E. Wildman.

ology that is discussed in detail in Chapter 12. Because water quality and irrigation are components of soil management, however, they are also discussed here.

WATER QUALITY

In evaluating the suitability of land for irrigated agriculture, both the availability and quality of water must be considered. Water used for irrigation almost always contains measurable quantities of dissolved substances that, in general, are soluble salts. These include small but important amounts of dissolved salts originating from the weathering of parent rocks. The value of water for irrigating crops is determined by the amount and kind of salts present.

Saline Water and Salinity

If irrigation water contains soluble salts in sufficient quantities to accumulate within the root zone and interfere with crop yields, a salinity problem arises (Fig. 6–40). The concentration of salts in the water can increase soil salinity to the point where the osmotic pressure leaves plants unable to extract sufficient water for growth. The plants show much the same symptoms they would in a drought—wilting, reduced growth, and in some plants a color change from bright green to bluish green. The wide range of salt tolerance among agricultural crops permits the use of some saline water for irrigation: water that is too saline for one crop can be used for a more salt-tolerant crop.

Permeability

Poor soil permeability adds to cropping difficulties by increasing soil crust formations, which interfere with seedling emergence. Water-logging of surface soil can also occur, increasing diseases, limiting gaseous diffusion, and causing nutritional problems.

The first step in evaluating soil permeability problems is to determine the total amount and kind of soluble salts in the water. Water with low salt content can result in poor soil permeability, just like water with an excessive salt content. Pure water brings into solution precipitated calcium and other soluble salts, resulting in dispersion of soil particles, temporarily blocking and reducing soil pore space. The second step is to consider the ratio of the sodium to calcium plus magnesium content [Na(Ca + Mg)] in the water. When the ratio is higher, dispersion is increased. Third, the carbonates and bicarbonates are evaluated because their presence also contributes to poor permeability.

Toxicity

High concentrations of some salts in water are toxic to some plants. The problem occurs when certain specific ions such as boron, chloride, or sodium are taken up by plants from the soil solution in sufficient quantities to reduce yield or stop growth.

Other Related Problems

Various other situations related to salt concentrations in water can arise. For example, excess nitrogen in irrigation water can cause lodging in grain, delayed maturity of some fruit and vegetable crops, and increased vegetative growth at the expense of root or tuber growth. Sometimes, calcium, magnesium carbonates, or bicarbonates deposit on the leaves and fruits of orchard crops, grapes, and vegetable crops after sprinkle irrigation. These white deposits, while not particularly harmful, do tend to lower quality by creating an unsightly product.

IRRIGATION

Farmers have irrigated crops for over 4,000 years. Records indicate that crops were irrigated along the Nile, Ganges, Tigris, and Euphrates rivers as early as 2600 B.C. It has been suggested that crop irrigation contributed to the founding of the great civilizations in these areas. It is interesting to note that early civilizations began alongside the rivers in arid or semiarid regions. Irrigation canals over 1,000 years old have been found along the Gila River in Arizona. But even more amazing, while irrigation has been practiced so long, modern practices date back less than 200 years, and even today farmers in many parts of the world lift water by treadmills or water screws.

The importance of irrigation is evident when precipitation patterns over the earth's total land area are considered. Seventy-five percent of the total land is semiarid to arid and receives, on average, less than 50 cm (20 in.) of rainfall annually; 20 percent receives less than 25 cm (10 in.).

By the end of the twentieth century, the total irrigated area in the world more than tripled since 1900. Most of the increases have occurred in China, India, Pakistan, the United States, and the former Soviet Union. The expansion of water management and irrigation is illustrated by the increased number of plans and proposals executed since the 1950s. China has built no fewer than forty-six dams, and India has almost doubled its irrigated acreage. Ambitious programs are in progress in many parts of the world, including Africa, Pakistan, Spain, Australia, Italy, and the United States.

With the completion of the Aswan High Dam in Egypt in 1967, the Nile River could no longer flood its banks. While this project permitted reclamation of large areas of desert, the settling of fertile silt in Lake Nasser behind the dam removes plant nutrients that formerly were responsible for the creation of the fertile Nile Delta. Also, with irrigation water being available to Egyptian farmers upon demand instead of annually as before the High Dam, intensive cropping is common. This practice is causing a multitude of soil problems, such as salinity and high water tables.

In 1957, a California water plan was adopted to better control, conserve, and utilize the state's water supplies. The plan provides for storage and transportation of water from water-rich northern California to water-deficient areas to the south through a system of aqueducts on both sides of the Central Valley (Fig. 6–41).

Currently a massive project is underway in China to move nearly 45 billion cubic meters of water annually from the Yangtze River in southern China to the Yellow River in northern China. The increased flow of the Yellow River will provide water for crop irrigation and human consumption in areas where water supply is low. The project is highly controversial however; many believe it is unlikely that the third phase, which would provide much of the water designated for irrigation, will be constructed.

Methods of Application

Selection of the proper water distribution system can save expensive labor and assure better crop yields as well as saving water. The method of application is important, especially if the cost of water is high. Some factors that determine the method and type of system used are: (1) climate, (2) type of crop, (3) cost of water, (4) slope of field, (5) physical properties of soil, (6) water quality, (7) water availability, (8) drainage capability, and (9) salinity or other problems.

Border or Flood Method

Flood irrigation is used where the topography is flat and level. This method is often used for drilled or broadcast crops, such as hay, pasture, and small cereal grains. Orchards and vineyards are also sometimes flood irrigated (Fig. 6–42).

The land must be graded and leveled for flood irrigation. The amount of grading needed depends on the topography, cropping system, and cost of grading. A uniform downslope of 0.1 to 0.4 percent is used for most soils and crops, with little or no cross-field slope. [A 0.1 percent downslope drops

Figure 6–41 The Tehama-Colusa canal shown here (*above*) is part of a water transport system through which water from a northern California reservoir (*below*) flows down the west side of the Central Valley to southern California. This and similar large lined canals effectively transport large volumes of water for agricultural as well as industrial and domestic use.

Figure 6–42 An orchard irrigated by flooding. The levees are contoured, causing the water between the levees to be the same depth over the entire field. This provides even water distribution and uniform water penetration into the root zone. In an older orchard, the root zone could be 2 or more m (6 or more ft) in depth. *Source:* USDA Soil Conservation Service.

Furrow Irrigation

Furrow irrigation is a modification of flooding—water is confined to furrows rather than wide checks. Water is used more efficiently with furrows than with flooding because the entire surface is not wetted, thus reducing evaporation losses.

Furrow irrigation is frequently used for row crops, orchards, and vineyards. The length of furrow varies from 30 m (100 ft) for small gardens to 450 m (1,500 ft) for field crops, but lengths of 90 to 180 m (300 to 600 ft) are more common. Long furrows cause greater loss of water because of deep percolation and excessive soil erosion at the head of the field.

Furrow spacing is determined by the plant row spacing. One irrigation furrow is generally provided for each crop row (Figs. 6–43 and 6–44). The furrow spacing can be 60 to 180 cm (2 to 6 ft), depending on the type of crop and wetting characteristics of the soil.

The depth of the furrow should be such that the water can be controlled. Water should flow in the furrow for sufficient time to allow it to percolate into and across the bed, although not directly wetting the surface of the bed. For most row crops and orchards, furrows from 20 to 30 cm (8 to 12 in.) deep provide the necessary control. Furrows from 10 to 15 cm (4 to 6 in.) in depth are better for small-seed crops.

Uniform crop maturity, necessary for mechanical harvesting, is easier to achieve with furrow than with flood irrigation. Furrow irrigation applies water more uniformly than flood irrigation and improves uniformity of crop development.

An increasing number of fields combine a type of flood irrigation with wetland management. These fields have

0.1 meter in elevation for each 100 m of field length (0.1 ft/100 ft).] Permanent or temporary levees are constructed running downslope, with a border disk forming ridges that divide the field into strips or checks, preferably not over 10 to 20 meters (33 to 66 ft) wide and 90 to 300 meters (300 to 990 ft) long.

Water from an irrigation pump or canal is turned into the supply, or head, ditch at the higher end of the field. It is released or siphoned into one or more checks and allowed to flow slowly downslope, spreading evenly and uniformly over each entire check as it advances toward the lower end. Ponding, excessive percolation, and inadequate wetting of the soil in different areas in the field should be avoided. Designing and operating such a system efficiently requires considerable experience, skill, and knowledge.

Figure 6–43 Pre-planting furrow application of water in process. The raised beds are reshaped just before or during planting to provide a smooth uniform surface. Note the penetration of water from the furrows into the beds.

Figure 6–44 A tomato field being furrow irrigated with water brought to the field in an open ditch, then siphoned over the ditch bank into the field. The water level in the ditch must be higher than that in the field. *Source: The Daily Democrat.* Woodland-Davis, California.

a drainage system that flows into a wetland. During periods of drought, some of the water from the wetland is pumped back into the field through the drainage pipes. The soil is flooded from below.

A variation of flood irrigation called ebb and flood or ebb and flow is used in many greenhouses. Plants are grown on benches designed to fill and drain mechanically. The process can be controlled manually or by a computer. The water level reaches approximately 0.75″ up the pot, so the pots absorb the water through capillary action. If the pots are dry when the beds are flooded, capillarity may not be established with the water in the bench. In that case, the pots have to be irrigated with a hose until there is enough moisture in the pot to establish capillarity. Liquid fertilizers can be applied through the system too. The water and nutrient solu-

tions can be recycled. When recycling is practiced, caution must be taken to prevent the spread of disease. Fertilizer concentrations must be closely monitored and adjustments made for losses due to plant uptake.

Sprinkler Irrigation

Sprinklers are often used when flood or furrow irrigation is impractical. Sprinklers can have some advantages over other irrigation methods; for example, through nozzle size selection, sprinklers are adaptable to high or low soil permeabilities. Through uniform wetting of the surface area, seed germination is more uniform; less total water may be used, and sprinklers are effective in washing salts away from salt-sensitive crops. Additionally, the surface need not be level; a properly designed system offers a method of adequately and accurately applying water, even on sloping land with grades up to 3 percent. Sprinklers are sometimes also used for frost control.

In some situations, sprinkler systems may require less labor, whereas with systems that are highly portable, more labor may be used. In general, equipment and energy costs are higher than with flood or furrow methods.

There are different types of sprinklers, each with certain advantages.

The hand set was the first type of sprinkler system developed. The main lines are either buried or portable. This system is used on many crops and is particularly useful for germinating small-seed crops, especially if the seedbed is rough. A disadvantage of the hand-set system is the added labor required to move the lateral lines as each set is completed. This can be overcome by having a solid-set system, one that is not moved. However, this requires more investment in equipment. The main line is located at the edge of the field, and the laterals are usually set perpendicular to the main line. With row crops, the laterals may be placed either parallel or perpendicular to the rows.

The permanent set type has all lines buried below the surface. In some systems, the lines are suspended above the surface. Because of the high investment cost, this type of system is usually restricted to orchards, vineyards, high-value crops, or recreation facilities (Fig. 6–45).

The wheel line system was designed to reduce labor by moving pipe across the field with a small gasoline engine. The sprinklers are mounted in the lateral pipe, which serves as the axle. To move to the next position, the gasoline engine turns the axle and propels the entire line across the field, saving labor and time (Fig. 6–46). Under windy conditions, application uniformity is poor.

The center pivot system is another labor-saving variation of sprinkler irrigation. The system is used more often in areas where land values are low or the availability of labor is low. The line, mounted on wheels driven by water pressure or electric motors, follows a circular path, pivoting around a fixed central point. One disadvantage of the system is that the circular irrigation pattern leaves the corners in rectangular fields unirrigated (Fig. 6–47). In some areas where land

Figure 6–45 It would be impractical to use methods other than sprinkling for irrigating this permanent pasture because of the uneven topography. This is a permanent set type of sprinkler system similar to those used for landscape golf courses and athletic fields. *Source:* USDA Soil Conservation Service.

Figure 6–46 A wheel line system of sprinkler irrigation.

Figure 6–47 Aerial view of grain fields in Nebraska being sprinkle-irrigated by central pivot systems. The white radius lines are the operating sprinklers. Modification of this equipment will enable the corners of these fields to be irrigated. *Source:* Valmont Industries.

and crop values justify the added costs, the system is modified by adding sprinklers that fold out as the corners of the field are approached, then close back as the line moves past the corner. Sometimes the corners are irrigated with a permanently buried auxiliary line of sprinklers, again increasing equipment costs.

Drip Irrigation

Drip or trickle irrigation is the latest development in irrigation systems. Small amounts of water are allowed to trickle slowly into the soil through mechanical devices called emitters, wetting the soil without runoff (Fig. 6–48). The emission rate of water ranges from about 2 to 8 liters per hr (0.5 to 2 gal/hr).

Figure 6–48 *Above:* One type of water emitter used in drip irrigation. The tiny plastic tubing is attached to a larger plastic tube from which the emitter receives the water. The number of emitters depends on the type and size of crop irrigated. *Below:* A less permanent but popular method of trickle irrigation utilizes thin wall plastic tubing having small perforations in the wall that allow water to trickle out. The advantages of this system are its low cost and ease of installation. It is operated at low pressures and easily repaired. Some disadvantages are its short life, a tendency to tear, and frequent plugging of the small perforations. It is well adapted for some annual crops, particularly those of high value, like the asparagus, shown here in fern.

Emitters are connected to a small plastic lateral tube, laid either on the soil surface or buried just beneath it for protection. Some systems have the emitters built into the lateral line or tube. The lateral lines are connected to a buried main line that receives water from a head source. The head source is the control station for the system. Here the water is filtered, may be treated with fertilizers, and is regulated for pressure and timing of application. Some advantages of drip irrigation are (1) the system need not be moved, (2) there is little interference with other cultural operations because much of the soil surface is not wetted, (3) there is less fluctuation of soil moisture in the root zone area because of the constant and slow drip application of water, and (4) less water is needed to grow a crop. The area of wetted soil can be as little as 10 percent of the total area of newly planted tree crops or up to 60 percent of the area of a mature crop. The amount of soil wetted depends on the soil's physical properties, the time of application, and the number of emitters used. Some objections are (1) expensive filtration equipment is needed to avoid frequently clogged emitters; (2) water distribution may be uneven on hilly land; (3) salts tend to concentrate on the soil surface and near the wetted area boundary, and because leaching with excess water does not occur; and (4) the distribution of roots may be restricted to the small volume of wetted soil.

Drip irrigation may not fit the needs of every crop or situation, but its use by orchardists, strawberry growers, or-

Figure 6–50 A five-year-old peach orchard is drip-irrigated with emitters placed around each tree. Only the soil around the root area receives water. As the trees grow and enlarge their root systems, more emitters are added to wet a larger volume of soil.

Figure 6–49 In the western United States, the use of drip irrigation on multi-row raised beds for strawberry production has increased and has largely replaced furrow irrigation practices. Strawberries such as these are often mulched with plastic film. The film is placed over the beds, and openings in the film are made where the plants are located. Thereafter, further foliage growth is above the plastic mulch. Clear film is used to warm the soil and the film prevents fruit from contacting the soil. Soil fumigation is used to control soil diseases and weeds.

namental nurseries, and for some high-yield field crops is rapidly increasing (Figs. 6–49 and 6–50). It is the most common form of irrigation in greenhouses and nurseries.

FERTILITY MANAGEMENT

Use of fertilizers has probably increased crop yields and reduced hunger more than any other single agricultural practice. In the United States, from 1970 to 1985, the annual consumption of chemical fertilizers has increased about 5 percent per year.

In addition to supplying nutrients to crops to increase yields, fertilizers can also cause marked changes in soil characteristics, some beneficial, some not. These secondary influences play an important role in the choice of fertilizer.

Sixteen chemical elements are known to be essential for the growth of most plants, and a few others are used by some plants under certain conditions. The essential elements are carbon (C), hydrogen (H), oxygen (O), nitrogen (N), phosphorus (P), potassium (K), calcium (Ca), magnesium (Mg), sulfur (S), iron (Fe), manganese (Mn), molybdenum (Mo), copper (Cu), boron (B), zinc (Zn), and chlorine (Cl). Some halophytes (plants that require salts) have been shown to need sodium, and some microorganisms that fix nitrogen symbiotically or nonsymbiotically require cobalt.

Mineral nutrients are divided into groups according to the quantity plants use. The primary **macronutrients**—mineral nutrients used in largest amounts—are nitrogen, phosphorus, and potassium; the secondary mineral nutrients—used in lesser amounts than primary—are calcium, magnesium, and sulfur; and the remaining used in the smallest amounts are the **micronutrients.** Micronutrients are sometimes referred to as trace elements. A complete discussion of mineral nutrition in plants can be found in Chapter 13.

Mineral nutrients are supplied to the soil by applying crop residues, animal manures, chemical fertilizers, or naturally occurring minerals. Other sources are the atmosphere, irrigation water, rainfall, and the elemental nutrients of the soil itself that can enter the water solution. The actual source of the nutrient (organic or inorganic) is unimportant to the plant as long as the mineral nutrients are available in sufficient quantity and can be easily assimilated. Nutrients are lost from the soil when any part of a crop is removed. Returning the crop residue to the soil does not replace all of the removed nutrients. To maintain or improve the soil's fertility, nutrients must be added in one form or another in amounts equal to or greater than those removed by crop harvest (Figs. 6–51 and 6–52). Generally, commercial fertilizers are easier to apply and manage than manures and crop residues, but the latter two should not be disregarded. They are especially beneficial in adding organic matter to help improve soil structure.

A complete fertilizer contains the three primary nutrients: nitrogen, phosphorus, and potassium. It may also contain some secondary or micronutrients. Each bag of commercial fertilizer carries a label stating the analysis of

Figure 6–52 Applying a dry fertilizer to a field before plowing. *Source:* Robert Mullen, The Ohio State University.

its contents. This analysis is represented by three figures; for example, 5–10–5. The first figure is the percentage nitrogen by weight; in this case, 5 kg of nitrogen per 100 kg (5 lbs/100 lbs) of fertilizer. The second figure represents phosphorus, specifically 10 percent phosphoric acid (P_2O_5); the third figure is potassium, specifically 5 percent potash (K_2O).[1] Thus, one metric ton (MT 1,000 kg) of a 5–10–5 fertilizer contains 50 kg (110 lb) of nitrogen, 100 kg (220 lb) of phosphoric acid (P_2O_5), and 50 kg (110 lb) of potash (K_2O), a total of 200 kg (440 lb) of nutrients. The remaining 800 kg (1,764 lb) consists of other chemicals in the formulation of filler.

While the label on the bag states the percentage of each primary nutrient, it may not indicate the compounds used to make up the fertilizer. For example, the nitrogen in the fertilizer might be supplied as urea, ammonium nitrate, ammonium sulfate, or sodium nitrate, and so on. The formulation is important because it informs the user of what compounds are used and their chemical form. It indicates the fertilizer's nutrient availability, effect on soil pH, ease of incorporating into the soil, and freedom from caking.

Assume that nitrogen in the form of ammonium sulfate ($(NH_4)_2SO_4$), phosphoric acid as monocalcium diphosphate [$Ca(H_2PO_4)_2$], and potash as potassium chloride (KCl) are used to formulate a complete fertilizer. Their atomic weights

Figure 6–51 Significant amounts of fertilizer are applied by air. Large areas of range land not readily accessible by other equipment, water-covered rice paddies, and large fields seeded to cereal grains are particularly suitable for aerial application of fertilizers and other chemicals.

---

[1]% P = % P_2O_5 × 0.43
% K = % K_2O × 0.83
% P_2O_5 = % P × 2.29
% K_2O = % K × 1.21

are used to calculate the weight of each ingredient needed to supply the proper amount of nutrients. We find that it requires 238, 165, and 79 kg/MT (476, 329, 158 lb/t) of ammonium sulfate, calcium dihydrogen phosphate, and potassium chloride, respectively, to produce a fertilizer with 5–10–5 percent of N, P_2O_5, K_2O. The three constituents total 482 kg/MT (963 lb/t). The balance needed to make up the weight is filler. The filler used most often is dolomitic limestone or gypsum, but other materials, even sand, can be used.

Fertilizer recommendations are often given as a ratio. A ratio differs from an analysis because it expresses the amount of one nutrient in relation to the other. For example, the ratio of nutrients in 5–10–5 fertilizer is 1:2:1. Thus, if the recommendation was for a 1:2:1 fertilizer, a product labeled 5–10–5 or 10–20–10, or 15–30–15 would be equally acceptable as long as the rate of application was adjusted accordingly.

The fertilizer's physical properties are worthy of consideration because a lumpy or caked fertilizer is difficult to apply evenly. Some constituents, such as ammonium nitrate (NH_4NO_3), calcium nitrate [$Ca(NO_3)_2$], sodium nitrate ($NaNO_3$), and urea [$CO(NH_2)_2$] absorb water from the air (i.e., they are hygroscopic) and thus must be protected from moisture. They are packaged in moisture-proof bags, and often a conditioning material is added to decrease moisture absorption and caking.

Fertilization principles in areas of intensive cultivation such as landscapes, sports and recreational fields, nurseries, and greenhouses are the same as those in the field. In these situations, however, the practice of fertigation is often used. **Fertigation** is the application of soluble fertilizer when irrigating. Because of the high frequency of application, fertilizer rates are relatively low and are usually expressed in parts per million (ppm) of the dilute fertilizer solution. A typical dose of a complete (N–P–K) fertilizer is 200–100–200 ppm.

Fertilizer injectors are usually used in fertigation programs. The injector allows a concentrated fertlizer solution to be prepared to save space, then the injector measures a precise amount of that concentrate into the irrigation water at a constant and reliable rate. Typical dilutions for injectors are 50:1, 100:1, and 200:1. Thus, when a 100:1 injector is used, the concentration in the fertilizer tank is 100 times the recommended rate. For example, 250 ppm would be mixed at 25,000 ppm. The injector injects 1 part of the concentrate into 100 parts of water so the rate coming out the end of the hose or irrigation line is 250 ppm. It also means that it would take a 10,000-gallon tank to hold the same amount of unconcentrated fertilizer that can be held in a 100-gallon tank of concentrated fertilizer. The space-savings is well worth the investment in an injector and the small amount of additional plumbing required. Although it might seem to make sense to have injectors with higher ratios to be even more space-efficient, at concentrations higher than 200:1, the fertilizer solution becomes saturated so that not all of the fertilizer can dissolve.

Any fertilization program requires regular and consistent **soil testing.** Soil testing allows a grower to monitor the fertility of the soil or growing media and adjust fertilizer formulations or application frequency if necessary. Tests should be made, at a minimum, at the beginning of each crop or growing season. More frequent testing is recommended if fertilization is done frequently, as with golf course and greenhouse or nursery management, or if problems develop. Fertility maps of fields can be created and incorporated into a global positioning system (GPS), thus enabling precision applications of appropriate fertilizers.

The availability of certain elements is regulated by the acidity or alkalinity of the soil (Fig. 6–53). In certain acid soils, aluminum solubility increases to toxic levels at low pH values, but its availability decreases at high pH values. Iron and zinc become less available to plants as the pH increases, but molybdenum is more available at higher pH levels. Phosphorus is more available at a soil pH of about 6.5 to 7.0 than at either higher or lower values. Calcium and magnesium applied as carbonate salts not only increase their availability, but also decrease soil acidity, raising the soil pH.

Primary Nutrients

Nitrogen

In the past, farmers "grew" much of their nitrogen fertilizer; that is, they plowed under legume crops that had been inoculated with bacteria (*Rhizobium*), which fix atmospheric nitrogen biologically. This remains a practice in many areas. Animal manures are also a nitrogen source, returning to the soil that amount taken out by plants used to feed the animals minus that used by the animal for its own growth. Electrolysis of atmospheric nitrogen by lightning during thunderstorms fixes nitrogen gas as oxides. In modern agriculture, however, the most important source of nitrogen is the synthetic fixation of atmospheric nitrogen gas. This is the Haber-Bosch process that occurs in fertilizer plants, where natural gas is combined with nitrogen gas under tremendous pressure and at high temperatures to form ammonia. The synthetic ammonia is then further processed into various forms in the manufacture of the commercial fertilizer.

Plants absorb nitrogen as inorganic nitrate ions (NO_3^-) and, in a few cases, as ammonium (NH_4^+) or amino (NH_2^+) ions. Most natural soil nitrogen is in the organic form—that is, combined in some manner with carbon. Organic nitrogen occurs in manures, decomposing organic matter, and urea [$CO(NH_2)_2$] and must be oxidized before most plants can use it.

The transformation of organic matter to the mineral or inorganic form—for example, organic nitrogen to NH_4^+,

Strongly Acid	Medium Acid	Slightly Acid	Very Slightly Acid	Very Slightly Alkaline	Slightly Alkaline	Medium Alkaline	Strongly Alkaline

Nitrogen

Phosphorus

Potassium

Sulfur

Calcium

Magnesium

Iron

Manganese

Boron

Copper and Zinc

Molybdenum

4.0   4.5   5.0   5.5   6.0   6.5   7.0   7.5   8.0   8.5   9.0   9.5   10.0

Figure 6–53   Availability of some essential nutrient elements as influenced by soil acidity or alkalinity.

NO_2^-, NO_3^-—by microorganisms is called **mineralization.** The conversion of the mineral form to the organic form is called **immobilization** (Fig. 6–54). When organic material with a carbon:nitrogen (C:N) ratio greater than 30 is applied to a soil, the nitrogen is immobilized during initial decomposition, then it is mineralized as decomposition proceeds.

The major role of plants in the processes illustrated in Figure 6–54 is the reduction of nitrate to ammonium; this is an extremely important process for all life on the earth and is a process that cannot be carried out by animals. The ammonium is then rapidly converted into a wide variety of important nitrogen-containing metabolites including amino acids, storage proteins, catalytic proteins (enzymes), and nucleic acids. Also, nitrogen is a key component of chlorophylls.

Continued use of nitrogen fertilizers can affect the pH of the soil (Table 6–6). Conversely, fertilizers can alter soil pH. Some fertilizers are residually acid forming, others residually basic, and some have little or no effect (see Fig. 6–53). Circumstances permitting, avoid acid-forming fertilizers on acid soils and basic fertilizers on alkaline soils. This suggestion is not always followed in practice. Fertilizers containing ammonium ions as the source of nitrogen are residually acid and may be used on alkaline soils. Fertilizers containing basic cations (Ca^{2+}, Na^+, or K^+) with nitrate (NO_3^-) anions as the source of nitrogen are residually basic and may be used on acid soils. In either case, the residual effect moves toward soil neutrality. Potassium sulfate (K_2SO_4) and potassium chloride (KCl) are residually neutral.

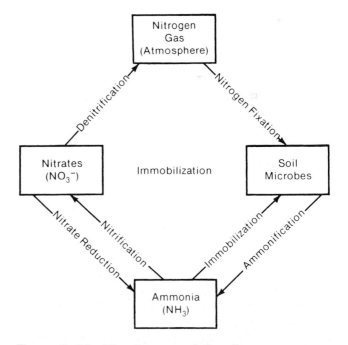

Figure 6–54   The nitrogen cycle in soil.

The kind of ion determines the mobility, solubility, and availability of nitrogen to the plant. For example, by ionic exchange, positively charged ammonium ions are held by negatively charged soil particles and are prevented from leaching out of the root zone. Under these conditions, the ammonium ion is considered to be relatively immobile. The

TABLE 6–6   Nitrogen Fertilizers, Their Composition, and Residual Effect on Soil pH

Organic	%N	Residual Effect	Inorganic	%N	Residual Effect
Dried blood	12	Acid	Anhydrous ammonia	82	Acid
Guano, Peruvian	12	Acid	Urea	46	Acid
Fish meal, dried	10	Acid	Ammonium nitrate	33	Acid
Tankage	8	Acid	Aqua ammonia	30	Acid
Soybean meal	7	Acid	Ammonium chloride	28	Neutral
Peanut meal	7	Acid	Calcium cyanamide	22	Basic
Cottonseed meal	7	Acid	Ammonium sulfate	20	Acid
Activated sludge	6	Acid	Diammonium phosphate	20	Acid
Bone meal, raw	4	Acid	Sodium nitrate	16	Basic
Garbage tankage	3	Basic	Calcium nitrate	16	Basic
			Potassium nitrate	14	Basic

nitrate ions are negatively charged and are not attracted to the soil particles; therefore, they move freely in the soil solution. More of the ammonium ions are retained in fine-textured soils than in coarse-textured soils because the fine-textured soils have more exchange sites. Conversion of ammonium to nitrate ions (nitrification) speeds up possible nutrient losses by leaching.

Nitrates, while essential for plant health, are detrimental to humans. Nitrate leaching from fields and other areas of crop production has resulted in groundwater contamination in several areas of the United States. As a result, in many areas, runoff from farms, greenhouses, and nurseries is clearly monitored and the amount of nitrate runoff is strictly limited. This has forced many crop producers to invest in water collection or filtration systems.

Nitrogen fertilizers are applied as solids, liquids, or gases. They can be applied as pre-plant fertilizers, top-dressed after the crop has emerged (Fig. 6–55), broadcast evenly over the field and mixed with the soil by disking, drilled or injected into the soil at desired depths, dripped into irrigation water, or injected as a gas or liquid into the soil (Fig. 6–56). These methods can be combined, depending upon the crop, or fertilizing can be combined with planting. Nitrogen fertilizers must not be placed in direct contact with the foliage because they burn the leaves. Ammonia is toxic to living tissue.

Many plants deficient in nitrogen show pale green to yellow leaves, but each crop has its own characteristic symptoms. Nitrogen deficiency generally causes the plant to grow slowly and restricts crop yields (Table 6–7).

Figure 6–56 Anhydrous ammonia fertilizer injected into the soil through the tubing attached to the rear of the shanks being drawn through the soil. The fertilizer is injected either in gaseous form, directly from the pressurized tank, or as liquid (aqua-ammonia), where the gas was previously dissolved in water. The placement can be made at the desired depth and location within the row in relation to the plants, making for more efficient use of fertilizer. *Source:* Robert Mullen, The Ohio State University.

Figure 6–55 Application of fertilizers as a liquid is a convenient and widely used method for providing supplemental plant nutrients to crop lands. In some situations, fertilizers are applied directly onto the leaves of some crop plants. These foliar applications usually contain secondary or micronutrients.

TABLE 6–7	Summary of Roles of Mineral Elements in Plant Nutrition			
Nutrient Element	**Function in Plants**	**Deficiency Symptoms**	**Losses from Soil**	**Fertilizer**
PRIMARY NUTRIENTS Nitrogen (N)	Synthesis of amino acids, proteins, chlorophyll, nucleic acids, and coenzymes.	Stunted growth, delayed maturity, light green leaves; lower leaves turn yellow and die.	Erosion, leaching, crop removal.	Inorganic salts of ammonia, calcium, sodium, potassium, urea; organic fertilizers; legume crops; animal manures, crop residues, animal waste.
Phosphorus (P)	Used in proteins, nucleo proteins, metabolic transfer processes, ATP, ADP, photosynthesis, and respiration. Component of phosphyolipids.	Purplish leaves, stems, and branches; reduced yields of seeds and fruits, stunted growth.	Crop removal, fixation in soil. Reversion to unavailable form in soil.	Superphosphate, treble superphosphate, ammonium phosphate, animal manures.
Potassium (K)	Sugar and starch formation, synthesis of proteins. Catalyst for enzyme reactions, stomate activity, growth of meristematic tissue.	Reduced yields; mottled, spotted, or curled older leaves; marginal burning of leaves; weak root system, weak stalks.	Crop removal. Soil fixation leaching.	Potassium sulfate, potassium chloride.
SECONDARY NUTRIENTS Calcium (Ca)	Cell walls, cell growth and division; nitrogen assimilation. Cofactor for some enzymes.	Deformed terminal leaves, reduced root growth. Some plants turn black, dead spots in midrib in some plants. Failure of terminal bud to grow.	Leaching, crop removal.	Calcium sulfate, calcium nitrate, calcium carbonate, dolomitic limestone.
Magnesium (Mg)	Essential in chlorophyll, formation of amino acids and vitamins. Neutralizes organic acids. Essential in formation of fats and sugars. Aids in seed germination.	Plants usually chlorotic (interveinal yellowing of older leaves); leaves may droop.	Leaching, plant removal, and erosion. Some losses by fixation to unavailable form in acid peaty soils.	Foliar sprays with magnesium sulfate, dolomitic limestone.
Sulfur (S)	Essential ingredient in amino acids and vitamins. Flavors in cruciferous plants and onions.	Light green leaves, reduced growth, yellowing of leaves. Weak stems. Similar to N deficiency.	Erosion, leaching, crop removal.	Ammonium sulfate, calcium sulfate, super phosphate, sulfuric acid, elemental sulfur.
MICRONUTRIENTS Boron (B)	Affects flowering, pollen germination, fruiting, cell division, nitrogen metabolism, water relations, hormone movement.	Terminal buds die, lateral branches begin to grow, then lateral buds die, branches form rosettes. Leaves thicken, curl, and become brittle.	Crop removal, leaching.	Sodium or calcium borate, animal manure, superphosphate.
Copper (Cu)	Constituent in enzymes, chlorophyll synthesis, catalyst for respiration, carbohydrate and protein metabolism.	Terminal leaf buds die. Chlorotic leaves. Stunted growth. Terminal leaves die.	Tied-up by highly organic soils and acid soils. Leaching.	Copper sulfate or other copper salts.
Chlorine (Cl)	Not too much known except that it aids in root and shoot growth. Required for growth and development.	Plants wilt. Chlorotic leaves. Some leaf necrosis. Bronzing in leaves.	Almost never deficient under field conditions.	Chloride salts.

(CONTINUED)

TABLE 6–7	(continued)			
Nutrient Element	**Function in Plants**	**Deficiency Symptoms**	**Losses from Soil**	**Fertilizer**
Iron (Fe)	Catalyst in synthesis of chlorophyll. Involved in formation of many compounds. Components in many enzymes.	Paling or yellowing of leaves, chlorosis between veins at first. Grasses develop alternate rows of yellowing and green stripes in leaves.	Crop removal, leaching, and erosion. Unavailable in alkaline soils.	Foliar applications of iron chelates, ferrous sulfate, or ferrous ammonium sulfate.
Manganese (Mn)	Chlorophyll synthesis, acts as coenzyme.	Network of green veins on light green background of intervenous tissue. Leaves later become white and abscise.	May be unavailable in alkaline soil. Toxic in acid soils.	Manganese sulfate as a foliar spray or as a soil application.
Molybdenum (Mo)	Essential in some enzyme systems that reduce nitrogen. Protein synthesis.	Plants may become nitrogen deficient. Pale green, rolled or cupped leaves, with yellow spots. Leaves of crucifers become narrow, cereal glumes do not fill out.	May have been lacking when soil was formed or become unavailable.	Solution of sodium molybdate sprayed on plants or soil. Also dusted on seeds before planting.
Zinc (Zn)	Used in formation of auxins, chloroplasts, and starch. Legumes need Zn for seed production.	Abnormal roots; mottled bronzed, or rosetted leaves. Intervenous chlorosis.	May not be available in alkaline soils; and toxic in acid soils. Crop removal.	Zinc sulfate as a foliar spray or added with other fertilizers.

(left column, rotated label) MICRONUTRIENTS (continued)

Phosphorous

Phosphorous is needed by plants in smaller amounts than nitrogen or potassium, but this does not reflect its true importance in plant nutrition. Phosphorus is a key element in the formation of AMP, ADP, and ATP (adenosine mono-, di-, and triphosphate), which play essential roles in photosynthesis and respiration. Phosphorus is a constituent of nucleic acid and phospholipids. Also, many intermediates in plant metabolism are phosphorylated compounds. Phosphorus is often associated with early crop maturity, increased root proliferation, and seed formation. Deficiencies result in stunted growth, accumulation of anthocyanin pigment (purpling the leaves of some plants), and reduced yields of seeds and fruits.

Phosphorus is absorbed mainly as orthophosphate ions $(H_2PO_4^-)$ and to a lesser extent as monohydrogen phosphate (HPO_4^{2-}). The quantity of either of these formed in the soil at any time is small, but the supply is constantly being renewed. The rate of renewal depends on soil pH. Phosphorus is most available to the majority of crops in the pH range of 5.5 to 7.0 (Fig. 6–53). Iron and aluminum phosphates precipitate in alkaline soils.

The primary source of phosphorus for fertilizers is mined apatite (rock phosphate), and much of the world's reserve is located in the United States. Apatite exists in several forms, all of which are more or less insoluble in water. To increase phosphorus availability to plants, apatite is treated with acids or heat when made into fertilizer. Rock phosphate treated with sulfuric acid yields superphosphate, the most

TABLE 6–8	Approximate Composition of Some Common Phosphate Fertilizers		
		Phosphorus (%)	
Source		**P**	**P_2O_5**
Ammonium phosphate		7.0	16
Superphosphate		7.0	16–20
Triple superphosphate		18–23	42–50
Dicalcium phosphate		23	52
Phosphoric acid		24	54
Potassium phosphate		18–22	42–50
Raw rock phosphate		10–17	25–30
Calcium metaphosphate		27	62

abundant phosphorus source, containing 16 to 20 percent phosphoric acid (P_2O_5) (Table 6–8). Liquid phosphoric acid (H_3PO_4) is made by treating apatite with sulfuric acid (a wet process that gives calcium sulfate as a precipitate). Ammonium phosphate results from treating phosphoric acid with ammonia. Superphosphate fertilizers do not appreciably affect soil pH, but phosphoric acid (H_3PO_4) is residually acidic. Another source includes basic slag, a by-product of the steel industry and an important phosphorus source in many European countries.

Potassium

Potassium is absorbed by plants in its ionic form (K^+). It plays roles in regulating the opening and closing of stomata and in water retention. It promotes the growth of meristematic tissue, activates some enzymatic reactions, aids in nitrogen metabolism and the synthesis of proteins, and aids in carbohydrate metabolism and translocation (Table 6–7). Potassium does not appear to be an integral part of plant constituents—protoplasm, fats, or carbohydrates—as do other nutrient elements. Plants absorb macroquantities of potassium.

The first visible deficiency symptom appears in the leaves. Weak stems (lodging in grains), decreased yield, lack of disease resistance, and crop quality have been associated with potassium deficiency.

Quantities of potassium salts are found in several areas of the world. Some lie below the earth's surface and some in dead lakes or seas. Potassium is found naturally in most soils. It comes from the decomposition of rocks and minerals, such as feldspars, muscovite, and biotite. West Germany, the Soviet Union, and Canada have the world's largest reserves, but the United States also has large deposits in New Mexico, Oklahoma, and Texas.

The most widely used potassium fertilizers are potassium chloride (KCl), potassium sulfate (K_2SO_4), and potassium nitrate (KNO_3), commonly known as muriate of potash, sulfate of potash, and saltpeter, respectively. The nitrate form is the most expensive and is mainly used on orchards, vegetables, or other high-value crops.

In the soil, potassium is more mobile than phosphorus but less mobile than are nitrates. It is readily leached from light sandy soils. In soils, potassium is bound by the negative charges on the clay colloids. In soils that are very high in clay and low in potassium, this binding of K^+ to the ion exchange sites may make potassium unavailable to plants, but this situation is rare in most agricultural soils. Potassium can be made more available to plants when other cations are added to the soil, for example, Ca^{++} added when the soil is limed.

Secondary Nutrients

Calcium

Calcium is absorbed by the plant in the ionic form (Ca^{2+}). It is essential to all higher plants. A deficiency kills terminal buds in shoots and apical tips in roots, reducing plant growth.

Calcium comes from dolomite, calcite, apatite, and some feldspars. Besides being an essential plant nutrient, calcium carbonate ($CaCO_3$), also called limestone, is used to correct soil acidity. Calcium sulfate (gypsum) is used to help reclaim sodic soils and also to improve soil structure and aggregation in saline soils. The best way to correct calcium deficiency in the soil is to add either limestone ($CaCO_3$) or gypsum ($CaSO_4$). The specific form depends on the soil's pH, and other soil characteristics.

Magnesium

Like calcium, magnesium is absorbed by the plant as an ion (Mg^{2+}). It is an essential nutrient, the central atom in the structure of the chlorophyll molecule. The magnesium salt of

ATP is the chemical form of ATP that actually participates in hundreds of biochemical reactions in plant cells. Magnesium is mobile and can translocate from older to younger leaves. Its deficiency causes an intervenous chlorosis (yellowing) in older leaves.

Magnesium is an exchangeable cation in the soil resulting from the decomposition of minerals such as biotite, dolomite, olivine, and serpentine. It also appears in the soil solution. In some arid areas of the world, magnesium occurs in such large quantities that it precipitates in the soil profile and, in rare instances, is so abundant that it is toxic.

A good source of magnesium (and calcium) for fertilizer is dolomitic limestone ($CaCO_3 \cdot MgCO_3$). Magnesium sulfate ($MgSO_4$) and potassium magnesium sulfate ($K_2SO_4 \cdot 2MgSO_4$) are also sources. Magnesium fertilizers are applied much the same as calcium.

Sulfur

Sulfur is absorbed by the plant as the sulfate ion (SO_4^{2-}). It is reduced to the disulfide (S–S) or sulf-hydryl (SH) group before being utilized in the plant. Sulfur is a component of several essential amino acids and several coenzymes, and is a key component of most proteins where it plays a role in maintaining the protein structure. Sulfur also contributes to the characteristic flavor compounds of some vegetables, such as cabbage and onions.

The deficiency symptoms of sulfur can be confused with those of nitrogen, except that sulfur is not as readily translocated from older to younger leaves. Therefore, the younger leaves of a plant suffering from sulfur deficiency appear yellow. Reduced plant growth, uniform yellowing of younger leaves, and weak stems are characteristic symptoms of sulfur-deficient plants.

The primary source of sulfur is the decomposition of metal sulfides in igneous rocks. Appreciable quantities also come from the atmosphere and some from irrigation waters. Sulfur is present in the soil as sulfates and sulfides and in organic combinations in the soil humus.

Sources of sulfur for fertilizers are sulfate salts of aluminum, ammonium, calcium, iron (ferrous), magnesium, manganese, potassium, sodium, and zinc. Other sources are lime sulfur, sulfuric acid, and elemental sulfur.

Micronutrients

Micronutrients are needed by plants in minute quantities, but this fact does not detract from their importance. Deficiency of a necessary micronutrient is as devastating to a plant as a deficient macronutrient. One characteristic common to all micronutrients is that, while they are essential in small quantities, they are toxic in large quantities (Table 6–9).

Boron

Boron occurs in most soils normally in quantities of 20 to 200 ppm. In some arid areas, it occurs in toxic amounts. In

TABLE 6–9	Approximate Range of Micronutrient Deficiency and Toxicity in Stems and Leaves (Dry Weight)		
Microelement	**Deficiency (ppm)**	**Normal (ppm)**	**Toxicity (ppm)**
Boron	5–30	30–75	75
Copper	4	4–15	20
Manganese	15	15–100	Depends on Fe: Mn ratio
Molybdenum	0.1	1–10	Low toxicity
Zinc	8	15–50	200

most humid regions, boron occurs as a borosilicate in the form of tourmaline. This material is quite insoluble. Boron-deficient soils occur along the Atlantic coast from Maine to Florida, in the Gulf coast states, the North Central states, and along the Pacific coast from Washington to California. Borax ($Na_2B_4O_7 \cdot 10 \; H_2O$) is water soluble, easily leached from sandy soils, and is a good source of boron.

Copper

Many areas throughout the world have copper-deficient soils. In the United States, deficient soils are found in the Great Lakes area, the West Coast, and Florida. Copper deficiencies appear more frequently in highly organic soils but have been found in mineral soils in some countries. The amount of organic matter, soil pH, and the presence of other metallic ions influence the availability of copper. Large quantities of copper in the soil can cause iron deficiency in some plants, and there have been reports of copper toxicity. Copper deficiencies are corrected by applying copper salts to the soil by foliar spraying with a solution of soluble copper salts. Copper sulfate ($CuSO_4 \cdot 5 \; H_2O$) or copper ammonium phosphate ($CuNH_4PO_4$) are often used.

Chlorine

Chlorine evaded detection as an essential nutrient for many years because it is so universally abundant in nature. Spray from ocean waves carrying huge amounts of sodium chloride and potassium chloride are blown many miles inland by the wind. Large quantities are also deposited in the soil by precipitation. Larger amounts of this nutrient are used by crop plants than any other micronutrient except iron.

Most chlorine exists in the soil as simple chloride salts and is absorbed by the plant as chloride ions (Cl^-). Because of their similarity, bromine can substitute for some chlorine in some plants. Chlorine deficiency is seldom seen in the field, but symptoms have been observed in tobacco, tomatoes, buckwheat, peas, cabbage, sugar beets, barley, corn, cotton, and potatoes. Deficiency symptoms of chlorine appear to be stunted root growth, leaf bronzing, and chlorotic leaves with some necrosis; often the plant wilts. Excess chlorine is toxic, especially to tobacco and potatoes.

Iron

Iron is more abundant in most soils than any other micronutrients, but often it is deficient because it is unavailable to plants because of its extremely low solubility. Iron deficiency has been noted in crops grown on alkaline or calcareous soils (Fig. 6–53) and on acid soils with high phosphate levels. Citrus, deciduous fruits, soybeans, strawberries, vegetable crops, and many ornamentals have shown chlorosis caused by iron deficiency. The deficiency appears first in the younger leaves as an intervenous yellowing that later progresses over the entire leaf; in severe cases, the leaves become almost white. Iron is essential in photosynthetic processes and also functions in several enzymatic reactions. The plant can absorb iron through its roots or leaves as either an ion (Fe^{2+}) or as a complex with soluble organic salts (chelated).

Manganese

Manganese is similar to iron in many ways. It is also a heavy metal and rather immobile in the plant. Manganese exists in the soil in several forms, depending on the soil environment. It is most available to plants if in the manganous state (Mn^{2+}, MnO) as an exchangeable cation in the soil. In this state, however, it is more subject to oxidation by microorganisms to its trivalent state (Mn^{3+}, Mn_2O_3); in well-aerated soils it can be further oxidized to its least soluble, four-valent state (Mn^{4+}, MnO_2). High soil pH and good oxidizing conditions encourage MnO_2 formation. Poorly aerated or waterlogged soils favor the reduced manganous (Mn^{2+}) state. Manganese is absorbed by the plant in the manganous ionic form (Mn^{2+}), and is often applied to plants as manganese sulfate ($MnSO_4$) or a chelated (complexed with some organic molecule) compound. Manganese may be applied as a foliar spray that is absorbed through the leaves.

Like iron, the first deficiency symptoms show intervenous chlorosis in the younger leaves. This nutrient participates in photosynthesis, activation of enzymes, carbohydrate metabolism, and phosphorylation. Manganese toxicity from large amounts of the micronutrient has been observed in cotton (crinkle leaf) and tobacco on high-acid soils. Liming the soil corrects the malady.

Molybdenum

Molybdenum is an essential nutrient, very noticeable when deficient in crops such as clovers, alfalfa, cereals, vegetables, soybeans, poinsettias, and forage grasses. A condition known as whiptail in cauliflower is caused by molybdenum deficiency. Soil environment affects the availability of this element to a large extent. It is unavailable to plants in strongly acid soil, where it reacts with iron and aluminum silicates to form insoluble compounds. Liming the soil usually increases the availability of molybdenum.

Phosphates seem to aid plants in the absorption of molybdenum, but sulfates tend to hinder its uptake. Symptoms of deficiency vary among crops, but intervenous chlorosis is often the first observable symptom. Legumes are stunted and the leaves turn yellow, as with nitrogen deficiency.

Zinc

Soil characteristics influence zinc availability. In alkaline soils, deficiencies are expected, and in strongly acid soils, toxicity is possible. Deficiencies occur in a wide range of soils but are most frequent in calcareous soils high in phosphorus. Zinc deficiencies have been observed in deciduous and citrus fruits, vegetables, and field crops such as corn, cotton, sorghum, and legumes. Zinc was one of the first micronutrients to be recognized as essential. Zinc attracted scientific interest early because of its importance in human nutrition, and considerable research was conducted in an endeavor to increase the concentration in plants. Zinc primarily acts as an enzyme activator in both plants and animals.

Zinc can be absorbed by the roots from the soil as the exchange cation (Zn^{2+}), or through the leaves when it is sprayed on the foliage as a $ZnSO_4$ solution or chelated compound. Deficiency symptoms first appear in the younger leaves as intervenous yellowing. Later, reduced shoot growth becomes evident. The midrib and margins of corn leaves remain green while a broad band of bleached tissue appears from the base to the tip.

Chelating Agents

The word **chelate** derives from a Greek word meaning "claw." A chelate is a large organic molecule that attracts and tightly holds specific cations like a chemical claw, preventing them from taking part in inorganic reactions but at the same time allowing them to be absorbed and used by plants. Chelates combine with metallic cations—iron, manganese, zinc, and copper—to prevent the cations from reacting with inorganic anions that would render them insoluble and unavailable to plants. For example, chelated iron cannot react with hydroxyl anions (OH^-) to form insoluble ferric hydroxide [$Fe_2(OH)_3$]. Chelated cations are more soluble at higher pH than are inorganic ions. Chelates are also known as **sequestering agents.**

Several important agricultural chelates or sequestering agents are commercially available: (1) **e**thylene-**d**iamine**t**et-ra**a**cetate (EDTA) sequesters copper, iron, manganese, and zinc; (2) **e**thylene**d**iamine**d**i-o-**h**y-droxyphenyl**a**cetic acid (EDDHA) sequesters iron; (3) **d**iethylene**t**riamine**p**enta **a**cetic acid (DTPA) sequesters iron; (4) **n**itrito**t**ri**a**cetic acid, (NTA) sequesters zinc; and (5) **h**ydrox**ye**thylethylene **d**iamine**t**etra**a**cetic acid (HEDTA) sequesters iron and zinc.

SUMMARY AND REVIEW

Soils form a complex ecosystem composed of biological organisms, living and dead; inorganic materials; water; and air. The interaction over time of many factors including parent material, weathering, types of organisms present, and the topography of the area influence the formation and properties of soil.

Physical soil properties include texture (percentage of sand, silt, and clay particles) and structure (the arrangement of the primary particles), Soils with good structure have enough pore (air) spaces that are large enough to transmit water and air without restriction and sufficient smaller pores that retain some water against the pull of gravity. Chemical soil properties include cation exchange capacity (CEC), soil acidity and alkalinity, and salinity. Cation exchange capacity is the ability of the soil to exchange cations it already holds with cations in the soil water. This factor is very important in fertility management. Soil acidity and alkalinity affect the availability of nutritional elements to the plant. Saline soils naturally contain large quantities of soluble salts. These salts may be harmful to plants if they are present in great enough quantities.

Soil organisms play a crucial role in determining soil characteristics. They create pore space by their tunneling activities. They add organic material when they die and decay. They often form beneficial and sometimes even necessary symbiotic relationships with the roots of plants growing in the soil. On the other hand, plant pathogen and herbivore organisms that can infest or feed on roots can also be a part of the soil ecosystem.

For several thousand years, crop production required the soil to be tilled before planting. Tilling was done to reduce residue from plants growing in the field, reduce weed and pest infestations, promote seed germination, and promote crop growth. Sophisticated tilling systems were developed and are still used today. Now, however, soil improvement for plant growth may involve leaving the soil alone for several years or tilling it as little as possible to let the soil ecosystem redevelop. The reduced tillage systems require paying closer attention to all phases of the production cycle; however, the benefits of reducing erosion is worth the extra attention needed.

Soilless potting media is often used in nurseries and greenhouses. The components used in these mixes are selected to provide the same desirable characteristics found in native soils. Issues of water-holding capability, degradation, compaction, CEC quality, and so on, are important considerations in the formulating and handling of potting media, just as they are in field soils.

The water in the soil is as important as the soil itself. The balance of soil particles, water, and air (primarily oxygen) determines how well a plant will grow. Water not only is necessary itself for the growth of the crops, but it is required for fertility practices too. Water carries the fertilizer elements to the surface of the soil particles, where it can be exchanged to the plant root.

When precipitation is not sufficient to promote acceptable crop growth, irrigation is practiced. There are many different irrigation systems, but they all have one main purpose—to get water to the roots of the plant. Irrigation systems can be adapted for almost every growing system from vast fields of agronomic crops to the smallest tray of seedlings in the greenhouse.

The nutritional elements required by the plant are almost all (with the exception of carbon, hydrogen, and oxygen) taken up from the soil by the roots. When these are in short supply naturally, they must be added to the soil. If they cannot be replenished by the use of manure, returning plant residue to the soil, or other organic method, fertilizer will most likely have to be used. Caution must be taken when fertilizing because over- or underdosing can occur, resulting in toxicity or continuing deficiency problems. Also, careless application of fertilization can have a negative effect on the environment, such as when too much nitrate is applied and the excess leaches into water supplies.

FOOD FOR THOUGHT

1. What do you think makes natural soil different from a pile of finely ground rocks?

2. Describe how the organic content of a soil might be different in soils formed in central Iowa and southern Florida.

3. Why is texture considered one of soil's most important characteristics?

4. Why is soil compaction considered a serious problem for anyone who grows plants for a living?

5. Describe how a maple leaf or turfgrass clipping is converted to humus.

6. Why was tillage a favorite form of land preparation for several thousand years? Explain why now no or low tillage is coming into favor.

7. Tensiometers are used to monitor and control irrigation in large fields. Why do you think that using a tensiometer to signal to a computer that a pot is dry has not worked very well for irrigating a large number of pots in a nursery or greenhouse?

8. Why do you think it is important to know the pH of your soil or potting mix when determining how to correct a nutrient deficiency? Why might some plants thrive at a low soil pH while others do not?

9. You've done a soil test on your golf course and find that the grass is yellow on the greens because of insufficient nitrogen, even though you have been fertilizing at the same time you've fertilized the fairway, which looks great. The greens have been constructed to use mostly sand for the soil, while the fairway turf grows in native soil. How might the difference in soil explain the differences in nutrient status?

SUPPLEMENTARY READING

BRADY, N. C., and R. R. WEIL. 1999. *The nature and properties of soils.* Twelfth Edition. Upper Saddle River, N.J.: Prentice Hall.

BROOKS, K. N., P. F. FOLLIOTT, H. M. GREGERSEN, and L. F. DEBANO. 1997. *Hydrology and the management of watersheds.* Second Edition. Ames, Iowa: Iowa State University Press.

HARPSTEAD, M. I. 1997. *Soil science simplified.* Third Edition. Ames, Iowa: Iowa State University Press.

HUDSON, N. 1995. *Soil conservation.* Third Edition. Ames, Iowa: Iowa State University Press.

MARSCHNER, H. 1995. *Mineral nutrition of higher plants.* Second Edition. New York: Academic Press.

MCLAREN, R. G., and K. C. CAMERON. 1996. *Soil science. Sustainable production and environmental protection.* Auckland, New Zealand: Oxford University Press.

PLASTER, E. 2003. *Soil science and management.* Thomson Delmar Learning. Clifton, N.J.

REED, D. W. (ed.). 1996. *Water, media, and nutrition for greenhouse crops.* Batavia, Ill.: Ball Publishing.

WILD, A. 2003. *Soils, land, and food.* Cambridge University Press. Cambridge, England.

CHAPTER

7

Integrated Management of Weeds, Insects, and Diseases

Michael Boehm, Jerron Schmoll, David Shetlar

KEY LEARNING CONCEPTS

After reading this chapter, you should be able to:

♦ Explain the foundational concepts of weed science, entomology, and plant pathology.
♦ List the major weeds, insect pests, and diseases that lead to plant losses.
♦ Discuss five major strategies for managing weeds, insects, and diseases and how they can be combined to develop an integrated pest management (IPM) program.

Since the beginning of agriculture, farmers have had to develop ways to manage weeds, insect pests, and diseases. Even today, with all the scientific research on crop protection, it is estimated that insects, diseases, weeds, and animal pests eliminate half the food produced in the world during the growing, transporting, and storing of crops. In the tropics, where heat and humidity favors the development of many diseases, two-thirds of some crops are lost, thereby aggravating the world hunger problem. Although not directly related to the world's supply of safe food, similar losses occur in timber, ornamental, floricultural, and turfgrass production systems. In economic terms, annual losses in food, fiber, and ornamental production systems caused by weeds, insects, and diseases are estimated in the hundreds of billions of dollars. Because of the significant impact of weeds, insects, and diseases—on both human and animal welfare as well as on the global, regional, and local economies—it is important for those interested in growing plants to develop a firm understanding of weed science, **entomology** (study of insects), and **plant pathology** (study of plant disease) and to learn how to eradicate, manage, or otherwise minimize plant losses caused by these important plant pests.

INTEGRATED PEST MANAGEMENT (IPM)

Whether you are managing a weed, insect pest, or disease-causing organism (also called a **pathogen**), most specialists interested in plant health recommend the use of a multipronged approach or strategy—commonly referred to as an integrated pest management (IPM) approach. Integrated pest management programs rely on the use of several methods rather than on a single means for avoiding or otherwise minimizing the impact of plant pests and pathogens (Fig. 7–1). Although sometimes called different names by weed scientists, entomologists, and plant pathologists, the methods for managing or eliminating plant pests fall into five distinct categories:

1. *Genetic host resistance*—use of genetically resistant plants to minimize or avoid losses caused by insect pests and/or pathogens.
2. *Cultural practices*—use of agronomic or horticultural practices that favor plant development and minimize pest or pathogen activity.
3. *Chemical applications*—use of pesticides such as herbicides (weeds), insecticides (insect pests), miticides (mites), antibiotics (bacteria), fungicides

(fungi), and nematicides (nematodes) to suppress or inhibit pest/pathogen activity.
4. *Biological control*—the use of beneficial or antagonistic organisms that kill or otherwise suppress plant pests or pathogens.
5. *Government regulatory measures*—the use of quarantines and pest-eradication programs to stop the introduction or spread of deleterious plant pests and/or pathogens.

How plant production specialists integrate or mix and match these individual pest/disease management approaches to develop an effective IPM strategy depends on many different considerations, such as personal preference, public perception, availability of effective options or tools in each management category, and profit margin. Each manager, regardless of cropping system, must consider the complexities of the unique system being managed to develop an IPM strategy that best meets his or her needs (Fig. 7–2). The use of IPM strategies is often perceived to be more environmentally sound compared to those that rely heavily on the use of pesticides. In response to recently mandated and self-imposed restrictions on pesticide use (based primarily on concerns over human, animal, and environmental health), many farmers, growers, and professional plant production specialists have either been forced or have made a conscious effort to rely less on the use of pesticides and more on IPM strategies. In some cases, the shift from a reliance on chemi-

Figure 7–1 Four of the five categories of pest/disease management approaches available to plant production specialists for use in developing effective integrated pest management programs. In an ideal situation, plant production specialists would have a pool of equally effective pest/disease management tools from which to choose. *Source:* Michael Boehm, The Ohio State University.

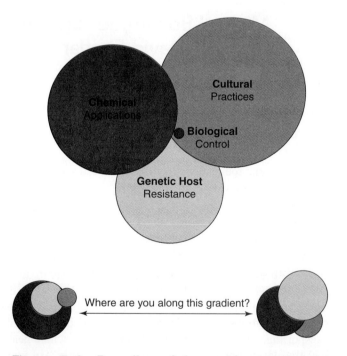

Figure 7–2 Regardless of the cropping system being managed, plant production specialists need to consider the complexities of their system to develop an IPM strategy that best meets their needs. *Source:* Michael Boehm, The Ohio State University.

cal control has been intensified by the development of pest populations resistant to previously effective pesticides.

Genetic Host Resistance

Genetic host resistance involves the use of genetically resistant plants to avoid or minimize plant losses caused by insect pests and/or pathogens. The use of genetically resistant plants is often recommended by entomologists and plant pathologists as the first line of defense against avoiding or minimizing plant damage caused by insects and pathogens. In some cropping systems, such as large-acreage field or row-crop agriculture (corn, soybeans, cereal crops, rice, cotton, etc.), the use of genetically resistant plants may be the only cost-effective means for managing a particular pest or disease. In some cases, the use of genetically resistant cultivars or varieties might be the only means of managing a disease or pest effectively, as is the case of managing plant diseases caused by viruses. The development of resistant plant types also reduces the need for costly and sometimes dangerous pesticides. Although genetic resistance should be considered when dealing with all insect pests and diseases, it is especially useful when dealing with annual cropping systems where new seed is sown each season, thereby providing an opportunity to introduce new cultivars or varieties with insect or pathogen resistance. Although important in perennial cropping systems such as orchards, timber production, and turfgrass swards, once the initial crop is planted, the introduction of resistant lines is limited due to the long-term nature of these crops.

In nature, plants susceptible to a prevalent disease or insect pest tend to disappear, while resistant types remain. The work of geneticists and plant breeders in purposely developing resistant cultivars of agriculturally important species has been one of their most useful pursuits, and the world's population owes much to them. The development of rust-resistant wheat cultivars, for example, has added food for untold millions of people. Genetically resistant plants may occur naturally, be derived through classical plant-breeding programs, or be derived through the use of modern molecular biology techniques. Plants derived from the latter are commonly referred to as **genetically modified organisms (GMOs).** Two examples of GMOs that are widely used in agriculture include the use of Round-Up resistant cotton, corn, and soybeans, and Bt corn that produces a toxin that affects Lepidoptera. In many instances, cultivars or varieties have been developed that yield plants in spite of heavy disease or insect pest pressure. (The ability of a plant to maintain its yield in spite of heavy disease or pest pressure is known as **tolerance.**) Ideally, all insect pests and diseases could be managed in this manner.

Cultural Practices

Cultural practices in pest management involve the use of agronomic or horticultural practices that favor plant development and minimize pest or pathogen activity. The second line of defense available to those interested in managing

plant pests is the use of cultural management practices that favor plant growth over pest or pathogen activity. As discussed in Chapters 4 and 6, all plants require light, water, and nutrients to grow. Knowing how to provide the correct balance of these essentials to maximize plant growth and yield—whether measured in bushels per acre or aesthetics—is the responsibility of the professional plant production specialist. By providing growing conditions that favor plant growth over insect and pathogen development or activity, it is possible to avoid or minimize losses caused by pests and diseases. Plant pathologists have developed a simple model called the disease triangle to illustrate this concept (Fig. 7–3). Several cultural practices can be used to change the environment in which plants are grown and they can severely influence pest and pathogen activity. These practices include (1) tillage practices, (2) water management, (3) fertility, (4) crop rotation, and (5) sanitation (cleaning or removal) of equipment and diseased or infested plant material.

In field agriculture, tillage—both plowing and cultivation—is a widely used agronomic practice used to reduce soil compaction, increase water infiltration, and reduce weed growth or bury weed seeds and survival structures of plant pathogens. Cultural practices that focus on water management—drainage, irrigation, and other specialized practices such as syringing on golf courses—is another group of cultural practices that can often be used to manage important pests and diseases. For example, management of *Phytophthora* and *Pythium,* fungal-like organisms that grow best in poorly drained soils, is directly related to avoidance of saturated soils and flooding. The use of a balanced fertility program

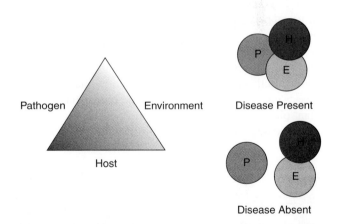

Figure 7–3 The disease triangle concept. A susceptible host (**H**), a virulent pathogen (**P**), and environmental conditions (**E**) that favor the pathogen must be present in the right mix to yield disease. If any one of these three components is missing or minimized, disease will not occur. Genetic host resistance, cultural practices, chemical applications, and biological control are tools used by professional plant production specialists to favor plant growing conditions and minimize pathogen activity or development. *Source:* Michael Boehm, The Ohio State University.

(discussed in Chapter 6) is another important requirement for growing healthy plants. When plants are either over- or underfertilized, however, they may become more or less susceptible to pathogen and insect attack. Knowing the diseases and insect pests for which a particular crop is predisposed is important and is often linked directly to the fertility status of the crop. Crop rotation is a simple but often effective means of controlling certain insect pests and diseases. If the same annual crop is grown year after year on a given plot of land, a particularly serious pest may keep increasing year after year, overwintering in crop residues, until it reaches such overwhelming populations that the crop cannot be produced on that piece of land. But by rotating the crop in a one- to several-year rotation with other nonsusceptible crops, the insect or disease pest, lacking a crop host for a long period, practically dies out. Cultural practices designed to achieve or maintain disease- or pest-free conditions on the farm or in the greenhouse or nursery play a key role in reducing plant losses by minimizing or eliminating sources of insect pests, weed seed, or pathogen inoculum. Practices such as the removal and destruction of infected plants or infested soil or potting mix, the use of certified pathogen-free or weed-free seed, and the use of clean tools and equipment is critical to reducing reservoirs and the spread of weeds, insects, and pathogens. Many other horticultural and agronomic cultural practices are also used routinely by farmers, growers, golf course superintendents, and other professional plant production specialists in the course of raising healthy plants that might also be modified to reduce pest and disease pressure.

Chemical Applications

The use of pesticides such as herbicides (weeds), insecticides (insect pests), miticides (mites), antibiotics (bacteria), fungicides (fungi and fungal-like organisms), and nematicides (nematodes) to suppress or inhibit pest/pathogen activity are included in this category. In an ideal world, plant pests and diseases could be managed effectively using genetically resistant plants and cultural management practices. Although some diseases and pests can be managed effectively using

these two approaches, many still require the use of pesticides to achieve commercially acceptable disease or pest thresholds (Figs. 7–4 and 7–5). Thus, the third line of defense available to those interested in managing plant pests is the use of pesticides to either kill or, as is more often the case, suppress plant pests and pathogens. Because of the concerns over the potential dangers of pesticides to humans, the environment, food products, farm and domestic animals, wildlife, beneficial insects, and the atmosphere, they are often considered the least desirable method of managing harmful insects and plant diseases (Fig. 7–6). As a result of these concerns, pesticide manufacturers, governmental regulatory agencies, and growers' groups have worked hard over the last forty years to reduce the use of broad-spectrum pesti-

Figure 7–5   Dusting grapes with sulfur dust to control powdery mildew caused by the fungus *Uncinula necator. Source:* Blue Anchor, Inc.

Figure 7–4   Certain insect and disease pests are controlled in commercial orchards by insecticides and fungicides applied by power air blast sprayers. *Source:* USDA.

Figure 7–6   When spraying pesticides to control insects and diseases in enclosed areas like greenhouses, workers should wear protective masks and clothing.

cides in favor of the development and use of pest/pathogen-specific compounds. Detailed information about each and the current status and use of herbicides, insecticides, antibiotics, fungicides, and nematicides is provided later in this chapter.

Biological Control

Biological control (sometimes referred to as biocontrol or biorational pest management) involves the use of beneficial or antagonistic organisms that kill or otherwise suppress plant pests or pathogens. The use of biological control is considered advantageous and environmentally sound because it provides an environment-friendly alternative to the use of pesticides. However, few biocontrol products provide consistent and commercially acceptable levels of pest or disease control.

Within all eco- and plant production systems, a balance usually develops among organisms, both plant and animal. Certain organisms are antagonistic to others and retard their expansion. Environmental or human-induced changes (such as those used by plant production specialists in the course of growing plants) that upset this balance by eliminating one of the organisms can lead to explosive proliferation of the others and to subsequent attacks on vulnerable crops. In a similar manner, if an organism is introduced into an environment but its antagonizing organism is not, the results can be devastating to a vulnerable crop plant. Biological control in such a situation would consist of introducing the antagonizing organism of the pest into the area, thus bringing the pest under control again. Biocontrol organisms kill or suppress pathogens and pests by (1) parasitizing the pest or pathogen directly, (2) competing with the pest or pathogen for space or nutrients, (3) producing toxins that kill or make the pest or pathogen sick, and (4) inducing a physiological or biochemical change in the host plant, thereby making it less susceptible to (more tolerant of) pest or pathogen attack. Specific examples of how biological control can be used to manage weed and insect pests and plant pathogens will be discussed later in this chapter. Several examples of how biological control has been used successfully to suppress important insect pests and disease include the use of parasitic wraps (introduced from Asia to control *Olive parlatoria* scale in California olive orchards) and the use of compost-amended potting mixes to suppress *Pythium* damping-off and root rot in greenhouses and nurseries (Fig. 7–7).

Government Regulatory Measures

The fifth and final means for managing plant pests and diseases is through the use of intensive inspection programs and quarantines. Although used by growers on a small-scale (i.e., the inspection and isolation of newly received plant materials prior to introduction), the use of quarantines is often implemented and regulated by local, state, regional, national, and international agencies and authorities. Whether dealing with people, animals, or plants, a shared concept in preventive medicine is that when pathogens and insect pests evolve with their hosts, the host will develop resistance or tolerance against the pathogen or pest. The result of such coevolution is that the pathogen or pest is kept in check. A practical corollary to this concept is that when an exotic pathogen or pest is introduced into a population that has never been exposed to

(A)

(B)

Figure 7–7 The use of compost-amended potting mixes by the floricultural and nursery industry for suppressing damping-off and root rot caused by *Pythium* and *Rhizoctonia* species. (A) The potting mix in both pots was inoculated with *Pythium ultimum*. The potting mix on the left (CP) was comprised of sphagnum peat and perlite and was severely stunted with extremely rotted roots. The potting mix on the right was the same as that on the left but was amended with composted pine bark (CPB) that contained beneficial biological control organisms (bacteria and fungi). (B) The top row shows the roots of five plants grown in CP. The bottom row shows roots from plants grown in CPB. The middle row is another type of potting mix without composted pine bark. *Source:* Michael Boehm, The Ohio State University.

or that did not coevolve with it (such populations are said to be naive), serious losses are likely to take place because of a lack of resistance in the host. Many serious plant disease and insect pests were not known to occur in certain countries until they were brought in by travelers or were shipped in, usually unknowingly, on contaminated plant material. The early colonists from Europe brought into North America certain insects and diseases of seeds and plants from their homelands. German soldiers hired by the British during the Revolutionary War brought the Hessian fly to North America with them in their straw bedding. This pest later moved westward through the United States, devastating wheat fields and ultimately causing far more problems for the new country than the soldiers did. Other historical examples of the devastating impact of exotic pests or pathogens on our native plant populations include the introduction of chestnut blight and Dutch elm disease fungi to North America in the early 1900s. Emerald ash borer, soybean rust, and sudden oak death are examples of recently introduced exotic pests and/or diseases that have caused or are capable of causing significant plant losses.

Strict government inspections and quarantines of imported plants, plant products, and soil have kept such pests out of many countries and many areas. However, modern jet travel, taking people and their belongings swiftly from country to country, greatly increases the possibility of introducing pests dangerous to people, plants, and animals. In addition to the passive or accidental means for moving pathogens and pests, the recent anthrax attacks in the United States highlight the potential for the intentional release of exotic pests and pathogens as agents of bioterrorism. Recent outbreaks of mad cow disease, avian influenza virus, foot and mouth, and soybean rust emphasize the continued need for strict government monitoring and enhanced reporting and response programs for key pests and diseases. The National Plant Diagnostic Network (NPDN), implemented following the events of September 11, 2001, is an excellent example of just such an effort. The NPDN was designed and implemented to enhance agricultural security within the United States by linking plant disease and pest specialists at land grant universities and government agencies so that high-consequence pests and pathogens could be quickly identified and an appropriate response coordinated. The identification and subsequent quarantining of imported geraniums infected with the bacterial pathogen *Ralstonia solanacearum* race 3 biovar II is an example of this effort. This biotype is not naturally found in the United States and could pose a threat to U.S. tomato and potato production.

In some cases, certain plants or plant products are forbidden entry into the country or areas of the country, and agricultural inspectors check luggage for such outlawed products. For example, mangoes, guavas, and passion fruit from Hawaii are not permitted entry into the U.S. mainland because they may be carrying the Mediterranean fruit fly, which is widespread in Hawaii but absent on the mainland. If introduced into California, for example, the fly could devastate the state's huge fruit industry. However, such plant products can be imported commercially if they are properly fumigated to kill the pests before shipping. Certain kinds of living plant material can be imported into the United States under permit or if it is quarantined after entry for two growing seasons to reveal any diseases. Travelers coming into the United States should not attempt to bring or send in agricultural materials such as fruits, vegetables, plants, bulbs, seeds, or cuttings unless advance arrangements have been made and a permit obtained. Any such plant materials being carried or imported through commercial channels must be reported to agricultural quarantine or customs officials upon arrival. The movement of certain crops among the contiguous forty-eight states is also prohibited or closely monitored.

Government eradication programs are conducted when a serious insect or disease pest breaks out. Often the trouble is eliminated before it has a chance to spread. Such programs require highly trained personnel who know the potential insect and disease problems and can recognize the pathogens and the symptoms of their activities. The dangerous Mediterranean fruit fly has been introduced accidentally into Florida on several occasions. It has been eradicated each time, but always at considerable expense. In California, too, discoveries of the Mediterranean fruit fly, probably brought in on illegal tourist importations of fruit from Hawaii, have also caused costly eradication procedures several times.

WEEDS

A **weed** can be defined as a plant out of place, growing where it is not wanted, that interferes with human activities or attempts to grow other plants. Wheat plants in a field of oats would be considered weeds, as would oat plants in a field of wheat. A weed can be any plant that competes with a desired plant for water, nutrients, or sunlight; interferes with harvesting a crop; reduces crop quality; reduces the aesthetic appeal of a crop; harbors undesirable insects, pathogens, or nematodes; or competes with the crop for pollinating insects. For the purpose of organization, the study of weeds is generally divided into two broad components: (1) weed biology and ecology and (2) weed control and management. However, all effective weed control strategies involve careful consideration and integration of both components.

Weed Biology and Ecology

Weed biology involves the study of the growth and reproduction of individual species of weeds. The classification (taxonomy) of a weed and its competitiveness, reproductive capabilities, and genetics all fall under the area of weed biology. Weed ecology is the study of the interaction of a weed species with its physical and biological environment. The physical components of the environment include variables such as soil structure, soil acidity, and the quantity of rainfall and sunlight available. The study of weeds' interactions with the biological environment requires developing an understanding of the dynamics involved in regulating how communities of plants, animals (such as insects and herbivorous mammals), and microorganisms (such as bacteria, viruses, and fungi) interact. Understanding the biology and ecology of a weed species is fundamental to the development of effective weed management strategies.

Classification

Plants can be classified in many different ways. Plants may be referred to by their habitat, for instance, tropical versus desert plants, or aquatic versus terrestrial plants. Broad classification of plants as either monocots (most of which are grass species) or dicots (broad leaves) is sometimes helpful when making general comparisons among plant species. Weed scientists have found that classification by life cycle (i.e., annuals, biennials, or perennials) is particularly useful because understanding weeds' life cycles is a key component of weed management. The term *invasive weed species* has recently been added to the vocabulary of weed scientists and is used to define species that are particularly aggressive, regardless of their life cycle.

Annuals

Annual plants reproduce by seed only and complete their life cycle within one year. Annuals may be subclassified as either summer annuals, obligate winter annuals, or facultative winter annuals. Summer annuals normally germinate in the spring or early summer, grow vegetatively throughout the remainder of the summer, and then set seed and die in the fall or early winter. The seeds of summer annuals typically lie dormant all winter and then germinate the following spring. Common summer annual weeds in the United States include crabgrasses (*Digitaria* spp.), foxtails (*Steraria* spp.), common lambsquarters (*Chenopodium album*), and pigweeds (*Amaranthus* spp.). Summer annual weeds tend to be troublesome in summer annual crops such as corn, soybeans, cotton, rice, peanut, and many vegetable crops (Fig. 7–8). Obligate winter annual weeds germinate in the fall, overwinter as immature plants, and then set seed and die in the spring or early summer. Facultative winter annuals germinate more frequently in the fall, but are capable of germinating in the early spring (in which case they follow the summer annual life cycle). The seeds of obligate winter and fall-germinating facultative annual weeds tend to lie dormant in the soil through the hot summer, and then the cycle continues with the germination of seed in the fall. Examples of winter annual weeds are annual bluegrass (*Poa annua*), common chickweed (*Stellaria media*), and most of the weed species in the mustard family, such as field pennycress (*Thlaspi arvense*) and the pepperweeds (*Lepidium* spp.). Like summer annuals, winter annuals tend to be most problematic in crops that share the winter annual life cycle, such as fall-planted nursery crops and small grains.

Figure 7–8 Foxtail (*Setaria* spp.), a summer annual weed, competing with soybean (*Glycine max*), a summer annual crop. *Source:* Kent Harrison, The Ohio State University.

Biennials

Biennial weeds live for two years and reproduce only by seed. In the first year of the life cycle, biennial plants grow vegetatively, producing a fleshy tap root and a rosette (a loose whorl of leaves attached to a highly compressed stem; Fig. 7–9). The tap root and rosette overwinter, and exposure to the cold vernalizes the plant and induces production of an elongated flower stalk the second year. After seed are produced, the plant usually dies but may persist for some additional time. Common burdock (*Arctium minus*), poison hemlock (*Conium maculatum*), and bull thistle (*Cirsium vulgare*) are examples of biennial weeds that can cause economic losses in pastures, orchards, and agronomic crops.

Perennials

Perennial weeds can live for an indefinite period of time and reproduce both by seed and vegetative propagules. Perennial weed species can be subclassified as simple perennials, herbaceous creeping perennials, and woody perennials. Simple perennials, such as common dandelion (*Taraxacum officinale*) and the plantains (*Plantago* spp.) (Fig. 7–10), have a nonspreading growth habit and regrow each year from adventitious buds located on the roots. Simple perennials may live for many years as single, individual plants. Creeping herbaceous perennials have a more aggressive growth habit and can spread vegetatively using modified root, stem, or leaf structures, such as rhizomes, stolons, tubers, bulbs, or creeping roots (Fig. 7–11). The herbaceous (nonwoody) shoot tissue normally dies back during the winter, and plants regrow in spring from the vegetative propagules that overwinter in soil. Over time, one individual creeping perennial seedling can grow into a large patch with an extensive root system and many shoots, which makes these species particularly difficult to control. In fact, many of the most troublesome weeds in the United States are creeping perennials (see

(A)

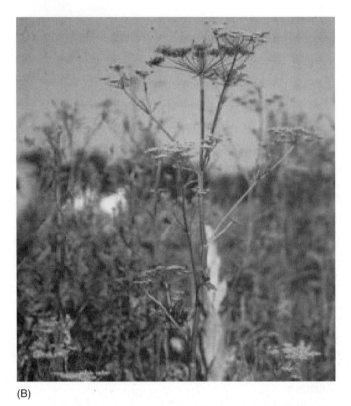

(B)

Figure 7–9   (A) Wild parsnip (*Pastinaca sativa*), young first-year rosette. (B) Second-flower stalk. *Source:* Kent Harrison, The Ohio State University.

Figure 7–10   Plantains (*Plantago* spp.) have a simple perennial life cycle. *Source:* Kent Harrison, The Ohio State University.

Table 7–1). These species tend to become established in crops that are grown under some form of reduced tillage, as well as in perennial crops such as pastures and hayfields. Once established, however, they can also persist in tilled environments for a long period of time because the tillage operations may fragment and spread the vegetative propagules to other areas of the field.

Woody perennial weeds are usually vine or shrub plants that regrow each year from persistent woody stems. These weeds are sometimes encountered in crop fields in reduced tillage systems; however they are most troublesome in low maintenance areas, such as pastures, forests, and rights-of-way. Examples of woody perennials are multiflora rose (*Rosa multiflora*), poison ivy (*Rhus radicans*), and bittersweet nightshade (*Solanum dulcamara*).

Invasive Weed Species

The U.S. National Invasive Species Council, established in 1999, defines an invasive species as "a species that is (1) nonnative (or alien) to the ecosystem under consideration and (2) whose introduction causes or is likely to cause economic or environmental harm or harm to human health. Invasive species can be plants, animals, and other organisms (e.g., microbes). Human actions are the primary means of invasive species introductions." Some weed scientists argue that some particularly aggressive and problematic native plants should also be considered invasive species. Regardless of origin or terminology, the hallmark of invasive weeds is their lack of natural enemies and the ability to infest new environments rapidly. These species are capable of aggressive growth under a variety of environmental conditions, which allows them to compete successfully with a wide range of native and cultivated plants. In North America, garlic mustard

Figure 7–11 Examples of vegetative propagules. (A) Johnsongrass (*Sorghum halepense*) rhizomes. (B) White clover (*Trifolium repens*) stolons. (C) Yellow nutsedge (*Cyperus esculentus*) tubers. (D) Wild garlic (*Allium vineale*) bulbs. (E) Honeyvine milkweed (*Ampelamus albidus*) creeping roots. *Source:* Kent Harrison, The Ohio State University.

TABLE 7–1	Examples of Troublesome Weed Species in the United States
Species	**Life Cycle**
Cogongrass (*Imperata cylindrical*)	Creeping perennial
Common lambsquarters (*Chenopodium album*)	Summer annual
Japanese honeysuckle (*Lonicera japonica*)	Woody perennial
Johnsongrass (*Sorghum halepense*)	Creeping perennial
Kudzu (*Pueraria lobata*)	Creeping perennial
Large crabgrass (*Digitaria sanguinalis*)	Summer annual
Leafy spurge (*Euphorbia escula*)	Creeping perennial
Purple loosestrife (*Lythrum salicaria*)	Creeping perennial
Red rice (*Oryza sativa*)	Summer annual
Purple nutsedge (*Cyperus rotundus*)	Creeping perennial

Source: Jerron Schmoll, Pioneer, Inc.

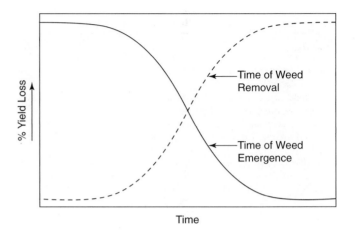

Figure 7–12   Effects of weed competition on crop yields. *Source:* Jerron Schmoll, Pioneer, Inc.

(*Alliaria petiolata,* a biennial) and Japanese honeysuckle (*Lonicera japonica,* a woody perennial) are examples of exotic (nonnative) invasive weed species.

Competition

Competition for sunlight, water, and mineral nutrients is often the first factor that comes to mind when one considers the negative effects of a weed on a crop. Indeed the time that a weed spends in contact with a crop plant can have a dramatic effect on yield. Figure 7–12 illustrates the effects of competition on crop yield.

The dashed line in Figure 7–12 shows what happens when weeds emerge with the crop and are allowed to compete for a period of time before they are removed. The weeds have very little effect on crop yield in the first few weeks primarily because there is sufficient light, water, and nutrients available for all plants in the field. As time passes and the crops and weeds grow larger, competition starts to have an increasingly negative impact on crop yields, which is indicated by the steep slope in the middle of the curve. At the top of the curve, weed competition is maximized and crop yields can be driven to zero by some of the most competitive weed species.

The solid line in Figure 7–12 shows that weeds emerging early in the season have a much greater impact on crop yield than those that emerge later in the season. Crop yields can be reduced drastically by a large population of weeds that emerges with the crop, but with later emerging weeds, the crop can use its head start to grow rapidly and to shade out the weeds. Once the crop canopy closes and the entire soil surface is shaded, weeds have a significantly reduced capacity to affect crop yield. Decreasing the time to canopy closure by maintaining a healthy, actively growing crop is an important goal for many crop producers.

The effects of competition are so pronounced that researchers have developed emergence prediction models for the more troublesome annual weed species in agronomic crops. These models usually draw on environmental factors, such as rainfall and temperature, to predict when emergence is likely to occur. From a conceptual standpoint, these models could then be used to predict when control measures, such as tillage or herbicide application, would be most effective. However, the complexity of weed seed dormancy and the fact that many weed species germinate sporadically over a long period of time have rendered most emergence models less than adequate for use in situations where multiple weed species are involved. Current models are most useful in situations where only one or a few weed species are of primary concern. As research in weed biology and ecology continues, emergence prediction models should help to improve the timing of weed control efforts, which would serve both to improve the efficacy of the control efforts and to reduce growers' costs.

Reproduction

Annual weeds rely on high seed production, seed longevity, seed dormancy, or a combination of these three factors for survival. Table 7–2 shows the variation in the number of seed produced per plant for a few common weeds and lists an estimate of the length of time that seeds from that plant may remain viable in the soil. The high numbers produced and the longevity of many common annual weed species allows a **seed bank** to build up in the soil. This seed bank serves as the primary source of seed for annual weed infestations in crop fields. While these species can be controlled effectively by tillage performed when seedlings are still small, tillage can also serve to bury ungerminated seeds deeper in the soil, where the cool, relatively stable temperatures, protection from seed-eating organisms; and lack of light provide an ideal environment for long-term survival of the seed.

Seed dormancy also helps to ensure the long-term survival of many weed species. Most of the seeds that are produced by annual plants will germinate after exposure to a

TABLE 7–2	Seed Production and Longevity		
Species	**Life Cycle**	**Seeds/Plant**	**Longevity (Years)**
Ambrosia trifida (giant ragweed)	Summer annual	5,000	21
Abutilon theophrasti (velvetleaf)	Summer annual	32,000	39
Capsella bursa-pastoris (shepherd's-purse)	Woody annual	38,500	35
Arctium minus (common burdock)	Biennial	31,600	—
Digitaria sanguinalis (large crabgrass)	Summer annual	150,000	50
Amaranthus retroflexus (redroot pigweed)	Summer annual	230,000	40
Chenopodium album (common lambsquarters)	Summer annual	500,000	39
Cirsium arvense (Canada thistle)	Creeping perennial	5,300	25

Source: Regnier, E. E. 1995. Teaching seed bank ecology in an undergraduate laboratory exercise. *Weed Technology:* 9:5–16.

prolonged cool, moist period (summer annuals) or a prolonged warm period (winter annuals). A small percentage of seeds will not germinate, however, even after this primary dormancy is broken and the seeds are exposed to conditions that are favorable for growth. These seeds are considered to be in a state of secondary (or conditional) dormancy and will not grow until the proper conditions are present to break this secondary dormancy. These conditions might include exposure to an additional cold (or warm) period, decay of the seed coat, exposure to light, leaching of growth-inhibiting hormones from the seed coat, or some combination of these factors. Thus, the seed production from a single year's annual weed infestation can provide enough seed to infest a field for many years. This predicament has led many growers to adopt the goal of zero seed return for annual weed species.

Annual weed species also produce seed that differ significantly in size. Larger seeded weeds, such as giant ragweed (*Ambrosia trifida*), produce relatively low numbers of seeds, but these species can rely on the large reserves of energy stored in their cotyledons to emerge from considerable depths in the soil profile. Smaller seeded weeds, such as redroot pigweed (*Amaranthus retroflexus*), produce many more seeds per plant but the seed can germinate only from shallow depths in soil. Both strategies are equally successful.

Biennial weeds also rely on seed longevity and seed dormancy for long-term survival. One additional survival mechanism that many biennials possess is that they have the capacity to behave as short-lived perennials. If the elongated flower stalk is removed, perhaps by mowing or grazing animals, the plant may overwinter a second year and produce seed in the third year.

Unlike annuals and biennials, perennials do not rely as heavily on seed production for establishment. While perennials do sometimes establish by seed, they are also capable of spreading via vegetative **propagule** fragments, which can easily be carried from one area to another by tillage equipment or other transport mechanisms (Fig. 7–12). The ample reserves of carbohydrates in these propagules allow the weeds to establish quickly when transported to a new area. In fact, understanding how the carbohydrate levels in the

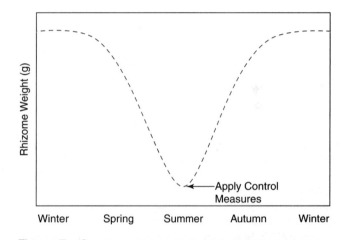

Figure 7–13 Seasonal carbohydrate distribution in rhizomatous creeping perennials. *Source:* Jerron Schmoll, Pioneer, Inc.

propagules change during the course of the year is a key component in the management of these species. The seasonal change in carbohydrate levels for a typical rhizomatous creeping perennial is shown in Figure 7–13. Note that carbohydrate storage is at its maximum in the winter months, which provides the plant with sufficient energy to produce new shoots aggressively in the spring. Carbohydrate levels are at their minimum levels in the summer months because they have been expended to support early spring shoot growth. At this minimum point, which usually corresponds to the early reproductive stage, the plant shoots have become large enough that they start to export sugars downward to replenish the vegetative propagules, which accumulate adequate levels of stored energy for winter survival and growth the following spring. The optimum time to apply some systemic herbicides (herbicides that are mobile within the plant) to creeping perennials is at this stage when the carbohydrate flow starts to move back to the vegetative propagules. It is also possible to deplete the carbohydrate supply of vegetative propagules gradually and eliminate perennials by repeated tillage, mowing, or other methods of shoot removal during the spring. Whatever method is chosen, it must be repeated frequently enough that the shoots are never allowed to reach the early reproductive stages where they begin to replenish the vegetative propagules.

Genetics

The study of weed genetics has gained more attention as the number of reported cases of herbicide resistance has steadily increased. Understanding the basic genetic concepts that drive herbicide resistance is key to maintaining the long-term utility of commonly used herbicides. Since the first reported case of herbicide resistance in common groundsel (*Senecio vulgaris*) in 1968, the number of herbicide resistance cases has increased to 174 species and 291 biotypes. A **biotype** refers to a subpopulation of plants within a species that differs only slightly from the norm of the general population. Occasionally a biotype of a weed species has some level of resistance to one particular herbicide or to a group of herbicides that have a similar mode of action. If a single resistant plant survives and reproduces, some of the pollen or seed resulting from that plant may carry the gene that imparts resistance. If the same herbicide is applied again in subsequent years, plants that carry the gene for resistance will survive and reproduce. If this cycle repeats, a population of herbicide-resistant biotypes may become prevalent within a relatively short period of time. Integration of several weed-management strategies, rather than strict reliance relying on chemical control, is a useful tool for controlling these especially problematic weeds.

Genetic variability within a weed species is not restricted to physiological differences. Many weed species also vary significantly in vegetative features such as leaf and seed morphology. For instance, in giant ragweed (*Ambrosia trifida*), the size and shape of the seeds produced varies dramatically from one individual to the next. Even individuals from closely related species may differ so much in their morphology that accurate identification becomes challenging. The pigweeds are a good example because there are several different species of pigweeds (*Amaranthus* spp.) that can be differentiated only by slight differences in seed-head shape and flower structure. Morphological variation likely plays a significant role in the adaptive survival and evolution of a species. For instance, a seed-boring insect might prefer to lay its eggs on the smaller seeds of a given species, while a seed-eating rodent might prefer to forage for larger seeds of the same species. Therefore, if variability exists in the size of seeds produced in a given year, the weed species may prove to be less susceptible to one particular predator, which increases the chance that some of the seed will survive until the following year.

Ecology

Ecology is the study of a species in both its physical and biological environment. Many weed species have special adaptations that allow them to compete effectively in specific physical environments. For instance, field horsetail (*Equisetum arvense*), reed canarygrass (*Phalaris arundinacea*), and the sedges (*Cyperus* spp.) thrive in damp areas. Red sorrel (*Rumex acetosella*) favors acidic soil conditions. Goosegrass (*Eleusine indica*) tolerates highly compacted soils. These adaptations give each species a distinct advantage over other species under specific environmental conditions. At the same time, these adaptations provide land managers with a powerful control strategy: removing the set of conditions to which a particular species is adapted will reduce the competitive ability of that species. For instance, sedges are rarely a problem in a field with a properly installed tile drainage system.

The study of the biological environment as it relates to a weedy species is a rich field of study. Numerous organisms use weeds as a food source, for shelter, and for nesting habitat. For instance, quackgrass (*Elytrigia repens*), while a troublesome weed in agronomic and vegetable crops, is a nutritious food source in pastures for grazing animals. Insect larvae such as the ragweed fruit fly (*Euaresta festiva*) use giant ragweed (*Ambrosia trifida*) seeds as a food source. Foxtail (*Setaria* spp.) seeds are eaten by many species of birds, including the mourning dove (*Zenaida macroura*). A wide variety of birds; including ring-necked pheasants (*Phasianus colchicus*), meadowlarks (*Sturnella* spp.), and sparrows (*Ammodramus* spp., *Chondestes grammacus,* and *Passerculus sandwichensis*), use grasslands (including weeds) as nesting habitat and foraging grounds.

The study of weed ecology may provide information that increases our capacity to use biological methods to control weeds. For instance, researchers have been studying the potential to reduce the number of weed seeds that enter the seed bank by optimizing habitats for seed-feeding insects, rodents, and birds. While practical recommendations are not yet available, researchers have determined that various seed feeders can consume a large number of weed seeds within days after the seed are deposited onto the soil surface. Integration of practices that increase the populations of seed feeders may become an important nonchemical weed-control method available to crop producers.

WEED MANAGEMENT
Preventive Weed Control

One of the first lines of defense against weeds is to prevent their initial establishment. Many different preventive measures are effective at keeping weeds from entering a field. Cleaning planting, tillage, and harvesting equipment when moving from one area to the next is essential to reducing the spread of weed seed and vegetative propagules. Using weed-free crop seed is another method of preventing weed establishment. Seed companies must comply with rigorous quality standards restricting weed seed content of planting seed, thus making the spread of weeds through contaminated seed much less problematic than in the past. Preventing seed from moving into a field from surrounding areas such as roadsides, ditches, and fencerows helps to control weed spread into a field. These areas should be regularly mowed, sprayed, or otherwise managed to keep weeds from spreading. Preventing weed reproduction in crop fields is another important preventive weed control measure, particularly for the control of annual weed species. Finally, many federal, state, and local laws require landowners to control noxious weed species on their property.

Mechanical Weed Control

Mechanical weed control is the term used to describe any physical weed-control measure. Hand weeding, tillage, mowing, mulching, and the use of fire are all examples of mechanical weed control. Prior to the 1940s, weed control in agronomic crops was accomplished primarily by hand removal or by some form of tillage. The first primitive tools used for mechanical weed control were probably sharpened sticks or metal objects, culminating with the common hoe. Even today in gardens, flower beds, and small crop fields, the hoe is often the principal means of eliminating weeds. With the advent of mechanized agriculture, tools were developed that accomplished the work of the hand hoe on a much larger scale. A variety of rotary hoes, cultivators, disks, and harrows are available to today's producers to accomplish the work of weed control in crop fields. In some areas, cultivating between the rows can remove most of the weeds, and only a small band of herbicides applied over the crop row is required for good weed control. While tillage is a highly effective control strategy for many weed species, concerns about soil erosion and compaction have reduced the utilization of this control measure in recent decades.

Mowing is another important method of weed control in many situations. Home lawns, roadsides, pastures, waterways, fallow ground, and hay fields are examples of areas in which mowing is the principal, if not the sole, source of weed control. Mowing at the correct time accomplishes two goals: (1) it eliminates the production of seed, which is of particular importance in the management of annual crops, and (2) it depletes the stores of carbohydrates in the roots and vegetative propagules of weedy plants, which is of importance in perennial crops.

Mulching is a common method of weed suppression in flower beds, vegetable crops, and nurseries. Many different types of mulch are available, and essentially any substance that can be used to cover the ground and smother weeds can be considered mulch. Wood chips, sawdust, gravel, newspaper, and straw are just a few of the materials that can be used. While these mulches can effectively suppress the germinating seeds of many annual weeds that may be present in the seed bank, they are often less effective against creeping perennials, whose shoots can grow through the mulch layer. Another issue is that over time, as organic mulches decay, they can serve as a fertile seedbed for annual weed seed that blow in from nearby areas; thus, new layers of organic mulch must be put down every year to maintain annual weed control. Plastic and fabric mulches have been developed to suppress the more aggressive creeping perennials species; they also prevent the establishment of newly deposited annual seed. However, plastic mulches often significantly alter soil temperatures and restrict the amount of rainfall that penetrates the ground near the roots of the crops; thus, use of these mulches can increase the amount of care that must be given to the crop.

Fire has been tried as a weed-control measure in almost every type of environment that humans have attempted to control. While burning ultimately has been found to have limited value in controlling weeds in most cropland, specialized flame weeders are sometimes used to burn off small annual weeds in organically grown vegetable fields. Controlled burning has been of considerable value in removing unwanted brush from forest areas, in prairie management and restoration, in warm-season grassland management, and in rangeland management.

Cultural Weed Control

Cultural weed control includes using methods of crop production that enhance the capacity of a crop to compete with weeds. Modification of crop planting date, row spacing, crop rotation, and flooding can all be used as components of a cultural weed-control program. Cultural weed control also encompasses the use of additional crops to help control weeds. The use of **cover crops**—crops grown for an objective other than harvest—is the most highly utilized of these practices.

While early planting dates are normally recommended to provide the crop with a longer growing season and thus higher potential yields, early planting also increases the length of time that the crop must compete with weeds. Many summer annual weeds also emerge early in the season, so planting a crop a few weeks later may allow the grower an opportunity to implement an additional mechanical weed-control measure such as disking or harrowing to remove these weeds. While some crop producers do utilize this strategy, the vast majority of agronomic crop producers rely on chemical weed-control measures to eliminate this early season flush of weeds.

Modification of crop row spacing is a more widespread practice that directly affects weed competitiveness. The goal in optimizing row spacing is to maximize crop yield potential while reducing the time required for the plant canopy to close, thereby effectively shading the soil surface and reducing the light available for late-emerging weeds to grow. A tremendous amount of research has been completed on the subject of optimum plant population and row spacing, and local recommendations can be obtained for most economically important crops.

Crop rotation can be an effective weed-control strategy in various cropping systems. Because the weeds that are most prevalent in a crop often have the same life cycle as the crop, implementing a perennial-annual crop rotation, such as a short-term hay crop with a row crop, can serve to disrupt weed life cycles and prevent the establishment of large weed populations. Weed populations also adapt and shift in response to tillage practices and herbicide applications, so additional weed-control advantages may be gained by using a similar rotational approach combined with these practices.

Flooding is a cultural weed-control practice that can be used in rice. Flooding allows the grower to regulate the germination of annual weed seeds by indirectly managing the amount of oxygen available in the soil. Flooded soils typically have very low oxygen content, which is required by most weed species for germination. Water management is an important method available for controlling red rice, which is closely related to cereal rice, in the southeastern United States.

The use of cover crops is another effective cultural weed-control measure. Cover crops such as cereal rye (*Secale cereale*) and hairy vetch (*Vicia villosa*) can suppress germination and compete effectively with existing weed species. These crops are often planted in the fall and killed in the spring prior to the establishment of the crop intended for harvest. Because they provide winter cover, this practice can also be an effective method of reducing soil erosion. Historically, legume cover crops were widely used as an important nitrogen source as well as for weed control. In addition to the effects of competition, **allelopathy** (the production by the cover crop of chemicals that are toxic to weeds) may play a role in the suppressive effect of cover crops on weeds. Leachates from crops such as cereal rye have been shown to inhibit germination and growth of susceptible weed species in laboratory and greenhouse experiments. Under field conditions, however, it has been difficult to ascertain the role of allelopathy in weed control because separating the effects of competition from allelopathic chemical inhibition is challenging. In the future, it may be possible to develop crop varieties that can defend themselves against weeds by allelopathy.

Biological Weed Control

Biological weed control involves introduction of a living organism, often a natural pest or disease, to control a weed species. There are two approaches to biological weed control: the classical (or inoculative) approach and the inundative (or augmentative) approach. The classical approach involves a one-time introduction of an organism to control weeds. In a successful introduction, the organism must establish itself in the ecosystem and reproduce, thereby providing ongoing control of the weed, usually for many years.

The inundative approach involves application of the biological organism on an as-needed basis. The organism is expected to provide short-term control of the target weed, but it is not expected to establish itself in the environment.

The control of prickly pear cactus (*Optunia* spp.) in Australia by a moth borer (*Cactoblastis cactorum*) from Argentina is perhaps the most frequently cited example of an effective classical biological weed-control effort. The cactus was introduced in the mid-1800s as an ornamental, but it escaped and spread rapidly throughout Australia, covering 60 million acres by 1925. It was eventually discovered that the *Cactoblastis* moth larvae fed exclusively on the cactus, so the moth was released in Australia. The feeding damage caused by the moth also allowed a fungus to establish in the cactus, which enhanced control. The moth populations grew quickly and reduced the cactus populations, to the extent that the moths ran out of food. This allowed a resurgence in the cactus population, which was soon followed by an increase in the moth population. This pattern is typical of classical biological weed control: a cyclic balance between predator and prey is often established. Therefore, weed control in this system is neither instantaneous nor complete. Other successes have followed, often involving an interaction between a weed and an insect. The approach has been most successful against single invasive species in relatively undisturbed, low-management areas, such as rangelands.

Successes have also been achieved with the inundative approach to biological weed control. *Colletotrichum gloeosporioides* is a fungus sold under the trade name Collego. It provides control of northern jointvetch (*Aeschynomene virginica*) in rice and soybeans. Another example is *Phytophthora palmivora*, a fungus sold as DeVine, it kills milkweed vine (*Morrenia odorata*) in citrus plantings.

Weed Control in Organic Crop Production Systems

Organic producers grow crops without the use of synthetic herbicides and fertilizers; therefore, successful weed control in these systems requires more careful integration of preventive, mechanical, and cultural weed-control methods than does conventional crop production, in which chemical weed control plays a major role. Because most of the research in recent decades has focused on chemical weed control, the most effective and economical weed-control strategies in annual crop production systems will almost invariably include chemical weed control. The result is that organic producers will normally have higher weed-management costs for less effective overall weed control than their counterparts in conventional production. Increasing interest in organic production has spurred a corresponding increase in weed-management research in organic systems, which should provide producers with better weed-control strategies in the future. In the meantime, organic crop producers must continue to receive premiums for their crops to compensate them for the increase in production costs. If consumer demand for organically grown produce continues to increase, so will the need for more effective weed-control strategies in organic systems.

Chemical Weed Control

Introduction

Sporadic research with chemicals as a way to control weeds started about 1910 in both Europe and the United States, but the basis for modern chemical weed control was laid out with the initial studies of auxins and plant hormone physiology in the 1930s, culminating with the discovery of the herbicidal properties of 2,4-dichlorophenoxyacetic acid (2,4-D) in 1942. Since that time, many more herbicides have been discovered and introduced into the market. Chemical weed control helped revolutionize crop production in the second half of the twentieth century, when crop producers integrated the use of herbicides with improved crop varieties, mechanized farm machinery, and advances in crop production practices to reduce labor costs and increase yields.

Herbicides have several characteristics that are important to understand so that they can be used effectively. These characteristics are the selectivity of the product, the timing of its application, its use rate, and its mode of action. Recently, crops that have been bred or genetically engineered for herbicide resistance have been introduced into the market. These products have greatly expanded the utility of several commonly used herbicides.

Selectivity

One of the key characteristics of a successful herbicide is selectivity. Selectivity means that the herbicide eliminates many different weed species, while the crop in which the herbicide is being used possesses a certain degree of innate tolerance to the herbicide. Herbicides are often broadly classified as grass killers, broad-leaf killers, or nonselective (kills both grasses and broad leaves), although some grass herbicides possess a certain degree of broad-leaf activity and some broad-leaf herbicides control certain grasses. The development of herbicide-resistant crops represents the latest important advance in herbicide selectivity.

Herbicide Application

Herbicides can be classified in terms of the timing of application. Burn-down, pre-plant, preemergence, and post-emergence applications are the four basic classifications. Herbicides that are applied to remove existing weeds in the field and thus prepare a clean seedbed for planting are considered **burn-down herbicides. Pre-plant herbicides** are applied before the crop is planted. These herbicides often require incorporation into the soil, either by mechanical means or by rainfall. **Preemergence herbicides** are applied to soil after crop planting but prior to the emergence of the weeds, the crops, or both. Many preemergence herbicides also require rainfall for maximum efficacy. **Post-emergence herbicides** are applied directly to foliage after weed emergence. Because herbicidal chemicals are often toxic to a broad number of plants, it is advised that the equipment used for the application of herbicides not be used for other purposes to avoid the possibility of accidentally harming plants.

To help reduce the chance of accidents, herbicide-application technology has evolved significantly in recent years. Utilization of **global positioning system (GPS)** and **geographic information system (GIS)** technology has allowed growers to apply herbicides with pinpoint accuracy and to vary the application rate across the field based on the history of weed pressure or on visual observations of high-density patches of weeds. While much of this technology is relatively new, adoption rates are increasing among the largest and most progressive growers. Current research is focusing on sensing technology that would identify weed species and their growth stage. This information could then be sent to a computer that would automatically vary the application rate of the herbicide as the sprayer moves across the field. Given the potential of such technology to reduce significantly the amount of herbicides that are needed, it is likely that further advances will be seen in the near future.

Herbicide Use Rates

Utilization of the proper herbicide rate is key to effective weed control. In general, smaller weeds of a given species can be controlled with a much lower use rate than larger weeds of the same species. Growers typically favor low use rates because it lowers production costs while simultaneously reducing any potential negative environmental impacts from leaching or runoff. However, the difficulty with using low rates is that larger weeds in the field may survive the herbicide application, mature, and produce seed. Once this seed enters the seed bank, the weed will likely be present in subsequent years, requiring more herbicide applications in the future. Therefore, an optimum use rate decision requires that the grower have a thorough knowledge of the size and species of weeds present in the field.

Herbicide Mode of Action

Herbicides can be classified into groups based on the plant biochemical processes with which they interfere. **Mode of action** is the term commonly used when discussing the action of herbicides on these biochemical processes. Table 7–3 lists nine of the most common modes of action. Each of these modes of action contains one or more groups of chemical families, and each chemical family contains one or more commonly used herbicides, examples of which are provided. Understanding the mode of action of an herbicide

TABLE 7–3	Herbicidal Modes of Action	
General Mode-of-Action Categories	**Chemical Family Examples**	**Specific Herbicide Examples**
Amino acid synthesis inhibitors	Glycine derivatives, imidazolinones, sulfonamides, sulfonylureas	Glyphosate, imazethapyr, cloransulam-methyl, chlorimuron
Ammonia assimilation inhibitors	Phosphonic acids	Glufosinate
Auxin transport inhibitors	Semicarbazones	Diflufenzopyr-Na
Cell membrane disruptors	Bipyridiliums, diphenylethers, triazolinones, thiadiazoles, N-phenylphthalimides	Paraquat, lactofen, sulfentrazone, fluthiacet, flumiclorac
Growth regulator (auxinic) herbicides	Phenoxy acids, benzoic acids, 2-pyridine carboxylic acids	2,4-D, dicamba, clopyralid
Lipid biosynthesis inhibitors	Aryloxyphenoxypropionoates, cyclohexanediones, thiocarbamates	Fenoxaprop, sethoxydim, butylate
Meristem mitotic inhibitors	Benzamides, chloroacetamides, dinitroanilines, oxyacetamides	Pronamide, acetochlor, pendimethalin, flufenacet
Photosynthesis inhibitors	Benzothiadiazole, nitriles, triazines, triazinones, uracils, ureas	Bentazon, bromoxynil, atrazine, metribuzin, terbacil, diuron
Pigment synthesis inhibitors	Isoxazolidinones, isoxazoles, callistemone	Clomazone, isoxaflutole, mesotrione

Source: Jerron Schmoll, Pioneer, Inc.

is important for three main reasons. First, knowing how the herbicide kills the plant allows a producer to predict how a particular herbicide will perform under different growing conditions. For instance, many herbicides work much more quickly on actively growing plants; thus, if such a herbicide is applied during cool conditions when plants are growing slowly, it may take much longer for injury symptoms to be observed on the target plant. Second, understanding mode of action helps a grower identify and assess the extent of injury symptoms on desired plants and on nontarget plants. Third, knowledge of mode of action is an important asset in helping producers manage herbicide-resistant weeds. Repeated applications of herbicides with the same mode of action increase the probability that a herbicide-resistant biotype of a weed species will be selected. Rotating herbicidal modes of action reduces the probability that a herbicide-resistant biotype will be selected for a given field. However, a few species of weeds such as rigid ryegrass (*Lolium rigidum*) and horseweed (*Conyza canadensis*) have developed resistance to multiple modes of action. Therefore, while rotating modes of action is important, relying strictly on herbicides to manage weeds is not a viable long-term weed-control strategy for some species.

Herbicide-Resistant Crops

One of the latest advances in weed management has been the development of crops that are resistant to broad-spectrum, commonly used herbicides. These crops have been developed using both conventional plant-breeding methods and genetic engineering technology. Since the release of the first herbicide-resistant crop in 1991, the acreage planted with herbicide-resistant crops has grown

dramatically. Herbicide resistance research has been done with a wide variety of crops and almost every mode of action. Herbicide resistant corn, soybean, cotton, and canola varieties are currently on the market in the United States. The level of weed control obtained with herbicide-resistant crops can be quite impressive, and this result has fueled the rapid adoption of these crops by U.S. producers. However, much of the rest of the world prohibits their use or has stipulated that herbicide-resistant crops be used only for animal feed. One of the major factors that has hindered the utilization of herbicide-resistant crops outside the United States has been opposition from groups who point out that the long-term effects of releasing these new crop biotypes into the environment are unknown. Proponents of herbicide-resistant crops point out the many potential benefits from the utilization of these crops, including the potential to reduce herbicide use and the opportunity to utilize less toxic herbicides. Many groups of academic and industry scientists are currently working to answer questions pertinent to both sides of the debate, but this issue seems unlikely to be fully resolved, at least in the near future.

INSECTS AND MITES

Plants and products are the only sources of food for a host of insects and mites. When we grow these plants as vast monocultures (e.g., agricultural crops, forestry, lawns) or store their products in large quantities, we often enter into a fierce competition with these insects and mites for these resources. This competition has gone on since humans first cultivated plants as agricultural crops (Fig. 7–14). Insects and mites also pose secondary hazards because they spread diseases among plants and animals.

Figure 7–14 Insect swarms can devastate crops, as in this stand of corn ruined by grasshoppers. Grasshoppers can be controlled biologically by dropping into infested areas wheat bran sprayed with spores of *Nosema locustate,* which causes a deadly disease of grasshoppers and crickets. This disease does not harm people, plants, or other animals; it affects only grasshoppers and Mormon crickets. *Source:* USDA.

The competition between humans and insects has intensified as more and more land has been conscripted into agricultural production, especially as our populations continue to expand at unchecked rates. The conflict between agricultural production and insect and mite pests took another turn shortly after World War II. During this period, synthetic organic pesticides were discovered and developed. These new chemicals were cheap to produce and they had an amazing ability to kill insects and mites. Scientists in the late 1940s and early 1950s regularly maintained that all pests, their damage, and disease spread would soon be a mere memory. We soon discovered that most pests adapted when facing extinction or adaptation. In short, pests soon became resistant to the pesticides that bombarded them constantly. To make matters worse, stronger pesticides had to be developed, which were applied more often and/or at higher rates. In the 1960s and 1970s, we also realized that extensive use of insecticides and miticides were causing major adverse environmental effects.

Many scientists, realizing that the pesticide–pest resistance cycle was escalating, began to investigate other ways of managing insect and mite pests. Biological controls and cul-

tural techniques were rediscovered and developed. Extensive research soon indicated that reliance on a single method such as chemical (pesticide), biological, or cultural controls would rarely keep pests at desired levels. However, IPM kept pest damage to a minimum while minimizing environmental contamination and development of pest resistance.

More recently, pest management researchers have begun to adopt the concept of plant health care (PHC). The underlying principle in this concept is that healthy plants are generally better equipped to deal with pest attack, while stressed plants commonly succumb to pest outbreaks. In essence, plants and their needs (e.g., nutrition, water, weather, etc.) should be the primary focus in agricultural production, but if pest problems arise, then the IPM approach should be followed to keep the plants (crops) healthy.

Some insects, however, help agriculture. Certain harmless insect species prey on other insect species that destroy plants or crops. The praying mantis and ladybird beetle, both larvae and adults, feed on aphids. The larvae of the green lacewing feeds on mealybugs, scale, and aphids. Predatory *Aphytus* wasps attack red and black scale. Starts of these beneficial insects can be purchased through mail-order catalogs or at nurseries or garden centers, packaged and ready for release in the garden. Once a biological balance is established between the host (harmful insects) and the predatory or parasitic insects, then no further chemical spraying is necessary. In fact, if spraying is done, the beneficial insects are likely to be killed along with the harmful ones.

The honeybee and certain other insects do a tremendous service in pollinating fruit trees and other crops, such as alfalfa. Some insects produce useful products, like honey and beeswax from the honeybee and silk from the silkworm. Certain insects are relished as food in some parts of the world. The *Drosophila* fly, because of its short life cycle, ease of handling, and the giant chromosomes of its salivary gland, has been of inestimable value to geneticists in their studies.

Nevertheless, innumerable kinds of insects and mites, if left uncontrolled, would soon reduce the world's food supply to a shambles. Control methods for harmful insects are of four types that can be used in integrated pest management.

1. *Biological control.* See Figures 7–15 and 7–16. This category entails the use of insect parasites, predators, and disease-causing pathogens. These are produced in culture for release into, or application to, infested fields. In the case of parasites and predators, another form of biological control involves the preservation of those that are naturally occurring through careful use of chemical or other forms of control. Most insect pathogens in use today are not capable of perpetuating themselves in the field and must be applied—usually as a spray—each time they are needed.

Figure 7–16 Natural biological control of a destructive insect: a hymenopterous parasite injects eggs into the body of a much larger larval insect (the large, hairy form with its head on the right). *Source:* USDA.

Figure 7–15 Biological control of destructive insects; the lady-bird beetle adult (*upper left*) and larva (*upper right*) feeding on aphids in various growth stages. *Source:* USDA.

2. *Cultural control.* Many cultural practices, such as crop rotation, destruction of crop debris after harvest, careful attention to planting or harvesting date, proper pruning, and the use of pest-resistant cultivars, are useful in preventing or mitigating serious infestations of certain insects. Proper fertilization and irrigation schedules are useful in maintaining tree crops in a state of vigor sufficient to prevent certain insects from exploiting them.

3. *Physical or mechanical control.* Leaving fields fallow during critical periods in an insect's life cycle, screening of seedbeds or ventilators on greenhouses, disking soil beneath tree and vine crops, and burying or chipping prunings are examples of physical practices that control specific insects. Hand picking insects, syringing plants with water to dislodge insects, and picking up and destroying fallen fruit are examples of physical controls useful in small-scale plantings.

4. *Chemical control.* When confronted by an imminent pest infestation, the farmer often has little recourse but to rely on the application of a chemical insecticide or miticide. For many pests, there are no good alternatives because biological, cultural, or physical controls are not effective against certain insects.

Insects and Their Relatives

Insects and their near-relatives are placed in the phylum Arthropoda. This phylum contains the Arachnida (spiders, mites, ticks, and scorpions), Crustacea (crabs, shrimp, isopods, etc.), Chilopoda (centipedes), Decapoda (millipedes), and the Insecta. The class Insecta is usually broken down into twenty-five or more orders that contain more than 1 million species: over 80 percent of all the species of animals known on the earth! Fortunately, most insects are of no economic importance; most estimate that less than 5 percent of the insect species have any direct importance to humans, and only about 1 percent regularly reach pest status. The insect orders with the most species are:

Coleoptera—beetles, weevils.

Lepidoptera—butterflies, moths, caterpillars.

Hymenoptera—bees, wasps (including many insect parasites), ants.

Diptera—true flies.

Heteroptera—true bugs and buglike insects (scales, whiteflies, leafhoppers, etc.).

Most entomologists emphasize insect metamorphosis (development) because this knowledge often assists in developing control strategies. Three main types of metamorphosis are generally recognized: no metamorphosis (ametabolous), gradual or incomplete metamorphosis (paurometabolous), and com-

To tell the difference among insects, mites, spiders, and ticks, consult the following lists.

INSECTS

The adults are characterized by:

1. Three definite body regions—head, thorax, and abdomen.
2. Three pairs of jointed legs.
3. One pair of feelers or antennas.
4. Eyes that are usually compound.
5. One or two pairs of wings.

MITES, SPIDERS, AND TICKS

These are characterized by:

1. Two main body regions—the cephalothorax (head and thorax fused together) and the abdomen.
2. Four pairs of jointed legs.
3. A lack of antennas or wings.
4. Simple eyes.

plete metamorphosis (holometabolous). The paurometabolous insects are often subdivided: those in which the nymphs look like the adults and generally live in the same habitat as the adults (gradual), and those with aquatic nymphs (called naiads) that may look quite a bit different than the adults that live out of water. These differences are illustrated in Figure 7–17.

The mode of feeding is also emphasized by entomologists. Insects all have mandibulate mouthparts, but these may be chewing or modified into sucking forms. Chewing insects (like caterpillars, beetles, and their larvae) devour plant foliage, shoots, flowers, and fruit, or they may burrow into the plants (borers) (Fig. 7–18). Sucking insects (and mites) consume plant juices or cell contents; their mouthparts probe into the interior of the plant (Figs. 7–19 and 7–20). Many sucking pests inject saliva that can clog the plant's vascular bundles, thereby causing parts of the plant to wilt or die. Other sucking pests commonly pick up plant diseases and transmit them when the insect moves to another plant. The mode of feeding is also important in determining what kind of pesticide should be used and what may be the most efficient method of application. Stomach poisons that are applied to the surface of a plant are readily ingested by foliage eating insects, but a systemic or contact poison would be more effective against a sucking pest.

Insect and Mite Pest Management

As has been previously described in this chapter, insect and mite pest management generally relies on the integration of cultural, biological, and chemical controls using the IPM approach. In summary, pest managers rarely use the term *pest control* (or *pest eradication*) because we have never been

able to control insect and mite pests permanently. At best, we can keep them at levels that allow us to obtain a reasonable return on our investment when producing a crop. Sometimes the cost of using pesticides exceeds the value of the crop realized. By using a matrix of cultural and chemical controls to conserve effective biological controls, most crops can be coaxed into providing adequate returns.

While cultural and biological controls are important in managing insect and mite pests, pesticides still dominate. Their characteristics will be discussed more extensively in the following subsections.

Pesticide Modes of Entry

Each pesticide has one or more ways in which it can get into an insect or mite. These modes of entry are important to understand in terms of making the best application for control of each type of pest.

Stomach poison refers to materials that enter the insect by mouth and kill by being absorbed into the body through the digestive tract. For most chewing insects, pesticide residues should be on the plant surface(s) being ingested. Some insecticides can be absorbed into the tissues of plants (translaminar systemic) and/or even transported through the vascular system of the plant (translocated systemic). Systemic pesticides are most useful for control of sucking pests, borers, and leafminers. Because borers and leafminers feed within the plant tissues, systemics are often the only way to reach these pests. Great care must be taken when using systemic insecticides because the residues can remain in the plant tissues eaten by humans. Such pesticides usually have a preharvest interval period, which is the time needed to degrade the pesticide residue below active levels, written on the label.

Contact poison refers to materials that enter the insect or mite body directly through the pest's cuticle. Some contact poisons actually have to be applied to the pest, while others have sufficient residual action that the residues are picked up as the pest walks across treated surfaces and sufficient material is absorbed to kill the pest.

Fumigants are inhaled as a vapor by the insect or mite, and the material is then absorbed through the respiratory system. Fumigants are generally used in enclosed spaces (e.g., greenhouses and grain storage bins), but some can also be used in the soil if the surface is sealed with water or an impermeable covering (e.g., plastic or tarp).

Pesticide Modes of Action

Each pesticide group tends to have a unique way (mode of action) of disrupting vital body functions of insects and mites. While the list of mode's of actions discussed in the following subsections is not complete, it covers many of the most common.

Neural Disruption Nervous systems are very complex, and pesticides can interfere with their normal function in

Insects *without metamorphosis* emerge from the egg looking just as they will when grown, except they are much smaller.	ORDERS Thysanura Collembola Mallophaga Anoplura	 Egg — Young — Adult
Insects with *gradual metamorphosis* change shape gradually. Wings develop and, as growth takes place, the nymphs look like the adults.	ORDERS Orthoptera Isoptera Corrodentia Thysanoptera Hemiptera Homoptera Dermaptera	 Egg — Nymphs — Adult
Insects with *incomplete metamorphosis* change shape gradually. They do not look like adults until shedding their last skin, when there is a quick change.	ORDERS Ephemerida Odonata Plecoptera	 Egg — Naiads — Adult
Insects with *complete metamorphosis* go through four stages of growth. None of the young looks like the adult, but there is a great change in shape when the adult emerges from the pupal stage.	ORDERS Neuroptera Coleoptera Strepsiptera Mecoptera Trichoptera Lepidoptera Diptera Siphonaptera Hymenoptera	 Egg — Larvae — Pupa — Adult

Figure 7–17 Four types of insect metamorphosis, from egg to adult.

numerous ways. The three major neural systems affected are sodium pump channels, neural transmitter systems, and nerve cell metabolism.

The sodium pump is the system used by nerve cells to transmit an electrical charge from one end of the nerve to the other end. As the electrical impulse moves, ions move in and out of the cell membrane, then the sodium pump system resets the electrical potential. Pesticides such as pyrethroids and chlorinated hydrocarbons interfere with this system.

Neural transmitter systems are located in the nerve synapsis—the microscopic gap between two nerves. In this gap, a neural transmitter compound is released by the sending nerve to stimulate the receiving nerve. This neural transmitter is rapidly degraded so that the receiving nerve doesn't continue to be stimulated. Organophosphate and carbamate insecticides block the neural transmitter–degrading chemical, thereby causing the receiving nerve to continue to fire uncontrollably. Neonicotinyl insecticides block the receiving

Figure 7–18 The cabbage looper, an insect with chewing mouthparts, feeds on soybean leaves. Stomach poison insecticides sprayed on the leaves control such pests. *Source:* USDA.

Figure 7–20 Cottony cushion scale (*Icerya purchasi*) feeding by sucking on a branch of an orange tree. *Source:* University of California Cooperative Extension.

Figure 7–19 Aphids are sucking insects. *Source:* Luis Cañas, The Ohio State University.

nerve at the synapsis, which means that the sending nerve sends the signal, but the receiving nerve doesn't fire. The result is that the insects stop their normal behavior or simply sit and do nothing.

Cellular Metabolic Disrupters Various chemicals can disrupt different metabolic pathways of a cell. Arsenicals are examples of this group. Some interfere with the energy pathways, some disrupt oxygen utilization, and still others interfere with gene functions. Most chemicals in this group are general toxins, with similar actions in all cells, whether insect, animal, or plant. Many of these chemicals can also be slow acting, so when they are used in baits, affected colonial insects such as termites and ants can eliminate the entire colony. Fipronil is one such insecticide.

Cell Membrane Disruption In the past, many maintained that mineral oils (e.g., dormant oils, horticultural oils), crop oils (e.g., soybean oil), and insecticidal soaps (i.e., fatty acid salts) have killed insects and mites through suffocation and were considered to be suffocants. The concept was that oils and soaps coat the insects and/or clog the spiracles (breathing holes). In reality, these materials disrupt the lipoprotein matrix of cells, causing the cells to lose their contents and cease functioning. Even though insect cuticle is considered to be an impermeable structure, in fact, it is filled with microtubules that lead to the underlying cells. When oils or soaps contact insects, they penetrate into the underlying cells, causing them to be destroyed. The same process occurs in insect egg cells, a common target of dormant oils. The protein toxin of *Bacillus thuringiensis* (BT), often considered to be a biological control, actually attaches to the cell wall of insect gut cells, poking a hole in them. The dead cells then allow the leakage of cell bacteria into the insect's body cavity, eventually causing death.

Hormone Disruption   Insects use various hormones to control their growth, development, and exoskeleton formation. Various pesticides can either mimic insect hormones or disrupt the normal function of hormones. Juvenile-hormone mimics cause insects to remain in a juvenile state. Chitin inhibitors disrupt the normal formation of chitin: it may not harden properly or molting is disrupted. Molt accelerators cause the insects to molt before the proper time.

Pheromones   Pheromones are the chemicals that insects normally use to communicate with each other. Most pheromones are used in insect traps to monitor activity, but some are used to disrupt mating. In essence, permeating the air with a sex pheromone causes males to have great difficulty locating unmated females.

Dessicants   Dessicants are chemicals or materials that can cause insects to lose water faster than they can replace it. Physical desiccants include diatomaceous earth, which makes microcuts in insect cuticle, thereby reducing the waterproofing properties of an intact cuticle. Other compounds, like silica aerogel, can pull water through insect cuticle when the dust adheres to the exoskeleton. Obviously, desiccants work best in dry environments, like house or building crawl spaces and wall voids.

Insecticides Classified by Chemistry

A great many chemicals have been used over the years for controlling insects. Some were found to be so potentially harmful to humans and other animal life, including beneficial insects, and to the environment that their use is no longer permitted. Use of all insecticides is tightly controlled in most countries. In the United States, insecticides must be approved and registered by the Environmental Protection Agency (EPA), and by state agencies also, before they can be used. Even then there are usually heavy restrictions that confine use to certain plants at certain times and at certain concentrations. Residues on food or feed crops exceeding fixed tolerances subject the product to seizure and destruction. Although no pesticide tolerances have been established for ornamental or forest crops, use of these chemicals on such commodities is nevertheless tightly regulated.

Insecticides can be classified as follows:

1. *Inorganic compounds.* These compounds include arsenic, fluorine, phosphorus, and sulfur compounds. These insecticides have been used little in recent times.
2. *Organic compounds.*
   a. *Plant derivatives.* These derivatives include materials such as pyrethrum (from the dried and powdered flowers of *Chrysanthemum cinerariaefolium,* which is a safe and effective insecticide; rotenone (from the roots of several plants in the pea family); nicotine from *Nicotiana* spp.; and from the neem tree. Many others could possibly be developed for commercial use if there were sufficient demand.
   b. *Synthetic organic chemicals.* After World War II, an entirely new concept in insect control emerged with the development of synthetic organic insecticides.
      i. *Chlorinated hydrocarbons.* DDT, first discovered in Germany in 1874, was found to have insecticidal properties. During and after World War II, it was used to control mosquitos, flies, fleas, and many agricultural insects. However, the buildup of DDT in the world's ecosystems to levels many scientists considered dangerous led to banning its use in several countries, including the United States. Other chlorinated hydrocarbons, closely related to DDT, have also been very effective in insect control. They are not easily biodegraded, however, tending to build up in plants and in the soil and to be transmitted into fish, fish-eating birds, meat, and milk products.
      ii. *Organic phosphates.* Compounds in this group were developed near the end of World War II and were found to have good insecticidal properties. They decompose more rapidly than the chlorinated hydrocarbons, but some of the materials can be very toxic to mammals and must be used with great care.
      iii. *Carbamates.* Carbaryl was the first chemical in this group to be widely used. It has low mammalian toxicity and its residual action is short-lived. It is effective against a wide range of both sucking and chewing insects, acting as a contact insecticide. Some carbonates are **systemic** and highly toxic.
      iv. *Pyrethroids.* This is a relatively new class of insecticides. They mimic properties of pyrethrins such as low mammalian toxicity. They kill a broad spectrum of insects and have a much longer residual life than pyrethrins.

v. *Neonicotinyls.* The neonicotinyls form a recent class of insecticides that block the synoptic transmission of insect nerves. Affected insects do not behave normally; stop feeding, thus causing the insects to be more vulnerable to predators, parasites, and diseases. These compounds have very low toxicity to vertebrates.

c. *Spray oils.* Long used to control scale insects and mites on fruit trees and ornamentals, spray oils are prepared by the distillation and chemical refining of mineral oils. A distillation range is chosen to give an oil fraction that is relatively nontoxic to plant tissue, yet lethal to insects. Spray oils are treated with hot sulfuric acid to remove many of the unsaturated molecules in the oil that cause plant injury. Botanical oils have similar properties.

d. *Microbial insecticides.* Some kinds of insects are susceptible to certain toxins produced by bacteria, fungi, and viruses. Often these pathogens cause the spectacular disappearance of insects. Such natural biological control of insects has been encouraged and used commercially. For example, *Heliothis* virus is used commercially to kill the tobacco budworm and the cotton bollworm. It infects only these two insect species. Several species of caterpillars *(Lepidoptera)* can be controlled by the toxins produced by the bacterium *Bacillus thuringiensis.* The toxins are applied in dust suspensions or sprays to foliage, which is then ingested by the caterpillars. These bacteria do not sustain themselves naturally and must be reapplied each time caterpillar control is needed. BT corn and other crops have been genetically engineered to produce the microbial toxin that affects pests.

Important Insect Pests of Agricultural Crops and Plant Products

There are innumerable insect pests of food crops, but certain pests are outstanding for the havoc they have wreaked over the years. Several species are discussed in the following subsections. When infestation of these or other pests occur, it is best to consult a local agricultural extension agent, pest control consultant, or garden supply center. Insecticide recommendations are continually changing.

Corn Ear Worm (*Heliothis zea*)

This insect, with three to five generations per year, occurs all over the world wherever corn is grown. Caterpillars feed on the corn silks and kernels, making the ears wormy. The worm also feeds on the other crops—beans, cotton, lettuce, tomato, alfalfa, clover, peanuts, and tobacco. It is best controlled by insecticides.

Codling Moth (*Laspeyresia pomonella*)

This insect, with two to three generations per year, attacks apples and pears wherever they are grown throughout the world. It is also a pest for walnuts. The larvae tunnel into the fruits, making them wormy. Unless insecticides are used, up to 90 percent of the fruits can become affected.

Peach Twig Borer (*Anarsia lineatella*)

This species occurs all through the peach-producing areas of the United States. It also attacks most other stone fruits. Overwintering larvae bore into buds and shoots as they start to grow in the spring. There are two to three broods a year, the later ones feeding directly on the fruit.

Lygus Bugs (*Lygus Hesperus* and *L. elisus*)

These insects principally attack alfalfa, Ladino clover, sugar beets, safflower, beans, cotton, and carrots. The sucking mouthparts are inserted into buds, flowers, and young fruits, so damaging the crops that they may be unmarketable. In some seed crops, such as alfalfa, no seeds may develop because of blasting of the flower buds by the lygus bugs.

San Jose Scale (*Quadraspidiotus perniciosus*)

This scale insect is well established throughout North America, Europe, and Asia, attacking most fruit crops and many ornamental trees and shrubs. Heavy infestations of scale can reduce tree vigor and cause death unless they are controlled. Scale spots on fruits reduce their market quality.

Green Peach Aphid (*Myzus persicae*)

This insect is found throughout the world. It is a sucking pest on many vegetable crops, all stone fruits, and many ornamentals. While its feeding reduces plant vigor, the chief source of damage is its transmission of viral diseases. Viral particles in its salivary fluid are injected into the host plant. Viruses known to be transmitted by the green peach aphid are sugar beet yellows, potato leaf roll, bean mosaic, lettuce mosaic, and cucumber mosaic.

Egyptian Alfalfa Weevil (*Hypera brunneipennis*)

This is a very destructive pest for alfalfa grown as a hay crop. Principal damage is done by the larvae, which feed on shoot tips, buds, and leaves.

Spider Mites (*Bryobia praetiosa, Panonychus ulmi,* and *Tetranychus urticae*)

These pests (which are not true insects) are widely distributed and feed on many species of host plants, including a wide array of vegetable and field crops, greenhouse and nursery plants, fruits, nuts, and ornamentals (see Fig. 7–21). A typical webbing sometimes appears as the mite population

Figure 7–21   Spider mites can be very damaging to many plants. The lily plant on the left has a heavy infestation of mites. The plant on the right is free of mites. *Source:* University of California Cooperative Extension.

increases, with a yellow stippling on leaf surfaces as defoliation begins, resulting from removal of plant fluids by the mites' piercing and sucking mouthparts. Six to ten generations per year can occur. Spider mites have developed resistance to many miticides.

Cereal Wireworms (*Agriotes* spp.)

These are the main insect pests of cereals such as wheat, barley, oats, and rye, especially in the northern growing regions. They are the larval stage of so-called click beetles, which themselves do no harm. Their life cycle spans five years, most of which is spent in the larval wireworm stage in the soil.

Colorado Potato Beetle (*Leptinotarsa decemlineata*)

This well-known insect, with chewing mouthparts, occurs from Colorado to the eastern United States in all potato-growing areas. It is established on the European continent but has been eradicated from the British Isles. Both adult and larval stages feed on the foliage of the potato plant, completely denuding it. It seldom feeds on other plants. In both stages the insect is large and easily seen.

Cotton Boll Weevil (*Anthonomus grandis*)

This insect attacks cotton plants in the United States, Mexico (where it originated), and Central America. Probably no insect, other than perhaps the codling moth, has had such an impact on agriculture and probably no other insect has received more study. The larvae develop inside the flower or boll, arising from eggs deposited by the female. They feed on the developing floral parts and fibers. Many generations occur in a single season. Losses are heavy unless insecticides are used.

Sugarcane Shoot Borer (*Diatraea saccharalis*)

This is the most important insect pest of sugarcane in the Caribbean. Caterpillars feed on the leaves, then enter the stalk and bore into the center, killing the central growing shoot. Later generations bore into side shoots. Such mechanical injury causes the stalks to break and fall over. Yield and quality of extracted juice drops. Control attempts include resistant cultivars, insecticides, and the release of parasites of the borer; none is very effective.

Mediterranean Fruit Fly (*Ceratitis capitata*)

This notorious insect is of the greatest importance on all fruit species in the Mediterranean countries, south and central

Africa (where it originated), western Australia, the west coast of South America, and throughout Central America. It has been kept out of the United States (except for Hawaii) by strict government inspection, quarantine, and eradication measures. The fly attacks peaches, apricots, apples, citrus, bananas, and many other fruits and vegetables. The fly is slightly smaller than the common housefly and is mostly yellow and brown in color, with black markings. It is controlled mainly by poisonous sprays containing attractants and by the release of sterile male flies. Both the Oriental (Dacus dorsalis) and Mexican (Anastrepha ludens) fruit flies (see Fig. 7–22) can also be devastating to fruit crops.

Grape Berry Moth (Endopiza viteana)

This is the principal insect pest on grapes in Europe, eastern North America, North Africa, and Japan. The larvae feed on developing fruit. Two or three generations can occur during the season. Insecticide spray is the only control measure.

Insects Attacking Dried Fruits

Several fruits are preserved by drying, either outdoors on trays in the sun or in forced hot-air dehydrators. Important dried fruits are raisins, prunes, dates, figs, apples, peaches, apricots, and pears. All these dried fruits are food for various insects, which are best controlled by fumigation treatments. Several types of insects feed on dried fruits:

Figure 7–22 Fruit flies, such as the Oriental, Mediterranean, and Mexican, attack many fruit crops and must be kept out of major fruit growing regions if at all possible. The insects in this photograph are adult Mexican fruit flies on an orange fruit. *Source:* USDA.

Beetles: dried fruit beetle, saw-toothed grain beetle, small darkling beetle, hairy fungus beetle, corn sap beetle, pineapple beetle, and date stone beetle.

Moths: raisin moth, Indian meal moth, almond moth, dried fruit moth, navel orange worm, dried prune moth, and dusky raisin moth.

Flies: vinegar fly *(Drosophila),* soldier fly, blowfly, housefly.

Insects Attacking Stored Grains

Conservative estimates maintain that insects destroy at least 5 percent of the world's production of cereal grains, amounting to about 15 million tons annually. Important insects that feed on stored grains are the sawtoothed grain beetle, lesser grain beetle, flat grain beetle, red flour beetle, foreign grain beetle, larger black flour beetle, Angoumois grain moth, hairy fungus beetle, granary weevil, and the rice weevil.

Control measures include prompt harvesting, drying the grain with heated air to a low moisture content (11 to 13 percent), and storage in tight, insect-free bins raised above ground. After two to six weeks, the grain is fumigated.

RODENTS AND VERTEBRATE WILDLIFE

Rodents, particularly Norway rats, roof rats, and house mice, cause great losses to food crops. Such losses occur mainly in stored grains and other food products in open storage, although rats also feed on unharvested fruits and vegetables. Sugarcane fields in Hawaii, for example, are often invaded by rats. Several million metric tons (MT) of grain are lost annually to rats.

The strategy in rodent control is, first, to remove all food and water available to them from the areas they inhabit, and second, to place bait traps containing an anticoagulant rodenticide such as Warfarin in their runways. Control should be done around granaries just before harvest begins. Some rat species, however, show increasing resistance to rodenticides, and in the future the chemicals may not be effective. Chemical sterilants, acting as oral contraceptives, reduce rodent populations.

Young fruit trees and fall-planted seeds in nurseries are often damaged by deer, mice, gophers, squirrels, and rabbits. Effective repellents of rodents and birds have been developed as coatings of forest seeds sown in logged- and burned-over areas.

Certain birds—particularly crows, ducks, geese, starlings, blackbirds, ravens, magpies, and scrub jays—are a major menace to grain crops and many fruit and nut crops, such as cherries, grapes, prunes, plums, strawberries, almonds, pecans, walnuts, and pistachios. Blackbirds, ducks, and starlings, in particular, can decimate grain crops such as wheat and field and sweet corn just ready for harvest. They can also cause considerable losses in peanut crops. Canadian geese cause extensive damage to golf course greens and fairways. The use of nonlethal chemical repellents is one

Figure 7–23   A nondestructive method of keeping birds from eating the grapes in a vineyard. Amplified bird distress calls are played at intervals during the day at harvest time. *Source:* Blue Anchor, Inc.

type of bird control. Scare devices—carbide explosives, shell crackers, and amplified recordings of bird distress calls (Fig. 7–23)—are also used with varying degrees of success. Research on chemical reproduction inhibitors may eventually provide the best method of controlling depredating bird populations.

PLANT DISEASES

All species of native and cultivated plants are susceptible to disease and prone to injury. **Disease** is defined as suboptimal plant growth brought about by a continuous irritant such as a pathogen (an organism or entity capable of inciting disease) or via chronic exposure to less than ideal growing conditions. In contrast, injury is loss of plant vigor resulting from an instantaneous event such as a lightning strike, hail damage, chemical burn, or mechanical damage. Because of the instantaneous and cause-and-effect nature of injuries (i.e., the plants were pruned yesterday and they are dying today), they are often easy to diagnose. In the case of disease, the source of continual irritation may be abiotic (nonliving) or biotic (caused by a pathogen). Abiotic diseases are also referred to as noninfectious diseases because they do not spread from plant to plant. In lay terms, they are not contagious.

Abiotic diseases are very common and should be considered the likely suspect when attempting to diagnose the

cause of decreased plant vigor or death. This approach is very important when working with intensively managed cropping systems in which a high degree of manipulation or handling takes place. Examples of abiotic plant diseases include damage caused by chronic exposure to air pollutants such as nitrogen dioxide (NO_2—automobile exhaust), sulfur dioxide (SO_2—stacks of factories), and ground-level ozone (O_3—a by-product of photochemical reactions in the atmosphere); nutritional deficiencies and toxicities; and growth under less than ideal light, moisture, or temperature conditions. The latter emphasizes why it is so important for plant production specialists to know and understand the complexities of their unique cropping system and the environmental growth requirements of the plant species being cultivated.

Biotic diseases are caused by pathogens and are often referred to as infectious diseases because they can move within and spread between plants. Plant pathogens are very similar to those that cause disease in humans and animals and include viruses, bacteria, spiroplasmas, phytoplasmas (formally called mycoplasmalike organisms), fungal-like organisms, fungi, nematodes, and parasitic higher plants. Pathogens may infect all types of plant tissues including leaves, shoots, stems, crowns, roots, tubers, fruit, seeds, and vascular tissue and can cause a wide variety of disease, ranging from root rots and rusts to cankers, blights, and wilts. Most plants are immune to most pathogens; however, all are susceptible to attack by at least one pathogen in each of the groups listed above. Some plants are susceptible to many. For example, wheat, rice, corn, and soybean are each susceptible to more than thirty different fungal pathogens. Similar levels of susceptibility exist in most other crops such as citrus, turfgrass, trees, and the vast diversity of solanaceous, cucurbit, floricultural, and ornamental plant species. Some pathogens like *Rhizoctonia, Pythium, Fusarium,* and *Sclerotinia* infect a broad range of plant species and yet others only a given species. For an infectious disease to develop, the following three conditions must exist: a susceptible host, a pathogen capable of inciting disease, and a favorable environment for pathogen development (Fig. 7–3). If any one of these factors is not present, disease will fail to develop. In the case of infectious plant diseases, any practice that favors plant growth and reduces either the amount of pathogen present or its development or activity will result in significantly less disease.

Regardless of the type of pathogen, the development of visual disease symptoms on a plant requires that the infectious agent must (1) come into contact with a susceptible host (referred to as inoculation), (2) gain entrance or penetrate the host through either a wound or a natural opening (stomata, lenticel, hydathode, nectarthode) or via direct penetration of the host, (3) establish itself within the host, (4) grow and reproduce within or on the host; and ultimately (5) be able to spread to other susceptible plants (referred to as dissemination). Successful pathogens must also (6) be able to survive prolonged periods of unfavorable environmental conditions and the absence of a susceptible plant host. Collectively,

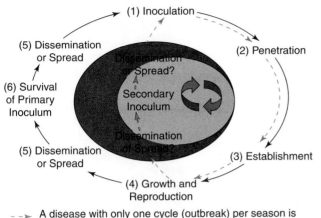

--→ A disease with only one cycle (outbreak) per season is called a MONOCYCLIC or "one-cycled" disease

——→ A disease with multiple cycles (outbreaks) per season is called a POLYCYCLIC or "many-cycled" disease

Figure 7–24 The six steps of a disease cycle: survival, inoculation, penetration, establishment, growth and reproduction, and dissemination or spread. A disease with only one cycle (outbreak) per season is called a monocyclic, or one-cycled, disease. A disease with multiple cycles (outbreaks) per season is called a polycyclic, or many-cycled, disease. Information about a pathogen's disease cycle can provide clues about how to intervene, disrupt, or break its cycle and thus reduce plant losses. *Source:* Michael Boehm, The Ohio State University.

Figure 7–25 Phosphorus deficiency in young corn growing in a drought-stressed field. *Source:* Michael Boehm, The Ohio State University.

these steps are referred to as the disease cycle (Fig. 7–24). If this cycle is disrupted, either naturally or via the concerted efforts of a grower, the disease will be less intense or fail to develop.

Disease Signs and Symptoms

Two terms used often when discussing plant disease and injury are *sign* and *symptom*. The term *sign* is used when the pathogen or part of the pathogen is observed in or on an infected plant. Examples include fungal hyphae or mycelium, spores, fruiting bodies, bacterial cells, virus particles, nematodes, and parasitic higher plants. Although many plant diseases can be diagnosed in the field based on the observation of key diagnostic signs, many require observation by trained specialists under laboratory conditions. *Symptoms* are visual or otherwise detectable reactions or alterations of a plant as the result of disease or injury. Symptoms of disease often change as the disease progresses. Initial symptoms are often invisible to the naked eye or very small and nondescript, and they may be quite different from those observed in the final stages of disease development. Symptoms can generally be placed in the following categories:

1. *Abnormal tissue coloration.* Leaf appearance commonly changes. Leaves may become chlorotic (yellowish) or necrotic (brown) or exhibit purpling, bronzing, and reddening. Mosaic or mottling patterns may ap-

pear, especially with virus diseases. Nitrogen-deficient plants often exhibit a generalized chlorosis or yellowing. Iron-deficient leaves often exhibit interveinal chlorosis. Phosphorus-deficient plants are often purple (Fig. 7–25).

2. *Wilting.* Drought stress causes wilt. If a pathogen interferes with the uptake of water by the host plant, part or all of the plant may die. Fungi belonging to the genera *Verticillium* and *Fusarium* and bacteria in the genus *Xanthomonas* are often associated with wilt diseases because they colonize the xylem of plants, leading to a lack of water transport.

3. *Tissue death.* Necrotic (dead) tissue can appear in leaves, stems, or root, either as spots or as entire organs. Decay of soft succulent tissue, as in damping-off in young seedlings, is common. Cankers caused by death of the underlying tissue sometimes appear as sunken, dead tissue on the trunks or limbs of woody plants.

4. *Defoliation.* As the infectious disease progresses, the plant may lose all its leaves and sometimes drop its fruit. Defoliation is a common symptom in sycamore anthracnose (caused by *Gnomonia veneta*) and apple scab (caused by *Venturia inaequalis*). Both are fungal diseases.

5. *Abnormal increase in tissue size.* Some diseases increase cell numbers or cell size in the plant tissues, twisting and curling the leaves or forming galls on stems or roots (Figs. 7–26 and 7–27).

6. *Dwarfing.* In some cases, the pathogenic organism reduces cell number or size, stunting parts or all of the host plant.

7. *Replacement of host plant tissue by tissue of the infectious organism.* This development occurs commonly when floral parts or fruits are involved. Examples include ergot on rye and other cereal crops caused by the

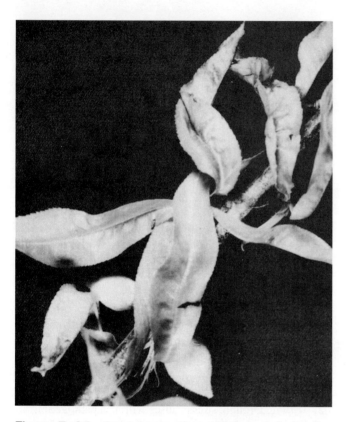

Figure 7–26   Peach leaves distorted by peach leaf curl, a disease caused by the fungus *Taphrina deformans*. The disease can be controlled by fungicidal sprays applied before the buds open in the spring.

Figure 7–27   Galls on a young fruit tree infected by crown gall bacteria (*Agrobacterium tumefaciens*).

fungus *Claviceps purpurea* and corn smut caused by the fungus *Ustilago maydis* (Fig. 7–28).

Classification of Infectious Plant Diseases

The pathogens responsible for causing most biotic plant diseases include viruses, bacteria, spiroplasmas, phytoplasmas (formally called mycoplasmalike organisms), fungi, fungallike organisms, nematodes, and parasitic higher plants. Each will be discussed in the following subsections.

Viruses

Viruses are intracellular (inside cells) pathogenic particles that infect other living organisms. Human diseases caused by viruses include chickenpox, herpes, influenza, rabies, smallpox, and acquired immunodeficiency syndrome (AIDS). Although most of us are familiar with these viruses, the first virus ever described and from which the term was eventually derived was tobacco mosaic virus (TMV). The term *virus* was derived from the original description of the causal agent of TMV—a "contagium vivum fluidum," or contagious living fluid. TMV was discovered by Martinus W. Beijerinck, a Dutch microbiologist, in 1898.

Virus particles are extremely small and can be seen only with an electron microscope. Most plant viruses are either rod-shaped or isometric (polyhedral). TMV, potato virus Y (PVY), and cucumber mosaic virus (CMV) are examples

---

KOCH'S POSTULATES: RULES FOR PROOF OF PATHOGENICITY

How do plant pathologists know that an organism isolated from a sick plant is actually the causal agent of a particular disease, especially the first time a new disease or set of symptoms is observed? Dr. Robert Koch, a German bacteriologist and a physician, asked this question in 1876 when studying anthrax in cattle. He developed a simple four-step process, commonly referred to today as Koch's postulates, to answer this question. Koch's postulates, or rules for proof of pathogenicity, have since been adopted as the principle means of confirming that an organism isolated from a sick or symptomatic plant, animal, or human is the pathogen responsible for a given disease. In simple terms, Koch's postulates state that (1) the microorganism in question must be present in every case of the disease, (2) the microorganism must be isolated from the diseased host and grown in pure culture, (3) the specific disease must be reproduced when a pure culture of the microorganism is inoculated into a healthy susceptible host, and (4) the introduced microorganism must be recovered from the experimentally infected host. If one can prove Koch's postulates, it is then possible to say conclusively that a particular organism is the causal agent of the disease in question.

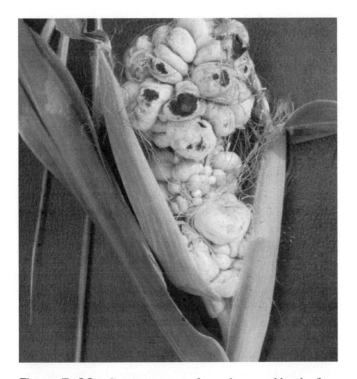

Figure 7–28 Common smut of corn is caused by the fungus *Ustilago maydis*. Sweet corn is considered more susceptible to smut than field corn. The disease occurs on leaves, stalks, tassels, and ears. The disease is easily recognized early in the season by the presence of silvery white tumorlike galls that, as the season progresses, mature, rupture, and release large masses of soot-colored smut spores (called teliospores) similar to those shown above. Young galls are considered a delicacy in certain parts of the world. *Source:* Michael Boehm, The Ohio State University.

of a short, rigid, rod-shaped; a long, flexuous, rod-shaped; and an isometric virus, respectively. Viruses consist of an inner core of nucleic acid [either ribonucleic acid (RNA) or deoxyribonucleic acid (DNA)] surrounded by an outer sheath or coat of protein referred to as the capsid. The capsid is also enclosed by a membrane in most human and animal viruses that helps the virus pass through the cell membrane in these types of cells. Because the cell membrane in plants is surrounded by a rigid cell wall, plant viruses require a wound for their initial entrance into a plant cell. Wounds in plants occur naturally, such as in the branching of lateral roots, or they may be the result of agronomic, horticultural practices or other mechanical means; fungal, nematode, or parasitic plant infections; or insects. In some cases, the organism creating the wound can also be carrying and can pass or transmit the virus. Organisms that transmit pathogens are called **vectors.** Mechanical and insect vector transmission are the two most important means by which plant viruses are spread. The activity of humans in propagating plants by budding and grafting or by cuttings is one of the chief ways viral diseases are spread. In fact, plant virologists use grafting-and-budding procedures to transmit and detect viruses in their studies. The seedling offspring of a virus-infected plant is usually, but not always, free of the virus, depending on the plant species and the kind of virus. Insect transmission is perhaps the most important means of virus transmission in the field. Insects in the Order Homoptera (such as aphids, planthoppers, leafhoppers, whiteflies, and mealy bugs) that have piercing, sucking mouthparts are the most common and economically important vectors of plant viruses (Fig. 7–19). Some plant viruses can also be transmitted in pollen grains or by seed.

Viruses are obligate parasites; that is, they require a living host to be able to grow and multiply. Once in a wounded cell, the virus particle sheds its protein coat, and the nucleic acid is then transcribed and/or translated, leading to the development of new virus particles. Cell-to-cell movement of plant viruses occurs through the cytoplasmic bridges between cells called **plasmodesmata,** and viruses move systemically throughout infected plants via the phloem. Although the details of plant virus replication are complex and beyond the scope of this chapter, the general idea is that plant viruses cause disease in part by causing a reallocation of photosynthates and a disruption of normal cellular processes as they replicate. Many kinds of plants are infected with viruses and show no symptoms. Such infections are referred to as being latent. Some viruses, such as cucumber mosaic virus (CMV) and cowpea mosaic virus (CPMV), occur as a complex of multiple component particles, each containing different nucleic acid cores. In multicomponent viruses, all components have to be present in a plant for infection and replication to take place.

Viruses are difficult to classify and, for want of anything better, they are given descriptive (and sometimes colorful) names based on the disease they cause—for example, tobacco ring spot, watermelon mosaic, barley yellow dwarf, potato mop top, citrus tristeza, sugar beet curly top, lettuce mosaic, maize dwarf mosaic, potato leaf roll, peach yellow bud mosaic, African cassava mosaic, carnation streak, and tomato spotted wilt. Many of these viruses also infect plants of other species. For example, tobacco ring spot virus causes a bud blight in soybeans; maize dwarf mosaic infects sorghum, Sudan grass, sugarcane, and johnsongrass in addition to corn, but it still retains its original name.

Once plants are infected, little can be done to free them from the virus. Because different cultivars and species show different degrees of resistance to some viruses, resistant types should be planted whenever they are available. Recent advances in plant cell molecular biology and virology have lead to the development of genetically modified plants with superior resistance to some viruses. For orchard, ornamental nursery, and floricultural crops, the best management approach is the planting of stock that has been propagated from known virus-clean or certified sources. The citrus industries in both Florida and California, for example, have set up certification and registration programs to ensure that citrus nursery stock is propagated from the most pathogen-free propagative materials available. Similar certification programs exist for seed potatoes. Another successful way to eliminate viruses, particularly from herbaceous plants, is to use meristematic tip-culturing techniques and tissue culturing to develop virus-free callus tissue that can then be used to generate new virus-free clones of the original plant. This procedure is based on the fact that virus is usually not present in the actively growing shoot tip of an infected plant. This procedure has been used to clear many herbaceous cultivars of viruses.

Numerous cultural practices can be used to reduce plant losses due to virus infection, including (1) scouting and the removal of symptomatic plants or known alternative weed or volunteer plants that may serve as a reservoir for a given virus, (2) the use of clean or sanitized tools and equipment, (3) hand washing, and (4) the use of disposable overgarments. Rotations to nonhost crops and geographic isolation of production facilities may also help avoid losses caused by plant viruses. The isolation of newly received plant material prior to its introduction into the rest of a production system can also minimize the unintentional introduction of pathogens. Some viruses are permanently inactivated by prolonged exposure of infected tissue to relatively high temperatures—for example, twenty to thirty days at 38°C (100°F). This procedure, called heat therapy, frees individual plants or cuttings of the virus. The clean tissue is then used as a propagative source, allowing large-scale production of virus-free plants. This approach has been taken with many cultivars of fruit and ornamental species. If insect vectors and infected plant material are kept out of the new virus-clean plantings, subsequent reinfection is unlikely, particularly if the planting is at a distance from virus-infected plantings. There are no chemical sprays or biological control approaches to eradicate viruses, although insecticides and biocontrol products can be used to control insect vectors. Management of insect vector populations in the field can be difficult to impossible unless it is coordinated on a regional basis, but it may be highly effective in closed production systems such as greenhouses or interiorscapes.

Bacteria

Bacteria are microscopic, single-celled prokaryotic organisms that reproduce asexually by binary fission (one cell splitting into two). They occur singly or in colonies of cells—in pairs, chains, or clusters. Bacteria are classified into four groups based on cell shape: (1) the spherical cocci, (2) the rod-shaped bacilli, (3) the spiral-shaped spirilli, and (4) a small group of filamentous forms—the actinomycetes. Only bacilli and actinomycetes are known to cause diseases in plants. Some types of bacilli and spirilli are motile—they have whiplike flagella that propel them through films of water. Bacteria multiply at alarming rates under suitable environmental conditions. Bacteria are divided into two groups based on their reaction when subjected to a relatively simple staining procedure called the Gram stain: Gram-negative bacteria stain red or pink, and Gram-positive

bacteria stain purple. The difference in color is directly related to the chemical composition and structure of the cell wall.

Although considered structurally simple, bacteria are extremely diverse from a metabolic standpoint and are found almost everywhere on earth in vast numbers—from living in jet fuel and on the rims of volcanoes to thriving in hydrothermal vents deep on the ocean floor. Beneficial bacteria are involved in diverse processes such as digestion in animals, nitrogen fixation in the roots of certain legumes, the decomposition of animal and plant remains, and sewage disposal systems. Pathogenic bacteria, on the other hand, cause severe and often fatal diseases in humans, animals, and plants. The first bacterial disease ever discovered was anthrax (caused by *Bacillus anthracis*) of cattle and sheep in 1876. The discovery of anthrax in cattle was immediately followed by the discovery of fireblight in pear and apple (caused by *Erwinia amylovora*) by T. J. Burrill from the University of Illinois (1877–1885). Strep throat, bacterial meningitis, infections of most skin wounds, botulism, and plague (both bubonic and pneumonic) are other examples of bacterial diseases that affect people. Table 7–4 gives some examples of important bacterial diseases of plants.

The taxonomy of plant-pathogenic bacteria is currently in flux based on recent advances in the classification of bacteria. Most plant-pathogenic bacteria belong to the following genera: *Erwinia, Pectobacterium, Pantoea, Agrobacterium, Rhizobium, Pseudomonas, Ralstonia, Burkholderia, Acidovorax, Xanthomonas, Clavibacter, Streptomyces,* and *Xylella.* Plant-pathogenic bacteria cause many different kinds of symptoms, including galls and overgrowths, wilts, leaf spots, specks and blights, soft rots, and scabs and cankers (Table 7–4). In contrast to viruses, which are intracellular pathogens, bacteria are considered intercellular (between cells) pathogens. The means by which plant-pathogenic bacteria cause disease are as varied as the types of symptoms they cause. Some plant-pathogenic bacteria produce toxins that lead to cell death or enzymes that break down key structural components of plant cells, such as the case of the soft-rotters that produce pectinase that degrades the middle lamella. Still others colonize the xylem and restrict water movement, resulting in wilt. Some even have the ability to modify or transform their host genetically and bring about the formation of galls and overgrowths such as the case of *Agrobacterium tumefaciens* (causal agent of crown gall).

Bacteria that cause plant diseases are spread in many ways—they can be splashed about by rains or carried by the wind, birds, or insects. People can unwittingly spread bacterial diseases, for instance, by pruning infected orchard trees during the rainy season. Water facilitates the entrance of bacteria carried on pruning tools into the pruning cuts. Propagation with bacteria-infected plant material is a major way pathogenic bacteria are moved over great distances. However bacterial pathogens are disseminated, they require a wound or natural opening to penetrate a plant host.

TABLE 7–4	Some Important Plant Diseases Caused by Bacteria		
Common Name of Disease	Pathogen	Hosts Attacked	Integrated Disease Management Measures
Crown gall	Agrobacterium tumefaciens	Woody ornamentals and tree fruits	Use of pathogen-free nursery stock; avoid planting in infested soils; avoid wounding; treat graft junctions and rootstock with biocontrol bacteria
Bacterial wilt of cucurbits	Erwinia tracheiphila	All cucurbits (cucumber, pumpkin, squash, and muskmelon)	Resistant varieties; control of insect vectors (spotted and striped cucumber beetles)
Stewart's wilt of corn	Pantoea stewartii (formerly Erwinia stewartii)	Corn (sweet and field hybrids)	Resistant varieties
Fireblight	Erwinia amylovora	Pome fruits (apple, pear, quince, and some ornamentals such as pyrcantha)	Moderately resistant varieties; prune out diseased tissues; maintain a balanced fertility program; chemical applications—copper sulfate, Bordeaux mixture, streptomycin or oxytetracycline; biological control—blight ban
Common blight of beans	Xanthomonas phaseoli	Beans (field or dry, garden or snap, lima, etc.)	Pathogen-free seed; sanitation; disposal of crop residue; three-year crop rotations
Bacterial canker	Pseudomonas syringe	Almonds, apricots, avocado, cherry, peach, plum	Resistant cultivars; use only pathogen-free budwood; prune out diseased tissues; chemical applications—copper, Bordeaux mixture
Citrus canker	Xanthomonas axonopodis	Citrus crops (oranges, grapefruits, lemons, limes, etc.)	Quarantines to exclude and restrict movement of pathogen and/or infected tissue; destruction of all trees in a location when disease is detected; chemical applications with copper
Soft rot of vegetables	Pectobacterium carotovora (formerly Erwinia carotovora)	Fleshy storage tissues of vegetables and ornamentals (potatoes, carrots, cucumber, lettuce, cabbage, onions, iris, etc.)—in the field, transit, and storage	Sanitary and cultural practices; avoid mechanical damage during harvest and transit; good ventilation during storage; storage under cool, dry conditions

As previously described, the means by which bacterial pathogens cause disease are varied. All bacterial plant pathogens replicate by means of binary fission. They overwinter or survive unfavorable environmental periods or the absence of a susceptible host by going dormant in infected tissue, infested soil or water, or an insect vector.

Bacterial diseases in plants are difficult to control. Integrated management measures for bacterial plant pathogens include using resistant varieties, cultivars or hybrids, bacteria-free seed or propagation materials, and sanitation and cultural practices that either eliminate or reduce sources of bacterial contamination; preventing surface wounds that permit the entrance of bacteria into the inner tissues; and propagating only bacteria-free nursery stock. Prolonged exposure to dry air, heat, and sunlight sometimes kills bacteria in plant material. Applications of copper-containing compounds or Bordeaux mixture (copper sulfate and lime) as well as the antibiotics streptomycin and/or oxytetracycline may also help kill or suppress plant-pathogenic bacteria. Similarly, the use of antagonistic or biological control products such as Blight Ban and Agrosin K84 may also be effective for managing bacterial diseases of plants.

Spiroplasmas and Phytoplasmas

Spiroplasmas and **phytoplasmas** are prokaryotic organisms that lack rigid cell walls and infect plants. Spiroplasmas are helical in shape (resembling a corkscrew), whereas phytoplasmas are round or ovoid. These organisms resemble bacteria in some ways but are very small (perhaps even submicroscopic) and are wall-less. Phytoplasmas were once called mycoplasmalike organisms because they resemble mycoplasmas—organisms that cause disease in humans and animals. About 200 different plant diseases have been shown to be caused by spiroplasmas and phytoplasmas. Some of the diseases they cause are aster yellows, western-X of peaches, cherry buckskin, pear decline, mulberry dwarf disease, corn stunt, and stubborn disease of citrus. As with viruses, a disease caused by phytoplasmas is named after the plant on which is was first studied, but it can also occur on many other plants. For example, aster yellows also affects other ornamentals—gladiolus and phlox, for example—and tomato, spinach, onion, lettuce, celery, carrots, strawberry, and many weeds. Spiroplasma- or phytoplasma-infected plants exhibit a wide variety of symptoms, similar to those typically associated with viral infections, including generalized stunting and decline, yellowing, excessive proliferation of shoots, sterility of flowers, the development of green flowers, dieback, and plant death. In fact, until the discovery that these organisms caused disease in plants in 1967, most diseases now known to be caused by spiroplasmas or phytoplasmas were believed to be caused by viruses. Unlike most bacterial pathogens, spiroplasmas and phytoplasmas live in the phloem and are vectored by sucking insects such as leafhoppers, planthoppers, and psyllids. Studies of corn stunt provide evidence that once the insect vectors establish the infective particles in their bodies, the insects retain the ability to transmit them the rest of their lives. One obvious method of controlling the spread of these diseases is an effective spray program that eliminates the insect vectors. As discussed earlier in this chapter, this type of program is often difficult or impossible in the field but may work well in enclosed production systems like greenhouses. Phytoplasmas and spiroplasmas are susceptible to certain antibiotics, particularly tetracycline, which has been used to treat pear trees with pear decline disease. Tetracycline must be injected into mature trees on a routine or therapeutic schedule to be effective, and even then it appears only to suppress the development of symptoms rather than curing the infected plant. Applications made during the early stages of infection tend to be more effective than those made in the later stages of disease development.

Fungi and Fungal-Like Organisms

Collectively, fungi and fungal-like organisms (FLOs) cause more plant diseases than any other group of plant-pathogenic organisms. Over 8,000 species of fungi and FLOs have been shown to cause disease in plants. Fungal-like organisms are organisms like *Pythium* and *Phytophthora* and those that cause downy mildew. Until recently, they were considered

fungi but, because of changes in fungal taxonomy, are now in the Kingdom Chromista (also called Stramenopila). Fungi and FLOs are heterotrophic eukaryotic organisms that lack chlorophyll and thus the ability to photosynthesize. Fungi and FLOs obtain nutrients by absorbing them through tiny threadlike filaments called **hyphae** that branch in all directions throughout a substrate. A collection of hyphae is referred to as mycelium (pl., mycelia). The hyphae are filled with protoplasm containing nuclei. Mycelia are the key diagnostic sign associated with diseases caused by fungi and FLOs. Most of us have seen mycelium growing on old bread or rotten fruit or vegetables and may have referred to these organisms collectively as molds or mildew.

Fungi and FLOs (indeed all pathogens) can be grouped into the following four categories based on their preference for surviving on dead or decaying organic matter versus living tissue:

1. Obligate saprophytes—always saprophytes. These organisms can only survive or are obliged to gain nourishment by colonizing dead or decaying organic matter. They are *not* parasites.
2. Obligate parasites—always parasites. They can grow only as parasites on or in a living host. They cannot survive as saprophytes or be cultured in the laboratory. This group of pathogens have a vested interest in prolonging the life of their host to increase their own viability. All viruses, downy mildews, powdery mildews, rusts, and smuts are obligate parasites.
3. Facultative parasites—usually survive as saprophytes but have the ability to parasitize and cause disease under certain conditions. Examples include *Pythium* species and many bacterial pathogens.
4. Facultative saprophytes—usually survive as parasites but have the ability to live on dead and decaying organic matter under the right conditions. Examples include *Phytophthora* and *Botrytis* species.

Some fungi and FLOs can live on only one host species, while others develop on many different kinds.

Fungi and FLOs, like bacteria, can be beneficial as well as pathogenic. Beneficial fungi participate in biological cycles, decaying dead animal and plant materials and thus converting them into plant nutrients that are absorbed by living plants. Some beneficial fungi grow in a symbiotic relationship with the root cells of higher green plants; this combination is termed a **mycorrhiza.** The roots of most cultivated plants—corn, soybeans, cotton, tobacco, peas, red clover, apples, citrus, pines, aspens, birches, turfgrass species, and others—have mycorrhizal relationships with soil fungi. The mycorrhizae appear to be highly beneficial, often necessary, for optimum growth of many plants. Interest and research in such fungi–root symbiotic relations is considerable because establishing proper mycorrhizal fungi with cultivated plants offers great potential for improved plant growth. Some beneficial fungi, such as those belonging to the genus *Trichoderma,* are effective biocontrol agents of plant pathogenic fungi, while others, like *Arthrobotrys dactyloides,* have been shown to trap and parasitize plant pathogenic nematodes.

Certain fungi produce useful antibiotics and enzymes. *Pencillium* fungi produce the famous penicillin G, which has prevented countless deaths from bacterial infection. It inhibits formation of the bacteria's cell wall. Many food-producing processes, such as the making of bread, wine, beer, and cheese, are based on the activities of fungi. Mushrooms, which are fungi, are important as food.

Most plant diseases are caused by fungi, and the food loss to fungal diseases is staggering (Fig. 7–29). Some of the world's great famines can be blamed on pathogenic fungi. Wheat crops of the Middle Ages were ruined when the grains became infected with a dark, dusty powder now known to be the spores of the fungus called bunt or stinking smut (*Tilletia* spp.) The potato blight in Ireland and northern Europe, rampant during two successive seasons (1845–1846 and 1846–1847), was caused by the fungal-like organism *Phytophthora infestans* (the genus *Phytophthora* was named by Anton de Bary in 1876 as "plant destroyer"). It resulted in the death of more than 1 million people by starvation and caused mass migration from Ireland. In the 1870s, an epidemic of downy mildew, caused by the fungus *Plasmopara viticola,* struck the grape vineyards of central Europe, causing great losses to grape growers and wine makers. In the United States alone, hundreds of millions of bushels of wheat have been lost in epidemic years to stem rust (*Puccinia graminis tritici*).

Because of the shear number of plant diseases caused by fungi and the huge diversity in how plant-pathogenic fungi cause disease, it is impossible and beyond the scope of this text to provide details about specific disease cycles and integrated fungal disease management strategies. For those interested in such information, an introductory level course or workshop in plant pathology is recommended. But like other groups of plant pathogens, fungal pathogens have developed ways to survive periods of unfavorable environmental conditions or in the absence of a susceptible host, spread, infect, and grow and reproduce on and within plants. The steps involved in a fungal or FLOs disease cycle are identical to those described previously for bacterial pathogens (Fig. 7–24). One important difference between fungi and FLOs, and bacteria and viruses is that fungi and FLOs not only penetrate a host via a wound or natural opening, but they can also penetrate actively via the production of specialized hyphal structures called appresoria (sing. appresorium). Appresoria are swollen tips of hyphae that allow the fungus to penetrate plant tissues directly through mechanical and enzymatic activity. The ability of fungi and FLOs to penetrate healthy plants is undoubtedly responsible for their place collectively as the most important group of plant pathogens.

More options are generally available to professional plant production specialists and growers to manage fungal and FLO diseases compared to viral and bacterial diseases. One of the most satisfactory methods of dealing with fungus diseases is strict sanitation to eliminate the pathogenic organism, starting with the initial stages of propagation and growth of the potential host plants.

Integrated management strategies for fungal and FLO diseases include the following.

1. Genetic host resistance:
   ♦ Using genetically resistant species, cultivars, varieties, and hybrids.
2. Cultural practices:
   ♦ Planting only disease-free certified seed.
   ♦ Maintaining a balanced fertility program that avoids excessive or inadequate levels of key plant-required nutrients.
   ♦ Maintaining an effective water management program—maintaining adequate soil drainage, monitoring irrigation practices, and adjusting accordingly, and so on.
   ♦ Ensuring proper lighting—both quality and quantity to optimize plant health. Especially important in turfgrass and floricultural and ornamental nursery production systems.
   ♦ Removing crop residues by burning or burying (plowing).

Figure 7–29 Fungal diseases can cause considerable damage to grain crops. Shown here is barley scald caused by the fungus *Rhynochosporium secalis.* The head on the left is only slightly affected, while the one on the right is so severely damaged that the lack of photosynthesis has prevented the grains from filling. *Source:* University of California Cooperative Extension.

◆ Implementing crop rotation strategies to reduce or eliminate the interaction of susceptible plants with pathogens.

◆ Growing crops in climates unsuitable for pathogenic fungi and FLOs.

◆ Careful handling of the crop (vegetables and fruits) to prevent cuts, bruises, and wounding during harvest, transit, and storage.

◆ Storing crop products at the proper temperatures.

◆ Soil pasteurization: moist heat at 82°C (180°F) for thirty minutes.

3. Chemical applications:

◆ Using pre-plant soil fumigants.

◆ Using fungicide drenches or seed treatments with fungicides.

◆ Using fungicides during the growth of the crop.

◆ Using post-harvest fungicides on fruits and vegetables.

4. Biological control:

◆ Using biological control organisms to suppress the activity of deleterious fungi and FLOs.

5. Government regulatory measures:

◆ Implementing strict quarantines that exclude or restrict the introduction or movement of fungal and FLO pathogens or infected plant material.

In many of the major crops, cultivars resistant to prevailing diseases are available, and more are continually being developed by plant breeders. As has been discussed with other types of diseases, the use of genetically resistant plants should be the first line of defense for diseases caused by fungi and FLOs, if they are available. Several examples of cultivars genetically resistant to fungal and FLO disease are notable. Certain hybrid potato cultivars are resistant to late blight (*Phytophthora infestans*). Soybean cultivars resistant to downy mildew (*Peronospora manshurica*) have been developed. In the United States, apple cultivars have been developed at the Indiana and the New York agricultural experiment stations that show high resistance or immunity to apple scab (*Venturia inaequalis*), a devastating disease of apples grown in cool humid climates with summer rainfall. In the cereal crops (oats, wheat, rye, barley), powdery mildew (*Erysiphe graminis*) can be controlled only by the use of resistant cultivars developed by plant breeders. Tomatoes can be grown in *Fusarium*-infested soils only if *Fusarium*-resistant cultivars are planted. Plant breeders are continuously breeding wheat cultivars resistant to stem rust (*Puccinia graminis tritici*), but the fungus continuously mutates, attacking the formerly resistant cultivars. Still newer types then have to be developed. The importance of resistant cultivars in controlling fungus diseases is shown in Table 7–5.

Although the use of resistant cultivars and eradication of the pathogen through the use of cultural practices are the most satisfactory ways of dealing with diseases caused by fungi and FLOs, these measures are not possible in many instances. Often the disease appears and its development must be slowed or stopped by whatever means are available. The use of fungicides provides an effective means for managing many plant diseases, and they are relied on heavily for certain fungal and FLO problems. Fungicide applications are often essential where there is a demand for plant health during environmental periods that favor pathogen growth. Fungicides are typically more effective when applied prior to the onset of disease symptoms (referred to as preventive applications). Some fungicides are effective when applied after the onset of symptoms and are said to have curative activity. In either case, fungicides must be delivered to the area of the plant where the pathogen is active to be effective. Many different types and chemical classes of fungicides are currently available. Numerous online extension-outreach and agrichemical company resources provide specific fungicide recommendations for nearly every major cropping system and pathogen. Always read and follow label recommendations when applying fungicides and pesticides.

Nematodes

Nematodes are simple, microscopic, multi-cellular animals—typically containing 1,000 cells or less. They are wormlike in appearance but are taxonomically distinct from earthworms, wireworms, or flatworms. They are microscopic soft-bodied (no skeleton), nonsegmented roundworms. The basic body plan of a nematode is a tube within a tube. Most nematodes are not pathogens but rather saprophytes. Some are serious human, animal, and plant pathogens. Those that attack animals or humans do not attack plants, and vice versa. Heartworm in dogs and cats and elephantiasis are examples of nematode diseases in animals and people, respectively. Plant parasitic nematodes may attack the roots, stem, foliage, and flowers of plants. All plant parasitic nematodes have piercing mouthparts called **stylets.** The presence of a stylet is the key diagnostic sign differentiating plant parasitic nematodes from all other types of nematodes. In other words, if a nematode has a stylet, it is a plant pathogen. The bacterial-feeding nematode, *Caenorhabditis elegans,* is one of the best-understood animals on earth. It was the first animal to be completely sequenced. The study of *C. elegans* has led to many new insights into animal development, neurobiology, and behavior.

Several genera and species of nematodes are highly damaging to a great range of hosts, including foliage plants, vegetable crops, fruit and nut trees, turfgrass, and forest trees. Some of the most damaging nematode species are root knot (*Meloidogyne* spp.), cyst (*Heterodera* spp.), root lesion (*Pratylenchus* spp.), spiral (*Helicotylenchus* spp.), burrowing (*Radopholus similis*), bulb and stem (*Ditylenchus dipsaci*), reniform (*Rotylenchulus reniformis*), dagger (*Xiphinema* spp.), and bud and leaf (*Aphelenchoides* spp.).

TABLE 7–5 | Some Important Plant Diseases Caused by Fungi and Fungal-Like Organisms

Common Name of Disease	Pathogen	Hosts Attacked	Integrated Disease Management Measures
Stem rust of wheat	*Puccinia graminis tritici*	Wheat	Resistant varieties
Corn smut	*Ustilago maydis*	Corn	Resistant varieties
Fusarium wilt	*Fusarium oxysporum*	Tomato, pea, celery, banana, cotton, watermelon, ornamentals	Resistant varieties
Powdery mildew	Many different species	All hosts	Resistant varieties; fungicide applications
Dollar spot	*Sclerotinia homoeocarpa*	Many turfgrass species	Moderately resistant species or cultivars; avoid excessive nitrogen fertility and prolonged leaf wetness; adequate soil moisture; fungicide applications
Rice blast	*Magnaporthe grisea*	Rice	Resistant cultivars; avoid excessive nitrogen; fungicide applications
Apple scab	*Venturia inaequalis*	Apples and crab apples	Resistant varieties; fungicide applications
Peach leaf curl	*Taphina deformans*	Peaches and nectarines	Fungicide applications
Verticillium wilt	*Verticillium albo-atrum* and *V. dahliae*	Potatoes and a wide range of woody fruit and ornamental species; many herbaceous plants	Resistant varieties; remove infected plant material; avoid infested soils; soil fumigation; fungicide applications
Fusarium head blight	*Fusarium graminearum*	Wheat and barley	Moderately resistant varieties; crop rotation; fungicide applications have limited effectiveness; biological control
Downy mildew of grape	*Plasmopara viticola*	Grape	Resistant varieties; fungicide applications
Late blight of potato	*Phytophthora infestans*	Potato and tomato	Resistant cultivars; destroy all culled potatoes; crop rotation; fungicide applications
Phytophthora root rot of soybean	*Phytophthora sojae*	Soybean	Resistant varieties; fungicide applications
Damping off	*Pythium* and *Rhizoctonia* species	Seedlings of most plant species	Fungicide seed treatments; biological control—use compost-amended mixes and antagonistic microorganisms
Soybean rust	*Phakopsora pachyrhizi*	Soybean and many alternative hosts	Quarantines to exclude/restrict movement of pathogen; destruction of infected plant material; fungicide applications

Typical root symptoms indicating nematode attack are root knots or galls (Fig. 7–30), root and tuber lesions (Fig. 7–31), excessive root branching, injured root tips, and stunted root systems. Symptoms on the above-ground plant parts indicating, root infection are a slow decline of the entire plant; wilting, even with ample soil moisture; foliage yellowing; and fewer and smaller leaves. In fact, these same symptoms appear in plants deprived of a properly functioning root system. Bulb and stem nematodes produce stem swellings and shortened internodes. Bud and leaf nematodes distort and kill bud and leaf tissue.

Parasitic nematodes are readily spread by any physical means that can move soil particles about—equipment, tools, shoes, birds, insects, dust, wind, and water. In addition, the movement of nematode-infested plants or plant parts spreads the parasites.

Various methods are available to reduce crop losses from nematodes.

Figure 7–30   Galls on tomato roots caused by root-knot nematodes.

Figure 7–31   Nematodes can severely damage white potato tubers. These microscopic roundworms drill into the tubers, damaging the skin, disfiguring the potatoes, and making them unmarketable. *Source:* University of California Cooperative Extension.

1. Genetic host resistance:
    ◆ Plant resistant species and cultivars. For example, in an area with soil heavily infested with the root-knot nematode, plant apricots, cherries, apples, pears, or plums, which are resistant, rather than peaches or nectarines, which are highly susceptible. A root-knot nematode-resistant peach rootstock called 'Nemagaurd,' developed by USDA plant breeders, is available, thus permitting peach production even on infested soils. Certain vegetable crops—sweet corn, asparagus, and cabbage—are resistant to root-knot nematodes, whereas radishes are susceptible. Resistant ornamentals include the African marigold, azalea, camellia, and oleander. In Long Island, New York, where the golden nematode is a serious problem for potato production, resistant cultivars are available. Similarly, soybean varieties resistant to soybean cyst nematode (Heterodera glycines) are also available.

2. Cultural practices:
    ◆ Use only nematode-free nursery stock for planting. In most countries, government nursery inspectors will condemn and destroy any nursery stock showing evidence of nematode infestation.
    ◆ In nursery operations, use benches raised off the ground and pot plants only in pasteurized soil mixes. Keep containers, bins, benches, and flats clean. Fumigate outdoor growing fields where nursery stock is grown.
    ◆ Rotate crops to control certain nematodes. Rotation is useful for types that have a narrow host range, such as sugar beets attacked by the cyst nematode. Where the crop value is too low to justify large-scale soil fumigation, crop rotation is the only practical method of nematode control.
    ◆ Use cover crops that reduce nematode damage. Cover crops can improve soil structure and fertility; decrease soil erosion; be used as animal feed; and suppress weeds, insects and pathogens. Examples of cover crops that have been shown to suppress nematodes include cowpea, rapeseed, velvet bean, and Sudan grass.

3. Chemical applications:
    ◆ Treat the soil area with fumigant before planting. Methyl bromide is often used to reduce the nematode population to levels not harmful to plants. Soil mixes for container-grown plants can either be treated with a fumigant or steam-pasteurized at 82°C (180°F) for about thirty minutes. This method is too expensive for field crops other than commercial strawberry fields. The impending loss of methyl bromide may seriously affect the crops where it is used.

♦ Use nematicides in certain cases. All nematicides are poisonous and must be used carefully, following the directions on the containers exactly. Most such materials will injure or kill plants if they are applied too close to their root zones. As the number of commercially available nematicides decreases, greater emphasis has been placed on the development of alternative IPM practices.

4. Biological control:
   ♦ Although not widely available, scientists have explored the use of antagonistic fungi like *Arthrobotrys dactyloides* to trap and parasitize plant pathogenic nematodes.

5. Government regulatory measures:
   ♦ Avoid importing soil (or plants with soil on their roots) from areas that could be loaded with a dangerous nematode species new to the area. United States plant importation regulations forbid the introduction of plants with soil on their roots from other countries.

Parasitic Higher Plants

Some higher plant forms live on the surface of or parasitize other plants and often cause harmful reactions in their hosts. These **higher plant parasites** can be placed in three groups: epiphytes, hemiparasites, and true parasites. The **epiphytes** do little or no harm to their host plants, using them merely for physical support and protection. Examples are Spanish moss and epiphytic orchids, which in their native habitat commonly grow on tree limbs (Fig. 7–32).

The hemiparasites, sometimes called water parasites, do injure their host plants, absorbing water and mineral nutrients from them. However, they possess chlorophyll and can manufacture their own carbohydrates by photosynthesis. Witchweed (*Striga asiatica*) is a hemiparasitic seed plant that severely damages sugarcane, corn, sorghum, many other grasses, and some broad-leaved plants. It attaches itself to the host's roots and utilizes most of the host's water and mineral nutrients, causing it to wilt, yellow, stunt, and die. The best control is to plant a crop, such as Sudan grass, that stimulates the witchweed seed to germinate, then plow under the entire field. Crops should be rotated, and susceptible crops should not be planted.

Mistletoe (*Phoradendron* spp.), another member of the hemiparasitic group, attacks many broad-leaved trees such as Modesto ash, silver maple, honeylocust, hackberry (see Fig. 7–33), cottonwood, walnut, oak, birch, and some conifers. The seeds germinate on the limbs of susceptible hosts, forming an attachment disk on the bark. The sticky berries are disseminated throughout the tree and from tree to tree by birds and wind. Although not very effective, the usual control is to cut out the mistletoe branches deep into the tree under the point of attachment. No good herbicidal control has been developed; however, a dormant-season application

Figure 7–32 The epiphytic plant, Spanish moss (*Tillandsia usneoides*), a member of the pineapple family, is commonly found attached to trees throughout the southern United States. The tree shown here is growing near Baton Rouge, Louisiana. Spanish moss absorbs most of its nutrients and water directly from the atmosphere and uses its host plant only as a place to obtain support and light for photosynthesis. It gets no food or water from the host plant, as do the parasitic plants. Similar appearing plants, which are called lichens (an algae and a fungus living in an intimate symbiotic relationship as a composite plant), are members of the *Usnea* genus and are sometimes called "old man's beard."

of ethephon, an ethylene-releasing material, is a possible control on some plants. The best control is to plant only tree species resistant to mistletoe attacks.

True parasites lack chlorophyll and depend on their hosts for all nourishment—carbohydrates, minerals, and water. Examples of this group are the dwarf mistletoe (*Arceuthobium* spp.) and dodder (*Cuscuta* spp.). Broomrape (*Orobanche* spp.) is a serious parasitic pest in Europe and has caused extensive damage to tomatoes in California. Dwarf mistletoe attacks many coniferous species in the western United States, reducing tree vigor and lowering lumber

Figure 7–33   Branch of a hackberry (*Celtis sinensis*) tree being killed by the growth of mistletoe.

Figure 7–34   Olive branches attacked by dodder.

quality. The sticky seeds (not the fruits) are forcibly ejected and can travel up to about 19 m (60 ft). This is the principal means of dissemination. Birds are known to carry the sticky seeds on their feathers; wind plays a very minimal role in dissemination. The best control is removal of infected trees.

Dodder has many species, but about six cause the major damage, attacking crops such as alfalfa, lespedeza, clover, flax, sugar beets, some vegetable crops, and ornamentals (see Fig. 7–34). Dodder seriously reduces yields and quality of crops. Strict regulations prohibit the sale of crop seed contaminated by dodder seed. Great effort should be taken to avoid planting seed that has dodder seed mixed with it. Patches of dodder in field crops or along fences or ditch banks should be eradicated by burning or with herbicides.

Practical Tips for Diagnosing Plant Diseases

Proper diagnosis is a critical step in the management of plant diseases. Without a solid diagnosis, it is impossible to suggest an adequate management approach. The more you know, the better equipped you will be to take corrective action. In the case of plant disease diagnosis, the more you know about the host, environment, and biotic and abiotic fac-

tors that cause disease (the disease triangle), the greater chance you have of making a correct diagnosis. The following five-step approach is just one of many approaches available for diagnosing plant diseases:

1. *Define the problem.* In other words, determine that a problem actually exists. Start by correctly identifying the host plant and by being familiar with its normal or healthy state and characteristics. Make sure to take seasonal effects into account. After all, a maple tree without leaves in December in Ohio should not trigger an alarm. A defoliated maple tree in June is another story. Know your hosts and how they change with the seasons. Only then can you determine that a problem exists.
2. *Examine the entire plant community.* Don't jump into examining the affected individual plant or area. Take stock of the entire plant community. For example, if you are making a trip to a golf course to examine a potentially diseased fairway, notice other golf courses along the way. Make a few calls to other golf course superintendents close to the course in question. Once at the course, observe other fairways to see how widespread the damage is. Once at the affected fairway, take a minute to look at the entire fairway.

Note light intensities, wind direction, slope of the land, air movement, and so on. Take time to develop an overview of the situation at hand because this approach often provides many valuable clues about the situation. Then focus your attention on the affected plant(s) or area. Even then, however, look at the entire plant first before jumping directly to any signs or symptoms that might be present. Check out the leaves, stems, roots, and fruits or flowers if appropriate.

3. *Look for patterns.* Is only a single plant affected? Is the potential disease restricted to a certain area or a single species? Are the symptoms randomly distributed, or can you see any distinct patterns or clear lines of demarcation between healthy and affected plants? Random patterns are often indicative of diseases caused by pathogens, whereas uniform damage such as streaks or lines or damage over a large area is indicative of an abiotic (chemical, physical, or mechanical) culprit.

4. *Consider how the damage developed over time.* Did the damage appear suddenly or over time? Has the damage spread or stayed in the same location? Progressive development and spread over time often indicates damage caused by pathogens. In contrast, damage that does not spread and which occurs suddenly is typically caused by an abiotic factor.

5. *Ask questions, gather information, and determine causes of plant damage.* Gather as many clues (i.e., as much information) as possible about the crop. Determine cultivar or variety, age of the stand (especially important when dealing with some perennial crops), recent fertilizer or pesticide applications, cultural practices implemented, recent weather trends, irrigation practices, the history of the site or stand, and how the damage progressed over time. Collect as much information as possible to help you develop a solid mental picture of what led up to the damage you are observing. Look for evidence of pathogen activity. Specifically, look for key diagnostic signs or symptoms indicative of plant pathogens or insects. For instance, the presence of large numbers of fruiting bodies or mycelium might lead you to suspect a fungal infection. After you have gathered sufficient background information and nothing strikes you as being obvious, such as a chemical misapplication or hail damage, and you have eliminated the possibility of pathogens and insect pests, retrace your steps and focus your diagnosis on abiotic factors. You may need to enlist the services of a plant pest or disease diagnosis laboratory or clinician to help narrow the range of probable causes. Consider contacting and coordinating with the clinician before sending samples. Whenever possible, include photographs or digital images to aid the diagnostician in the task.

THE SAFE USE OF AGRICULTURAL CHEMICALS—HERBICIDES, INSECTICIDES, FUNGICIDES, MITICIDES, AND NEMATICIDES

The application of agricultural chemicals to food-producing plants must not create a health hazard. Many countries have elaborate procedures for determining whether agricultural chemicals are reasonably safe before they can be registered for sale to growers of agricultural crops. In the United States, the Environmental Protection Agency (EPA)[1] is responsible for determining the safety of agricultural chemicals; state and local government agencies can also add their own safety requirements.

Before the EPA grants approval for the sale of agricultural chemicals, exhaustive tests are conducted to show:

1. That the product, at the recommended application rate, has low toxicity levels (both acute and chronic) as determined by experiments with test animals.
2. An absence of residues in food or feed crops—or, if there is a detectable residue, that is no more than the tolerance level established as safe.
3. The fate of residues and breakdown products in the environment—in the soil, runoff water, groundwater, or wildlife.
4. Whether the product affects the environment by inducing changes in the natural populations of higher plants and animals or of microorganisms.

It is estimated that an agricultural chemical company developing a new pesticide spends $50 million and that six to ten years of research are required to develop information sufficient to satisfy EPA standards.

When the EPA registers a pesticide for use, the label lists very specific restrictions on the product. It is registered for use on a certain crop or crops, to be applied at specific times and at specific concentrations. The reentry interval (REI) specifies the amount of time that must pass after chemical application and before reentry in the area of application is allowed. While a certain herbicide, for example, may be known to control a given weed species, it may not be permissible to use the herbicide to control such weeds if they are growing in a crop for which the herbicide is not registered. When using pesticides, including herbicides, the warnings given in the box below should be carefully read and followed.

---

[1]"The Environmental Protection Agency is charged by the United States Congress to protect the nation's land, air, and water systems. Under a mandate of national environmental laws focused on air and water quality, soil waste management, and the control of toxic substances, pesticides, noise and radiation, the Agency strives to formulate and implement actions which lead to a compatible balance between human activities and the ability of natural systems to support and nurture life." From the *EPA Journal.*

The term LD_{50} may be seen on labels of agricultural chemicals. LD means "lethal dose," and refers to the chemical's toxicity. Oral LD_{50} is the dose that will kill 50 percent of test animals ingesting the chemical by mouth. LD_{50} is expressed in milligrams of the chemical per kilogram of body weight of the test animal. The higher the LD_{50} value, the safer the chemical. According to EPA toxicology guidelines a chemical with an oral LD_{50} of 50 or less must be labeled "DANGER—POISON (FATAL)"; one with an LD_{50} between 50 and 500 is labeled "WARNING (MAY BE FATAL)"; an LD_{50} from 500 to 5,000, "CAUTION", and one with an LD_{50} over 5,000 is also labeled "CAUTION." All labels must also state "KEEP OUT OF REACH OF CHILDREN."

PESTICIDE EFFECTS ON THE ENVIRONMENT

Insecticides, miticides, fungicides, nematicides, and herbicides can be thought of as necessary evils. Without these chemicals, large-scale agriculture and our standard of living today would not exist. Too little food would be produced to feed the world's 6 billion people, and mass starvation would result.

However, chemical applications in agriculture do not always do just what they are supposed to do and nothing else.

This problem is recognized more and more now, and greater and greater precautions are being taken to avoid unwanted side effects from these chemicals. Government regulations on pesticides have become steadily tighter. Pesticides that are chemically stable and persist in ecosystems are the ones largely responsible for environmental contamination. The law has ordered the replacement of persistent chemicals such as DDT, DDD, dieldrin, aldrin, chlordane, BHC, and heptaclor with the low-persistence organophosphates (malathion, diazinon) and the carbamates (carbaryl and methomyl), which break down rapidly and are not taken up in food chains. There is the risk, however, that large-scale applications of pesticides, particularly herbicides, to agricultural crops, followed by irrigation or heavy rains, can result in leaching of the chemicals into the underground water supply and eventually into drinking water. In fact, traces of several pesticides have been found in several underground water supplies in many states in the United States.

It was pointed out earlier in this chapter that many insect pests are held in check very well by their own natural enemies—often other insect predators. Reducing populations of one serious primary insect pest with insecticides may, at the same time, so reduce the numbers of insect pred-

WARNINGS ON THE USE OF PESTICIDE CHEMICALS AND SUGGESTIONS FOR THEIR PROPER USE

Pesticides are poisonous and should always be used with caution. The following suggestions for using and handling pesticides help minimize the likelihood of injury from exposure to such chemicals to humans, animals, and crops other than the pest species to be destroyed:

1. Always read and exactly follow all precautionary directions on container labels before using sprays or dusts. Read all warnings and cautions before opening the container. Repeat this process every time you use the pesticide—regardless of how often you use it or how familiar you think you are with the directions. Apply materials only in amounts and at times specified.

2. Keep sprays and dusts out of reach of children unauthorized persons, pets, and livestock. Store all pesticides outside the house in a locked cabinet or shed and away from food and feed.

3. Always store sprays and dusts in their original labeled containers and keep them tightly closed. Never store them in anything but the original container.

4. Never smoke, eat, or chew anything while spraying or dusting.

5. Avoid inhaling sprays or dusts. When directed on the label wear protective clothing and a proper mask.

6. Remove contaminated clothing immediately and wash the contaminated skin thoroughly if liquid concentrates are accidentally spilled on the skin or clothing.

7. Always bathe and change into clean clothing after spraying or dusting. If this is not possible, wash hands and face thoroughly and change clothes. Wash clothing after applying pesticides, never reuse before laundering. Launder this clothing separately from the family wash.

8. Cover food and water sources when treating around livestock or pet areas. Do not contaminate fishponds, streams, or lakes.

9. Always dispose of empty containers so that they pose no hazard to humans, animals, valuable plants or wildlife. Never burn pesticide containers, especially aerosol cans.

10. Read label directions and follow recommendations to keep residues on edible portions of plants within the limits permitted by law.

11. Call a physician or get the patient to a hospital immediately if symptoms of illness occur during or shortly after dusting or spraying. Be sure to take the container of the label of the pesticide used to the physician.

12. Do not use the mouth to siphon liquids from containers or to blow out clogged lines, nozzles, and so on.

13. Do not spray with leaking hoses or connections.

14. Do not work in the drift of a spray or dust.

15. Confine chemicals to the property being treated and avoid drift by stopping treatment if the weather conditions are not favorable.

16. Protect nearby evergreen trees and shrubs from the dormant sprays used on fruit trees.

17. Do not use household preparations of pesticides on plants because they contain solvents that can injure plants.

Source: Division of Agricultural Sciences, University of California.

ators feeding on a secondary insect that the secondary pest increases explosively. For example, insecticidal control of codling moth, pear thrips, and pear psylla in pear orchards may be followed by large increases in spider mite populations because the spray applications have reduced other predaceous mites that feed on the spider mite.

Production of several major food crops relies on pollination of the flowers by honeybees and other bees. Many nut and fruit tree species—almonds, apples, plums, and sweet cherries—as well as certain vegetables, such as muskmelons and honeydew melons, and forage crops such as seed alfalfa and seed clover require thorough working of the flowers by bees during bloom. Elimination of bees by haphazard insecticidal applications and drift cannot be tolerated. Some states impose strict legal requirements wherever honeybees could be involved.

Insecticides should not be applied in areas where bees are working, particularly with chemicals highly toxic to bees such as diazinon, Guthion, malathion, parathion, and carbaryl. Some insecticides (Aramite, ethion, methoxychlor, Omite, pyrethrins, rotenone, and Tedion) are relatively nontoxic to honey bees.

The state of California requires that beekeepers post their names and telephone numbers on all hives. Anyone planning to apply pesticides in the vicinity must inform the beekeepers of upcoming spray applications. If potentially hazardous insecticides are to be used, spray applicators must notify all beekeepers within a one-mile radius and allow them forty-eight hours to move their hives.

Effects on Wildlife

There is no doubt that pesticides have harmed wildlife, even though such effects may be difficult to document. Some reported losses of fish and fish-eating birds have resulted from the improper or illegal use of pesticides and sometimes from legal use. Certain pesticides most lethal to wildlife, such as DDT, have been withdrawn from general use in the United States.

Some of the adverse effects of pesticides on wildlife have been indirect. For example, in some areas the pheasant population has declined when the weed cover, which had offered protection and nesting places, was cleared out along ditch banks, fencerows, and fallow lands by herbicides.

SUMMARY AND REVIEW

The management of weeds, insects and their relatives, and disease pathogens is important to anyone growing plants for a living. In the past, we tried to rely mainly on chemicals to control these pests, but today we realize that we have to take an integrated pest management approach and use many different methods. In general, these methods are genetic host resistance, chemical applications, cultural practices, biological controls, and government regulatory measures. Using these methods in the proper way at the proper time results in effective management of the pests and at the same time protects the environment and workers who grow and handle the plants.

Weed biology and ecology is the study of weeds and their interactions in the physical and biological environment. Classifying weeds by their life cycles (annual, biennial, and perennial) is a useful component in weed management, as is the broader classification of monocot or dicot. Invasive weeds are those species that lack natural enemies and have the ability to infest an area rapidly. Annual and biennial weeds rely heavily on seed reproduction. The seed bank is an important factor in the infestation of annual weeds in a field. Biennials also rely on their seed bank. Perennials rely somewhat on a seed bank, but they are more likely to persist year after year by the spread of propagules. Weed management includes preventive, mechanical, cultural, biological, and chemical (herbicide) control. Herbicides are classified by their time of application and their mode of action.

The metamorphosis of insects and their relatives is an important factor in the control of these plant pests. Vulnerability to a control method varies with life stage. Mode of feeding is also an important factor in the control of insect pests and also in the type of damage that pests inflict. The type of feeding determines which control measures are most effective. Some types of feeding create more damage by the transmission of pathogens than in the actual physical damage to a plant. Pesticides are classified by their mode of entry and their mode of action. Both factors are important to consider when choosing a pesticide. Pesticides are also classified by their chemistry. Some classifications tend to be more toxic to nontarget organisms than others.

Plant disease can be biotic (caused by a living organism) or abiotic (caused by environmental factors). Biotic diseases are considered to be infectious because they can be spread from plant to plant. Abiotic diseases are noninfectious. Disease symptoms include abnormal tissue coloration, wilting, tissue death, defoliation, abnormal increase in tissue size, dwarfing, and replacement of host tissue by tissue of the infectious organism. Pathogens responsible for causing most biotic diseases include viruses, bacteria, spiroplasmas, phytoplasmas, fungi, fungal-like organisms, nematodes, and parasitic higher plants. Viruses are difficult to control once a plant has become infected. Prevention methods include purchasing material that is pathogen-free, using resistant varieties and cultivars, eradicating infected plants, and sanitation of the facility and workers. Bacterial diseases are also difficult to control once plants become infected, so the same methods used to prevent viral infections can be used for bacterial infections.

Not all bacteria are harmful to plants; some can be beneficial. Spiroplasmas and phytoplasmas are vectored by sucking insects. Controlling the insects is the most effective way to control these pathogens. Fungi and FLOs are the main causes of most plant diseases. One reason for this is their ability to penetrate a healthy plant without the aid of a wound or natural opening. These pathogens can be controlled in part with the use of chemicals, but the other control mechanisms of an integrated pest management program are equally important and are often more effective. Nematodes are microscopic multicellular organisms. The plant parasitic nematodes are distinguished by their stylets used for feeding. Genetic host resistance is an important method of control of some nematodes, as is the government regulation that forbids the importing of soil or plants with soil on their roots into the United States. Other control measures include chemical fumigation or steam pasteurization of the soil, and the use of some ne-

maticides. Higher parasitic plants include the hemiparasites, such as witchweed and mistletoe, and true parasites, such as dwarf mistletoe and dodder. Hemiparasites possess chlorophyll and can photosynthesize, but they take water and mineral elements from their host. True parasites lack chlorophyll and depend on the host for all nourishment. It is difficult to kill these pests with chemicals; genetic host resistance and cultural methods such as physical eradication of host and pest work best.

The use of herbicides, insecticides, fungicides, and other pest-management chemicals is highly regulated by the government. Their safe use is determined by the EPA and by other state and local government agencies. Legal use of these chemicals requires compliance with the label instructions. In addition to keeping the use legal, compliance helps to ensure that those handling the chemicals (and the plants treated with them) and the enviroment are not harmed.

FOOD FOR THOUGHT

1. What are the methods used in integrated pest management?

2. Compare the cultural control methods for weeds, insects, and diseases. How are they similar? How do they differ?

3. Which IPM method do you think is the most labor intensive? Explain your answer.

4. How does resistance to herbicides and other pesticides become prevalent in the environment?

5. Describe the effects of weed competition on crop yield when weeds emerge with the crop.

6. Explain the role of carbohydrate storage in terms of creeping perennial weed management.

7. List the five types of weed-control measures. Describe one advantage and one disadvantage associated with each type of control measure.

8. The development of herbicide-resistant crops was a major breakthrough for agricultural science in the late twentieth century. What is your opinion of the utility of these crops?

9. Why are the life cyles of weeds and insects important to know when planning a control strategy?

10. Fungi and FLOs possess a characteristic that probably makes them the major pathogen in plant diseases. What is this factor? How does it give fungi an advantage over bacteria and viruses?

11. What is mode of action? How does it influence a grower's choice about what chemical to use to control a pest?

12. Why is compliance with label instructions important when using a chemical control measure?

13. List and briefly describe the five disease management strategies that comprise an integrated disease management program.

14. What are the two most important environmental conditions that influence the development of plant disease?

SUPPLEMENTAL READING

AGRIOS, G. N. 2005. *Plant pathology,* Fifth Edition. New York: Elsevier Academic Press.

HOROWITZ, A. R., and I. ISHAAYA (Eds.). 2004. *Insect pest management: Field and protected crops.* New York: Springer.

LIEBMAN, M., C. L. MOHLER, and C. P. STAVER. 2001. *Ecological management of agricultural weeds.* Cambridge, England: Cambridge University Press.

MONACO, T. J., S. C. WELLER, and F. M. ASHTON. 2001. *Weed science.* Hoboken, N.J.: John Wiley and Sons.

SCHUMANN, G. L. 1991. *Plant diseases: Their biology and social impact.* St. Paul, Minn.: American Phytopathological Society Press.

http://www.weedscience.com/

http://www.apsnet.org/ (The American Phytopathological Society)

http://www.invasivespecies.gov/ (United States Invasive Species Council)

UNIT
II

PLANT STRUCTURE, CHEMISTRY, GROWTH AND DEVELOPMENT, GENETICS, AND BIODIVERSITY

CHAPTER

Structure of Higher Plants

![leaf icon]

KEY LEARNING CONCEPTS

After reading this chapter, you should be able to:

♦ Describe and recognize parts of the plant cell, plant tissues, and plant organs.
♦ Understand the basic functions of cells, tissues, and organs.

Every day we can identify familiar plants in our immediate surroundings. Some plants and their features can be identified and appreciated from their external structure, but their internal structure and function are often overlooked. The beauty of an orchid blossom is greatly admired, but just as impressive are the parts of a cell as recorded with a scanning electron microscope. The purpose of this chapter is to develop an understanding of the internal and external structures of the higher plants.

An approach that capitalizes on what is already known is to follow a plant from seed germination to full size and then to observe the formation of fruits and seeds. We can then appreciate how the plant grows and develops and, at the same time, acquire the vocabulary necessary to understand the growth processes of plants. This approach allows us to study both the external form of the plant, or its **morphology,** and its internal structure, or anatomy and **histology** (microscopic features).

Our major food, fiber, wood, and ornamental plants belong to two main classes—the **gymnosperms,** represented mainly by the narrow-leaved, evergreen trees; and the **angiosperms,** usually broad-leaved, flowering plants. In the temperate zone, angiosperms are by far the most common in everyday life. Angiosperms are divided into two subclasses: the **monocotyledons,** which have an embryo with one cotyledon, and the **dicotyledons,** which have an embryo with two cotyledons. These names are often shortened to **monocot** and **dicot.** We begin by examining the life cycle of a common monocot, corn (Fig. 8–1), and a common dicot, the bean (Fig. 8–4).

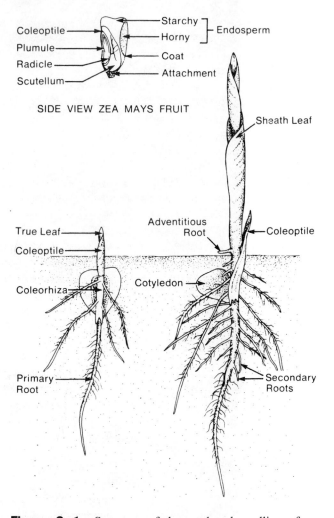

Figure 8–1   Structure of the seed and seedling of corn (*Zea mays*), a monocotyledonous plant.

THE LIFE CYCLE OF A CORN PLANT (A MONOCOT)

When a corn seed is planted in moist soil, it imbibes (absorbs) water from the soil. Germination begins with the emergence of the **radicle** (the primary root) and the **plumule** (the primary shoot). These two enlarging axes form the primary body of the plant.

The radicle grows downward through a protective sheath, the **coleorhiza**, from which the primary root develops and the secondary roots branch. A mature corn plant can develop roots 2 m (6.1 ft) long. **Adventitious roots** (roots other than those that develop from the radicle) grow from the shoot axis just at or above the soil surface (Fig. 8–2): These roots, also called anchor, brace, or prop roots, branch out in the soil to give added support to the plant.

The emerging plumule is protected by a sheathlike leaf, the **coleoptile**, that envelops the main stem as it grows upward through the soil. As the true foliage leaves develop, the main stem continues to produce sheathing leaves that encircle the stems at each node.

Figure 8–2   Adventitious or anchor roots develop above the soil line on the lower stem of a corn plant. These structures add support to the plant.

When the corn plant has reached a given size, producing a set number of leaves, female flowers, known as **pistillate flowers** or ears, appear at the base (axil) of one or more sheath leaves. Later, the male flowers, known as **staminate flowers** or tassels, develop at the top of the plant. Figure 8–3 shows both kinds of flowers. Blown by the wind, pollen grains from the tassels fall on and pollinate the long pistillate filaments (silks) and subsequently fertilize the ovaries, which become the individual corn kernels borne on a stalk (cob). Each ovary develops into a fruit, called a **caryopsis,** that encloses the true seed. After the kernels mature and dry, the fruits (containing the seeds) are harvested and stored over the winter. The seeds can be sown when weather conditions are favorable for germination, and the life cycle repeats itself.

Figure 8–3   *Left:* A pollen-bearing corn tassel (staminate flower). *Right:* The ear (pistillate flower), showing the "silks" that intercept the wind-blown pollen grains.

THE LIFE CYCLE OF A BEAN PLANT (A DICOT)

After a bean seed has been sown in moist soil, it imbibes water and swells. The seed coat bursts and the radicle emerges (Fig. 8–4). The radicle grows downward and the hook of the bean, known as the **hypocotyl,** emerges above the soil, carrying the two cotyledons with it. Between the cotyledons lies a growing point (**apical** or **shoot meristem**) flanked by two opposite primary foliage leaves. The stem region just above the cotyledons and the first trifoliate leaves is called the **epicotyl** (Fig. 8–4). Under favorable conditions the shoot apical meristem rapidly produces two trifoliate leaves opposite each other on the stem. The cotyledons, which have been supplying much of the reserve food for this initial growth, shrivel and abscise (drop off). The plant's green leaves are now capable of manufacturing food for future growth of the seedling. The bean plant produces trifoliate leaves, and flowers begin to develop in the axils of about the fourth set of leaves and in each succeeding set. These flowers are self-pollinated; thus, fruits (pods) develop as long as environmental conditions are favorable. The seeds mature and dry within the pod, and they can be sown at once to produce another generation of bean plants. The difference in emergence of the growing points of beans and corn from beneath the soil affects the tolerance of each

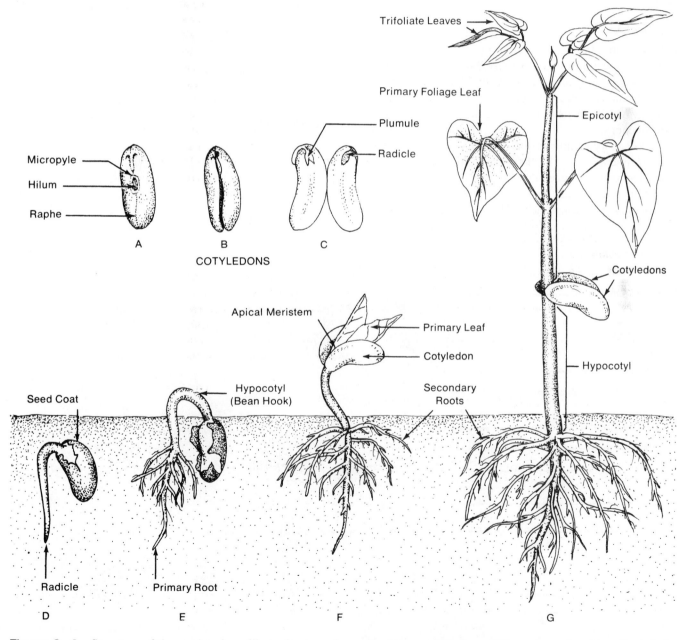

Figure 8–4 Structure of the seed and seedling—in several growth stages—of the bean (*Phaseolus vulgaris*), a dicotyledonous plant.

crop to light frosts. A late frost would be more likely to severely damage a newly emerged bean seedling than a newly emerged corn seedling because the growing point of corn is below the soil and protected. The leaves of the corn plant may be damaged, but the growing point will survive to generate new leaves. However, the bean plant's growing point would be severely injured or killed and unable to generate new growth.

The life cycles of plants like the corn and bean are more or less familiar to most of us from our own observations, but to enlarge our knowledge about higher plants, we must consider the largely unfamiliar areas of plant anatomy and histology.

THE CELL

The plant **cell** is the basic structural and physiological unit of plants, in which most reactions characteristic of life occur. The tissues of the plant develop through an orderly process of cell division and differentiation. **Cytology** is the branch of biology involved in the study of the components of cells and their functions.

Cells vary greatly in size. The smallest must be measured in micrometers (1/1,000 of a millimeter), but some wood fiber cells are several centimeters long.

Early cytological studies were conducted with light microscopes, which can demonstrate general cellular fea-

tures but which cannot resolve all the fine details within cells. Electron microscopes and enhanced light microscopes have revealed that living cells are not empty chambers but highly organized complexes of subcellular compartments with specialized metabolic functions. In the living cell, these complexes are distributed through a dynamic and orderly flow of materials within the **cytoplasm.**

CELL STRUCTURE

There are two types of cells. **Prokaryotic cells** have no separate subcellular units; for example, nuclear material is not enclosed in a membrane. These cells, considered primitive, are found in bacteria and blue-green algae. **Eukaryotic cells** are made up of compartments bounded by membranes, with specialized structures and functions. These units, called **organelles,** include the **nucleus, mitochondria, plastids,** microbodies, **vacuoles,** dictyosomes, and **endoplasmic reticulum** (Fig. 8–5). Plant cells are eukaryotic cells.

The Protoplast

The organelles of the plant cell are contained within a membrane-bounded **protoplast,** which in turn is encased within a cell wall. The major features of the protoplast are the outer membrane or plasma membrane, the cytoplasm, the nucleus, and the vacuole.

Mitochondrion
Golgi Body

Cell Wall

Plasmalemma

Cytoplasm (Ground Substance)

Vacuole Membrane (Tonoplast)

Vacuole

Plastid (Chloroplast)

Nucleus

Nucleolus

Endoplasmic Reticulum

Nuclear Membrane

Plasmodesma

Figure 8–5 Photomicrograph of a plant cell showing the various parts and organelles, × 6,000. *Source:* Keith Weinstock.

Plasma Membrane

The plasma membrane, also called the **plasmalemma,** is a lipid bilayer surrounding the cytoplasm. This membrane is important in maintaining a surface area for selective absorption and secretion by the cell, and plays a role in generating energy as well. Proteins embedded within the bilayer can function as enzymes; as surface receptors, both on the inside surface and in communication with the cell environment; or as channels for the uptake and efflux of ions.

Cytoplasm

The **cytoplasm** is a viscous fluid composed of matrix proteins, bounded by the semipermeable plasma membrane. The flow of organelles within the cytoplasmic matrix, called *cytoplasmic streaming,* is clearly visible in active leaf cells under a light microscope. Also within the cytoplasm is a very important network of membranes, the **endoplasmic reticulum (ER).** Proteins are synthesized on the surfaces of the ER throughout the cell, on small discrete structures called **ribosomes.** Proteins may be further processed inside the ER and transported to destinations that will be sites of activity within the cell.

   Plastids of several types are located within the cytoplasm. The colorless leucoplasts serve as storage bodies for oil, starch, and proteins. Chromoplasts contain the various plant pigments, including **chlorophyll.** Chromoplasts with chlorophyll are called **chloroplasts** and are responsible for photosynthesis in leaves and in some stems. Enclosed by a double membrane, most chloroplasts also contain other **pigments,** large quantities of proteins and **lipids,** and some stored **starch.** Within the chloroplast, light energy is first harvested by pigments bound to stacked membranes called grana and then converted into chemical energy in the form of sugars.

   Mitochondria are cytoplasmic bodies that are smaller than plastids. Like the chloroplasts, they are surrounded by a double membrane, and contain a specialized inner membrane system. The mitochondria are sites of respiration and are also involved in protein synthesis. They produce energy-rich compounds such as **adenosine triphosphate (ATP).**

The Nucleus

The **nucleus** is a prominent organelle within the cell, enclosed by a double membrane and containing one or more bodies called nucleoli. Within the nucleus are the **chromosomes,** long lengths of **deoxyribonucleic acid (DNA)** and associated proteins that contain the genetic information coding for all cell functions, for differentiation of the organism, and for reproduction. During cell division, the chromosomes replicate and condense into discrete rod-shaped bodies; a set of chromosomes is passed on to each of the daughter cells (see Chapter 14), thus ensuring continuity of genetic information from old to new cells. **Genetic codes** are transcribed from the DNA in the nucleus and translated into proteins on the **ribosomes.**

   DNA is also found outside the nucleus in the mitochondria and in the chloroplasts, thereby giving these bodies a role in heredity independent of the nucleus. Unlike nuclear DNA, mitochondrial DNA and chloroplastic DNA are inherited only from the female parent. There is no sexual segregation of genetic traits. This has proven useful in determining the relationship between species or individuals.

Vacuoles

Vacuoles may occupy a major portion of the interior of plant cells. In actively dividing cells, vacuoles are very small, but they can account for up to 90 percent of the volume of mature cells. The membrane surrounding the vacuole is called the **tonoplast,** and it serves an important role in regulating ion flow within the cell, maintaining cell turgor, and other functions. Vacuoles contain a watery solution of dissolved materials, including inorganic salts, blue or red pigments (**anthocyanins**), sugars, organic acids, and various inclusions of crystals. The vacuole serves as a storage reserve for water and salts, as well as for toxic products.

The Cell Wall

The **cell wall** protects the protoplast, provides an external structure, and in some tissues (e.g., bark, wood) may act as a strong support for the plant. The cell wall is nonliving, made up of **cellulose,** pectic substances, and lignins. Between cells lies an intercellular layer called the **middle lamella,** which contains many of the mucilaginous pectic compounds that hold adjacent cell walls together. Adjacent to the middle lamella is the primary wall, which is composed mostly of cellulose. This elastic but strong material is the chief constituent of most plant cell walls.

   The secondary wall layer, which lies within the primary wall and is laid down only after the primary wall is complete, is usually thicker than the primary wall when fully developed. The secondary wall is also composed of cellulose, but in some cells and tissues it may contain **lignins, suberins,** or **cutins.** Lignins are closely associated with the cellulose and give it added strength, as is well demonstrated by wood fibers. Large quantities of water are contained and transferred in cellulosic walls, which act as wicks. In some specialized cells, (for example, cork), water loss or flow is prevented by the presence of the waxy material, suberin, in the walls.

   Individual cells in a tissue are connected to one another via strands of cytoplasmic material, called **plasmodesmata,** which extend through the plasma membrane. The surrounding cell wall forms channels around the plasmodesmata, called **pits.** As the wall grows thicker, these pits are preserved and can become quite long, clearly visible in the light microscope. Water and dissolved materials can move from cell to cell through these connections.

PLANT TISSUES

Large tracts of organized cells of similar structure that perform a collective function are referred to as **tissues.** Tissues of various types combine to form complex plant organs such as leaves, flowers, fruits, stems, and roots.

In all plants, both young and mature, two basic kinds of tissues can be distinguished. One kind is the **meristem,** or **meristematic tissue,** which is comprised of actively dividing cells that develop and differentiate into yet other tissues and organs. Cells in the meristematic tissues have thin walls and dense protoplasts. Meristematic tissues are found in the root and shoot tips, just above the nodes (intercalary meristems) and in woody perennials, as cylinders in the shoots and roots (the **cambium layer**).

The second kind of tissue is that which develops from the meristems and has differentiated fully. This is the permanent tissue, of which there are two kinds: the simple, which includes the epidermis, parenchyma, schlerenchyma, and collenchyma; and the complex, which includes the xylem and phloem.

Meristematic Tissues

The common categories of meristematic tissues are:

> Apical meristems
>> Shoot
>> Root
> Subapical meristems
> Intercalary meristems
> Lateral meristems
>> Vascular cambium
>> Cork cambium

Apical Meristems

Shoot meristems, frequently referred to as shoot apical meristems, are the termini of the above-ground portions of the plant (see Fig. 8–6). They are responsible for producing new buds and leaves in a uniform pattern at the terminus of the stem and laterally along stems. The pattern of leaves and lateral buds that form from the shoot meristems vary with the species of plant. For example, in the maples (*Acer*), ashes (*Fraxinus*), and in members of the mint (*Amiaceae*) and olive (*Oleaceae*) families, the leaves and buds are opposite—at a 180° angle—to one another. On the other hand, in the oaks (*Quercus*) and walnuts (*Juglans*), for example, the leaves and buds alternate from one side of the stem to the other, and in the pines (*Pinus*) they form a spiral pattern.

The shoot apical meristem produces epidermis, cortex, primary xylem and phloem, and the central pith, tissues that form the primary structure of the stem. The shoot apex may eventually develop terminal inflorescences (floral groupings) instead of continuing to produce leaves and lateral buds as, for example, in the chrysanthemum, poinsettia, and sunflower.

Some shoot meristems always remain vegetative and continue to produce leaves and lateral buds, as in many vines such as the grape. In such cases, flowers or inflorescences are

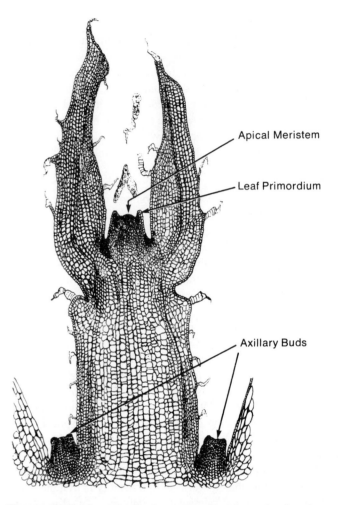

Figure 8–6 Longitudinal section of a shoot tip showing the apical meristem. Cell division in this region, along with cell elongation, is responsible for shoot growth. Buds in the axils of leaves also have meristematic regions at their tips that can develop into shoots. Should the apical meristem die or be broken off, an axillary bud can become the dominant shoot tip and continue shoot growth.

borne in the axils of lateral leaves somewhat behind the terminal growing point. Many trees also have this growth pattern; the dominant meristem (central leader) enables the plant to grow upright and gain height. In this case, individual flowers or inflorescences are usually borne on side branches in the axils of leaves at some distance below the apex. These growth and flowering characteristics are known for most cultivated plants. Gardeners, orchardists, and foresters use this knowledge to prune trees and other plants to direct the growth in the manner that they desire.

Root meristems, located at the various termini of the roots, are the growing points for the root system. Some plants have a dominant **tap root,** which develops downward, together with limited lateral root growth (see Fig. 8–7). Examples of plants with tap roots are carrots, beets, and turnips, all well-known root crops. Other species with tap roots include

Figure 8–7 Two types of root systems. *Left:* Fibrous root system of a cereal plant. *Right:* Tap root system as developed by the carrot plant (*Daucus carota*).

oaks, pecans, alfalfa, and cotton. Many plants, however, do not have a dominant tap root. Instead, the roots branch in many directions creating a **fibrous root** system (Fig. 8–7). Examples are the grasses, grain crops, and many kinds of shallow-rooted trees.

The root meristem lies just behind the root cap, which protects the meristem as the root grows through the soil. These root cap cells are constantly being destroyed, but the apical root meristem produces more to replace them. The root meristem produces the primary tissues—such as protoderm,

ground meristem, and procambium—that later become the epidermis, cortex, and vascular cylinder of the mature root.

Subapical Meristems

The **subapical meristem** produces new cells in the region a few micrometers behind an active shoot or apical meristem. The subapical meristematic region has long been thought of as a region where cells only elongate and expand. Cells do, indeed, expand in this region and thus increase internode length, thus adding to the growth in height of the plant. However, because new cells also form in this subapical region, it is a true meristem. The activity of the subapical meristem can be seen particularly in certain plants that lack tall stems when they are first producing leaves and that grow as a rosette. Examples are beets, carrots, China asters, lettuce, mustard, and turnips. These plants form rosettes of leaves on very short internodes. Later, when the shoot apical meristem initiates flowers, the stem below the flower elongates rapidly (bolts) because of the activity of the subapical meristem. During the period of fast growth, cells divide as well as elongate. The division and elongation together account for the rapid stem growth below the terminal flower buds.

Intercalary Meristems

The **intercalary meristems** are active tissues that have been separated from the apical meristem by regions of more mature or developed tissues. The separation occurs at an early stage of development, and therefore the separated cells retain their ability to divide. The best examples of intercalary meristems are found in monocots and especially in the grasses. The active meristematic cells just above the nodes in the lower region of the leaf sheath divide, and those cells develop (expand and elongate) rapidly. Grass leaves elongate at the base in a like manner.

Lateral Meristems

The lateral meristems, which produce secondary growth, are cylinders of actively dividing cells starting somewhat below the apical or subapical meristems and continuing through the plant axis. These meristems are the **vascular cambium**—which produces new xylem (water and mineral conducting elements) and new phloem (photosynthate conducting elements)—and the **cork cambium**—which chiefly produces bark, the protective covering of old stems and roots (Fig. 8–8). Stem girth of woody perennial plants and trees increases mainly by the activity of these lateral meristems. The number of growth rings indicates the tree's age (Fig. 8–9). Measuring the width of the annual growth rings in the stems is one way to determine the rapidity of lateral growth of a tree. Long-lived trees, such as the redwood (*Sequoia*), and certain pines, such as the bristle cone pine (*Pinus aristata*), increase greatly in size by lateral growth, but short-lived summer plants (annuals) such as marigolds (*Tagetes*), tomatoes (*Lycopersicon*), and peppers

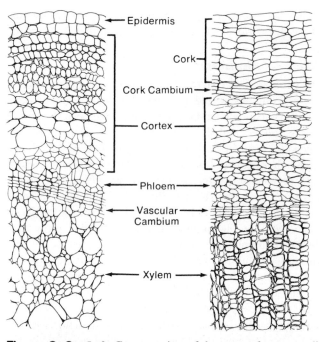

Figure 8–8 *Left:* Cross section of the stem of a young dicotyledonous plant showing the tissues of the meristematic region, the vascular cambium. The epidermis lies on the outside. The dividing cells of the vascular cambium layer, producing phloem to the outside and xylem to the inside, account for the thickening of the stem as it grows older. *Right:* Cross section of an older dicot stem where a cork cambium has developed. This meristematic layer produces the cork cells (bark), which protects the inner, more tender tissues.

(*Capsicum*) develop only a limited girth of the lower stem before the frost destroys them. Tomatoes, however, are perennials if they are grown in the frost-free tropics. Vascular and cork cambial growth turns the lower stem into a small trunk if growth continues.

Permanent Tissues

Permanent tissues can be classified into simple and complex tissues. The simple tissues are uniform, composed of only one type of cell. Examples are epidermis, parenchyma, sclerenchyma, collenchyma, and cork. Complex tissues are mixed, containing different kinds of cells. Examples are xylem and phloem.

Simple Tissues

The **epidermis** is a single exterior layer of cells that protects stems, leaves, flowers, and roots. The outside surface of epidermal cells is usually covered with a waxy substance called cutin, which reduces water loss. The epidermis of leaves is usually colorless except for the **guard cells** of the **stomata** (Fig. 8–10), which contain chlorophyll and are green. Some leaf epidermal cells are elongated into hairs and are called trichomes (Fig. 8–11). The root epidermis lacks cutin. It develops protuberances called root hairs, which actively absorb water from the soil.

Parenchyma tissue is made up of living thin-walled cells with large vacuoles and many flattened sides. This is the principal tissue of the cylindrical zone under the epidermis

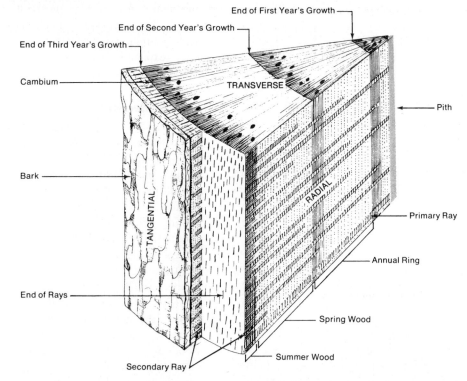

Figure 8–9 Section of a three-year-old stem of pine (*Pinus*) showing the annual rings by the end of the third summer. The porous, fast-growing spring wood is followed by the more dense, slower-growing summer wood.

Figure 8–15 Long files of parenchyma cells surround the vessel. Pitting in lateral vessel wall in red maple *(Acer rubrum)*. *Source:* Dr. Irving B. Sachs, Forest Products Laboratory, USDA Forest Service.

Roots

Roots are responsible for absorbing and conducting water and mineral nutrients and for anchoring and supporting the plant. In addition, some roots, as in sugar beets and carrots, act as storage organs for photosynthesized food. The roots of some plants develop secondary xylem from cambial activity and an abundance of parenchyma cells, which are able to store photosynthates and water. Dissolved mineral nutrients and water required for growth are absorbed by the root hairs, which are extensions of the epidermal cells (Fig. 8–18).

A few kinds of trees develop aerial roots from the underside of branches. Once these roots reach the soil and penetrate it, they become functional as ground roots. Good examples of this are the strangler fig *(Ficus aurea)* and the banyan tree *(Ficus benghalensis)* of the tropics. Some of the strangest roots are those in certain tropical orchids *(Cattleya, Phalaenopsis, Aerides,* and *Vanda)*. The roots contain chlorophyll for photosynthesizing food; they cling to rocks or tree surfaces and are fully exposed to receive light. Frequent rains and mist supply the moisture and nutrients necessary for growth.

The root system is a significant portion of the entire dry weight of any plant, about one-quarter to one-third of the total, depending on the storage or fibrous nature of the root. Measuring the total root system of a single mature rye plant showed that the plant had about 600 km (380 miles) of roots!

Figure 8–16 Sieve cells, sieve-tube elements, and companion cells in side view and cross section, showing detail structure of sieve plates. (A, B): From Canadian hemlock *(Tsuga canadensis)*; only one-third of cell shown. (C, D, E): From tulip tree *(Liriodendron tulipifera)*. C, D, with companion cells attached; E, detail of sieve plate. (F, G, H): From apple *(Malus pumila)*; H, detail of sieve plate. (I, J, K): From black walnut *(Juglans nigra)*; K, part of sieve plate in detail. (L, M, N): From black locust *(Robinia pseudo-acacia)*, with companion cells attached; N, detail of sieve plate. *Source:* Eames, A. J., and L. H. McDaniels. 1947. An introduction to plant anatomy. New York: McGraw-Hill.

Many of the functional roots of woody plants extend only into the top 1 m (3 ft) of soil. The depth that tree roots penetrate depends largely on the species of tree and on the structure and water status of the soil.

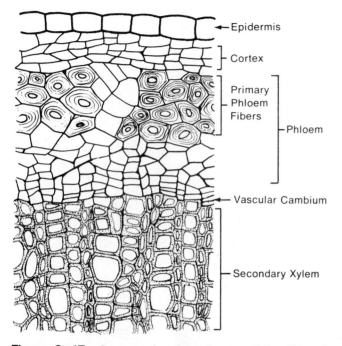

Figure 8–17 Cross section through stem of flax (*Linum*) showing thick-walled strengthening phloem fibers. *Source:* Esau, K. 1965. *Plant anatomy.* 2nd ed. New York: John Wiley.

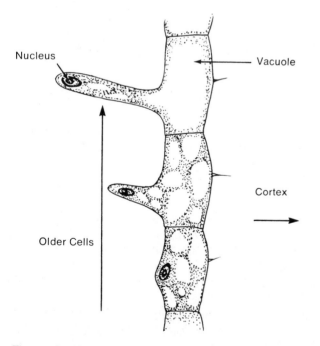

Figure 8–18 Section of epidermis of a young root showing three stages (bottom to top) in the development of root hairs.

After the radicle or primary root emerges from the seed, it can continue to grow principally as a tap root, or it may develop branch roots and form a fibrous root system (Fig. 8–7). The tap root usually grows downward, and the branch roots grow downward or horizontally. The tap root can be encouraged to branch at an early stage by removing or breaking the apical root meristem, which often happens when some seedlings are transplanted into the garden; for example, the tap roots of tomatoes are broken when the young plants are removed from the container, and the roots become fibrous. When young trees are transplanted from the nursery, the tap root is cut and the roots branch after the tree is transplanted into the home garden or commercial orchard.

The meristematic region of a root is composed of small, thin-walled cells with dense protoplasm that produce primary tissue at a rapid rate. Behind this active meristematic region lies the zone of elongation. Here the cells expand, especially in length; new protoplasm forms; and the size of the vacuoles increases. The apical meristem and the region of elongation take up only a few millimeters of each root. Behind the region of elongation is the region of maturation, where the enlarged cells differentiate into the tissues of the primary body. In the epidermis of this young region, the cells protrude and elongate and begin to form root hairs (Fig. 8–19). New root hairs arise in the newly developed region to replace old root hairs destroyed as the roots penetrate the soil.

The root apical meristem produces tissues different from those produced by the shoot apical meristem. The root meristem gives rise to the **root cap, epidermis, cortex,** and central vascular cylinder. The root cap is a thimble-shaped group of cells that protect the actively dividing meristem as it penetrates the soil (Fig. 8–19). These moist cells are sloughed off as the root comes in contact with sharp soil particles; the meristem forms new cells on the inner part of the cap to replace the damaged or lost cells. The meristem produces long rows of cells under the root cap. One of these becomes the protoderm that gives rise to the epidermis, or outer layer, of the root. The ground meristem, the tissue layer that gives rise to the cortex just below the epidermis, is usually thicker than the ground meristem found in stems. The cortical region is mainly composed of storage parenchyma cells, which have large intercellular spaces. A single layer of inner cortical cells forms the **endodermis,** a tissue found only in the root and not the stem. Each thin-walled endodermal cell is completely encircled by a narrow, thickened band of waterproofed material known as the **Casparian strip.** The solution of water and nutrients entering the root from the soil cannot penetrate the Casparian strip. For the soil solution to enter the inner tissue (pericycle) of the root, it must pass through the permeable endodermal cell walls and the protoplast (Fig. 8–20).

The Procambium Layer

The procambium layer gives rise to various tissues of the vascular cylinder. These include the **pericycle,** which is the outermost layer of cells of the central core and lies just inside the endodermis. The pericycle develops from a single

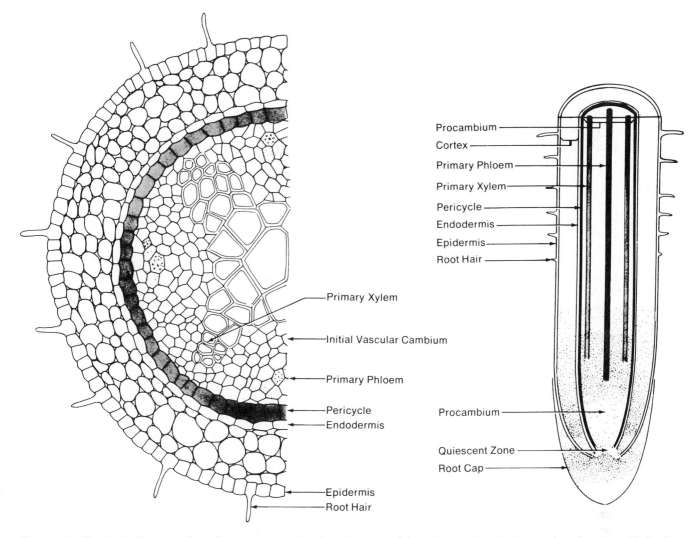

Figure 8–19   *Left:* Cross section of a young root showing the parts of the primary plant body and their location. *Right:* Developmental occurrences in the root tip, showing the various components and their relative location. *Source:* Weier, T. E., C. R. Stocking, M. G. Barbour, and T. L. Rost. 1982. *Botany, an introduction to plant biology.* 6th ed. New York: John Wiley.

parenchyma cell layer on the outer portion of the procambium. The pericycle is a meristematic region producing lateral (branch) roots that grow outwardly through the cortex and epidermis. It may also give rise to vascular and cork cambium. The procambium layer also produces a vascular cylinder of primary phloem and xylem, vascular cambium, and, in some species, pith. The pericycle and the vascular cylinder are collectively called the **stele.** The primary xylem is a central mass of tissue that may extend as arms beyond the primary phloem (Fig. 8–19). A layer of procambium cells separates these two primary tissues; this layer is the meristematic region for any new vascular tissue that subsequently forms.

As the root grows in girth and the plant matures, a continuous ring of secondary phloem forms outside the vascular cambium, and the primary phloem becomes less important than at earlier growth stages. The cambium layer

develops from the procambium and from pericycle cells outside the primary xylem. The primary xylem with its extending arms remains, but it is encircled by secondary xylem formed by the adjacent vascular cambium. Annual rings develop in the secondary plant body of the root as it grows in girth, much as in stems except for the star-shaped primary xylem at the core of the root. The cork cambium formed from pericycle produces a corky layer outwardly from the vascular system.

Adventitious roots form at any place on plant tissue other than the radicle of a germinating seed and its extensions. Adventitious roots arise from meristematic cells adjacent to vascular bundles (in herbaceous dicot stems) or from cambium or young phloem cells in young stems of woody perennials. This production of adventitious roots is the basis for propagation by stem cuttings (see Chapter 14). Adventitious roots can arise from plant parts other than

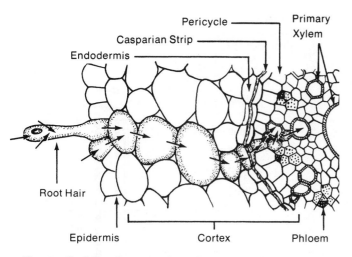

Figure 8–20 Cross section of a young wheat (*Triticum*) root showing the path of entrance of soil solution into the root from a root hair to the tracheary elements of the xylem. *Source:* Esau, K. 1965. *Plant anatomy.* 2nd ed. New York: John Wiley.

stems, such as from leaf petioles or leaf blades or even from old root pieces. Plant parts to be rooted are usually detached from the parent plant and placed under favorable environmental conditions for rooting. In the propagation procedure known as layering, adventitious roots are induced to form on plant parts still attached to the parent plant. Adventitious roots also develop on intact plants, as in the corn plant (see Fig. 8–2).

The ability of some plants to readily form adventitious roots allows seedlings and transplants to be planted deep in the ground if the seedling or transplant has become tall and spindly. By planting deep and having adventitious roots form, the young plant is much less likely to be damaged by wind or rain. For example, tomato transplants that have gotten tall and spindly are often planted deep in the field or home garden. Easter lily bulbs are planted in tall pots to allow adventitious roots to form on the stem and help support the plant.

Stems

The main stem and its branches are the scaffold of the plant, supporting the leaves, flowers, and fruits. The leaves and herbaceous green stems manufacture food, which is transported to the roots, flowers, and fruits through the phloem. Figure 8–21 illustrates the complexity of the primary vascular system. The greater part of the vascular system consists of xylem and phloem. The secondary xylem also serves as the major structural support in woody perennial plants.

The stem develops from three primary tissues produced by the apical meristem: the protoderm, the ground meristem, and the procambium. These give rise to the epidermis, cortex, and vascular cambium, respectively.

Figure 8–21 Three-dimensional skeletal view through a potato stem showing the primary vascular system extending through the stem with branches into the cutaway leaf petioles. *Source:* Eames, A. J., and L. H. McDaniels. *An introduction to plant anatomy.* 1947. New York: McGraw-Hill.

The epidermis, which is usually a single layer of surface cells, protects the stem. Epidermal cells are usually cutinized on their outer surface to retard desiccation. The epidermis of leaves and young stems has pores, the stomata (Fig. 8–10), that allow for the exchange of gases.

The cortex lies just beneath the epidermis and encircles the inner core of the vascular tissue. The cortex comprises parenchyma, collenchyma, sclerenchyma, and secretory cells, with parenchyma cells the most numerous. Some parenchyma cells have chloroplasts and are called chlorenchyma. Parenchyma cells have the ability to divide and form new tissue when wounded, thus providing a protective mechanism for the stem. Collenchyma is the outer cell layer of the cortex adjacent to the epidermal layer (Fig. 8–22). These cells may be thickened at the corners, and their walls contain cellulose, hemicellulose, and pectin. This tissue, therefore, adds strength to the stem. Sclerenchyma cells have thick lignified walls. They can form long fibers, which are the source of strength in mature stems. Secretory cells produce resinous substances and are commonly found, for example, in the resin ducts of pine trees.

The **vascular system** of seed-bearing plants consists of the pericycle, phloem, vascular cambium, xylem, pith rays, and pith. The arrangement of these complex tissues in the vascular system differs among three broad groups of

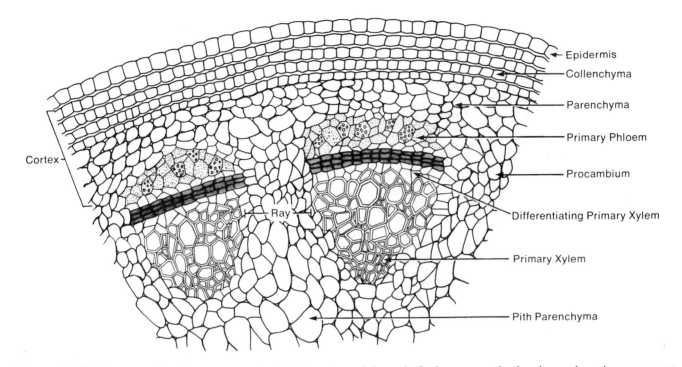

Figure 8–22 Cross section of a young woody plant stem toward the end of primary growth, showing various tissues present.

plants: (1) the gymnosperms and the woody dicotyledonous angiosperm perennials (which live for long periods), such as trees and shrubs; (2) the herbaceous dicotyledonous plants, such as potato, petunia, and phlox; and (3) the monocotyledonous plants, such as corn and date palms. These three groups are discussed separately to distinguish among their different internal stem structures and growth patterns.

Woody Perennials (Dicotyledonous Angiosperms and the Gymnosperms)

All the cells and tissues originate from a terminal shoot meristem that forms protophloem and protoxylem (primary tissue) (Fig. 8–22). As the stem grows in length, the secondary tissues form from the vascular cambium. The secondary phloem develops toward the outside of the stem and the secondary xylem forms inwardly from the vascular cambium. These growing secondary and permanent tissues crush the primary tissues until they become difficult to see. Secondary xylem is actively produced by the vascular cambium in the early spring and less actively in late summer. This xylem tissue becomes the early (porous) and late (dense) wood that form the annual growth rings in trees (Figs. 8–9 and 8–23). The vessels or tracheids formed during the spring flush of growth are larger than those formed during the summer (Figs. 8–21, 8–23, and 8–24). The narrow-leaved evergreen trees belonging to the gymnosperms are usually referred to as the softwoods or non-porous wood trees (Figs. 8–24 and 8–25). The xylem of

Figure 8–23 A cross section through the xylem of Douglas fir (*Pseudotsuga menziesii*) (× 200). The large pores to the left are the spring wood and the more dense cells were formed in the summer of the same year. The very large pore at the top is a resin duct. *Source:* Dr. Irving B. Sachs, Forest Products Laboratory, USDA Forest Service.

Figure 8–24 Three-dimensional view of the wood of a softwood forest species: (1) cross-sectional face; (2) radial face; (3) tangential face; (4) annual ring; (5) early wood; (6) late wood; (7) wood ray; (8) fusiform ray; (9) vertical resin duct; (10) horizontal resin duct; (11) bordered pit; (12) simple pit. *Source:* U.S. Forest Products Laboratory, Madison, Wis.

Figure 8–25 End-wall perforations in the ray cells of Douglas fir (*Pseudotsuga menziesii*). *Source:* Dr. Irving B. Sachs, Forest Products Laboratory, USDA Forest Service.

gymnosperms consists mainly of tracheids (Fig. 8–13). The broad-leaved angiosperm trees are called hardwoods or porous wood trees (Fig. 8–26); the xylem tissue is made up mostly of vessel elements (Figs. 8–12 and 8–13).

Both the gymnosperms and the woody perennial angiosperms grow in girth each year when the cells of the vascular cambium divide, forming annual rings of xylem. A stem nearing the end of its first season of growth is mostly xylem. The phloem, as it is crushed by the expanding xylem, is constantly being renewed by the vascular cambium. The phloem is a relatively thin layer of complex tissue protected by the bark or cork layer.

The cork cambium (**phellogen**), which is a meristematic tissue, provides cells that grow both outward and inward. The outward cells become cork cells; the inward, phelloderm. The cork cells become suberized and are, therefore, resistant to entry or loss of water. The cork cells soon die but retain their ability to resist desiccation, disease, insects, and extreme temperatures. The unusually thick bark of the cork oak (*Quercus suber*) is stripped for a multitude of commercial uses such as corks and insulation (Fig. 8–27).

In the young twigs and small trunks of many kinds of trees and shrubs, pore openings (**lenticels**) allow the inward and outward diffusion of gases (Fig. 8–28).

Figure 8–26 Three-dimensional view of the wood of a hardwood forest species: (1) cross-sectional face; (2) radial face; (3) tangential face; (4) annual ring; (5) early wood; (6) late wood; (7) wood ray; (8) vessel; (9) perforation plate. *Source:* U.S. Forest Products Laboratory, Madison, Wis.

Figure 8–27 The rough, thick bark of the cork oak (*Quercus suber*), which is commercially stripped for cork products.

Figure 8–29 labels: Primary Phloem, Primary Xylem, Vascular Ray, Epidermis, Pericycle Fibers, Pith Parenchyma, Procambium, Pericycle Fibers, Vascular Ray (Cortex), Vascular Bundles, Collenchyma, Epidermis

Figure 8–29 Three-dimensional cutaway view of an herbaceous dicot stem, showing the vascular bundles.

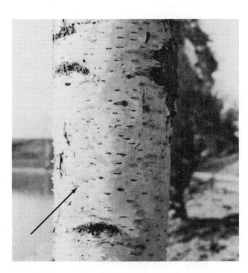

Figure 8–28 The bark of a birch tree (*Betula verrucosa*) showing the lenticels.

Herbaceous Dicotyledonous Plants

The early stem growth of plants in this category is much like the early growth of woody dicot stems. The vascular bundles (fascicles) of a herbaceous dicot usually remain separated and distinct; they are arranged in a single circle in the stem (Fig. 8–29). A larger proportion of the herbaceous stem is cortex and pith rather than xylem or phloem. Stem strength comes from the pericycle fibers adjacent to the phloem or from collenchyma or sclerenchyma tissue just beneath the epidermis (Fig. 8–29).

Herbaceous Monocotyledonous Plants

Stem growth originates from an apical meristem that produces vascular bundles scattered throughout the parenchyma (Fig. 8–30). The vascular bundles form most frequently near the epidermis. The sclerenchyma cells near the epidermis and thick-walled cells surrounding the bundles provide the principal support in monocot stems. Monocots (see Fig. 8–30) have no continuous cambium and, therefore, lack secondary growth. Stem diameter from the base to the apex is usually more uniform in monocot stems than in dicot stems with secondary vascular growth.

Woody Perennial Monocotyledonous Plants

In trees such as date and coconut palms (*Arecaceae*), the thickness at the stem apex increases by the activity of a primary thickening meristem. In the trunk below the terminal

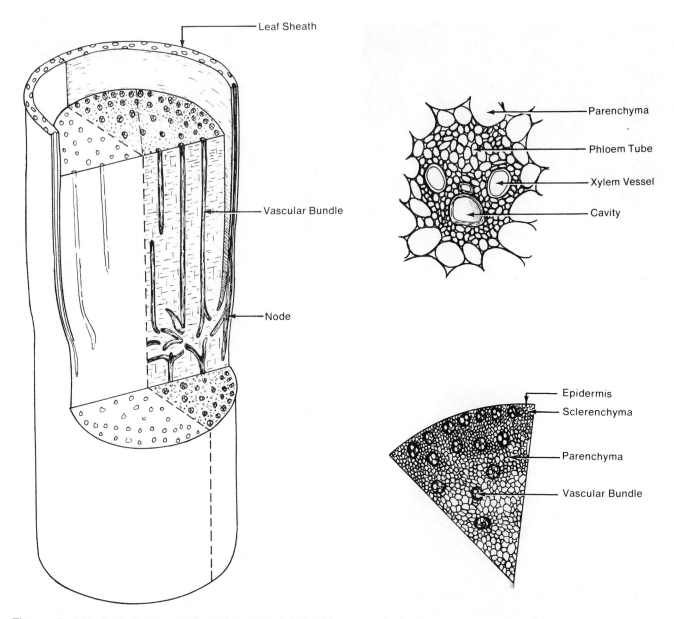

Figure 8–30 *Left:* Three-dimensional cutaway view of the stem of a herbaceous monocot (solid stem), showing the scattered vascular bundles. *Upper right:* Enlarged view of vascular bundle. *Lower right:* Enlarged cross-section area.

growing point, parenchyma cells continue to divide and enlarge, thus allowing for lateral stem enlargement. This is termed diffuse secondary growth because no actual lateral meristem is involved.

Stem Forms

When most people think of the stem of a plant, they envision the upright portion that bears branches, leaves, flowers, and fruits. Stems come in other forms, too. For example, certain fruit trees, such as apples, cherries, plums, and pears, bear flowers and fruits each spring on persistent shortened stems called spurs. Stems can also grow horizontally, as in a pumpkin or cucumber vine. Some species of plants have under-

ground stems; only a small portion of the stem shows above ground for a relatively short period in the spring. These are the so-called bulbous plants. The white (Irish) potato plant (*Solanum tuberosum*), ready for harvest, exemplifies two kinds of stems: the above-ground stem that bears the leaves and flowers, and an underground stem whose terminal portion swells into a tuber as it accumulates starches and sugars from photosynthesis in the leaves (Fig. 8–31). Just like other stems, the white potato tuber has buds (eyes) that sprout, when planted, to form new above-ground stems.

A **rhizome** is an underground stem that grows horizontally. Examples of plants with rhizomatous stems are bananas, cannas, certain irises, certain bamboos, and some grasses, such as quack grass, Johnson grass, and Bermuda grass.

Figure 8–31 A potato plant at time of harvest. The tubers are stem structures growing below ground whose terminals have stored sufficient photosynthate to swell into tubers.

Stolons are stems that grow horizontally above ground. Sometimes called runners, stolons can develop roots in the soil at every node or at every other node (strawberry). Examples of species with stolons are ajuga, Bermuda grass, and some ferns.

Corms are thickened compressed stems that grow underground (Fig. 8–32). Buds on corms sprout to produce upright stems, which bear leaves and flowers. Gladiolus, crocus, freesia, and ixia are some examples.

Bulbs are highly compressed underground stems to which numerous storage leaves (scales) are attached. These highly developed stems provide a means for some species to survive the cold of winter and the dry soil of summer. One or more buds on the bulb sprout in the spring to produce an elongated stem with leaves and flowers. Through photosynthesis, the leaves produce carbohydrates, which are translocated to the bulb for storage. The storage of food in bulbs is a mechanism to enable the species to survive through periods of unfavorable climatic conditions. Hyacinths, lilies, onions, and tulips are examples of bulbous plants.

Stem **tubers** are the enlarged, fleshy, terminal portions of underground stems. The white potato, as discussed previously, is a good example (Fig. 8–31).

Leaves

Leaves develop in a complex series of events closely associated with stem development. Some of these developmental events are still not clearly understood. Leaves are initiated by

Figure 8–32 A gladiolus corm, which is thickened compressed stem tissue, toward the end of the growing season. The spring-planted corm (below) has shriveled with a new corm (above) forming above the old corm. New small cormels are forming at the base of the new corm.

the apical shoot meristem. Their prescribed pattern, position, and shape are influenced to some extent by their environment. For example, the leaves of cacti are adapted to growth in the desert, whereas leaves of ferns are adapted for growth in a rainforest. Most monocots, such as the grasses and palm trees, have strap-shaped leaves with parallel veins and interveinous connections between major veins. The veins contain sheaths of vascular bundles including xylem and phloem elements. **Palisade** and **spongy mesophyll parenchyma cells,** containing chlorophyll for photosynthesis, surround these veins (Figs. 8–33 and 8–34).

The leaves of dicotyledonous plants vary considerably in size and shape (Fig. 8–35). Practically all have veins arranged in the shape of nets. The large primary veins divide into smaller secondary veins. The veins are made up of xylem and phloem connected to all segments of the leaf. The spongy **mesophyll** parenchyma (Figs. 8–33 and 8–34) contains the intercellular spaces through which carbon dioxide, oxygen, and water pass. The outside layer or skin of the leaf is largely made up of epidermal cells. This epidermal layer contains openings or pores called **stomata,** each surrounded by two **guard cells.** There are generally more stomata in the lower epidermal layer of the leaf than in the upper epidermal layer. Water lilies, however, are an exception to this pattern.

The primary function of leaves is photosynthesis; a secondary function is transpiration. The guard cells, which occur in pairs on both sides of the stomata, control the opening and closing of the stomata through which carbon dioxide, one of the raw materials for photosynthesis, enters the plant, and oxygen, a product of photosynthesis, is released. Water also enters or escapes through the stomata. The loss of water from the leaf by evaporation is a process called **transpiration.** Transpiration helps regulate leaf temperature and provides the force that draws water into and through the xylem. Some plants have modified leaf surfaces that affect the rate

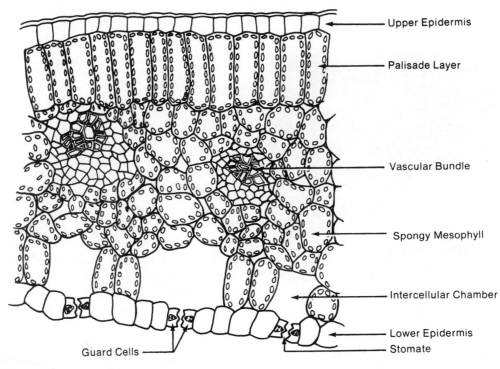

Figure 8–33 Cross section through a lily leaf showing the tissues involved in photosynthesis, transpiration, and translocation.

Figure 8–34 Three-dimensional cutaway view of an apple leaf showing the relation of cells in the various tissues. *Source:* Eames, A. J., and L. H. MacDaniels. 1947. *An introduction to plant anatomy.* 2nd ed. New York: McGraw-Hill.

Figure 8–35 Examples of leaf patterns, with descriptive names. The top two rows are the shapes of simple leaves; the center rows are simple leaves cut to depict leaf tips or bases; one row up from the bottom illustrates examples of leaf attachments or leaf edges; the bottom row shows examples of compound leaves. These characteristics are used to describe leaves in plant identification keys.

of transpiration. The leaves of some plants, such as cabbage, have a thick waxy surface (**cuticle**) that greatly reduces water loss. In the leaves of other plants, the epidermal cells produce elongated hairs that reduce the wind velocity at the leaf surface, thus reducing the transpiration rate. Some kinds of plants, especially those native to hot dry climates, such as the olive tree, minimize water loss by having stomata sunken deep in the epidermal layer.

Plants often have leaves modified to perform functions other than photosynthesis and transpiration. For example, the leaf stipules of the black locust (*Robinia pseudoacacia*) are modified to become thorns that aid in protecting the plant. Some viny plants, like the grape, have leaves modified in the form of tendrils that help support the vine when trained on trellises.

In most dicotyledonous plants, the leaf is made up of the **blade,** the flat thin part; the stemlike **petiole,** which attaches the blade to the stem; and, in some plants, the **stipules** at the base of the petiole. Some leaf blades are attached directly to the stem and lack a petiole or stipules. These are termed **sessile** leaves.

Leaf shapes among the many plant species vary greatly, and a special morphological terminology describes the leaf shape, margin, tip, and base. Leaves are usually classified (see Fig. 8–35) as **simple** (a single leaf) or **compound** (one with three or more leaflets). Distinguishing between simple and compound leaves can be difficult. The best test is to examine the base of the petiole. A true leaf has a bud in this location; a leaflet does not. A compound leaf resembling a feather is termed **pinnate;** one resembling the palm of a hand, **palmate. A trifoliate** compound leaf has three leaflets, as in the bean plant.

The shapes of simple leaves are described as linear, oblong, elliptical, lanceolate, deltoid, and so forth. The shapes of the tips and bases of simple leaves are also categorized as cordate (heart-shaped), sagittate (arrow-shaped), auriculate (ear-lobed), and sheathing, as in the grasses (Fig. 24–2). Leaf edges or margins range from entire (smooth), dentate (tooth like), serrate (sawlike) to lobed (rounded edges). These leaf characteristics aid in plant identification.

All gymnosperms native to North America and Europe have needlelike leaves and are evergreen. Some gymnosperms native to Australia, South Africa, and other parts of the world are broad-leaved plants.

Buds

Plant stems generally produce buds in the axils of leaves at the nodes or terminally on shoots. Buds usually do not occur on roots. A **bud** can be defined as an undeveloped shoot or flower, largely composed of meristematic tissue, and generally protected by modified leaf scales. Buds include: (1) vegetative buds, which develop into a shoot; (2) flower buds, which open to produce a flower or flowers; and (3) mixed buds, which open to produce both shoots and flowers. Cut-

ting through a bud longitudinally reveals the miniature parts of either a stem growing point or, in a flower bud, all the miniaturized parts of a flower.

Buds are especially prominent in winter on deciduous plants when the leaves have fallen. A leaf scar, where the leaf petiole was attached, is visible just below each bud. Buds may occur opposite each other on a stem or in an alternate arrangement around the stem.

Buds are initiated by terminal growing points as shoots elongate during the growing season. Some buds continue to grow after they are formed, developing into shoots. The growing points in other buds remain dormant until the following spring. Some buds may remain latent for long periods of time and become embedded in enlarging stem tissue; these become latent buds.

Adventitious buds can develop in places where buds generally do not form, such as buds arising on root pieces when root cuttings are made.

Buds of deciduous woody species usually go into a physiological resting or quiescent state shortly after they are formed in the summer and stay that way until they are subjected to low-temperature winter chilling, which overcomes their resting condition and enables them to resume growth the following spring. Buds of tropical and subtropical plants, however, generally do not develop such a resting condition.

Flowers

Flower buds form by the differentiation of vegetative buds into flower parts. In the angiosperms, specialized floral leaves borne and arranged on the stem are adapted for sexual reproduction; these are the **flowers.** After fertilization portions of the flower develop into a **fruit,** which bears the **seed(s).**

The flowers may be borne at the apex of a stem, as in the sunflower, rose, and poinsettia, or in the axils of the leaves lower down on the stem, as in the tomato, fuchsia, and many of the temperate zone fruit trees. Flowers or **inflorescences** vary in shape and form among the species, a fact that aids in identifying a plant's **species, genus,** and **family.** As with stems, botanists classify flowers in a specialized morphological terminology (see Figs. 8–36 and 8–37).

Complete Flowers

Complete flowers usually have four parts—sepals, petals, stamens, and pistil—which are usually borne on a receptacle (Fig. 8–36). The **sepals** are the leaflike scales that encircle the other flower parts, as in the carnation and rose. Most often the sepals are green, but sometimes they are the same color as the petals, as in tulips and lilies. The sepals may fold back (rose) or remain upright (carnation) as the petals grow and emerge. The sepals collectively are called the **calyx.** The **petals** are the next whorl of floral leaves inward from the sepals. The collective term for petals is **corolla.** The petals are usually brightly colored with some yellow, and they often contain **nectaries** that secrete nectar to attract insects,

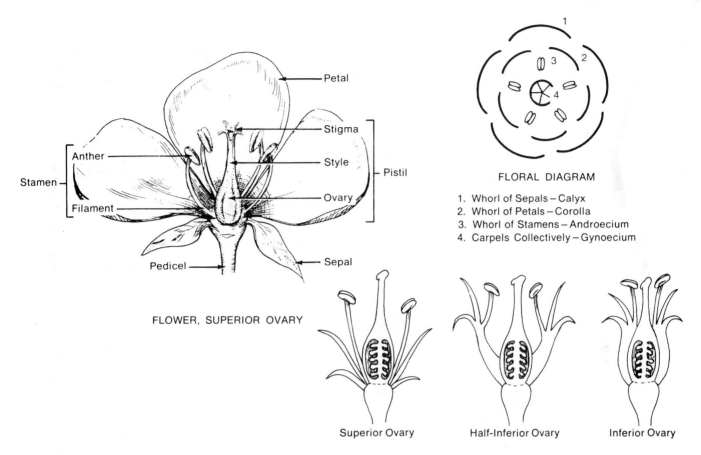

FLORAL DIAGRAM

1. Whorl of Sepals – Calyx
2. Whorl of Petals – Corolla
3. Whorl of Stamens – Androecium
4. Carpels Collectively – Gynoecium

FLOWER, SUPERIOR OVARY

Superior Ovary   Half-Inferior Ovary   Inferior Ovary

Figure 8–36 Types of flower patterns in angiosperms, showing the relative position of the ovary in relation to accessory structures.

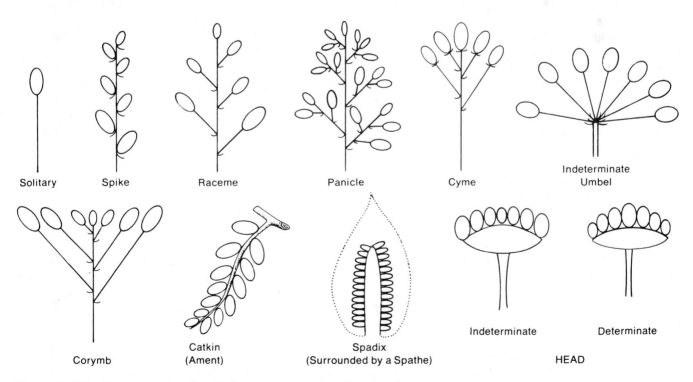

Solitary   Spike   Raceme   Panicle   Cyme   Indeterminate Umbel

Corymb   Catkin (Ament)   Spadix (Surrounded by a Spathe)   Indeterminate   Determinate   HEAD

Figure 8–37 Some typical stylized inflorescences used to describe plants.

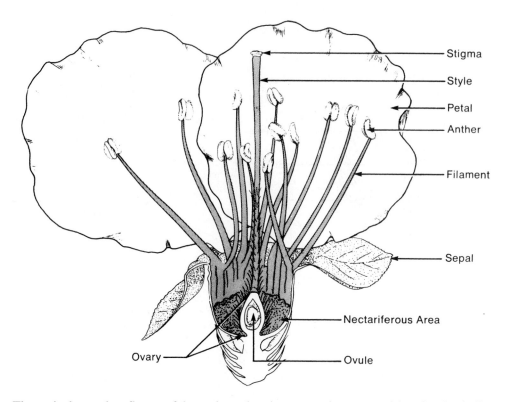

Figure 8–38   The typical complete flower of the apricot, showing a superior ovary with a simple pistil. *Source:* USDA.

which pollinate the flowers. Sepals and petals collectively are called the **perianth,** as in bulbous plants.

The next whorl of floral organs in a complete flower is the male part, or **stamen.** Each stamen consists of a **filament** and an **anther;** the anther produces the **pollen.** A group or whorl of stamens is the **androecium.**

The **pistil,** the central female component of the flower, is composed of three parts: the **stigma,** the receptive surface that receives the pollen; the **style,** a tube connected to the stigma; and the **ovary,** attached to the lower end of the style. The ovary contains undeveloped **ovules** that are attached to a **placenta;** the ovules develop into seeds after pollination and fertilization. The pistil can be simple (i.e., has but one carpel) or compound (i.e., has two or more fused carpels). Collectively, the carpels are known as the **gynoecium.** The apricot and the apple (Figs. 8–38 and 8–39) are examples of complete flowers with a simple and a compound pistil, respectively.

The stamens and pistils are considered the essential parts of the complete flower for sexual reproduction. The sepals (calyx) and the petals (corolla) are accessory flower parts.

Incomplete Flowers

Incomplete flowers lack one or more of the four parts: sepals, petals, stamens, or pistil. Flowers with both stamens and pistils are called **perfect flowers.** Flowers with stamens only and no pistils are called **staminate flowers;** those with pistils but no stamens are called **pistillate flowers.** Staminate or pistillate flowers are by definition **imperfect flowers.** Plants

having both staminate and pistillate flowers borne on the same plant are termed **monoecious** (e.g., alder, corn, walnut) (Fig. 8–40). If the pistillate and staminate flowers are borne on separate individual plants (male and female plants), the species is called **dioecious.** Examples are date palm, kiwifruit, gingko, pistachio, and asparagus.

Some flowers, like the tulip, are borne singly on a stalk and are called solitary, but others are arranged in multiples or in clusters known as **inflorescences** (see Fig. 8–37). The **corymb** is a short, flat-topped flower with an indeterminate cluster that continues to produce flowers until conditions become unfavorable; the lower flowers open first. An example is the cherry. The **cyme** resembles the corymb, except that the central or topmost flower is the first to open. Examples are chickweed and strawberry. The **raceme** is a single elongated indeterminate arrangement of stalked flowers, found in the mustard and cole crops (*Brassicaceae*), for example. The **spike** is an elongated, simple, indeterminate inflorescence with sessile (no stalk) flowers, as in wheat, oats, and gladiolus. The **catkin** is a spike with only pistillate or staminate flowers exemplified by alder, poplar, walnut, and willow. The **panicle** is an indeterminate branching raceme found in many of the grasses. The **umbel** is an indeterminate, often flat-topped, cluster of flowers that are of equal length and arise from a common point, as in carrots, dill, and onions. A **head** is a short dense spike; daisies and sunflowers have heads. A **spadix** is a complete densely flowered structure surrounded by a spathe (calla lily in the *Araceae* family).

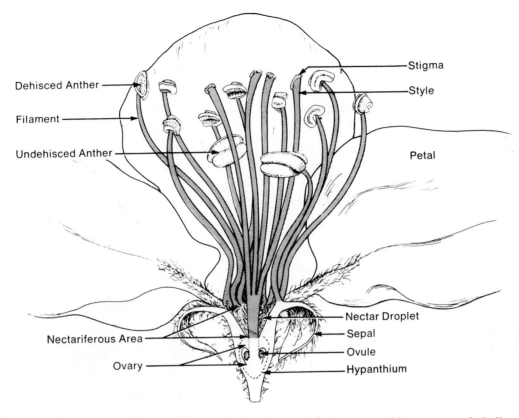

Figure 8–39 The typical complete flower of the apple, showing an inferior ovary with a compound pistil. *Source:* USDA.

Figure 8–40 Flowers of a monoecious species, the walnut. *Left:* Female flowers. *Right:* Male flowers, or catkins, are borne in structures separate from the female flowers on the same plant. The female flowers are wind pollinated.

Fruits

A fruit is a matured ovary plus associated parts; it is generally a seed-bearing organ, but there are parthenocarpic fruits that are seedless. The fruit protects the seed in some plants and helps disseminate it. For example, the seeds of fleshy fruits like peaches or plums occur inside their pits, which are discarded by animals or humans eating the fruit. The seeds in the discarded pits may eventually germinate and grow. One might tend to think of all fruits as fleshy organs, but there are many dry fruits such as nuts, capsules, legume pods, and grains. Fruits develop after pollination and fertilization. Flowers are self-pollinated or cross-pollinated by wind or insects. The pollen grows from a pollen grain on the stigma through the style and fertilizes the egg, causing the fruit to develop. Fruits may consist of a single carpel, as in beans and peas, or a combination of several carpels, as in apples (Fig. 8–39) and tomatoes. The fruit matures quickly, in a matter of weeks in the case of some summer annuals and strawberries, or it requires as long as nine months, as with oranges. The ovary wall, which is called the **pericarp,** can develop into different structures. The peel of an orange is part of the pericarp. The pod of a pea, or the shell of a sunflower seed, and the skin, flesh, and pit of a peach are all derived from the ovary wall or pericarp.

Simple Fruits

Simple fruits have a single ovary formed from one flower. The most common classification of simple fruits categorizes them as fleshy, semifleshy, or dry by the texture of the mature pericarp.

Fleshy Fruits   The entire pericarp and accessory parts develop into succulent tissue.

Berry. A pulpy fruit from one or more carpels that develops few to many seeds. Examples are bananas, dates, grapes, peppers, tomatoes, and papayas.

Hesperidium. A fruit with several carpels with inner pulp juice sacs or vesicles enclosed in a leathery rind; for example, orange, lemon, lime, and grapefruit.

Pepo. A fruit formed from an **inferior ovary** that develops from multiple carpels bearing many seeds. The pericarp is a thick and usually hard rind. Cucumbers, melons, squashes, and watermelons are examples.

Dry-Fleshy Fruits   Some parts of the pericarp become dry and the other portions remain succulent.

Drupe. This is a simple fruit derived from a single carpel. The **exocarp** (the outer layer) becomes the thin skin; the **mesocarp** (the middle layer) becomes thick and fleshy; the **endocarp** (the inner layer) becomes hard and stony and is often referred to as the pit (and often erroneously as the seed). Peaches, plums, cherries, apricots (Fig. 8–38), almonds, and olives are examples of drupe fruits.

Pome. This is a simple fruit made up of several carpels. The outer (and edible) portion forms from an accessory structure, the hypanthium of the flower, which surrounds the multiple **carpels.** Apple (Fig. 8–39), pear, and quince are examples.

Dry Fruits (Papery or Stony)   The entire pericarp is dry at maturity.

Dehiscent fruits. These fruits split at maturity to expose the seeds.

Legume or pod. A fruit from a single carpel which usually dehisces along both sutures. This fruit is typical of the pea family (*Fabaceae*), such as peas and beans.

Capsule. The fruits form from two or more carpels, each of which produces many seeds. Splitting can occur in several different ways. Iris, poppy, and jimson weed are examples.

Follicle. The fruits form from a single carpel that splits along one suture. *Delphinium* and *Helleborus* are examples.

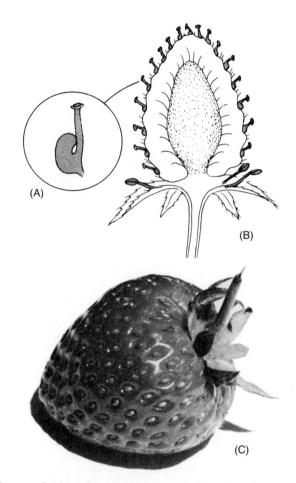

Figure 8–41 Strawberry "fruit." The true strawberry fruits on the left (A) are the small achenes borne on the surface of the large fleshy receptacle (B). The complete edible part is shown in (C).

Silique. Fruits form from two carpels with a septum between. The two halves separate longitudinally, exposing the seeds on a central membrane. Mustard, *Lunaria,* and stocks are examples.

Indehiscent fruits. These fruits do not split open when mature.

Achene. Simple, one-seeded, thin-walled fruit attached to an ovary wall. Achenes are very often mistaken for seeds as in the case of strawberry "fruits" (Fig. 8–41), the so-called seeds of the rose hip, and sunflower fruits.

Caryopsis (grain). A one-seeded fruit with a thin pericarp surrounding and adhering tightly to the true seed. Corn, rice, wheat, and barley are examples.

Nut. A one-seeded fruit with a thick, hard, stony pericarp. Oak (acorn), chestnut, filbert, walnut, and hickory are examples.

Samara. A one-seeded (elm) or a two-seeded (maple) fruit with a winglike structure formed from the ovary wall. Ash is another example.

Schizocarp. A fruit formed from two or more carpels that split at maturity to yield two one-seeded halves. Carrots, dill, caraway, and parsley are examples.

Aggregate and Multiple Fruits

Aggregate and **multiple fruits** form from several ovaries. The true fruits are attached to, or contained within, a receptacle or an accessory structure.

Aggregate fruits develop from many ovaries on a single flower. The strawberry, for example, has many achenes (true fruits), each attached to a single fleshy receptacle (Fig. 8–41). The many achenes of the rose hip develop inside the receptacle. The blackberry and raspberry are similar to the strawberry except that individual small drupes, instead of achenes, are attached to the fleshy receptacle.

Multiple fruits develop from many individual ovaries fused into a single structure borne on a common stalk. The fig "fruit" we eat is made up of small drupes (the true fruits) contained inside a fleshy receptacle. The whole structure is termed a syconium. The pineapple is a large accessory structure covered with seedless (parthenocarpic) berries. Mulberries are multiple drupelets borne on a fleshy receptacle.

Seeds

Seeds vary considerably in size, shape, structure, and mode of dissemination. The seeds contained in samaras take to the air on their small wings. The downy tufts of milkweed and dandelion seeds enable them to be carried great distances on wind currents. The coconut is known to have floated to new land masses on ocean currents. Some seeds are attached to barbs or hooks or are contained within burrs or siliques that can catch in clothing or animal fur and be transported great distances. Squirrels and other rodents bury nuts, many of which germinate later. Birds and other animals disseminate seeds by eating the fruit and passing resistant seeds through their digestive tract. Some kinds of seeds are forcefully expelled from dehiscing dried fruits as they mature. These varied means of seed dissemination aid in species survival by promoting plant growth in new locations with possibly different environments.

A seed is a mature ovule. The three basic parts are the embryo, the food storage tissue (endosperm, cotyledons, or perisperm), and the seed coats.

The **embryo** is a miniature plantlet formed within the seed from the union of the male and female gametes during fertilization. Basically, the embryo has two growing points: the **radicle,** which is the embryonic root, and the **plumule,** which is the embryonic shoot. One or two **cotyledons** are located between these two growing points on the root-shoot axis.

Food can be stored in the endosperm, cotyledons, or perisperm in the form of starch, fats, or proteins. Stored food in the cereal grains is largely endosperm. Seeds having a large portion of their food stored as endosperm are called **albuminous seeds.** Those seeds with no endosperm or only a thin layer surrounding the embryo are called **exalbuminous seeds.** Such seeds store food either in fleshy cotyledons, as in beans, or occasionally in the perisperm (developed nucellus), as in beets.

The seed coverings are usually tough, preventing damage to the enclosed embryo. They are also relatively impervious to water to save the embryo from desiccation, but again there are exceptions. There may be one or two **seed coats** (**testa**), which form from the **integuments,** the outer layers of the ovule. In dry fruits such as achenes, samaras, and schizocarps, the fruit adheres closely to the seed coat; consequently the true fruit is usually called a seed. The pits or stones of such drupe fruits as peaches and apricots are often called seeds; however, they are actually the hardened inner wall (endocarp) of the pericarp that surrounds the seed.

The scar that remains after breaking the seed from the stalk is called the **hilum,** and the small opening near the hilum is the **micropyle.** The ridge on the seed is the **raphe** (Fig. 8–4).

SUMMARY AND REVIEW

Plants are made up of cells and subcellular organelles including, but not limited to, the nucleus, mitochondria, chloroplasts, vacuoles, and endoplasmic reticulum. The organelles of the plant cell and the surrounding fluid matrix called the cytoplasm are contained within the plasmalemma. Surrounding the plasmalemma is the cell wall.

Groups of cells that perform a collective function are called tissues. Some of the major tissues that make up a plant include meristems, pith, epidermis, cambium, xylem, phloem, palisade layer, and spongy mesophyll. Tissues can be grouped together to form complex systems of organs such as roots, stems, leaves, flowers, and seeds.

Each cell, tissue, and organ has a specific function. Cells are the site of DNA and RNA synthesis, respiration, photosynthesis, protein synthesis, and other biochemical processes.

Tissues are collections of cells that have a specific function. Tissues can be simple or complex. Simple tissues are the meristems, epidermis, and parenchyma. Meristematic tissues, including apical, subapical, intercalary, and vascular cambium, are the regions of cell division (mitosis) and contribute to the growth of the plant by adding more cells. The epidermis is the exterior layer of living cells that protects all parts of the plant and includes the guard cells of the stomata. Some epidermal cells produce hairlike projections called trichomes. Parenchyma tissue is made up of thin-walled cells that are found in all parts of the plant, and they perform different functions depending on their location. These cells can become meristematic and regenerate the same or different tissues when a plant is injured.

Complex tissues are the vascular tissues, xylem and phloem, which move liquids and dissolved solids throughout the plant. Xylem tissue is nonliving and can be composed of vessels, tracheids, and fibers. Phloem tissue is living and is composed of sieve tubes, sieve tube members, companion cells, fibers, and parenchyma.

Plant organs are groups of tissues organized to perform very complex functions in the plant. They are the roots, stems, leaves, flowers, fruits, and seeds.

Roots can be tap or fibrous in form. They anchor the plant and provide the mechanism to bring water and nutrients into the plant. They are often the storage site of photosynthesized food. Parts of the root include the root cap, epidermis, root tip, cortex, endodermis, procambium layer, and stele (comprised of the vascular cylinder and pericycle).

The stem makes up the support system of the plant and supports the leaves, flowers, and fruit. Stem tissues vary from one plant to the next, but most, if not all, stems have an epidermis, cortex, vascular system of xylem and phloem and other tissues, and vascular cambium (not found in monocots). Stems can have different forms besides that usually associated with the aerial portion of a plant. These forms include rhizomes, stolons, corms, bulbs, and stem tubers.

Leaves are the site for most of the photosynthetic activity in a plant. Leaves also transpire large quantities of water to help regulate plant temperature and provide the force for carrying xylem contents through the plant. Leaves are comprised of epidermis, vascular bundles, palisade, and spongy mesophyll cells. Quite often, leaf shape is a distinguishing characteristic of a plant.

Flowers are the sexual reproductive structures of angiosperms. Flower parts include sepals, petals, stamens, and pistils. When a flower has all four parts, it is called complete. Flowers missing one or more parts are called incomplete. The stamens and pistils are the male and female reproductive organs, respectively. Stamens are composed of anthers (site of pollen production) and filaments. Pistils are composed of stigma, style, and ovary (site of ovule production). Perfect flowers have both stamens and pistils. Imperfect flowers lack one or the other. Monoecious plants have both male and female flowers on the same plant. Dioecious plants have female flowers on one plant and male flowers on another. Flowers can occur singly on a stalk (solitary) or there can be multiple flowers per stem (inflorescence).

Fruits are matured ovaries plus associated parts. Generally fruits are seed-bearing, but some fruits are parthenocarpic and are seedless. The ovary wall (pericarp) can develop in many different ways and often is a significant portion of the fruit. Fruits can be simple (developing from one ovary), such as berries (grapes and tomatoes), drupes (peaches and cherries), pomes (apples and pears), pods (beans), achenes (strawberries), caryopsis (grains), nuts, and others. Multiple and aggregate fruits form from several ovaries.

FOOD FOR THOUGHT

1. What is the main plant part used in each of the following: corn flakes, cotton T-shirt, wooden two-by-fours, spinach salad, banana pudding, baked potato, pickled beets, ornamental cabbage, strawberry jam?

2. How does an intercalary meristem explain why we can mow grass but not stop the growth of the leaf?

3. The age of trees is often determined by counting the annual rings. Explain why annual rings are made up of xylem and not phloem tissue.

4. Explain why you would or would not expect plants that evolved in wet climates, such as tropical or temperate rainforests, to have leaves with sunken stomates and a thick layer of cutin.

5. Leaf and bud shapes are often two of the most distinguishing features of plants and are key factors in their identification. What other anatomical features do you think might be useful in distinguishing one type of plant from another?

SUPPLEMENTARY READING

RAVEN, P. H., R. F. EVERT, and S. E. EICHORN. 1998. *Biology of plants,* Sixth Edition. New York: Freeman.

TAIZ, L., and E. ZEIGER. 2002. *Plant physiology.* Third Edition. Sunderland, Mass.: Sinauer.

CHAPTER

9

Stages of Growth and Development

James D. Metzger

KEY LEARNING CONCEPTS

After reading this chapter, you should be able to:

♦ Recognize definitions and measurements of plant growth and development.
♦ Understand factors that affect plant growth.
♦ Recognize the categories of plant hormones and understand the role of hormones in plant growth and development.

VEGETATIVE GROWTH AND DEVELOPMENT

Shoot and Root Systems

In living plants, we see primarily the shoot system, in all its diverse patterns—from the mosses to the magnificent towering redwoods, oaks, and pines. In many crop plants grown for livestock forage—alfalfa, corn for silage, and pasture grasses—the entire shoot system is harvested. A beautiful lawn results from the shoot and leaf growth of millions of small grass plants. A timber tree cut for lumber is the product of many years of shoot growth. The much-admired potted foliage plants in our homes are the shoot and leaf systems of particular kinds of plants brought into cultivation many years ago after their discovery by plant explorers in tropical lands.

In growing most plants—with the exceptions, perhaps, of bonsai[1] plants and container-grown ornamentals—we are interested in obtaining as much

---

[1] *Bonsai* is a Japanese term describing the art of dwarfing and shaping trees, shrubs, or vines grown in containers by careful pruning of roots and tops.

vigorous **vegetative growth** as we can as quickly as we can. This is particularly true for crop plants, where the difference between a profitable year and a loss for the farmer can be the amount of shoot growth harvested or the amount of leaf area produced to nourish the developing flowers, fruits, grains, or seeds.

The roots of plants are largely unseen and tend to be forgotten, but in the higher plants they are essential for growth. The four principal functions of roots in higher plants are:

1. Anchoring plants in the soil.
2. Absorbing water and mineral nutrients.
3. Conducting water and dissolved minerals, as well as organic materials.
4. Storing food materials, for example, in plants such as sweet potatoes, sugar beets, and carrots.

The roots of some plants can also function in vegetative reproduction, as in root cuttings.

The root system and the shoot system tend to maintain a balance. As the top of the plant grows larger and larger, the leaf area increases and water loss through transpiration increases. This increased water loss is made up by water absorption from an increasing root system. The enlarging shoot system also requires greater amounts of minerals that are absorbed by the increasing root system.

Definitions and Measurements

What is plant growth and how is it measured? We generally think of growth as an irreversible increase in volume or dry weight. The swelling of wood after it becomes wet is not growth because the wood will shrink upon drying. Growth can be measured as increases in fresh weight or dry weight, or in volume, length, height, or surface area. As its gross size increases, a plant's form and shape change as directed by genetic factors. Plant growth is a product of living cells, with all their myriad metabolic processes. A definition of plant growth is as follows: size increase by cell division and enlargement, including synthesis of new cellular material and organization of subcellular organelles.

Plant shoot growth can be classified as **determinate** or **indeterminate.** In the case of determinate growth, after a certain period of vegetative growth, flower bud clusters form at the shoot terminals so that most shoot elongation stops. Many vegetable species and cultivars grow this way, remaining small bush plants; bush snap beans are an example of a determinate type. Plants with indeterminate growth bear the flower clusters laterally along the stems in the axils of the leaves so that the shoot terminals remain vegetative and the shoot continues to grow until it is stopped by senescence or some environmental influence. Trailing pole beans, grapevines, and forest trees are examples of plants with an indeterminate growth pattern. The determinate, bush-type plants produce much less vegetative growth than do the indeterminate type.

Shoot Growth Patterns: Annuals, Biennials, and Perennials

Annuals, which are herbaceous (nonwoody) plants, complete their life cycle (seed to seed) in one growing season. Shoot growth commences after seed germination and continues in a fairly uniform pattern—provided no environmental influences are limiting—until growth is stopped by frost or some senescence-inducing factor. Flowering, followed by fruit and seed production, occurs at intervals through the summer. General growth curves for the annuals are shown in Figure 9–1, a detailed growth curve for barley, an herbaceous annual, is shown in Figure 9–2.

Figure 9–3 shows various events in the life cycle of a typical **angiosperm** annual plant. All these events occur during a single summer growing season.

The **biennials,** which are herbaceous plants, require two growing seasons (not necessarily two years) to complete their life cycle (seed to seed). As shown in Figure 9–4, stem growth is limited during the first growing season, the shoot system remaining mostly as a rosette. The plants remain alive but dormant through the winter. Exposure to chilling temperatures triggers hormonal changes leading to stem elongation, flowering, fruit formation, and seed set during the second growing season. Senescence and death of the plant follows shortly thereafter. Examples of herbaceous

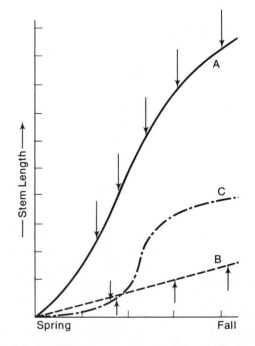

Figure 9–1 Vegetative growth patterns of annual plants. (A) Indeterminate vine-type plants. (B) Determinate, bush-type plants. (C) Terminal-flowering plants, such as cereals and grasses. Arrows indicate times of flower initiation. *Source:* Adapted from Rappaport, L., and R. M. Sachs. 1976. *Physiology of cultivated plants.* Davis, Calif.: UCD Bookstore.

Figure 9–2 Growth curve of a field-grown barley plant from leaf emergence to grain maturity. ○ = plant height. ● = dry weight of plant minus grain weight. ◉ = dry weight of plant plus grain weight. *Source:* Adapted from Noggle, G. R., and G. J. Fritz. 1976. *Introductory plant physiology.* Englewood Cliffs, N.J.: Prentice-Hall.

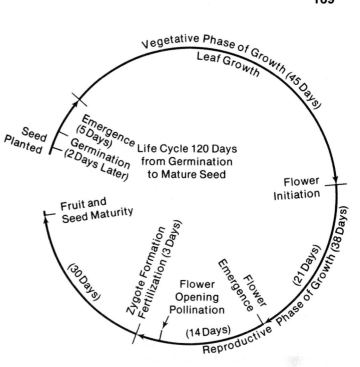

Figure 9–3 Events in the life cycle of a typical annual plant—from seed planting to seed maturity—accomplished in four months. *Source:* Adapted from Noggle, G. R., and G. J. Fritz. 1976. *Introductory plant physiology.* Englewood Cliffs, N.J.: Prentice-Hall.

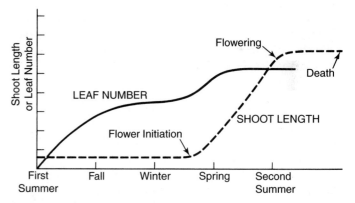

Figure 9–4 Growth curves of biennial plants during the first growing season (vegetative growth only), a required winter chilling-period, and a second growing season (flowering, fruiting, and seed production). *Source:* Adapted from Rappaport, L., and R. M. Sachs. 1976. *Physiology of cultivated plants.* Davis, Calif.: UCD Bookstore.

biennials are celery, Swiss chard, beets, and cole crops such as cabbage and Brussels sprouts.

Most annual and biennial plants flower and fruit only once before dying. The production of the flowers and fruits or, perhaps, just the flowering stimulus itself apparently causes the plants to senesce and die. In such plants, continued removal of flowers and fruits often delays senescence.

Perennials are either herbaceous or woody. In herbaceous perennials, the roots and shoots can remain alive indefinitely but the shoot system may be killed by frosts in cold-winter regions or by senescence-inducing factors. Shoot growth resumes each spring from latent or **adventitious** buds at the crown of the plant. Figure 9–5 shows typical growth patterns for two types of herbaceous perennials. Many perennial ornamental plants (e.g., pelargo-

niums) are grown as annuals in areas with severe winters, such as the Midwest region of the United States. When grown in areas with mild winters such as parts of the South and West (United States), these plants exhibit their perennial characteristics.

In woody perennial plants, both the shoot and root system remain alive indefinitely, each growing to the ultimate

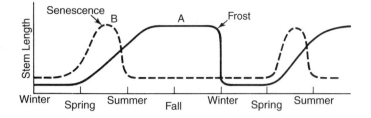

Figure 9–5 Shoot growth patterns of herbaceous perennials over a two-year period. (A) Plants such as garden (outdoor) chrysanthemums, peony, and phlox whose growth is stopped by cold weather in the fall. (B) Bulbous plants such as tulips, narcissus, and hyacinths, whose growth is terminated after spring flowering. *Source:* Adapted from Rappaport, L., and R. M. Sachs. 1976. *Physiology of cultivated plants.* Davis, Calif.: UCD Bookstore.

size for the particular plant as programmed by its gene complement. Shoot growth of temperate zone plants takes place annually during the growing season, as indicated by Figure 9–6, adding to the growth accumulated in previous seasons. The magnitude of growth can vary considerably from season to season under the influence of several environmental factors.

In the tropics some of the evergreen tree species continually grow vegetatively throughout the year, but other species exhibit intermittent growth, sometimes correlated with changes in weather, especially rainfall. Vegetative growth of coffee trees, for example, is much greater in the wet season than in the dry.

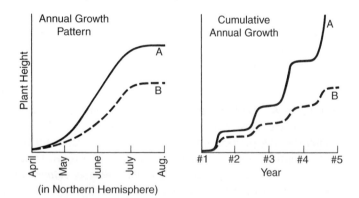

(in Northern Hemisphere)

Figure 9–6 (*Left*) Growth patterns for temperate zone woody perennials in the northern hemisphere during one growing season, and (*Right*) over a period of several years. (A) Curve for a rapid-growing species such as poplar. (B) Curve for a slow-growing species such as oak. *Source:* Adapted from Rappaport, L., and R. M. Sachs. 1976. *Physiology of cultivated plants.* Davis, Calif.: UCD Bookstore.

In the temperate zones all trees show intermittent annual vegetative growth. These growth patterns can be placed into four categories:

1. A single flush of terminal shoot growth during the growing season followed by the formation of a resting terminal bud; oaks, hickories, many conifers, and most fruit tree species are examples.
2. Recurrent flushes of terminal growth with terminal bud formation at the end of each flush; examples include the pines of the southeastern United States (*Pinus taeda* and *P. virginiana*) and certain subtropical evergreen and deciduous tree species such as citrus and Persian walnut.
3. A flush of growth followed by shoot tip abortion at the end of the season, with the shoot the following season starting up from a lateral bud. This gives rise to a zigzag pattern of shoot development, exemplified by the elm, birch, willow, beech, and honey locust.
4. A sustained growth flush for varying periods, producing new growing points that develop as late-season leaves, and ending the season with the formation of a distinct terminal bud; examples are the tulip tree (*Liriodendron tulipifera*) and sweet gum (*Liquidambar styraciflua*).

There is a pronounced diurnal variation in shoot growth of woody perennials. Shoots tend to grow more rapidly at night than in the day provided temperatures are favorable. This difference is probably due to a lower water stress at night than during the day. Some species show twice the growth at night compared with day growth.

Root Growth Patterns

Various studies of root growth patterns conclude that the roots of mature deciduous woody perennials start to grow in the spring before bud break and, in some species, stop growth at various times in the summer, but in others continue to grow until after leaf drop in the autumn. Growth peaks in the spring and again in late summer or early autumn. Roots of some species, however, continue to grow even during the winter months whenever soil temperatures and moisture are favorable. Not all roots of a tree may be growing at any one time. While some are actively growing, others may be quiescent. The spring flush of root growth results from the accumulated foods stored in the tree the previous year. When this source is depleted, root growth slows but, following gradual accumulation of carbohydrates from photosynthesis through the summer, root growth again increases in the autumn. It does not appear that the growing points of the roots are under the same control as the shoot buds, which in many woody perennials go into a resting period in late summer and require chilling through the winter to be reactivated.

How the Plant Grows

The importance of meristems in the growth of plants must be clearly understood. In dicotyledonous plants, vegetative buds at the shoot tips (**apical meristems**) and in the axils of leaves contain meristematic cells that are capable of dividing and redividing, by mitosis and cytokinesis,[2] producing millions of cells along a longitudinal axis. Cell division, together with cell elongation, causes the shoots to grow. A growing point at a shoot tip, with its meristematic region producing new cells year after year for many hundreds of years, can account for the towering height of the redwood, Douglas fir, and other forest trees. Similar meristematic cells are also located in the root tip, just behind the root cap.

The thickening of the trunks of growing trees and other woody perennials is due to secondary growth produced by another meristematic region, the vascular cambium, a cell layer that lies between the xylem and the phloem and encircles the tree from the roots to almost the top of the shoot. Cells of this vascular cambium also have the capability of dividing, producing both the permanent woody xylem tissues toward the inside that give the tree its girth and mechanical strength and the more fragile transitory phloem cells to the outside. External to the vascular cambium is another meristematic region, the cork cambium, which produces the cork in the bark layer.

Genetic Factors Affecting Plant Growth and Development[3]

One of the marvels of biology is the manner in which the offspring of plants and animals resemble, yet differ from, their parents. A peach tree, not a cherry tree, grows from a germinated peach seed. In a field planted with Marquis wheat seed, Marquis wheat plants develop, not oats or barley. The organism developing by cell division and elongation from the fertilized egg—the zygote—in every case is under the genetic control of the genes inherited from the parents at the time of fertilization. The genes, which direct the form and shape of the organism, are passed on to the daughter cells. All of them are on the chromosomes in the nucleus of each cell of the organism.

As the plant enlarges from the fertilized egg or zygote to its mature size, many developmental processes take place. Genes, which are discreet segments of DNA located on chromosomes, direct the synthesis of proteins through the process of transcription and translation using the information encoded by the specific sequences of DNA bases. (For detailed information, see Chapter 14.) Genes that encode for enzymes or structural proteins such as the mitotic spindles are termed **structural genes,** while **regulatory genes** encode for special proteins called **transcription factors** that regulate the activity of structural genes and other regulatory genes. The protein products of some regulatory genes are also known to regulate the activity of enzymes directly. At any given time, some of the organism's genes are transcriptionally active, while others are silent. The control of gene activity depends on the cell type, environmental conditions or the particular stage of development. As development proceeds, some genes become activated-switched on, whereas other genes are inactivated. The control of development through selective gene activation and deactivation is mediated mostly (but not entirely) through the action of transcription factors that turn on or turn off structural genes.

What are the signals that trigger the action of the regulatory genes? Although not clearly understood, they are believed to include growth regulators, certain inorganic ions, coenzymes, and other metabolites. Environmental factors such as temperature or light can also function as signals during certain developmental stages. Thus, the particular combination of genes directs the form and size that each plant is finally to assume, as altered by stresses from environmental influences, either beneficial or deleterious.

Environmental Factors Influencing Plant Growth and Development

Light   The sun is the source of energy for photosynthesis and other plant processes, but a substantial amount of radiation is lost because of absorption and refraction as it passes through the atmosphere. The gases of the atmosphere absorb radiation in a wavelength-dependent manner. Ozone in the upper atmosphere serves as a protective shield for living organisms by absorbing most of the harmful ultraviolet radiation. Other atmospheric gases such as carbon dioxide and water vapor absorb long-wave or thermal radiation. These gases play a pivotal role in the heat balance of the earth and its atmosphere, which in turn drives climate and weather. The atmospheric gases do not absorb very much of the sun's radiation with wavelengths of 400 to 700 nm; this is notable because this band of radiation, which is commonly referred to as **visible light** or **photosynthetic active radiation (PAR)**, is the most important to life on earth.

Radiation passing through the earth's atmosphere can be refracted, or scattered, resulting in significant reflection of light back into outer space and thus reducing the amount of light available for photosynthesis. One type of light scattering is due to an interaction of light waves with individual molecules of the atmosphere. This type of scattering is wavelength dependent and is the basis for the blue color of the sky. Blue light is more likely to be scattered than yellow or red light because it has a shorter wavelength. As a result of the increased scattering, much of the blue light reaches the earth's surface via an indirect route, so it appears that blue light is coming from a source other than the sun.

---

[2]Cell division occurs in two parts: (1) mitosis, where one nucleus is divided into two nuclei, and (2) cytokinesis, which is the division of the resulting binucleate cell into two uninucleate cells. The latter is accomplished by the formation of the cell plate, a thin layer of polysaccharide materials that develops across the spindle between the daughter nuclei toward the end of mitosis, thus dividing the cytoplasmic constituents of the mother cell.

[3]For a discussion of some basic concepts, see Chapter 14.

There is another type of scattering that is due to a combination of airborne particles such as dust, water droplets, and ice crystals in clouds. This type of scattering is wavelength independent, which means that all wavelengths are scattered equally well. Thus, clouds are typically white, unless they contain dust particles that absorb more light than is scattered. Scattering due to clouds substantially reduces light intensities at the earth's surface.

In addition to absorption and scattering, the amount of light reaching the earth's surface depends on the **angle of incidence** of the sunlight, that is, the angle that a beam of sunlight, makes with the earth's surface. Another term for the angle of incidence of sunlight is the sun's altitude. The lower the sun is in the sky (i.e., the lower its altitude), the lower the intensity of the sunlight. The sun's altitude is determined by the time of the day, the day of the year, and the latitude. On any given day, the intensity of sunlight begins to increase at dawn, when the altitude is zero; it reaches a maximum when the sun is at its highest point at solar noon; and then mirrors the decline in solar altitude as dusk approaches. In addition, the intensity at any time of the day varies with the day of the year. Because the earth is tilted so that its axis of rotation (the line through the earth connecting both poles) makes an angle of 23.5° with the plane of the earth's orbit around the sun, the sun's rays strike the earth more directly in the summer than in the winter. In other words, at any given time of the day, the solar altitude, and therefore the intensity of sunlight, is highest during the summer months.

The angle of incidence of sunlight is also affected by the geographical position relative to the equator. The higher the latitude above either the Tropic of Cancer (23.5°N) or the Tropic of Capricorn (23.5°S), the lower the sun's altitude is at any time on any given day. Thus, even on June 21, the summer solstice in the northern hemisphere, when the sun's rays are the most direct, the light intensity steadily declines as one travels north from Orlando, Florida, to Fairbanks, Alaska.

The angle of incidence of sunlight has two major effects on crop growth. First, light intensity is directly related to crop productivity. As light intensity increases, so does photosynthetic activity until the photosynthetic machinery is saturated. Typically photosynthetic activity is fairly tightly correlated with the daily change in the sun's altitude; that is, photosynthetic rates are low at dawn, they peak around solar noon, and they gradually decline as the day progresses toward sunset.

Not only does the angle of incidence affect the intensity of light, it also affects the amount of light entering the leaf and available for absorption by chlorophyll. Light striking a surface has two possible fates: it may penetrate the surface, where it is either absorbed by or transmitted through the material, or it can be reflected. The proportion of incident light that is either absorbed or transmitted by (or reflected from) a surface varies with the angle of incidence. As the angle of incidence decreases, a greater proportion of the incident light is reflected, which explains why so much more sunlight is reflected around sunrise and sunset than at midday.

Light that is reflected from the surface of a leaf is lost for photosynthesis. Some plants such as soybean can compensate for the reduced light intensity and increased reflection at low solar angles through **heliotropic movements,** in which leaf angles are adjusted so that the sun's rays are normal, or perpendicular, *to the leaf* during most of the day. Heliotropic movements are also used to lower the amount of light absorbed by reducing the angle of incidence of sunlight on the leaf when overheating might occur.

Another characteristic of light that is important for plant processes is quality. **Light quality** does not refer to how beneficial the light is for the plant, but rather to the relative quantity of individual component wavelengths contained in an incident beam of light. Better terms for light quality are *spectral composition* or *spectral distribution,* but these have not yet gained widespread acceptance by plant scientists. The band of wavelengths that affect plant processes range from about 380 nm to 800 nm, but individual processes such as photosynthesis have much more narrow requirements for the specific wavelengths that are most effective. For example, blue and red light at 440 and 650 nm, respectively, are much more effective in driving photosynthesis than is green light. Other light-controlled processes have different spectral requirements.

Light also affects plant growth and development through photomorphogenesis. **Photomorphogenesis** describes several highly integrated processes that are regulated to produce the shape or form of the plant, and they include seed germination in light-sensitive seeds, de-etiolation (the greening of young seedlings when they emerge from the soil), and stem growth in plants competing for light with other plants. A characteristic that distinguishes most photomorphogenic responses from photosynthesis is the relative insensitivity to light intensity. In fact, most photomorphogenic responses are fully induced by light intensities that are below the compensation point for photosynthesis. In addition, photomorphogenic responses typically have more specific requirements for the spectral composition of the incident light than does photosynthesis.

Most photomorphogenic responses are regulated by the phytochrome pigment system. Phytochrome is a pigment that has two interconvertible forms: a red light absorbing form and a far-red light absorbing form. Far-red light is the region of the spectrum with wavelengths between 700 and 1,000 nm, but the far-red absorbing form of phytochrome absorbs maximally in the region of 720 to 740 nm. As shown in Figure 9–7, a phytochrome molecule in the red absorbing form is converted to the far-red absorbing form following irradiation with red light. Irradiation with far-red light is required for the phytochrome molecule to be converted back to the red absorbing form. As a consequence, the relative amounts of the far-red absorbing form compared to the red absorbing form are proportional to the ratio of red to far-red

Figure 9–7 How red and far-red light affect phytochrome. Phytochrome$_{red}$ can absorb only red light, while phytochrome$_{far-red}$ can absorb only far-red light. The wavelength of light that is most effectively absorbed by the phytochrome molecule is signified by λ_{max}. Slightly longer and shorter wavelengths are also absorbed, but less effectively.

(R:FR) light in the environment. During most of the day, the R:FR ratio is about 1.0 to 1.2, so roughly speaking, about two-thirds of the total pool of phytochrome is in the far-red absorbing form. This is important because the far-red absorbing form resulting from irradiation with red light leads to biological action, while the red absorbing form is inactive. Thus, plants have a highly sensitive system for sensing changes in the light environment by continuously measuring the R:FR ratio. The R:FR ratio declines dramatically as light penetrates through leaf canopies because red light is efficiently absorbed by chlorophyll, while the far-red light is either transmitted or reflected. In some dense canopies, for example in a corn field, R:FR ratios can be as low as 0.05. The R:FR ratio of light impinging on an individual plant is highly dependent on how close and, of course, how big are the neighboring plants. The information contained in the R:FR ratio provides an accurate means to sense neighboring plants and to respond so that it can compete for resources, especially light.

Plants compete for light by redirecting growth and development so that they produce leaves that are above the leaves of their neighbors. It is no surprise then that plant architecture is highly dependent on the R:FR ratio under which the plant grows. Plants grown in light conditions in which the R:FR ratio is high (for example, open sunny areas) are typically compact, with dark green leaves and more branches and tillers. Plants respond to a decline in the R:FR ratio with increased height, reduced branching, and smaller stem diameters. Chlorophyll synthesis declines when the R:FR ratio is low, so plants appear chlorotic even though soil fertility is adequate. These changes in plant architecture result in tall, spindly plants that are more prone to lodge (fall over) under typical field conditions and are more easily damaged during shipment. Such plants are also more susceptible to disease and environmental stress. Reduced branching and tillering under low R:FR ratios also results in lower yields. The practical implication of the neighbor detection system is that a maximum number of plants can be grown in an area; any number above the maximum means that yield or quality are adversely affected.

Another pigment system mediating photomorphogenic responses is the **cryptochrome** system. In contrast to phytochrome, this pigment is a blue light photoreceptor. Not much is known about cryptochrome, but it appears to complement the functions of phytochrome in providing the plant with more complete information about the light environment in which it is growing.

Phototropism is a photomorphogenic response of plants to the direction of light. For the most part, aboveground organs such as stems, leaves, and flowers bend toward a unidirectional beam of light and are said to exhibit a positive phototropic response. A blue light receptor, different than cryptochrome, called **phototropin,** is responsible for sensing the direction of light. The bending in positive phototropic responses is due to increased cell growth on the side away from the light source. It is believed that the plant hormone auxin accumulates on the shaded side, promoting cell expansion. In some species, the roots exhibit a negative phototropic response; that is, they grow away from the light source.

Another characteristic of light that is important to plant growth and development is duration. **Photoperiodism** is the photomorphogenic response to seasonal variations in the amount of daylight. Numerous aspects of plant growth and development are controlled by photoperiod including flowering (which will be discussed in more detail later in this chapter); induction of bud dormancy in woody species; and the formation of vegetative propagules such as bulbs, tubers, corms, and runners (stolons).

All photoperiodically controlled processes can be categorized into three basic response types: **long-day plants (LDPs); short-day plants (SDPs);** and **day-neutral plants (DNPs);** which are photoperiodically insensitive (there are actually more response types, but the three described here represent the vast majority of species). The designation as a long- or short-day plant is not based on the absolute length of the day, but rather if the photoperiodically controlled process is induced only at daylengths longer or shorter than specific daylength, called the **critical daylength (CDL).** A plant with photoperiodically controlled process that is induced only when the days are *longer* than the critical daylength is considered an LDP for that process, while an SDP represents the inverse situation: the process is induced only when the daylength is shorter than the CDL. It is important to understand that no direct relationship exists between the response type and the absolute length of the CDL. For example, consider the flowering response of a typical LDP and an SDP. Red clover is an LDP with a CDL of twelve hours, meaning that it will not flower unless the daylength is longer than twelve hours. In contrast, the CDL for the garden chrysanthemum, which is an SDP, is fifteen hours, considerably longer than red clover, a so-called long-day plant. In general, whether or not a plant is an SDP or LDP determines when the process is induced relative to the summer solstice. In other words, long-day responses are initiated prior to the longest day of the year, while short-day responses are initiated when the days begin to shorten, so that the shorter the CDL, the later in the summer or early fall the process is induced.

Next to flowering, the collection of integrated processes known as the autumn syndrome observed in many woody plants is the most visible photoperiodic process. The autumn syndrome describes a series of processes that occur in plants in temperate climates at the end of the summer in preparation for winter; these processes include acquisition of freeze tolerance; dormancy of buds; and, in deciduous trees, leaf fall. Dormancy is a temporary cessation of growth and is often accompanied by the formation of bud scales, modified leaves that protect the delicate shoot tips from desiccation during winter.

Dormancy is induced by short days in several familiar trees and shrubs, including red maple (*Acer rubrum*), redbud (*Cercis canadensis*), American elm (*Ulmus americana*), eastern hemlock (*Tsuga canadensis*), and weigela (*Weigela florida*). However, there are some notable exceptions in which dormancy and other aspects of the autumn syndrome are not controlled by photoperiod. Some examples include ash (*Fraxinus* spp.), mountain ash (*Sorbus* spp.), and common fruit trees such as apple (*Malus*) and pear (*Pyrus*).

The critical daylength for dormancy induction can vary by several hours among members of a species, especially when individuals from different latitudes are compared. This variation maximizes the number of geographical areas a species can colonize. Table 9–1 provides a comparison of the approximate critical daylengths for the induction of bud dormancy in various woody plants at three different latitudes. Typically, the higher the latitude, the longer the CDL for the induction of bud dormancy (and other short-day responses as well). The longer CDL at higher latitudes results in increasingly earlier dates at which buds begin to go dormant, and this process is directly related to earlier onset of winter.

In addition to flowering and bud dormancy, other developmental processes are controlled by photoperiod. Many of these processes are related to strategies for survival in unfavorable environmental conditions such as the cold temperatures of winter or the hot, dry conditions of summer. Many herbaceous perennials survive winter with the formation of tubers that are protected by being buried underground; some examples of plants in which tuber formation

is a photoperiodically controlled process are potatoes, dahlias, yams, Jerusalem artichokes, and tuberous begonias. In all these cases, tuber formation is a short-day process that is initiated at the end of the summer. The formation of vegetative rosettes at the base of certain herbaceous perennials and adventitious crown or root buds are often short-day induced processes that are specially adapted to survive winter.

In contrast, processes relating to vegetative reproduction are often induced in late spring and early summer by long days to take full advantage of the growing season. Examples include runner formation and growth in the strawberry plant and bulb formation in onion, garlic, and related species. (Note that bulb and corm formation in spring blooming ornamental species like tulip, hyacinth, and crocus are *not* under photoperiodic control.) In commercial onion production, it is important to match the critical daylength of a cultivar with the latitude in which it will be grown. Table 9–2 provides a comparison of the critical daylength for bulb formation in different onion cultivars. It is obvious that some of these cultivars (e.g., Yellow Zittau) would exhibit poor bulb formation if grown, say, in Orlando, Florida (28°N), in which the longest day of the year is about fourteen hours.

Temperature   The seasonal variation of light intensity is responsible for the temperature changes from summer to winter in the various temperature zones. Farmers depend on climatic records in their areas to predict the last day of frost in the spring and the number of available growing days before the first killing frost in the fall. The greater the distance from the equator, the fewer the number of available growing days to mature crops. It is possible to grow temperature-sensitive tomatoes in Alaska or northern Europe, but precautions must be taken to protect the seedlings from frost in early spring. But once summer arrives in these northern latitudes, the days are so long and the temperatures are warm enough that plants grow, flower, and set fruit rapidly. The growing season may be short, but plants develop quickly.

TABLE 9–1	Approximate Critical Daylength and Day of the Year with That Daylength for the Induction of Bud Dormancy in Various Woody Plants Collected at Different Latitudes in Scandinavia

56°N Latitude		63°N Latitude		70°N Latitude	
CDL	**Date**	**CDL**	**Date**	**CDL**	**Date**
14 hours	Sept. 1	16 hours	Aug. 16	20 hours	Aug. 5

Source: Adapted from Håbjørg, A. (1988). *Horticultural Science 23*(3): 539. American Society of Horticultural Science, Alexandria, Va.

TABLE 9–2	The Effect of Daylength on Bulb Formation in Different Cultivars of Onion*

Cultivar	Approximate Critical Daylength for Bulb Formation
Yellow Bermuda	12 hours
California Early Red	13 hours
Ohio Yellow Globe	13.5 hours
Italian Red	14 hours
Sweet Spanish	15 hours
Yellow Zittau	16 hours

* Cultivars bred to be grown at lower latitudes have a shorter critical daylength because maximum daylengths in the summer are shorter than at higher latitudes.

Source: Magruder, R., and H. A. Allard. 1937. *Journal of Agricultural Research.* USDA-ARS.

All plants have optimal temperatures for maximum vegetative growth and flowering, as noted for plants discussed as crops in subsequent chapters. Most temperate-region plants grow between temperatures of 4°C (39°F) and 50°C (122°F), but these are generally the limits of plant growth. The high temperatures destroy the protoplasm of most cells; however, some spores and seeds can withstand the temperature of boiling water for short periods. At the low temperatures, most plants just fail to grow because of a lack of cell activity. However, there are some arctic or mountain plants that function near freezing, but these are rare exceptions.

Plant parts are injured by very high temperatures, even if the exposure is short. Leaves may be **solarized** or sunburned when exposed to high light intensities. In the leaf, light energy converts to heat, which destroys the cells. Young trees in orchards are prone to sunscald, which kills the cambium layer just under the thin bark of the trunk and limbs. Injury can be prevented by whitewashing the bark to reflect the heat. Occasionally, plants suffer heat damage when the relative humidity drops suddenly because of hot, desiccating winds. Heat damage in greenhouse-grown plants can be prevented if the relative humidity is increased by misting in the immediate vicinity of the plant.

The most common low-temperature injury is evident after the night of the first killing frost in the fall. Plant surfaces may be coated with frost depending on the dew point temperature, but soon after the sun shines on the leaves, the damage is evident in blackened leaves. The contents of the cells are damaged by the ice and, on thawing, the water in the protoplasm moves to the intercellular spaces, the protoplasm dehydrates, and death occurs. Frost damage to plants due to heat loss at night by radiation can be avoided to some extent by placing a shield opaque to thermal radiation (3,000 to 10,000 nm) between the plant and the clear sky on a calm night. This shield prevents radiation from the leaves to the cold sky. Citrus and grape growers sometimes use smudge pots burning oil to create heat that radiates to the trees. In locations where there are temperature inversions above the plants (slightly warmer temperatures in a layer some distance above the soil than at ground level), wind machines may be useful. The power-driven propellers mix the warm air above with the cool air below, thus preventing low-temperature injury.

Many woody perennials grown in locations with severe winters enter into a rest period—brought on by shortening days in fall—and become resistant to low winter temperatures. A covering of snow helps the plants withstand the winter because it acts as a good insulator at extremely low temperatures.

Some plants of tropical origin may be injured at temperatures just above the freezing point. This is referred to as a chilling injury. The leaves may wilt and never recover and developing fruits may not mature; some examples are avocado, banana, mango, okra, and tomato. Unripe tomatoes, which show pink color, are injured and do not finish the ripening process if refrigerated at 4°C (39°F) or less. Many ornamental foliage plants are susceptible to chilling injury. These plants need protection when being shipped during the winter from warm production areas such as Florida to colder areas. They should also be kept away from cold areas in homes and office buildings.

While we usually think of low temperatures as being injurious or as negatively affecting plant processes, low, nonfreezing temperatures are sometimes used by plants as cues to coordinate growth and development with the changing seasons. Temperatures for these cold-induced processes are usually in the range of 0°C (32°F) to 10°C (50°F). Examples of cold-induced processes include:

1. *Seed germination.* Some seeds require a period of time during which the seeds are imbibed at low temperatures (**stratification**) before germination is possible. The seeds must be imbibed because dry seeds cannot sense the cold temperatures.
2. *Flowering.* The cold induction of flowering is called **vernalization** and will be discussed in more detail later in this chapter.
3. *Dormancy breakage.* As spring approaches, bud dormancy and other processes (such as freeze tolerance induced in the fall in many woody plants) are gradually lost, while the ability to resume growth is regained. The environmental cue initiating these changes is often cold temperature. The duration required for complete loss of dormancy is called the **chilling requirement.** The length of the chilling requirement varies among species and even among cultivars of species. Plants with a short chilling requirement have an increased risk of damage if growth is initiated before the possibility of freezing temperatures occurring is gone. In fruit crops, matching the duration of the chilling requirement with local climate is an important consideration in cultivar selection.
4. *Acquisition of cold and freeze tolerance.* For many herbaceous perennials, the low, nonfreezing temperatures (0°C to 10°C) that frequently occur during autumn nights induce the physiological processes responsible for the ability to survive the freezing temperatures of winter. The freeze tolerance acquired in the autumn is quickly lost when temperature rises the following spring. Unlike the first three cold-induced processes, the duration of the cold period required to induce maximum freeze tolerance is relatively short—about one to two weeks.

Water Most growing plants contain about 90 percent water. It is stored in various plant tissues and is used as one of the raw materials for photosynthesis. A corn plant near harvest may contain as much as two liters (2.1 qt) of water, but during its growth from a seedling to a mature plant with ears, it may extract 100 times this amount from the soil. Some of this water is used in growth and development, but most of it is lost through the leaf stomates in the transpiration process.

Therefore, to produce a mature crop of corn or tomatoes, the equivalent of 30 to 60 cm (12 to 24 in.) of water must be applied to the soil surface as rain or irrigation. The quantity of water necessary depends on the crop (some plants use much more than others) as well as the available sunshine during the season. When the light from the sun is strong, the leaves tend to lose large quantities of water by transpiration and, additionally, the soil loses water by evaporation. Substantially more soil water is lost from transpiration by plants than by evaporation from the soil. The total soil water loss by both means is called **evapotranspiration.** The rate of evapotranspiration is known in many farming regions and is a guide to the farmer concerned with how much irrigation water must be applied to compensate for the soil water loss.

The plant obtains water from the soil by forces in the transpiration process. The mesophyll cells of the leaves, filled with water, are connected to the intercellular spaces that lead to the stomates; the stomates, when open, lose moisture to the air. Thus, water is pulled through the plant via the conducting tissue (xylem). If more water is lost through the stomates than the roots can supply, the plant wilts. Herbaceous plants may wilt slightly at midday or later on a bright sunny day but they usually recover during the night. Deciduous fruit plants often fail to recover from this water loss. Wilted plants eventually die if they cannot recover enough soil water to regain their turgidity. While the plant is wilted, the stomates are closed, cutting off the intake of CO_2 for photosynthesis, and thus reducing carbohydrate manufacture.

The quality of soil water, as determined by the quantity of minerals and salts dissolved in the water, is very important to the growth of plants. If there is too much dissolved salt, or if one salt, such as sodium chloride, predominates, the roots and shoots become damaged and the top of the plant suffers from lack of water. The plant may suffer from both lack of water and the toxic effects of specific ions, for example, sodium (Na) and chloride (C1). Some desert plants are resistant to high salts in water and manage to survive. In contrast, most domesticated plants require good quality water.

Agricultural production is influenced by both the quality and quantity of water available in a region of favorable temperatures. The quality of water is no problem when there is adequate rainfall. However, if the rain must be stored as runoff water in reservoirs, poor quality may result if the water accumulates too much salt before it is collected and used in irrigation.

Gases The two gases most important to the growth of green plants are oxygen (O_2) and carbon dioxide (CO_2). Green land plants and the extensive phytoplankton of the oceans help keep a balance of O_2 and CO_2 in the atmosphere throughout the world. Green plants use CO_2 for photosynthesis and return O_2 to the atmosphere. Plants and animals use O_2 and give off CO_2 during respiration. Ocean and freshwater algae account for about 90 percent of the photosynthesis in the world! These plants do most of the work to keep our atmo-

sphere in a favorable balance. Of the land plants, the forests of the world account for most photosynthetic activity.

Stomatal opening and closing is regulated in part by the CO_2 level, which is influenced strongly by photosynthesis. When the CO_2 concentration in the leaf cells behind the stomata is lower than that found in the atmosphere, stomata open. The use of CO_2 during photosynthesis keeps the concentration lower than atmospheric concentration and the stomata remain open if other conditions are favorable. (However, the whole process regulating stomatal opening and closing is affected by many factors, not just CO_2 level.)

Oxygen is important in the respiration of all plant parts. Respiration is the release of energy captured and stored in the carbohydrates (sugars) synthesized during photosynthesis. The released energy is used to drive the complex biochemical reactions needed for the growth and development of not only plants but also all living organisms.

Soils saturated or waterlogged for long periods do not contain adequate O_2 for root respiration activity. Roots cannot absorb minerals from the soil if the soil is lacking O_2 due to low respiration rates. When plants are grown with their roots in water (hydroponics), air is bubbled through the water to supply O_2 in constant small quantities so that the roots can respire.

Atmospheric pollution is a by-product of our technical society. We burn fossil fuels at a high rate for heat, transportation, and manufacturing. The burning releases the energy and CO_2 that plants (millions of years ago) stored as carbohydrates through photosynthesis. While increasing CO_2 may increase the photosynthetic productivity of crops, more CO_2 is being released than can be absorbed by plants, resulting in an increase in atmospheric CO_2, which has been associated with global warming. The loss of millions of acres of forestland has significantly reduced the amount of CO_2 that land plants can absorb. Reforestation on a large scale is being considered as a means to reduce CO_2 buildup in the atmosphere.

Other atmospheric pollutant by-products of burning include ozone, nitrogen oxide, sulfur dioxide, fluorides, and ethylene, all of which can be harmful to plants. Ozone is formed in a photooxidation process in the presence of strong sunlight. It is the major pollutant in smog and is usually found in large cities that have many automobiles. Exhausts from cars, in the presence of nitrogen oxide and abundant sunlight, produce ozone and peroxyacetyl nitrates (PAN), which harm both plants and animals. Plants affected by pollutants may have a leaf margin burn, as in some grasses, or the leaves—of petunias, lettuce, and beans—may lose chlorophyll in the center. This area of the leaf then becomes bleached or dies. Automobile exhausts also emit ethylene into the air. Ethylene is a growth regulator that can cause unusual plant growth patterns, such as leaf distortions. Sulfur dioxide results from the combustion of fossil fuels (except most natural gases) and is particularly evident in smelting processes using sulfide ores. Sulfur dioxide can cause pro-

Carbon dioxide is the third most abundant gas in the atmosphere, behind nitrogen (N_2) and O_2. In relative amounts, CO_2 is very small. Nitrogen is approximately 78 percent ($\approx$ 780,000 ppm), O_2 is approximately 21 percent ($\approx$ 210,000 ppm), and CO_2 is approximately 0.035 percent ($\approx$ 350 ppm). Atmospheric CO_2 rose 20 to 30 ppm during the 1900s. This elevation is believed to be enough to contribute significantly to global warming. Global warming results from atmospheric gases, especially CO_2, which trap heat at the earth's surface. The phenomenon is sometimes referred to as the greenhouse effect. Gases that contribute to global warming are often misleadingly referred to as greenhouse gases. As a result, greenhouses are often looked upon by the public as being contributors to global warming. Ironically, because of the plants growing in them, greenhouses most likely reduce the greenhouse effect.

TABLE 9–3	Comparison of the Duration of the Juvenile Phase in Various Species	
Species (Common Name)		**Duration of Juvenile Phase**
Chenopodium rubrum (coast blite)		0
Pharbitis nil (Japanese morning glory)		0
Perilla crispa (perilla)		1–2 months
Bryophyllum daigremontianum (bryophyllum)		1–2 years
Malus pumila (apple)		6–8 years
Citrus sinensis (orange)		6–7 years
Citrus paradisi (grapefruit)		6–8 years
Pinus sylvestris (Scotch pine)		5–10 years
Betula pubescens (birch)		5–10 years
Pyrus communis (pear)		8–12 years
Larix decidua (European larch)		10–15 years
Pseudotsuga menziesii (Douglas fir)		15–20 years
Fraxinus excelsia (ash)		15–20 years
Acer pseudoplatanus (sycamore maple)		15–20 years
Picea abies (Norway spruce)		20–25 years
Abies alba (white fir)		25–30 years
Quercus robur (English oak)		25–30 years
Fagus sylvatica (European beech)		30–40 years

Source: Metzger, J. D. 1995. "Hormones in reproductive development." In P. J. Davies (ed.), *Plant hormones,* second edition. Springer-Verlag, Germany. Used with kind permission of Springer Science and Business Media.

nounced areas of damage on leaves of sensitive plants. Fluorides can be emitted into the atmosphere by the manufacture of cement, glass, ceramics, bricks, aluminum, and phosphate fertilizers. Fluoride damage is usually seen along the margins or tips of leaves, especially in monocots.

Phase Change: Juvenility, Maturation, Senescence

A newly emerged seedling undergoes a phasic development throughout its life that is essentially the same as in animals. It will pass through embryonic growth; juvenility; a transition stage, which in plants is called phase change; maturity or **adult phase;** senescence; and death. The **juvenile phase** is characterized by the inability to reproduce sexually; in angiosperms, plants at this stage cannot flower, even though conditions are permissive. Flowering occurs only in plants that have reached the adult phase following phase change. Adult plants that have not flowered because the conditions are not proper are said to be ripe to flower. The duration of the juvenile phase varies from a week or two up to thirty or forty years in some tree species. It should be noted, however, that plants do not measure time in terms of calendar days, but rather with some factor related to an increase in plant size, probably number of nodes (leaves) produced. From a practical point of view, juvenility is a serious obstacle in breeding programs for economically important fruit and forest trees. A comparison of the duration of the juvenile phase in various plant species is shown in Table 9–3.

The morphology of juvenile and adult plants is often quite different. For example, juvenile eucalyptus trees have opposite leaves that are broad and lack a petiole. In the upper portions of adult plants, the leaves assume a different appearance: they become alternate, are narrower, and have a distinct petiole. Juvenile citrus seedlings are thorny, but thorns are produced when the plant undergoes phase change (Fig. 9–8).

The growth form of juvenile and adult plants may also be drastically different. The familiar English ivy (*Hedera helix*) in its juvenile stage has three or five leaves and a hairy vinelike stem (plagiotropic growth). Branches formed after

Figure 9–8 *Acacia melanoxylon* seedling showing phase change from juvenile to mature form. Lower, juvenile leaves have compound bipinnate structure. Upper, mature leaves are actually expanded petioles (phyllodia). Transition stages are evident in between. *Source:* Kester, D. E., and H. T. Hudson. *Plant Propagation: Principles & Practices,* 4th Edition, © 1983. Adapted by permission of Pearson Education, Inc., Upper Saddle River, NJ.

phase change exhibit a much more upright growth habit (orthotropic growth) and are hairless; leaves are nonlobed ovate and opposite. In all of these cases, the adult plant is actually a chimera, with the adult portion located with the youngest portion of the plant, while the juvenile portion is found at the base and oldest part of the plant. A summary comparing morphological features of juvenile and adult plants in various species can be found in Table 9–4. Rooting ability of cuttings is one of these features that is of significant economic importance. Plant propagators would like to maintain the juvenile stage in stock plants as long as possible so that cuttings taken from them will root properly and in high percentages.

Aging and Senescence

The life spans of the different kinds of flowering plants differ greatly, ranging from a few months to thousands of years. Some of the coniferous evergreen forest trees are the earth's oldest living organisms. Some of the giant California redwoods (*Sequoia sempervirens*) are known to be over 3,000 years old. Olive trees with huge trunks found in the eastern Mediterranean area are believed to be several thousand years old (see Fig. 9–9).

Figure 9–9 Ancient olive tree (*Olea europaea*) growing on the Mount of Olives in Jerusalem. Trees of this species are known to live for several thousand years.

There seems to be no reason why a clone could not continue to exist indefinitely if it was protected from diseases (especially viruses) and other environmental stresses and repropagated frequently. (A **clone** is genetically uniform

TABLE 9–4	Comparison of Morphological Characteristics Associated with the Juvenile and Adult States of Various Species		
Characteristic	**Species**	**Juvenile Form**	**Adult Form**
Growth habit	*Hedera helix*	Plagiotropic	Orthotropic
	Ficus punula	Plagiotropic	Orthotropic
	Euonymus radicans	Plagiotropic	Orthotropic
Leaf shape	*Cupressus* spp.	Acicular	Scalelike
	Acacia spp.	Pinnate	Phyllodes
	Eucalyptus spp.	Oval, sessile	Lanceolate with petioles
	Pinus spp.	Flat, glaucous	Scale- and bractlike
	Hedera helix	Palmate	Ovate, entire
Phyllotaxis	*Eucalyptus* spp.	Opposite	Alternate
	Hedera helix	Alternate	Spiral
Anthocyanin pigmentation in leaves	*Malus pumila*	Present	Absent
	Carya illinoisiensis	Present	Absent
	Acer rubrum	Present	Absent
	Hedera helix	Present	Absent
Thorniness	*Robinia pseudoacacia*	Thorns	No thorns
	Citrus spp.	Thorns	No thorns
Autumn leaf	*Fagus sylvatica*	Keep leaves	
Abscission in deciduous trees	*Quercus* spp.	Keep leaves	Abscise
	Robinia pseudoacacia	Keep leaves	Abscise
	Carpinus spp.	Keep leaves	Abscise
Rooting ability of cuttings	*Hedera helix*	Will root	Will not root
	Quercus spp.	Will root	Will not root
	Fagus sylvatica	Will root	Will not root
	Pinus spp.	Will root	Will not root
	Pyrus malus	Will root	Will not root

Source: Metzger, J. D. 1995. "Hormones in reproductive development." In P. J. Davies (ed.), *Plant hormones,* second edition. Springer-Verlag, Germany. Used with kind permission of Springer Science and Business Media.

material derived from a single individual and propagated exclusively by vegetative means, such as cuttings, divisions, or grafts.) Some clones are known to be very old—the Winter Pearmain apple, cultivated in England, was being grown there about 1200 A.D. The Reine Claude plum was also grown in England as long ago as 1500 A.D. The Black Corinth grape has been grown in Greece for thousands of years. Other present-day clones of plants that are easily propagated by vegetative means, such as the fig, olive, and date, were being grown in Biblical times.

Senescence is considered to be a terminal, irreversible deteriorative change in living organisms, leading to cellular and tissue breakdown and death. It is a conspicuous period of physical decline, particularly evident toward the end of the life cycle of annual plants (population senescence) and of individual plants (whole plant senescence), but it can also occur in leaves, seeds, flowers, or fruits (organ senescence). Plants exhibit senescence in different ways. In annuals, the entire plant dies at the end of one growing season, after and probably because of fruit and seed production. In herbaceous perennials, the tops of the plants die at the end of the growing season, perhaps killed by frost, but the shoot grows again in the spring and the roots can live for many years. In deciduous woody perennials, the leaves senesce, die, and fall off each year but the shoot and root systems remain alive, often for a great many years.

Senescence is usually considered to be due to inherent physiological changes in the plant, but it can also be caused by pathogenic attack or environmental stress. As individual trees age, for example, they are more and more vulnerable to lethal attacks by fungi, bacteria, and viruses. The long-lived trees mentioned above characteristically have very durable heartwood, containing high levels of resins and phenolic compounds that resist decay. As large trees get older, the ratio of foliage area to the surface area of roots, trunk, and limbs often diminishes progressively, thus lessening the amount of food the tree produces in relation to the amount it consumes in respiration. Eventually the leaves are unable to supply adequate amounts of food to nourish the tree. The reduced foliage cover for the trunk and large limbs as the trees age can also lead to sunburn damage and the death of exposed tissues.

Considerable study has been given to senescence in plants, particularly in regard to leaves and their abscission. During leaf senescence, DNA, RNA, proteins, chlorophyll, photosynthesis, starch, auxins, and gibberellins decrease, sometimes drastically. Senescence is not entirely degradation, however; particular mRNAs and proteins are synthesized only in senescing tissues.

A decline in photosynthetic activity of many determinate annuals such as wheat plants following flowering; the decline in photosynthesis, of course, soon leads to senescence and death. Plant senescence is hastened, too, by the transfer of stored nutrients to the reproductive parts—the flowers, fruits, and seeds—as they develop and mature, at the expense of the root and shoot systems. As a result, senescence can be postponed in many plants by picking off the flowers before seeds start to form. In sweet peas, for example, removing the flowers once they start to wither and before seeds form prolongs the blooming period. Just as plant hormones are involved in many plant functions, they are involved, too, in senescence. For example, ethylene plays a major role in fruit ripening and deterioration.

REPRODUCTIVE GROWTH AND DEVELOPMENT

Fruit and seed production involves several phases:

1. Flower induction and initiation
2. Flower differentiation and development
3. Pollination
4. Fertilization
5. Fruit set and seed formation
6. Growth and maturation of fruit and seed
7. Fruit senescence

Flower Induction and Initiation

Some annuals mature and can flower in only a few days or weeks after the seeds sprout; some forest and fruit trees require years before flowering. Once mature, the plant can be induced to flower by becoming sensitive to the conditions of its environment. What brings about the formation of flowers? In some species, it is daylength (photoperiodic effect) and/or low temperatures (vernalization), although in most trees neither daylength nor cold temperatures induces flowering.

Photoperiodism (Daylength)

Earlier in this chapter, we discussed the general features exhibited by photoperiodically controlled processes. In this section, we will consider in more detail the photoperiodic control of flowering. The influence of the photoperiod on the flowering of several plant species was first studied in detail by the United States Department of Agriculture (USDA) at Beltsville, Maryland, and the results were published by W. W. Garner and H. A. Allard in 1920. They grew plants in containers that could be wheeled into dark sheds at the end of the workday and returned to the sunlight in the morning. They found that Maryland Mammoth tobacco and certain cultivars of soybeans and cosmos required short days for flower induction. A set number of successive short days was required to complete differentiation (change from vegetative to reproductive growing points or shoot terminals). Once induced by short days, the plants could be moved to a long daylength without interfering with the flowering process. These plants, such as strawberry, poinsettia, and soybean, were called short-day plants. Later studies by many other workers found that long days were necessary to induce flowering of some plants, such as spinach, sugar beets, and winter barley. These were classified as long-day plants. A third group of plants, including tomato, corn, fruit trees, and cucumber, were those in which flowering was not affected by daylength and were called day-neutral plants. This phenomenon whereby daylength controls certain plant processes, as noted above, was termed photoperiodism. Other workers in later experiments found this phenomenon to be more complicated and that many plants did not fit nicely into these three categories because of interactions of daylength with temperatures.

Further studies have shown that there is a daily fluctuation in the sensitivity to light such that, depending on the time of day, light either inhibits or promotes flowering. For SDPs, light promotes flowering if it is provided between dawn and the critical daylength; light after the critical daylength prevents flowering. Long-day plants are the inverse. Light during the early part of the day actually inhibits flowering; for flowering to occur, light must be provided after the critical daylength has been reached. This cycle of sensitivity to light is one of many circadian rhythms exhibited by plants and animals alike. Circadian rhythms are biological rhythms that complete one cycle in approximately 24 hours; they are typically initiated, or entrained, by transitions between darkness and light, as occurs at dawn. (Some circadian rhythms are entrained by the changes in temperature that occur between day and night.)

The circadian rhythms of light sensitivity in SDPs and LDPs are qualitatively identical, but they differ in time of day when the promotive and inhibitory stages are maximally expressed. Figure 9–10 demonstrates how a flash of light (or night break) of sufficient intensity or duration inhibits flowering of a short-day plant (long-night plant) but may induce flowering of a long-day plant (LDP). This information has been useful to commercial chrysanthemum growers who grow these short-day plants on a year-round schedule. When they want the young plants to reach a size adequate for flowering, the growers use incandescent or fluorescent lamps over the chrysanthemum plants (near midnight), each night for one to four hours, depending on time of year and latitude. This night break inhibits flowering until the plants reach the desired height. Conversely, when the natural daylight of summer is too long for chrysanthemum plants to flower, they cover the plants of proper size with black cloth or plastic each evening about 6 P.M. and remove it in the morning at about 8 A.M. This shortens the plant's day (lengthens the night) enough to induce and fully develop the flowers. These manipulations enable chrysanthemum plants to be flowered every day of the year. Poinsettias flower the same way and could be produced year-round. However, because of the association with Christmas, the public will purchase poinsettias only for Christmas.

As is the case for other photoperiodically regulated processes, the pigment system responsible for sensing light is phytochrome. The phytochrome system is located in the leaf,

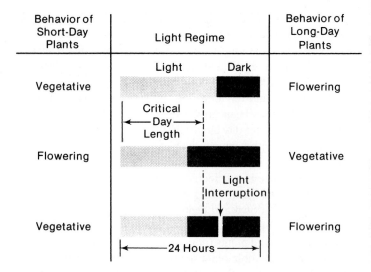

Figure 9–10   The effect of light interruption of the dark period on flowering in short-day and long-day plants. *Source:* Galston, A. W., and P. J. Davies. *Control Mechanisms in Plant Development,* 1st Edition, © 1970. Reprinted by permission of Pearson Education, Inc., Upper Saddle River, NJ.

yet it is the apical meristem that must go through a transition to produce floral organs instead of leaves. Thus, both positive and negative signals are transported via the phloem from the exposed leaves to the apical meristem, the timing of which is determined by the response type and the amount of time that has elapsed after the start of the day. Most of the attention has been focused on the positive signal, which promotes flowering. This signal is sometimes called **florigen,** or the floral stimulus. The existence of the floral stimulus was discovered by partial leaf removal when plants were placed under the inductive photoperiod for flower formation, then failed to flower. The SDP *Xanthium* (cocklebur) exposed to long noninductive days initiated flowers if light was blocked from a single leaf. Some experiments in which flowering plants (donor) were grafted to nonflowering plants (receptor) caused the latter group to flower. These and other experiments gave rise to the theory that a flowering hormone, which might be similar in all plants, was responsible for flower induction. However, the hormone will remain a hypothetical compound unless it can be isolated and characterized. Even less is known about the nature of the negative or inhibitory signal, except that its existence was postulated from grafting experiments.

There has been considerable documentation to categorize plants as short-day, long-day, or day-neutral. Although not complete, one such list appears in Table 9–5. The critical daylength of many of the species shown in the table may be changed by a slight shift in temperature above or below the optimum for that species.

Understanding photoperiodic response allows crop producers to select species and cultivars that flower and seed at the right time for their geographic location or market window.

TABLE 9–5	A Partial List of Long-Day, Short-Day and Day-Neutral Plants		
Long-Day Plants	**Length of Daily Light Period Necessary for Flowering**	**Short-Day Plants**	**Length of Daily Light Period Necessary for Flowering**
Althea (*Hibiscus syriacus*)	More than 12 hours	Bryophyllum (*Bryophyllum pinnatum*)	Less than 12 hours
Baby's breath (*Gypsophila paniculata*)	16 hours	Chrysanthemum (*Chrysanthemum × morifolium*)	15 hours
Barley, winter (*Hordeum vulgare*)	12 hours	Cocklebur (*Xanthium strumarium*)	15.6 hours
Bentgrass (*Agrostis palustris*)	16 hours	Cosmos, Klondyke (*Cosmus sulphureus*)	14 hours
Canary-grass (*Phalaris arundinacea*)	12.5 hours	Cotton, Upland (*Gossypium hirsutum*)	14 hours
Chrysanthemum frutescens	12 hours	Kalanchoe (*Kalanchoe blossfeldiana*)	12 hours
Chrysanthemum maximum	12 hours	Orchid (*Cattleya trianae*)	9 hours
Clover, red (*Trifolium pratense*)	12 hours	Perilla, Common (*Perilla crispa*)	14 hours
Coneflower (*Rudbeckia bicolor*)	10 hours	Poinsettia (*Euphorbia pulcherrima*)	12.5 hours
Dill (*Anethum graveolens*)	11 hours	Rice, winter (*Oryza sativa*)	12 hours
Fuchsia hybrida	12 hours	Soybean (*Glycine max*)	12 hours
Henbane, annula (*Hyoscyamus niger*)	10 hours	Strawberry (*Fragaria × Ananasia*)	10 hours
Oat (*Avena sativa*)	9 hours	Tobacco, Maryland Mammoth (*Nicotiana tabacum*)	14 hours
Orchardgrass (*Dactylis glomerata*)	12 hours	Violet (*Viola papilionacea*)	11 hours
Ryegrass, early perennial (*Lolium perenne*)	9 hours		
Ryegrass, Italian (*Lolium italicum*)	11 hours		
Ryegrass, late perennial (*Lolium perenne*)	13 hours		
Sedum (*Sedum spectabile*)	13 hours		
Spinach (*Spinacia oleracea*)	13 hours		
Timothy, hay (*Phleum pratensis*)	12 hours		
Timothy, pasture (*Phleum nodosum*)	14.5 hours		
Wheat, winter (*Triticum aestivum*)	12 hours		
Wheatgrass (*Agropyron smithii*)	10 hours		

Day-Neutral Plants

Balsam (*Impatiens balsamina*)
Bluegrass, annual (*Poa annua*)
Buckwheat (*Fagopyrum tataricum*)
Cape jasmine (*Gardenia jasminoides*)
Corn (maize) (*Zea mays*)
Cucumber (*Cucumis sativus*)
English holly (*Ilex aquifolium*)
Euphorbia (*Euphorbia peplus*)
Fruit and nut tree species
Globe-amaranth (*Gomphrina globosa*)
Grapes
Honesty (*Lunaria annua*)
Kidney bean (*Phaseolus vulgaris*)
Pea (*Pisum sativum*)
Scrofularia (*Scrofularia peregrina*)
Senecio (*Senecio vulgaris*)
Strawberry, everbearing (*Fragaria × Ananasia*)
Tomato (*Lycopersicon lycopersicum*)
Viburnum (*Viburnum* spp.)

Source: Modified from *Plant Physiology*, 3rd edition by Salisbury/Ross. © 1985. Reprinted with permission of Brooks/Cole, a division of Thomson Learning: *www.thomsonrights.com*. fax: 800-730-2215.

Low Temperature Induction

Some plants, including many of the biennials, require low temperature for flower induction. The term for this is vernalization, which means "making ready for spring." It was first observed in winter wheat over a century ago. Vernalization is any temperature treatment that induces or promotes flowering. The temperatures required to vernalize a given plant and the length of the vernalization period vary among species and may even differ among cultivars of the same species. Broadly speaking, however, vernalization temperatures range between 0° and 10°C (32° and 50°F). Some of the biennials that require vernalization are beets, Brussels sprouts, carrots, celery, and some garden flowers such as Canterbury bells and foxglove. Winter annuals—such as the cereal crops, barley, oats, rye, and wheat—also respond to cold by flowering. Some plants, such as lettuce, peas, and spinach, can be induced to flower earlier with vernalization, but vernalization is not an absolute requirement; they will eventually flower without it. Some species can be vernalized as seeds (beet and kohlrabi), but most plants must reach a minimum size or produce a certain number of leaves to be sensitized by the cold.

Garden perennials, plants with corms or tubers, and many flowering shrubs and fruit trees require low temperatures to overcome the rest period, but few require low temperatures for flower induction. The true bulb plants, such as the hyacinth, narcissus, and tulip, do not require a vernalization period to break rest, but low temperatures are required to promote flower development once the flower has formed within the bulb. The olive tree (*Olea europaea*) does need chilling temperatures for induction of flower parts. In the kiwifruit (*Actinidia deliciosa*), too, there is no evidence of reproductive structures in the bud until after chilling temperatures occur.

It is important to note that many plants do not respond with flower induction to changed daylength or low temperature. In fact, the majority of agricultural plants are self-inductive for flowering; that is, they initiate or form flowers at almost any photoperiod and without vernalization. Many of the garden annuals are good examples of plants that flower when they reach a certain morphological maturity. Most fruit trees, shrubs, woody plants, garden perennials (roses, carnations, gerbera) and vegetable crops (beans, peas, tomatoes, peppers, cucumbers) have self-induced flowering.

Flower Development

When an apparent floral stimulus is transmitted from the leaves to the apical meristems, a change from a vegetative to a flowering state takes place. In the case of some short-day plants, such as *Xanthium,* only one inductive short day (long night) is required, whereas chrysanthemums, poinsettias, and kalanchoe require three to four consecutive short days for induction. Once the apex has changed to a flower primordium, the process is not reversible. The floral apex may abort, however, if the subsequent environmental conditions are not favorable for full flower development. In such a case, the axillary buds below the aborted floral apex usually grow vegetatively until daylength or temperature conditions once again are favorable for flower induction.

The number of days from flower initiation to anthesis (time of flower opening) depends on the species and the cultivar. These periods of development can be modified somewhat by raising the temperatures slightly in the last third to half of the development phase. Very high temperatures in the early developmental stages, however, may cause flower abortion.

Pollination

In the production of most floral crops and flowering shrubs— for example, carnations, petunias, chrysanthemums, roses,

and camellias—the flower itself is the desired product. There is little interest in any resulting fruits and seed, except in the case of the plant breeder working with such species. But in the food crops—the cereals, fruits, and many vegetable species— the postflowering structures are the desired products. It is the grains, fruits, and seeds that are harvested.

In angiosperms, **pollination** is defined as the transfer of pollen from an anther to a stigma. The anther and stigma may be in the same flower (self-pollination), in different flowers on the same plant (self-pollination), in different flowers on different plants of the same cultivar (self-pollination), or in different flowers on plants of different cultivars (cross-pollination).

Pollen grains come in many sizes and shapes and, while essential for sexual plant reproduction, can be devastating to many people as allergy producers. See Figure 9–11.

Figure 9–12 shows the various parts of a simple flower dependent upon pollination for fruit set.

If a plant is **self-fertile,** it produces fruit and seed with its own pollen, without the transfer of pollen from another cultivar. If it is **self-sterile,** it cannot set fruit and seed with its own pollen, but instead requires pollen from another cultivar. Often this is due to **incompatibility,** where a plant's own pollen will not grow through the style into its embryo sac (see Figs. 9–12 and 9–13). Sometimes, too, cross-pollination between two particular cultivars is ineffective be-

Figure 9–11   Scanning electron micrograph of pollen grains produced in a male cone of red pine (*Pinus resinosa*). *Source:* USDA.

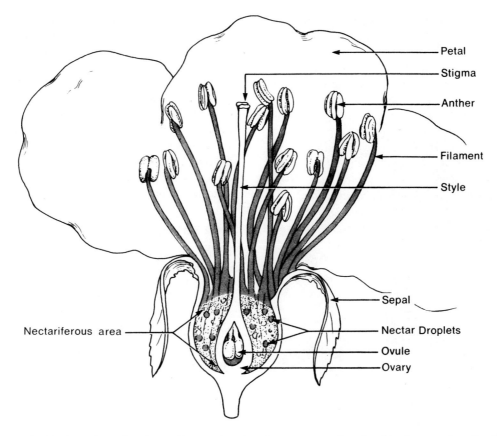

Figure 9–12   Longitudinal section of a cherry flower showing the structures involved in the transfer of pollen from anthers to stigma (pollination). *Source:* USDA.

cause of incompatibility, which is believed to be due to factors that inhibit pollen tube germination or elongation.

Pollen transfer from the anthers to the stigmas is principally by:

1. Insects, chiefly honeybees (Fig. 9–14). Insect pollination is common among cultivars with white or brightly colored flower parts and attractive nectar. Most fruit crops, many vegetables, and legume forage crops are pollinated by insects.

   Adequate pollination is so important in many crops that considerable efforts are made to aid the bees in their pollen distribution. Commercially prepared pollen is collected and tested for vigor and compatibility and applied to the crop flowers by various wind-generating devices. Also pollen inserts are placed at the entrances of hives to coat bees with pollen as they enter and exit.

2. Wind. This is the main pollinating agent for plants with inconspicuous flowers—the grasses, cereal grain crops, and forest tree species, as well as some fruit and nut crops such as the olive, walnut, pistachio, and pecan.

Other minor pollinating agents are water, snails, slugs, birds, and bats.

Figure 9–13 shows a longitudinal section through the pistil of a flower following pollination. Note the elongated

Figure 9–13   A longitudinal section through the carpel of a flower following pollination and just before fertilization.

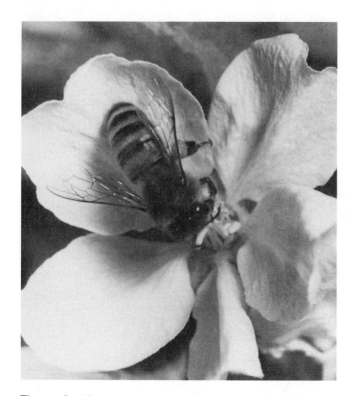

Figure 9–14   Honeybees, collecting nectar from the flowers, also cause pollination by distributing pollen from the anthers to the stigma. Bees perform a great service in the culture of many crops by their pollination activities. *Source: USDA.*

pollen tube. A pollen grain that germinated the sticky surface of the stigma has grown down through the style carrying the male gametes to the embryo sac in the ovary.

Parthenocarpy

If pollination—and subsequent fertilization—do not occur, fruit and seed rarely develop. One important exception, however, is fruit that sets parthenocarpically. **Parthenocarpy** is the formation of fruit without the stimulation of pollination and fertilization. Without fertilization, no seeds are produced; therefore, parthenocarpic fruits are seedless. There are many examples of parthenocarpic fruits such as the Washington Navel orange, the Cavendish banana, the oriental persimmon, and many fig cultivars. (Not all seedless fruits are parthenocarpic, however; sometimes, as in certain seedless grapes, pollination and fertilization occur and the fruit forms but the embryo aborts, thus no viable seed is produced.)

Fertilization

In the angiosperms the pollen tube grows through the micropyle opening in the ovule into the embryo sac and discharges two sperm nuclei (1N each). One unites with the egg (1N) to form the zygote (2N), which will become the embryo

and eventually the new plant. The other sperm nucleus unites with the two polar nuclei (1N each) in the embryo sac to form the endosperm (3N), which develops into food storage tissue. This process is termed **double fertilization.** The elapsed time between pollination and fertilization in most angiosperms is about twenty-four to forty-eight hours.

In the cone-bearing plants of the gymnosperms, the entire process of pollination and fertilization is different than in the angiosperms. Pollen-producing cones are produced on the tree separately from the ovulate cones. There is no wall enclosing the ovaries as in the angiosperms, thus producing naked seeds on the cone scales.

Fruit Setting

Following formation of the zygote—at which time the genetic makeup (the genotype) of the new seedling plant is determined—many significant changes occur, leading to the formation of the fruit and (usually) seeds within the fruit.

Accessory tissues in the flower are often involved in fruit formation, such as the enlarged fleshy receptacle surrounding the ovary wall in the apple and pear. Botanically, however, the true fruit is just the enlarged ovary.

Figure 9–15 shows various stages in the development of the different tissues in a lettuce fruit.

In many plants only a small percentage of the flowers develop into fruits. This is particularly true in fruit crops where a tree could not possibly mature as many fruits as there are flowers. Many of the flowers drop without fertilization of the egg, and many of the flowers with a fertilized egg abort at the zygote stage or later. When the zygote fails to develop and no seed forms, the immature fruit usually drops. In some seedless grapes, such as Thompson Seedless, the embryo starts to develop, then aborts; the seed fails to develop, and the fruit does not abscise but grows to full size.

Certain of the plant hormones appear to be involved in fruit setting, but the actual physiological mechanisms are largely unknown. In fruits of some species—tomatoes, peppers, eggplants, and figs—applied auxin can replace the stimulus of pollination and/or fertilization. Fruit set can also be induced in grapes, certain stone fruits—apricots, for example—and apples and pears by gibberellin sprays. Cytokinins also stimulate fruit set in grapes.

One of the chief problems in fruit production is obtaining the optimal level of fruit setting. Too low a fruit set gives a light, unprofitable crop. Too heavy a set leads to undesirable small, poor-quality fruits that mature late, possibly exhausting the tree's food supply and often resulting in little or no crop the following year. To overcome excessive fruit set, half—or more—of the fruits are removed at a very early stage, either by hand thinning, machine shaking, or chemical sprays. Interestingly, fruits of some species (the Washington Navel orange, for example) are self-thinning. Most of the tiny fruits originally forming drop, leaving an optimum number to develop to maturity.

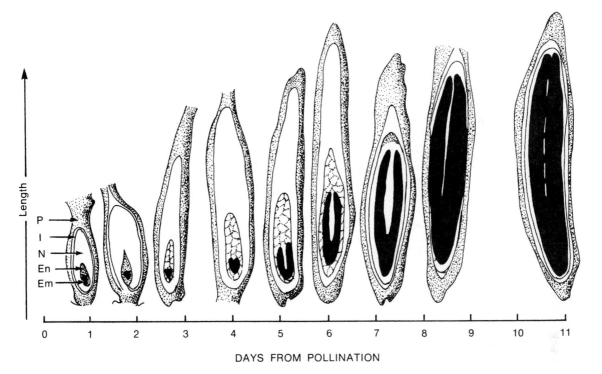

Figure 9–15   Developmental pattern of the tissues in a lettuce fruit, from the fertilized egg to the mature fruit. The ovary wall (the pericarp) is firmly attached to the seed coat (integument), so the structure is correctly considered a fruit and not a seed. P = pericarp; I = integument; N = nucellus; En = endosperm; Em = embryo. *Source:* Adapted by Hartmann, H. T., and D. E. Kester. 1983. *Plant propagation,* 4th ed. Englewood Cliffs, N.J.: Prentice-Hall. From Jones, H. A. 1927. Pollination and life history studies of the lettuce (*Latuca sativa*). *Hilgardia* 2:425–79.

As might be expected, temperature strongly influences fruit set. Temperatures that are too low or too high at this critical period are often responsible for crop failures. Low light intensity and lack of adequate soil moisture can also adversely affect fruit set.

Fruit Growth and Development

Once fruit has set, the true fruit and, sometimes, various associated tissues begin to grow. Food materials move from other parts of the plant into these developing tissues. Hormonal substances, such as the auxins, gibberellins, ethylene, and cytokinins, may be involved in some phases of fruit growth just as they are in fruit set. These materials originate in both the developing seeds and fruit, although it is significant to note the parthenocarpic fruits (without seeds) continue to grow to full size.

An interesting relationship between fruit growth and the presence of auxin has been observed in strawberry fruits. Removing some of the achenes ("seeds") from the surface of the strawberry at an early growth stage causes it to be lopsided; the strawberry fails to develop under the section where the achenes were removed. The stimulatory effect of some mobile material originating in the achenes is lost. Presumably this material is an auxin because application of auxin

paste to the area where the achenes were removed allows the strawberry to develop normally.

Evidence of the participation of gibberellins in fruit growth has been shown in the grape. Application of gibberellin to Thompson Seedless grape clusters at an early stage of berry development markedly increases the ultimate fruit size. The size increase is so pronounced that almost all table grapes of this cultivar grown in California are now treated with gibberellin. This effect on size also holds true for certain other grape cultivars.

While various plant hormones may be involved in fruit growth, the basic mechanisms are still barely understood. During flower development and in the early stages of fruit growth, there is considerable cell division. Following this period of intense cell division, most fruits increase in size because of cell enlargement.

Fruits have two basic patterns of growth, as shown in Figure 9–16. One is the simple sigmoid growth curve—typical of fruits such as the orange, apple, pear, pineapple, olive, almond, tomato, and strawberry—in which there is a slow start followed by a period of rapid size increase, then a decrease in growth rate near fruit maturity. The second pattern is a double sigmoid growth curve, in which the single sigmoid growth curve is repeated. Near the center of the

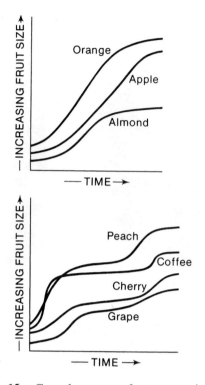

Figure 9–16   Growth curves of representative kinds of fruits showing the two characteristic types. *Top:* The sigmoid growth curve. *Bottom:* The double sigmoid growth curve. *Source:* Kester, D. E., and H. T. Hudson. *Plant Propagation: Principles & Practices,* 4th Edition, © 1983. Adapted by permission of Pearson Education, Inc., Upper Saddle River, NJ.

growth period, the growth curve is flat; the fruit increases little, if at all, in size. The stone fruits—peach, apricot, plum, and cherry—as well as the grape and fig show a double sigmoid growth pattern. In the stone fruits, which have a hard endocarp or pit, the pit hardens during the second phase of fruit development. In addition, some important changes take place in seed development within the pit, as illustrated in Figure 9–17.

PLANT GROWTH REGULATORS

In plants, as in animals, many of the behavioral patterns and functions are controlled by hormones. Hormones are produced in minute amounts at one site in the plant and translocated to other sites where they can alter growth and development. The natural hormones and other materials are essentially chemical messengers, influencing the many patterns of plant development.

A distinction must be made between the terms *plant hormone* and *plant growth regulator.* A plant hormone is a natural substance (produced by the plant itself) that acts to control plant activities. Plant hormones that are chemically synthesized can initiate reactions in the plant similar to those caused by the natural hormones. **Plant growth regulators,**

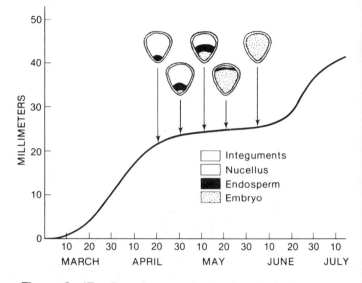

Figure 9–17   Growth curve of an apricot fruit through the growing season. During the second growth period the pit (endocarp) hardens and the seed within the pit develops mostly from nutritive tissue (nucellus and endosperm) to finally consist entirely of the embryo. *Source:* Adapted from Crane, J. C., and P. Punsri. Comparative growth of the endosperm and the embryo in unsprayed and 2,4,5-trichlorophenoxyacetic acid sprayed royal and Tilton apricots. *Proc. Amer. Soc. Hort. Sci.* 68: 96–104. 1956.

on the other hand, include plant hormones—natural and synthetic—but also other, nonnutrient chemicals not found naturally in plants but that, when applied to plants, influence their growth and development.

There are five traditionally recognized groups of natural plant hormones: **auxins, gibberellins, cytokinins, ethylene,** and **abscisic acid** (Fig. 9–18). Recently additional hormones or hormone classes, including the Brassinolides, salicylic acid, and jasmonates, have been identified. The discovery and subsequent study of plant hormones is one of the most exciting and fascinating chapters in the history of plant physiology. Despite considerable study of hormones, however, the mechanism of their actions in the plant is still not completely understood.

In addition to natural hormones, synthetic growth regulators have been developed to allow growers to manipulate plant growth and development. Synthetic growth regulators are used to promote rooting, reduce stem elongation, encourage branching, regulate flowering, and influence other aspects of growth.

Auxins

Auxins were the first group of plant hormones to be discovered. The discovery came in the mid-1930s, and for many years thereafter auxins and their activities in plants were studied intensely throughout the world.

The auxins, both natural and synthetic, influence plant growth in many ways, including cell enlargement or

Figure 9–18 Structural formulas of some natural and some synthetic plant growth regulators.

elongation, photo- and geotropism, apical dominance, abscission of plant parts, flower initiation and development, root initiation, fruit set and growth, cambial activity, tuber and bulb formation, and seed germination. Auxins operate at the cellular level, affecting activities such as protoplasmic streaming and enzyme activity. Auxins are related to many other chemical control mechanisms and are readily transported throughout the plant, principally in an apex-to-base direction (basipetally).

The natural auxins originate in meristems and enlarging tissues, such as actively growing terminal and lateral buds, lengthening internodes, and developing embryos in the seed. Auxins are produced in relatively high amounts in the shoot tip or terminal growing point of the plant and move down the plant through the vascular tissues, causing the phenomenon known as **apical dominance** (blockage of the growth of lateral buds by the presence of terminal buds). High levels of auxin in the stem just above the lateral buds block their growth. If the shoot tip supplying this auxin is broken or cut off, the auxin level behind the lateral buds is reduced and the lateral buds begin to grow. This is part of the reason why, when a shoot tip is removed, many new shoots arise from buds down along the stem.

One of the most widespread auxins that occurs naturally in plants is indoleacetic acid (IAA) (see Fig. 9–18). Several other natural auxins have also been identified, and there are others whose chemical structure is yet unknown. Many synthetic auxins induce the same effects as natural auxin. Some of these are indolebutyric acid (IBA), naphthaleneacetic acid (NAA), and 2,4-dichlorophenoxyacetic acid (2,4-D). (see Fig. 9–18).

Some important commercial uses of these synthetic auxins are:

1. *Adventitious root initiation.* One of the first responses attributed to auxins was the stimulation of root formation in stem cuttings. Two synthetic auxins, indolebutyric acid and naphthaleneacetic acid, are now widely used commercially in treating the bases of stem cutting to stimulate the initiation of adventitious roots.
2. *Weed control.* The synthetic auxin, 2,4-dichlorophenoxyacetic acid, is in widespread commercial use as a selective weedkiller that eliminates broad-leaved weeds in grass or cereal fields.
3. *Inhibition of stem sprouting.* Many kinds of woody ornamental trees produce masses of vigorous sprouts from the base of the trunk that, if not removed, would transform the tree into a bush. Continual removal of these sprouts by hand is costly and time consuming. It has been found that treatment of the tree trunks with the auxin naphthaleneacetic acid at about 10,000 ppm (1.0 percent) strongly inhibits the development of such sprouts.
4. *Tissue culture.* The initiation of roots and shoots on small pieces of plant tissue cultured under aseptic conditions has become a standard method of micropropagation of some plant species. Often an auxin, such as IAA or 2,4-D, has to be included in the culture medium for roots to initiate.

Gibberellins (GAs)

The gibberellins are a group of natural plant hormones with many powerful regulatory functions. The most obvious is the stimulation of stem growth dramatically, far more than auxins can. Gibberellins may stimulate cell division, cell elongation, or both, and they can control enzyme secretion.

In some plants, GA is involved in flower initiation and sex expression (male or female flower parts). Fruit set as well as fruit growth, maturation, and ripening seem to be controlled by gibberellin in some species. Senescence of plant parts, particularly leaves, is also affected by GA. Certain dwarf cultivars of peas and corn, if treated with GA, grow to a normal height, indicating that the dwarfed plants lack a normal level of gibberellin (Fig. 9–19).

Gibberellins are also involved in overcoming dormancy in seeds and in buds. Their role in the germination of barley seed has received much study (see Fig. 9–20). After the seed has been moistened and placed at room temperature, a natural gibberellin produced in the embryo translocates to the aleurone layer surrounding the endosperm. Triggered by the GA, cells in the aleurone layer synthesize enzymes such as amylases, proteases, and lipases. These enzymes then diffuse throughout the endosperm, hydrolyzing starches and proteins into sugars and amino acids that then become available to the embryo for its growth and development.

Figure 9–19   Overcoming dwarfness in corn by spraying with gibberellin. *Left:* Untreated, genetically dwarf corn plants. *Center:* Nondwarf corn sprayed with gibberellin. *Right:* Genetic dwarf corn sprayed with gibberellin. Photographs taken six weeks after spraying. *Source: From Plant Growth Substances in Agriculture* by Robert J. Weaver. W. H. Freeman and Company. Copyright © 1972.

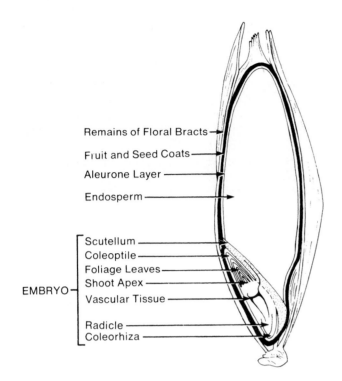

Figure 9–20 Longitudinal section of a barley seed. *Source:* From a drawing by Peter J. Davies, In: *The Life of the Green Plant* (3rd edition), by Arthur W. Galston, Peter J. Davies, and Ruth L. Satter. Prentice-Hall. Copyright 1970 and 1980.

The molecular structure of the gibberellins is well known; a typical one is shown in Figure 9–18. By 2000, more than 100 different gibberellins had been discovered in tissues of various plants. Some common ones are GA_1, GA_3 (gibberellic acid), GA_4, and GA_7.

Gibberellins were first discovered in 1926 by Japanese researchers studying a disease of rice plants caused by the fungus *Gibberella fujikuroi*. Plants infected with the fungus grew excessively and abnormally. Extracts from this fungus applied to noninfected plants stimulated the same abnormal growth. By 1939, the active ingredient was extracted from the fungus, crystallized, and named gibberellin. This early work with gibberellin in Japan went unnoticed in the Western world until the early 1950s when a great surge of gibberellin research began, particularly in the United States and England. This research led to the isolation of many different forms of gibberellin extracted from the *Gibberella* fungus and from higher plants.

Gibberellins are synthesized in the shoot apex of the plant, particularly in new leaf primordial. They are also found in embryos and cotyledons of immature seeds and in fruit tissue. In addition, the root system synthesizes large quantities of gibberellin, which moves upward throughout the plant. GA translocates easily in the plant in both directions, unlike auxin, which moves largely in an apex-to-base direction.

Pharmaceutical companies produce crystalline GA_3 as the acid or the potassium salt for research studies and certain commercial applications. These preparations are all obtained from growth of the *Gibberella* fungus in a process similar to that used to produce antibiotics.

Even though it has been demonstrated that gibberellins occur naturally in many of the higher plants, little is known of the physiological mechanisms of gibberellin action or transport.

Although gibberellins are a powerful and important group of plant hormones involved in many of the plant functions, only a few agricultural uses have been found for them. Some are:

1. *Increasing fruit size of seedless grapes.* This is the principal commercial application of gibberellin. Practically all vines of the Thompson Seedless grape grown for table use in California are sprayed each year. Berry size of other grape cultivars, such as Black Corinth, is also increased by gibberellin sprays, as shown in Figure 9–21.

2. *Stimulating seed germination and seedling growth.* Several cases have been reported where soaking seeds in solutions of gibberellic acid before germination greatly stimulates seedling emergence and growth. Such responses have been obtained with barley, rice, peas, beans, avocado, orange, grape, camellia, apple, peach, and cherry. Figure 9–22 shows the stimulation obtained with grape seedlings by gibberellin treatment.

Figure 9–21 Effect of gibberellin sprays on growth of Black Corinth grapes. (A) Untreated control. (B) Stem girdling control. (C) Plants sprayed at an early growth stage with gibberellin at 5 ppm. (D) At 20 ppm. Photos taken 59 days after spraying. *Source:* From *Plant Growth Substances in Agriculture* by Robert J. Weaver, W.H. Freeman and Company. Copyright © 1972.

Figure 9–22   Effect of gibberellin on germination and growth of Tokay grape seeds. Seeds soaked (before planting) at 0, 100, 1,000, or 8,000 ppm in potassium gibberellate solution for 20 hr. *Source:* From *Plant Growth Substances in Agriculture* by Robert J. Weaver, W.H. Freeman and Company. Copyright © 1972.

3. *Promoting male flowers in cucumbers.* When pollen is wanted for hybrid seed production, a single application of GA$_3$ to the leaves stimulates maleness of a cucumber. This has proved an important discovery for hybridizers.

4. *Overcoming the cold requirement for some plants.* Azalea plants require six weeks of cool temperatures (8°C or 46°F) to develop flower buds. Several leaf applications of 1,000 ppm gibberellin completely or partially replace this cold requirement for flower bud development. It has been shown experimentally that gibberellins applied to biennial plants that require a cold period before they flower causes early flowering. Most of these treatments, however, have limited commercial value.

Cytokinins

This group of plant hormones primarily promotes cell division but they also participate in a great many aspects of plant growth and development, such as cell enlargement, tissue differentiation, dormancy, different phases of flowering and fruiting, and in retardation of leaf senescence.

Cytokinins interact with auxins to influence differentiation of tissues. As shown in Figure 9–23, externally applied cytokinin alone stimulates bud formation in tobacco stem segments; auxin applied alone causes roots to develop, but when cytokinin plus auxin are applied together, there is a canceling effect—only masses of undifferentiated callus form.

There are both natural cytokinins, such as zeatin, and synthetic forms, such as kinetin and benzyladenine (BA) (see Fig. 9–18). There are over 100 known natural and synthetic cytokinins. Cytokinins occur in many plant tissues as both the free hormonal material and as a component of transfer RNA. They are found in abundance in embryos and germinating seeds and in young developing fruits—all tissues with considerable cell division. Roots supply cytokinins upward to the shoots.

Figure 9–23   Effects of a cytokinin and an auxin on growth and organ formation in tobacco stem segments. (A) Control, no treatment. (B) Cytokinin—bud formation but no root formation. (C) Auxin—root formation with prevention of bud development. (D) Cytokinin plus auxin—stimulation of callus growth but no organ formation. *Source:* Kester, D. E., and H. T. Hudson. *Plant Propagation: Principles & Practices,* 4th Edition, © 1983. Adapted by permission of Pearson Education, Inc., Upper Saddle River, NJ.

The mechanism of cytokinin action in the plant is not clear. Cytokinins indirectly increase enzyme activity and increase the DNA produced in some tissues. Their regulatory effects seem to result from interactions with other hormones in the plant.

Cytokinins were discovered when scientists at the University of Wisconsin in 1955 used a synthetic material—kinetin, later named a cytokinin—to cause cell division of tobacco stem pith. After many interesting physiological activities of kinetin became apparent, plants were examined for possible natural similar materials. In 1964 such a material was isolated from young corn seeds by researchers in New Zealand and was named zeatin. An active promoter of cell division known to exist in coconut milk was finally determined to be a zeatin-riboside.

Even though cytokinins are strongly involved in plant growth regulation, no important agricultural uses have been developed for them. In media for aseptic tissue culture, however, cytokinin usually must be added to induce shoot development. Applications of cytokinins to green tissue have been shown to delay senescence. These materials have also been used experimentally, and in limited commercial applications, on greenhouse roses and potted chrysanthemum plants to stimulate growth of axillary buds by overcoming natural bud inhibitors.

Ethylene

It has been well established for many years that ethylene gas (see Fig. 9–18) evokes many varied responses in plants. As long ago as 1924 it was found the ethylene could induce fruit ripening; in 1925, it was determined that ethylene could overcome bud dormancy in potato tubers; in 1931, that it could induce leaf abscission; in 1932, that it could induce flowering in pineapple plants; in 1933, that it could cause roots to form on stem cuttings. By the mid-1930s, it was determined that ethylene was itself a plant product, and arguments arose among plant scientists about whether ethylene, a gas, should be considered a plant hormone.

Little further attention was paid to possible roles of ethylene as a natural growth regulator until the development in the 1960s of gas chromatographic techniques that permit the detection of ethylene in concentrations as low as one part per billion. Vast amounts of research of ethylene physiology in the 1960s and 1970s established ethylene as a plant hormone.

Ethylene itself is a tiny molecule ($CH_2 = CH_2$), compared with the other plant hormones. The pathway for ethylene biosynthesis in plants has been fairly well elucidated. Ethylene, as a gas, diffuses readily throughout the plant, moving much like carbon dioxide, and it can exert its influence in minute quantities. Its solubility in water also enhances its movement through the plant. The cuticular coatings on external cell surfaces tend to prevent losses from the plant. Ethylene is apparently produced in actively growing meristems of the plant, in ripening and senescing fruits, in senescing flowers, in germinating seeds, and in certain plant tissues as a response to bending, wounding, or bruising. Synthetic ethylene, from ethephon, applied to plant tissues can cause a great burst of natural ethylene production—an autocatalytic effect.

Just how ethylene exerts its regulatory effects is no better known than the basic mechanisms involved in the action of the other plant hormones. One theory is that ethylene regulates some aspect of DNA transcription or RNA translation, thus changing RNA-directed protein synthesis and, consequently, enzyme patterns. But many other mechanisms are also likely to be in operation.

The possible commercial uses of ethylene were greatly increased with the development in the 1960s of ethylene-releasing compounds such as ethephon (2-chloroethyl) phosphonic acid. This compound applied as an agricultural spray gradually releases ethylene into plant tissues. In contrast to some of the other plant hormones, ethylene and ethylene-releasing chemical have several valuable commercial applications:

1. *Fruit ripening.* Ethylene gas, injected into airtight storage rooms, is used commercially to ripen bananas, honeydew melons, and tomatoes. The ethylene-releasing chemical ethephon also ripens tomatoes that are green but horticulturally mature. To harvest canning tomatoes by machine harvesting equipment, where the entire crop is picked at one time, it is important that most of the fruits be ripe and fully colored at the time of harvest. Spraying the field with an ethylene-releasing material before harvest promotes uniform red color development of the green fruits. Ethephon is used as a preharvest spray to promote uniform ripening of apples, cherries, figs, blueberries, coffee, and pineapple.

2. *Flower initiation.* Ethylene gas released from ethephon has initiated flowers in several ornamental bromeliad species, including *Ananas* spp., *Aechmea fasciata, Neoregelia* spp., *Billbergia* spp., and *Vriesia splendens.* Ethephon has been widely used to promote uniform flowering in the cultivated banana.

3. *Changing sex expression.* Ethylene application to certain plants, such as cucumbers and pumpkins, can dramatically increase the production of female flowers. Some cucumber cultivars produce both female and nonfruiting male flowers on the same plant. Spraying the vines with ethephon causes all flowers to be female, which develop into fruits and thus increase yields. This practice gives results similar to previous studies where auxins were used.

4. *Degreening oranges, lemons, and grapefruit.* Sometimes the rind of maturing oranges and grapefruits remains green owing to high chlorophyll levels, even though the eating quality, juice content, and ratios of soluble solids to acid are high enough to meet grade standards for harvest. Citrus packers can treat such fruits with ethylene at about 20 ppm for twelve to seventy-two hours. This breaks down the chlorophyll and allows the orange and yellow carotenoid pigments to show.

5. *Harvest aids.* Certain fruit and nut crops, such as sour cherries and walnuts, are harvested by mechanical tree shakers that shake the trees until the crop falls into catching frames or onto the ground to be picked up later. Often this practice is not completely successful because the fruits or nuts are so tightly attached that the tree shakers do not remove them. However, by spraying the trees about a week before harvest with an ethylene-releasing compound—ethephon, for example—the abscission-inducing effects of the ethylene result in a much higher percentage of crops removed.

6. *Growth regulation.* One of the physiological effects of ethylene application to plants is a reduction in the growth of stems and leaves. The greenhouse industry recently made use of this fact in controlling excessive growth in floriculture crops. Florel® is a special formulation of ethephon registered for this purpose. Great care must be exercised when using ethylene as a growth regulator because high concentrations of this hormone can cause leaf abscission and other deleterious effects.

Ethylene can also harm plants. It can cause **unwanted** leaf abscission and can hasten senescence of most flowers. The introduction of a few parts per billion of ethylene into the surrounding air causes carnation flowers to close, rose buds to expand prematurely, and orchid flower petals and sepals to develop a water-soaked appearance. The pollination of an orchid flower can generate sufficient ethylene to cause injury to the flower parts. Ethylene can cause flower bud abortion of bulbs during shipment. A few diseased tulip bulbs give off enough ethylene in a packing crate to stop further development of the flower buds within the bulbs. In certain commercial flowers such as carnations and geraniums, the deleterious action of ethylene may be blocked with the application of methylcyclopropene (MCP).

Abscisic Acid (ABA)

ABA was originally identified as a component of a complex of inhibitory substances associated with the dormant buds of ash trees and with substances that accumulated in abscising leaves. In fact, the original names for ABA were *dormin* and *abscisin II* because researchers believed there was a primary function for ABA in these two processes. Later, when the chemical structure (Fig. 9–18) was elucidated, dormin and abscisin II were shown to be identical compounds, and the name abscisic acid was adopted. Ironically, further research showed that ABA has no direct role in either bud dormancy or abscission. Today, ABA is recognized to have two major roles in the life of a plant. First is the regulation of processes in seed development, including the accumulation of seed proteins and the prevention of precocious seed germination, that is, germination while on the mother plant. The second role is one of a mobile stress hormone in which ABA action initiates plants' responses to cold and water stress. One of the most important stress responses mediated by ABA is the closure of stomates when the loss of water from the plant reaches a critical value. Once this critical value is reached, plant cells begin to synthesize large amounts of ABA, which is then transported to the guard cells signaling them to reduce the stomatal aperture, thus minimizing further water loss by transpiration.

ABA is synthesized in both leaves and roots. This characteristic provides the plant with a mechanism to adjust the amount of water loss through transpiration in response to the water status of not only the leaves but also the roots and surrounding soil. For example, during a hot, sunny day, leaves may begin to wilt despite adequate moisture in the soil because more water is lost than can be replaced via the xylem transport system. The reduced leaf water content results in the production of ABA by the mesophyll cells, which signals the guard cells to close the stomates until leaf turgor is restored. On the other hand, as soil moisture reserves are gradually depleted through direct evaporation to the atmosphere and transpiration, the water content of the roots declines commensurately. When a threshold level of water loss is reached, root cells produce ABA, which is transported to the leaves via the xylem, closing the stomates. Thus, plants have a very elegant system for adjusting stomatal apertures to match soil moisture content.

At present, no commercial uses of ABA exist in crop production, although considerable effort was made to see if it could be used as an antitranspirant to minimize transplant shock. The biggest obstacle for such use, besides cost, is the fact that the plant rapidly deactivates ABA, so any effect on depressing transpiration is transient.

Additional Hormones or Hormone Classes

In recent years, four additional hormones or hormone classes have been discovered, indicating that other likely plant hormones are waiting to be discovered. **Brassinolides** are steroids, closely related in structure to animal steroid hormones such as estrogen and testosterone. In plants, brassinolides appear to function in the regulation of cell division and elongation. Plants lacking brassinolides are dwarf and exhibit weak growth. A second new hormone is **salicylic acid,** which coincidentally is the biologically active component of aspirin. This hormone is an important component of plants' response to pathogen attack; it serves as a signal to activate genes involved in pathogen defense. Through a relatively minor chemical modification, plants use a derivative of salicylic acid as a form of interplant communication in an early warning system that a nearby plant is under attack by a pathogen. This modified version is the methyl ester of salicylic acid, which is wintergreen oil. Methyl salicylate is much more volatile than salicylic acid and readily evaporates from the leaves. The vaporized molecules diffuse through the atmosphere and can be absorbed by the leaves of neighboring plants. Once inside living cells, methyl salicylate is easily hydrolyzed to re-form salicylic acid, which can then induce plant defense systems without the plant actually being infected.

While salicylic acid has a major role in pathogen defense, the **jasmonates** represent a group of compounds involved in systems that defend plants against herbivores. Jasmonates are derived from fatty acids and are similar in structure to the class of animal hormones known as prostaglandins. Jasmonates are also volatile and are the major component of the fragrance associated with jasmine tea and gardenia flowers. The volatile nature of jasmonates provides a mechanism similar to that of methyl salicylate for interplant signaling of attacks by herbivores such as insects.

The last hormone is also involved in herbivore defense. **Systemin** is unique among plant hormones because it is the only one known to be a peptide (there are several animal peptide hormones—insulin is a well-known example) composed of eighteen amino acids. It is produced in tissue wounded by herbivores and is transported to remote tissues and organs, where it induces defense genes.

Synthetic Growth Retardants

A rather diverse group of growth retardants developed since about 1950 has several important commercial uses regarding ornamental plants, principally in obtaining compact, dwarf-type plants. These materials generally act by slowing, but not stopping, cell division and elongation in subapical meristems, usually without causing stem or leaf malformations.

Figure 9–24 Growth retardation in chrysanthemum plants when treated with a growth retardant, *Chrysanthemum x morifolium.* 'Circus' plants sprayed with daminozide (B-Nine) to retard shoot growth; *left:* control, no daminozide; *center:* 2,500 ppm; *right:* 2,500 ppm sprayed Aug. 14 and again on Aug. 21.

Five cuttings were planted in 6-in. (15 cm) pots on July 31, pinched (apex removed) Aug. 7, sprayed Aug. 14, given short days under black cloth on Aug. 18 to initiate flowers; the plants flowered about Oct. 15, when the photo was taken. Average heights above the soil at flowering time were: *left,* 37 cm; *center,* 31 cm; *right,* 26 cm.

The primary effect of these materials is the opposite of gibberellin, often converting a tall-growing plant into a rosette. Most act by blocking gibberellin synthesis. Plants treated with these growth retardants have a compact, scaled-down appearance, which is often more attractive than larger, untreated plants with a loose, open growth. The treated plants also often have darker, more attractive foliage and more flowers than untreated plants (see Fig. 9–24). Some of the better known synthetic growth retardants are described below:

Daminozide (succinic acid-2, 2-dimenthyl hydrazide, Alar, B-Nine). Tests have shown that daminozide effectively retards growth and stimulates flowering of several kinds of herbaceous and woody ornamental plants and enhances the size and color of various fruit species. Those plants that respond well include chrysanthemums (Fig. 9–24), various bedding plants (2,500 to 5,000 ppm), and azaleas (2,500 ppm). Sometimes two applications two or three weeks apart are required to maintain the desired dwarf form.

Chlormequat [(2-chloroethyl) trimethylammonium chloride, Cycocel, CCC]. Chlormequat is effective in retarding the height of some ornamental plants. The height of poinsettias may be controlled if chlormequat is applied as a drench to the soil or as a spray to the stems and foliage.

Ancymidol (A-Rest), α-chloropropyl- (P-methoxyphenyl)- 5-pyrimidine-methanol = ancymidol. This growth retardant is very effective for reducing the height of some bulbous and other potted ornamental crops.

Paclobutrazol [2RS, 3RS]-1-[4-chlorophenyl] 4-4-dimethyl-2-1, 4-triazol-yl-pentan-3-ol] Bonzi®. paclobutrazol and its close chemical relative uniconazole (Sumagic®) are very potent growth-retarding chemicals that effectively control the height of many herbaceous and woody ornamentals. The rates used for these chemicals are much lower than for other growth retardants.

Trinexapac-ethyl [4-(cyclopropyl-α-hydroxy-methylene)-3,5-dioxocycloheanecarboxylic acid ethyl ester] Primo®. Trinexapac-ethyl is a relatively new growth retardant registered for use in turfgrass. It is relatively specific for monocots.

> **Caution: Chemical use on plants must be in accordance with the law.** Before using any chemical, check the label to be certain that the crop you are treating, the rate you are using, and the intended use, along with any other considerations, are in agreement with the label.

> ### CHEMICAL REGISTRATION
> In the United States, the application of chemicals to plants for commercial use is strictly controlled by Environmental Protection Agency (EPA) regulations. Before the chemicals can be used legally, an EPA registration must be obtained for each crop, stating the dosage allowable and the time of year that application is permissible. Application for registration is usually made by the chemical company manufacturing the material after a patent has been obtained. Regulations for obtaining a registration for use of chemicals to be applied to food crop plants are much stricter than are those for ornamentals.

SUMMARY AND REVIEW

Growth is the increase in size of a plant by cell division and enlargment. Shoot growth can be determinate (shoot elongation ceases with the formation of reproductive structures) and indeterminate (bearing clusters of fruits and flowers along the stem). Plant growth patterns include annual (complete life cyle in one growing season), biennial (life cycle covers two growing seasons), and perennial (growth resumes each growing season for several years). Shoot growth often occurs in flushes, that is, periods of growth followed by inactive periods during a growing season. Vegetative growth is the growth of the roots and nonflowering or fruiting top growth. Reproductive growth is the growth and development of flowers, fruit, and seed.

Factors that affect plant growth can be genetic, environmental, and stage of development. Genetic factors include the overall genetic (DNA) composition of the plant and the active or inactive state of genes at any particular time. Environmental factors include light, heat, water, dissolved nutritional minerals, and atmospheric gases. Light provides the energy for photosynthesis and the accumulation of carbohydrates needed for growth. Changes in light quality or duration direct the shape of the growing plant, including the flowering of photoperiodic plants. Heat determines how fast most plants grow and develop. Most plants have an optimum growing temperature. Temperatures much below or above the optimum slow growth rate and may be detrimental or even lethal. However, low temperatures can serve as cues for the plant to coordinate growth with the changing seasons. Water is stored in cells and used in cell processes. Water and solutes are carried throughout the plant through the vascular system. Most water entering the plant leaves the plant through the transpirational stream. Water evaporating out of the leaves cools the plant through the process of evapotranspiration. The most important atmospheric gases for plants are CO_2 (needed for photosynthesis) and O_2 (needed for respiration). Stages of development include juvenile, mature, and senescing. During each period, specific types of growth and development can occur that may not occur in other stages.

There are five traditionally recognized classes of plant hormones: auxins, gibberellins (GAs), cytokinins, ethylene, and abscisic acid (ABA). Endogenous auxins influence cell enlargement, photo- and geotropism, apical dominance, and other growth traits. Auxins and related compounds are used to promote the rooting of cuttings, to kill weeds, and to promote rooting in tissue culture systems. Gibberellins can promote flower initiation and stem elongation, and can overcome dormancy in seeds and buds. Agricultural uses of GAs include increasing the size of seedless grapes and overcoming or reducing cold requirements of some plants. Most chemical growth retardants act by blocking synthesis of GAs in the plant. Cytokinins promote cell division and slow leaf senescence and influence other physiological processes. They are used to promote shoot development in tissue culture. Ethylene is the only known gaseous hormone. It promotes senescence and fruit ripening in many species. Ethylene is used to ripen fruit on the field or after harvest; to defoliate cotton and tomatoes to aid in harvest; to initiate flowering in some species; and to increase the number of female flowers, and subsequently fruit number, in cucumbers, melons, and pumpkins. Unwanted exposure to ethylene can cause severe damage to plants. Some fruits give off ethylene as they ripen. The ethylene can harm any ethylene-sensitive produce, including flowers, stored in the same area. Ethylene can build up to toxic levels in greenhouses as a result of incomplete combustion in and improper venting of flame-burning heaters. Some fungal pathogens, especially those found in bulb crops, generate ethylene, causing damage to infected plants and nearby plants. Abscisic acid (ABA) is involved in seed development. ABA also promotes the closing of stomates when plants are stressed by low water availability. In some plants, ABA also promotes leaf abscission and/or dormancy. In recent years, four more classes of hormones have been discovered and more are likely to be discovered. Chemical growth regulators generally promote or inhibit the influence of hormones. In some cases, such as with GA and ethylene, the growth regulator is the hormone.

FOOD FOR THOUGHT

1. Plant growth may be continuous for some plants, such as those growing in tropical rainforests, or it may be cyclic, depending on changes in environmental conditions. Crop production often involves directing plant growth by regulating environmental conditions. Name some environmental conditions that we can control and describe how controlling them can influence plant growth.

2. Why is the angle at which the sun strikes the earth such an important factor in how much light is available for photosynthesis? What are heliotropic leaves?

3. Ethylene is the hormone that causes senescence (aging and death) in plants. Why then do we sometimes use ethylene as a growth regulator on plants we want to continue to grow and develop?

4. One of the concerns with genetically modified crops (GMOs) is that pollen from modified crops may escape and fertilize native plants in the area, thereby introducing a new gene into the native population. Can you think of any ways that the risk of pollination of wild plants by GMO plants can be reduced?

SUPPLEMENTARY READING

HALEVY, A. H., ed. 1985–1987. *Handbook of flowering.* Six vols. Boca Raton, Fla.: CRC Press.

SALISURY, F. B., and C. W. ROSS. 1992. *Plant physiology,* Fourth Edition. Belmont, Calif.: Wadsworth.

TAIZ, L., and ZEIGER, E. 2002. *Plant physiology,* Third Edition. Sunderland, Mass.: Sinauer.

WILKINSON, R. E., ed. 1994. *Plant-environment interactions.* New York, N.Y.: Dekker.

CHAPTER

10

Plant Chemistry and Metabolism

John Streeter

KEY LEARNING CONCEPTS

After reading this chapter, you should be able to:

- Know the main biochemicals found in plants.
- Know how those chemicals are formed and some of their uses.
- Understand how relatively few elements (carbon, hydrogen, oxygen, nitrogen, phosphorus, sulfur) are combined in nearly innumerable ways to create the structures or perform the functions required for plant growth and development.

INTRODUCTION

In this chapter, a very brief introduction to the structure and interchange of plant metabolites is provided. The student should understand that a vast amount of additional information is available in other sources. Advanced students will want to study plant physiology and plant biochemistry to gain a better understanding of these subjects. Some additional information can be found in the references listed at the end of this chapter.

CARBOHYDRATES

The creation and metabolism of carbohydrates or sugars begins with photosynthesis. This process, involving the conversion of light energy into chemical energy, is the beginning point for all the compounds discussed here. Thus, carbohydrates are the compounds from which all other plant biochemistry is derived. Relative to other organisms, plants are carbohydrate-rich organisms, and they depend on carbohydrates for much of their structure and energy storage.

The general composition of carbohydrates is just what their name implies—hydrates of carbon. Thus, the general formula for carbohydrates is $C(H_2O)$, that is, one carbon atom, two hydrogen atoms, and one oxygen atom. (There can be minor deviations from this general formula.) These three elements are assembled in various combinations to form many different structures.

Carbohydrates are classified into different structural groups, and these distinctions are important in understanding the function of these compounds:

> *Monosaccharides.* Single carbohydrate molecules.
>
> *Disaccharides.* Two carbohydrate molecules linked together.
>
> *Oligosaccharides.* More than two molecules but less than seven molecules linked together.
>
> *Polysaccharides.* Combinations of seven or more molecules linked together.

Monosaccharides fall into two general categories: aldoses, which include an aldehyde group (Fig. 10–1a), and ketoses, which include a keto group (Fig. 10–1b). The smallest molecules considered to be carbohydrates are three-carbon molecules such as glyceraldehyde. It is important to understand a little about the structural diversity of these compounds, and this information will be presented using the most common and important of the six-carbon sugars, namely, glucose. The glucose molecule exists in three structural forms (Fig.10–2). In solution, these structural forms of glucose all exist and are in equilibrium (represented by the double arrows). The carbon atoms in the molecule are numbered to help us keep track of the different structures, and vertical lines in the ring structures represent the –OH groups. Note that (1) glucose is an aldose (aldehyde group at carbon 1 in the linear structure), (2) the position of the –OH groups is important in determining structure and chemical properties, (3) the ring structures differ only by the position of the –OH at carbon 1, and (4) the Greek letters alpha (α) and beta (β) are used to indicate the position of the OH group at carbon 1. The ring structures are intended to represent planar configurations with OH groups either above or below the plane of the ring. The ring structures are the molecules involved in metabolism, so it is important to understand their structure and diversity.

If the position of one of the OH groups at carbons 2, 3, or 4 is changed, compounds with different chemical properties and functions are created. Examples of other hexoses important in plant metabolism and similar in structure to glucose are galactose and mannose. Fructose, another very common sugar in nature, is a six-carbon ketose where the carbon at position 1 is $-CH_2OH$ and the carbon at position 2 is $-C=O$; the structure of fructose is otherwise identical to the structure of glucose. Smaller monosaccharides, pentoses (five carbons) and tetroses (four carbons), are also important in plant structure and function. The pentose ribose is particularly important; ribose and its deoxy derivative, which is an important component of nucleic acids (discussed later in this chapter), are shown in Figure 10–3. The *deoxy* terminology refers to the fact that carbon 2 in the molecule has the structure

Other carbohydrates that are important in plant structure and metabolism include polyols, in which all carbons have a hydroxyl group (e.g., sorbitol, the main form of carbohydrate transported in apple plants), amines (e.g., glucosamine), and uronic acids (e.g., galacturonic acid, the major component of pectin). The ring structures of these compounds are shown in Figure 10–4.

Figure 10–1   Example of an aldehyde group (a) and a keto group (b).

Figure 10–2   Illustration of the ring and planar structures of glucose. Note the difference between the α and β forms.

Figure 10–3   Oxygenated and deoxygenated forms of the sugar, ribose.

Figure 10–4 Ring structures of sorbitol, α-glucosamine, and α-galacturonic acid.

The crucial importance of these structural details in defining the function of carbohydrates can be illustrated by the comparison of the structures of starch and cellulose, two common plant polysaccharides. The general structure of starch is:

—glucose- α1-4—glucose- α1-4—glucose- α1-4— glucose- α1-4 - . . .

There may also be branch points in the structure but, here, only the linear chain is considered. The 1–4 notation means that the linkage between the two glucose molecules is from carbon 1 of one molecule to carbon 4 of the next molecule. The α notation means that the OH group at position 1 of the ring structure is down (Fig. 10–2).

Only one simple change in the structure of starch is needed to make the structure of cellulose:

—glucose- β1-4—glucose- β1-4—glucose- β1-4—glucose- β1-4 — . . .

Thus, the two structures differ only in the position of the OH group at carbon 1. But the chemical nature of the two polymers is very different. Starch is slightly soluble in water and can be broken easily down by various organisms (including humans and plants) to the simple glucose molecule. Thus, starch is the main form of carbohydrate stored in plants and is very important in animal nutrition. In contrast, cellulose is essentially insoluble in water and, when assembled into fibrils, provides the rigidity required for cell walls. Cellulose is composed of long linear chains of up to 2,000,000 glucose units. Cellulose is the main structural component in plants and is essentially inert to higher organisms (plant fiber passes through the human gut without hydrolysis) and can be utilized for energy only by microorganisms—bacteria and fungi. Examples of substances that are important in our everyday lives and are essentially pure cellulose are paper products and cotton cloth.

Another common carbohydrate in plants is sucrose, the most common form of carbohydrate used in carbon transport in plants. Sucrose is a dissaccharide comprised of a

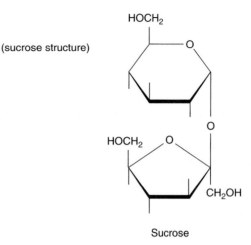

(sucrose structure)

Sucrose

Figure 10–5 Ring structure of sucrose. Sucrose is made up of a glucose sugar attached to a fructose sugar.

glucose and a fructose molecule (Fig. 10–5). Sucrose, which is obtained from sugar cane and sugar beets, is another carbohydrate that plays an important role in our everyday lives.

LIPIDS

Lipids are a group of compounds that are derivatives of glycerol, a simple 3-carbon polyol:

$$CH_2 — CH — CH_2$$
$$OH \quad OH \quad OH$$

The complexity of this group of compounds lies in the coupling of the hydroxyl groups to fatty acids and other molecules. Plants and animals have a wide variety of fatty acids, which have the following general structure:

$$OH$$
$$O = C – (CH_2)_n — CH_3$$

The subscript n after the parenthesis means that the CH_2 group is present a certain number of times, depending on the fatty acid; this number is generally between 10 and 22. Some examples are:

Fatty Acid	n
Lauric acid	10
Palmitic acid	14
Stearic acid	16

In addition to having different chain lengths, fatty acids may be either saturated or unsaturated. The term *unsaturated* refers to the presence of C=C double bonds instead of C–C single bonds. The above listed acids are all saturated, that is, linear chains of nothing but CH_2 groups.

Examples of unsaturated fatty acids that are common in plant lipids are listed below:

Oleic acid	$CH_3(CH_2)_7CH=CH(CH_2)_7COOH$
Linoleic acid	$CH_3(CH_2)_4(CH=CHCH_2)_2(CH_2)_6COOH$
Linolenic acid	$CH_3CH_2(CH=CHCH_2)_3(CH_2)_6COOH$

It is not important to memorize the composition of these compounds, only to understand in a general way what they are and how diverse they can be.

To complete the structure of the fundamental lipid molecule, linkage of fatty acids to the glycerol molecule is required to give a triglyceride molecule (Fig. 10–6). The fatty acids have been written out to convey their true molecular size and complexity. Depending on the molecular weight of the fatty acid chains and their degree of unsaturation, a wide variety of molecules with different chemical properties can be constructed. Thus, materials like fats, waxes, and oils—materials with very different physical properties—all belong to this class of compounds.

Another type of lipid is an important part of membrane structure (see Chapter 13). The membrane lipids are described as phospholipids, which refers to the fact that one of the OH groups on the glycerol molecule is linked to a phosphate group, not a fatty acid (Fig. 10–7).

In addition, another chemical group is almost always attached to the phosphate group. Here, a compound named choline is used to illustrate the point. The name of this structure is phosphatidyl choline (Fig. 10–8). Note that the phosphate carries a negative charge and the choline carries a positive charge. Because a wide variety of other chemical groups can be attached to the phosphate, this diversity, coupled with the presence of different fatty acids, means that the variety of structural and chemical properties of these phospholipid molecules is enormous.

In the actual structure of the membrane, the phosphate and attached group are rotated 180° in the other direction. This orientation gives these molecules two very different chemical surfaces; the phosphate and other attached group on the left-hand side are charged and hydrophilic (water-loving), while the long hydrocarbon chains of the fatty acids on the right are hydrophobic (water-hating). When mole-

$$O^-$$
$$|$$
$$CH_2O-P-OH$$
$$|$$
$$OH$$
$$CHO-(C=O)-(CH_2)_5\ CH=CH\ (CH_2)_4\ CH_3$$
$$|$$
$$CH_2O-(C=O)-(CH_2)_6\ CH=CH\ CH_2\ CH=CH\ CH_2\ CH=CH\ (CH_2)_2\ CH_3$$

Figure 10–7   A phospholid molecule is comprised of two fatty acids and a phosphate group attached to a glycerol molecule.

$$O^-$$
$$|$$
$$CH_2O-P-OCH_2CH_2N^+(CH_3)_3$$
$$|$$
$$OH$$
$$CHO-(C=O)-(CH_2)_5\ CH=CH\ (CH_2)_4\ CH_3$$
$$|$$
$$CH_2O-(C=O)-(CH_2)_6\ CH=CH\ CH_2\ CH=CH\ CH_2\ CH=CH\ (CH_2)_2\ CH_3$$

Figure 10–8   Structure of phosphatidyl choline.

cules with these two different chemical properties are combined, they position themselves so that the hydrophobic groups are in close contact with other hydrophobic groups, and the hydrophilic groups are in contact with other hydrophilic groups. In addition, this single layer of lipid molecules are arranged into a double layer because of the mutual attraction among the hydrophobic hydrocarbon chains. This arrangement gives the basic structure of a membrane, as illustrated in Figure 13–3.

This double-layered structure would be very unstable were it not for the fact that the negative charges on the phosphate groups are connected with a divalent cation, usually Ca^{++}. In addition, the two different regions of the membrane attract proteins that sit on the surface of the membrane, or are

Figure 10–6   A triglyceride molecule is comprised of three fatty acids attached to a glycerol molecule.

imbedded in the membrane, or transverse the entire membrane structure, thus giving it more stability. The orientation of proteins in the membranes depends on the surface properties of the proteins (see next section). A sketch of a bilayered phospholipid membrane with imbedded proteins is shown in Figure 13–3.

One other feature of membrane structure should be mentioned—a group of molecules called sterols. These molecules also have hydrophobic and hydrophilic regions, but they are planar and rigid relative to the fluid nature of the pure phospholipid bilayer. When sterols are inserted into the membrane, they increase the stability of the membrane structure; the sterol content of some plant membranes reaches as high as 50 percent. Of course, too much sterol in the membranes renders them too inflexible; there is much concern about high cholesterol and its relation to cardiovascular diseases in humans. Cholesterol is the main sterol in animal membranes. Sitosterol is the common sterol in plant membranes, which contain very little cholesterol. Note the similarity in the structure of cholesterol and sitosterol and the planar nature of these molecules in Figure 10–9. Positions in the structures represent carbon (C) and hydrogen (H) unless otherwise indicated; the hydrophilic region of the molecules is the OH group at one end.

PROTEINS

Proteins are chains of amino acids that have complex three-dimensional structures. As discussed in Chapter 13, ammonium becomes available for amino acid synthesis via nitrate reductase and nitrite reductase or via nitrogen fixation. The reactions involved in the assimilation of ammonium in plants are so important that they will be considered in detail (see Fig. 10–10).

Figure 10–9 Structures of cholesterol and β-sitosterol.

Figure 10–10 Incorporation of ammonium into the amino acid glutamate via two reactions. Ammonium may be the product of more than one process, but incorporation into biologically important compounds is almost entirely via this route.

The key point of entry for ammonium into combinations with carbon compounds is the enzyme glutamine synthetase, which catalyzes a reaction involving glutamate plus ammonium. Energy is required in the form of adenosine triphosphate (defined later). The second enzyme has several names—technically, glutamate oxo-glutarate amino transferase, or glutamate synthase, or GOGAT. This reaction transfers the ammonium assimilated in the first step to 2-oxoglutarate from the tricarboxylic acid (TCA) cycle (discussed in Chapter 11). Note that this second reaction results in the synthesis of two glutamate molecules; one is recycled to pick up another ammonium molecule. The other glutamate molecule is available for other chemical reactions. Of special importance here is the process called transamination (Fig. 10–10). Via transamination, the amino group of glutamate can be passed to various organic acids to form the complete array of amino acids required for protein synthesis.

All amino acids have a common sequence of atoms:

$$NH_2 - \underset{\underset{NH_2}{|}}{CH} - \underset{\underset{OH}{|}}{C}=O$$

Under physiological pH conditions, the COOH group loses its H atom to form the charged group COO^-. The R in the structure above represents a side group of some type (Table 10–1); there are twenty different amino acids in proteins and, because of the diversity of these side groups, it is possible to obtain an enormous variety of proteins.

Note that some of the amino acids are called essential amino acids. These eight amino acids cannot be synthesized by animals; thus, it is essential that animals (including humans) obtain these amino acids from plant protein, either directly or through the consumption of meat or fish. It is not important to memorize the chemical groups in the various amino acids, but it is important to appreciate the wide chemical diversity in this group of compounds.

In the primary structure of proteins, amino acids are linked together between the carbonyl (C=O) carbon of one amino acid and the amino group of another amino acid to form what is known as the peptide bond (Fig. 10–11). After the series of peptide bonds is completed, the chain can be folded to give two general types of secondary structure—an α helix or a β pleated sheet (Figs. 10–12b and 10–12c). The α helix and β pleated sheet structures fold again to form what is called the tertiary structure of the protein (Fig. 10–12d). Certain amino acids play an important role in determining the protein structure. For example, proline leads to a peptide bond with a different angle than the standard peptide bond because of its special amino group. Cysteine has a terminal sulfhydryl (SH) group (Fig. 10–13). The SH on a cysteine in one part of the chain may form a disulfide bond with a second cysteine in an adjacent chain to stabilize the tertiary structure, that is:

$$-CH_2-SH + -CH_2-SH \Leftrightarrow -CH_2-S-S-CH_2$$

Enzymes are protein catalysts. They do not create chemical reactions; they simply facilitate or accelerate chemical reactions. Many enzymes have a fourth level of structure wherein two or more of the tertiary structures (or subunits) combine in a specific way to form a quaternary structure (Fig. 10–12e). All enzymes have an active site where the substrate molecule is bound and altered to some other chemical structure (Fig. 10–14). Living cells have hundreds of different enzymes, one for each of the many metabolic reactions that are needed to sustain life and growth.

Storage of N represents another critical role of proteins. A young seedling requires inputs of nitrogen (N) and other nutrients before it can begin active assimilation of N by

TABLE 10–1	Alpha Amino Acids Involved in Protein Synthesis		
Name	**Number of N Atoms**	**% N**	**Unique Chemical Group**
Glycine	1	18.7	None
Alanine	1	15.7	α-methyl
Serine	1	13.3	Hydroxyl
Cysteine	1	11.6	Sulfhydryl
Methionine*	1	9.4	Sulfide
Valine*	1	12.0	
Leucine*	1	10.7	Aliphatic carbon chains
Isoleucine*	1	10.7	
Threonine*	1	11.8	Hydroxyl
Aspartic acid	1	10.5	Second carboxyl
Asparagine	2	21.2	Amide
Glutamic acid	1	9.5	Second carboxyl
Glutamine	2	19.2	Amide
Phenylalanine*	1	8.5	Phenyl
Tyrosine	1	7.7	Hydroxyphenyl
Tryptophan*	2	13.7	Indole
Histidine	3	27.1	Imidazole
Proline	1	12.2	Secondary amino
Arginine	4	32.2	Guanidino
Lysine*	2	19.2	ε-amino

* Essential amino acid.

Figure 10–11 Linkage of one amino acid to another to provide the primary structure or proteins. R1, R2, R3, and so on, indicate side groups on different amino acids (see Table 10–1).

(A) Primary structure

(B) Secondary structure (α helix)
(R groups not shown)

(C) Secondary structure (β pleated sheet)
(R groups not shown)

(D) Tertiary structure

(E) Quaternary structure

Figure 10–12   Schematic picture of the secondary, tertiary, and quaternary structure of proteins. Follow the shaded portions in the diagrams to see the progression of structural complexity. (A) The peptide bond linkage shown in Figure 10–2 but with correct atomic configuration. (B) and (C) Helix and sheet structures formed by the attraction of one peptide chain to another. (D) Assembly of helices and sheets into a globular protein molecule (tertiary structure). Most of this assembly occurs because of the attraction of various side groups (Table 10–1) to each other. (E) Combination of several identical proteins (termed subunits) to form the functional protein.

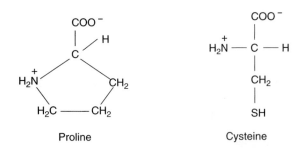

Proline   Cysteine

Figure 10–13   Structures of proline and cysteine.

Figure 10–14   Schematic representation of an enzymatic protein and an active site on the enzyme. The active site is the point at which the substrate is bound and the chemical reaction takes place to form the product.

itself. Seeds contain various storage protein structures, most of which are known. Protein concentration in seeds varies widely but is usually between 5 and 25 percent of dry weight. Seeds containing high levels of protein are important in agriculture because they serve as sources of protein for animal feeds. Depending on the balance of amino acids in a seed storage protein, it may be a more or less desirable source of protein for animal feeds. In particular, the balance of essential amino acids, which animals cannot synthesize by themselves, is critical for nutritional value, and it is not uncommon for supplemental methionine or lysine to be required in the diet because the content of these amino acids in the seed storage protein is too low.

NUCLEOSIDES, NUCLEOTIDES, AND NUCLEIC ACIDS

Nucleosides are comprised of the sugar ribose coupled to an organic base; the organic bases that are important in metabolism are derivatives of compounds know as purines or pyrimidines. The structure of purine and pyrimidine, along with one example of each type of base, is shown in Figure 10–15. The nucleoside adenosine (ribose + adenine) is not important in plant metabolism until the molecule is coupled to a phosphate group to form a nucleotide.

The structure of adenosine-3-monophosphate (AMP) is shown in Figure 10–16, alongside the structure of adenosine triphosphate (ATP), one of the most important molecules in all of metabolism: The importance of ATP lies in the energy stored in the $-P-O-P-O-PO(OH)_2$ bonds. In particular, the terminal $P-O-P$ bond, when hydrolyzed, yields about 8,000 calories/mole—a very large energy yield. In reactions involving ATP energy transfer, the magnesium salt of ATP binds to the active site of an enzyme and, when the terminal phosphate bond is hydrolyzed, the energy released can be utilized in the conversion of substrate to product. The other important feature of ATP is that it is very mobile within a cell. Thus, following ATP synthe-

(A) Purine   (B) Adenine   (C) Pyrimidine   (D) Cytosine
                    (6-amino purine)   (2-oxy-4-amino
                                                                   pyrimidine)

Figure 10–15   Structures of the organic base purine (A), adenine, a type of purine (B), the organic base pyrimidine (C) and cytosine, a type of pyrimidine (D).

Adenosine-3'-monophosphate (AMP)
(A)

Adenosine triphosphate (ATP)
(B)

Figure 10–16 Structures of (A) adenosine-3'-monophosphate (AMP) and (B) adenosine triphosphate (ATP).

sis in metabolic reactions or (mainly) in respiration in mitochondria (see Chapter 11), ATP moves readily to other sites to participate in a wide variety of enzymatic reactions. Glutamine synthesis (Fig. 10–10) is one example.

Perhaps an even more important role of these organic bases is their function in the nucleic acids—deoxyribonucleic acid (DNA) and ribonucleic acid (RNA). RNA is a chainlike molecule with a ribose + phosphate backbone (Fig. 10–17). Note that the phosphate group links carbon 3 one ribose to carbon 5 of the adjacent ribose. In DNA, the ribose sugar is deoxy at carbon 2 (see Fig. 10–3).

In DNA, one of four organic bases (adenine, guanine, thymidine, or cytosine) is attached to each ribose at carbon 1. What is remarkable is that this simple four-letter code can be organized to contain all of the information needed to construct all of the proteins in a cell. Sequences of three bases code for a particular amino acid, and each sequence for a particular protein contains many three-letter codons flanked by a transcription start signal and a transcription stop signal. To provide enough information to build and maintain a plant requires between 10^7 and 10^{12} bases in a precise sequence. With this many bases and multi-lettered messages, it is possible to see how instructions for building and operating an organism can be included in this remarkable molecule. Description of how the nucleic acid code is transcribed and translated into a protein structure is beyond the scope of this book. References cited at the end of the chapter provide additional detail.

SECONDARY PRODUCTS

The term *secondary* is used because these compounds are not part of the main processes of metabolism in plants; that is, they are not part of primary metabolism, including carbohydrate metabolism, energy production, protein production, and so on. The story of these compounds is difficult to present because of their chemical complexity and diversity. However, these metabolites are an extremely important feature of plant chemistry. Animals do not synthesize these compounds, but humans have exploited some of these chemicals for a wide variety of uses.

Alkaloids

This class of compounds includes secondary metabolites that should be familiar to students as a result of daily life and general knowledge:

♦ *Morphine.* This first alkaloid to be identified (in the early nineteenth century) is synthesized and extracted from the opium poppy (*Papaver somniferum*). It has long been used in medicine as a pain killer, but continued use of the compound can lead to addiction.
♦ *Cocaine.* An alkaloid produced by the cocoa plant (*Erythroxylum cocoa*), which grows in the Andes mountains of South America. It is relatively harmless in small doses (used by the natives as a stimulant), but purified, concentrated cocaine is strongly addictive and dangerous.

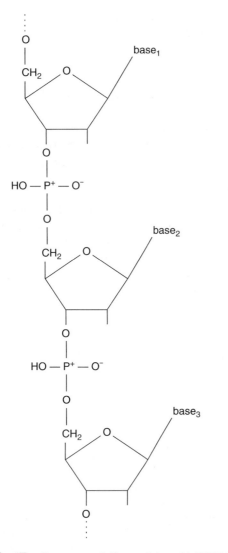

Figure 10–17   Structure of ribonucleic acid (RNA) showing its chain-like configuration.

♦ *Nicotine.* This alkaloid is obtained from the tobacco plant (*Nicotiana tabacum*) and is very toxic when consumed in large quantities. It acts to constrict the blood vessels when consumed in small doses via tobacco smoke.

♦ *Caffeine.* This compound is obtained from the coffee plant (*Coffea arabica*) and is common in many popular beverages, carbonated soft drinks in particular. It acts as a stimulant in warm-blooded animals.

So why should plants synthesize these noxious compounds? Most of the alkaloids are formed rapidly from precursors upon wounding of the plant. This formation occurs when the plant is attacked by a chewing insect or is grazed by an animal. Most of these compounds have a bitter taste and act as deterrents or protective chemicals. Because the plant is a carbohydrate-rich organism and cannot move to

Phenol

Figure 10–18   Structure of phenol.

avoid attackers, it is only natural that compounds of this type are produced by plants.

Phenolics

Another very important group of secondary metabolites are termed phenolics. This term results from the presence of the phenol molecule in some form in all of these compounds (Fig. 10–18). By far the most important phenolic in plants is lignin, the most abundant compound in plants after cellulose. The structure of lignin molecules is difficult to determine because of their high molecular weight and complex structure. Most lignin molecules probably contain twenty to twenty-four phenol units linked together in many different ways. Lignin is deposited in many cell walls to provide additional rigidity and strength. A good example is the lignification of secondary xylem tissues in woody plants; that is, the polymer cellulose is rigid, but it is insufficiently strong by itself to permit long-distance upright growth. Woody plants have highly lignified secondary xylem tissues to stand tall for the acquisition of light energy.

Another important group of phenolics are the flavonoids. Probably the most important role of the flavonoids is acting as pigments produced in flowers to attract pollinating insects and birds. Most of these pigments are shades of red, but blue and purple pigments are also in this class. One of the most intriguing stories regarding flower pigmentation has emerged only recently: flowers produce flavonoid pigments that absorb ultraviolet (UV) light. This phenomenon went unnoticed for many years because humans cannot see in the ultraviolet range; however, honeybees can. Thus, many flowers appear different to a bee than they do to a human (Fig. 10–19). The presence of UV-absorbing pigments is assumed to present a more easily recognized target for the beneficial insects.

Terpenoids

This largest of all the classes of secondary metabolites are all polymers of the hydrocarbon isoprene (Fig. 10–20). One of the most important group of terpenoids are the carotenoids—the yellow pigments in leaves. These pigments serve a protective role in chloroplasts but have no further use in senescing leaves, leading to the yellow leaf colors we see in the autumn. Other terpenoids exude from insect-damaged trees and appear to play a protective role against insect inva-

Figure 10–19 Two photographs of the golden eye flower (*Viguiera*). The photo on the left was taken with regular film, which records an image that can be seen by the human eye. The photo on the right was taken using a different light source and film that is sensitive to ultraviolet (UV) light. Because honey bees can see light in the UV range, the picture seen by the bees (right) is different from that seen by humans (left). The pattern on the right is due to the presence of UV-absorbing flavonoids in the flower petals.

Isoprene (C_5H_8)

Figure 10–20 Structure of isoprene.

sion. Some terpenoids are used in flavoring food; this group includes menthol, a compound that provides a strong aroma in many products. Perhaps the most exploited terpenoid is rubber, a very high molecular weight compound obtained

from latex. The latex is obtained from the tropical tree *Hevea brasiliensis.*

Finally, some terpenoids are poisonous to many insects and are clearly deterrents to feeding. However, one insect, the Monarch butterfly in the caterpillar and adult stages, has taken advantage of the toxic flavonoids in the milkweed plant (*Asclepia* species) by ingesting sufficient quantities and storing them inside its body. When ingested by birds, these insects provide a bitter taste and will upset the bird's stomach. Birds soon learn to leave these particular insects alone.

Other secondary products have been described, but the short discussion above serves as an introduction to this plant-specific group of compounds. Much more detail is available in the references cited below.

SUMMARY AND REVIEW

Carbohydrates are the product of photosynthesis, they contain only three elements—C, H, and O—and are the beginning point for nearly all other plant metabolism. Most carbohydrates are five or six-carbon compounds and are known as monosaccharides. The hundreds of carbohydrates found in plants consist of various combinations of monosaccarides. Depending on the dgree of polymerization and the boding between monosaccharides, the carbohydrates have a wide variety of chemical properties. Some plant carbohydrates like sucrose and starch are very important in animal nutrition.

Lipids are compounds that are present in fats, waxes and oils. They are constructed with glycerol plus other compounds known as fatty acids, and these long chain acids have a wide variety of different chemical properties. Some lipids are used as storage compounds (e.g., corn oil, soybean oil) but the most important role of lipids is in the formation of cell membranes. The lipids in membranes also usually contain a phosphate molecule and, in addition, a variety of other compounds may be appended to the phosphate.

Proteins are long chains of amino acids linked together by a unique chemical bond—the peptide bond. All amino acids have a common sequence of N, H, C, and O atoms but amino acids differ in the wide variety of chemical groups attached to the common structure. Plants have many amino acids, but only twenty are involved in protein synthesis. Protein molecules are very complex structures, and their shape plays an important role in their function, especially as enzymes. Enzymes are proteins that are biochemical catalysts and facilitate most, if not all, biochemical reactions. Other proteins serve as food reserves for seeds. Examples of protein-rich seeds are peanuts and soybeans.

Nucleosides, nucleotides, and nucleic acids form the basis for many important molecules. Adenosine triphosphate (ATP) is one of the most important metabolic molecules because of the high energy levels it carries in its phosphate bonds. Genes consist of long chains of nucleic acids known as DNA. Transcription and translation of the coded message in the DNA leads to formation of all of the proteins that make up an organism.

So called secondary plant chemicals include the alkaloids, phenolics, and terpenoids, whose specific functions and characteristics provide protection, structural strength, and coloration along with other features for the plant. Humans have long used many of these products for industrial, pharmaceutical, and recreational uses.

FOOD FOR THOUGHT

1. What are the structural differences between carbohydrates and lipids?

2. Carbohydrates in plants include sugars, starches, and cellulose. Lipids are fats, waxes, and oils. What products that we commonly use likely come from carbohydrates? Which ones likely come from lipids?

3. What are the six chemical elements found in the largest quantity in plant molecules? Of these, which three are most common?

4. One of the concerns with genetically modified (GMO) food crops is that some individuals may exhibit allergic reactions to the new food. Can you explain why this is a concern considering that all DNA in all organisms is comprised of the same four nucleic acids (though arranged in different sequences) and we consume DNA in relatively large quantities with every bite of food we take?

SUPPLEMENTARY READING

BUCHANAN, B. B., W. GRUISSEM, and R. L. JONES. 2000. *Biochemistry and molecular biology of plants.* Rockville, Md.: American Society of Plant Physiologists.

RAVEN, P. H., R. F. EVERT, and S. E. EICHHORN. 1992. *Biology of plants*, Sixth Edition. New York: W. H. Freeman & Co.

TAIZ, L., and E. ZEIGER. 2002. *Plant physiology*, Third Edition. Sunderland, Mass: Sinauer Associates, Inc.

CHAPTER

11

Photosynthesis and Respiration

John Streeter

KEY LEARNING CONCEPTS

After reading this chapter, you should be able to:

♦ Understand the process of photosynthesis and where in a cell each part of the process takes place.
♦ Understand the process of respiration and where in the cell each part of the process takes place.
♦ How photosynthesis and respiration are related and what each process does with the energy that originated from the sun or other light source.

INTRODUCTION

Two fundamental concepts of chemistry underlie an understanding of photosynthesis and respiration (and metabolism in general). The first is the uniqueness of the carbon atom. Carbon has six protons, six neutrons, and six electrons (atomic weight of 12). The two inner electrons (K shell) cannot be shared, but the four outer electrons (L shell) can be shared. Thus, carbon can form up to four bonds, each bond representing two electrons; we indicate shared pairs of electrons as short lines, as in ethane:

$$
\begin{array}{ccc}
& H & H \\
& | & | \\
H - & C - C & - H \\
& | & | \\
& H & H
\end{array}
$$

These four electrons are held just weakly enough to the nucleus of the carbon atom so that they form shared pairs relatively easily. At the same time, bonds (shared electron pairs) formed by carbon with other elements have sufficient strength to make them relatively stable. The strength of the C–C bond is 82.6 kcal/mole (an estimate of how much energy is needed to break the bond). Compared to the bond strength of N–N (39 kcal/mole), S–S (54 kcal/mole), and O–O (35 kcal/mole), the C–C bond is relatively strong. Many shared pairs of electrons lead to much greater bond strengths, for example, the N≡N bond discussed in Chapter 13 in relation to N_2 fixation has a strength of 226 kcal/mole. Other examples of carbon bonds are C–O at 85.5 kcal/mole and C–H at 98.7 kcal/mole, just in the right range for a combination of sharability and stability. *No other element has this ideal combination of sharability of electrons plus stability of the bonds formed.*

The second fundamental concept is oxidation/reduction. A compound that is reduced is a good electron donor relative to a compound that is oxidized. Another way to say this is that reduced compounds have electrons ready to donate to other compounds, whereas oxidized molecules seek to obtain more electrons. When a reduced compound is oxidized, it shares its electrons with the oxidized compound, and the originally oxidized molecule becomes more reduced. Again, carbon is at the center of this process in living organisms because it is so versatile in sharing electrons. *This transfer of electrons from relatively reduced to relatively oxidized carbon compounds drives all of metabolism and underlies life itself.* Thus, the one way to look at photosynthesis is to say that it is the process of converting oxidized carbon (CO_2) to reduced carbon [carbohydrate or $C(H_2O)$]. The conversion of reduced carbon to oxidized carbon in metabolism is used to generate other forms of energy that are central to the function of living organisms.

A nonbiological example of oxidation/reduction involving carbon is when we mix methane (natural gas, fully reduced carbon) with oxygen and light a match to start the reaction. As the methane is oxidized to carbon dioxide, the process is accompanied by the generation of heat energy:

$$4\ CH_3 + 7\ O_2 \rightarrow 4\ CO_2 + 6\ H_2O + HEAT$$

It is important to qualify this example by noting that, in the conversion of reduced carbon to oxidized carbon in plant metabolism, molecular oxygen may not be the oxidizing agent and heat is rarely released. Thus, in plant metabolism, the conversion of one form of chemical energy into a different form of chemical energy is generally driven by the oxidation of reduced carbon compounds. An example of a high-energy compound produced in metabolism is adenosine triphosphate (ATP), where the terminal phosphate bond has a lot of stored chemical energy (Chapter 10). We will return to these subjects later in this chapter.

PHOTOSYNTHESIS: LIGHT ENERGY → REDUCING POWER

There are several ways to define photosynthesis. Probably the best way is to say that photosynthesis is the process of conversion of light (radiant) energy to chemical energy in the form of reduced carbon compounds. The simplest chemical reaction that can be written to represent photosynthesis is:

$$CO_2 + H_2O \xrightarrow{\text{light}} C(H_2O) + O_2$$

Note that this process involves the conversion of carbon dioxide (oxidized carbon) to carbohydrate (CH_2O, relatively reduced carbon) and molecular oxygen. However, this equation is a gross oversimplification of what really happens in photosynthesis. Also, this process is easy to write down on paper but is exceedingly difficult to accomplish. So bacteria and plants have very complex systems to accomplish this complicated chemical process.

In plants, photosynthesis occurs in chloroplasts. These organelles were described in a general way in Chapter 8. An electron microscope picture conveys a considerably specialized structure (Fig. 11–1). The chloroplast contains a complex array of membranes called thylakoids, and many enzymatic proteins are imbedded in the thylakoid membranes (Fig. 11–2). In this chapter, a detailed discussion of the biochemistry of Photosystem I, Photosystem II, and so on, (shown in the figure) will not be undertaken. The essential function of these photosystems in the chloroplast is to utilize light energy to obtain electrons from water:

$$2H_2O \rightarrow O_2 + 4H^+ + 4e^-$$

Note that the oxygen produced in photosynthesis comes from water, not carbon dioxide. The photosystems also use light energy to raise the electrons to a very high energetic state so that they can be used to start the cascade of reduction reactions. The protons (H^+) generated in the water-splitting reaction also set up a proton (H^+) gradient across the thylakoid membranes, and this gradient is used to drive ATP synthesis.

Only about 5 percent of the radiant energy provided by the sun is used directly in photosynthesis (Fig. 11–3) mainly because the photochemistry in chloroplasts requires only a small portion of the radiant energy spectrum. (A related issue is that some of the energy received is converted to heat, which presents a problem to the plant; most of this heat energy is dissipated through transpiration. See Chapter 12.) The portion of the light energy that can be used for photosynthesis is absorbed by the chlorophyll molecules in the thylakoid membranes. The reception of a photon by a chlorophyll molecule is not sufficient to energize an electron. Instead, multiple photons of light of the desired wavelength range must be received by multiple chlorophyll molecules. The analogy of chlorophyll as an antenna molecule is appropriate. So,

Figure 11–1 (A) Sketch of a chloroplast (*left*) with emphasis on membrane structures (thylakoids) involved in photosynthesis. More detailed sketch (*right*) of "grana," or stacks of membranous structure, in which are imbedded many catalytic and transport proteins. (B) Electron micrographs showing a whole chloroplast (*left*) and the membrane networks at higher magnification (*right*). The circles in the right-hand micrograph are lipid bodies. *Source:* (A) Buchanan, W. B., W. Gruissem, and R. L. Jones. 2000. *Biochemistry and molecular biology of plants.* Rockville, Md: American Society of Plant Biologists. (B) Stachlin and Van der Staay. 1996. Ort and Yocum, Oxygenic Photosynthesis: The light reactions. Kluwer Academic Press, Dordrecht, The Netherlands.Used with kind permission of Springer Science and Business Media.

Figure 11–2 Detailed sketch of the structure of grana. Photosystem II (large circles) is located in the inside portion of the grana, whereas photosystem I (squares) and the enzymes for ATP synthesis are located mainly in the thylakoid membranes between grana (called stroma). Other components of the electron-energizing system (cytochromes, triangles) are found in both locations. *Source:* Taiz, L., and E. Zeiger. 1991. *Plant physiology.* Redwood City, Calif.: Benjamin Cummings. Used with permission.

THE NATURE OF LIGHT

The sun's energy is derived from thermonuclear reactions that convert hydrogen atoms to helium atoms with the release of tremendous amounts of energy. In the sun, four hydrogen atoms, each weighing 1.008 units of atomic mass, combine to form one helium atom weighing 4.003 units of atomic mass. Thus, 4.032 atomic mass units of hydrogen are converted to 4.003 atomic mass units of helium with the conversion of 0.029 units of atomic mass to energy. The amount of energy is calculated from the equation $E = mc^2$, where the mass (m, 0.029 atomic mass units in this case) is multiplied times c^2 (c is the speed of light, 3×10^{10} cm/sec). The product of these equals 8.7×10^8 ergs. Researchers have estimated that the sun's mass is being coverted to energy at the rate of 120 million tons per minute. This energy travels through space to the earth as electromagnetic radiation waves at the speed of light, about 300,000 Km/s (186,000 mi/s). The wavelengths of this radiation vary from very long radio waves of more than 1 kilometer (0.62 mi) to very short cosmic rays of less than 10^{-4} hm (3.9×10^{-10} in). Visible light is that portion of the electromagnetic spectrum between 400 and 760 mn; however, plants respond to the somewhat wider spectrum of about 300 to 800 nm (Fig. 1). Light is transmitted to the earth in discrete bundles of energy called photons or quanta. The number of waves passing a given point in space per second is called its frequency. Light traveling in shorter wavelengths at higher frequencies is more energetic than light at longer wavelengths and lower frequencies. Figure 2 shows the relationship between wavelength and frequency.

The rate of photosynthesis can be determined by measuring the rate at which carbon dioxide is absorbed by a leaf. A leaf is placed in a sealed glass chamber in the light, and the loss of carbon dioxide from inside the chamber is measured.

The wave nature of light. The wavelength (A) is twice that of (B) and the wave frequency in (A) is half that in (B). The short wavelength and high frequency make (B) more energetic than the long wavelength and low frequency of (A).

But while carbon dioxide is being used by photosynthesis, some is also being evolved by the leaf in either light or darkness because of cell respiration. The above method of analysis measures **net photosynthesis** because some of the photosynthesis taking place is reusing the carbon dioxide evolved from respiration. **Total photosynthesis** is determined by measuring the carbon dioxide absorbed in photosynthesis and adding to this value the carbon dioxide evolved during respiration in the light. The relationship is: rate of total photosynthesis = rate of net photosynthesis + rate of respiration in light.

The electromagnetic spectrum of radiant energy with the expansion of the photosynthetic active radiation (PAR) spectrum.

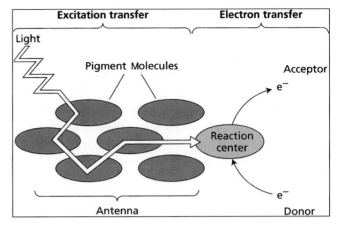

Figure 11−4 Sketch of the photosynthetic apparatus emphasizing the antennalike properties of the chlorophyll molecules. Multiple photons of light are transferred to a reaction center, where an electron is obtained from water or converted to a high-energy electron. *Source:* Taiz, L., and E. Zeiger. 1991. *Plant physiology.* Redwood City, Calif.: Benjamin Cummings. Used with permission.

Figure 11−3 Fate of light energy reaching the earth from the sun. Most of the energy is reflected by the earth's atmosphere and, of the energy that actually reaches the earth's surface, much is lost in reflection or in the generation of heat. Because of wavelength restriction of the light reception pigments, only a small portion of the light energy reaching the earth is used in photosynthesis. *Source:* Taiz, L., and E. Zeiger. 1991. *Plant physiology.* Redwood City, Calif.: Benjamin Cummings. Used with permission.

multiple photons are received and shunted to a reaction center where the electrons from water are energized (Fig. 11–4).

In reality, there are many biochemical steps between the release of electrons from water and the formation of a chemical reductant. More detail on these steps is available in the references cited at the end of this chapter. The first chemically stable reductant formed is called reduced nicotinamide dinucleotide phosphate (NADPH). This molecule contains ribose and the nucleoside adenine (discussed in Chapter 10) plus nicotinamide: nicotine with an amide group attached. The oxidized and reduced forms of NADP are shown in Fig. 11–5, but only the structure of nicotinamide is detailed. Note that, in accepting two electrons, the nicotine ring structure changes from having three double bonds to two double bonds. Thus, NADPH is the electron-rich version of the

Figure 11−5 Electrons from water are finally utilized in the reduction of the nicotinamide ring of NADP to form the first chemically stable reductant. In a sense, NADPH is at the interface between the photochemical reactions of photosynthesis and the use of the electrons to reduce CO_2 to carbohydrate. *Source:* Taiz, L., and E. Zeiger. 1991. *Plant physiology.* Redwood City, Calif.: Benjamin Cummings. Used with permission.

molecule, and NADPH is a strong reductant that is used in subsequent biochemical steps.

PHOTOSYNTHESIS: CARBON DIOXIDE → CARBOHYDRATE

So far, the discussion has shown that the light-requiring reactions of photosynthesis result in the conversion of water to O_2, and in the generation of NADPH and the energy-rich compound adenosine triphosphate (ATP; see Chapter 10). The first step in the generation of carbohydrate is the reaction of CO_2 with the compound ribulose 1,5 bisphosphate. Ribulose is a five-carbon sugar very similar to ribose (discussed in Chapter 10). The reaction, one of the most important in all of biochemistry, is catalyzed by the enzyme ribulose bisphosphate carboxylase, usually referred to as RUBISCO. Because of the importance of this reaction, it is shown in detail here:

Ribulose 1;5-bisphosphate   2 carboxy-3-ketoarabinitol 1,5-bisphosphate (a transient, unstable enzyme-bound intermediate)   3-phosphoglycerate

Note that an unstable intermediate is formed first, and that this intermediate is spontaneously hydrolyzed to two molecules of 3-phosphoglycerate.

At this point, the NADPH reductant and energy-rich ATP generated in the light reactions have still not been used. The next two reactions do require ATP and reductant. They are presented here in word form only:

Phosphoglycerate + ATP → 1,3-diphosphoglycerate + ADP

1,3-diphosphoglycerate + NADPH → glyceraldehyde 3-phosphate + NADP + inorganic phosphate

Note that glyceraldehyde 3-phosphate is a reduced carbon compound relative to phosphoglycerate.

Two more biochemical processes are required to complete the story. The first is the regeneration of the beginning substrate, ribulose 1,5-bisphosphate. This process requires many chemical steps and requires several ATP molecules. It is known as the photosynthetic carbon reduction cycle, or Calvin cycle. The term *cycle* is appropriate because the product of the RUBISO reaction is used to regenerate the starting compound. Because a new carbon atom has been obtained from the CO_2 in the RUBISCO reaction, a totally new molecule of glyceraldehyde 3-phosphate can be spun off with every three revolutions of the cycle.

The other requirement is for the incorporation of glyceraldehyde 3-phosphate into the sugar, glucose. Glucose is a stable, and translocatable form of reduced carbon. Glucose can also combine with sugar molecules to form compounds such as sucrose, starch, and cellulose (see Chapter 10).

In both sucrose and starch synthesis, the fundamental unit is hexose (6-carbon sugar). Sucrose is a disaccharide comprised of two hexoses, one fructose and one glucose, and starch is a polymer of glucose (see Chapter 10). The new glyceraldehyde 3-phosphate molecule is combined with another 3-carbon phosphorylated sugar to give fructose-1, 6-bisphosphate (reaction 1 below). The phosphate at carbon 1 of fructose 1,6-bisphosphate is lost, and the product is fructose-6-phosphate (reaction 2). The fructose 6-phosphate is then converted to glucose 6-phosphate (reaction 3) and, in turn, the phosphate on carbon 6 of glucose is moved to carbon 1 (reaction 4):

1. glyceraldehyde 3-phosphate + dihydroxyacetone 3-phosphate → fructose-1,6-bisphosphate
2. fructose-1,6-bisphosphate → fructose 6-phosphate + inorganic phosphate
3. fructose 6-phosphate → glucose 6-phosphate
4. glucose 6-phosphate → glucose 1-phosphate
5. Net: glyceraldehyde 3-phosphate + dihydroxyacetone 3-phosphate → glucose 1-phosphate + PO_4^{3-}

It is not important to memorize these reactions. Instead, remember two important concepts. The first is to appreciate how reduced carbon compounds are shuffled around to generate the compound required for a specific purpose. Second, phosphorylated compounds are often involved in metabolism (new concept) because the phosphate group attached to a carbohydrate ($-CH_2OPO_3^{2-}$) makes the shared electrons (bonds) at that carbon more likely to break apart and form new bonds; that is, phosphorylated versions of the sugars are more reactive than the nonphosphorylated forms.

The compounds shown in reactions 3 and 4 above participate in many other reactions involved in carbohydrate metabolism. Two of the most important examples are sucrose and starch synthesis. At the point of glucose 1-phosphate, the

sucrose and starch synthesis pathways diverge. However, the remaining steps are simple, except for the fact that an even more energized form of the sugar is required to make the final transition. This form is obtained by coupling the sugar to a phosphorylated base. In the case of starch synthesis, the base is adenine and the phosphorylated compound is adenosine diphosphate (ADP; see Chapter 10). This highly reactive form of glucose is coupled to a chain of other glucose molecules, with the release of ADP:

Starch

In the case of sucrose synthesis, the organic base is uridine, a derivative of pyrimidine and closely related to cytosine (see Chapter 10). The active form of glucose in

this instance is uridine diphosphoglucose (UDP-glucose). This highly reactive form of glucose reacts with fructose 6-phosphate:

The phosphate is hydrolyzed from the sucrose 6-phosphate to complete the synthesis of sucrose.

Why is sucrose and not some other simpler carbohydrate used for long-distance transport in plants? In general, dead-end metabolites are used for long-distance transport. By "dead-end," we mean metabolites that do not participate in many other reactions. The simpler carbohydrates (fructose and glucose) participate in many reactions; these reactions have not been considered in this general overview. In other words, these monosaccharides are metabolically active, whereas sucrose is not. In fact, sucrose is required for only a single function, namely, sucrose breakdown to glucose and fructose. Thus, sucrose can be moved from leaves to other plant parts where reduced carbon is required for synthesis of new cells or other functions, and there is little chance of sucrose metabolism between the origin and destination of the sucrose. Loading of sucrose into the phloem and unloading of the sucrose at the destination need to be controlled, but this control can be accomplished relatively easily.

This same general rule applies to other compounds. For example, a common nitrogen transport molecule is the amino acid asparagine. Whereas many other amino acids are involved in many reactions, asparagine is involved in only two: hydrolysis to yield ammonium plus aspartic acid, and in the synthesis of protein. Like sucrose, asparagine is, relatively speaking, a dead-end metabolite and a logical candidate for long-distance transport.

FACTORS AFFECTING THE RATE OF PHOTOSYNTHESIS IN HIGHER PLANTS

Environmental and plant growth factors that affect the rate of photosynthesis are:

1. Light quality (wavelength).
2. Light intensity (the amount of incident light energy absorbed by the leaf).
3. Carbon dioxide concentration.
4. Heat.
5. Water availability.
6. Plant development and source-sink relationships.

Light Quality

The energy level of different colors of light differs. Blue light, with its shorter wavelength and higher frequency, is about 1.8 times more energetic than the same number of photons of red light.

The plant must absorb light to keep the photosynthetic machinery running. Photosynthesis occurs in special structures, or organelles, called chloroplasts found within some plant cells. The chloroplasts contain pigments capable of intercepting light and converting electromagnetic energy into the chemical energy necessary to drive the photosynthetic processes. When these pigments (chlorophyll a, chlorophyll b, and some carotenoids) are irradiated with light containing all visible wavelengths, they absorb mostly from the red and blue portions of the spectrum and reflect the green portion. Light quality is particularly important when plants are grown under artificial light. Ample radiation from the red and blue wavelengths must be assured for photosynthesis.

Light Intensity

Wavelengths of light that are active in photosynthesis are not uniform across the entire visible spectrum. Blue and red wavelengths are more photosynthetically active than green. The light that is active in photosynthesis is referred to as photosynthetic photon flux (PPF). PPF is usually expressed as "photon flux density," which has units of micromoles of photons per m^2 per second.[1] Sensors that accurately measure PPF over the 400 to 700 nm range are used in plant science research. Until recently their cost prohibited their use by growers; however, models are now available that growers can afford. The PPF of full sunlight is approximately 2,000 μmol $\times$ m^{-2} $\times$ sec.

Light intensity affects plant growth, in part by influencing the rate of photosynthetic activity. The effect varies with different plants. Some species require high light intensities to grow well and are often termed sun-loving plants; examples are corn, potatoes, sugarcane, many turfgrasses, and some fruit trees. On the other hand, plant species that do not grow well in high light intensities are sometimes referred to as shade-loving plants. Many such plants grow in the dense shade of the forest floor, and some of them are useful as house ornamentals. Other plant species are intermediate between the sun-loving and shade-loving types and grow well in moderately intense light.

Light intensity distinctly affects the size and shape of leaves. Generally, the leaves of a given plant species grow thinner and larger in area at the low light intensity than leaves grown at the high light intensity. Also, leaves of plants grown in high light intensities tend to be darker green than those grown in low light intensities. These responses highlight the role of leaf blades in collecting solar energy.

In a given plant, there is a light intensity at which photosynthesis and respiration rates are equal and net gas exchange is zero. This intensity is the **light compensation point** (Fig. 11–6). At the light compensation point, steady-state equilibrium exists between respiration and photosynthesis, but CO_2 is still exchanged. The gas is evolved in respiration at the same rate that it is used in photosynthesis. The plant is said to be **light saturated** when further increases in light intensity increase photosynthesis little or not at all. At very high light intensities, the rate at which CO_2 is available to the plant could limit the photosynthetic rate. The light intensity at which saturation occurs increases as the CO_2 concentration surrounding the plant rises (Fig. 11–7).

---

[1] Units of footcandles and lux are sometimes used to express light "intensity," but these measurements and units refer to the entire visible spectrum. Although these light measurements are useful for some activities (e.g., photography), they are not very relevant to plant growth. These units of light are no longer accepted in the scientific literature relating to photosynthesis.

Figure 11–6 The effect of light intensity on net photosynthetic activity.

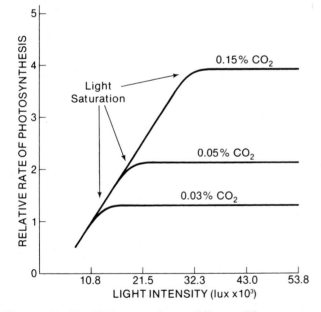

Figure 11–7 Light saturation at different CO_2 concentrations, and at a constant temperature of 25°C (77°F). Increasing CO_2 concentration increases photosynthetic activity. Note that the point of light saturation is also increased at higher CO_2 concentration.

Carbon Dioxide

When a plant is placed in a closed system, CO_2 is assimilated until the CO_2 concentration reaches equilibrium at some low value. The carbon dioxide concentration at equilibrium is

such that the amount of carbon dioxide evolved in respiration exactly equals the amount consumed in photosynthesis. This CO_2 concentration is called the **carbon dioxide compensation point.** It is reached when a plant is growing at a constant light intensity higher than the light compensation point. At the CO_2 compensation point, no further assimilation (photosynthesis) occurs until an equal amount of CO_2 is evolved by respiration.

The concentration of CO_2 in the air surrounding the leaves markedly affects photosynthesis (Fig. 11–7). Normally the atmosphere contains an average of about 0.03 percent CO_2 and 21 percent O_2. Plant physiologists have found that increasing the CO_2 concentration in a closed system, such as a sealed greenhouse or an acrylic plastic chamber, to about 0.10 percent approximately doubles the photosynthetic rate of certain crops such as wheat, rice, soybeans, some vegetables, and fruits. Many greenhouse crops, such as carnations, orchids, and roses, are grown commercially in a CO_2-rich atmosphere. Increasing the CO_2 concentration is feasible in the greenhouse or laboratory, but it is not possible to markedly increase the CO_2 concentration in the air above a corn or wheat field.

However, there is now good evidence to indicate that the CO_2 concentration in the atmosphere is steadily increasing due to the enormous consumption of fossil fuels related to human activity on the earth. In fact, there is presently considerable research on the impact of higher CO_2 concentrations on the growth and reproduction of horticultural and agronomic crops.

A few practices can have an impact on CO_2 contraction in some field situations. For example, the density of the crop and the height of the crop canopy can sometimes be altered to increase the diffusion rate of CO_2, which in turn increases its concentration in the vicinity of the leaves. Applications of organic matter in the form of crop residues or green manure crops to the soil tends to increase CO_2 levels in the atmosphere above the soil. On a warm, sunny day after a rain and with no air movement, some plants grow so rapidly that CO_2 availability at the leaf surfaces become limiting. Under such conditions, wind machines can increase the available CO_2 by circulating and mixing the air. In greenhouses, the use of horizontal airflow fans (HAFs) increases airflow around plants and exposes the leaves to a constant supply of CO_2. The use of HAFs may eliminate the need for supplemental CO_2 in most greenhouses.

Heat

At low light levels, temperature has little effect on the rate of photosynthesis because light is the limiting factor. As a general rule, however, if light is not limiting, the rate of photosynthetic activity approximately doubles for each 10°C (18°F) increase in temperature for many plant species in the temperate climates. The effect of temperature varies with species; plants adapted to tropical conditions require a higher temperature for maximum photosynthesis than those adapted to colder regions. However, excessively high temperatures reduce the photosynthetic rate of some plants not accustomed to such high temperatures by causing the stomata of the leaves to close. A reduced rate of photosynthesis, together with the increased respiration rate at high temperatures, lowers the sugar content of some fruits (for example, the cantaloupe) grown under these conditions.

Water

Under conditions of drought (low soil moisture and hot, drying winds), plants often lose water through transpiration faster than their roots can absorb it. The rapid loss of water causes the stomata to close and the leaves to wilt temporarily. When this occurs, the exchange of CO_2 and O_2 is restricted, resulting in a dramatic drop in photosynthesis. Thus, water deficit reduces photosynthesis, in part, by markedly reducing gas exchange.

Excessive soil moisture sometimes creates an anaerobic condition (lack of oxygen) around the roots, reducing root respiration and mineral uptake and transport of water and minerals to the leaves—thus indirectly depressing photosynthesis in the leaves.

Different Photosynthetic Mechanisms

Photosynthesis requires the leaf to take in CO_2. Usually this is accompanied by the loss of water vapor (transpiration). When sufficient water is available, transpiration and photosynthesis work together. During times of drought and high light conditions, however, the two may be in conflict. The stomates need to open to allow CO_2 in for maximum photosynthesis. But excessive water loss through the stomates may threaten plant survival.

Most native temperate zone plants receive enough water during the growing season that compromise between CO_2 intake and water loss is not an issue. However, plants that evolved in more arid regions have adaptations that help conserve water but do not inhibit photosynthesis during dry periods. Plants with these adaptations are called C4 and crassulacean acid metabolism (CAM) plants.

The C4 mechanism is found in tropical grasses such as corn, sorghum, warm-season turfgrasses, and other species. These plants have a special anatomy that concentrates CO_2 in special areas of the leaf that allow CO_2 to be fixed more efficiently into glucose. They are called C4 plants because CO_2 entering the leaf is temporarily attached to a three-carbon organic acid making a four-carbon organic acid. The acid is shuttled to the special areas and the CO_2 is released at the areas where it can be used efficiently. The 3-C acid goes back to the stomatal opening to get another CO_2 molecule as it enters. Because of the increased efficiency, not as much CO_2 has to enter the leaf, so the stomates can be closed more to reduce water loss.

The CAM plants (named for the species, *Crassula*, where the mechanism was first discovered) are water-storing desert plants such as succulents and cacti. Some orchids also use this mechanism. As with C4, the CAM process also attaches CO_2 to a three-carbon organic acid, but this time it happens at night when the stomates can open without extreme water loss. The CO_2 is held on the acid until light is available for photosynthesis, then it is released from the acid.

Adaptations for both C4 and CAM plants is expensive in terms of the energy required to construct and maintain them. But because they grow in areas where light is plentiful, the plants can afford them. The problem arises when we try to grow these plants in areas where light is not sufficient to support their growth. This is part of the reason why corn cannot be grown as far north as other grasses, and why warm-season turfgrasses do not do well in the middle and northern latitudes of the United States.

Plant Development and the Source-Sink Relationship

Provided that all other environmental factors are constant, the growth of the plant also significantly influences the rate of net and total photosynthesis, both in single leaves and in the total canopy of leaves on a plant. As the leaves expand and grow larger, chloroplast development and replication proceed in the new cells until the leaf has fully expanded and has reached maturity.

As single leaves on a plant develop, their photosynthetic rate rises in step with the expansion of the leaf until the leaf has matured. For example, when a soybean or spinach plant leaf begins expansion, its chlorophyll content per square decimeter of leaf area and its rate of net photosynthesis per square decimeter of leaf area both rise to maximum values just after the leaf reaches full expansion. When the leaf reaches full expansion, it is often called a **source leaf** because the carbohydrate synthesized in that leaf is in excess of local requirements and is exported to other parts of the plant that are actively growing. These sites of active growth and metabolism are often called **sink tissues.** Examples of sinks are roots and reproductive organs (i.e., fruits and seeds). Note that juvenile leaves are also sinks until their photosynthetic capacity provides carbohydrate above amounts required for local growth.

Considerable research has established that sink tissues can place a demand on source leaves so that there can be an actual increase in the activity of photosynthetic CO_2 assimilatory enzymes. The result is reflected in a sustained increase in photosynthetic rate in the source leaves.

In many plant species, the photosynthesis rates in all the mature leaves begin to decline drastically when the plants flower and envelop fruits and seeds. Actually, proteins in the leaves, and specially photosynthetic enzyme proteins in the chloroplasts, are degraded to amino acids as the leaves begin the process called **senescence.** The amino acids in the leaves are transported to the developing seeds, and subsequently seed storage protein is synthesized from the amino acids derived from protein degradation in the leaves. In the case of plants such as soybean and wheat, when the seeds are totally mature, the foliar photosynthetic activity declines to zero, followed by death of the foliage.

RESPIRATION

All organisms require the reduced carbon formed in photosynthesis to survive. But the building and maintenance of cells and tissues require not only carbohydrates, but also proteins, lipids, nucleic acids, and so forth. So it is necessary to convert the energy stored in the reduced carbon compounds into other chemical forms to accomplish the thousands of metabolic tasks required for survival.

A simple way to think about respiration is to consider the process as the conversion of carbohydrate into energy-rich ATP:

$$C_6H_{12}O_6 + 6O_2 + 36\ ADP + 36\ PO_4^{3-}$$
$$\rightarrow 6CO_2 + 6H_2O + 36\ ATP^{3-}$$

So, in a sense, respiration is the reverse of photosynthesis with respect to carbon dioxide $\leftrightarrow$ carbohydrate. The above reaction pertains to all organisms, including plants, and the number of ATPs produced (36) is a theoretical number, as will be explained in subsequent paragraphs. Even though plants can use light energy and animals cannot, plants must also respire to support the synthesis of all of the compounds needed for building and maintenance of structure—that is, for growth and reproduction.

The *first step of respiration* is a process called glycolysis—the conversion of carbohydrate to a three-carbon organic acid named pyruvic acid (or pyruvate as the anion). Starting with glucose, the carbohydrate is phosphorylated and then split into two three-carbon pieces—namely, glyceraldehyde 3-phosphate, the same compound involved in the photosynthetic cycle. Further transitions lead to the generation of a few high-energy ATPs and two reduced pyridine nucleotides—NADH. NADH is very similar to NADPH (as already mentioned), except that the latter compound has a phosphate group on the ribose (Fig. 11–5). The overall process of glycolysis results in the following:

$$C_6H_{12}O_6 + 2\ ADP + 2\ PO_4^{3-} + 2\ NAD^+$$
$$\rightarrow 2\ C_3H_4O_3^- + 2ATP + 2NADH + H_2O$$

Note that pyruvate is oxidized relative to the starting material, glucose, but only two ATPs have been produced. The initial equation indicated that thiry-six ATPs should result, so thirty-four more are still possible.

The *second step of respiration* is called the tricarboxylic acid (TCA) cycle because several of the reactions involve acids with three carboxyl groups. Before entry into the cycle, the pyruvate is first converted to acetyl coenzyme A, with the loss of one CO_2 and the production of another NADH (Fig. 11–8). The reduced carbon enters the cycle as an acetyl group and is degraded to two more CO_2 molecules coupled to the generation of three more NADH molecules. A new type of electron acceptor—flavin adenine dinucleotide (FAD)—is also involved here. Also, one ATP is gained with each turn of the cycle. Note that a complete turn of the cycle leads to the formation of a four-carbon acid that serves as the receptor for the next acetyl group.

For every glucose that is processed through glycolysis, two pyruvate molecules are formed. With two ATPs formed in two rounds of the TCA cycle, the tally of ATPs formed is now up to four, still a long way to go to reach thirty-six. But much of the reducing power in the original glucose molecule has been retained in the NADH and $FADH_2$ molecules formed: two NADH in glycolysis + eight in the formation of acetyl CoA and in the TCA cycle + two $FADH_2$.

These reduced nucleotide molecules are oxidized to generate ATP in the *third step of respiration*—called the electron transport system. The system consists of a series of cyclic reactions (Fig. 11–9). One way to describe this process is to imagine that the electrons in NADH are passed down a cascade of stairs, with energy being released at each step. During this process, those valuable electrons in NADH are cashed in for ATP production, one NADH for three ATPs. $FADH_2$ enters the cascade after the first ATP synthesis step, so it yields only two ATPs. Note that molecular oxygen is the final electron acceptor (O_2 is relatively easily reduced), and that H_2O is the final product. Carbon dioxide—the substrate for photosynthesis—is also released in respiration, but this occurs in the operation of the TCA cycle (Fig. 11–8).

Before making the final tally for recovery of ATPs from the reducing power in glucose, localization of the three steps of respiration needs to be considered. Enzymes for glycolysis, the first step, are located in the cell cytoplasm. The remaining two steps are localized in mitochondria. Thus, pyruvate—the product of glycolysis—must be transported into the mitochondrion (Fig. 11–10). Note in Figure 11–10 that an intermediate of glycolysis—phosphoenol pyruvate (PEP)—is used to produce a four-carbon acid that is also transported into the mitochondrion. This is important and necessary because certain acids are drawn off from the cycle (Fig. 11–8) for synthesis of other compounds (not shown). An example would be the use of ketoglutaric acid (oxoglutaric acid) in the synthesis of glutamate via GOGAT (Chapter 15). Because all of the 2-carbon pieces that enter the cycle at each turn are converted to CO_2, additional carbon must be put into the cycle to keep it going when organic acids are being drawn off.

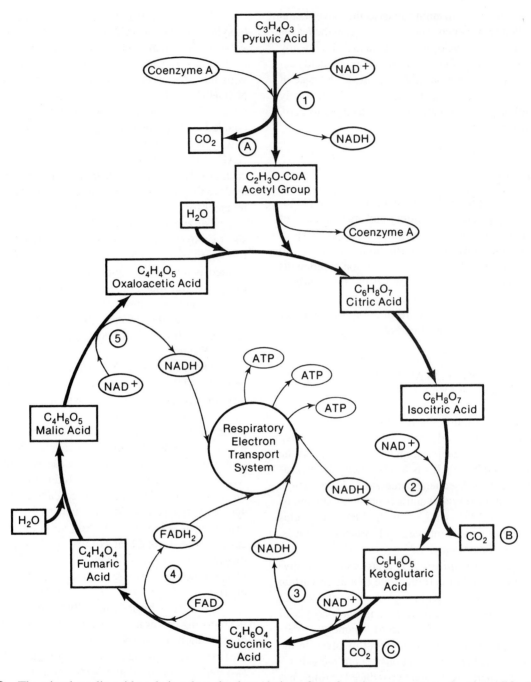

Figure 11–8 The tricarboxylic acid cycle involves the degradation of one 3-carbon pyruvate molecule to CO_2. The reducing equivalents in pyruvate are transferred to NAD and to flavin adenine dinucleotide (FAD).

The enzymes for the electron transport system are localized in the mitochondrial inner membranes—the site of NADH consumption and ATP generation. Note that the NAD^+ generated in electron transport can be recycled effi-

ciently to supply the TCA cycle because both of the last two steps of respiration are in this compartment. To calculate the theoretical final number of ATP molecules, it is important to note that, while ATP can pass through membranes without

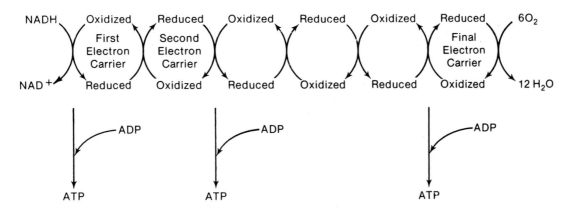

Figure 11–9 A sketch of the electron transport system. Electrons from NADH pass through a cascade of enzymatic electron transfer reactions, and the high energy compound ATP is synthesized at three points in the cascade. The final step involves the reduction of molecular oxygen to form water.

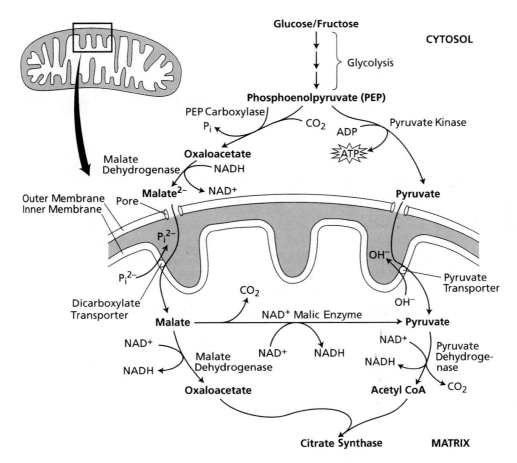

Figure 11–10 Coupling of glycolysis in the cytoplasm to the TCA cycle inside mitochondria. Note that, to keep the cycle operating, the 3-carbon acid (pyruvate) is converted completely to CO_2 at every turn of the cycle (Fig. 11–8). To enable the spin-off of acids from the cycle for other biosynthetic purposes, it is also necessary to synthesize and supply 4-carbon acid (oxalacetate) to the cycle. Additional 3-carbon fuel for the cycle can also be provided by the conversion of malate to pyruvate + CO_2.
Source: Taiz, L., and E. Zeiger. 1991. *Plant physiology.* Redwood City, Calif.: Benjamin Cummings. Used with permission.

cost, there is a cost for the transport of NADH, namely, one ATP per NADH transported into the mitochondrion. So each NADH produced in the cytoplasm yields only two ATPs.

With all of the above information, it is finally possible to calculate the theoretical yield of energy-rich ATP molecules from one glucose:

Step	NADH Produced	ATP Synthesis			Total ATP Production
		Direct	From NADH	From FADH$_2$	
Glycolysis	2	2	4*	—	6
TCA cycle + electron transport	8	2	24	4	30
				Grand total	36

*Six are produced but two must be subtracted to support the transport of NADH into the mitochondrion.

The number 36 in the table above is a theoretical number. When actual ATP production by isolated mitochondria is measured in the laboratory, the number is slightly less, probably thirty-two ATP/glucose. Even so, it is estimated that nearly 60 percent of the energy in a glucose molecule is conserved in ATP production.

One might ask: Why not oxidize glucose more directly rather than going through all these metabolic gyrations? The answer is in the above table; that is, metabolism has evolved to get the absolute most out of the reducing power of a molecule of carbohydrate. All these many reactions represent the most efficient possible conversion of one form of chemical energy to another, without the production of other energy forms (e.g., heat, light) that cells cannot use for growth and reproduction. ATP is the energy currency of living organisms. It is transported to other places in the cell to be used in a wide variety of other chemical reactions. The respiratory steps are well tuned to achieve maximum ATP production.

SUMMARY AND REVIEW

Photosynthesis takes place in the chloroplast and is the conversion of light (radiant) energy to chemical energy in the form of reduced carbon compounds. It is a complex process that starts with the synthesis of ATP and NADPH in the light reaction. These energy-rich compounds provide the energy for the generation of the carbohydrate (sugar) glucose and the shuffling and recombining of carbon and compounds in the Calvin cycle. The Calvin cycle is the regeneration of the compound ribulose 1,5-bisphosphate, the molecule to which the enzyme RUBISCO attaches CO_2. Glucose can leave the chloropast. It can combine with another sugar, fructose, to form sucrose, the most commonly transported carbohydrate in plants. Glucose can also be joined with other glucose molecules to form starch, the most common form of stored reduced carbon in plants.

Several factors can affect photosynthesis. The different wavelengths (light quality) of photosynthetically active light affect photosynthetic efficiency unequally. Blue and red are very efficient; yellow and green are less efficient. Light intensity affects photosynthetic productivity, with higher intensities generally producing more photosynthesis. The light environment in which plants evolved can determine whether plants grow well in low-, moderate- or high-light conditions. Carbon dioxide availability is necessary for maximum photosynthesis. Heat influences photosynthetic rate. Generally if light is not limiting, photosynthesis doubles for each 10°C increase in temperature. However, temperatures that are too low or high can slow photosynthesis. Sufficient water must be available for the plant to keep its stomates open to allow CO_2 to enter the leaf. C4 and CAM plants have special adaptations that let them conduct photosynthesis relatively efficiently when the water supply is limited.

The source-sink relationship describes the relationship between photosynthesizing leaves (source) and the developing tissues and organs that use photosynthetic products tissues.

Respiration, or the release of energy stored in the reduced carbon compounds, takes place in three complex steps. The first step is glycolysis, and it occurs in the cell cytoplasm. In glycolysis, glucose is oxidized into two molecules of pyruvate. Two NADHs and two ATPs are also produced. The pyruvate is converted to acetyl coenzyme A and a CO_2 molecule is given off. The two carbon acetyl groups enter the TCA cycle, which is the second step of respiration and occurs in the mitochondria. In the TCA cycle, NADH, FADH, and ATP are produced and more CO_2 is given off. The third respiration step is the electron transport system, which occurs on the inner membrane of the mitochondria. Here the energy from NADH and FADH is transferred to ATP, and H_2O is generated. In all, thirty-six ATPs are the net energy product of respiration.

FOOD FOR THOUGHT

1. As a professional landscaper, you are asked to talk to a garden club about why light is important to the growth of plants in a landscape. What are the main points about photosynthesis that you want to make in your talk?

2. You are a farmer (or a golf course superintendent) and you have observed that the soybeans/Bermuda grass growing most of the day in the shade of a row of trees do not look as healthy as the rest of the field/fairway. You know the problem is low light. What do you think is happening to the plant in terms of photosynthesis and respiration?

3. How do you think the density of plants in a field or on a greenhouse bench influences the amount of light reaching each plant and consequently the amount of photosynthetic activity in each plant? Would plant spacing be an important part of a crop production plan?

SUPPLEMENTARY READING

BUCHANAN, B. B., W. GRUISSEM, and R. L. JONES. 2000. *Biochemistry and molecular biology of plants.* Rockville, Md.: American Society of Plant Physiologists.

RAVEN, P. H., R. F. EVERT, and S. E. EICHHORN. 1992. *Biology of plants*, Sixth Edition. New York: W. H. Freeman & Co.

TAIZ, L., and E. ZEIGER. 2002. *Plant Physiology*, Third Edition. Sunderland, Mass.: Sinauer Associates, Inc.

CHAPTER 12

Soil and Plant Water Relations

James D. Metzger and Margaret McMahon

KEY LEARNING CONCEPTS

After reading this chapter, you should be able to:

♦ List the uses of water in plants.
♦ Understand the forces that move and hold water in the soil.
♦ Understand the forces that move water from the soil, through the plant, and into the air.
♦ Explain how these forces influence plant growth and irrigation practices.

WATER

Uses in Plants

Water makes up approximately 90 percent of a plant's mass and performs many functions in plants. It is required for seed germination, serves as part of the plant's structure, carries minerals into and through the plant, transports photosynthates and other biochemicals through the plant, and cools the plant by evaporation. Without sufficient water, plant growth and development is inhibited and if the insufficiency is great enough, the plant dies. Almost all the water a plant takes in comes from the soil through the roots. A small amount may be absorbed by other organs and tissues, but usually only under special circumstances, such as during the propagation of stem and leaf cuttings.

Characteristics

Water is the universal solvent; it dissolves more substances than any other liquid. The characteristics of water arise from its unique structure, which also accounts for its remarkable stability. Water is one of nature's most stable

Figure 12–1 Two hydrogen atoms join one oxygen atom at an angle of 105°, giving the water molecule an asymmetrical configuration that accounts for many of its unique properties. Note the electrical polarity of the molecule.

compounds, so much so that for centuries it was considered a single element, not the compound it is. The water molecule in fact comprises two hydrogen atoms attached to an oxygen. The water molecule is not symmetrical. This lack of symmetry creates a dipole with the molecule having more positive electrical charge and the opposite end a more negative charge (Fig. 12–1). This phenomenon of polarity creates an attraction between water molecules: the positive end of one molecule attracts the negative end of an adjacent molecule. Water molecules can also attract or be attracted by cations, such as Na^+, K^+, and Ca^{++}, or by anions or clay colloids in the soil. Although we write the chemical designation for ions as in the last sentence, they are, in fact, hydrated in the real world, giving them unique chemical and physical properties.

Surface tension is that physical property of water in the liquid state that is due to the intermolecular attraction (hydrogen bonding) between the water molecules. The molecules in the surface film in water are inwardly attracted, resulting in a strong surface tension. Were the water's surface tension not so strong, soil would hold little water, water could not reach the top of a tall tree, and blood would not flow through our bodies. This strength of attraction between water molecules is illustrated by the fact that a steel needle can float on the surface of water.

In the liquid state, the attraction among water molecules is chaotic and random, but as the liquid freezes, a rigid symmetrical lattice with an open porous structure forms. This arrangement accounts for the reduction in density as the water solidifies. Water also has an unusually high **specific heat,** which allows it to absorb a large amount of heat without a large increase in temperature. As a result, water can moderate the temperatures of nearby areas.

SOIL WATER

Availability

Even though water is present in the soil, it sometimes is not available to the plant for reasons that will be discussed later in this chapter. Water held by and moving within the soil sup-

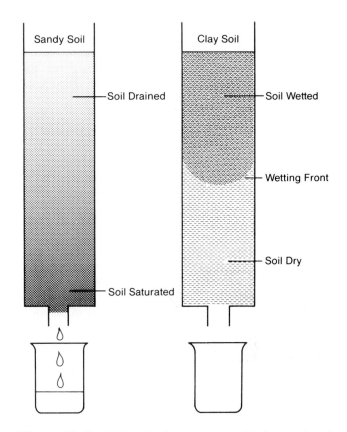

Figure 12–2 Fifty ml of water were added to each column. After one hour, water has drained to the bottom and some dripped out of the sandy soil (*left*), while the water remained in the top half of the clay soil (*right*). Obviously, clay soil can hold more water.

plies the plant with mineral nutrients and oxygen, as well as water. It moves through the soil pore spaces in the liquid or gaseous state. The pore spaces are always filled with water, air, or a mixture of both. When the pore spaces are filled with water, the soil is said to be saturated. Saturation is an unhealthy condition for plants if it lasts too long because the oxygen needed for respiration is missing. On the other hand, when the pore spaces are filled mostly with air, the soil is too dry for good plant growth. The number and size of the soil pores vary with the soil's texture and structure. Clay soils have smaller but more numerous pores than sandy soils. Thus, an equal volume of clay soil holds more water than a sandy soil when the pores are filled (Fig. 12–2). The ability of the soil to retain water is called its water-holding capacity. The capacities of different textured soils for holding water are plotted in Fig. 12–3.

Water Movement and Retention in Soil

Three forces—gravity, adhesion, and cohesion—are responsible for water movement within the soil. Gravity causes water to move downward and is the principal force when a soil is saturated. **Adhesion** is the force of attraction between

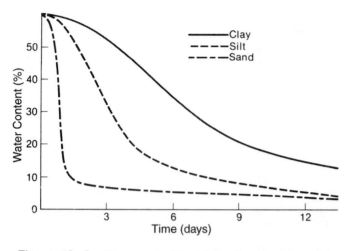

Figure 12–3 The speed with which soil water moves into or out of a soil is determined by the soil's texture. For example, assume that a sandy soil (coarse texture), a silt soil (medium texture), and a clay soil (fine texture) are saturated at the same time ($t = 0$) and that the moisture contents of the three soils are determined and plotted with time. The graph shows that the sandy soil loses moisture very quickly, the silt loses water moderately, and the clay soil loses it slowly. Thus, sandy soils have to be irrigated more frequently than do silt or clay soils.

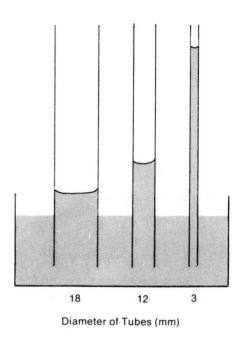

Figure 12–4 Capillary rise: water ascends in capillaries, reaching higher levels in smaller tubes. Doubling the diameter of the tube doubles the area for the water molecules to adhere to—but it also quadruples the weight of the water to be pulled up. Therefore, water does not rise as high in the columns with larger diameters.

unlike molecules (soil particles and water). **Cohesion** is the force of attraction between like molecules (water and water). The latter two forces can cause water to move by capillarity in any direction—upward, downward, or laterally—and are the principal forces that move water in an unsaturated soil.

The upward movement of water, called capillary rise, is responsible for the loss of water from the soil surface by evaporation. Capillary rise can be demonstrated with one end of a strip of blotting paper inserted partway into water or with capillary tubes (Fig. 12–4).

As soil dries, the water film surrounding each soil particle thins. Consequently, the adhesive and cohesive forces of attraction increase rapidly, making it more difficult for the plant to extract water that is held tightly in soil particles. The forces that act on soil water create changes in energy level, called water potential, symbolized by the Greek letter psi (Ψ). Water potential determines the direction in which water flows. Water flows from high to low water potential. Think of a waterfall, where water flows from a high to a low level (potential), or a water balloon, where pressure inside the balloon creates a very high water potential, forcing water out of the opening of the balloon if you let go of it. Conversely, tension lowers water potential, which draws water into an area. An example is sucking on a straw in a glass of iced tea. The greater the tension, the lower the water potential. Water potential is measured in units of pressure, most commonly atmosphere (atm) and MPa (megapascal). One atm = 0.101

MPa. Saturated soil with the atmosphere at 100 percent relative humidity has a water potential of zero. As water drains away, soil water potential decreases (becomes a negative number). When the negative potential is stronger than gravitational pull, water no longer drains from the soil. Negative potential, or pressure, is called tension. *Tension* is the term often used by soil scientists to describe soil water potential. The conversion between water potential and tension can be done by changing the sign of the values, for example, a tension of 1.0 MPa equals -1.0 MPa Ψ.

Soil water potential is determined by several factors: (1) gravitation, due to the earth's gravity; (2) matric, from the adhesion and cohesion between soil particles and water, resulting in adsorbed water and capillary water; and (3) osmotic, caused by dissolved salts in the soil water.

Imagine an irrigation canal in the soil surface. Water is applied at a constant rate to the soil. The water immediately begins to wet the soil by moving downward through the profile as a result of the force of gravity. As water is added to the channel, it also moves slowly both laterally and downward. Gravity does not move water laterally, so a different energy source must be responsible. The lateral movement of water is under the influence of matric potential resulting from the attraction between soil particles and water molecules (cohesive forces) (Fig. 12–5a and b).

When irrigating, a grower has to remember that water will move laterally only if there is enough matric potential to

Figure 12–5 Water from an irrigation source has to move from that source into the soil. In the case of an irrigation canal system (A), water moves down because of gravity and laterally because of the matric potential of the soil. The same forces work on a drip irrigation system (B). A wetting problem can occur when using drip irrigation in pots containing a high percentage of peat moss. If the medium is very dry, the water does not move laterally (C), and the irrigation water is not distributed evenly. Only the area under the tube becomes wet. *Source:* Margaret McMahon, The Ohio State University.

Figure 12–6 Effect of osmotic energy. Even though the plants in both beakers started with the same quantity of water, it was not equally available. In beaker (A), salt was added, creating an osmotic pressure too great for the plant to overcome. Since water was limited in its availability, the plant wilted. Beaker (B) contained pure water, which was readily available to the plant. Remember that fertilizer is a salt. High fertilizer concentrations can create an osmotic potential in soil, which in turn can inhibit water uptake by plants.

pull the water in that direction. This is usually more of a problem with drip irrigation systems, where the water comes in at a single point at the same time the media in the pot that contains a high percentage of peat moss. When the peat moss gets very dry, there may not be enough cohesive force necessary to move water laterally. In that case, the water will run straight down, and the medium to the sides is not wet (Fig. 12–5b and c). A thorough surface watering using a hose or other overhead irrigation system is needed to rewet the medium and establish the cohesive forces. Salts are present in soil water. The salts create osmotic energy, and if the salts are present in a sufficiently high concentration, the osmotic energy prevents water movement into the plant (Fig. 12–6).

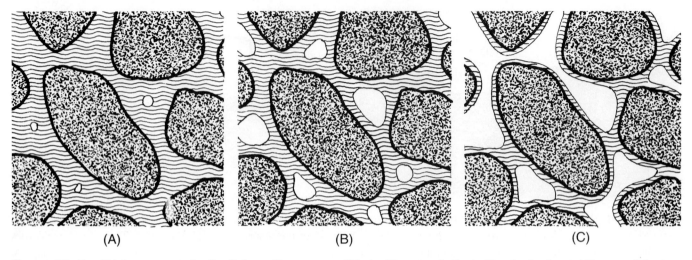

(A)   (B)   (C)

Figure 12–7   (A) In a saturated soil, all the void spaces are filled with water, indicated by the horizontal lines, and the tension (negative pressure) needed to remove the water from the soil particles is zero. (B) As the soil drains, the air:water ratio becomes larger and air occupies some of the void spaces. The tension at field capacity (i.e., when drainage ceases) is about 0.33 atm. (C) If the soil continues to dry, then the water film becomes so thin that extracting the water requires a tension of 15 atm (15.2 bars). This is more than most plants can exert. Therefore, the plants wilt permanently.

Fertilizer is a salt and, in high concentrations, can create an osmotic potential that inhibits water uptake by the plant. Container-grown plants that receive fertilizer in the irrigation water (fertigation) can be especially susceptible to this problem because the containers can become dry between fertigations. It is good practice to apply enough clear water shortly ahead of fertigating to make sure the soil or medium is moistened before applying the fertilizer solution.

After a prolonged rain or irrigation, the air in the soil pores is displaced with water. As mentioned earlier, in this condition, the soil is saturated, the soil moisture potential (tension or suction) is zero, and no energy is required to remove water from the soil particles (Fig. 12–7). This state will prevail as long as water is applied at a rate equal to the rate of water loss by drainage, by plant use, and evaporation. When no more water is added, losses continue, first from the larger macropores and then from the smaller micropores. Loss of water continues until the adhesive and cohesive forces equal gravity. This usually takes two to three days in a loam soil. At this moisture content, the soil is said to be at **field capacity.** The water has drained from the macropores but the micropores still contain water. If plants are growing in the soil, water loss continues, and if no water is added, eventually the soil reaches a moisture content that does not sustain plant life and the plants permanently wilt. The soil moisture content at which a plant wilts (the sunflower is often used as a reference) and cannot recover when placed in an environment of 100 percent relative humidity is termed the **permanent wilting point (PWP).** Soil texture determines the relationship between soil water content and soil moisture tension (Fig. 12–8).

Note in Figure 12–8 that the permanent wilting point of approximately 15 atmospheres of tension occurs at differ-

Figure 12–8   Soil moisture release curves showing the relationship between water content and soil moisture tension with three different soil textures. For example, compare the 10 percent loss of water from the clay soil (30 percent – 20 percent) and the accompanying slight increase in soil moisture tension (from a to b), with the 3 percent loss of water from the sandy soil (4 percent – 1 percent) and the accompanying large increase in soil moisture tension (from *c* to *d*). A similar analysis can be made on this graph with the same textured soil in different soil moisture ranges. In the wet range (high soil moisture percentage), a large loss of water produces a small increase in soil moisture tension, but in the dry soil moisture range, a small loss of water produces an increase in soil moisture tension that could easily go from wet enough to sustain plant growth to the permanent wilting point.

Figure 12–9 Relationship between soil moisture tension and the moisture coefficients. Water available to plants is defined as the water between field capacity and PWP. Note that soil scientists often use the term *tension* rather than water potential when describing soil water. To convert between water potential and tension, change the sign of the values from the positive to negative, and vice versa.

ent soil moisture contents, depending on the soil texture. For an average loam soil at the PWP, all macropores and all but the smallest micropores are emptied of water. If the soil moisture depletes further, the soil becomes air dry. In this condition, all liquid water is gone except that held as a thin, tightly bound layer around the soil particles. This adsorbed water is called **hygroscopic water,** and the soil moisture tension is far beyond the availability range for plant use (Fig. 12–9). Another form of water, even less available to plants and called the **crystal lattice water,** is held in the soil's crystalline structure. This water can be removed only by applying sufficient heat to destroy the crystalline structure.

Anyone who grows plants for a living is interested mainly in the water available for plant growth. **Available water (AW)** is defined as the soil moisture between field capacity (FC) and the permanent wilting point (PWP):

$$AW = FC - PWP$$

Water between FC and saturation is not considered available because it is lost through drainage. Water at tensions greater than PWP is held too tightly by the soil for plants to remove. The definition of AW could imply that the water between FC and PWP is equally available to all plants, but this is not necessarily true. Water near PWP is not as available to most plants as that near field capacity. Also, the PWP is not the same for all plant species. Deep-rooted crops may have an advantage when soil closer to the surface is dry because soil moisture in the area of the deeper roots may be adequate for growth. Water in pots and planters is subject to the same influences as that found in field soils, although on a much smaller scale.

PLANT WATER

The role of water in the absorption and transport of raw materials used for photosynthesis in the green parts of the plant and the translocation of the products of photosynthesis to ar-

eas of storage or consumption is important for an understanding of plant growth and development.

Absorption and Conduction of Water

Water is absorbed by the roots from the soil and distributed throughout the plant, even to its highest leaves—perhaps 60 m (200 ft) or more above the source. This formidable task requires considerable energy, estimated to equal about 16 kg/cm^2 (225 lb/in^2) of pressure. The energy to move the water comes from the water potential gradient that develops in the soil, plant, and air continuum. In most cases, the water potential of moist soils is greater than that of the roots, so water flows into the roots (some anatomical features of the root aid or deter this process). Once inside the xylem of the root, however, the flow of water is almost strictly pressure dependent, with the pressure decreasing from the root to the stem, to the leaves, to the air (Fig. 12–10). The pressure, along with the cohesive (hydrogen bonding) properties of water, create a chain of water molecules that is pulled through the plant and out the stomates to the air surrounding the leaf. Not all water leaves the leaf. Some water is part of the plant's structures or held in the cytoplasm, some is used for biochemical processes, and some is stored in the tonoplast. The pressure of water inside a cell creates turgor pressure, which gives plants rigidity. When there is insufficient water to create turgor, the plant wilts.

The upward movement of water occurs mainly in the xylem. The xylem is the thin, hollow tube that extends from the root through the stem and other parts of the plant. Minerals dissolved in the water are also carried to leaves, stems, and fruits via the xylem stream. (Remember that xylem tubes are very thin, much like capillary tubes.) Examples of important minerals are nitrogen used for protein synthesis, phosphates used to make ATP during photosynthesis, and magnesium that becomes part of the chlorophyll molecule.

The upward movement of water from the roots through the xylem in the stems to the uppermost leaves is sometimes

	Ψ (MPa)
Air	−95
Leaves	−0.8
Stem	−0.7
Roots	−0.6
Soil	−0.4

(A)

(B)

Figure 12–10   (A) Decreasing water potential (Ψ) from the soil through the plant to the atmosphere creates tension that pulls the cohesive chain of water molecules (B) from the soil into the roots, through the xylem of the stem and leaves, and out into the atmosphere. *Source:* Margaret McMahon, The Ohio Sate University.

called the transpiration stream because transpiration is the primary cause of this movement. The upward transpiration pull in the leaves is started by the evaporation of water molecules from the outer surfaces of the mesophyll cells. As this water is lost, the mesophyll cells become water deficient, and a potential difference is created between the dry mesophyll

cells and the walls of adjacent moist cells. Because of water's cohesive properties, water from the wetter adjacent cell walls begins to diffuse into the less hydrated cells, thus relieving the pressure difference. Continued loss of water molecules at the leaf surface establishes a flow of water throughout the plant from cell to cell, from the roots to the leaves. In addition, a pressure difference exists between the outer cells of the root tops and the soil moisture, creating a driving force that moves water into the roots. Part of the water is absorbed into the roots by osmosis. **Osmosis** is the flow or diffusion that takes place through a semipermeable membrane (as in a living cell) that typically separates either a solvent (water) and a solution, or a dilute solution and concentrated solution. These concentration gradient differences create flow conditions through the semipermeable membrane until equilibrium is established. At this equilibrium concentration, osmosis ceases. The proportion taken in by osmosis compared with the amount taken in by the transpiration stream depends on the rate of transpiration. Normally, however, the amount entering by osmosis is relatively small.

Water absorption and translocation by the plant are of primary importance. Of all materials taken in by plants, water is absorbed in the largest quantities, but only a fraction of the water taken in is used in metabolic processes. Most of the water is lost from the leaves by transpiration. Water loss in the plant is regulated to a certain extent by the opening and closing of stomata on the leaf surfaces. The flow of water out of the leaves is a critical component of the plant's cooling mechanism.

Absorption and Transport of Mineral Nutrients

In higher plants, the minerals initially needed to start growth of a new plant are normally provided by the seed or stored in the propagating tissue. But as these are used, additional nutrients for the plant's continued growth must be absorbed from the soil.

Chemical analyses of the sap in root cells for various mineral nutrients reveal that these cells have concentrations 500 to 10,000 times higher than those of the same element in the soil solution. If simple diffusion were the only mechanism involved in taking up soil nutrients, the mineral nutrients could not move into the roots against such a high concentration gradient. In addition, the plasma membranes of cells are largely impermeable to the movement of ions. Thus, energy is required to move ions against the concentration gradient and through the impermeable membranes. In the root cells, this energy is obtained from the respiration of starches and sugars that originated in the photosynthetic processes.

A plant absorbs mineral nutrients against a high concentration gradient by a process called **active transport.** Active transport is facilitated by special proteins present in the plasma membranes of cells. Various models for these "carrier proteins" have been proposed, and one model is shown in Figure 12–11. In this model, a potassium ion, which has moved through the cell wall, is bound by the carrier at the

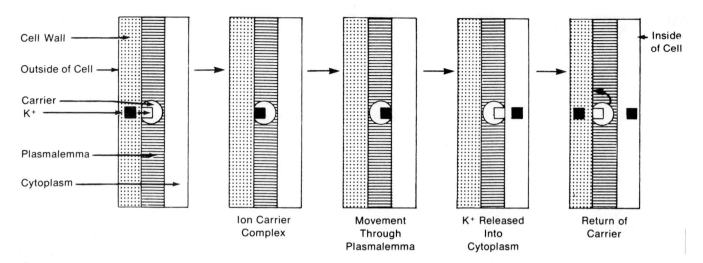

Figure 12–11 A schematic diagram of carrier ion transport. *Source:* Figure drawn using data from Weier, T. E., C. R. Stocking, M. G. Barbour, and T. Rost. 1982. *Botany: An introduction to plant biology.* 6th ed. New York: John Wiley.

outer surface of the plasma membrane. In the energy-requiring step, the protein moves or changes shape so that the K^+ ion is moved to the inside of the membrane. Following release of the ion, the protein resumes its original shape or position.

Translocation of Sugars

Sugars that are synthesized during photosynthesis move throughout the plant, primarily in the phloem tissues. The movement is mostly downward from leaves to roots, but lateral or even upward movement from leaves to fruits or buds or other storage organs also occurs. In woody perennials, the phloem tissue is just underneath the bark following development of secondary vascular tissue. Thus, when this phloem tissue is severed by girdling, the tissue above the cut proliferates whereas the tissue below the cut is starved for photosynthates (Fig. 12–12). Accidental or unintentional girdling, which can be caused by deer feeding on bark during the winter, constricting wires or ties left around a tree trunk, or roots circling around a tree trunk below the ground, can severely stunt and even kill a tree by starving the roots for photosynthates. In some cases, trees are intentionally girdled to kill them.

The rate of translocation of sugars in the phloem is, in some instances, more than a thousand times faster than simple diffusion of sugar through water. The rate of translocation has been measured in many plants, and average values of 1 to 6 g/cm²/hr have been found in developing fruits and tubers. The forces that create these high rates of movement are called active transport, as already mentioned.

To conclude the discussion of transport, it is useful to note that metabolites, ions, and water move through the plant by various processes, some relatively slow and some relatively fast:

Figure 12–12 Grape shoot girdled by removing the bark and the phloem and leaving the xylem intact causes sugar and other carbohydrates to concentrate above the girdle. Grape shoots are often girdled commercially to increase fruit size.

1. Diffusion, which transports ions and molecules slowly.
2. Cytoplasmic streaming, which transports molecules and ions within the cytoplasm at a considerably faster rate than diffusion.
3. Downward mass flow translocation of material in the phloem.
4. Very rapid upward movement of water and mineral nutrients through the xylem.
5. Lateral transport of materials along the vascular rays radially from sieve tubes into the cambium tissue and xylem.

SUMMARY AND REVIEW

The importance of water in so many plant functions and processes means that water uptake and distribution in the plant is a very important concept for anyone who grows plants to understand. The first thing to understand is that water is a dipole, meaning one side of the molecule has a negative charge and the opposite side has a positive charge. The positive side of one water molecule attracts the negative side of another water molecule, creating a cohesive force among water molecules. In soil, water is influenced by several forces that exert pressure or tension on water. The energy from these forces is called water potential (symbolized by the Greek letter, ψ). Water potential determines the direction in which water moves. Water moves from high to low water potential. A saturated soil has zero water potential when the relative humidity is 100 percent.

In soil, water potential is influenced by many factors, including pore size, gravity, electrical charges on soil particles, and ions present in the soil. Each determines how much water the soil can hold. The ability of the soil to hold water is called its water-holding capacity. The first force to act on saturated soil is gravity, which drains away water until the forces holding the water are stronger than gravitational pull. The water remaining after gravity has removed some is called the field capacity of the soil. Evaporation from the soil surface and uptake by plants removes more soil water. Eventually enough water is removed that plants can no longer take up enough to prevent wilting, even at 100 percent relative humidity. This soil condition is called the permanent wilting point. The water that exists in the soil between field capacity and the permanent wilting point is that available for plants.

Water moves laterally through the soil by matric potential. Lateral movement is very important in some irrigation systems such as drip. If the soil is too dry, the matric potential may not be great enough to move water laterally away from the water flow, so only the soil or potting medium directly below the drip tube opening gets wet.

Water enters the plant through the roots and enters the xylem. The force initially pulling water into the plant is mainly from the difference in water potential between the soil and the root, with the root being somewhat lower than the soil. Inside the root, both active and passive mechanisms move the water into the xylem, where it moves up the plant stem, into the leaves, and then out through the stomates into the atmosphere, forced again by a water potential differential that decreases along the way. Some water is held by the plant for its growth and development, but most water leaves the plant through evaporation. As long as the soil water potential allows a continuous flow of water molecules, creating a cohesively linked chain, into the plant, water readily moves into and out of the plant.

As water moves into and through the plant, it carries mineral nutrition elements from the soil with it. The concentrations of these elements in the plant's water-carrying system is much greater than in the soil. The uptake and movement of these elements requires an active transport system along with the flow of water. The products of photosynthesis, such as sugars and other soluble metabolites, are translocated through the plant along with water through the phloem tissue. Active transport of these compounds is required.

FOOD FOR THOUGHT

1. In each of the following, place an arrowhead on the line to indicate the direction in which water will flow.

   a. 10 MPa —— 2 MPa

   b. −10 MPa —— −2 MPa

   c. −0.1 MPa —— −2 MPa

2. In terms of soil texture, why does soil that is 75 percent sand and 25 percent clay drain faster than a soil that is 50 percent sand and 50 percent clay?

3. In terms of water potential, why does an application of a high concentration of liquid fertilizer on dry soil or potting mix likely create a serious water uptake problem for plants growing in that soil or mix?

4. What are some ingredients you might add to an artificial growing mix to make it drain fairly rapidly? To make it retain water?

5. Drip irrigation systems now sometimes use rings with several openings that go around a plant, instead of the single tube used in the past. What do you think is the reasoning behind this technology?

6. Why are the narrow tube structure of xylem, decreasing water potential, and the adhesive and cohesive forces of water required to move water into and through a plant?

SUPPLEMENTARY READING

Brooks, K. N., P. F. Folliott, H. M. Gregorson, and L. E. DeBano. 1997. *Hydrology and watershed management,* Second Edition. Ames, Iowa: Iowa State University Press.

Reed., D. W. (ed.). 1996. *Water, media, and nutrition for greenhouse crops.* Batavia, Ill.: Ball Publishing.

Taiz, L., and E. Zeiger. 2002. *Plant physiology.* Sunderland, Mass.: Sinauer Associates, Inc., Publishers.

CHAPTER 13

Plant Nutrients

John Streeter

KEY LEARNING CONCEPTS

After reading this chapter, you should be able to:

♦ Understand how nutritional elements are taken into the plant through the roots.
♦ Know the fourteen essential nutritional elements and the relative amounts of each that is required for plant growth and development.
♦ Know the sources of nitrogen, phosphorus, potassium, calcium, magnesium and how each is used by the plant for growth, development, and reproduction.
♦ Understand the concepts of nutrient deficiency and remobilization.

HOW DO PLANTS ABSORB NUTRIENTS FROM THE SOIL?

Before discussing plant nutrients and their function, some understanding of how these nutrients get into the plant is needed. Nutrients do not simply flow into the cytoplasm of a plant cell in the water absorbed. Instead, there are significant barriers to the movement of nutrients inside the plant, and the presence of these barriers means that systems are required to move nutrients around. These systems for nutrient movement have a strong attraction for a single type of nutrient (e.g, nitrogen, potassium), so the uptake and movement of individual nutrients is efficient and is controlled by the plant.

First, it is important to emphasize that the plant nutrients are charged forms (ions) of the elements. Some ions may have one or more different charges and chemical forms (Table 13–1). In addition, most of the ions in the soil solution are hydrated; that is, they are surrounded by water molecules.

Long-distance transport in plants is accomplished by two tissues: xylem and phloem (Figs. 13–1 and 13–2). In general, the xylem is the vascular tissue

TABLE 13–1	Summary of Mineral Elements and Typical Concentrations in Plants		
Element	**mg/kg g dry wt (ppm)**	**Relative Number of Atoms**	**Typical Forms in Plants**
Nickel	Trace	<1	Ni^{2+}
Molybdenum	0.1	1	Mo^{3+}
Copper	6	100	Cu^{2+}
Zinc	20	300	Zn^{2+}
Manganese	55	1,000	Mn^{2+}
Iron	112	2,000	Fe^{2+}, Fe^{3+}, heme
Boron	22	2,000	BO_3^{3-}
Chlorine	106	3,000	Cl^-
Sulfur	960	30,000	SO_4^{2-}, SO_3^{2-}, SH^-
Phosphorus	1,860	60,000	PO_4^{3-}, $P_2O_7^{4-}$
Magnesium	1,940	80,000	Mg^{2+}
Calcium	—	125,000	Ca^{2+}
Potassium	—	250,000	K^+
Nitrogen	—	1,000,000	NO_3^-, NO_2^-, NH_4^+

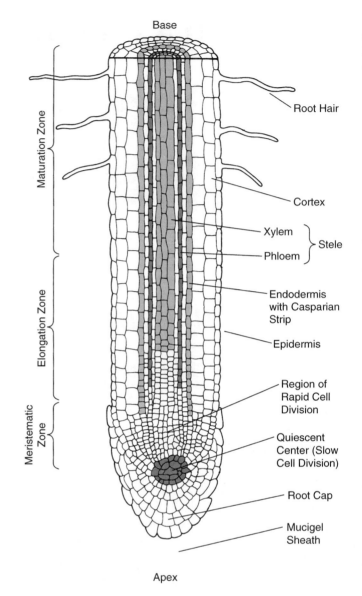

Figure 13–1 Longitudinal view of a growing plant root. The dotted lines for xylem and phloem indicate the regions of the root where differentiation occurs. Note that the phloem, which delivers nutrients to the growing root, extends farther down than the xylem, which transports nutrients from the soil solution to other plant tissues. Casparian strips are the waxy deposits in the cell walls that prevent movement of water and ions into the xylem via the apoplast (see Fig. 13–2). The formation of endodermis complete with Casparian strips corresponds to the point where differentiation of xylem is complete. *Source:* Esau, K. 1953. Plant Anatomy. Reprinted with permission of John Wiley and Sons, Inc.

that provides upward movement of water and ions to the shoots, whereas the phloem is the vascular tissue that provides downward or lateral movement of nutrients from leaves to other parts of the plant (including roots). Thus, the question that forms the heading of this section can be asked in a more specific way: how do ions get from the soil solution to the xylem tissue in the center of a root?

Plant cell walls are composed of high molecular weight polysaccharides—cellulose, hemicellulose—and pectin. You can think of the cell wall as very porous—much like a paper towel. That is, there are spaces in the cellulose fibers and in the pectin between cell walls that are large enough to allow relatively unimpeded flow of ions and water into the root. Even though the ions are hydrated, they are still several times smaller than the pores in a cell wall. Thus, the ions have a free ride until they reach the endodermis. Flow of ions via this route is termed apoplastic movement (Fig. 13–2).

The endodermis is composed of cells that have cell walls impregnated with a waxy substance called suberin. Suberin provides an impenetrable barrier for both water and ions, so that transit through the cell wall and intercellular spaces is no longer possible and access to the xylem is blocked. The ions must pass through the plasma membrane (or plasmalemma) at this point or at some place further out in the cortex of the root. If an ion passes the plasma membrane further out in the root, it can move to the xylem by going from cell to cell through the cytoplasm. This transit route is termed symplastic (Fig. 13–2). Regardless of their route, once the ions are past the endodermis, they can move to the xylem vessels relatively easily.

The major control point for movement of ions is at the plasma membrane. The plasma membrane is composed of a lipid bilayer, that is, a double layer of lipid molecules. The important concept to remember is that this lipid bilayer, although readily permeable to water, is essentially imperme-

able to ions. As an example, the plasma membrane is about 1 million times less permeable to a K^+ ion than it is to a water molecule. (Gases like O_2 and N_2 move through membranes even more easily than water.) This impenetrability is based mainly on the fact that ions are charged and hydrated.

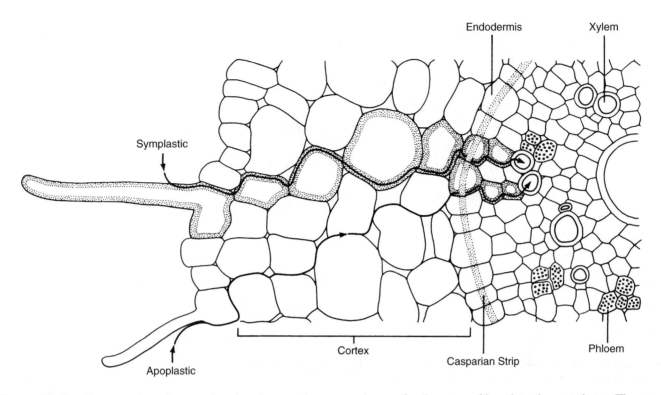

Figure 13–2 Cross section of a root showing the two alternate pathways for the entry of ions into the root tissue. The presence of Casparian strips in the endodermis prevents apoplastic movement of ions into the vascular tissue. Thus, at some point, ions must cross the plasma membrane of root cells and move through the cytoplasm.

Because the interior of the lipid bilayer is neutral in charge and is hydrophobic, it repels the diffusion of ions.

So how do ions get through the plasma membrane? The answer is that proteins are embedded in the membrane (Fig. 13–3), and these proteins serve as channels or carriers for the ions. The exact mechanisms by which ions are transported through membranes are beyond the scope of this discussion. But with these transport proteins, the specificity and transport activity for ion movement are established. To repeat, ions do not move freely in the plant tissue; their movement is strictly controlled and facilitated by transport mechanisms in membranes.

Before concluding this section, the growth and structural features of roots that allow for absorption of nutrients in the soil must be considered. First, it is important to recognize that roots are constantly growing during the spring, summer, and fall months. One can think of roots as exploring the soil; that is, as nutrients are depleted at one area of the soil, it is necessary for the plant to search for ions in new locations. Second, it is important to emphasize the role of root hairs (Fig. 13–1). These structures, which emerge about 1 mm behind the root tip, are extensions of epidermal cells that increase the surface area of the growing root many, many times. As the root grows, these hairs are eventually sloughed

off. Thus, continued growth is also necessary to promote continuous root-hair development. Finally, as one might guess, root hairs are the site of much of the active ion uptake because of their enormous surface area.

ELEMENTS THAT PLANTS NEED FOR GROWTH AND REPRODUCTION

Plants are mainly composed of the elements carbon (C), hydrogen (H), and oxygen (O). In fact, if the water is removed from plant tissues, these three elements comprise greater than 95 percent of the dry weight of these tissues. Relative to the amount of nitrogen shown in Table 13–1, the amount of oxygen in a typical plant tissue is thirty times greater, carbon is forty times greater, and hydrogen is sixty times greater. These three elements are abundant on the earth in various forms: water, CO_2, and O_2, for example.

In this chapter, the emphasis is on the other elements required for the growth of plants. To begin, note the relative concentrations of various elements in plant tissues (Table 13–1). Of course, these are very rough approximations; the actual numbers vary widely from plant to plant and from tissue to tissue. The objective here is to understand a little about why these other elements are important and to learn

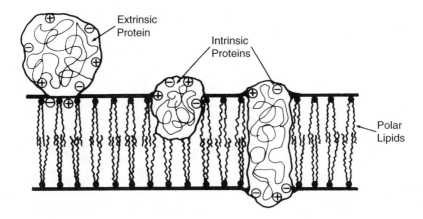

Figure 13–3   General features of proteins associated with the plasma membrane. Surface charges on the proteins allow them to be associated with the surface of the membrane or be partially or completely embedded in the membrane. Proteins that span the entire membrane can act as channels or carriers of ions.

something about where the elements come from and how the plant obtains them. Most of the emphasis will be on the elements that most often limit plant growth in crop production systems—nitrogen (N), phosphorus (P), and potassium (K).

Nitrogen

Nitrogen is the element that most often limits plant growth. There are two reasons that plants are relatively starved for nitrogen. First, although there are enormous amounts of nitrogen in the atmosphere, dissolved in the oceans, and trapped in rocks, this nitrogen is in a chemical form that cannot be used by plants, namely, nitrogen gas (N_2). In contrast, the biologically useful forms of N are extremely scarce.

Second, although there is also a huge amount of N in most soils, again, the N is in a chemical form that plants cannot use directly. This soil N is in the form of organic matter—a term that includes a broad group of materials and compounds, mostly resulting from the decay of plant or animal material. For example, many agricultural soils contain about 3,000 kg of N per hectare in the organic matter fraction. However, only about 2 to 3 percent of the organic matter N is converted to usable forms of N in a given year. This conversion is the result of microbial growth and activity. The availability of 60 to 90 kg per hectare of N from soil organic matter is enough to support minimal crop growth, but it is not nearly enough to support profitable crop growth. For example, a good crop of field corn requires at least 300 kg of N per hectare just for seed production.

Another natural source of N for plant growth comes from the conversion of molecular N_2 in the atmosphere to nitrate during lightning strikes. This source of N is small and is highly variable between locations and years. Thus, while natural vegetation (forests, wild grasslands) can get by on N

from organic matter and lightning strikes, these sources of N are not nearly sufficient to supply the N needs of crop plants. Note that in natural systems, nitrogen is recycled through the death and decomposition of plant material. In crop systems, much of the nitrogen is removed during the harvest.

Why is nitrogen so critical for plant growth? Nitrogen is the key element in proteins, and cells cannot function without a wide variety of proteins. Because of their relatively high protein content, plant seeds are important food sources for animals, including humans. Animals and humans cannot live without this source of plant proteins.

Nitrogen exists in several forms in nature. A sketch of the relationship between various natural forms of nitrogen is provided in Figure 13–4. In the figure, we see that plants have access to several forms of nitrogen. Plants cannot use nitrogen gas (N_2), the most abundant form of N on the earth. Also, nitrite ion (NO_2^-) is toxic to plants, leaving only two forms of N that plants can absorb and utilize for the construction of protein: ammonium (NH_4^-) and nitrate (NO_3^-).

These two forms of N are common in soils, but nitrate is by far the dominant form in almost all agricultural soils because soil microorganisms can get energy from ammonium and rapidly oxidize the ammonium to nitrate. Thus, if a farmer applies anhydrous ammonia or urea fertilizers, these compounds are converted to nitrate within a few days, and nitrate becomes the form of N that is available to plants. This process of nitrification presents a problem in agriculture because nitrate N is not bound to soil particles and can be leached into the groundwater by rain. Consumption of too much nitrate by humans is detrimental to health, so contamination of drinking water with nitrates from agricultural runoff is a problem.

Nitrate is absorbed by roots by way of a nitrate-specific transport protein in the plasma membranes of the

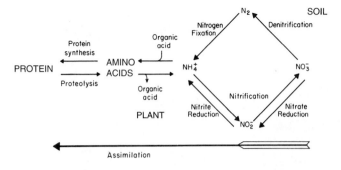

Figure 13–4 The biological nitrogen cycle. The incorporation of single elements or ions (e.g., ammonium) into organic compounds is termed assimilation. When plant tissues die and disintegrate, the reverse process takes place, making the elements and ions available to other organisms or available for remobilization (see Fig. 13–10).

root cells. Because of the very large N requirement of plants and the relative scarcity of N in soils, it is not surprising that this nitrate transporter has a high affinity for the NO_3^- ion and that this transporter is present with high activity in all roots. Once inside the cytoplasm of a root cell, the nitrate has two possible fates. First, it can move to the xylem and be transported as nitrate ion to the shoot for further processing. A second possibility is that the nitrate may be reduced to ammonium in the root, and the ammonium may be assimilated into amino acids and then transported to the shoot as amino nitrogen (see Fig. 13–4). Note that ammonium, like nitrite, can be toxic to plant cells. Thus, while the nitrate content of plant tissues is relatively easy to measure, the concentrations of nitrite and ammonium are usually below detectable levels. When ammonium is absorbed from the soil solution, it is rapidly incorporated into amino acids.

As already stated, the main importance of N is its requirement for incorporation into proteins. More than 95 percent of the N in a plant leaf is in the form of protein. Other N pools include nucleic acids (DNA, RNA), free amino acids, membrane components, pigments (especially chlorophyll), and hormones. Protein content of seeds is generally high relative to other plant tissues; for example, the protein concentration of soybean seeds is about 38 percent of dry weight. Protein is stored in the seed to provide a source of N and other nutrients to the seedling after germination. Of course, humans take advantage of this seed protein by collecting the seeds and feeding them to animals or eating them directly.

Besides nitrate, some plants have access to another source of N by way of a process called nitrogen fixation. This process, which involves the conversion of N_2 gas from the atmosphere into ammonium (Fig. 13–4), can be carried out by several groups of bacteria. The bacteria that are of greatest agricultural importance are named rhizobia. Within this group are several genera and many species. These bacteria infect legumes such as the vegetable crops peas and beans and the agronomic crops soybean and alfalfa (also legume trees such as the locust). They enter the root through root hairs and form growths on the root called nodules (Fig. 13–5, left).

When one slices open a root nodule, several different tissues can be seen (Fig. 13–5, center). The outer portion of the nodule is called the cortex, and vascular bundles in the cortex connect the nodule and the root. The darker interior region of the nodule contains cells that house millions of nitrogen-fixing bacteria (Fig. 13–5, right). All the cells in the nodule are plant cells that have developed to accomplish a specific task. The plant supplies the bacteria with organic nutrients (e.g., carbohydrates), and in return the bacteria supply the plant with ammonium. For the growth of most legumes, application of nitrogen fertilizer is not needed or recommended. Some soil nitrogen is needed to supplement fixed nitrogen, but the supply from organic matter or carried over from the previous crop is sufficient.

The chemistry of biological nitrogen fixation is similar to what occurs in the industrial manufacture of ammonia. In fertilizer manufacture, the conversion of N_2 to NH_3 occurs at a very high temperature and at great pressure in the presence of a catalyst (Fig. 13–6). Methane is used as a source of energy and CO_2 is a by-product. The process is quite energy intensive and expensive. In the biological process, carbohydrates supplied by the plant are the source of energy; the product, ammonia, is the same; and CO_2 is again a by-product (Fig. 13–7). The remarkable thing about the biological process is that it occurs at 1 atmosphere of pressure and at ambient temperature. Thus, the catalyst in the bacteria is much more efficient than anything chemists have been able to devise.

In general, a species of rhizobia infects only one species of legume. Thus, pea rhizobia do not form nodules on bean plants and soybean rhizobia do not form nodules on alfalfa. When a legume is first grown in the field, it is very important to supply an appropriate rhizobial inoculant, and these inoculants are commercially available at feed and fertilizer dealers. Once a host plant has been inoculated and grown in a field, the rhizobia for that host can survive in the soil for several years. Survival varies among species so, if the time between legume crops is more than a few years, re-inoculation at the next planting is recommended. Inoculants are a minor portion of total crop production costs, and the benefits from nitrogen fixation are great. In general, inoculants are applied to the seeds at the time of planting; both dry and liquid formulations are available. Inoculation at planting time is important to allow the bacteria to become established in the soil before the first root hairs appear.

Phosphorus

After nitrogen, phosphorus is generally the second most limiting element for plant growth because much of the P in the soil is tied up in organic matter and because many salts of

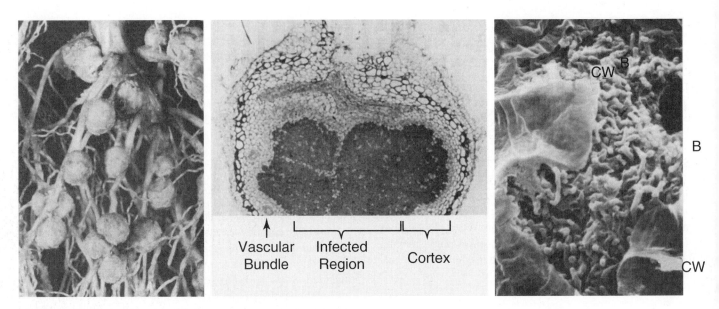

Figure 13–5 *Left:* Mature nodules on the root of a soybean plant. *Center:* Cross section through a mature nodule. Note the location of vascular bundles in the cortex and the connection of these bundles to the root, which was located at the top of the section. *Right:* Scanning electron micrograph of an infected cell of a soybean nodule showing nitrogen-fixing bacteria. CW = cell wall; B = bacteria.

Figure 13–6 Some features of the manufacture of ammonia. Note the very high pressure and temperature required for the conversion of N_2 gas to ammonia. Other common nitrogenous fertilizers—urea, ammonium nitrate, ammonium sulfate—can be synthesized from the anhydrous ammonia. *Source:* John Streeter, The Ohio State University.

phosphate are only slightly soluble. In most agricultural soils, the concentration of available P is on the order of 1/4,000 the concentration of nitrate and 1/500 the concentration of potassium. Thus, although the P requirement is only about 1/20 the requirement for N and only about one-fourth

the requirement for K (Table 13–1), achieving a sufficient supply of P for plant growth and economic yield is often a challenge for the growing plant.

The availability of P is most severely limited in both acid and alkaline soils, and a soil pH of between 6 and 8 is

0

Figure 13–7 The conversion of N_2 gas to ammonia in legume nodules. In this case, sugar from the plant provides the energy, but much of the chemistry is the same as the industrial synthesis of ammonia. *Source:* John Streeter, The Ohio State University.

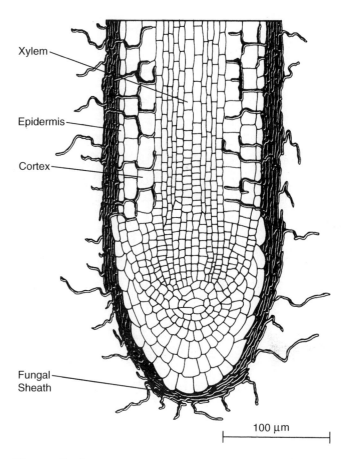

Figure 13–8 A plant root infected with an ectotrophic mycorrhizal fungus. With this type of fungus, the mycelium forms a sheath around the root and also enters the intercellular spaces in the root cortex. *Source:* Taiz, L. and E. Zeiger. 1991. *Plant Physiology,* 2nd edition. Sunderland, Mass. Sinauer Assoc.

needed to maximize P availability. Thus, one strategy for supplying adequate P is simply to check soil pH routinely by testing the soil and then applying lime when necessary. A second strategy is to add P to the soil in the form of fertilizer when a soil test indicates the need for additional P to support profitable crop yield. Phosphorus fertilizer is usually incorporated into the soil at the time of planting because P is relatively immobile in the soil. P from agricultural runoff becomes a problem usually because soil particles wash into rivers and lakes.

A third possible way that plants obtain P from the soil is with the help of mycorrhizal fungi. Some types of mycorrhizal fungi form a sheath around the root surface, and extensions of the hyphae grow between the cells of the root cortex (Fig. 13–8). The hyphae of a second type of mycorrhizal fungus actually enter the cells of the root cortex (Fig. 13–9). In both cases, the function of the fungus is much like that of root hairs. That is, the fungal mycelium allows for exploration or a much larger volume of soil, and the fungus associates with the roots even after the root hairs are gone. Also, the fungus can absorb P from the soil more efficiently than the plant root; that is, the fungus has a P transporter in its plasma membrane that has higher affinity for P than the transport system in the plant root.

These fungi associate with the roots of a wide variety of plants, and new mycorrhizal/plant associations are continuously being discovered. There is currently great interest in exploiting these fungi in crop production. At some time in the future, it may be possible to buy a mycorrhizal inoculant that would increase the production of a crop due to the greater efficiency of P acquisition.

Phosphorus plays many vital roles in the plant cell. One role is the presence of this element in energy-rich compounds such as adenosine triphosphate (ATP), which was discussed in more detail in Chapter 10. A second critical role is

as a component of membranes; phosphate groups on the surface of membranes help to stabilize the structure and to bind other membrane components such as proteins. A third role is the presence of P in nucleic acids (DNA, RNA); the backbone of these molecules is a chain of sugars linked together by phosphate groups.

Potassium, Calcium, and Magnesium

These three elements are grouped together because, in general, the amounts and availability of these elements are not so problematic compared to N and P. Although addition of K in fertilizer is usually required to provide maximum crop yields, K^+ is retained in the soil because of its positive charge and is not easily leached. Also, sources of K on the earth are more plentiful than are sources of N and P. Potassium fertilizers are also cheaper to produce because the mineral sources of K can be used directly. Potassium is highly mobile in the plant and, in fact, one of its roles is to provide charge balance

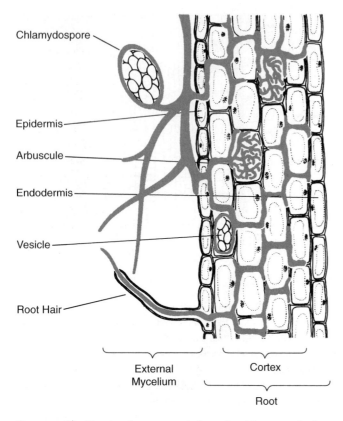

Chlamydospore

Epidermis

Arbuscule

Endodermis

Vesicle

Root Hair

External Mycelium

Cortex

Root

Figure 13–9 A plant root infected with a vesicular-arbuscular mycorrhizal fungus. With this type of fungus, a sheath around the root is not formed, but the fungal mycelium actually enters the cortical cells. The nomenclature of these fungi is based on the formation of sacklike structures called vesicles and branched structures called arbuscles in the root cells. *Source:* Taiz, L. and E. Zeiger. 1991. *Plant Physiology,* 2nd edition. Sunderland, Mass. Sinauer Assoc.

in the translocation of negatively charged metabolites and anions like nitrate and phosphate. Potassium is critical for the osmotic control of stomates and thus it is important in the processes of photosynthesis and transpiration. Finally, potassium is required for the activation of many enzymes.

The size and charges on the Ca^{++} ion make it ideal for several structural roles. Pectin is the polymer that acts as a glue between the walls of adjacent cells. It is a polymer of galacturonic acid, which is a compound that carries a negative charge. Calcium ions bind to adjacent negative charges on the galacturonic molecules, giving the pectin glue a stable structure. This role of Ca^{++} is quite apparent because severe Ca^{++} deficiency usually results in plant tissues collapsing and becoming soft and rotted. This problem is fairly common in tomato plants, where there is often insufficient transport of Ca^{++} into the rapidly growing fruit. As a result, the fruit becomes unmarketable.

A second major role of calcium is similar to the first—stabilization of structure. That is, Ca^{++} preferentially binds to the phosphate groups on the surface of membranes. The solid circles at the surface of the membrane shown in Fig. 13–3 represent head groups, which vary widely in chemical composition but which almost always include a phosphate group. The membrane structure is inherently unstable, and binding of Ca^{++} to the phosphate (PO_4^{3-}) groups serves to stabilize the structure. Because of its size and hydration, the Ca^{++} ion is uniquely suited for this role. This function of Ca^{++} can be demonstrated by growing plants in solution culture with very low Ca^{++} supply. Under these conditions, the roots leak metabolites such as carbohydrates; however, plant tissues rarely become leaky under natural growing conditions because the supply of calcium in the soil is usually adequate.

In fact, the supply of Ca^{++} can become a problem to the plant cell because this ion forms salts of phosphate that are only very slightly soluble. The Ca^{++} concentration in the cell cytoplasm must be kept very low, and this level is accomplished by transporting excess Ca^{++} into the vacuole. Plant cells have directional pumps for Ca^{++} that allow for maintenance of appropriate Ca^{++} levels in various cellular compartments.

Magnesium plays several roles in plants, but two stand out. First, Mg^{++} is an essential component of chlorophyll. Chlorophyll is one of the most abundant molecules in nature and plays a key role in reception of light energy, so the availability of Mg^{++} is crucial. The second critical role of magnesium is as a cation balancing the phosphate charges in adenosine triphosphate (ATP). ATP is involved in hundreds of essential metabolic reactions, and the reactive form in ATP is the magnesium salt. Thus, although we often write:

$$\text{substrate} + \text{ATP} \rightarrow \text{product} + \text{ADP} + PO_4^{-3}$$

the biochemically correct way to write the reaction is:

$$\text{substrate} + \text{MgATP} \rightarrow \text{product} + \text{MgADP} + PO_4^{-3}$$

The Mg salt of ATP is the reactive form because the Mg atom is required for binding of ATP at the active site on enzymes. (The function of ATP was discussed in more detail in Chapter 10.)

MINOR ELEMENTS

The limited space in this textbook does not allow for a detailed discussion of the other elements required for plant growth. The remaining required elements are listed in Table 13–1. Elements other than sulfur (often considered one of the major elements) are needed in relatively very small amounts and are usually termed minor nutrients. Most soils supply adequate amounts of these elements, although inclusion of them in nutrient solutions in greenhouse growth systems is often required. Some of the roles of these minor elements are listed in

TABLE 13–2 Minor Elements Essential for Plant Growth and Some Examples of Their Roles

Element	Example of Roles in the Plant
Sulfur	A component of some amino acids and essential for the maintenance of structure for some proteins. A component of factors involved in many metabolic reactions. Plays a role in flavor in some vegetables such as cabbage.
Chlorine	Mainly involved in ion balance; that is, as a plant absorbs more cations than anions, negative charges are required to avoid development of a charge. Required for one of the photosynthetic reactions.
Boron	Required in nucleic acid biosynthesis and in maintaining membrane structure.
Iron	Required for chlorophyll biosynthesis. A component of cytochromes and some reactions involved in electron transfer.
Manganese	Required for the catalytic activity of several enzymes. Also required in one of the key photosynthetic reactions.
Zinc	Also required for the catalytic activity of several enzymes.
Copper	Same as for zinc.
Molybdenum	Required for the activity of nitrate reductase, a key component in the assimilation of nitrogen (see Fig. 13–4).
Nickel	A component of the enzyme urease.

Table 13–2. A common theme in the table is the importance of these minor elements in plant metabolism.

NUTRIENT DEFICIENCIES

Many books list typical nutrient deficiency symptoms. Although they can sometimes be helpful, reliance on so-called typical nutrient deficiencies can be dangerous because deficiency symptoms vary slightly from plant to plant, deficiency of more than a single element may be involved, and many deficiency symptoms are confounded by reactions of the plant to disease. Thus, to examine plant leaves and decide that the plant needs more nitrogen or more iron or more sulfur is difficult because all of these deficiencies lead to chlorosis of leaves.

In recent years, several informative websites have appeared showing colored pictures of deficiencies of elements on various horticultural and agronomic crops. Two such websites are listed in the reference section, and you are urged to visit them to see what information they provide and to witness the difficulty associated with visual analysis of deficiencies. Other sites continue to appear and can be found through web searches.

A more reliable way to establish the cause of an apparent deficiency is to send samples of the soil or medium in which the plant is growing and plant tissue to a laboratory for analysis. Many such laboratories are maintained by agricultural supply companies and can be found through the local agricultural extension agent. These laboratories maintain lists of normal concentrations of elements in a wide variety of crop plants. Acceptable nutrient levels have been established through many years of research at agricultural universities. Generally, the testing lab supplies an analysis of all major and most minor elements and provides warning if the concentration of one or more elements is outside the normal range for that crop. Although some cost is involved, this approach is clearly the most reliable way to establish the cause of a nutrient deficiency.

REMOBILIZATION

Reproductive fecundity is so important in the survival of a species that mechanisms have evolved for plants to preserve precious nutrients for future generations. In practical terms, this adaptation means that seeds of crop plants are usually a rich source of nutrients relative to other plant tissues. This generality is true for perennial plants as well, although long-lived perennials like trees store much of their critical nutrients in roots during the winter for reuse in the following growing season.

As you might guess, the main nutrients that are important for storage in seeds are those that are relatively scarce in the environment: K and especially N and P. During the growing season, the plant expends considerable energy importing these nutrients from the soil solution. All three of these nutrients are relatively mobile, and large quantities of these elements are moved from vegetative tissues to reproductive tissues during reproductive development and the onset of senescence. This process is known as remobilization.

An illustration of N, P, and K remobilization is shown in Fig. 13–10. Note that the vertical scale on this figure is percentage, with 100 percent being shown near the end of dry-matter accumulation. Thus, the actual amounts of nutrients translocated from vegetative to reproductive tissues are not shown. However, this presentation emphasizes the relative depletion of nutrients from vegetative tissues during reproductive development. Note that accumulation of N, P, and K continues during early reproductive development but that, at roughly eighty-five to ninety days after emergence, nutrient accumulation stops. Beyond this time, the nutrient content of leaf blades, stems, and petioles starts to decline. Between eighty-five and 120 days after emergence, a massive expansion of the proportion of nutrients occurs in the beans. Some of the nutrients are lost in fallen leaf blades and petioles but by the time of leaf drop, much of the N, P, and K has already been withdrawn and remobilized to the seeds.

The same process occurs in all plants that undergo annual senescence of some vegetative tissues, although for

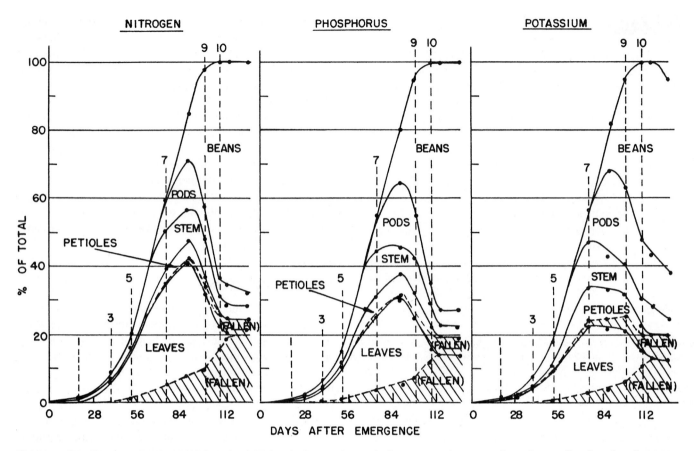

Figure 13-10 Distribution of N, P, and K in soybean plants during vegetative growth and reproductive development. "Fallen" refers to leaf blades and petioles that have senesced and fallen from the plant. Note that the vertical axis of the figure is in units of percent, so that actual amounts of nutrients are not shown, only the relative distribution of elements in the plants. *Source:* Agronomy Journal, Vol. 63, page 408. 1971. Madison, Wisc. American Society of Agronomy.

some plants, the destination for nutrients is the roots and not the seeds. Thus, by harvest time, the N, P, and K content of corn stalks or grape leaves or apple leaves is very low—too low for these organs to be functional. They are largely composed of cellulose and other polymers of C, H, and O. As such, these expendable plant tissues provide organic material when returned to the soil, but they do not constitute a supply of N, P, and K that is adequate to grow the subsequent crop.

SUMMARY AND REVIEW

Nutritional elements move into the plant root as charged ions through a complex system that involves free movement (apoplastic) and selective transport (symplastic). Once inside the root, ions are transported relatively easily through the xylem to other parts of the plant, where once again selective transport directs where each specific ion is allocated. The plasma membrane is the sight of most selective control.

The absorption of minerals into the root depends on the growth of the root and its structural features. One can think of growing roots as exploring the soil to search for ions in new locations. The root hairs are the main site of nutrient absorption and in most healthy roots, they have an enormous amount of surface area.

Nitrogen is the element that most often limits plant growth. It is not available to plants as N_2; it has to be con-

verted to NO_3^- or NH_4^+. Much of the N in the soil is in an organic form, also unavailable to plants until converted to a usable form during decomposition. Nitrogen is essential for plant growth for many reasons, but its role in protein structure accounts for the greatest allocation of the element in plants. Some plants develop a relationship with rhizobia bacteria. These bacteria form nodules on the plant root where they convert N_2 to ammonium in a process called nitrogen fixation. The plant can use the ammonium and in turn, the bacteria can feed on the carbohydrates produced by the plants.

Most crop plants have a need for N that is greater than the supply in the soil and thus require N fertilization. N fertilizers can be applied as ammonia/urea, NH_4^+, and NO_3^-. NO_3^- is the form most readily taken up by plants, but it is also the form that is most easily leached from the soil into the groundwater drinking supply, creating health hazards for humans.

Phosphorus is the second most limiting element for plant growth. Although plentiful in most soils, it is in unavailable or only slightly available forms. Soil pH influences P availability, with a pH of 6 to 8 being the ideal. Phosphorus availability can often be improved by adjusting soil pH. If fertilization is necessary, it is often done at the time of planting (or when mixing artificial media) because it is not easily leached from the soil. When it is a problem in water, it is because of surface erosion, not leaching. Mycorrhizal fungi can play a role in helping plants obtain P. These fungi are associated with many different plants and form a sheath around the plant root. Hyphae from the fungi spread out and extract P from areas distant from the root. In a sense, they act like root hairs. The role of phosphorus in plant metabolism is in energy-rich compounds such as ATP. Phosphorus is also a critical element in the structure and function of membranes. A third role is as part of the sugar-phosphate backbone in DNA and RNA molecules.

Potassium, calcium, and magnesium availability is not as problematic in crop production as is N and P, although K is usually provided to improve crop yield. These elements are positively charged and do not leach out of the soil very easily. Potassium is critical for the function of stomates in regulating transpiration and CO_2 uptake into the leaf. Calcium has two roles as a structurally stabilizing molecule. The first is as part of pectin, which acts as a glue between cell walls. The inability of plants to transport Ca into rapidly developing tissues and organs such as fruit results in the tissues becoming soft and rotten. The second is in membrane stabilization. Magnesium has several roles in plant metabolism, but two stand out. The first is as an essential component of chlorophyll and the second as the cation that balances the anion phosphate charge in ATP.

Except for S, the rest of the essential elements are minor and are needed in small amounts. Each has one or more critical functions in plants. Most soils provide these elements in adequate quantities, but plants grown in artificial mixes require their addition to fertilizer formulations.

Whether an element is major or minor, if it is deficient in the plant, deficiency symptoms will develop, and the growth, development, and reproduction of the plant will be compromised. Deficiency symptoms are sometimes visible and characteristic of a particular nutrient, but most often the visual symptoms are hard to distinguish between elements, or more than one element may be deficient. Soil and tissue testing are necessary to determine how much of an element is available and how much is actually being taken up and incorporated into plant tissues. Deficiency symptoms for mobile ions usually appear first in old leaves. Immobile ion deficiencies usually appear in young leaves.

Remobilization is the process plants use to reallocate nutrients from one part of the plant to another. Usually the allocation is from old tissue to reproductive tissue. Some elements are more readily mobilized than others. N, P, and K are readily mobile. Others such as iron are relatively immobile.

FOOD FOR THOUGHT

1. Farmers, landscapers, golf course superintendents, and anyone else who makes a living from growing plants are always as concerned about the health of the roots of their plants as much as the above-ground portions. In terms of nutrient uptake, describe what would happen if a plant lost a large portion of its roots to insect feeding or disease infection. What symptoms would likely appear?

2. Two classes of microorganisms (rhizobia and mycorrhiza) play important roles in nutrient uptake. How do each of these fungi aid in nutrient uptake?

3. Although N is one of the most abundant elements in the soil and air, it is often one of the most limiting in crop growth. Explain how something so abundant can be so limiting.

4. The terms *major* (macro-) and *minor* (micro-) *nutrients* seems to imply that one group is more important than the other. Explain why each group is equally important to crop growth and development.

5. What is remobilization and how does it affect reproduction in plants? What characteristic of an element determines if it is remobilized or not?

SUPPLEMENTARY READING

EPSTEIN, E., and A. J. BLOOM. 2004. *Mineral nutrition of plants: Principles and perspectives,* Second Edition. Sunderland, Mass.: Sinaur Associates, Inc.

MARSCHNER, H. 1995. *Mineral nutrition of higher plants,* Second Edition. London: Academic Press.

TAIZ, L., and E. ZEIGER. 2002. *Plant physiology,* Third Edition. Sunderland, Mass.: Sinauer Associates, Inc.

Websites for nutrient deficiency symptoms:
http://www.luminet.net/~wenonah/min-def/tomatoes.htm
http://www.nrs.mcgill.ca/whalen/nutrient/Calcium/ BigPixCa.html#Tomato01Ca

CHAPTER

14

Genetics and Propagation

KEY LEARNING CONCEPTS

After reading this chapter, you should be able to:

♦ Understand how the basic concepts of genetics relate to the production and utilization of plants.

♦ Explain the common methods of plant breeding and sexual and asexual propagation.

♦ Understand how genetic engineering is used to introduce genetic traits into plants from unrelated or distantly related organisms.

BASIC GENETIC CONCEPTS IN PLANT SCIENCE

The plants we cultivate for our survival and pleasure all originated from wild plants. However, most of our domesticated plants appear very different from their wild relatives. These differences have come about mainly through the recombination of genes and the redistribution of heritable traits through the generations separating the wild types from the domestic plants. Manipulating natural genetic processes (discussed in Chapter 15) has been the basis for crop improvement since the time of the first plant scientists. Understanding basic genetic principles is crucial to understanding the principles of the traditional methods of plant breeding and plant propagation and the most recent tool for crop improvement: bioengineering.

Chromosomes

Some plants consist entirely of living cells, whereas others, such as woody perennials, are made up of both living and dead cells. The living cell consists of a **cell wall** containing a fluid, the **cytoplasm,** in which the **nucleus** is suspended, along with many other structures. The nucleus contains the **chromosomes,** which carry most of the genetic information and transmit that information from one generation of cells to the next. The number of chromosomes is the same in all vegetative cells of an entire plant species. This number is usually the **2n,** or **diploid,** chromosome number (although higher numbers—tetraploid or octaploid, for example—also occur). In the sex cells—the egg and sperm—the number is reduced by half and is termed the **haploid,** or **1n,** chromosome number.

Although usually constant for a given species, the number, size, and appearance of chromosomes vary considerably between different plant species. The chromosomes can be counted by microscopic examination of the nucleus at a stage just before cell division. The chromosome numbers are known for most plant species. For example, the diploid chromosome number for alfalfa (*Medicago sativa*) is 32; for barley (*Hordeum vulgare*), 14; for corn (*Zea mays*), 20; and for sugar beet (*Beta vulgaris*), 18.

Chromosomes change in appearance during cellular development and division, but basically each is a long, threadlike structure consisting of **deoxyribonucleic acid (DNA)** plus associated **ribonucleic acids (RNA)** and various proteins. DNA can replicate itself and it can transmit information to other parts of the cell. DNA is a polymer—a very large molecule made up of many repeating units, but the repeating units called **nucleotides** can vary. DNA, the active genetic material, is a very large molecule composed of two spiral strands (see Fig. 14–1). The "backbone" of these strands is composed of sugar residues (S) linked by phosphates (P) on each side. A sugar residue on one strand is connected with a sugar residue on the other by two bases that are linked to each other by hydrogen bonds (Fig. 14–1). These bases are cytosine (C), guanine (G), adenine (A), and thymine (T). The sugar residues in each strand are held tightly together by the phosphate radicals, but the two spiral strands are bound together more loosely by the hydrogen bonds.

One of the characteristics of DNA that makes it possible for the chromosomes to transmit genetic information from one cell generation to the next is its ability to replicate itself. This ability arises from the double-strand structure of the molecule and the properties of the four bases. Adenine and thymine are held together by two hydrogen bonds (A = T). Their molecular structure is such that only they can join together. In the same manner, only cytosine and guanine can join together, and they are held by three hydrogen bonds (C = G). When the chromosome divides during cell division, the two spiral strands of the DNA molecule unravel and separate at the position of these hydrogen bonds. Every base (cytosine, guanine, adenine, or thymine) attached to each strand attracts its complementary base, so that each single strand immediately becomes a new double strand exactly the same as the original double strand (see Fig. 14–1).

Although DNA includes the self-perpetuating **genetic code,** a related substance—ribonucleic acid (RNA)—actually controls the growth processes in the cell. DNA and RNA have certain differences. DNA is double-stranded, whereas RNA is a single strand. RNA sugars have one more oxygen atom than DNA sugars. RNA has uracil (U) as a base in place of thymine (T). The DNA molecule acts as a template from which a complementary strand of RNA is formed (in the same manner by which a complementary strand of DNA forms from a single strand).

The form of RNA that carries the genetic instructions as a complementary copy of the DNA series of bases is called **messenger RNA.** Still another form of RNA—**transfer RNA**—brings the amino acids to the ribosomes to construct the proteins. Proteins may be structural or they may be enzymes that act as catalysts in biochemical reactions. For all organisms, the genetic code is the same, meaning that a DNA sequence is translated into the same protein in every organism. This universal concept is what enables genetic engineers to take the DNA from one species and put it into another and get the same protein in the second organism. The proteins then perform the same function in the new plant as they did in the old, as long as anything influencing that protein's function is the same. One of the difficulties in genetic engineering is that often the new plant does not have the same factors influencing the new protein that the original plant had.

Genes

Structurally, a gene[1] is a sequence of triplet organic bases (cytosine, guanine, adenine, and thymine) along a DNA molecule. The gene is the ultimate hereditary unit that functions as a certain part of a chromosome determining the development of a particular characteristic in an organism. Thus, one gene (or several interacting genes) may determine plant height, leaf shape, flower color, or fruit size. Genes are much too small to be seen, even with an electron microscope. There are, of course, a great many genes in each cell of the higher plants; they number in the thousands. Any individual gene may have a large effect or a small effect. Some genes act independently, whereas others act only in conjunction with other genes.

Because genes are arranged along the chromosome, genes on the same chromosome are linked—that is, genes on the same chromosome move from one cell generation to the next as a unit. Linkage is not perfect, however; sometimes during meiosis (reduction cell division), chromosomes break and exchange parts, as we will soon study in more detail.

---

[1]Virus particles are similar to genes. Because viruses can be isolated and easily studied, much knowledge of gene structure is obtained from the study of viruses.

Figure 14–1   *Left:* Schematic diagram of the DNA molecule showing the helical structure and the base pairing. Lower detailed view shows the alternate attachment of sugars (S) and phosphates (P) on each strand. Connecting the two strands are the base pairs adenine (A) —thymine (T) guanine (G) —cytosine (C), which are held together by hydrogen bonds and attached to the sugars on each strand. *Right:* During cell division and chromosome splitting, the DNA molecule replicates itself by unraveling, separating at the hydrogen bonds. Then each strand quickly becomes a new double strand just like the original, with each base attracting to itself its complementary base (A = T and G = C). *Source: Left:* McElroy, William D; Swanson, Carl P. *Modern Cell Biology.,* 2nd Edition, © 1976. Reprinted by permission of Pearson Education, Inc., Upper Saddle River, NJ. *Right:* Wright, J. W. 1976. *Introduction to forest genetics.* New York: Academic Press.

Homologous Chromosomes

In each vegetative cell are pairs of each individual chromosome. These pairs are called **homologous chromosomes** and are properly defined as such if they each have the same gene or genes affecting the same traits at corresponding positions. Genes are termed **alleles** to each other if they occupy the same position on homologous chromosomes and affect the same trait. If a plant had the genetic constitution ABCDEFG/abcdefg on a certain pair of homologous chromosomes, the genes A and a would be alleles, B and b would be alleles, and so forth.

Allelic genes can be dominant or recessive to each other. A **dominant gene,** A, is one that causes a certain characteristic to be expressed whether the plant is **homozygous,** AA—both alleles the same—or **heterozygous,** Aa—the two

alleles different. A **recessive gene** causes the character it controls to be expressed only if both alleles are recessive, aa. (By convention, capital letters express dominant genes; lowercase letters express recessive genes.)

Mitosis

Cell division in the shoot and root tips, axillary buds, leaf primordia, and the vascular cambium—all of which increases plant size—is called **mitosis.** These vegetative cells usually contain two sets of homologous chromosomes—the 2n or diploid number. During cell division, the chromosomes split longitudinally, replicating to produce two chromosomes that are identical to each other. One of each pair goes to one daughter cell, and one to the other. An equatorial plate, as it is called, develops between them to form a new cell wall and thus two new cells, each with its full complement of chromosomes—and genes. So each daughter cell has a genotype identical to that of the mother cell.

Meiosis and Fertilization

Meiosis refers to the type of cell division, sometimes called **reduction division,** that occurs in the flower—in the angiosperms—to form the cells from which the pollen grains and the embryo sac (which contains the egg) develop (see Figs. 14–2 and 14–3). In this type of cell division, the homologous chromosomes separate from each other without replicating, one going to one daughter cell and one to the other, thus reducing the number of chromosomes—the 1n or haploid number. During pairing of the two sets of homologous chromosomes, **crossing over** can occur. The chromosomes may break at the same locus on each so that they may rejoin after exchanging segments. Thus, if the original gene sequence on the homologous chromosomes was ABCDEFGHI/abcdefghi and crossing over occurred between E and F, then the gene sequence for this particular chromosome appearing in the pollen grain or egg cell would be ABCDEfghi or abcdeFGHI. This gene alteration would, of course, be expressed in altered characteristics of the new plants.

In **fertilization** in angiosperms, one male gamete (1n) from the pollen grain unites with a female gamete—the egg (1n)—to form the zygote (2n), which develops into the embryo and finally the new plant (see Figs. 14–2, 14–3, and 14–4). The manner in which these gametes can segregate and reunite to form new combinations is illustrated in Figure 14–4 for both a monohybrid and a dihybrid cross. Also during fertilization in the angiosperms, one male gamete (1n) unites with the two polar nuclei (1n each) in the embryo sac to form a food storage tissue, the endosperm (3n), that can serve as a nutritive material for the developing embryo. The seed coats, developing by mitosis from the female parent, cover both the endosperm and the embryo. In gymnosperms, the endosperm tissue—more correctly termed the female gametophyte—is 1n tissue, rather than 3n, as in the angiosperms (see Fig. 14–3).

Mutations

As plants grow, hundreds of thousands of cells may be dividing constantly. At each cell division, DNA must replicate itself—the double strand must untwist, the millions of nucleotide triplets must be reproduced exactly on new single strands, and the single strands must then twist around each other to form new double strands. It is a marvel that this complicated process takes place on such a grand scale. With so many cells involved, however, errors can and do occur during replication. When they do, they are called **mutations,** and the altered genes may possibly result in changes in the characteristics of the plant—although most mutations have such slight effects that they go unnoticed. The great majority of mutations are deleterious, but some are not and they provide a source of variability that aids the plant breeder in developing new cultivars. Mutation rates can be vastly increased by treatment with ionizing radiation and by certain chemicals. Mutations that occur during the formation of pollen grains and egg cells and appear in the plant's seedling offspring are particularly important to the plant breeder.

In addition to gene mutations, hereditary modifications can also be caused by gross changes in chromosome number or structure. This can involve the doubling of chromosome numbers, the addition or subtraction of an entire chromosome, or some other structural change in a chromosome. Such gross chromosome changes may cause pronounced changes in the plant's characteristics.

Polyploidy

Polyploidy is a condition in which individual plants have more than two sets of homologous chromosomes in their somatic (vegetative) cells. Beyond the normal diploid (2n) number, plants may be triploid (3n), tetraploid (4n), pentaploid (5n), hexaploid (6n), and so forth. Polyploid plants may arise by duplication of the chromosome sets from a single species—autoploidy—or by a combination of chromosome sets from two or more species—alloploidy. The latter is the more common type of polyploidy in nature. Many of the cultivated crop species evolved in nature as polyploids, as shown in Table 14–1 for oats, wheat, and tobacco.

Cytoplasmic Inheritance

While most inherited characteristics are transmitted by the genes in the nucleus, certain characteristics in some herbaceous plants can be controlled by cytoplasmic factors, which are contributed only by the female parent. Male sterility in corn is due, in part, to a cytoplasmic factor and is used to produce hybrid seed without the laborious procedure of hand detasseling.

Genotype and Phenotype

The term **genotype** refers to the genetic makeup of the plant—its genetic constitution—the kinds of genes it has on the chromosomes and the order in which they are situated. **Phenotype** refers to the plant's appearance, behavior, and chemical and physical properties. The phenotype expressed is also influenced by the environment. A plant may have genes for very vigorous growth but when grown under a deficiency of soil nitrogen, for example, its inherent vigor is not expressed. However, the genes still control the plant's characteristics even if an environmental factor drastically alters the phenotype. Therefore, if seeds are taken from stunted plants grown under deficient soil nitrogen conditions and planted in rich fertile soil, the plants again show the strong, vigorous growth called for by their genotype.

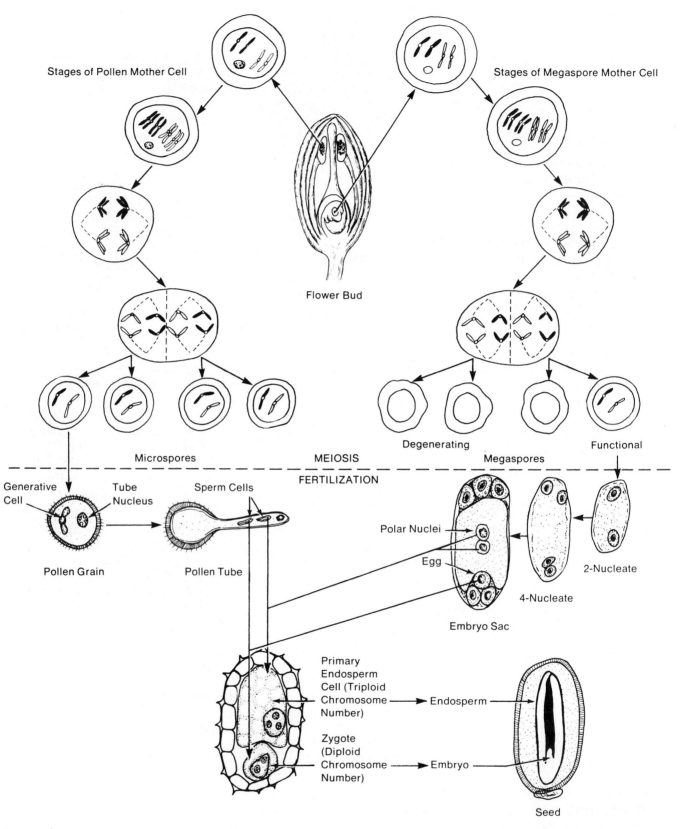

Stages of Pollen Mother Cell

Stages of Megaspore Mother Cell

Flower Bud

Microspores

MEIOSIS

Degenerating

Megaspores

Functional

FERTILIZATION

Generative Cell

Tube Nucleus

Sperm Cells

Polar Nuclei

Egg

2-Nucleate

4-Nucleate

Pollen Grain

Pollen Tube

Embryo Sac

Primary Endosperm Cell (Triploid Chromosome Number) ⟶ Endosperm ⟶ Endosperm

Zygote (Diploid Chromosome Number) ⟶ Embryo ⟶ Embryo

Seed

Figure 14–2 Diagrammatical representation of the sexual cycle in angiosperms. Meiosis occurs in the flower bud in the anther (male) and the pistil (female) during the bud stage. During this process, the pollen mother cells and the megaspore mother cells, both diploid, undergo a reduction division in which homologous chromosomes segregate to different cells. This is immediately followed by a mitotic division, which produces four daughter cells, each with half the chromosomes of the mother cells. In fertilization, a male gamete unites with the egg to produce a zygote, in which the diploid chromosome number is restored. A second male gamete unites with the polar nuclei to produce the endosperm. *Source:* Kester, Dale E., Hartmann, Hudson T.: *Plant Propagation: Principles & Practices,* 4th Edition © 1983. Adapted by permission of Pearson Education, Inc., Upper Saddle River, NJ.

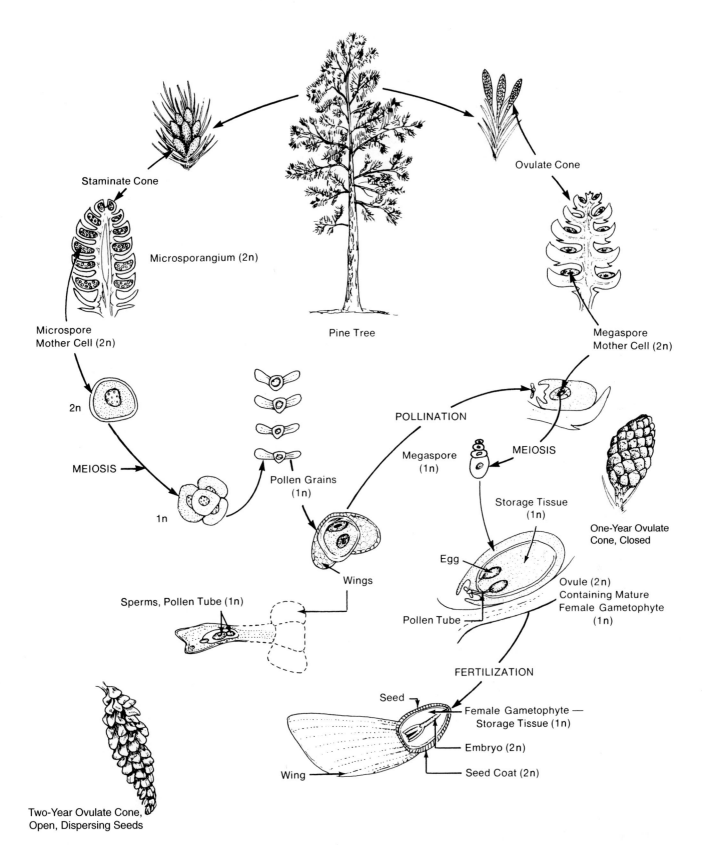

Figure 14–3 Diagrammatical representation of the sexual cycle in a gymnosperm (pine), showing meiosis and fertilization. *Source:* Kester, Dale E., Hartmann, Hudson T.: *Plant Propagation: Principles & Practices,* 4th Education © 1983. Adapted by permission of Pearson Education, Inc., Upper Saddle River, NJ.

Figure 14–4 *Top:* Inheritance involving a single gene pair in a monohybrid cross (one characteristic involved). In garden peas, tallness (D) is dominant over dwarfness (d). A tall pea plant is either homozygous (DD) or heterozygous (Dd). Segregation occurs in the F_2 generation to produce three genotypes (DD, Dd, dd) and two phenotypes (tall and dwarf). *Bottom:* Inheritance in a dihybrid cross (two characteristics involved) of the peach (*Prunus persica*). Fuzzy skin (G) of a peach is dominant over the glabrous (i.e., smooth) skin of the nectarine (g). White flesh color (Y) is dominant over yellow flesh color (y). In the example shown, the phenotype of the F_1 generation is different from either parent. Segregation in the F_2 generation produces nine genotypes and four phenotypes. *Source:* Kester, Dale E., Hartmann, Hudson T.: *Plant Propagation: Principles & Practices,* 4th Edition © 1983. Adapted by permission of Pearson Education, Inc., Upper Saddle River, NJ.

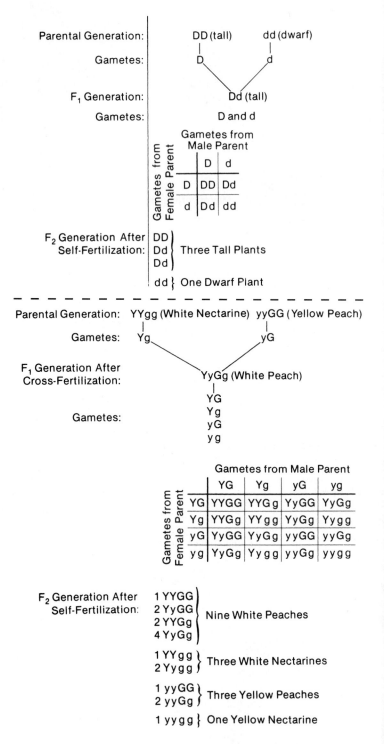

PROPAGATION OF PLANTS

Once we have a plant with desired characteristics, we have to increase the numbers of that plant. The process of increasing plant numbers is called **propagation.** Plants are propagated by either sexual (seed) or asexual (vegetative) methods. Some kinds of plants are almost always propa-

gated by one method or the other; other kinds can be propagated successfully either way. The various ways of propagating plants are outlined in Table 14–2. A successful propagation method is one that transmits all the desirable characteristics of the original plant to all the progeny. If the characteristics of the original plant are lost or changed dur-

TABLE 14–1	Polyploidism in Oats, Wheat, and Tobacco	
Common Name	**Species**	**Somatic Chromosome Number**
Sand oats	*Avena strigosa*	14
Slender wild oats	*Avena barbata*	28
Cultivated oats	*Avena sativa*	42
Einkorn	*Triticum monococcum*	14
Emmer	*Triticum dicoccum*	28
Common wheat	*Triticum aestivum*	42
Wild tobacco	*Nicotiana sylvestris*	24
Cultivated tobacco	*Nicotiana tabacum*	48

TABLE 14–2 Methods of Propagating Plants, with Typical Examples

I. Sexual
  A. Propagation by seed
II. Asexual (vegetative)
  A. Apomictic embryos—citrus mango
  B. Cuttings—geraniums
  C. Grafting—dwarf fruit trees
  D. Budding
  E. Layering
  F. Runners—strawberry
  G. Suckers—red raspberry, blackberry
  H. Separation
  I. Division
    1. Stem tubers—white potato
    2. Tuberous roots—sweet potato, dahlia
    3. Rhizomes—iris, canna
  J. Micropropagation—orchid, fern

ing the propagation procedures, that particular method is unsatisfactory for that type of plant. It is then necessary to use another propagation procedure that will preserve these characteristics.

It is important to realize that the kinds of agricultural and ornamental plants being grown today are plants with particularly desirable characteristics. Such plants originated from a plant or plants found growing in the wild, in cultivated plant populations, from mutations, or from breeding programs conducted by government agencies or private plant breeders. With the appropriate propagation procedures, such plants can become the starting point for populations of many millions of individuals, all just like the original mother plant. The groups of plants that people have developed and cultivated have been given cultivar names; for example, Redhaven peach, Bing cherry, Pawnee wheat, Imperial Blue pansy, Peace rose, Ranger alfalfa, and so on. The propagation methods used for increasing the populations of these groups of plants must do so without changing their characteristics.

There are several kinds of cultivars. If the plant group reproduces "true" by seeds—with no characteristics changed—the cultivar is termed a **line.** A line is homozygous[2] and, if self-pollinated (or if cross-pollination is prevented), seed propagation yields progeny like the original plant. Many of the world's leading economic plants—the cereals, the vegetables, and the garden flowers—are made up of groups of these lines. They are seed propagated and faithfully maintain their characteristics when propagated in this manner. Many forest species, although seed propagated, are not considered lines because of their higher level of variability. They are known by their species names, such as Douglas fir (*Pseudotsuga menziesii*), ponderosa pine (*Pinus ponderosa*), or coast redwood (*Sequoia sempervirens*).

In addition to lines, there are other types of seed-propagated cultivars, such as **inbred lines**—used to produce hybrid cultivars—and **hybrids**—as in hybrid corn (see Fig. 14–5).

Many groups of plants are heterozygous[3] rather than homozygous. These include fruit and nut species, many forage crops, and woody ornamentals. These plants have many dissimilar genes controlling their characteristics. During meiosis and fertilization leading to embryo formation in the seed, these genes segregate and recombine in a great many different ways so that the resulting plants differ from each other and from their parent. With these plants, seed propagation cannot maintain the characteristics of the female parent and cannot be used as a successful propagation procedure. With such heterozygous plants, vegetative (asexual) propagation is usually used. A piece of vegetative tissue—a section of a stem, root, or leaf—is placed in a suitable environment, such as a warm, humid rooting frame in the greenhouse. In time, the piece of tissue may regenerate the missing part: a stem piece forms roots, a root piece forms shoots, or a leaf forms both shoots and roots. By these means, new plants form that are exactly the same genetically as the plant from which the piece of vegetative tissue was taken; therefore, the new plant has all the same characteristics as the parent. The flower is not involved in vegetative propagation, allowing no opportunity for genetic change (unless, perhaps, a mutation has occurred, which does happen, but rarely). When such vegetative propagation is used, cultivars, even though heterozygous, can be perpetuated generation after generation, involving hundreds of thousands or more individual plants. Cultivars originating from a single plant or plant part and maintained in this manner by vegetative propagation are called **clones.**

The vegetative procedures just described—where a piece of tissue regenerates the missing part—is termed cutting propagation. There are a number of other, somewhat more complicated vegetative propagation methods, such as grafting, budding, layering, and runners.

---

[2]Having similar genes of a Mendelian pair present in the same cell as, for example, a dwarf pea plant with genes (tt) for dwarfness only.

[3]Having different genes of a Mendelian pair in the same cell as, for example, a tall pea plant with genes for tallness (T) and genes for dwarfness (t).

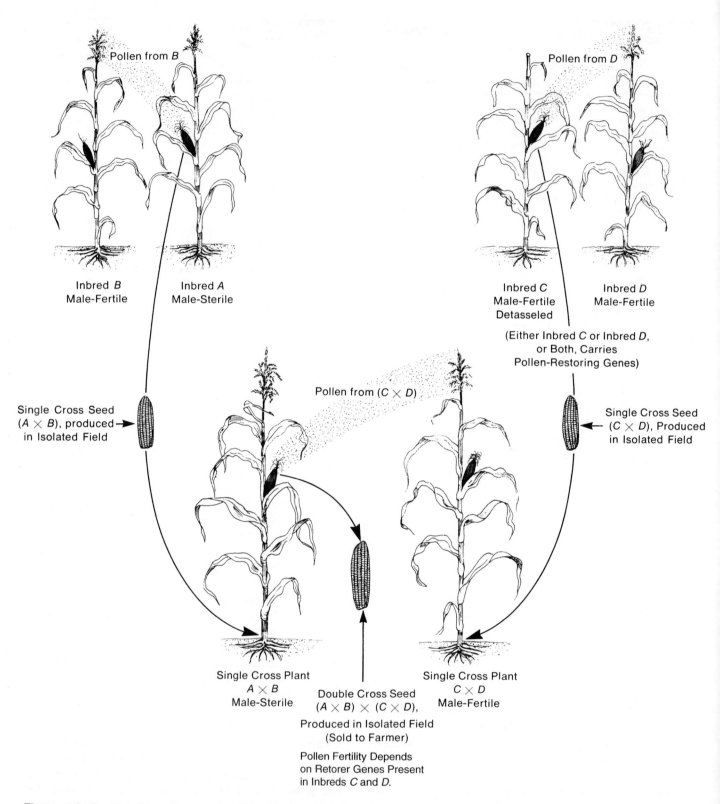

Pollen from B

Pollen from D

Inbred B
Male-Fertile

Inbred A
Male-Sterile

Inbred C
Male-Fertile
Detasseled

Inbred D
Male-Fertile

(Either Inbred C or Inbred D,
or Both, Carries
Pollen-Restoring Genes)

Single Cross Seed
(A × B), produced →
in Isolated Field

Pollen from (C × D)

Single Cross Seed
← (C × D), Produced
in Isolated Field

Single Cross Plant
A × B
Male-Sterile

Double Cross Seed
(A × B) × (C × D),

Single Cross Plant
C × D
Male-Fertile

Produced in Isolated Field
(Sold to Farmer)

Pollen Fertility Depends
on Retorer Genes Present
in Inbreds C and D.

Figure 14–5 Hybrid seed corn production by the utilization of cytoplasmic male sterility in the production of single cross and double cross. In this example, only one inbred, A, is male sterile. The cytoplasmic male sterility is transmitted to the single cross A × B. Pollen-restoring genes carried by the inbreds C or D give pollen fertility to the double cross (A × B) × (C × D) also. *Source:* USDA.

250

SEXUAL PROPAGATION

In the sexual reproduction of plants, a seed must be produced in a flower. Seed formation is preceded by a type of cell division termed **meiosis** or reduction division, in which the number of chromosomes in the cells is reduced by half to form the male sperm cell and the female egg. The egg and sperm combine during fertilization in the ovule to form the zygote that develops into the embryo.

Seed Production

If a cultivar—Great Lakes head lettuce, Calrose rice, or Kombar barley, for example—is to be maintained by seed propagation, careful control of the seed source is essential. If the cultivar is homozygous and self-pollinated, seed purity is generally assured. If the cultivar is homozygous but cross-pollinates with other cultivars or with other species, then the plants used to produce the seeds must be separated by considerable distances from other plants with which they are likely to cross to prevent pollen contamination and loss of genetic purity.

With certain plants propagated by seed, some variability may be tolerated, or it may even be desirable. Examples are reforestation projects, tree and shrub plantings in shelter belts, plantings for wildlife cover, and plant breeding projects.

Most states in the United States and many other countries have established seed certification programs to protect and maintain the genetic quality of cultivars. Government agencies set standards for seed, particularly for field and forage crops such as soybeans, rice, wheat, oats, alfalfa, and clovers. Isolation standards for the seed-producing fields are established, as are requirements for roguing or culling out off-types, diseased plants, and weeds. Field inspections and final seed testing are included in such programs. Thorough cleaning of seed harvesting equipment between seed lots is required.

Seed certification programs in the United States recognize four classes of seeds in agronomic crops, such as cotton, alfalfa, soybeans, and cereal grains:

1. *Breeder seed.* This is produced only in small amounts and is under the control of the plant breeder. It is planted to produce foundation seed. Breeder seed is labeled with a white tag.
2. *Foundation seed.* This is multiplied from breeder seed; it is available only in limited amounts and is planted to produce registered seed. It is controlled by public or private foundation seed stock organizations. Foundation seed is also labeled with a white tag.
3. *Registered seed.* This is the seed source for growers of certified seed and is under the control of the registered seed producers. It is the progeny of either breeder or foundation seed. It is labeled with a purple tag.
4. *Certified seed.* This seed is available in large quantities and is sold to farmers for general crop production. It is labeled with a blue tag. Certified seed is of known genetic identity and purity.

Genetic quality of vegetable and flower seeds, however, is largely regulated by the seed companies, which maintain careful control over their seed plantings and continually test seed purity in special test gardens.

The production of plants from **hybrid seeds** has been one of the outstanding scientific breakthroughs in agricultural history. The use of hybrid seed has more than doubled the yield of both sweet corn and field corn. In the United States, almost all corn is now produced from hybrid seed. This method of producing hybrid corn seed was developed in 1918, but it was not used commercially until about 1935 (Fig. 14–1). Hybrid barley and alfalfa cultivars have also been released, and almost all grain sorghum produced in the United States comes from hybrid seed. Considerable research and millions of dollars have been expended by public and private agencies in trying to develop practical methods of producing hybrid seeds of wheat to increase yields, but these are still without the spectacular success achieved with corn. Wheat flowers are not structured for easy cross pollination, as are the flowers of corn.

The use of F_1[4] hybrid seeds of several of the herbaceous bedding plants—petunias, zinnias, pansies, and marigolds—has made possible superior flower types and increased plant vigor, and has given a great boost to the bedding plant industry. Hybrid seeds also give greatly increased yield and quality of vegetable crops such as tomato, squash, cucumber, muskmelon, cauliflower, and broccoli.

A seed production industry has developed since about 1935, and it has made outstanding contributions to the increased production of world crops. Working with government and private plant breeders and seed certification agencies, the seed industry has supplied the channels for rapid seed increase and distribution while safeguarding genetic cultivar purity. Farmers in North America have become accustomed to using certified seed obtained from recognized seed producers rather than relying on their own or a neighbor's "bin-run" seed of dubious quality.

Seed Formation

Seeds originate as the final product of a plant's sexual reproduction system. A seed has three essential parts:

1. The embryo, which develops into the new plant.
2. Food storage material, which is available to nourish the embryonic plant. This may be either endosperm tissue or the fleshly cotyledon(s), part of the embryo itself.
3. Seed coverings, which are usually the two seed coats but may include other parts of the ovary wall. Coverings protect the seed and may help control seed dormancy.

---

[4]F_1 hybrid is the first generation offspring of a cross between two individuals differing in one or more genes.

The parts of the seed as they develop from the parts of the flower are:

Ovary → fruit (sometimes composed of more than one ovary, plus additional tissues)

Ovule → seed (sometimes coalesces with the fruit)

Integuments → seed coats (two)

Nucellus → perisperm (usually absent or reduced but sometimes develops into storage tissue)

2 polar nuclei
+ 1 sperm nucleus → endosperm (triploid—3n)

Egg nucleus
+ 1 sperm nucleus → zygote → embryo (diploid—2n)

After pollination and fertilization are completed, many changes occur in the flower to produce the fruit and the seed. Fruits and seeds appear in innumerable forms, depending on the species, but they all contain the same essential parts listed above.

Seed Storage and Viability Testing

For a seed to germinate, the embryo must be alive. In some plants, such as the willow, maple, and elm, the embryos are very short-lived, remaining viable for only a few days or months. In others, such as the hard-seeded legumes, the embryo generally remains alive for a great many years. Seeds of other kinds of plants range between these extremes, the length of embryo viability often depending on seed storage conditions. For example, seeds stored in a sealed container under refrigeration at 0°C to 4°C (32°F to 40°F) and at low relative humidity (e.g., 15 percent, which would give a seed moisture level of about 4 to 7 percent) generally retain viability considerably longer than seeds sorted at room temperature and high relative humidity. Seeds of certain plants, however, soon become desiccated if stored under dry conditions, and the embryos die. Examples are citrus, maple, oak, hickory, walnut, and most tropical species.

It is often advisable before planting seeds, especially those of woody perennials and those that may have been stored for several years, to test the viability of a representative sample of the seed lot to be planted. There are several seed viability tests.

One easy means of determining the possible germinability of a seed lot is a cut test. Seeds of a representative sample are simply cut in half to see whether there is an embryo inside. Often the embryo has aborted or has been eaten by insects and, of course, the seed would not germinate. The mere presence of an embryo, however, does not mean it is alive.

Another simple test is to float the seeds in water. Quite often the "floaters" are empty seeds and can be skimmed off. The sound, full seeds sink and are the ones to be planted.

X-ray photographs of seeds do essentially the same thing as a cut test and are used in some seed laboratories to determine if the seed is empty or the embryo is shrunken. X-ray tests can also be used to determine the optimum time to harvest seeds by observing when the embryos have completely filled the seed.

These tests are not, strictly speaking, viability tests but are useful to rule out seeds that have no possibility of germinating. They still do not give the viability status of seeds with full-sized, apparently sound embryos.

Germination Test

The germination test is useful for seeds that have no dormancy problems, such as flower, vegetable, and grain seeds. They can be tested for germinability by several methods, such as rolling them in several layers of moist paper toweling, or actually planting the seeds in flats of a suitable growing medium in a greenhouse. Seed-testing laboratories use elaborate seed germination boxes with controlled lighting and temperature. After several days or weeks, viability is calculated as the percentage of seedlings developing from the number of seeds planted.

Tetrazolium Test

The chemical 2,3,5-triphenyl tetrazolium chloride, which is colorless when dissolved in water, changes to the red-colored chemical, triphenyl formazan, whenever it contacts living, respiring tissue. In living tissues, enzymes change the tetrazolium salt to formazan. In dead, nonrespiring tissues, these enzymes are not active. In this test, the seeds are usually soaked in water to allow them to become completely hydrated, then cut in half lengthwise to expose the embryo. They are then placed in a 0.1 to 0.5 percent tetrazolium solution and held at room temperature or somewhat higher for several hours to several days, the exact time depending on the species of seed. The parts of the seed that are living (and respiring) become red; the nonliving parts remain white. If the embryo turns red, the seeds are viable; if the embryo remains white, the seeds are nonviable. Sometimes only a portion of the embryo becomes red, making it difficult to interpret the results of the test. This difficulty in interpretation is the principal weakness of this test, which is nevertheless widely used.

Excised Embryo Test

The embryos in the seeds of many woody plant species have profound dormancy conditions (see page 253) and do not respond in a direct germination test. However, if the embryos are carefully excised from the seed and placed on moist paper in a covered dish, viable embryos will show some activity—possibly a greening and separation of the cotyledons with definite indications of life. Nonviable embryos remain white and succumb quickly to fungus and bacterial attacks. Although this method takes time and requires skill in removing the embryo, it is routinely used by some seed laboratories with good results.

Seed Dormancy

Seeds of many plant species, especially woody perennials, do not germinate when they are extracted from the mature fruit and planted, even though all temperature, light, and moisture conditions favor germination. This is an important survival mechanism for the species and a result of evolutionary development. These species have survived because their seeds have not germinated just before adverse weather conditions that would kill the young, tender seedlings. Thus, in nature, these dormancy factors prevent seed germination of woody perennials in the autumn, allowing the embryonic plant within the seed to overwinter in a very cold-resistant form. Any plant species whose seeds did germinate in the fall in an area with severe winters would likely not survive in that region. Often, the causes for dormancy can persist indefinitely in the seed and require specific treatments to overcome them before germination can take place. This poses problems for the propagator and requires a knowledge of seed dormancy and how to overcome it. Seed dormancy can result from structural or physiological conditions in the seed coverings, particularly the seed coats, or in the embryo itself, or both.

Seed Coat Dormancy

Seed coats or other tissues covering the embryo may be impermeable to water and gases, particularly oxygen, which therefore cannot penetrate to the embryo and initiate the physiological processes of germination. This situation usually occurs in species whose seeds have hard seed coats, such as alfalfa, clover, and other legumes as well as in some pine, birch, and ash species. In nature, continued weathering, the action of microorganisms, or passage through the digestive tract of animals can soften the seed coverings sufficiently so that they do become permeable and germination can proceed. In some species, the seed coats are apparently permeable to water and gases but have such high mechanical resistance to embryo expansion that germination does not occur unless the seed coats are softened in some manner.

Various artificial methods of softening seed coats are widely used to enhance germination. Three principal procedures are:

Scarification The surface of the seed is mechanically scratched or ruptured. This is often done by rubbing the seed between sheets of sandpaper.

Heat treatment In many kinds of seed, exposure to heat, usually boiling water, for a short time sufficiently disrupts the seed coat to permit passage of water and gases.

Acid scarification Soaking seeds with impervious coverings in concentrated sulfuric acid for the proper length of time etches their coats enough for germination.

Embryo Dormancy

Embryo dormancy is very common in seeds of woody perennial plants. It is due to physiological conditions or germina-

tion blocks in the embryo itself that prevent it from resuming active growth even though all environmental conditions (temperature, water, oxygen, light) are favorable.

It has been known for hundreds of years that dormant seeds will germinate readily in the spring if they are allowed to winter outdoors in regions with cold climates so that they received some chilling while being kept moist. From this arose the practice known as **stratification,** in which boxes are filled with alternate layers of moist sand and seed and set outdoors in a protected shady place to overwinter. The following spring, the seeds are removed from the box and planted. The critical conditions in seed stratification are:

1. *Chilling temperatures.* From about 1°C to 7°C (34°F to 45°F).
2. *Moisture.* The seeds should be soaked in water to start, then kept moist.
3. *Adequate oxygen.* The seeds should have adequate air and not be kept in an airtight container.
4. *Period of time.* The optimum stratification time varies considerably among species. Seeds of the American plum (*Prunus americana*), for example, require at least ninety days chilling while, at the other extreme, apricot seeds (*Prunus armeniaca*) require only twenty to thirty days.

More precise stratification treatments can be given if controlled refrigeration is used rather than natural outdoor winter cold, which can fluctuate considerably. Polyethylene plastic bags are suitable containers for stratifying seeds. The seeds are mixed with a slightly moist medium—sand, vermiculite, or peat moss—and placed in the bag. Polyethylene allows sufficient oxygen to pass through for the seed's requirements but slows water loss. It is advisable to soak seeds in water for twenty-four hours to saturate the tissues thoroughly before the chilling treatment begins.

Seeds of some species germinate better if they are given a warm (24°C; 75°F) moist stratification period for several weeks just before the cold stratification period.

In nursery practice many propagators sow seeds having embryo dormancy in outdoor nursery beds in the fall, allowing the natural winter chilling to satisfy the embryo's chilling requirement.

There is evidence that during the stratification treatment, growth-promoting hormones (e.g., gibberellins and cytokinins) in the seeds increase while the level of growth-inhibiting hormones (e.g., abscisic acid) decreases, thus permitting germination.

The term *after-ripening* is often used to describe the physiological changes in the seed that allow germination to take place.

Double Dormancy

Seeds of some species have both seed coat and embryo dormancy. An example is redbud (*Cercis occidentalis*). To obtain good germination of such seeds, they should be first treated in some manner as described above to soften the seed coats, then given a cold stratification treatment to overcome the embryo dormancy.

Rudimentary Embryos

Some plants shed their fruits before the embryo within the seed has matured enough to germinate. A period of time—several weeks to several months—after harvest is required for the embryo to develop to the point where it can continue growth. This process can take place either while the seed is in storage or after planting. Seed dormancy due to rudimentary embryos occurs in genera such as *Fraxinus, Ilex, Magnolia, Pinus, Ranunculus,* and *Viburnum.*

Chemical Inhibitors

In many species, the seeds contain one or more chemicals that can block essential steps in the germination process. Sometimes—for example, in iris seeds—the inhibitor is not in the embryo but in the endosperm tissue. If the embryo is excised from the seed, it starts to grow readily in a sterile nutrient culture. Chemical inhibitors can also occur in seed coats or in the pericarp (ovary wall). Often, leaching such seeds in running water for several hours removes the inhibitors and permits germination. Seeds of some desert plants contain chemical germination inhibitors that are leached out by heavy soaking rains (but not by light showers). The heavy rains soak the soil sufficiently so that the seedlings can become established before the soil dries out. This interesting evolutionary phenomenon permits the continuation of these species by allowing survival of the seedling offspring in a difficult environment.

Some of these germination inhibitors are well-known chemicals such as coumarin and caffeic acid. Seeds of certain fleshy fruits—tomatoes, lemons, strawberries—do not germinate while still attached to the fruit because of certain of these germination inhibitors in the fruits.

Secondary Dormancy

Seeds that are ready to germinate after all germination blocks are removed can become dormant again because of exposure to some environmental condition. For example, seeds of some woody perennial plants, after undergoing stratification to overcome embryo dormancy, become dormant again if the germination temperatures reach 26°C to 32°C (79°F to 90°F). Exposing winter barley or spring wheat seeds to certain unfavorable conditions, such as high temperatures or high moisture levels, can also induce a secondary dormancy.

Seed Germination

If the seeds have viable embryos, have all germination blocks removed, and are placed under proper environmental conditions of moisture, temperature, and (sometimes) light, the quiescent embryos in the seeds can resume their growth. The nutrients stored in the endosperm or cotyledons of the seed nourish the developing embryo until the new shoot rises above ground, develops leaves, and produces its own food by photosynthesis.

Germination can proceed in several ways, depending on the species (see Figs. 14–6 and 14–7). Sometimes the cotyledons are pushed above ground (**epigeal germination**) and sometimes they remain below ground (**hypogeal germination**). The sequence of events during seed germination is as follows:

1. *Imbibition of water by the seeds.* The colloidal properties of seed tissues give them great water-absorbing properties. Moist seeds may swell to a size much larger

Figure 14–6   Seed germination in a monocotyledonous plant, barley.

Figure 14–7 Seed germination in dicotyledonous plants. *Above:* Epigeous germination as shown in the cherry. The cotyledons are above ground. *Below:* Hypogeous germination as shown in the peach. The cotyledons remain below ground. *Source:* Kester, Dale E., Hartmann, Hudson T.: *Plant Propagation: Principles & Practices,* 4th Edition © 1983. Adapted by permission of Pearson Education, Inc., Upper Saddle River, NJ.

than the dry seeds. The cells become turgid and the seed coverings soften and rupture, permitting easy passage of oxygen and carbon dioxide.

2. *Activation of hormones and enzymes.* After water is absorbed, various enzyme systems are activated or synthesized, often as a result of stimulation by hormones. The enzymes convert complex food storage molecules into simpler food materials that can be readily translocated and used for growth. Other enzymes are involved in the respiratory processes, which release energy for cell division and growth. Food materials are translocated to root and shoot growing points.

3. *Embryo growth and development.* The root-shoot axis (plumule, epicotyl, hypocotyl, and radicle) grows by cell division and enlargement. At the same time, food materials translocate to the growing points from the storage tissues, which gradually become depleted. The seed coats rupture and photosynthetic tissue (green leaves and shoots) emerge into the light to carry on photosynthesis. In addition, the embryonic root (radicle) emerges and grows into moist soil to supply the newly developed leafy tissues with water—which will be lost through transpiration. By this time, if no unfavorable environmental influences interfere, the seedling has become established and can exist as an independent plant.

Environmental Factors Influencing Seed Germination

For successful seed germination and seedling growth, certain environmental conditions are required:

1. Adequate moisture
2. Proper temperature
3. Good aeration
4. Light (in some cases)
5. Freedom from pathogenic organisms
6. Freedom from toxic amounts of salts

Moisture   It is essential that water be available in adequate amounts to initiate the physiological and biochemical processes in the seed that result in reactivation of embryo growth. Germination usually takes place satisfactorily at moisture levels between field capacity and permanent wilting percentage, although seeds of some species (lettuce, peas, rice, beets, celery) germinate best at high soil-moisture levels, whereas those of others (spinach) do best with low moisture.

Temperature   The temperature can strongly influence the percentage and rate of seed germination, varying with the kind of seed. Seeds of the cool-season crops germinate best at relatively low temperatures of 0°C to 10°C (32°F to 50°F). Examples are peas, lettuce, and celery. Seeds of warm-season

crops germinate best at temperatures ranging from 15°C to 26°C (59°F to 79°F); examples are soybeans, beans, squash, and cotton. Seeds of many other species germinate over a wide temperature range. In addition, many kinds of seeds germinate much better when the temperature fluctuates daily about 10°C (18°F) between maximum and minimum.

Aeration   Respiration rates are high in germinating seeds, which thus require adequate oxygen. The usual amount in the air is 20 percent. If this concentration is decreased, germination rate and percentage germination of most kinds of seeds is retarded.

In seedbeds that are overwatered and poorly drained, the soil pore spaces may be so filled with water that the amount of oxygen available to the seeds becomes limiting and germination of most kinds of seeds is retarded or prevented.

Light   Light is essential to the germination of some kinds of seeds, such as lettuce, celery, most grasses, and many herbaceous garden flower plants. Such seeds should be planted very shallow for good germination. However, seeds of other plants—onion, amaranth, nigella, and phlox—are inhibited by light and do not germinate unless planted deep enough to avoid light. The light requirement for seed germination is very complex, depending on the age of the seed, degree of seed imbibition with water, temperature, day length, and certain germination-stimulating chemicals. A pigment in seeds—phytochrome—is involved in the controlling mechanism.

Pathogenic Organisms   **Damping off** describes the situation in which the seedlings die during or shortly after germination. Damping off is caused primarily by attacks of certain universally present and very destructive fungi—*Pythium ultimum* and *Rhizoctonia solani,* and to a lesser extent, *Botrytis* spp. and *Phytophthora* spp. Mycelia of these organisms and the spores of *Pythium* and *Phytophthora* are often found in the germination medium, on the seed surfaces, in the water, or on tools. The best control methods are fumigation or heat pasteurization of the germination medium, surface treatment of the seeds with fungicides before planting, and good sanitation procedures.

Salinity Problems   If the germination medium is watered lightly but frequently after the seeds have been planted, evaporation of water from the surface leaves salt deposits. If this situation continues, the salinity can increase to such a level that it will injure or kill the seedlings as they germinate. This is a particular problem with small, shallow-planted seeds that may dry out quickly, and in areas having high salt concentrations in the water. Such salts may originate in the germination medium, the irrigation water, or added fertilizers. Salinity damage to seedlings often looks like damping-off injury. This problem can be prevented by using soil mixtures and water low in salts, withholding fertilizers, and by irrigating more copiously but less frequently so that excess salts are leached out.

VEGETATIVE PROPAGATION

Vegetative or asexual propagation is accomplished entirely through **mitosis,** the same nonreductive cell division process by which the plant grows. Each daughter cell is an exact replica of its mother cell. Chromosome numbers and composition do not change during cell division. Mitotic cell division, as illustrated in Figure 14–8, produces the adventitious roots and shoots as well as the callus (parenchyma cells required for healing of a graft union) necessary for successful vegetative propagation (see Fig. 14–9). **Adventitious shoots** are those appearing any place on the plant other than from the shoot terminals or in the axils of leaves. **Adventitious roots** appear any place on the plant other than from the radicle (root tip) of the seed or its branches. **Callus** is a mass of undifferentiated and proliferating parenchyma cells.

Vegetative propagation is used primarily for woody perennial plants that are highly heterozygous; that is, those that do not "breed true" from seed. In these plants, the mother plant's desirable characteristics are lost if seed propagation is used. To maintain the genetic identity of the mother plant in such cases, it is necessary to avoid use of the flower and seed altogether in reproducing the plant. Vegetative tissues (stem, root, or leaf) are used to develop new plants by inducing the formation of adventitious shoots, roots, or both. If pieces of stem tissue will not produce adventitious organs (roots, shoots, or both), or if a particular kind of rootstock is required, it is necessary to graft or bud two pieces of tissue together, one to become the top part of the plant and one to become the root system.

As noted previously, a cultivar that must be reproduced by asexual methods to maintain its characteristics is termed a *clone,* as distinguished from a *line,* which maintains its characteristics without change by seed propagation. Almost all fruit and nut cultivars and many ornamental cultivars are clones. All plants that are members of the same clone have the same genetic makeup and are, in reality, exact descendants of the mother plant from which the clone originated, although the original mother plant may have died many years earlier. Some clones, such as the Thompson Seedless grape, have been in existence since ancient times and consist of many millions of individual plants scattered all over the world.

Cultivated clones originate in two ways. The first, and usual, way is as seedling plants that some person recognizes as having some superior qualities and proceeds to propagate vegetatively. For example, the world-famed Golden Delicious apple originated as a seedling tree that A. H. Mullins found growing on his farm in Clay County, West Virginia. Mullins recognized this apple's value. Later it came to the attention of Stark Brothers Nurseries of Louisiana, Missouri, who in 1912 paid Mullins for the propagation rights to the tree. They named it Golden Delicious and introduced nursery trees for orchard planting in 1916. Although the original tree died long ago, millions of Golden Delicious apple trees are now growing throughout the world, all with the same genetic makeup as the one original tree and all producing the same kind of apples. Although many clones have originated as chance seedlings found growing in the wild, they can also originate from controlled crosses made by plant breeders.

A second way clones originate is from mutations (bud sports). A single bud on a plant may have its genetic makeup altered during cell division so that, as the bud grows and develops into a branch, one or more of its characteristics differs from those of the rest of the plant. Many mutations are minor or inferior or go unnoticed. But occasionally some strikingly superior characteristic appears; someone notices it and propagates new plants taken from shoots on this mutated branch. This, then, becomes the start of a new clone. The desirable pink-fleshed Ruby grapefruit originated in 1929 as a mutated branch on the Thompson Pink grapefruit, which in turn had originated in 1913 as a mutated branch on a white Marsh grapefruit tree.

The primary advantage of clones is the uniformity of the member plants. All members have the same genetic makeup (**genotype**) and potentially they can all be exactly alike. However, environmental factors can so modify the appearance and behavior of the plant (**phenotype**) that individual plants can differ strikingly. For example, a vineyard of Concord grapevines, pruned, irrigated, sprayed, and fertilized properly, would appear totally different from an adjacent abandoned vineyard of the same cultivar, yet the vines in both would be genetically identical.

The mutation process can also cause undesirable genetic changes in clones. This happens frequently in many species and must be avoided to prevent deterioration of the clone.

Some mutations, called **chimeras,** genetically change only a portion rather than the entire shoot. Some of the variegated leaf patterns found in the foliage of certain plants such as *Pelargonium* or citrus are due to chimeras (Fig. 14–10). The word *chimera* comes from Greek mythology. The chimera was a fire-breathing beast with the head of a lion, the body of a goat, and the tail of a dragon. Chimeras can often be the source of desirable traits such as thornlessness, reduced fuzziness in peaches, and variegated colors in leaves and flowers. However, chimeras can also be very difficult to propagate true to form. Because many of our fruit and ornamental crops are chimeras, a more detailed explanation of them is presented here.[5]

No discussion of the origin of chimeras would be complete without a review of the organization of the shoot apex. The pattern of cell division, frequency of cell division, and layered organization of the cells in the apex interact in determining the type of chimera that is produced and the stability of the pattern that results. The apical meristem of a shoot is the location where most of the cells that produce the plant body are formed. Cell division occurs at a very rapid rate in

---

[5]The following discussion is adapted with permission from Dr. Dan Lineberger, Texas A&M University.

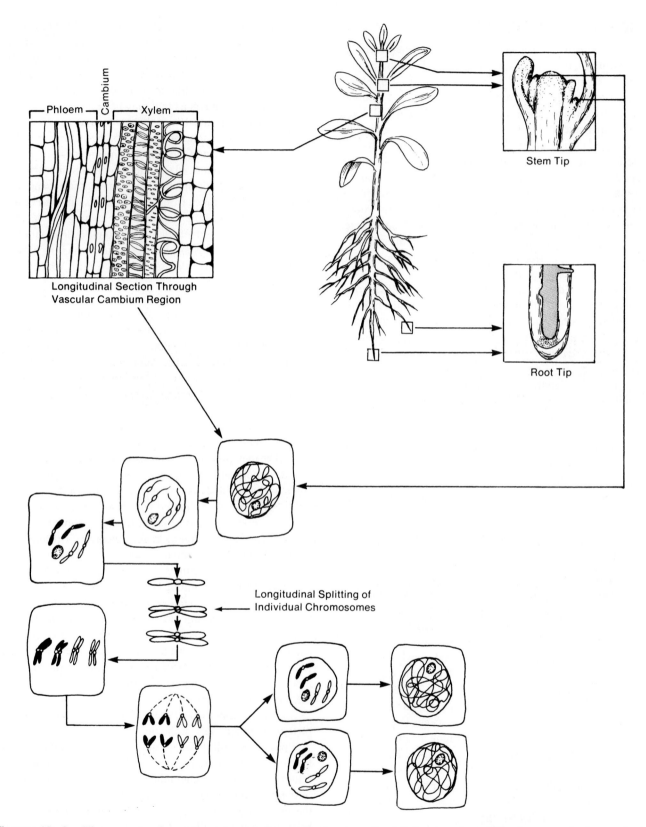

Figure 14–8 The process of growth and asexual reproduction in a dicotyledonous plant. Mitosis occurs in three principal growing regions of the plant: the stem tip, the root tip of primary and secondary roots, and the cambium. A meristematic cell is shown dividing into two daughter cells whose chromosomes are (usually) identical with those of the original cell. *Source:* Kester, Dale E., Hartmann, Hudson T.: *Plant Propagation: Principles & Practices,* 4th Edition © 1983. Adapted by permission of Pearson Education, Inc., Upper Saddle River, NJ.

Figure 14–9 Regeneration in asexual propagation. *Left:* Adventitious shoots growing from a root cutting. *Center:* Adventitious roots developing from the base of a stem cutting. *Right:* Callus tissue produced to give healing of a graft union. *Source:* Kester, Dale E., Hartmann, Hudson T.: *Plant Propagation: Principles & Practices,* 4th Edition © 1983. Adapted by permission of Pearson Education, Inc., Upper Saddle River, NJ.

Figure 14–10 A variegated pink Eureka lemon, a type of chimera.

an actively growing shoot, and these cells in turn elongate or expand, resulting in lengthening of the shoot. Figure 14–11 is a sketch of a longitudinal (or lengthwise) section through the apical meristem of a typical woody or herbaceous dicot shoot. Leaf primordia arise on the sides of the apical dome, and lateral buds develop in the axils of these young leaves.

The apex is organized into a layered region (the tunica) and a region where layering is not evident (the corpus). The controlled pattern of cell divisions in the tunica results in the maintenance of discrete layers, with the number of layers varying somewhat among the different species. Note that the layers retain their organization into the region where the leaves and lateral buds are developed.

The derivatives (or progeny, if you will) of the outermost layer (L.I) give rise to the epidermis. The epidermal layer is continuous as an outer covering over all tissues of the leaf, stem, flower petals, etc. Derivatives of layer II (L.II) give rise to several layers within the stem and a large proportion of the cells in the leaf blade. Derivatives of layer III (L.III) give rise to most of the internal tissue of the stem and a number of cells around the veins within the leaf. The significance of the cell layers and their resulting progeny will be discussed in more detail.

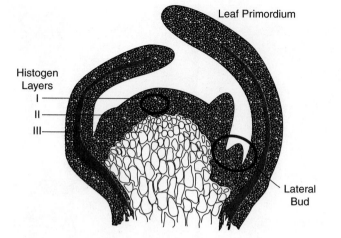

Figure 14–11   Typical angiosperm meristem showing the histogen layers: I, II, and III. Note the lateral bud in the axil of the immature leaf. *Source:* Courtesy of Dr. Dan Lineberger, Texas A&M University.

Chimeras arise when a cell undergoes mutation. This mutation may be spontaneous or it may be induced by irradiation or treatment with chemical mutagens. If the cell that mutates is located near the crest of the apical dome, then all other cells that are produced by division from it will also be the mutated type. The result will be cells of different genotypes growing adjacent in a plant tissue, the definition of a chimera.

If the location of the cell at the time of mutation is in a region where little further cell division will occur, then the likelihood of detecting this mutation by visual inspection of the whole plant is low. Furthermore, if the mutation results in a genotype that is not very different morphologically from the rest of the plant, then the likelihood of identifying the plant as a chimera is also low. A mutation that results in colorless rather than green cells (variegation) is easily detectable, whereas a mutation that results in greater sugar accumulation in the cells cannot be observed.

Chimeral plants can be categorized on the basis of the location and relative proportion of mutated to nonmutated cells in the apical meristem (Fig. 14–12). **Periclinal chimeras** are the most important category because they are relatively stable and can be vegetatively propagated. A mutation produces a periclinal chimera if the affected cell is positioned near the apical dome so that the cells produced by subsequent divisions form an entire layer of the mutated type. The resulting meristem contains one layer that is genetically different from the remainder of the meristem. If, for example, the mutation occurs in L.I, then the epidermal layer of the shoot that is produced after the mutation is the new genetic type.

A classical example of an L.I periclinal chimera is the thornless blackberry. The epidermal layer of this type produces no thorns (the modified epidermal cells are correctly called prickles). The thornless epidermis covers a stem whose cells

contain the information for the thorny genotype. This can be demonstrated by taking root cuttings. The adventitious shoots that differentiate on the root cuttings are not chimeral and therefore revert to the thorny genotype. Other common periclinal chimeras we see in cultivated crops involve loss of epidermal appendages (thornlessness in blackberries, "fuzzless" peaches), alteration in bract color in poinsettia, and various petal or flower color patterns in carnation and chrysanthemum.

Mericlinal **chimeras** and **sectorial chimeras** result in the development of an epidermis that is not entirely chimeric. Further cell division and development can give rise to shoots and leaves that are not chimeras. Reliable propagation of these types of chimeras is therefore nearly impossible. For that reason, they will not be discussed here.

Although periclinal chimeras cannot be reproduced by sexual propagation, the most commonly used techniques for asexual plant propagation (described later in this chapter) result in the formation of true-to-type periclinal chimeras. However, important departures from the rule are found in cases where the propagules differentiate from adventitious shoots. A case discussed previously that illustrates this point is the thornless blackberry. Adventitious roots that form on a blackberry stem cutting or tip layer originate in the subepidermal tissues of the stem (L.II and L.III). If root cuttings are then taken from these plants, the adventitious shoots do not contain the layer with the thornless genotype (L.I) and the propagules are of the thorny type. Likewise, the pinwheel-flowering African violets similarly cannot be propagated in a true-to-type fashion from leaf cuttings. Shoots that originate on the leaf cuttings are either a single color or irregularly mottled bicolors without the characteristic flowering pattern. Propagation of the pinwheel-flowering African violets is currently achieved by separating suckers, which originate from lateral buds. Because microprogation generates plants from a small number or even single cells, segregation of plantlets with and without the chimeric trait is likely to occur, making microprogation unsuitable for use with chimeras.

Apomixis

Apomixis is an interesting phenomenon in which the genetic identity of the mother plant is transmitted to daughter plants that develop by seed formation and germination. Apomixis is a form of asexual propagation because there is no union of male and female gametes before seedling production. There are several types but a common one is found in citrus seeds where, in addition to the sexual embryo formed through the usual pollination and fertilization processes, embryos also arise in the nucellar tissue (nucellar budding). The nucellar tissue enclosing the embryo sac has not undergone reduction division and has the same genetic makeup as the female parent. So the nucellar embryos, although developing in a seed, are exactly the same genetically as the mother plant and thus maintain the clone.

Figure 14–12 Development of mericlinal, periclinal, and sectorial chimeras. The location of the mutated cell determines the type of chimera that will develop. *Source:* Courtesy of Dr. Dan Lineberger, Texas A&M University.

Such seeds can contain several nucellar embryos in addition to the sexual embryo. Thus, several seedlings are obtained from one seed, a situation known as polyembryony.

Even though plants arising by apomixis from nucellar embryos maintain the clone, they go through the juvenile to mature transition stages just as any woody plant seedling would, taking several years to flower and fruit.

Disease Problems in Clones

Propagating plants vegetatively with clonal material has one great disadvantage: clones can become infected with systemic viruses and mycoplasmalike organisms that are passed along to the daughter plants during asexual propagation procedures. In time, all clonal members may become infected with viruses. Some viruses are latent in particular nonsusceptible clonal material, but if this material is used in a graft combination where the virus can move through the graft union to the graft partner—which is susceptible—then the entire grafted plant will be killed by the virus. On the other hand, virus-free seedlings can be obtained in many species by seed propagation because the virus is not transmitted through the embryo.

Some viruses can be removed from clonal material by heat treatment. The virus-infected plant material, perhaps a small nursery tree growing in a container, is held at 37°C to 38°C (98°F to 100°F) for two to four weeks or longer. This combination of time and temperature eliminates the virus. After treatment, cuttings can be taken for rooting, or buds may be taken for budding into virus-clean seedling root stocks.

Another procedure to eliminate viruses from clones is shoot-tip culture. In virus-infected plants, the terminal growing point is often free of the virus. By excising this shoot apex aseptically and growing it on a sterile medium, roots will often develop, producing a new plant free of the virus. Here again, a starting point becomes available for continued propagation of the clone but without the virus. This method has succeeded with many herbaceous plants such as carnation, chrysanthemum, hops, garlic, rhubarb, orchid, and strawberry. Sometimes, in strawberry plants, for example, both procedures are required—heat treatment of the plant, followed by excision and culture of the shoot tip—to free the plant of known viruses.

In recent years certification programs have been established by government agencies in many states in the United States and in other countries to provide nurseries with sources of true-to-name, pathogen-free propagation material. Elaborate programs, for example, have been established for citrus in Florida and California and for deciduous tree fruits, grapes, strawberries, potatoes, and certain ornamentals in many states.

In such programs, mature flowering or fruiting plants known to be true-to-name and true-to-type are selected as mother plants. These are "indexed" by certain grafting procedures or other tests such as enzyme-linked immunoassays (ELISAs) to be sure that no known viruses or other diseases are present. If no pathogen-free source plants can be located, then procedures like those described above for eliminating viruses must be used to obtain pathogen-free plants. Once a "clean" source is obtained, it must then be maintained under isolated and sanitary conditions, with periodic inspection and testing to ensure that it does not again become infected. Sometimes it is necessary to grow the plants in insect-proof screenhouses or greenhouses or in isolated areas far from commercial production fields.

Distribution systems from these foundation plantings are necessary, sometimes requiring plots of land, "increase blocks," to grow a greater amount of propagating material that may be needed. This material can be termed "certified stock" if it is grown under the supervision of a legally designated agency with prescribed regulations designed to maintain certain standards of cleanliness and clonal identity.

Many of the major agricultural industries, based on crops that are susceptible to various pathogens, particularly viruses, could not exist without such certification programs. These crops include citrus, grapes, potatoes, roses, cherries, peaches, and strawberries.

Propagation by Cuttings

Cutting propagation is a vegetative method widely used for propagating herbaceous and woody ornamental plants and, to a much lesser extent, fruit species. A cutting is essentially a piece of vegetative tissue that, when placed under the proper environmental conditions, regenerates the missing parts—roots, shoots, or both—and develops into a self-sustaining plant.

Cuttings can be classified according to the part of the plant from which they are obtained:

> Stem cuttings
>> Hardwood
>>> Deciduous
>>> Narrow-leaved evergreen
>> Semihardwood
>> Softwood
>> Herbaceous
> Leaf cuttings
> Leaf-bud cuttings
> Root cuttings

Stem cuttings already have terminal or axillary buds (potentially a new shoot system), but new roots must develop at the base of the cutting before a new plant will be formed. Stem cuttings can be prepared to include the shoot tip, or cuttings can be made from the more basal parts of the shoot that have only axillary buds. Leaf cuttings have neither buds nor roots, so both must form. Leaf-bud cuttings have a bud at the base of the petiole—for the new shoot system—so only new roots must form. Root cuttings must produce a new adventitious shoot and continue growth of the existing root piece, or develop roots from the base of the new shoot. Figure 14–13 illustrates types of cuttings.

Plant species and cultivars vary markedly in their ability to develop adventitious roots. Cuttings from some kinds of plants root easily even when the simplest procedures are used. Cuttings of others root only if the influencing rooting factors are carefully observed. Cuttings of still other kinds of plants have never been rooted, or rooted only rarely and in meager numbers, despite great efforts and much research. The basic reasons for such differences in rooting ability among different kinds of plants are not well understood.

Stem Cuttings

Deciduous Hardwood Stem Cuttings
Such cuttings are made in late winter or early spring, using leafless shoots that grew the previous summer. The shoots are cut into lengths of 15 to 30 cm (6 to 12 in.) and propagated outdoors in nursery beds. Many plants can be propagated in this manner, including some deciduous trees, such as willow and poplar; many deciduous shrubs, such as forsythia, weigela, privet, spiraea, honeysuckle, and roses; and several fruit species, such as grape, mulberry, currant, gooseberry, quince, olive, fig, and pomegranate.

Narrow-Leaved Evergreen Hardwood Stem Cuttings
Many conifers can be propagated by stem cuttings. Cuttings are best made in early winter. Needles are removed from the lower half of the cutting but left on the upper. The cuttings are usually inserted into flats or directly into greenhouse benches filled with a rooting medium of sand or of equal parts perlite and peat moss. The cuttings are best rooted in a cool, humid greenhouse or cold frame, preferably with high light intensity, and kept lightly watered or misted until they root. Rooting may take several weeks or even months, depending on the species.

Semihardwood Stem Cuttings
Most broad-leaved evergreen ornamentals—rhododendron, camellia, pittosporum, holly, evergreen azalea, escallonia, euonymus, and boxwood—as well as some fruit species—citrus and olive, for example—can be propagated by this type of cutting. Cuttings are best taken in midsummer, following the flush of spring growth.

The cuttings are inserted into flats or greenhouse benches. A porous rooting medium, such as equal parts of perlite and peatmoss or of perlite and vermiculite, is used. The cuttings are placed in a well-lighted, humid location to minimize water loss from the leaves. Closed rooting frames or flats covered with polyethylene plastic sheeting are suitable for keeping the humidity around the cuttings high. Most commercial nurseries root this type of cutting in mist propagating beds, where a fine mist is sprayed over the cuttings intermittently during the day.

Figure 14–13 Types of cuttings. (A) Hardwood stem cutting. (B) Leafy stem cutting. (C) Leaf cutting. (D) Leaf-bud cutting. (E) Root cutting.

Figure 14–14 Mist propagation bed operating in a green-house. The mist is on only a few seconds—just long enough to wet the leaves—then turns off. When the leaves start to dry, the mist is turned on again. The mist is generally off all night.

Softwood Stem Cuttings These are similar to semi-hardwood cuttings except they are prepared from young leafy shoots arising in the spring from deciduous trees or shrubs.

Herbaceous Stem Cuttings This type of cutting is used in propagating plants such as coleus, carnation, geranium, chrysanthemum, and many tropical house plants, all of which root easily. Typically a stem is harvested and the basal end is placed in a rooting medium in a warm (75°F to 80°F), draft-free environment and misted (see Fig. 14–14) until new roots develop. Rooting time depends on species and cultivar, often taking several days to several weeks. Heating the root zone area to 75°F decreases the time of root formation and makes rooting more uniform. Misting frequency should decrease as callus and roots form to reduce the risk of disease and in-crease hardiness of the cutting.

Leaf Cuttings

There are various types of leaf cuttings, as shown in Figure 14–15. A common one consists of a single leaf blade and petiole, as might be taken from an African violet. As with other types of leafy cuttings, the humidity must be kept very high, preferably by using a closed rooting frame, as for the herbaceous cuttings, or by the use of a humid greenhouse. Roots and shoots generally develop from the same point at the base of the petiole and grow into a plant independent of the leaf blade, which functions to nourish the new plant. African violet, peperomia, begonia, and sansevieria are ex-amples of plants commonly started by leaf cuttings.

Leaf-Bud Cuttings

Leaf cuttings of some species form roots at the base of the petiole but do not develop a shoot, resulting only in a rooted leaf that may stay alive for months (or years). To avoid this, a leaf-bud cutting can be prepared. This cutting consists of a short piece of stem with an attached leaf and a bud in the axil of the leaf, as shown in Figure 14–16. Such cuttings are rooted under high humidity, as described for semihardwood or herbaceous cuttings. The axillary bud develops into the new shoot system. Leaf-bud cuttings are useful as substitutes

Figure 14–15 Types of leaf cuttings. (A) New plants arising from base of petiole in African violet (*Saintpaulia*). (B) New plants arising from cuts in veins of a begonia leaf. (C) New plant arising from base of leaf blade in sanse-vieria. (D) New plants growing from notches of leaf in *Kalanchoe (Bryophyllum)*.

for stem cuttings in obtaining as many plants as possible from scarce propagating material. Leaf-bud cuttings give one and perhaps two (if the plant has opposite leaves) new plants from each node, whereas each stem cutting generally re-quires a minimum of two nodes.

Root Cuttings

Many plant species can be propagated by cutting the small, young roots into pieces and planting them horizontally in soil about 1.3 cm (0.5 in.) deep or vertically with the upper end (nearest the crown of the plant) just below the soil level. One or more new adventitious shoots form along the root piece,

Figure 14–16 Leaf-bud cuttings of peperomia. Each cutting consists of a leaf blade, petiole, axillary bud, and a piece of stem. Arrows show axillary buds starting to grow.

and either this shoot forms roots or the root piece itself develops new branch roots, thus producing a new plant. The best time to obtain the root pieces from the mother plant and to prepare and plant the root cuttings is late winter or early spring. The roots contain the highest quantity of stored foods at that time and the cuttings start to grow at the beginning of the growing season. Cuttings can be planted in an outdoor nursery or in flats of soil in a greenhouse or cold frame.

Origin of Adventitious Roots in Stem Cuttings

In stem cuttings of herbaceous plants, adventitious roots generally originate laterally and adjacent to the vascular bundles, whereas in cuttings of woody perennials, the roots originate in the region of the vascular cambium, often in young phloem parenchyma (Fig. 14–17). In each case, the new roots are in a position to establish a vascular connection with the conducting tissues of the xylem and phloem in the cutting.

Much study has been given to the mechanisms that lead to the initiation of adventitious roots in cuttings. There is convincing evidence that auxin, one of the natural growth hormones, is essential. It has long been known that the leaves on cuttings strongly promote root initiation. Materials originating in the leaves, called rooting cofactors, are believed to be essential to rooting; they combine with auxin to form a complex that directs RNA to activate enzymes that cause root initials to form. The composition of these cofactors is not clear but some, at least, are likely to be phenolic compounds. Other natural hormones, such as gibberellins and abscisic acid, also influence adventitious root formation. Controlled studies under aseptic conditions have shown, too, that sugar, nitrogen, calcium, and other mineral nutrients must be present for roots to form.

Factors Influencing the Rooting of Cuttings

There is a large group of plants whose cuttings root only with considerable difficulty. It is necessary to consider carefully the factors described below to satisfactorily root cuttings in this group.

Source of Cutting Material

Generally, the cuttings most likely to root come from stock plants that are growing in full sun at only a moderate rate and that have thus accumulated carbohydrates in their tissues. If the cutting material for woody plant species can be taken from the young, nonflowering plants only a few years away from a germinated seed, rooting will be much better than when the cuttings are taken from old, mature flowering and fruiting plants. The juvenility influence in the young material can sometimes be retained by keeping the stock plants cut back heavily each year to force new shoot growth out from the lower part of the plants near ground level.

Time of Year the Cutting Material Is Taken

In woody perennial plants, cutting material can be taken at any time of the year. In some species, the time the material is taken can dramatically influence rooting. Hardwood cuttings often root best if the material is gathered in late winter, while softwood cuttings usually root best if taken in the spring shortly after new shoot growth has attained a length of 10 to 15 cm (4 to 6 in.). Semihardwood cuttings are best taken in midsummer after the spring flush of growth has matured somewhat. Herbaceous cuttings can be easily rooted any time of the year, especially when succulent.

Etiolation

It has long been known that stem tissue developing in complete darkness is more likely to initiate adventitious roots than tissue exposed to light. Thus, if the basal part of shoots that are later to be made into cuttings can be kept in darkness, they are likely to form roots. Such techniques are used successfully in rooting cuttings of difficult species, such as the avocado. The mechanism for etiolation in promotion of rooting is not clear, but the harmful effects of light on rooting may be due to photoinactivation of one or more natural rooting factors in the stem tissues. (**Etiolation** is the growth of shoots in the absence of light or in low light, causing them to be abnormally elongated and colored yellow or white due to the absence of chlorophyll.)

Treatment of Cuttings with Auxins

In the mid-1930s, it was discovered that one of the natural plant hormones, auxin (indoleacetic acid [IAA]), stimulated the initiation of adventitious roots on stem cuttings. Synthetic IAA was just as effective as the natural material. But it was soon discovered that other closely related synthetic auxins—indolebutyric acid (IBA) and naphthaleneacetic acid (NAA)—were even more effective. This knowledge was quickly picked up by plant propagators, who now routinely treat the base of cuttings with one of these materials, particularly IBA, just before the cuttings are stuck in the rooting medium. Commercial preparations of IBA dispersed in talc are available; the lower end of the cutting is swirled around in the mixture to coat it with the powder. The propagator can also prepare solutions from the pure chemicals, then dip the base of the cuttings in the solution for about five seconds just before the cuttings are stuck in the rooting medium. The optimum concentration for promoting rooting varies with the species but ranges from about 2,000 to 10,000 parts per million (ppm).

Misting

Misting is an essential factor in the propagation of most cuttings, especially herbaceous cuttings. The frequency of

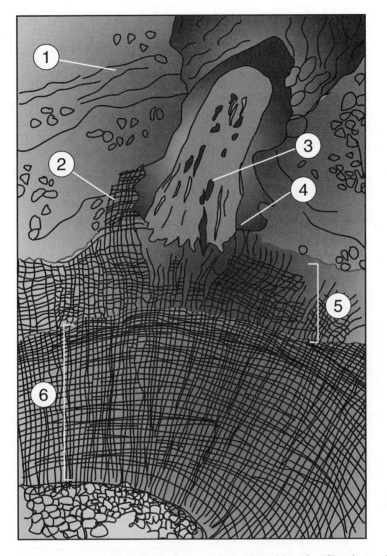

Figure 14–17 *Left:* Cross section of euonymus shoot before rooting: (1) epidermis, (2) primary bark, (3) phloem, (4) cambium, (5) xylem, (6) medullary rays, (7) perimeduler zone, (8) pith. *Right:* Cross section of euonymus shoot (cutting) after an adventitious root has formed: (1) primary bark, (2) phloem, (3) newly formed adventitious root, (4) cambium, (5) xylem formed after rooting, (6) primary xylem. *Source:* Prof. Bojinov Bogdanov and *Proceedings International Plant Propagators' Society,* Vol. 35, pp. 449–53. 1985.

misting is most critical during the early stages of rooting, before callus formation. As rooting takes place, frequency tapers off. The primary purpose of misting is to prevent cuttings from dehydrating by keeping relative humidity near 100 percent around the cutting. Misting also helps to keep the cutting cool, which reduces transpirational loss of water. Frequency is determined not only by root development, but by environmental factors such as light intensity, relative humidity, and air temperature. Low light, high relative humidity, and cool temperatures reduce the need for mist. Timers can be used to control the frequency of misting but are not the best mechanism. Timers do not sense changes in environment and do not make adjustments as conditions change. Using light or relative humidity sensors to set misting frequency allows misting frequency to change as environmental conditions that affect the need for mist change.

Bottom Heat in the Cutting Beds To force rooting at the base of cuttings before shoot growth starts, it is advisable to maintain temperatures at the base of the cuttings at about 24°C (75°F)—or about 6°C (10°F) higher than that at the tops of the cuttings, 18.5°C (65°F). This is best done by providing bottom heat from thermostatically controlled electric heating cables or hot water pipes under the rooting frames. Bottom heat under the cuttings often greatly stimulates rooting.

Propagation by Grafting and Budding

Grafting and budding are vegetative methods used to propagate plants of a clone whose cuttings are difficult to root. These methods are also used to make use of a particular rootstock rather than having the plant on its own roots. Certain rootstocks are often utilized to obtain a dwarfed or invigorated plant or to give resistance to soil-borne pests.

Grafting

Grafting can be defined as the art of joining parts of plants together so that they will unite and continue their growth as one plant. The **scion** is that part of the graft combination that is to become the upper or top portion of the plant. Usually the scion is a piece of stem tissue several inches long with two to four buds. If this piece is reduced in size so there is just one bud, with a thin layer of bark and wood under it, then the operation is termed **budding.** The **rootstock** (or **understock** or **stock**) is the lower part of the graft combination, the part that is to become the root system. Root grafting is a common method of propagation in which a scion is grafted directly onto a short piece of root and the combination is then planted. Sometimes grafting is used to change the fruiting cultivar in a fruit tree or grapevine to a different one (**top-grafting** or **top-working**). Grafting may be used to repair the damaged trunk of a tree (bridge-grafting) or to replace an injured root system (inarching). Grafting, or budding, is sometimes used to study the transmission of viral diseases (indexing). The indexing test involves inserting a bud from a plant suspected of carrying a virus into another indicator plant by T-budding. A definite visual response results from the presence of a virus, such as gumming around the inserted bud.

Several standard methods of grafting and budding have been widely used over the world for a great many years.

Whip Grafting

The whip graft (Fig. 14–18) is useful in grafting together material about 0.6 to 1.3 cm (0.25 to 0.5 in.) in diameter. It is often used to make root grafts in late winter. A small stem piece of the scion cultivar is grafted onto a root piece. The completed grafts are buried for two or three weeks in a moist material, such as wood shavings, at about 21°C (70°F) to encourage callusing or healing of the union. After the union has healed, the graft can be planted in the nursery.

Various grafting machines can be used as a substitute for the whip graft and are considerably faster than hand grafting. Such machines have been used mostly in grape grafting but are also satisfactory for making fruit tree root grafts.

Cleft Grafting

The cleft graft (Fig. 14–19) is mostly used in top-grafting, where scions 0.6 to 1.3 cm (0.25 to 0.5 in.) in diameter are inserted into stubs of older limbs that are 8 to 10 cm (3 to 4 in.) in diameter after they have been cut off a foot or so out from the trunk of the tree. It is very important

in cleft grafting to match the cambium layer of the scion as closely as possible with the cambium layer of the stock so that the two pieces heal together. Cleft grafting is usually done in late winter or early spring.

Bark Grafting

The bark graft is also used for top-grafting. It can be done only when the bark separates easily from the wood along the cambium layer. Thus, bark grafting is usually done in early to mid-spring after new growth is well underway. Bark grafting is easy to do and is widely used, especially for species considered somewhat difficult to graft successfully.

Budding

T-budding

This technique is widely used in propagating fruit trees and roses. Buds taken from budsticks are inserted under the bark of small seedling rootstock plants a few inches above ground level. Buds are then inserted and tied into place with budding rubbers, but the tops of the seedling rootstocks are not cut off above the inserted cultivar bud until just before growth starts the next spring. For fruit trees, budding is usually done in late summer. Roses are usually budded in the spring, with the top of the rootstock broken over above the bud after about two weeks following budding. The break forces the bud to grow, and after the shoot has grown 10 to 20 cm (4 to 8 in.) from the bud, the rootstock is completely cut off above the inserted bud. By late fall, a sizable rose plant is ready to dig.

Patch-Budding

Patch-budding is mostly used for plants such as walnuts, pecans, and other species that are difficult to T-bud because of their thick bark. Patch-budding must be done during the growing season when the bark on both the budstick and the rootstock is slipping easily (separating from

---

In all types of grafting and budding, the two parts must be held together very tightly and securely by wedging, tying, nailing, or wrapping with string, rubber bands, or with plastic or cloth tape. The graft union must also be completely covered by grafting wax to prevent the cut surfaces from drying out. Polarity must be observed: the scions must be inserted so that the buds point upward.

The proper selection of scionwood and budwood is very important in all types of grafting and budding. Scionwood and budwood should be taken from source trees true-to-type for the cultivar to be propagated. They should be free of known viruses and any other diseases. Some American states and some other countries have programs to supply propagating material of many fruit and woody ornamental species certified as produced under disease-free conditions.

For all types of grafting and budding, the buds on the scionwood and budwood must be dormant when the grafting or budding is done. For the whip and cleft graft, where the grafting is done during the dormant season, the scionwood can be collected from dormant source trees and used immediately in the grafting operation. For the bark graft, made later in the spring after vegetative growth has started, it is necessary to gather the scionwood earlier during the dormant season, wrap it in polyethylene bags with some slightly damp peatmoss, then hold it under refrigeration, preferably at about 0°C (32°F), until grafting. For broad-leaved evergreens, such as citrus or olives, scionwood can be obtained from the tree at the time the bark grafting is to be done, using two-year-old shoots having dormant latent buds.

PREPARING THE STOCK

A long, sloping cut 2.5 to 6.4 cm (1 to 2½ in.) long is made at the top of the stock.

A second downward cut is made starting one-third of the distance from the tip to the base of the first cut.

Pulled apart it looks like this.

PREPARING THE SCION

A long, sloping cut is made at the base of the scion the same length as the cut on the stock.

A second cut is made under the first, just as for the stock.

The stock and scion are slipped together, the tongues interlocking.

The graft is then tied and waxed.

Figure 14–18   The whip, or tongue, graft. This method is widely used in grafting small plant material and is especially valuable in making root grafts, as illustrated here. *Source:* Kester, Dale E., Hartmann, Hudson T.: *Plant Propagation: Principles & Practices,* 4th Edition © 1983. Adapted by permission of Pearson Education, Inc., Upper Saddle River, NJ.

PREPARING THE STOCK

The stub is split several cm (in.)

A smooth straight-grained section should be used so the split will be even.

PREPARING THE SCION

The scion is made by cutting a long, gradually tapering wedge.

The outside edge of the wedge should be slightly thicker than the inside.

INSERTING THE SCIONS INTO THE STOCK

The split in the stock is held open by a wedge for insertion of the scions.

Two scions are inserted in a stub, one at each end of the split. The scions must be carefully placed so the cambium layers match.

After the scions are properly placed, the wedge is withdrawn. The entire union, including the tips of the scions, is then thoroughly covered with grafting wax.

Figure 14–19   Steps in making the cleft graft. This method is very widely used and is quite successful if the scions are inserted so that the cambium layers of stock and scion match properly. *Source:* Kester, Dale E., Hartmann, Hudson T.: *Plant Propagation: Principles & Practices,* 4th Edition © 1983. Adapted by permission of Pearson Education, Inc., Upper Saddle River, NJ.

the wood along the cambium layer). For propagating nursery trees, patch-budding is generally done in mid to late summer; the subsequent handling is the same as for T-budding.

Healing of the Graft and Bud Union

In preparing a graft or bud combination, the two parts are joined by one of the methods just described so that the cambial layers of stock and scion exposed by the grafting cuts are brought into intimate contact. They are held in place by wedging, nailing, or wrapping so that the parts cannot move about or become dislodged. Then the graft union is thoroughly covered with plastic or cloth tape or, better, by grafting wax to keep out air. The union heals by callus production from young tissues near the cambium layers of both stock and scion. Temperature

levels must be conducive to cellular activity—generally from about 10°C to 30°C (50°F to 86°F)—and no dry air must contact the cut surfaces because it would desiccate the tissues.

The steps in healing of a graft union are illustrated in Figure 14–20. Healing usually takes about two weeks and must be completed, with vascular connections made for translocation of water through the xylem, before the buds on the scion start to grow and develop leaves. The transpiring leaves would soon desiccate the scion unless the graft union has healed by this time.

Limits of Grafting

There are certain limits to the combinations that can be successfully established by grafting or budding. The partners

Cleft Graft (Top View)
Just After Being Made

Scion   Stock

Enlarged View of Cambial
Region Where Healing Occurs

Cambium →   ← Cambium

First step—production of callus
tissue (parenchyma cells) by both
graft components (stock and scion)
in the cambium region.

Second step—intermingling and
interlocking of parenchyma cells.

Third step—differentiation of certain
parenchyma cells of the callus into
new cambium cells connecting with
the original cambium in the stock
and scion.

New
Xylem

New
Phloem

Fourth step—production of new
vascular tissues by the new cambium,
permitting passage of nutrients and
water between the stock and scion.

Figure 14–20   Developmental sequence during the healing of a graft union as illustrated by the cleft graft. *Source:* Kester, Dale
E., Hartmann, Hudson T.: *Plant Propagation: Principles & Practices,* 4th Edition © 1983. Adapted by permission of Pearson Education, Inc.,
Upper Saddle River, NJ.

(stock and scion) in the combination must have some degree of botanical relationship—the closer the better. For instance, woody perennial plants in different botanical families have never, as far as is known, been successfully grafted together. It is useless to attempt to graft a scion taken from a grapevine (VITACEAE) onto an apple tree (ROSACEAE).

In only a few cases have completely successful graft combinations been made between plants in the same family but different genera. For example, the thorny, deciduous large shrub trifoliate orange (*Poncirus trifoliata*) is widely used commercially as a rootstock for the common sweet orange (*Citrus sinesis*), a large evergreen tree. Both are in the family RUTACEAE, but different genera—*Poncirus* and *Citrus.*

If the two graft partners are in the same genus but different species, then the chances of success are greatly improved. Still, many such graft combinations will not unite. Plant propagation books should be consulted for information that has been accumulated by trial and error over the years. For example, in the family ROSACEAE and the genus *Prunus,* it is well established that almonds (*P. dulcis*), apricots (*P. armeniaca*), European plums (*P. domestica*), and Japanese plums (*P. salicina*) can all be successfully grafted onto peach seedlings (*P. persica*) as a rootstock. But two other members of the genus *Prunus,* sweet cherry (*P. avium*) and sour cherry (*P. cerasus*) fail when grafted onto peach roots.

If the two partners are different cultivars (clones) within a species, the chances are almost 100 percent that the graft combination will succeed. For instance, if you have a Jonathan apple tree (*Malus pumila*), you could successfully top-graft on it any other apple cultivar, for example, the Golden Delicious (*Malus pumila*).

Graft Incompatibility

There are many puzzling, unexplained situations among various graft combinations. Some pear cultivars are commercially grafted onto quince roots (an intergeneric combination), but scions of other pear cultivars grafted on quince roots soon die. The reverse, quince on pear roots, always fails. Plums can be successfully grafted on peach roots but peaches on plum roots are a failure. Japanese plums (*Prunus salicina*) can be grafted on European plum (*P. domestica*) roots, but in the reverse combination, the trees soon die.

There are many examples of incompatible graft combinations. Compatibility in grafting is the ability of two different plants, grafted together, to produce a successful union and to develop satisfactorily into one composite plant. The causes for **graft incompatibility** are little understood in spite of many years of research into the problem. There is some evidence, however, that in certain graft combinations one partner (scion or stock) produces chemicals that are toxic to the other, killing the entire plant.

Effect of Rootstock on Growth and Development of the Scion Cultivar

Tree fruit growers often select a certain rootstock for a particular fruiting cultivar because it will dwarf the tree to some

extent and thus make harvesting easier. This is particularly true in apples, where an entire series of clonal rootstocks is available to produce apple trees with any desired degree of dwarfness. Quince roots will dwarf pear trees. Trifoliate orange roots will dwarf sweet orange trees.

These dwarfing influences extend to the tree only—not to the fruits produced by the trees. But in some species, particularly citrus, the kind of rootstock used can also strongly influence the quality of fruit produced by the scion cultivar. For instance, when sweet orange seedlings are used as the rootstock for orange trees, the fruits will be of much higher quality than when Rough lemon is used as the rootstock.

What mechanisms are involved in such influences? Why are trees dwarfed by certain rootstocks? Considerable research has been conducted by plant physiologists and horticulturists over the years to try to explain these results, but convincing arguments are few. Probably the best hypothesis is that certain dwarfing rootstocks produce relatively large amounts of growth inhibitors, such as abscisic acid (see p. 192) that translocate through the graft union to the top fruiting cultivar and slow its growth. In addition, such dwarfing stocks produce low amounts of growth promoters, such as the gibberellins.

Layering

A simple and highly successful method of propagating plants is layering. Layering is similar to propagation by cuttings except that, instead of severing the part to be rooted from the mother plant, it is left attached and receives water and nutrients from the mother plant. After the stem piece (layer) has rooted—no matter how long it takes—it is cut from the mother plant and transplanted to grow independently. Various layering procedures can be used for many different kinds of plants (Fig. 14–21).

Tip Layering

This is a natural means of vegetative propagation, shown by black raspberries and the trailing blackberries. The tips of the long canes, if buried a few inches deep in the soil toward the end of their first summer's growth, root and produce a shoot that grows upward through the soil, forming a new plant. The cane can be cut from the parent bush and the new plant dug and replanted.

Simple Layering

This method can be used with plants that produce long shoots arising from the plant at ground level. In early spring the ends of these shoots can be bent over, placed in a hole in the soil several inches deep, and recurved so that the shoot tip is exposed above ground. The curved section to be buried should be nicked or twisted slightly to retard translocation of food materials through the stem; this promotes rooting. The hole is then filled in with loose, moist soil tamped firmly in place. Sometimes it is necessary to drive a stake with a hook on it into the soil beside the layer to hold it in place. After the

Tip Layering

Simple Layering

Mound Layering

Air Layering

Figure 14–21 Steps in preparing four kinds of layers. See text for details.

first summer's growth, the layer will usually have rooted and can be cut from the parent plant, dug, and transplanted. Simple layering is used to propagate filberts commercially.

Mound Layering

This method, also called **stooling,** is widely used commercially to propagate apple and plum clonal rootstocks as well as currant and gooseberry cultivars. In late winter, the mother plants in the stool bed are cut back almost to the ground. As the new shoots start to grow in the spring, moist soil—or sometimes a mixture of soil and wood shavings or sawdust—is placed around the base of the shoots. As the shoots continue to grow, the mixture is mounded higher. The stool beds are kept continually moist, usually by overhead sprinklers. Roots develop from the base of these shoots, which by the end of the summer are well rooted. In early spring, the soil is pulled away, and the rooted layers are cut off and transplanted to the nursery row for another season's growth. The stool bed is then handled again in the same manner during the next season.

Air Layering

In this method, the rooting medium is brought up to the stem to be rooted, rather than bringing the stem down to the soil. It is used for propagating the India rubber plant (*Ficus elastica*) or for *Dieffenbachia,* as well as many stiff-stemmed tropical plants, such as the litchi (*Litchi chinensis*) and Jackfruit (*Artocarpus heterophyllus*). The leaves of the branch to be air-layered are removed just below a good clump of foliage. The stem is girdled by cutting the bark away down to the wood for a width of 2.5 cm (1 in.) and a root-promoting auxin powder is rubbed into this cut. A ball (about two handfuls) of moist (not wet) sphagnum moss is wrapped around the girdling cut, then a sheet of polyethylene plastic or aluminum foil is wrapped snugly around the sphagnum moss and tied tightly above and below the ball. No further watering is needed because the moss absorbs moisture from the plant itself and the covering retards water loss. After several weeks, roots start developing. When a heavy root system has formed, the layer can be cut off just below the root ball, removed from the plastic, and, without disturbing the roots, planted in a large pot of soil. A few leaves should be removed to reduce water loss and the plant should be put in a cool, humid, and shady place until it becomes well established on its own new roots system.

Other Plant Structures Providing Natural Propagation Methods

Runners

Some plants, such as the strawberry (Fig. 14–22), grow as a rosette crown, with runners (stolons) arising from the crown. New plants arise from nodes at intervals along these runners. From these runner plants additional new runners arise, thus developing a natural clonal multiplication system. The runner plants must have favorable moist soil conditions to root. The strawberry produces runners in the summer in response to long days, and stops producing runners as the days shorten in the fall. Then the strawberry runner plants can be dug, packed in polyethylene-lined boxes, and placed in cold storage (−2°C; 28°F) for planting later, usually the following spring. Other plants with

Figure 14–22 Strawberry plant with new plants developing from nodes on the runners (stolons).

this natural asexual reproduction system are the ground cover (*Duchesnea indica*) and the strawberry geranium (*Saxifraga stolonifera*).

Suckers

Some plants, such as the blackberry and red raspberry, produce **adventitious** shoots—or suckers—from their horizontal root system, which eventually spread to form a dense thicket of new plants. These individual shoots with a piece of the old root attached can be dug and replanted. This is a simple, highly successful asexual propagation method.

Crowns

Many perennial plants exist as a single unit, becoming larger each year as new shoots arise from the crown (root-shoot junction) of the plant. Vegetative propagation of such plants consists of crown division—cutting the crown into pieces, each having roots and shoots, and transplanting to a new location (Fig. 14–23).

Propagation Using Specialized Stems and Roots

A number of herbaceous perennial plants have structures such as **bulbs, corms, tubers, tuberous roots,** or **rhizomes.** These structures function as food storage organs during the plant's annual dormant period and also as vegetative propa-

Figure 14–23 Crown of a herbaceous perennial, the Shasta daisy. Lateral shoots develop from the underground portion of older stems and root. These rooted shoots can be cut from the mother plant and replanted. Underneath is a rooted shoot cut from the mother plant.

gation structures. In bulbs and corms, the newly formed plants break away from the mother plant naturally. This type of propagation is termed **separation.** The remaining structures—tubers, tuberous roots, and rhizomes—must be cut apart; this is termed propagation by **division.**

Bulbs

These are short, underground organs having a basal plate of stem tissue with fleshy leaf scales surrounding a growing point or flower primordium. In the axils of some of these leaf scales, new miniature bulbs develop that eventually grow and split off from the parent bulb to form a new plant. There are two types of bulbs: (1) the tunicate bulb has a solid tight structure with fleshy scales arranged in concentric layers and covered by a dry membranous protective layer; examples are onion, daffodil, tulip, and hyacinth; (2) the scaly bulb, such as the lily, has loose, separate scales with no protective layer to prevent desiccation. The scales can be removed from the bulb; placing them under moist, humid conditions promotes the development of one or more new miniature bulbs at the base of each scale.

Corms

A **corm** is the swollen underground base of a stem axis; it has nodes and internodes and is enclosed by dry scalelike leaves. The gladiolus, freesia, and crocus are examples of plants having a corm structure. Following bloom in gladiolus, one or more new corms develop just above the old corm, which disintegrates. In addition, several new small corms called cormels are produced just below each new corm. These may be detached and grown separately for one or two years to reach flowering stage.

Stem Tubers

The edible Irish (white) potato is a good example of a plant having a tuber structure. The underground tuber consists of swollen stem tissue with nodes and internodes. When a potato tuber is cut into sections and planted, shoots arise from the buds (eyes) on the potato piece, which itself serves as a food supply for the developing shoot. From the lower portion of such shoots, adventitious roots develop, along with several underground horizontal shoots that are stem tissue. The terminal portions of these horizontal shoots enlarge greatly to form the fleshy potato tuber. Potato cultivars are clones and are propagated by dividing each tuber into several sections. The crop eventually developing from each section usually consists of three to six new tubers.

Tuberous Roots

Tuberous roots are root tissue. The dahlia, tuberous-rooted begonia, and the sweet potato (Fig. 14–24) are examples. Sweet potatoes are propagated vegetatively by placing the tuberous roots in beds so they are covered with about 5 cm (2 in.) of soil. Adventitious shoots (slips) develop from these root tubers; adventitious roots form from the base of the shoots. These rooted slips are then pulled from the tuberous root and planted. As the sweet potato vines grow, some of their roots swell to form the familiar edible sweet potato.

Figure 14–24 *Left:* Sweet potato tuberous root, producing rooted adventitious shoots. *Right:* Two detached shoots (slips) ready for planting.

Rhizomes

Certain plants, such as the German iris and bamboo (Fig. 14–25), have the main stem axis growing horizontally just at slightly below the soil surface. This stem is a rhizome. Some plants have thick and fleshy rhizomes, while others have thin, slender rhizomes. Like any stem, rhizomes have nodes and internodes. Leaves and flower stalks and adventitious roots develop from the nodes. Propagating plants with rhizomes is easy. At a time of year when the plant is not actively growing, the rhizome can be cut into pieces several inches in length and transplanted. Noxious weeds, such as Johnson grass, that have a rhizome structure cannot be controlled by cultivation because it merely breaks up the rhizomes and spreads the pieces about, each piece developing a new plant. Many important economic plants are propagated from rhizomes; examples are banana, ferns, ginger, and many grasses. Sometimes the above-ground horizontal stems are termed stolons, particularly in grasses such as Bermuda grass.

MICROPROPAGATION (TISSUE CULTURE)

A major advance in plant propagation involves the use of very small pieces of plant tissue grown on sterile nutrient media under aseptic conditions in small glass containers. These small pieces of tissue, called **explants,** are used to regenerate new shoot systems, which can be separated for rooting and growing into full-size plants. Some of the pieces of tissue are retained for further regeneration. The increase is geometrical, giving rise to fantastically large numbers of new individual plants in a short time. Some nurseries produce millions of plantlets a year by tissue culture methods, which require the use of highly trained technicians working under sterile conditions.

Figure 14–25 Rhizome of bamboo with lateral buds and adventitious roots arising at nodes. Plants with rhizomes are propagated simply by cutting the rhizome into pieces and planting them.

Micropropagation was used at first with herbaceous plants, such as ferns, orchids, gerberas, carnations, tobacco, chrysanthemum, asparagus, gladiolus, gloxinia, strawberry, and many others. Later, with modifications of the media, it was found that many woody plants, such as rhododendrons, kalmias, deciduous azaleas, roses, plums, apples, and many others could also be successfully propagated by tissue culture methods (Fig. 14–26).

Different parts of the plant can be taken as the explant. Entire seeds themselves can be used; for example, very tiny orchid seeds have been commercially germinated in sterile culture for many years. Embryos can be extracted from seeds and grown on a sterile nutrient medium. Shoot-tip culture involves excision of the growing point, which increases in size when it is kept in a nutrient medium and is divided over and over, greatly increasing plant numbers. This method has been successful with orchids, ferns, apples, and carnations. In other cases, tissue culture involves, for example, the excision of a piece of stem tissue and placing it on a nutrient medium. The explant then develops masses of **callus** by continuous cell division. From these callus clumps, roots and shoots may differentiate to form new plants. This method has been used with carrot, tobacco, asparagus, endive, aspen, Dutch iris, and citrus.

Single pollen grains of some plants, such as tobacco, have been germinated in sterile culture, when taken at just the right stage, to form **haploid** (1n) plants. Doubling the chromosome number with colchicine treatments yields diploid homozygous plants.

In addition to propagation, these aseptic tissue cultures procedures are used for pathogen elimination and germplasm preservation. The nutrient media used for micropropagation are also favorable substrates for the growth of bacteria, fungi, and yeasts, so the prepared nutrient medium and its containers must be sterilized. The sealed containers are placed in an autoclave or pressure cooker for 20 to 30 minutes at 120°C (248°F). The plant tissue itself underneath the epidermis is sterile, but surface sterilization of the explants with a material such as a 10 percent Clorox solution is necessary, followed by rinsing with sterile water. Excision and insertion of the plant tissues onto the nutrient medium in the sterilized containers must be done with laboratory skill and procedures to prevent recontamination.

The nutrient media on which the excised plant parts are grown include mineral salts, sugar, vitamins, growth regulators (auxins and cytokinins), and sometimes certain organic complexes such as coconut milk, yeast extract, or banana puree. Many different media have been developed for obtaining shoot and root proliferation on explants of various species. The constituents of one widely used medium are given in Table 14–3. A mixture of micronutrient elements, including iron, is often added.

Figure 14–26 Small plantlets of plum trees being propagated in a sterile nutrient medium in test tubes (10 cm rule). *Source:* F. Loreti and S. Morini.

TABLE 14–3 Components of the Murashige and Skoog Medium for Growing Tissue Explants Under Sterile Culture

NH_4NO_3	400 mg/l	Indoleacetic acid	2.0 mg/l
$Ca(NO_3)_2 \cdot 4H_2O$	144	Kinetin	0.04–0.2
KNO_3	80	Thiamin	0.1
KH_2PO_4	12.5	Nicotinic acid	0.5
$MgSO_4 \cdot 7H_2O$	72	Pyridoxine	0.5
KCl	65	Glycine	2.0
NaFe—EDTA	25	Myo-inositol	100
H_3BO_3	1.6	Casein	
$MnSO_4 \cdot 4H_2O$	6.5	hydrolysate	1,000
$ZnSO_4 \cdot 7H_2O$	2.7	Sucrose	2%
KI	0.75	Powdered purified agar	1%

Micronutrient elements may or may not be required but are usually added routinely. The following stock solution provides the required materials. One milliliter of this solution is added per liter of culture medium.

$MnSO_4 \cdot 4H_2O$	1.81 g	$CuSO_4 \cdot 5H_2O$	0.08 g
H_3BO_3	2.86 g	$(NH_4)_2MoO_4$	0.09 g
$ZnSO_4 \cdot 7H_2O$	0.22 g	Distilled water	995 ml

Iron is usually essential and can be supplied in several ways: iron tartrate (1 ml of a 1 percent stock solution), inorganic iron ($FeCl_3 \cdot 6H_2O$, 1 mg/l) or $FeSO_4$, 2.5 mg/l; or chelated iron. Chelated iron can be supplied as NaFeEDTA, 25 mg/l, or by mixing Na_2EDTA and $FeSO_4 \cdot 7H_2O$ in equimolar concentrations to give 0.1 mM Fe.

BIOTECHNOLOGY

Biotechnology, in a broad sense, can be defined as the management of biological systems for the benefit of humanity. By this definition, it is obvious that farmers have been practicing biotechnology for many years in the production of their crops. But the word *biotechnology,* as generally used in the late twentieth century, means the *new* biotechnology, developed by means of molecular biology and molecular genetics using advanced genetic engineering techniques.

Genetic engineering differs from traditional breeding and genetics because union of pollen/sperm and egg or a mutation does not have to occur to have different DNA incorporated into an organism. Genetic engineering complements but does not replace traditional breeding techniques. Because the genetic code is the same for all DNA-carrying organisms, DNA from one organism can be utilized by another, even though the two organisms may be biologically very different. Scientists from many different disciplines have worked since the 1970s to perfect methods of identifying genes, separating them from the host DNA, and transfering them to new hosts. It is a complicated process and it takes many years to introduce a plant or animal that has the foreign DNA in it. However, improvements in the process have enabled plant scientists to introduce many plants with foreign DNA that gives resistance to insects, heribicides, certain environmental stresses, or other traits. Such organisms are commonly called genetically modified organisms (GMOs). When you think about it, any organism except a clone is genetically modified, but the term refers to those that have been altered through the introduction of DNA by artificial means.

Most genetic engineering starts with the identification of a gene that imparts a desired trait such as insect resistance or more nutrititive content. The gene is cut from the chromosome of cells taken from the host organism. If insufficient numbers of the gene can be extracted, the gene is amplified, meaning it is increased in great numbers, usually by a process called polymerase chain reaction (PCR). The gene is then inserted either by a bacterium or gene gun into the nucleus of cells of the plant being transformed. The plants that develop from these cells are tested to be sure they carry the gene. Then through traditional breeding or asexual propagation, large numbers of the transformed plant are produced. To date, no entirely new genes have been created. Gene modification and transfer has been done with genes that exist naturally. To make a gene from scratch would be an extremely difficult task.

When a gene is inserted into a cell, it then can be induced to produce the protein or proteins that confer the desired trait. Sometimes the gene is attached to a promoter that has the gene turned on constantly. In other situations, the promoter turns the gene on only under specific conditions or in certain tissues or organs. Keep in mind, however, that even though a protein is generated, its function may or may not be the same as in the original organisms, depending on what other factors are influencing the protein. As mentioned earlier in the chapter, the inability to influence a protein the same in the modified organism is a challenge to genetic engineers.

Two of the most commonly found GMO crops are those designated as Bt® and Round-Up Ready® Bt® means the plant carries a gene from the *Bacillus thuringenis* bacterium. The gene generates a protein that is poisonous to some insects that feed on it. Round-up Ready® plants are resistant to the herbicide, Round-Up®. By using Bt® and Round-Up Ready® plants, farmers have been able to reduce the use of chemical pesticides and implement tilling practices that reduce soil erosion and improve soil quality. A type of rice has been developed that carries the genetic information to produce beta-carotene in the rice grains. Beta-carotene can help prevent blindness in children and reduce the incidence of childhood blindness in underdeveloped countries where rice is a dietary staple.

Although many GMO plants are now commercially available, there is considerable resistance from the general public to buying and using products from the crops. For instance, there is fear that the Bt® toxin would be poisonous to humans or that Round-Up® resistance will be transferred by natural breeding to native weed species in an area. The plant science community must work with other segments of society to educate people and offer the scientifically sound assurance that GMO plants are safe to use.

SUMMARY AND REVIEW

Crops are domesticated by selecting and propagating plants with superior characteristics. The propagation can be sexual (seed) or asexual (vegetative). With many crops, using seeds is the most practical method of propagation. As long as the seeds produce plants that maintain the desired characteristics of the parent(s) (true to type), the seeds can be saved from one crop to produce the next. When seeds do not produce true to type, then hybrid seeds that are obtained from specific parental crosses must be obtained for each crop. The development of hybrid seed has been one of the more important breakthroughs in crop improvement. Reliable production of crops from seeds requires careful control of the seed source. Seed certification programs are designed to maintain the genetic quality of each generation of seed.

In many cases, asexual propagation allows desirable traits to be passed easily from one generation to the next. Asexual propagation involves the production of an entire plant from a cell, tissue, organ, or other part of another plant. Except in the case of mutation or chimera, the new plant has the same genetic makeup as the original plant. The technique is very useful when genetic integrity cannot be maintained by seed production or when seed production is inefficient.

Cutting production involves removing a part of a stem or root of a plant and regenerating the missing organs and tissues by providing an environment that favors the regeneration process.

Grafting is the attaching of a twig called the scion from a plant with one genetic complement to the rootstock from a plant with a different genetic complement. The scion and rootstock must be related closely enough for grafting compatibility. There are several different kinds of grafts, including whip and cleft. Budding is similar to grafting except the scion is a bud.

Layering is similar to cutting except the part to be propagated is not removed from the parent plant until the missing stem or roots have regenerated. Layering can be simple, mound, or air.

Other plant structures that can be used for propagation are plantlets that can be separated from the parent plant, or crowns, bulbs, corms, and tubers that can be divided. Plantlets can come from runners, which extend from the main portion of the plant or form naturally in some species along leaf margins and at leaf bases. Care must be taken to ensure that, over time, the new generations have not acquired undesirable mutations or become infected with pathogens.

Crop improvement has progressed from its early stages, when superior plants were selected by a farmer and the seeds of those plants used for the next crop, to modern methods that involve very rigid and competitive breeding and genetic engineering programs throughout the world. The potential for crop improvement through traditional breeding and genetic engineering is inestimable. No longer is the genetic information in one plant limited to what can be obtained through compatible sexual crossing. The ability to impart resistance, productivity, or nutrition in plants appears to be limitless from a scientific basis. However, public sentiment does not completely support GMOs. The agricultural community is being forced to find ways to show the public that genetic engineering is scientifically sound and environmentally safe.

FOOD FOR THOUGHT

1. Indicate which of the following are homozygous and which are heterozygous: AABB Aabb aaBB AaBb

   Which can produce all true-to-type seed if they were self-pollinated?

2. When seed companies first introduced hybrid seed, farmers were very skeptical and thought that the seed companies were just trying to get them to buy new seed every year. Why would farmers think that way? What do you think changed their minds after a few years and made them overwhelmingly favor buying the hybrid seed each year?

3. Grafting is often a difficult and time-consuming process. What are the advantages of grafting that makes the method worth using for many crops?

4. What are adventitious shoots, adventitious roots, and callus? Why are they important factors in asexual propagation?

5. Make a list of four traits you would like to see in the ideal plant. For each trait, list an existing plant or other organism that carries that trait. Determine if that trait can be passed to your ideal plant by traditional breeding or if it would require genetic engineering.

6. Chrysanthemums are often chimeras. Explain why this characteristic requires that most chrysanthemum cultivars be propagated asexually rather than by seed.

7. Check the daily newspapers or find a GMO site on the Internet. Determine how much scientific data are presented by both those attacking and those defending the use of GMOs. Evaluate the validity of the data presented.

SUPPLEMENTARY READING

ANONYMOUS. 1992. *New crops, new uses, new markets: Industrial and commercial products from U.S. agriculture.* Washington, D.C.: Office of Publishing and Visual Communication, U.S. Department of Agriculture.

COOK, G. 2004. *Genetically modified language: The discourse of arguments for GM crops and food.* New York: Routledge.

HARTMANN, H. T., D. E. KESTER, F. T. DAVIES, and R. L. GENEVE. 2002. *Plant propagation: Principles and practices,* Seventh Edition. Upper Saddle River, N.J.: Prentice Hall.

JENSEN, N. F. 1988. *Plant breeding methodology.* New York: Wiley.

RICE, L. W. and R. P. RICE, Jr. 2006. *Practical horticulture,* Sixth Edition. Upper Saddle River, N.J.: Prentice Hall.

CHAPTER
15

Crop Biodiversity: Naming, Classifying, Origin, and Germplasm Preservation

David Tay

KEY LEARNING CONCEPTS

After reading this chapter, you should be able to:

♦ Understand how plants are named and classified.
♦ Use the nomenclature and system of taxonomic classification to identify plants and their relationship to each other.
♦ Explain how several crops originated and where they were domesticated.
♦ Understand the importance of saving germplasm from extinction and the global system created to preserve germplasm.

There are over 500,000 different kinds of plants, and for humans to be able to communicate about them, some method of classifying and naming them had to be developed. The process of developing plant names started over 4,000 years ago, but the first recorded names were attributed to Theophrastus (370–285 B.C.). Naming plants, no doubt, began with simple names that referred to the plant's use, growing habit, or other visible attribute. One example is the milk weed—so named because its sap (latex) is milky in appearance. One difficulty with such names is that often they are only used locally. People in one place know the plant by one name, while elsewhere the same plant is known by a different name. The weed *Tribulus terrestris* is known as the puncture vine in some areas because the seeds have sharp spines that puncture tires or bare feet; another common name for the same plant in other areas is goat-head because the shape of the seed resembles a goat's head. As plant knowledge expanded and the exchange of this knowledge became desirable, it was obvious that a uniform and internationally acceptable system was needed to name and classify plants.

There are many ways to classify plants, and any system depends on how the classification is to be used. Some classifications relate directly to specific environmental requirements of the plant for satisfactory growth. For example, such a classification could categorize plants according to their climatic requirements.

CLIMATIC AND RELATED CLASSIFICATIONS

Farmers obviously have to be able to identify and name crops. But this alone is not enough. They also have to distinguish which of the many crops suit their climate. Most farmers in the United States are working in the temperate zone. For example, some fruit and nut crops grown in the temperate zone are almond, apple, apricot, cherry, peach, pear, pecan, and plum. Fruit growers in a tropical region would have an entirely different choice of crops, such as cocao, cashew and macadamia nuts, banana, mango, papaya, and pineapple. The fruits of the subtropical region, between the temperate zone and the tropics, cannot withstand the severe winters of the temperate zone but may need some winter chilling. Some of these subtropical plants are citrus, date, fig, olive, and pomegranate. Thus, by using climate as a criterion, plants can be classified into distinct groups.

Agronomic crops like grains, forage, fiber, and oil crops and the vegetable and ornamental plants can also be classified by their temperature requirements. For example, some annuals have specific climatic requirements for growth and flowering and are distinguished as winter or summer annuals. Winter annuals are planted in the fall and bloom early the following spring. Summer annuals are planted in the spring and bloom through the summer and fall.

Some crops grow best in certain seasons and are thus classified. For example, warm-season plants such as corn, beans, tomatoes, peppers, watermelons, petunias, marigolds, zinnias, and Bermuda grass grow best where monthly temperatures average 18°C to 27°C (65°F to 80°F), while broccoli, cabbage, lettuce, peas, flowering bulbous plants, snapdragons, cyclamen, and bluegrass are cool-season crops growing best at average monthly temperatures of 15°C to 18°C (60°F to 65°F). Plants can be classified by the seasons in which they are most likely to flower and fruit or when the quality of the product can be expected to be at its maximum.

Numerous flower and vegetable **cultivars** can be classified as early, midseason, or late maturing.

Vegetables are classified into groups according to their edible parts. Some are grown for fruits and seeds, such as the tomato, bell pepper, string bean, pea, and corn. Many are grown for their shoots or leafy parts, such as asparagus, celery, spinach, lettuce, and cabbage. Others are grown for their underground parts (either roots or tubers), such as the carrot, beet, turnip, and potato.

Ornamentals are sometimes classified by use—that is, as houseplants, greenhouse plants, garden plants, street trees, and various classes of landscape plants. Houseplants, which have become very popular, are often classified according to their foliage, flowers, or growth habits. Common foliage plants, which stay green all year, are the philodendrons, dieffenbachias, and ferns, to name a few. Blooming of some flowering plants may be only seasonal, as in the lily and poinsettia, but the African violet and chrysanthemum are available in flower year-round. Plants outside the home are typically classified by use, for example, bedding plants such as petunias, marigolds, and zinnias, and landscape plants like trees and shrubs.

The forester classifies trees into two broad groups: the hardwoods and softwoods. Some hardwood types are oaks (*Quercus*), maples (*Acer*), birch (*Betula*), and beech (*Fagus*). Some softwood trees are pines (*Pinus*) (Fig. 15–1), firs

Figure 15–1 Jeffrey pine (*Pinus jeffreyi* Grev. and Balf.) tree growing in the Sierra Nevada mountains of northern California near Lake Tahoe. This tree is about 2 m (6.5 ft) in diameter at breast height (DBH), the point where lumber trees are measured. *Source:* Robert A. H. Legro.

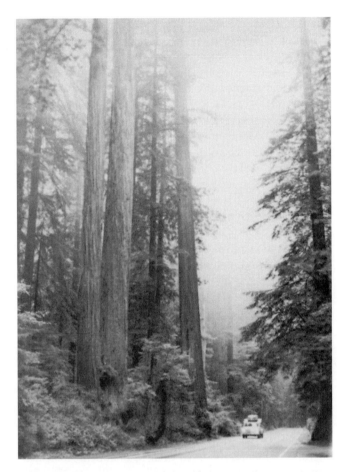

Figure 15–2 A grove of coast redwoods (*Sequoia sempervirens* D. Don, Endl.) along Highway 101 on "The Avenue of the Giants" in northern California. The camper truck is dwarfed by these majestic trees.

(*Abies*), redwood (*Sequoia*) (Fig. 15–2), cedars (*Cedrus*), and spruce (*Picea*). Trees are also classified according to the hardiness zones in which they can survive. Some can withstand very low temperatures during the winter, whereas others are subject to frost damage and therefore must be grown in a subtropical climate.

COMMON AND BOTANICAL NAMES

Most plants are generally known by their common names because common names are often easier to remember, pronounce, and use. Maple or elm trees growing along the streets are referred to by common names in everyday conversation. Common names often evolve because of certain plant characteristics.

A common name has value in conversation only if both persons know exactly what plant is being discussed. This is most likely when persons are from the same community and the common name of the plant cannot be mistaken for another. But take the case of the jasmine, a plant known all over the world and prized for its fragrance, flavoring (tea), and landscaping. Many common plant names contain the word *jasmine*, but such plants do not resemble one another and may not even be closely related botanically. Some of the so-called jasmines are listed below, with the botanical (italicized) name following the common name:

Star jasmine	(*Jasminum gracillimum*)
Star jasmine	(*Trachelospermum jasminoides*)
Blue jasmine	(*Clematis crispa*)
Cape jasmine	(*Gardenia jasminoides*)
Crape jasmine	(*Tabernaemontana divaricata*)
Night jasmine	(*Cestrum nocturnum*)
Night jasmine	(*Nyctanthes arbor-tristis*)

It is obvious from these examples that common names have their limitations for universal written or verbal communication. There are too many and they are too variable to serve most scientific purposes.

DEVELOPMENT OF BOTANICAL CLASSIFICATIONS

Theophrastus (370–285 B.C.), a student of Aristotle, classified plants by their texture or form. He also classified many as herbs, shrubs, and trees. He noted the annual, biennial, and perennial growth habits of certain plants, and described differences in flower parts that enabled him to group plants for purposes of discussion. He is known as the father of botany for these significant contributions.

Carl von Linné (1707–1778), better known as Carolus Linnaeus, devised a system of categorizing plants that led to the modern taxonomy or nomenclature of plants.

Scientific Classification

The scientific system of classification has all living things divided into groups called **taxa** (sing. taxon) based on physical characteristics. The first taxon, called **Domain,** divides all living things into two Domains: Prokaryotes (cells having no separate subcellular units) and Eukaryotes (cells having subcellular units). The Eukaryote Domain is divided into the four Kingdoms of Fungi, Protista, **Plantae,** and Animalia. The Plantae Kingdom is divided into two groups: **bryophytes** (includes mosses and liverworts) and **vascular plants.** The vascular plants are divided into two subgroups: seedless and seeded. Seedless and seeded plants are further classified by **Phyla.** Seedless phyla include the Pterophyta (ferns). Seeded phyla include Cycadophyta (cycads), Ginkgophyta (ginkgo), Coniferophyta (conifers), and Anthophyta (angiosperms, which are subdivided into monocotyledon and dicotyledon). Almost all commercially important crop plants are in the seeded group. After phylum, plants are classified in descending rank by **class, order, family, genus,** and **species.** Each descending rank more closely defines the physical characteristics common to members of that rank.

PLANT IDENTIFICATION AND NOMENCLATURE

The family is usually the highest taxon commonly included in plant identification or study. Students of plant science are usually required to learn the family, genus, and species of some plants as well as their common names.

Since the early Christian era, naturalists wrote their books in Latin, which was the language of all educated people in Europe. Thus, Linnaeus used names of Latin form. Most of the names he gave, which describe morphological characteristics of the plants, came from Latin words, although some were derived from Greek and Arabic.

The names are usually phonetic and often give a clue to the plant's characteristics, its native habitat, or for whom it is named. Such derivatives are numerous. Names that refer to leaves include *folius, phyllon,* or *phylla,* usually as suffixes. The names can also have prefixes, such as *macro* or *micro.* Thus, words are created, such as *macrophylla* (large leaf), *microfolius* or *microphylla* (small leaf), *illicifolius* (holly leaf), and *salicifolius* (willow leaf). The Latin for flower is *flora;* add the prefix *grand* and it becomes *grandiflora* (large flower), as in *Magnolia grandiflora* L., the southern magnolia (Fig. 15–3). Shapes or growing habits of plants can be described with *altus* or *alta* (tall), *arboreus* (treelike), *compactus* (dense), *nanus* or *pumilus* (dwarf), *repens* or *reptans* (creeping), and *scandens* (climbing). Names based on flower or foliage color include *albus* or *leuco* (white), *argentus* (silver), *aureus* or *chryso*

(gold), *rubra, rubens,* or *coccineus* (red), and *croceus, flavus,* or *luteus* (yellow). Species names sometimes reflect the plant's place of origin. Examples are *australis* (southern), *borealis* (northern), *canadensis* (from Canada), *chinensis* or *sinensis* (from China), *chilensis* or *chileonsis* (from Chile), *japonica, nipponica,* or *nipponicus* (from Japan), *campestris* (field), *insularis* (island), and *montanus* (mountain). Sometimes Linnaeus took names, like *Narcissus,* from classical mythology, or devised names to honor other scientists, like *Rudbeckia,* honoring Linnaeus' botany professor, Rudbeck.

Each plant has a two-word, or **binomial,** name[1] given in Latin. The first name refers to the plant's genus, the second to its species. The Latin binomial name is international and understood universally.

Complete Linnaean names have a third element—the **authority,** or the abbreviated name of the scientist who named the species. Consider the name for the common white (Irish) potato, *Solanum tuberosum* L. The "L." means Linnaeus, and his initial appears commonly, because Linnaeus named so many species. Books and journals often omit the authority for brevity and simplicity, but it is important in determining which taxon is being referred to in situations where different botanists have used different binomials for the same plant. In such cases, the botanical or scientific name that was published first takes precedence.

Wild or naturally occurring plants are named under the rules of the *International Code of Botanical Nomenclature.* Cultivated plants are named according to the same principles but are covered by the *International Code of Nomenclature Cultivated Plants.* Some of the basic rules of nomenclature follow.

The generic name always begins with a capital letter; it is underlined when written by hand or typewriter and italicized in print. Thus, the genus name for potato is *Solanum.* The specific epithet *tuberosum* is likewise underlined or italicized. The specific epithet usually begins with a lowercase letter, but it *may be* capitalized if it is a person's name; it is always correct if written entirely in lowercase. Many of the original species names are frequently capitalized (e.g., *Pinus Jeffreyi* Grev. and Balf. or *Pinus jeffreyi* Grev. and Balf.).

To complete the binomial name, the authority for describing and naming the plant is given after the genus and species; thus, *Solanum tuberosum* L. In this text, unless otherwise specified, both the binomials and the authorships agree

Figure 15–3   The southern magnolia (*Magnolia grandiflora* L.). The name *grandiflora* is more than justified; the flower measures 15 cm (6 in.) across.

---

[1]The binomial is a binary combination of the name of the genus followed by a single specific epithet. If an epithet consists of two or more words, these are to be united or hyphenated [Article 23 *International Code of Botanical Nomenclature* (2)]. Although the second word of the binomial is technically the "specific epithet," many persons refer to that second name as the "species." This shorter and therefore more convenient form is the one used in this text.

with those given in *Hortus Third,* which is a widely recognized compilation of the cultivated plants for the United States and Canada. These authority names are often abbreviated. Each taxonomy book has a list of the full names of these authorities.

When several plants in the same genus are listed, the genus name is given in full for the first plant, then shortened to the first initial (which is always capitalized) for the other plants in the list. As an example, the apricot, the European plum, and the peach are all members of the genus *Prunus.* Listed as scientific binomials, they would be *Prunus armeniaca, P. domestica,* and *P. persica,* respectively. This procedure should not be used if there is any chance of confusion with another genus with the same first initial.

The singular and plural spelling of *species* is the same. Occasionally the plant genus is known but the exact species is not known because it is difficult or impossible to identify. In such a case, the genus name is given and followed by the lowercase letters "sp." for species (singular) and "spp." for species (plural). An example is *Prunus* spp. The "spp." usually refers to all of the many species in the genus. However, the "sp." refers to a definite plant whose specific epithet is not known. The "sp." or "spp." is never underlined or italicized.

Subspecific Categories

Sometimes a botanical binomial is not sufficient to identify a species—wild or cultivated. Botanists and horticulturists may form subspecific categories, such as botanical variety, cultivar, and group.

Botanical Variety

A plant group can be so different in the wild from the general species described originally that it warrants a botanical variety classification below that of species. An example of this is *Buxus microphylla* Sieb. and Zucc. var. *japonica* Rehd. and Wils., which is native to Japan. The "var." stands for *varietas,* Latin for "variety." Another botanical variety originated in Korea; it is *Buxus microphylla* var. *koreana* Nakai. These botanical varieties are sufficiently different to warrant unique names and authorities to distinguish them from one another. In this case, the name of the variety *japonica* or *koreana* is underlined or italicized. When a varietas epithet is formed from a surname, it may or may not be capitalized depending on the personal preference of the author. However, the trend is to not capitalize them, as recommended by the International Codes.

Cultivar

Many kinds of plants that are valuable in agriculture must be propagated with little or no genetic change in the offspring. In agriculture and horticulture, there are cultivated varieties that remain genetically true. These cultivated varieties may

be different from botanical varieties and are called **cultivars,** a contraction of *culti*vated *var*iety. There are two main categories of cultivars—the clones and the lines. If propagated by vegetative methods, they are called *clones*; if by seeds (under certain specified conditions), they are called *lines.* The word *cultivar* is abbreviated "cv." and the plural is "cvs." A cultivar is often a distinct variant selected by someone who believed it was uniquely different from any plant already in cultivation. The flower color may have changed from red to white because of a mutation, as in some carnations. Perhaps a plant has fewer spines or thorns than does the ordinary species; an example is a Chinese holly (*Ilex cornuta* Lindl. and Paxt.) found to have few or no spines. It was named *Ilex cornuta* cv. Burfordii. The cultivar name is always capitalized but never underlined or italicized. The term "cv." after *cornuta* may be dropped in favor of single quotes around the cultivar name. Either way of expressing the cultivar name is acceptable, and both can be used in the same article. Either single quotes or the term "cv." are used, but *never both.* Tables or lists usually use "cultivar" or "cv." in the heading to avoid single quotes around each cultivar name.

Many annual flowers, vegetables, grains, and forage crops are cultivars that are propagated by seed. Others are F_1 hybrids, uniform and nonuniform assemblages. An example is *Petunia* $\times$ *hybrida* Hort. Vilm.-Andr.—the hybrid garden petunia. A breeder may develop a new strain that is believed to warrant a cultivar name such as 'Fire Chief' or 'Pink Cascade.' The parent plants can be maintained and crossed to produce the same F_1 hybrid cultivar year after year. Many vegetables and flowering annuals are maintained as cultivars in this manner, with the parents maintained to produce new crops of seed each year for planting. Cultivars of fruit trees, grapes, and woody ornamentals are usually maintained as true-to-type **clones** by vegetative propagation methods.

Group

The **group** category is used for some vegetables and some ornamentals such as lilies, orchids, roses, and tulips. It is a category below the species and not used as frequently as the cultivar category. A group includes more than one cultivar of a particular kind of plant. For example, when there are evident differences among plants of the same species, they can be further categorized by a group name. When a species has many cultivars, cultivars that are similar are categorized into groups. For example, cultivars of *Brassica oleraceae* can be grouped into the Acephala Group, the Alboglabra Group, the Botrytis Group, or the Capitata Group, depending on their morphological characteristics. These groups have the same botanical name—*Brassica oleracea.* The name of the group is written within parentheses between the species name and the cultivar name, as *Brassica oleraceae* (Capitata) 'King

Figure 15–4 Side and front views of potato (*upper*) and tomato (*lower*) flowers. These views show the similarities of the flowers of plants in the same family, SOLANACEAE. *Source:* Moira Tanaka.

Cole,' and the group name is always capitalized but not enclosed in single quotes.

Family

The family is a group of closely related genera. The relationship can be based on certain plant structures or on chemical characteristics, such as the presence of latex in the milkweed family ASCLEPIADACEAE, but flower structure is the usual basis for association. The nightshade family SOLANACEAE contains not only *Solanum* (potato) but also *Lycopersicon* (tomato), *Capsicum* (pepper), *Nicotiana* (tobacco), *Datura* (deadly nightshade), *Petunia,* and many others. This is a large family (about ninety genera and more than 2,000 species), most of which are native to the tropics. All species in this family have similar flower structures; the similarities between a tobacco, a tomato, and a potato flower, for instance (see Fig. 15–4), are readily seen.

The first letter of family names is always capitalized and the names are sometimes underlined or italicized. The family names may be written entirely in capital (upper case) letters, the method used in this text. Most families' names end with -*aceae* (pronounced ace-ay-ee) attached to a genus name; for example, SOLANACEAE, ROSACEAE, AMARYLLIDACEAE,

LILACEAE, and MAGNOLIACEAE. Eight families, however, did not follow this standard rule. For the sake of uniformity, new names have been adopted for these families. The old names appear in parentheses following the new names:

> ASTERACEAE (COMPOSITAE)
> BRASSICACEAE (CRUCIFERAE)
> POACEAE (GRAMINAE)
> CLUSIACEAE (GUTTIFEREAE)
> LAMIACIACE (LABIATE)
> FAVACEAE (LEGUMINOSAE)
> ARECACEAE (PALMAE)
> APIACEAE (UMBELLIFERAE)

Plant classifications can be studied in detail in various plant biology or taxonomy books and references.

PLANT IDENTIFICATION KEY

See Table 15–1 for a simplified yoked (indented) key used to identify some commonly known seed-bearing plants (Spermatophyta). To use a key you need to know the vocabulary of plant structure. Examine the plant in question to decide if its characteristics fit in one category or the other offered by the key. Keying plants is a process of elimination by making yes or no decisions to characteristics offered in the key—rejecting those that do not apply (dichotomy). Keys must be written to include the unknown plant to ensure accuracy, so that there is no question about whether one should pursue one branch of the dichotomous key or its alternative.

To use a simplified key (Table 15–1), eliminate the alternative that does not pertain to the plant in question, and then proceed to the next pair of numbers directly under the proper choice. As an example, the first choice in this key is to determine whether the seeds of the plant being identified are borne naked, as in the cone-bearing plants of the gymnosperms (Fig. 15–5), or enclosed in an ovary, as in the angiosperms. Once this yes or no decision has been made, the next step is to compare the next pair of descriptions directly under the previously chosen characteristic. If the seeds are enclosed within an ovary, the plant in question is an angiosperm and the next pair of numbers to compare is 8. Note that the two 8s are separated by several pairs of subsidiary descriptions. A choice between the two 8s must be made before proceeding further. If the plant in question has parallel-veined leaves and the seeds have one cotyledon, the plant belongs to the monocotyledon subclass. This procedure is followed until the plant to be identified "fits" a given set of plant characteristics.

A slight variation of this yoked key appears in Chapter 24 for the identification of turfgrasses; the number at the end of the phrase in that case indicates the next number to pursue.

TABLE 15–1	A Simplified Yoked (Indented) Key for Identifying Some Seed-Bearing Plants (Spermatophyta)

1. Ovules and seeds borne naked on scales in cones without typical flowers (Fig. 15–5); trees or shrubs, often evergreen. Gymnospermae
    2. Plant foliage palmlike. CYCADACEAE (Fig. 15–7)
    2. Plant foliate not palmlike
        3. One seed in a cup-shaped, drupelike fruit. TAXACEAE (Fig. 15–8)
        3. Many seeds in a dry woody cone
            4. Leaves alternate and single
            4. Leaves alternate and in clusters; needle-shaped
                5. Cone-scale without bracts with two to nine seeds
                5. Cone-scales in axils of bracts, flattened, with two seeds. PINACEAE
                    6. Cones upright on top of branchlets. *Abies*
                    6. Cones not upright on branchlets. *Pinus*
                        7. Twigs not grooved. White pines or soft pines
                        7. Twigs grooved. Pitch or hard pines
1. Plants with seeds borne in an ovary (base of pistil) with typical flowers. Herbs, trees, and shrubs. Angiospermae
    8. Leaves usually parallel veined, flower parts usually in multiples of three. One seed-leaf or cotyledon. Do not form annual rings when increasing in stem girth. Monocotyledonae
        9. Plant with palmlike leaves. PALMAE
            10. Leaves fanlike
            10. Leaves featherlike. Feather and fishtail palms
                11. Lower feathery leaves not spinelike
                11. Lower feathery leaves spinelike, fruit fleshy with long grooved seed. *Phoenix* spp.
                    12. Plant is a tree with shoots at base, trunk about 50 cm (20 in.) in diameter. Fruit edible. *Phoenix dactylifera,* date palm
                    12. Plants not as above. Other *Phoenix* spp.
        9. Plants without palmlike leaves
            13. Perianth none or rudimentary
                14. Stems solid. CYPERACEAE
                14. Stems mostly hollow. GRAMINEAE
                    15. Plants woody, bamboolike. Bamboos
                    15. Plants herbaceous, not bamboolike
                      16. Grasses that produce sugar. *Saccharum* spp.
                      16. Grasses that produce little sugar
                        17. Small grains and their kin (rice, wheat, etc.)
                        17. Cornlike plants and their kin
                          18. Plants monoecious. *Zea mays,* corn
                          18. Plants not monoecious. *Sorghum* spp.
             13. Perianth present
                19. Pistils several, not united. APONOGETONACEAE
                19. Pistils one, carpets united, ovary and fruit superior. AMARYLLIDACEAE
                    20. Anthers six, stem a fibrous rhizome. *Agapanthus* spp.
                    20. Anthers six, stem a corm or bulb *Allium* spp.
                      21. Leaves large, usually hollow and cylindrical. Bulb rounded and large. *Allium cepa,* onion
                      21. Leaves large, usually hollow and cylindrical. Bulb slightly thicker than neck. *Allium fistulosum*
    8. Leaves usually without parallel venation, two cotyledons. Herbs, trees, and shrubs with stems increasing in thickness with cambium cells, which form annual rings in woody plants. Dicotyledonae
        22. Corolla absent or not apparent, calyx present or lacking
        22. Corolla present, calyx usually forming two series of calyxlike bracts
            23. Petals united
            23. Petals separate
                24. Ovary inferior or partly so
                24. Ovary superior
                    25. Stamens few, not more than twice as many as petals
                    25. Stamens numerous, more than twice as many as petals
                      26. Habit aquatic. NYMPHAEACEAE, water lilies
                      26. Habit terrestrial
                        27. Pistils more than one, filaments of stamens united into a tube. MALVACEAE
                        28. Style—branches slender, spreading at maturity, seeds kidney-shaped. *Hibiscus* spp.
                        28. Styles united, ovary several carpels, calyx deciduous, seed angular. *Gossypium* spp.

(CONTINUED)

TABLE 15–1	**(continued)**

29. Staminal column long, anthers compactly arranged on short filaments. *G. barbadense,* sea-island cotton

29. Staminal column short, anthers loosely arranged and of varying lengths. *G. hirsutum,* upland cotton (Fig. 15–9)

  27. Pistil more than one, filaments not united into a tube. ROSACEAE

    30. Ovaries superior, fruit not a pome

      31. Pistils one, leaves simple and entire. *Prunus* spp.

        32. Fruit soft and pulpy. *P. armeniaca,* apricot

        32. Fruit dry and hard. *P. mume,* Japanese apricot

      31. Pistils two to many, leaves compound (at least basal leaves)

        33. Plants woody shrubs. *Rosa* spp.

          34. Styles not extended beyond mouth of hip. Stamens about one-half as long as styles. *R. odorata*

          34. Styles extend beyond mouth of hip, stamens about as long as styles. *R. multiflora*

        33. Plants herbaceous. *Fragaria* spp.

          35. Underside of leaves are bluish white. *F. chiloensis,* wild strawberry (Fig. 15–10)

          35. Underside of leaves are green. Other *Fragaria* spp.

    30. Ovaries inferior, fruit a pome

      36. Fruit with stone cells. *Pyrus* spp., pears

      36. Fruit without stone cells. *Malus* spp., apples

Figure 15–5 Cones of the sugar pine (*Pinus lambertiana* Dougl.) in closed and open conditions. When the cones dry, the winged seeds, shown on the 17.8-cm (7-in.) ruler, are released and can be disseminated by the wind.

Figure 15–6 An inflorescence of the pear (*Pyrus communis* L.) showing the five petals and the many stamens typical of the ROSACEAE. Mature fruits do not develop for another four months.

Certain plant parts might not be available because they may not be in season—when needed to identify the plant. An example of this occurs in the simplified key in Table 15–1. In the family ROSACEAE (the second item), both flowers and fruit are needed to be certain of making the correct choice. In the case of strawberries, both fresh flowers and fruits can be obtained at the same time; however, it is not possible to have flowers and fruits at the same time for the apricot or the pear (Fig. 15–6). In such cases, it might be necessary to preserve

flowers for future examination or simply describe them thoroughly in the spring when they are abundant, and then wait for the fruit to develop to accurately determine its characteristics. See Figures 15–7 through 15–10 for additional examples of hard-to-classify plants.

This simple key illustrates how choices are made between two characteristics—eliminating alternatives that are not pertinent to finally identify the plant in question. Much of the upper portion of the key can be ignored by a trained tax-

Figure 15–7 The inflorescence of a cycad (*Dioon edule* Lindl.) native to Mexico. The inflorescence is about 23 cm (9 in.) tall. The pinnate leaves might erroneously lead one to call this plant a palm.

Figure 15–8 The fruits of *Taxus baccata* L. 'Lutea'. The single seed (arrow) is within the cuplike fruit, which is poisonous. *Source:* Robert A. H. Legro.

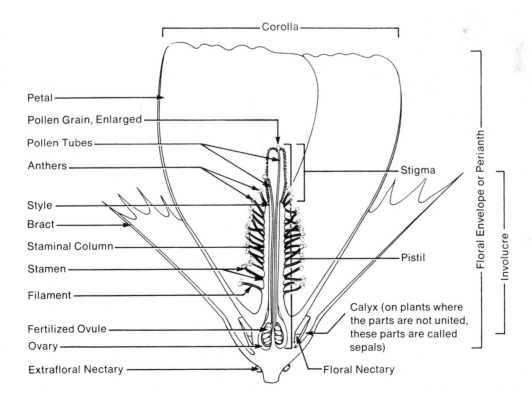

Figure 15–9 Cotton (*Gossypium hirsutum* L.) flower in a longitudinal section. *Source:* USDA.

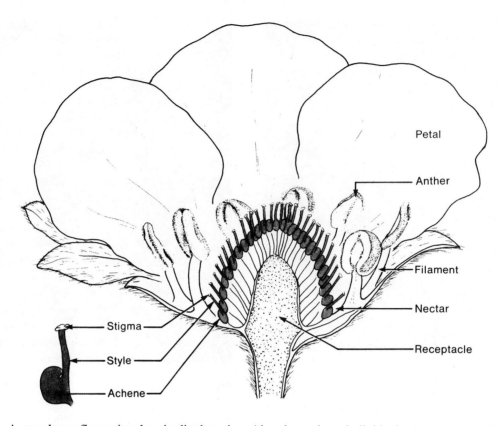

Figure 15–10   A strawberry flower in a longitudinal section. Also shown is an individual achene. *Source:* USDA.

onomist or a good botany student because, to them, it is easy to identify a gymnosperm or an angiosperm. The family characteristics are determined by observing and studying certain flower characteristics. Sometimes it is difficult but necessary to distinguish between an inferior or superior ovary (the characteristic that divides the LILIACEAE from the AMARYLLIDACEAE families; see Fig. 8–36), but one finds these determinations easier after some experience is gained. Once the family is known, the correct genus and species are identified by referring to taxonomic books with detailed keys.

Obviously one must be familiar with plant parts and structures (Chapter 8) to use a botanical key and determine the identity of the plant. Various plant parts and shapes are useful in illustrating plant identification (see Figs. 8–35, 8–36, and 8–37).

ORIGIN OF CULTIVATED PLANTS

Most of the crop plants important today were cultivated in a primitive way long before recorded history. Agriculture began about 10,000 years ago when ancient peoples selected certain plant types they found growing about them and thus enlarged their food sources beyond hunting and fishing. Most of these early food plants are still cultivated but in

much improved forms. Others that were unknown to early humans have been added over the centuries to make up our present-day selection of food plants. Then, much later, as people became more concerned about the aesthetics of their environment, they also added shade trees, shrubs, flowering plants, and lawn grasses to the list of cultivated plants.

Several botanists have brought together fascinating information on the subject of where and when the crops that now feed the world's peoples, their livestock, and other domestic animals originated. The Swiss botanist Alphonse De-Candolle wrote *Origin of Cultivated Plants,* first published in 1833. Later, the famous Russian plant geneticist Nikolai Vavilov also studied the origin of cultivated plants. The results of his studies were published in *The Origin, Variation, Immunity, and Breeding of Cultivated Plants,* translated from the Russian and published in English in 1951. Vavilov concluded from his studies that the various cultivated plants originated in eight independent centers: (1) central China, (2) India, (3) Indochina and the Malay Archipelago, (4) the Turkey-Iran region, (5) the Mediterranean area, (6) the Ethiopia-Somaliland area of east Africa, (7) Mexico and Central America, and (8) the Peru-Ecuador-Bolivia and the Brazil-Paraguay area of South America. Many archeological studies carried out in later years generally confirm Vavilov's locations for these centers.

More recent studies and theories on the origins and movements of the world's agricultural crops have been summarized by Carl Sauer in *Agricultural Origins and Dispersals,* by Jack Harlan in *Crops and Man,* by Jack Hawkes in *The Diversity of Crop Plants,* by Barbara Bender in *Farming in Prehistory,* and by Norman Simmonds in *Evolution of Crop Plants.*

DOMESTICATION OF PLANTS

Toward the end of the Ice Age, technically the Pleistocene epoch of the Cenozoic era, about 11,000 to 15,000 years ago, when the glaciers were in full retreat and early humans were wandering about the earth, the stage was being set for the initiation of food production (Table 15–2). Before this time, tool-using hunter-gatherers had been on earth for about 4 million years. Then—only about 400 generations ago—there was a gradual transition to food-producing activities. This transition phase suggests several interesting questions, such as why it took so long for humans to become food producers and why such activities began in various parts of the world— southwest and southeast Asia, middle America, and western South America—at about the same time.

There is definite evidence from archeological sites that agricultural villages existed about 8000 to 9000 B.C. in the area of southwest Asia known as the Fertile Crescent. It extends from the alluvial plains of Mesopotamia (now Iraq) across Syria and down the eastern coast of the Mediterranean sea to the Nile Valley of Egypt. Radiocarbon dating suggests that plants and animals were being domesticated at several places in this area by at least 5000 to 6000 B.C. There is evidence from site excavations that einkorn wheat, emmer wheat, barley, lentil, chickpeas, oats, and vetch were being cultivated, as well as dates, grapes, olives, almonds, figs, and pomegranates. Excavations at Jarmo in Iraqi Kurdistan, at ancient Jericho in the Jordan Valley, and many other sites in the Fertile Crescent have supplied much of the evidence to support these conclusions.

An indigenous savanna type of agriculture apparently was developing from domesticated native plants about 4000 B.C. in a belt across central Africa (see Fig. 15–11). This area was the first home of the human race, as we understand it now. The genus *Homo* originated here, where most of human evolution subsequently occurred. Some important world crops brought under cultivation in this area include coffee, sorghum, millet, cowpeas, yams, and oil palm.

The Chinese center of agricultural origins became important about 4000 B.C. according to radiocarbon dating. Crops domesticated include millet, chestnuts, hazelnuts, peaches, apricots, mulberries, soybeans, and rice. The West did not learn of the all-important rice plant until about 350 B.C., and the peach was unknown outside China until about 200 A.D.

A farming culture in southeast Asia and what is now Indonesia apparently had domesticated rice around 6000 B.C.

Other important crops appearing later under cultivation in this area were sugar cane, coconut, banana, mango, citrus, breadfruit, yams, and taro.

In the New World, evidence from archeological sites shows agricultural beginnings in two areas. One is present-day southern Mexico and Central America, where plant cultivation began about 5000 to 7000 B.C. The plants grown were early forms of maize (corn), sweet potato, tomato, cotton, pumpkin, peppers, squash, runner beans, papaya, avocado, and pineapple.

The second American region is a broad "noncenter" of agricultural origins stretching from Chile northward to the Atlantic Ocean and eastward into Brazil. There is evidence that both the snap and lima beans were cultivated here by about 6000 B.C. or earlier. Other important cultivated crops from this region are the potato, peanut, cacao, pineapple, cashew, papaya, avocado, Brazil nut, peppers, tobacco, guava, tomato, yam, cassava (manioc), and squash.

No major cultivated crop originated in the area of the present-day United States. Agriculture here relies in a large measure on introduced crops. There are, however, many minor native American fruit and nut crops, such as the American grapes and plums, the pecan, chestnut, hickory nut, hazelnut, black walnut, persimmon, blueberry, raspberry, blackberry, and cranberry. The sunflower (which has long been an important oil crop in the former Soviet Union and Eastern Europe) originated in the United States, as did hops, the tepary bean, Jerusalem artichoke, and some grasses. Many plants now used as ornamentals, mostly in improved forms, and many of the world's great timber tree species also originated in the United States.

The only food crop to originate in the entire continent of Australia was the macadamia, or Queensland nut.

Methods of Plant Domestication

When ancient peoples started to domesticate plants—that is, propagating and growing plants under their own control— they would have had the choice of vegetative propagation or seed propagation. Both methods have advantages and disadvantages in crop improvement, but because of genetic segregation between generations, the offspring from sexual propagation are often different (not true to type) from parents, and desired traits may not be present in the offspring. For this reason, vegetative propagation is the more immediate and direct approach to selecting and maintaining superior plants.

Vegetative or Asexual Propagation Methods

It is probably no coincidence that some of the oldest cultivated woody plants are the easiest to propagate by vegetative methods such as hardwood cuttings. These plants include the grape, fig, olive, mulberry, pomegranate, and quince.

An observant person in prehistoric days might have noticed among a wild stand of seedling grapevines, for example,

TABLE 15–2 Beginnings of Agricultural Development in the Old and New World

The Postglacial Time Scale / Climatic Changes Date	Old World — Climatic Changes in Northern Europe	Old World Cultural Stages	New World Cultural Stages
10,000 B.C.	Last glacial stage (Würm-Wisconsin ice)	Late Paleolithic hunting cultures (Cro-Magnon, etc.)	
9000 B.C.	Retreat of the glaciers (Pre-boreal period, cold dry)	Mesolithic fishing, hunting, collecting cultures	
8000 B.C.			Hunting cultures established (Folsom Man, etc.)
7000 B.C.		Agricultural beginnings	
6000 B.C.			
5000 B.C.	Boreal period (Warm dry)		
	Atlantic period (warm moist)	Neolithic agriculture established and spreading	
4000 B.C.		Beginnings of civilization (Egyptian-Sumerian)	
3000 B.C.		Neolithic agriculture in northern Europe	
2000 B.C.			American agricultural beginnings
	Subboreal period (colder dry)	Babylonian Empire Invasions: Aryans to India; Medes and Persians to southwest Asia and Mesopotamia	
1000 B.C.		Rise and flowering of Greek civilization	
B.C.–A.D.	Sub-Atlantic period (cool moist)		Early Mexican and Mayan civilizations
		Roman empire	
		Invasions: Goths, Huns	Decline of Mayans
		Rise of Islam	
1000 A.D.		Norsemen to America	
		Mongol and Tartar invasions	Aztecs and Incas
		Voyages of discovery and colonization by Europe	
		Industrial revolution and modern period	
2000 A.D.			

Source: Dasmann, R. F. 1984. *Environmental conservation,* Fifth Edition. New York: Wiley.

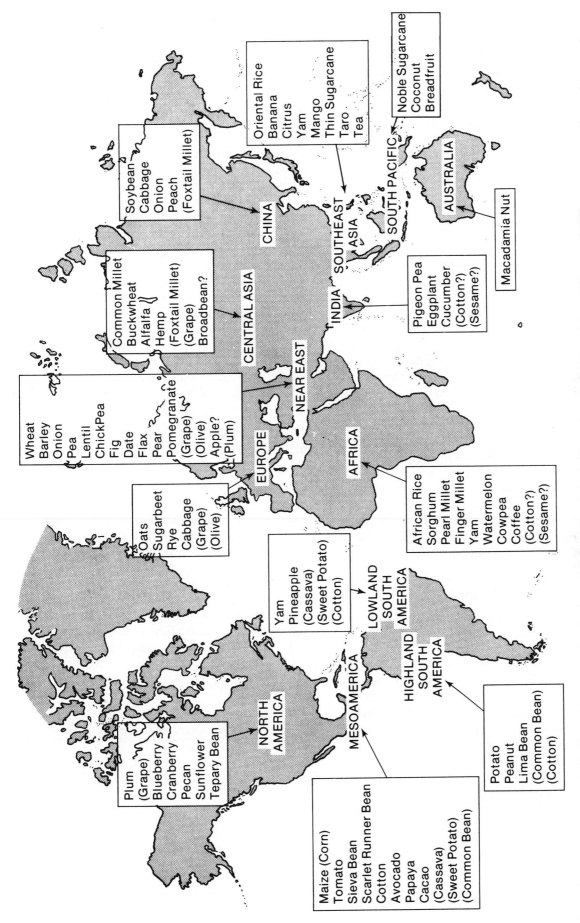

Figure 15–11 Regions of the world where major food crops were domesticated. Crops that apparently originated in several different areas are shown in parentheses. Question marks after the name indicate doubt about the location of origin. *Source:* Harlan, J. R. 1976. The plants and animals that nourish man. *Sci. Am.* 235(3):88–97. © September 1976 by Scientific American, Inc. All rights reserved.

291

one single plant standing out from the others in its heavy production of large, high-quality fruits. From past experience, too, the person might have known that a piece of the grape cane stuck into the ground in early spring would take root and produce a new plant just like the original. The next logical step, of course, would be to break off and plant many cane pieces (cuttings) from this one very desirable grape plant, then to protect them against animals and the competition of weeds. This action would establish a vineyard based on cultivating a superior seedling and disregarding all seedlings with inferior characteristics.

Many of the tree fruit species—fig, almond, quince, apple, pear, cherry, pomegranate, and walnut—are native to areas of the Near East on the southern slopes of the Caucasus Mountains. Today, there are still forests in the area consisting entirely of seedling wild trees of these species, showing great variability in many characteristics. One may also see trees in which the local farmers have grafted the superior wild types onto inferior seedlings, making use of them only as rootstocks. Also, while clearing the brush and forests to make space for grain fields, the farmers have left superior individual wild apple, pear, and cherry seedlings in the fields for harvesting.

Many types of ancient plants brought under domestication by early people could have been—and probably were—easily maintained and increased by unsophisticated vegetative methods. Such plants would have included the potato and sweet potato, propagated by dividing the tubers; banana, bamboo, and ginger, propagated by cutting up rhizomes; pomegranate, quince, fig, olive, and grape, propagated by stem cuttings; filbert, propagated by layering; and pineapple and date, propagated by a form of suckering.

Seed or Sexual Propagation Methods

Domestication of seed-propagated plants, such as the cereal crops, by the ancient agriculturists probably began as the purposeful harvesting of wild grass seeds, some of which were sown to produce the next year's crop. This established two population types for the next harvest—one that was not harvested and reseeded itself, and one that was sown from the harvested seed. In cereals, this procedure immediately starts to separate the "shattering" types (the seed separates from the head) from the "nonshattering" types. Most of the seeds that shatter fall to the ground, while nonshattering seeds are harvested and can be resown. Thus, seed with the desirable nonshattering trait is easily obtained without intentional selection. The mere practice of harvesting sets up a selection pressure for improved forms.

Planting the harvested seeds close together in a cultivated plot kept free of competing weeds automatically selects for the stronger, more vigorous plants. In some species, seedlings derived from the larger seeds—those having a considerable amount of stored foods—are likely to germinate and

grow rapidly, thus suppressing seedlings arising from the smaller seeds with low vitality. The harvested crop would, therefore, come from the more vigorous plants producing the desirable large grains, and the plants producing the smaller, inferior grains would be eliminated. Thus, seedling competition sets up an automatic selection pressure for an improved form.

By these procedures, early peoples unconsciously developed superior forms of their cereal crops. Other desirable characteristics arising from selection pressures were loss of seed dormancy, increased flower numbers and larger inflorescences, and a trend toward determinate rather than indeterminate growth.[2] These superior characteristics in the cultivated races over the wild types can usually be obtained with only small genetic differences between the two.

Plant size, productivity, seed output, and ecological adaptations are often complex quantitative traits and involve many small genetic changes. Their evolution is a slow, continuous process. Often some other characteristics, such as nonshattering seeds, resistance to natural enemies, or novel fruit or seed color and shape might be controlled by one or very few genes. Such traits can evolve rapidly by early humans' observation and by harvesting separately.

EXAMPLES OF IMPROVEMENT IN SOME IMPORTANT CROP PLANTS[3]

The following examples illustrate patterns in plant improvement, beginning with: (1) the harvest of crops from wild plants by primitive humans, followed by (2) selection of superior types from prehistoric eras to the present time, and going on to (3) modern methods of plant breeding that can dramatically increase crop yield and quality by applying genetic principles and gene transfer to developing improved cultivars. It should be realized that the present, improved cultivars of important crops such as wheat, corn, rice, potato, sweet potato, and all fruit crops have been so adapted to conform to human cultural practices that they all now completely depend on our care for their continued existence.

Grains and Vegetable Crops

> **Wheat (*Triticum aestivum* and *T. turgidum* Durum group)**

Wheat is the most widely cultivated plant in the world today, the chief cereal, and is used worldwide for making bread.

---

[2]In determinate growth, shoot elongation stops when flowers form on the shoot terminals. In indeterminate growth, elongation continues after flowering.

[3]For a discussion of the genetic terminology used in this section, see Chapter 14.

Present wheats evolved from wild wheatlike grass plants found and cultivated by ancient humans in the Near East region about 7000 B.C. Native wheatlike plants can still be found in this area. Even in early prehistoric times, wheat probably improved naturally by spontaneous hybridizations, by chromosome doubling, and by mutations to increase fertility. Wheat species occur in a series with increasing chromosome numbers (polyploidy): First is the small primitive diploid einkorn wheat (7 pairs of chromosomes); second is the much larger tetraploid emmer wheat (14 pairs of chromosomes); and third is the hexaploid bread wheats (21 pairs of chromosomes), the ones grown today. Humans probably had no role in this early advancement of wheat. From the Near East, these early wheat forms were taken into ancient Egypt, the Balkans, and central Europe. The Spanish brought wheat to the Americas; eventually, the United States, Canada, and Argentina became the world's largest wheat producers. A chance introduction of 'Turkey Red' wheat into central Kansas by a small group of Mennonite immigrants from Russia in 1873 established the basis for the tremendous hard red winter wheat industry of the central Great Plains area of the United States.

The two major wheat species today are *Triticum aestivum,* used for flour in making breads and pastries, and *T. turgidum* (Durum group), used for products such as macaroni, spaghetti, and noodles. In the twentieth century, plant breeders, through hybridization and selection, have produced perhaps a thousand cultivars of bread wheat alone designed for certain climates, high productivity, special milling properties, and particularly disease resistance. Wheat cultivars resistant to the devastating stem rust disease must continually be developed to cope with the mutating stem rust pathogen.

Today's plant breeders, too, have developed dwarf forms of wheat, which can produce high seed yields without falling over (lodging) when heavily fertilized. Some of these new dwarf cultivars were developed by Norman E. Borlaug, working at the Rockefeller Foundation's International Maize and Wheat Improvement Center in Mexico. Borlaug was awarded the Nobel Peace Prize in 1970 for his work. Certain of these cultivars have been so successful with fertilizers and irrigation in Pakistan and India that both countries have become exporters rather than importers of wheat, a result of the so-called green revolution.

Wheat plants are self-pollinated, allowing farmers to save their seeds for future planting. F_1 hybrid wheat for increased plant vigor and yields has not been developed to the extent it has with corn, owing to difficulties arising from the wheat's flower structure. Wheat has perfect (bisexual) flowers, making cross pollination difficult, whereas corn has the male and female flowers separate on the same plant and is cross-pollinated easily.

Corn (*Zea mays*)

Corn (or maize) originated in the New World about 5000 to 6000 B.C., but its earliest history is still a mystery. Maize is

Figure 15–12 Improvement in corn (maize) from a primitive type (*left*) still growing on the eastern slope of the Andes mountains in South America, to modern hybrid corn (*right*). The primitive type corn contains a valuable trait—multiple-aleurone layers in the grain—that is being transferred by plant breeders to modern dent corn. This example illustrates the need for maintaining the germplasm of seemingly worthless plant types. *Source:* USDA.

known only as a domesticated plant. There is no wild form except, apparently, teosinte, a close relative. The economic life of the ancient American civilizations—the Aztecs, Mayas, and Incas—depended on corn. At the time of Columbus's expeditions to the New World, corn was being grown by Indian tribes from Canada to Chile. Corn probably originated in several places in both Mexico and South America. An early form of corn is still growing in South America (see Fig. 15–12).

Several hypotheses have been advanced to explain the origin of corn: (1) it developed from "pod corn," a type in which each individual kernel is enclosed in floral bracts—as in the other cereals; (2) it originated from teosinte, corn's closest relative, by gradual selection under the influence of harvesting by humans; (3) corn, teosinte, and *Tripsacum* (a perennial grass) descended along independent lines directly from a common ancestor; or (4) there is the tripartite theory that (a) cultivated corn originated from pod corn, (b) teosinte is a derivative of a hybrid of corn and *Tripsacum,* (c) the majority of modern corn cultivars are the product of an admixture with teosinte or *Tripsacum,* or both.

Corn, even today, is an extremely variable species, from the color of the grains to the size and shape of the grains and ears (Fig. 15–12). Corn mutates easily, forming new types. The Indians of America must have made considerable conscious selection of corn for so many types to be introduced to European agriculture following the New World explorations.

The development of hybrid corn in the 1930s is one of the outstanding achievements of modern agriculture (see Fig. 14–5). High-yielding F_1 corn hybrids were developed for different climatic zones. In 1935, 1 percent of the corn planted in the United States was of the hybrid type. By 1970, almost all corn produced in the United States was of the hybrid type. The planting of hybrid cultivars, along with fertilizers, irrigation, and mechanization, dramatically improved production. Corn is now one of the world's chief food crops for both humans and domesticated animals, but it would not survive in nature without human attention and cultivation.

The latest development in corn improvement is biotechnology germplasm, including the Bt® corn for corn borer resistance, and the Roundup Ready® and Liberty Link® corn for herbicide tolerance. In 2004, 46 percent of US corn plantings (79 million acres) was projected to be biotechnology cultivars. On the other hand, specialty corns—such as white, waxy, hard endosperm food grade, high oil, nutritionally enhanced, high amylase and high extractable starch corn—are being developed. Ethanol production from high-starch corn is increasing in use as a replacement for some petroleum fuels.

Rice (*Oryza sativa*)

Rice is the basic food for more than half the world's population and one of the oldest cultivated crops. It is believed to have originated in southeast Asia about 5,000 years ago, or even earlier, and it spread to Europe and Japan by the second century B.C.

There are about twenty-five species of *Oryza,* but *O. perennis,* widely distributed throughout the tropics, is probably the one from which cultivated rice was developed. Primitive humans most likely collected seeds and cultivated the wild types. During cultivation of rice, mutations, plus hybridization with other *Oryza* species, probably occurred, leading to improved forms with larger grains and nonshattering fruit stalks. Rice was first cultivated in America along the coast of South Carolina about 1685. Four rice experiment stations were established by the US government in the early 1900s. Breeding work at US government rice experiment stations has resulted in many superior cultivars being introduced to the rice-growing areas of the United States.

Much of the rice harvested in Asia was from old, unproductive types grown under primitive conditions. The Ford and Rockefeller Foundations established the International Rice Research Institution (IRRI) in the Philippines in 1962, bringing together scientists from eastern and western nations for the purpose of improving rice culture. This group of researchers developed new early, high-yielding dwarf cultivars by hybridization that did not fall over (lodge) when heavily fertilized and up to three crops can be grown in the same field each year. These cultivars dramatically increased yields, and some are highly resistant to insects and disease pathogens native to the Far East, South America, and Africa. The recent trend in Asia is the adoption of hybrid rice. In China, about 50 percent of the entire rice crop is planted with high-stability-yield hybrid cultivars. Other countries, like Vietnam and India, are pursuing this technology.

Today, a genetically modified rice produces beta-carotene in its kernels, giving the rice a yellow color. Adding beta-carotene, a precursor for vitamin A (which is essential for human health), to rice increases the nutritional value of the rice significantly. It is predicted that the modified rice will help prevent blindness in millions of children in developing countries where rice is a dietary staple. However, with public sentiment strongly opposed to genetically modified foodstuffs, it is not known if the rice will ever be available to these children.

Soybean (*Glycine max*)

Sometimes called the Cinderella crop, the soybean has risen spectacularly to prominence in the United States in recent years with, for example, an increase in production from 135,000 metric tons (5 million bushels) in 1925 to about 75 million tons (3 billion bushels) in 2000. This great increase is due, in part, to the availability of more productive, disease-resistant cultivars. From about 1910 to 1950, large numbers of new strains and seed lots of soybeans were introduced into the United States from the Orient, its native home, largely by United States Department of Agriculture (USDA) plant explorers. From these diverse sources of germplasm, hybridization programs developed many superior cultivars. Much of this work was done at the USDA Regional Soybean Laboratory at Urbana, Illinois, in cooperation with various midwestern agricultural experiment stations. These cultivars give high yields, proper bean maturity for the particular area, strong erect plants that hold their seeds until harvest, and high disease resistance and bean quality. The soybean is particularly well adapted to the United States' Midwest corn belt and the southeastern states, which together account for more than 26 percent of the world's soybean production.

Sugar Beet (*Beta vulgaris*)

The modern sugar beet is a plant developed entirely by human efforts in plant breeding. It is our only major food crop that was not grown in some primitive form in ancient times. It was developed only about 250 years ago in Europe as a source of sugar to compete with the then very expensive cane sugar. The German chemist Andreas Marggraf found in 1747 that the kind of sugar in cultivated garden beets was the same chemically as that in cane sugar. Breeding and selection increased the sugar content in the root from about 2 percent to about 16 to 20 percent.

In the early part of the nineteenth century, Napoleon encouraged the development and production of the sugar beet industry in France to free that country from the British monopoly on cane sugar. By the end of the nineteenth century, sugar beets were being grown in North America, and they have now become an important temperate zone crop in many areas of the United States and southern Canada, and even more so in other parts of the world. All production phases are now completely mechanized, permitting the sugar beet to compete favorably with the tropical sugar cane plant. To maintain profitable production, however, plant breeders have had to continue to develop sugar beet cultivars resistant to virus and fungus diseases.

Potato (*Solanum tuberosum*)

Potatoes are one of the big four crops that feed the world's population. Wild potato species are widespread in South America, particularly in the Andes Mountain region. Potatoes were probably cultivated by primitive peoples in this area more than 4,000 years ago, but they became a more productive crop over the years with selection of superior types. The potato was cultivated over the length of the Andes Mountain area at the time the Spanish explorers arrived. About 1575, they carried it back to the European continent, where today 90 percent of the world's potato crop is grown.

The potato plant produces pink, white, or blue flowers that develop into small green berries containing seeds. The seeds, when planted, produce new types of potato plants much different from the parent plant and from each other in many respects. In the early days of potato culture, the South American Indians undoubtedly selected superior plant types resulting from natural crosses. Once one single such superior plant was obtained, it could be perpetuated and increased in great numbers as a clone by tuber division. Some of these superior selections propagated by vegetative methods over the years became commercially successful cultivars. It was noted, however, that certain of these cultivars would degenerate after many generations of such asexual propagation and yield only weak unproductive plants. It was observed, too, that sowing seeds from such plants gave progeny plants with changed characteristics, including renewed vigor and productivity. It is now known that these clonal cultivars had become infected with viruses that passed along through the tubers to the new plants generation after generation. Such viruses did not pass through the

seed to the new seedling plant, so that its growth was no longer inhibited by the virus.

In modern commercial potato growing, the planting stock ("seed" tubers) is produced under carefully observed conditions in regions where any viruses can readily be detected and the infected stock discarded. Growers who plant only "certified" stock can be reasonably sure that their fields will not be infected with viruses or other pathogens and will be true to type. In the United States, certified "seed" potatoes are produced by commercial growers in the northern states and are strictly inspected by state government agencies. This is also true for many other, but not all, countries.

Many potato cultivars now being grown have been developed by plant breeders who have introduced superior characteristics into an existing cultivar. An example is resistance to *Phytophthora infestans* (late blight)—from *Solanum demissum,* a wild potato species, obtained from collections in Mexico. Many cultivars are being grown today for different purposes and for different climatic zones. New cultivars are constantly being introduced by plant breeders and older ones are being discarded.

Tomato (*Lycopersicon esculentum*)

The cultivated tomato originated from wild forms in the Peru-Ecuador-Bolivia area of the Andes Mountains in South America. Prehistoric Indians carried it to Central America and Mexico. Early explorers introduced the tomato to Europe about 1550, and it was brought back west, to the Carolinas in North America, about 1710. Thomas Jefferson grew tomatoes on his plantation around 1780. In those days, most people considered the tomato poisonous, but in the United States it started gaining acceptance as a food plant about 1825. The early Indians undoubtedly improved the tomato by planting seeds taken only from the best fruits on the most productive plants. This selection process continued in the early days of tomato culture in Europe and the United States. Early cultivars introduced by US seed companies were Stone and Globe in 1870. In the early 1900s, the USDA and the state experiment stations began breeding tomato cultivars to include specific characteristics. Marglobe was introduced by the USDA in 1925 and Rutgers by New Jersey in 1934. Some tomato cultivars, more recently introduced, carry resistance to fusarium wilt, verticillium wilt, and nematodes. Cultivars developed especially for machine harvesting have firm-fleshed fruits that all ripen at one time. Vigorous, highly productive F_1 hybrids, marketed both as seeds and as bedding plants, are recent developments by seed companies.

Fruit Crops

The major fruit crops are all heterozygous. They do not "come true" when propagated by seed, so vegetative propagation must be used to maintain an improved seedling selection.

In ancient times, most kinds of fruit—other than those very easily propagated vegetatively, such as bananas, grapes, figs, pomegranates, and olives—were probably seed propagated and the variability in offspring accepted by farmers. Later, however, as more sophisticated methods of vegetative propagation were developed, such as budding and grafting, it would have been found that certain superior individual fruit plants could be maintained and increased by these methods. Nevertheless, considerable seed propagation of fruit species were undoubtedly practiced in the early days of fruit growing.

Apple (*Malus pumila*)

The early US colonists planted many seedling apple trees probably because it was easier to bring seeds taken from their favorite apple trees in their native homes rather than material for grafting. As a result, they continued to increase their apple orchards by planting seedlings. In these early days, much of the apple crop was preserved as cider, for which fruit from seedling trees was quite satisfactory. Certain individual seedling trees, no doubt, were much superior to the others and formed the starting point for vegetative propagation and the origin of the many hundreds of apple cultivars grown in the United States up to the early part of the twentieth century. These numbers dwindled, however, until the 1980s, when only fifteen cultivars accounted for over 90 percent of the apples produced in the United States. Important cultivars are 'Delicious,' 'Golden Delicious,' 'McIntosh,' 'Rome Beauty,' 'Jonathan, 'Winesap,' 'York,' 'Stayman,' 'Cortland,' and Granny Smith.' Although apple breeding programs have been conducted by the USDA and by some state agricultural experiment stations, all the major apple cultivars now being grown originated as chance seedlings many years ago. For example, the 'McIntosh' apple was found growing as a seedling tree near Dundela, Ontario, Canada by John McIntosh in 1796. The 'Delicious' apple started as a single chance seedling near Peru, Iowa, about 1870. The 'Golden Delicious' also originated as a chance seedling in Cass County, West Virginia, about 1910. However, more recent cultivars released from breeding programs at the New York Agricultural Experiment Station—such as 'Empire,' 'Macoun,' and 'Jonagold'—are starting to find grower and consumer acceptance.

Pear (*Pyrus communis*)

In pears, too, cultivars originating as chance seedlings have dominated the markets. The 'Bartlett' (Williams Bon Chretien) originated in England as a chance seedling in 1796 and has been the world's leading pear cultivar ever since. Other leading pear cultivars in the United States, such as 'Beurre d'Anjou' and 'Beurre Bosc,' originated in Belgium as open-pollinated seedlings of cultivars then grown there. No pear cultivar from a controlled breeding program in the United States has become commercially important, although European pear breeders have developed several superior cultivars that produce well in France, Italy, and Belgium.

Peach and Nectarine (*Prunis persica*)

In contrast to the apples and pears, the important peach and nectarine cultivars grown today are the products of public and private plant breeding programs. Plant breeders have provided the consuming public with truly outstanding peaches and nectarines, in contrast to the small-fruited, non-productive cultivars of earlier days that got their start as chance seedlings.

Strawberry (*Fragaria* × *Ananassa*)

Today's garden strawberry first originated in France about 1720 as a natural hybrid between two native American *Fragaria* species. From this and subsequent hybridizations, a number of cultivar selections were made and maintained vegetatively by runners in the early days of strawberry culture in Europe. However, many of these early cultivars were susceptible to viruses, verticillium wilt, and other diseases and were low in productivity, so that around 1945 the entire US strawberry industry was falling into a precarious position.

Since World War II, a parade of new strawberry cultivars has been replacing older ones, coming from USDA and several state strawberry breeding programs and from similar programs in other countries. New cultivars have been developed for specific climatic regions and for characteristics such as adaptability of fruits to be used for freezing or for fresh shipping, resistance of plants to viruses and fungi and to winter cold, and fruit appearance and flavor.

PLANT IMPROVEMENT PROGRAMS

From the time of Neolithic humans up to the early 1900s, sexual plant improvement mostly consisted of selecting seeds from those individual plants in a mixed population that had the desired characteristics. The seed was planted and from that population seeds were again taken from the most desirable plants, and so on. While improvements resulted, this method offered no way to transfer desirable characteristics from one line to another.

Evolution and Darwinism

Charles Darwin, the great English naturalist working in the middle of the nineteenth century, provided scientific explanations of how evolution occurred, published in his monumental work (1859), *On the Origin of Species by Means of Natural Selection, or the Preservation of Favoured Races in the Struggle for Life*. Darwin's concept of evolution, generally accepted today, is:

1. Variation exists in an initial population of plants or animals.

2. Environmental stresses place certain individuals at an advantage.

3. Because certain individuals survive and reproduce more successfully, they leave more offspring, which then carry the same genetic traits.

4. The abundance of the advantageous traits increases in this way in every generation, but variation still persists.

In Darwin's time, nothing was scientifically known about heredity. Darwin, in his proposals, had great difficulty in accounting for a sufficient supply of variation. But the discoveries by the Austrian monk Gregor Mendel, in the 1860s, demonstrated the genetic mode of plant inheritance and developed the foundation for the science of genetics. His published works lay unappreciated in an obscure journal for thirty-four years, but in 1900 his papers were discovered independently by three European botanists. The discoveries of Mendel provided exactly the mechanisms needed by Darwin, and removed his difficulties in explaining variation. The integration of Darwinian selection and Mendelian genetics are now generally accepted as the proper explanation of evolution.

Many crop scientists have discovered patterns of evolution in various crop genera, primarily from two points of view: (1) relationships between crops and their related wild and weedy species, and (2) adaptive variation in geographical races. Plant breeders have made fascinating and useful applications of this knowledge in terms of collecting and utilizing many germplasm resources. In fact, all of the processes of evolution in native plants have direct analogues in breeding methods. For example, induced mutations and hybridization follow natural sources of novel variation; selection and cycles of recombination in hybrid materials provide the genetic changes in both natural and breeders' populations.

Based on Mendel's work and utilizing the expertise of trained geneticists, modern breeding programs have produced an array of new cultivars for many crops. These have been bred for characteristics such as resistance to disease, insects, and cold, and for productivity, flavor, and nutritive value. Today's breeding programs have been called directed and accelerated evolution.

Products of the twentieth century, public and private plant breeding programs have had a tremendous beneficial impact on our food supply and range of ornamental plants. These programs have produced a great many new superior cultivars for almost all cereal, vegetable, forage, fruit, and ornamental crops. The USDA and most state agricultural experiment stations in the United States maintain such programs. Most other countries also have plant breeding programs, often specializing in certain crops. Private plant breeding has mostly been done by seed companies producing new agronomic, vegetable, and flower cultivars.

Innovations by plant breeders include the development of F_1 hybrid corn and new vegetable and flower cultivars. Most of the new vegetable and flower lines are far superior

THE MECHANISM OF EVOLUTION

Evolution, as accepted today, can be explained as follows:

1. Genes, in the chromosomes, are largely responsible for the structure, metabolism, and development of plants and animals.

2. The complement of genes does not remain constant because mutations occur that modify the metabolism and structure of the individuals that contain them.

3. Mutations, which may cause considerable change, may kill or greatly weaken the plant because they can upset the equilibriums that exist between the plant and its environment.

4. Hybrids differ from their parents because of the resulting new combinations of genes.

5. If the variants, as developing from mutations or by hybridization, are better adapted to the existing environment than the parent plants, then the parent plants may be replaced by the new forms.

6. As the earth's surface and climate change, or as plants' habitat may be changed in other ways, those plants best adapted to the new environment replace those poorly adapted.

7. Evolution thus results from slow changes in the environment, variations occurring in plants and animals, and adjustments taking place between the changes in the environment and changes in the living organisms.

to previous cultivars in vigor and in insect and disease resistance; the new vegetable cultivars are also superior in flavor, appearance, and productivity.

By developing plants that show strong resistance to insects and disease, plant breeders are reducing the need for insecticides and fungicides. This, in the long run, would be the best method of pest control. For example, potato cultivars have been developed that are resistant to the late blight disease (*Phytophthora infestans*), which was responsible for the nineteenth-century Irish potato famine. Other potato cultivars have been developed that are resistant to the golden nematode, permitting potatoes to be produced in soils infested by these worms without expensive soil fumigation. Wheat breeders must continually develop new wheat cultivars resistant to stem rust (*Puccinia graminis tritici*) because this fungus continually changes to attack formerly resistant cultivars.

Plant breeders have a useful procedure for obtaining improved plant forms by spontaneous or induced mutations resulting from chromosome or gene changes. These changes can be induced by chemical treatment with colchicine or by irradiation with gamma or X-rays.

In one instance, plant breeders have gone beyond just improving native plants. They have created a new human-made cereal, triticale, by hybridizing the ancient grains, wheat and rye. The name *triticale* derives from the generic names of these two grains: *Tritium* and *Secale.* The hybrid combines the high yield and protein content of wheat with the winter hardiness of rye. The triticale plant is disease-resistant and thrives in some unfavorable soils and climates.

SEARCHING FOR AND MAINTAINING NEW GERMPLASM

There is no assurance that we are at present cultivating all the useful food and ornamental plants in existence on earth, or that all the germplasm containing useful genes has been found. **Germplasm** is the protoplasm of the sexual reproductive cells containing the units of heredity (chromosomes and genes). Plant explorers have long roamed the world and they continue their searches. They have found many plants that have subsequently made a major impact on the world's agriculture, often in different parts of the world from the plants' native regions. Accessible plants have always been moved about over the world by explorers on land and sea, armies, immigrants, and travelers. Plants moved into a new region often perform much better than they did in their original home. For example, the coffee plant (*Coffea arabica*) is native to the Ethiopian area of eastern Africa. It never developed into much of a crop there, but when moved to Brazil and Colombia in South America, the coffee tree prospered so well that these countries now produce the bulk of the world's coffee supply.

Several prominent plant explorers have contributed much to the wealth of available plant materials. Many of the early plant collecting trips were less than successful because the plants did not survive the long trip back. A London physician who was an amateur horticulturist, Nathanial Ward, invented the Wardian case early in the 1800s. The case was a small glass-enclosed box containing soil in the bottom. Plants kept in the Wardian case could survive long sea voyages, permitting the importation of species never before received alive. Large, magnificent, ornate glasshouses were built in England about this time to house the many tropical and subtropical plants brought back by the plant hunters.

Many plant explorers from the United States brought back plants that would later be the foundation for several crops grown in the United States. Colonel Agoston Haraszthy (wine grapes), N. E. Hansen (cold-resistant fruit and cereal plants), Mark Carleton (hard red winter wheat), David Fairchild, Frank Meyer, and others brought back many different species that have become the genetic basis for many of our edible and ornamental crops.

Plant-collecting trips to the native homes of certain desirable plant types are still being made by plant explorers. Plant explorers may be looking for an entirely new plant species to serve as a crop in a particular climatic region of their own country. Safflower came to the northern Great Plains of the United States this way, and it has proven most profitable. Plant explorers may also be searching for new germplasm for existing crops. Closely related types can be used by plant breeders to introduce, for example, genes for insect or disease resistance or improved vigor or quality into cultivars already being grown.

International Conventions Relating to Plant Genetic Resources, Conservation, and Utilization

Currently, two international conventions relating to the promotion of conservation and protection of biological diversity are in effect. The Convention on International Trade in Endangered Species of Wild Fauna and Flora (CITES), which came into operation in 1975, helps to ensure that international trade in specimens of wild animals and plants does not threaten their survival. CITES has about 28,500 species of plants listed under protection, and this list is growing as more and more plant habitats are damaged. For example, many of the orchids are on this list. Further information on CITES can be found at **http://www.cites.org/**. The second is the Convention on Biological Diversity (CBD) which is an outcome of the 1992 United Nations Environment Programme's Earth Summit in Rio de Janeiro, Brazil, and became effective in 1993. Its objectives are

> the conservation of bilogical diversity, the sustainable use of its components and the fair and equitable sharing of the benefits arising out of the utilization of genetic resources, including appropriate access to genetic resources and appropriate transfer of relevant technologies, taking into account all rights over those resources and to technologies, and appropriate funding.

This also means that plant germplasm becomes a country's natural resource, and the use of a germplasm by another country now requires prior official consents, collection permits, and specific material transfer agreements, including benefit sharing when a germplasm is commercialized. Annually, an estimated $500 to $800 billion of genetic resources derived products is traded globally. In the past, plant germplasm was shared and exploited globally as common properties of humankind. Thus, CBD can be seen as a response of germplasm-rich countries to the intellectual property rights protection systems in place in the world to protect developed germplasm. The International Treaty on Plant Genetic Resources for Food and Agriculture (ITPGRFA), a Food and Agriculture Organization (FAO) of the United Nations initiative, is the first endeavor to create a common multilateral agreement to ensure the conservation of sixty-four food and feed crops in harmony with CBD for sustainable agriculture and food security. It covers approaches in germplasm exploration and conservation, measures to induce sustainable germplasm uses, recommendations on international cooperation, measures to uphold farmers' rights, sovereign rights of contracting states, a standard material transfer agreement, benefit sharing, and so on. To implement ITPGRFA, a $260 million endowment—the Global Crop Diversity Trust—was instituted to provide the necessary funds for protecting and conserving the most threatened and valuable collections of crop diversity. Detailed information on CBD is available at **http://www.biodiv.org/**.

Moving plant materials about the world can also introduce devastating insect or disease pests into a country where they have never before appeared. For example, the chestnut blight fungus (*Endothia parasitica*) was inadvertently introduced on imported plant material to the New York area from the Orient in the late 1800s. By about 1935, this fungus had practically eliminated all varieties of the beautiful native American chestnut trees (*Castanea dentata*) from the eastern United States. To guard against the introduction of such pests, most countries have set up elaborate inspection, fumigation, and quarantine procedures. The 1997 International Plant Protection Convention (**http://www.ippc.int/IPP/En/default.htm**), an international treaty under the auspice of FAO and signed by 127 governments in 2004, helps to harmonize and standardize the rules and procedures, and agreements in plant quarantine to prevent the spread and introduction of pests of plants and plant products, and to promote appropriate measures for their control in the world. Where appropriate, storage places, packaging, conveyances, containers, soil, and any other organism, object, or material capable of harboring or spreading plant pests are included, particularly where international transportation is involved. As a result, regional plant protection organizations, for example, the North American Plant Protection Organizations (NAPPO), were established to promote the development and use of relevant international standards for phytosanitary measures, and to encourage interregional cooperation in promoting harmonized phytosanitary measures for controlling pests and in preventing their spread and/or introduction.

Plant material can be introduced as seeds, bulbs or corms, rooted cuttings, or as scions or budwood for grafting or budding onto related growing plants. Seeds are the easiest to ship and pose the least danger of carrying pathogens. Rooted plant parts with soil particles around the roots are particularly suspect because the soil may contain nematodes or soil-borne diseases. Vegetative plant material coming into the United States is usually fumigated, then held under postentry quarantine for as long as two years before distribution to nurseries is permitted.

Shipment of plant material is now easy and highly successful because of specialized packing materials, refrigeration, and frequent worldwide air flights. This contrasts to earlier days when only slow ship transport, often through hot tropical seas, was possible.

Preservation of Desirable Germplasm

A concerted worldwide effort is needed to ensure the survival of the earth's endangered plant species and thus preserve genetic diversity. International germplasm networks operate under the auspices of the Consultative Group on International Agricultural Research (CGIAR) and the World Conservation Union (IUCN). These networks interact with, and are supplemented by, many national seed banks and agricultural centers.

In 1998, FAO estimated that 6.1 million accessions of primary food, forage, and industrial crops had been collected and conserved globally. Such gene pools are needed for developing improved crops in the future, for introducing beneficial genes into existing crops from close relatives, for maintaining attractive plants and trees valued for their aesthetic purposes, and for keeping plants intact as part of ecosystems where their presence is necessary to the survival of other plant and animal species. The genetic diversity of plants, as well as animals and microbes, is of fundamental importance to our survival on earth. Food and other agricultural crops are derived from the genetic diversity of natural plant populations.

To help prevent eradication of many plant species, the US Congress in 1973 passed the Endangered Species Act, which directed the Smithsonian Institution to prepare a list of endangered plant species and to recommend measures for saving them. As many as one in ten plant species is now extinct or endangered because of the encroachment of agricultural operations, the removal of rare plants by plant collectors, and the general destruction of vegetation from various causes. Such endangered germplasm can be saved and stored for future use as seeds or as living plants in special protected locations.

The CGIAR international research centers such as the International Rice Research Institute (IRRI) in the Philippines and the International Maize and Wheat Improvement Center (CIMMYT) in Mexico have assembled one of the most comprehensive collections of our major food and forage crops (Table 15–3). This endeavor is coordinated by its specialized plant germplasm conservation center, the International Plant Genetic Resources Institute (IPGRI). IPGRI also provides technical leadership to national gene bank programs. The IUCN has the mission "to influence, encourage and assist societies throughout the world to conserve the integrity and diversity of nature and to ensure that any use of natural resources is equitable and ecologically sustainable." The focus of IUCN is the conservation of the whole ecosystem and the total biodiversity. It monitors the state of the world's species in the IUCN Red List of threatened and endangered species (**http://www.redlist.org/**). This list is adopted globally by national governments, nongovernmental organizations, and scientific institutions in their biodiversity conservation efforts. An estimated 60,000 to 100,000 plant species, representing about one-third of the world's plants, are currently threatened or facing extinction in their native habitats. In 2002, the Global Strategy for Plant Conservation (**http://www.plants2010.org/global_strategy.pdf**) was adopted by CBD and in 2004, the Global Partnership for Plant Conservation (**http://www.bgci.org.uk/files/2/786/GlobalPartnershipstatement.doc**), consisting of international and national agencies and organizations active in plant conservation, was established to support the worldwide implementation of the strategy. The Botanic Gardens Conservation International (**http://www.bgci.org.uk/**), founded in 1987, is one of such organizations that is taking the forefront in this attempt. Over 500 member institutions in 112 countries

TABLE 15–3   Collections of Food, Forage, and Forestry Crops Held in Trust for the World Community at the Eleven CGIAR Centers in 2002

Center	Crop(s)	Number of Accessions
International Center for Tropical Agriculture (CIAT)	Cassava	15,728
Cali, Colombia	Forages	18,138
	Bean	31,718
International Maize and Wheat Improvement Center (CIMMYT)	Maize	20,411
Mexico	Wheat	95,113
International Potato Center (CIP)	Andean roots and tubers	1,112
Lima, Peru	Sweet potato	6,413
	Potato	5,057
International Center for Agriculture in the Dry Areas (ICARDA)	Barley	24,218
Aleppo, Syria	Chickpea	9,116
	Faba bean	9,074
	Wheat	30,270
	Forages	24,581
	Lentil	7,827
International Crops Research Institute for the Semi-Arid Tropics (ICRISAT)	Chickpea	16,961
Patancheru, India	Groundnut	14,357
	Pearl millet	21,250
	Pigeonpea	12,698
	Sorghum	35,780
	Minor millets	9,050
International Institute for Tropical Agriculture (IITA)	Bambara groundnut	2,029
Ibadan, Nigeria	Cassava	2,158
	Cowpea	15,001
	Soybean	1,909
	Wild *Vigna*	1,634
	Yam	2,878
International Livestock Research Institute (ILRI)	Forages	11,537
Nairobi, Kenya		
International Plant Genetic Resources Institute (IPGRI)	*Musa*	931
Maccarese, Italy		
International Rice Research Institute (IRRI)	Rice	80,617
Los Banos, Philippines		
West Africa Rice Development Association (WARDA)	Rice	14,917
Bouaké, Cote d'Ivoire		
World Agroforesty Center	Sesbania	25
Nairobi, Kenya		
TOTAL		532,508

Source: http://www.cgiar.org/research/res_accessions.html.

are working together to implement a worldwide botanic gardens conservation strategy for plant conservation. For example, the Royal Botanic Gardens, Kew, England, has established the collaborative Millennium Seed Bank Project to collect and conserve 24,000 plant species from around the world and protect them against extinction.

In the United States, the National Plant Germplasm System (NPGS) is a coordinated network of four main plant introduction stations in four geographic regions of the country, a long-term storage gene bank in Fort Collins, Colorado, and about twenty-three repositories and active collection sites. The functions of the repositories and active sites are to collect, introduce, maintain, characterize, evaluate, catalog, and distribute plant germplasm. Financial support comes from the USDA and from state agricultural experiment stations as well as from commercial plant breeding and seed trade organizations. The

general mission of the system is to provide plant scientists with the germplasm needed to carry out their work, for example, in breeding new cultivars resistant to certain insects, diseases, smog, or high soil salinity. In 2002, more than 450,000 accessions are in conservation (Table 15–4).

The activities of the National Plant Germplasm System are:

1. Introduction of plant materials into useful scientific channels is done by planned foreign and domestic exploration trips, by exchanges with foreign agencies, or by traveling scientists. There may also be useful domestic germplasm that should be maintained—for example, mutations, species hybrids, or germplasm resistant to a certain insect or disease—and may be valuable for future crop development. These, along

TABLE 15–4	The Number of Accessions Maintained at the NPGS Locations in 2002			
Site	**Accession**	**Country**	**Genus**	**Species**
Barley Genetic Stocks Center	3,044	3	1	1
Clover collection	246	30	1	118
Cotton collection	9,308	124	1	41
Desert Legume Program	2,585	56	198	1279
Maize Genetic Stock Center	4,710	2	1	1
National Arboretum	1,909	51	257	861
National Arctic Plant Genetic Resources Unit	493	31	50	145
National Arid Land Plant Genetic Resources Unit	961	32	12	124
National Center for Genetic Resources Preservation	23,299	106	198	493
National Small Grains Collection	126,563	170	15	148
National Germplasm Repository—Brownwood	881	3	2	23
National Germplasm Repository—Corvallis	11,687	92	56	757
National Germplasm Repository—Davis	5,105	79	19	202
National Germplasm Repository—Geneva	5,136	58	6	91
National Germplasm Repository—Hilo	675	41	22	76
National Germplasm Repository—Mayaguez	560	40	137	229
National Germplasm Repository—Miami	4,606	90	213	527
National Germplasm Repository—Riverside	1,167	30	38	152
North Central Regional PI Station	47,032	177	319	1767
Northeast Regional PI Station	11,730	126	32	196
Ornamental Plant Germplasm Center	967	58	62	287
Pea Genetic Stock Collection	501	3	1	2
Plant Germplasm Quarantine Office	4,827	59	19	76
Potato Germplasm Introduction Station	5,503	38	1	168
Southern Regional PI Station	82,579	184	246	1433
Soybean Collection	20,415	91	1	16
Tobacco Collection	2,106	67	1	65
Tomato Genetic Stock Center	3,287	18	2	22
Western Regional PI Station	69,946	162	368	2423
Wheat Genetic Stocks Center	334	1	1	1
Total	452,162	N.A.		

Note: The twenty-six NPGS germplasm stations, centers, and repositories listed on the NPGS website (http://www.ars-grin.gov/npgs) are as follows:

1. Barley Genetic Stock Center (GSHO), Aberdeen, Idaho—http://www.ars-grin.gov/ars/PacWest/Aberdeen/hang.html
2. C. M. Rick Tomato Genetics Resource Center. Davis, California—http://tgrc.ucdavis.edu/
3. Desert Legume Program. Tucson, Arizona—http://ag.arizona.edu/bta/bta20.html
4. Maize Genetics Cooperation—Stock Center (GSZE), Urbana, Illinois—http://w3.aces.uiuc.edu/maize-coop/
5. G. A. Marx Pea Genetic Stock Center (GSPI), Pullman, Washington—http://www.ars-grin.gov/ars/PacWest/Pullman/GenStock/pea/MyHome.html
6. National Clonal Germplasm Repository (COR), Corvallis, Oregon—http://www.ars-grin.gov/ars/PacWest/Corvallis/ncgr/ncgr.html
7. National Clonal Germplasm Repository for Citrus and Dates, Riverside, California—http://www.ars-grin.gov/ars/PacWest/Riverside/homepg1.htm
8. National Clonal Germplasm Respository for Tree Fruit/Nut Crops and Grapes (DAV), Davis California—http://www.ars-grin.gov/ars/PacWest/Davis/
9. National Germplasm Resources Laboratory (NGRL), Beltsville, Maryland—http://www.barc.usda.gov/psi/ngrl/ngrl.html
10. National Center for Genetic Resources Preservation (NSSL), Fort Collins, Colorado—http://www.ars-grin.gov/ars/NoPlains/FtCollins/nsslmain.html
11. National Small Grains Collection (NSGC), Aberdeen, Idaho—http://www.ars-grin.gov/ars/PacWest/Aberdeen/nsgc.html
12. National Temperate Forage Legume Genetic Resources Unit, Prosser, Washington—http://www.forage.prosser.wsu.edu/
13. North Central Regional Plant Introduction Station (NC7), Ames, Iowa—http://www.ars-grin.gov/ars/MidWest/Ames/
14. Ornamental Plant Germplasm Center (OPGC), Columbus, Ohio—http://opgc.osu.edu
15. Pecan Breeding & Genetics, Brownwood and Somerville, Texas—http://extension-horticulture.tamu.edu/carya/
16. Plant Genetic Resources Conservation Unit (S9), Griffin, Georgia—http://www.ars-grin.gov/ars/SoAtlantic/Griffin/pgrcu/
17. Plant Genetic Resources Unit (NE9), Geneva, New York—http://www.ars-grin.gov/ars/NoAtlantic/Geneva/
18. Plant Germplasm Quarantine Office (PGQO), Beltsville, Maryland—http://www.barc.usda.gov/psi/fl/pgqo.html
19. Soybean/Maize Germplasm, Pathology, and Genetics Research Unit, Urbana, Illinois—http://www.life.uiuc.edu/plantbio/ars/ppgru.html
20. Subtropical Horticulture Research Station (MIA), Miami, Florida—http://ars-grin.gov/ars/SoAtlantic/Miami/homeshrs.html
21. Tropical Agriculture Research Station, Mayagüez, Puerto Rico—http://www.ars-grin.gov/ars/SoAtlantic/Mayaguez/mayaguez.html
22. Tropical Plant Genetic Resource Management Unit (HILO), Hilo, Hawaii—http://pbarc.ars.usda.gov/pages/research/tpgrmu/germplasm.shtml
23. United States Potato Genebank—NRSP-6, Sturgeon Bay, Wisconsin—http://www.ars-grin.gov/ars/MidWest/NR6/
24. Western Regional Plant Introduction Station (W6), Pullman, Washington—http://www.ars-grin.gov/ars/PacWest/Pullman/
25. Wheat Genetic Stock Center (GSTR), Aberdeen, Idaho—http://www.ars-grin.gov/ars/PacWest/Aberdeen/hang.html
26. Woody Landscape Plant Germplasm Repository, Washington, D.C.—http://www.usna.usda.gov/Research/wlpgr.html

with introduced foreign material, are eligible to enter the National Germplasm System.

2. Maintenance of this potentially valuable germplasm for future research programs is the responsibility of the regional and interregional plant introduction stations, the National Center for Genetic Resources Preservation, and the curators of collections of specific crops.

3. Characterization and evaluation of the plant genetic resources is done by initial screening and subsequent tests in the field, greenhouse, and laboratory by cooperating state, federal, and private scientists.

4. Distribution of plant germplasm is made free of charge to all qualified scientists and institutions requesting it, in sufficient amounts to enable them to initiate their research program.

The decentralized system is held together by a common web-based database management system, the Germplasm Resources Information Network (GRIN), which helps NPGS curators to handle both the plant and store inventory data. Another notable component of NPGS is about forty crop germplasm committees, which are made up of crop specific experts for each crop from USDA, universities, public institutions, and industry to advise on policy and coordinating activities to meet the immediate and long-term national goals of agriculture in the United States.

BROADENING THE BASE OF AGRICULTURAL PRODUCTION

The world's peoples are largely fed today by only about twenty crops. Reliance on so few crops could lead to a catastrophic famine if but a few of them were obliterated by insect or disease attacks or by climatic changes. The ravage of U.S. corn plantings by the corn blight disease in 1970 is an example of such a possibility.

In an attempt to broaden the base of agricultural plants in the tropics and to promote interest in neglected but seemingly useful tropical plants with economic potential, the U.S. National Academy of Science promoted a compilation of plants nominated by plant scientists around the world and published an account of thirty-six plants selected from the 400 proposed. Each plant was described along with its special requirements, research needs, selected readings, research contacts, and germplasm sources.

All thirty-six plants were thought to have considerable potential, but most have not been cultivated out of their own limited region of origin. Among the cereals, for example, the report cited an almost completely neglected grain species in the genus *Amaranthus,* native to Central America, that has very high levels of protein and the essential amino acid lysine, which is usually deficient in plant proteins. Among the vegetables studied, the wax gourd (*Benicasa hispida*) gives three crops a year of a large melonlike fruit that can be stored for twelve months without refrigeration. The mangosteen (*Garcinia mangostana*) is perhaps the world's best-tasting fruit, but it is little known outside the very humid tropics of southeast Asia.

Since 1990, crop scientists have met regularly in the United States to share information on new crops and the potential for new crop development. These scientists are looking for solutions to global problems such as hunger, malnutrition, deforestation, desertization, and agricultural sustainability. A series of symposia through these years has produced several volumes of accumulated information on new crops, with the aim of broadening crop diversity in agriculture. These volumes include *Advances in New Crops* (1990), published by Timber Press; *New Crops* (1993), published by Wiley; and *Progress in New Crops* (1996), *Perspectives in New Crops and New Uses* (1999), and *Trends in New Crops and New Uses* (2002), published by ASHS Press.

Other crops, native to the Americas, which have undeveloped but great potential are jojoba (*Simmondsia chinensis*), guayule (*Parthenium argentatum*), groundnut (*Apios americana*), and the tepary bean (*Phaseolus acutifolius*), as well as *Cuphea* species, whose seeds contain valuable oils. In recent years, an ANNONACEAE, pawpaw (*Asimina triloba*), the largest tree fruit native to the United States, is gaining interest among researchers and growers. It is the only temperate annona in the world and has tremendous potential to become an important nutritious fruit in North America and other temperate regions of the world. The leguminous tree *Leucaena leucocephala,* native to Central America, already has its own success story. It is fast growing, with high-protein leaves, and is used in India for feeding dairy cattle. In Africa, the leaves are used as a much-needed fertilizer, while in Sumatra and the Philippines, it is used as a commercial regenerative timber source.

Numerous exotic tropical fruits are found in America's supermarkets, many being grown in southern Florida. Others arrive by plane from various tropical regions.

SUMMARY AND REVIEW

Plants are classified in many different ways. They may be classified by taxonomy (common physical characteristics), climate (temperate, tropical, desert, etc.), season (warm, cool), temperature or light requirements, edible or usable parts, or other criteria.

Plants usually have two different names: common and scientific. The common name is usually in the language of the region where the plant is growing. The same plant can have many different common names. Likewise, different plants can have the same common name. To avoid confusion, every species of plant is now given a scientific name based on its taxonomic ranking. Taxonomic ranking goes from general to very specific common characteristics. The most general category is Domain. The most specific is species. The scientific name of a species is composed of the genus and specific epithet. Assigning a plant a scientific name is a formal process that ensures each species has a name that is distinct from any other species. Most scientific names are in Latin, although Greek and Arabic words are sometimes used.

The use of some species of plants for agricultural purposes (domestication) goes back many thousands of years. Although theories on how species came to be domesticated vary, in general, everyone agrees that certain regions gave rise to crops that developed from plants native to those regions. Some crops originated in more than one region. Few major crops originated in the United States, but several minor crops can be traced back to that country.

Crops are domesticated and improved by continually selecting and propagating plants with superior characteristics. Crop improvement has evolved from simple selection by a farmer to a complex system using selective breeding and bioengineering techniques. When plants are domesticated, the result is often a loss of genetic traits that at the time of selection were not considered important. Many times, however, these traits were what allowed the plant to survive in the wild. With time and changes in where and how the crop is grown, these traits regain their importance, but the genetic information is no longer available in the domesticated lines used for crop improvement. We have come to understand the importance of preserving the genetic information that was found in the ancestors of our crops. To facilitate the collection and preservation of germplasm, centers have been established around the world.

FOOD FOR THOUGHT

1. Look at the ingredients of your favorite cereal. Decide which ones are plant-derived. Then using the tables and figures provided in the chapter:

   a. Determine the likely region of origin of the plant-derived ingredients.

   b. Determine where the germplasm for those ingredients is most likely to be kept in preservation.

2. Find a landscape plant in your area for which you do not know the common or scientific name. Using a plant identification key (found in most libraries or on the Internet: key words: taxonomic key plant identification) determine the identity of that plant. From that identification, list closely related plants. Why do you think it is useful when growing plants to know closely related species?

SUPPLEMENTARY READING

BAILEY, L. H., E. Z. BAILEY, and staff of the L. H. Bailey Hortorium. 1976. *Hortus third.* New York: MacMillan.

BRICKELL, C. D. 1980. International code of nomenclature for cultivated plants. *Regnum Vegetabile* 104:7–32.

Convention on Biological Diversity (CBD). 2002. CBD—Text of the convention. **http://www.biodiv.org/convention/articles.asp**. Secretariat of the CBD, United Nations Environmental Programme, Montreal, Canada.

Convention on International Trade in Endangered Species of Wild Fauna and Flora (CITES). 1979. Text of the convention. Geneva: CITES Secretariat. **http://www.cites.org/eng/disc/text.shtml**.

Food and Agriculture Organization of the United Nations. 1998. The state of the world's plant genetic resources for food and agriculture. Rome: FAO, p. 98.

Food and Agriculture Organization of the United Nations. 2003. *International treaty on plant genetic resources for food and agriculture.* Rome: FAO. **ftp://ext-ftp.fao.org/ag/cgrfa/it/ITPGRe.pdf**.

GRIFFITHS, M. 1994. *Index of garden plants.* Portland, Ore.: Timber Press.

HARLAN, J. R. 1975. *Plants and man.* St. Paul, Minn.: American Society of Agronomy.

IUCN (The World Conservation Union). 2004. *IUCN Red List of threatened species.* **http://Species.://www.iucn.org/themes/ssc/red_list_2004/main_EN.htm**.

LANJOUW, J. (Ed.). 1966. *International code of botanical nomenclature.* Utrecht, Netherlands: International Bureau for Plant Taxonomy and Nomenclature of the International Association for Plant Taxonomy.

MOONEY, P. 1983. The law of the seed: Another development and plant genetic resources. *Development Dialogue* 1983(1–2):1–172.

RAVEN, P. H., R. F. EVERT, and S. E. EICHORN. 1999. *Biology of plants.* New York: Freeman.

RAY, D., J. JANICK, D. DIERIG, R. MYERS, and C. BAILEY. 2002. Preface, pp. xvi–xvii. In J. JANICK and A. WHIPKEY (Eds.), *Trends in new crops and new uses.* Alexandria, Va.: ASHS Press.

VAVILOV, N. I. 1951. The origin, variation, immunity and breeding of cultivated plants (trans by K. S. Chester). New York: Ronald Press.

WOODLAND, D. W. 2000. Contemporary plant systemics, third edition. Berrien Springs, Md.: Andrews University Press.

UNIT
III

CROPS: THEIR PRODUCTION AND UTILIZATION SYSTEMS

CHAPTER

16

Field Crops Grown for Food, Fiber, Fuel, and Other Industrial Uses

Peter Thomison

KEY LEARNING CONCEPTS

After reading this chapter you should be able to:

- ♦ Understand the cultural practices common to nearly all field crops and the reasons behind those practices.
- ♦ Know the major field crops grown for food, fiber, fuel, and other industrial uses.
- ♦ Understand the specific cultural practices used for growing many of those crops.

INTRODUCTION

Table 16–1 lists the world's major field crops and the principal uses of each. Successful crop production requires an understanding of the various management practices and environmental conditions affecting crop performance. Planting date, seeding rates, variety selection, tillage, fertilization, and pest control all influence crop performance. A crop's response to a given cultural practice is often influenced by one or more other practices. The keys to developing a successful production system are (1) recognizing and understanding the types of interactions that occur among production factors, as well as various yield limiting factors; and (2) developing management systems that maximize the beneficial aspect of each interaction. Knowledge of crop growth and

TABLE 16–1	Field Crops Grown for Food, Fiber, and Fuel		
Crop	Food	Fiber	Fuel and Other Industrial Uses
Agave	X	X	
Barley	X		
Beans	X		
Castor bean			X
Coconut	X		
Corn	X	X	X
Cotton		X	
Flax		X	
Hemp	X	X	
Jute		X	
Kenaf	X	X	
Oats	X		
Olive	X		
Palm oil	X		
Peanut	X		X
Ramie		X	
Rape	X		
Rice	X		
Rye	X		
Safflower	X		
Sesame	X		
Soybean	X		X
Sorghum	X		
Sugar beets	X		
Sugarcane	X		
Sunflower	X		
Wheat	X		
Wild rice	X		

development is also essential to using cultural practices more efficiently to obtain higher yields and profits.

TILLAGE

The major objective of tillage is to create a seedbed that will promote germination and early seedling growth, control weeds, and modify the soil environment to make it as favorable as possible for crop growth while minimizing soil erosion and fuel costs. Tillage affects soil structure, texture, temperature, moisture, air, and nutrient availability. The appropriate tillage practices needed for seedbed preparation depend on location and soil conditions. The success of a tillage system depends on how well it is adapted to the soil and climatic conditions of a region.

Tillage in crop production is often described as either primary (involving the first and generally major tillage operations) or secondary (usually harrowing, firming, and smoothing). As new tillage implements have been developed, differences between these two types of tillage have become less pronounced. Many different tillage systems are currently used in crop production for seedbed preparation. The most common systems are based on a moldboard plow, chisel plow, disk plow, rotary tiller, or large disk harrow for primary tillage; secondary tillage operations involve the use of the disk har-

row, spring-tooth harrow, and spike tooth harrow. Such systems are frequently referred to as conventional tillage.

Since the 1970s, tillage systems that minimize soil erosion have been adopted widely. These conservation tillage systems leave more crop residue on the soil surface, which is effective in reducing soil erosion, thereby protecting the valuable resources of soil and water. Crop residues on the surface during the fallow season increase rainfall infiltration into the soil and reduce surface runoff, thereby reducing soil erosion. Higher rates of residue returned to the soil increase the organic matter content that make soil more productive.

Conservation tillage systems range from those in which only a small slit is opened in the soil ahead of the planter unit (no-till) to near full-width tillage. The success of reduced tillage has been due mostly to the effective use of herbicides. In addition to helping to reduce erosion and enhance soil quality, no-till farming may also offer a solution to global warming. Soil loses most of its carbon content during plowing, which releases carbon dioxide (CO_2) gas into the atmosphere. Increased levels of CO_2 in the atmosphere have been associated with global climate change. No-till farming helps soils retain carbon, thus promoting carbon sequestration.

When no-till farming was introduced, some crop scientists were concerned that leaving crop residues on the soil surface would result in more pest problems. Crop residues harbor insects and diseases, which can be reduced by burying the residues through tillage. For the most part, concerns about no-tillage leading to greater pest problems have not been realized, but problems have occurred where no-till continuous cropping has been practiced. In the United States gray leaf spot (GLS), caused by the fungus *Cercospora zeae-maydis,* was a relatively minor disease problem of corn until the advent of no-till crop production, However, following widespread adoption of no-till cropping in continuous corn production in the southern United States in the early 1970s, GLS became a serious, and eventually a major foliage disease problem of corn. Many midwestern states were affected by a major GLS epidemic in 1995.

CROPPING SEQUENCES

Crops are produced using different cropping sequences based on agronomic and economic considerations. Some of these cropping sequences are complicated and require a high level of expertise and management to be successful, whereas others are fairly simple. The most popular cropping sequence in the US corn belt is the corn-soybean rotation, which involves alternating crops of corn and soybean. Crop rotations may also include a small grain, usually wheat, or forage crops. The corn-soybean rotation offers several advantages over growing either crop continuously. Benefits to growing corn in rotation with soybean include more weed control options, fewer difficult weed problems, less disease and insect buildup, and less nitrogen fertilizer use. Corn grown following soybeans typically yields about 10 percent more than growing continuous corn. No-till cropping systems are more likely to succeed on poorly drained soils if corn follows soybean or meadow rather than corn or a small grain, such as wheat. The yield advantage to growing corn following soybean is often much more pronounced when drought occurs during the growing season.

In contrast to crop rotation, monoculture is growing the same crop in a field continuously over a period of years. In the US corn belt, continuous corn production is the most common monoculture practiced. Farmers have used monoculture to take advantage of favorable markets or government support programs, an environment that strongly favors producing a high-value crop, or to reduce the need for diverse machinery or facilities. The ratoon cropping system, another form of monoculture, is used in sugarcane production. After the first harvest is completed, the crop is allowed to regrow and a second harvest is obtained from the regrowth prior to destroying it in the fall.

Double cropping involves the production of two crops, one following another, during one growing season. The most common double-cropping system in the corn belt is following winter wheat with soybeans. Intercropping consists of simultaneously producing two or more crops in the same field. Intercropping is most advantageous when crops of different characteristics are planted. Examples of intercropping include growing rows of corn and soybeans, or growing cassava and beans or peas in tropical systems. Relay intercropping consists of seeding a second crop into an initial crop prior to harvest of the initial crop. This provides a longer growing season for the second crop. Seeding soybeans into winter wheat prior to wheat harvest in the spring is an example of relay intercropping.

VARIETY SELECTION

Selecting varieties for planting is a key step in designing a successful crop production system. To stay competitive, growers must evaluate and select the most adapted varieties for their locale on an annual basis.

Growers need to choose the highest yielding hybrids that are maturity adapted to their growing region. Maturity, yield potential, and disease resistance are important factors to consider in variety selection for all crops. For the cereal crops (including wheat, barley, oats, rye, spelt, and triticale), winter hardiness, straw stiffness, and nonshattering characteristics are also important cultivar selection criteria. High-quality seed is also important to ensure varietal purity, high germination, and minimum contamination by weed seed and other impurities.

Maturity is a key characteristic in choosing a variety. Varieties that use most of the growing season to mature generally have greater yield potential than those that mature more quickly. However growers often plant varieties of different maturities. This practice may help limit damage to the crop from a period of stress during the growing season and may allow the harvest period to begin earlier. Planting varieties with

different maturities also ensures some degree of genetic diversity, which may help buffer the crop from various biotic and abiotic stresses. Most crops are characterized by a maturity designation. Soybeans are assigned to ten different maturity groups from 00 (grown in short season environments, e.g., Canada) to VIII (grown in semitropical environments). Winter and spring designations for small grain crops are associated with their planting date within a region.

In corn, the most common maturity rating system is the days-to-maturity system. This system does not reflect actual calendar days between planting and maturity—108-day hybrid, for example, does not actually mature 108 days after planting. A days-to-maturity rating is based on relative differences within a group of hybrids for grain moisture at harvest. A one-day maturity difference between two hybrids is typically equal to a ½ to ¾ percentage point difference in grain moisture. For example, a 108-day hybrid would be, on average, 3 to 4.5 points drier than a fuller season 114-day hybrid if they were planted the same day (6 days multiplied by 0.5 or 0.75). The days-to-maturity–moisture relationship is usually dependable when comparing hybrid maturities within a single seed company. However, because there are no industry standards for the days-to-maturity rating system, grain moisture comparisons of similar hybrid maturities from different seed companies may vary considerably. Days-to-maturity ratings are satisfactory for pre-season hybrid maturity selection when the length of the growing season is usually not an issue. For delayed planting or replanting hybrid selection needs, growers need more absolute descriptions of a hybrid's maturity requirements to manage the risk of a killing fall frost to late-planted corn.

The **growing degree day (GDD)** maturity rating system is more accurate in determining hybrid maturity than the days-to-maturity system because growth of the corn plant is directly related to the accumulation of heat over time rather than the number of calendar days from planting. The GDD system has several advantages over the days-to-maturity system. The GDD system provides information for choosing hybrids that will mature reliably, given a location and planting date; allows the grower to follow the progress of the crop through the growing season; and aids in planning harvest schedules.

The GDD calculation method most commonly used for corn in the United States is the 86/50 cutoff method. Growing degree days are calculated as the average daily temperature minus 50:

$$GDD = \frac{T_{max} + T_{min}}{2} - 50$$

If the maximum daily temperature (T_{max}) is greater than 86°F, 86 is used to determine the daily average. Similarly, if the minimum daily temperature (T_{min}) is less than 50°F, 50 is used to determine the daily average. Growing degree days are calculated daily and summed over time to define thermal time for a given period. Each corn hybrid requires a certain number of accumulated GDDs to reach maturity, and seed corn dealers have information on specific hybrids. As with any system, the GDD system has several shortcomings. Under certain delayed planting situations and stress conditions, GDD requirements for maturity may be reduced significantly.

Because weather conditions are unpredictable, the most reliable way to select superior varieties is to consider their performance during the last growing season and previous year over a wide range of locations and climatic conditions. In many countries, including the United States, state universities and/or government agricultural agencies conduct crop evaluations of private and public varieties to provide farmers with objective, unbiased performance information. Two to three years of data from several test locations is usually adequate when using results of performance trials. Test summaries for more than three years may exclude recently released varieties with better performance potential.

Resistance to lodging is an important trait in the selection of many crops (see Fig. 16–1). Lodging is the breaking or collapse of stalks (e.g., for corn, sorghum) or stems (e.g., for soybean, wheat, rice) usually at or near maturity. Lodging causes serious crop loss of grain because it makes harvesting difficult. Strong winds often contribute to the severity of lodging problems. In corn, lodging is promoted by stalk rots and harvest delays. In some wheat varieties, excessive nitrogen fertilization causes the heads to become too heavy with grain and may result in lodging.

The number of hectares, soil type, tillage practices, desired harvest moisture, and pest problems determine the need for traits such as disease resistance, early plant vigor, plant height, and so on. End uses of the crop should also be considered. For example, with corn, will it be sold directly to an elevator as shelled grain or used on the farm? On-farm capacity to harvest, dry, and store grain is another consideration. Premiums may be available for the contract production of specialty hybrids or cultivars with nutritionally enhanced grain characteristics.

The development of transgenic crops has given farmers new options for pest control and additional traits to consider when selecting hybrids to plant. Since the mid-1990s, transgenic (also referred to as GM or genetically modified) grain and fiber crops have been commercialized and widely produced. Transgenic corn, soybean, cotton, and canola account for most of the hectarage planted to transgenic crops. In the United States as of 2003, transgenic soybean cultivars were planted on more than 80 percent of the soybean hectarage, transgenic cotton cultivars were planted on more than 70 percent of the cotton hectarage, and transgenic corn hybrids were planted on nearly 40 percent of the corn acreage. Transgenic insect resistant corn hybrids contain a gene from bacteria that produces the insecticide known as Bt. Planting Bt hybrids may eliminate the need for postemergent insecticide applications for European corn borer and corn rootworm that are less effective and potentially harmful to nontarget beneficial insects. Planting crops tolerant or resistant to herbicides, like Roundup®, offers growers new weed-control techniques that involve fewer herbicide applications and use of more environmentally benign chemicals. However, consumer protests against GM foods, combined with marketing restrictions in export markets on grain from crops containing the Bt® and Roundup® resistance traits, have limited widespread adoption of some transgenic varieties. The efficacy of transgenic crops may also be short-lived if targeted insects and weeds develop resistance to Bt® and Roundup®.

PLANTING AND CROP ESTABLISHMENT

The planting operation is a key step in crop establishment. Soil and residue conditions should be evaluated before planting. Soil should be slightly moist and should crumble when squeezed. Planting in excessively wet soil may result in poor seed-soil contact or seed furrows that reopen upon drying. Wet residues may be jammed into the seed slot rather than being cut by the coulter, thus causing poor seed-soil contact and germination. Planting in dry soil may cause penetration problems, too shallow planting, and failure of the seed slot to close. All of these factors affect plant stands, either by reducing stand or causing uneven emergence of the crop and thus result in uneven development.

Many types of equipment are available for planting crops. Row crop planters are used to plant corns and soybeans in row spacings of 0.38 to 1.02 m (15 to 40 in.). Most soybeans and cereal crops are usually broadcast or planted with a seed drill (box drill) and air seeder. Row crop planters provide more effective seed placement that results in more accurate crop plant populations. To achieve this accuracy, the planting units on row crop planters take up more space than drill, thereby limiting the row spacings to less than 0.76 m (30 in.). Row crop planters are also designed to accommodate attachments for applying liquid or dry fertilizer, insecticides, and herbicides during the planting operation.

Figure 16–1 Lodging, as shown here, is the collapse of stems, causing the plant to fall over. This condition causes serious crop loss of grain in grain crops such as barley, corn, rice, and wheat because it makes harvesting difficult. In wheat, excessive nitrogen fertilization causes the heads to become too heavy with grain. Lodging is aggravated by strong winds. Lodging is also promoted by stalk rots in corn.

Rice is also planted with grain drills, but in some areas it is seeded into flooded fields using aircraft. Precision planters capable of metering seed at desired spacings are widely used in sugar beet production to achieve uniform emergence. Pelleted seed used in conjunction with these precision planters enhances the uniformity of seed spacing.

Planting Date

In temperate regions, the major environmental factors determining planting date are soil temperatures and moisture. Early plantings in the US corn belt are often associated with excessive soil moisture and low temperatures, which may delay emergence. These environmental conditions are conducive to infection of seed and seedlings by various fungal and bacterial pathogens, which may reduce emergence or cause postemergence seedling blights, further reducing the plant population. In the warmer tropical and subtropical climates, the planting date is often determined by the onset of rains (i.e., the monsoons in India). In such regions, high soil temperatures exceeding 35°C (95°F) and low soil moisture availability may limit germination of cool season crop species.

In the United States and Canada, cereal crops are planted in either the late fall or early to mid-spring months (Table 16–2). Winter and spring types exist for most of the cereal crops. Winter wheat, barley, and oats are generally planted in fall or winter plantings in the more southern latitudes. There is only a limited period when fieldwork can oc-

cur throughout much of the US corn belt because of its wet spring weather. This indicates the need to begin planting as soon as field conditions allow (Table 16–3).

Plant Population and Seeding Rate

The seeding rate of a crop is determined by the plant population (or plant density) required to optimize yields in a particular environment. For many field crops, a strong relationship exists between plant population and yield potential. Crop yields can be optimized only when water, nutrients, and light are not limiting, and temperatures and soil conditions are conducive for growth. The relative availability of water and nutrients are key factors determining the optimal plant population. In production environments with low rainfall and reduced fertility, the optimal crop population is usually lower than in environments associated with high fertility, adequate rainfall, and soils with good moisture holding capacity (Table 16–4). Yields for grain crops generally rise to a maximum as plant density increases and then fall or level off with further increases in plant density.

Plant population can also affect other economically important crop traits. Increases in plant population may increase plant height and reduce stem and stalk strength, thereby increasing the risk of lodging. Increased plant height may be beneficial in some crops such as soybean and cotton because the height of the lowest pod or boll may be increased, thereby facilitating harvest and producing a cleaner crop. The cutter

---

TABLE 16–2 Suggested Rates and Date of Seeding for Important Field Crops of the Eastern US Corn Belt at Normal Planting Times

	Rate to Plant		Date to Plant	
Crop	**Kg/ha***	**Seeds/ha†**	**Northern**	**Southern**
Barley, winter	100.8–134.4 (90–120)	3.0–3.7 million (1.2–1.5)	9/15–9/25	9/15–10/5
Corn, dent	137.9–20.2 (12–18)	54,000–79,000 (22,000–32,000) 61,700–84,000	4/28–5/10	4/15–5/10
Corn, pop	6.7–11.2 (6–10)	(25,000–34,000) 3.0–3.7 million	4/25–5/10	4/15–5/10
Oats, spring	84–112 (75–100)	(1.2–1.5 million) 272,200–395,500	3/15–4/15	3/5–4/5
Soybeans		(110,000–160,000) 2.5 million–3.7 million	5/1–5/20§	4/20–5/10§
Wheat	84–134.4 (75–120)	1 million to 1.5 million	9/22–10/13	9/28–10/18

* Pounds per acre in parentheses.
† Seeds per acre in parentheses.
§ Double cropped soybeans planted as late as early July.
Source: Adapted from Ohio Agronomy Guide, 13th ed. 1995. Bulletin 472. The Ohio State University Extension

---

TABLE 16–3 Days Available to Plow, Till, or Plant in Ohio, March 21–June 30, 4 Out of 5 Years (7-Day Week)

Soil Drainage Types	**March 21–30**	**April 1–30**	**May 1–31**	**June 1–30**
Poorly drained	0 or more	4 or more	8 or more	9 or more
Poorly drained with drainage improvements	1 or more	6 or more	11 or more	11 or more
Well drained	2 or more	8 or more	13 or more	13 or more

Source: Adapted from Nolte et al. 1976. Bulletin 605. The Ohio State University Extension.

TABLE 16–4   Corn Yield Response to Plant Population Across Various Environmental Conditions*

Number of Environments	Harvest Population (Plants/A)			
	18,000	22,000	26,000	30,000
	Yield (Bu/A)			
5	86	84	83	82
14	109	114	118	116
27	125	132	135	135
46	139	149	155	158
35	157	168	174	177
14	177	189	199	203

* Environmental conditions can be favorable or unfavorable.

Source: Adapted from Ohio Agronomy Guide, 13th ed. 1995. Bulletin 472. The Ohio State University Extension.

bar of the combine used for harvest does not have to be set as low to the soil surface to remove soybean pods and cotton bolls, thereby minimizing the potential damage and interference of the harvest operation caused by rocks and residue.

Seeding rates vary greatly among crops (Table 16–2). As the plant population increases, tillering and branching is reduced. Grain crops such as wheat can compensate with tillers to some extent for lower-than-optimal plant density. For other crops such as soybeans, which produce branches, specific plant populations are less critical, with crops producing optimal yields across a broader range of plant populations.

An inverse relationship usually exists between plant population and the ear or seed head. In some crops, this effect may be advantageous because smaller seed heads dry more quickly and may facilitate the harvesting operation. However, when drought conditions limit nutrient and water availability in corn, plant populations above the optimum may result in ears with few or no kernels.

Seed germinability and the vigor and timing of planting are major factors that need to be considered when determining the seeding rate required to achieve a target plant population. When planting is delayed in some crops, including small grains and soybeans, seeding rates are increased to compensate for a reduced period in which tillering, branching, and flowering can occur. In other crops, like corn, the seeding rates may be reduced because plant mortality associated with emergence declines as soils become drier and warmer later in the spring.

The number of plants/hectare at harvest is always less than the number of seeds planted. Planting date, tillage practices, pest problems, chemical injury, planter performance, and seed quality affect final populations obtained in the field. To compensate for these losses, a grower needs to plant more seed than the desired population at harvest. To determine an appropriate seeding rate, use the following formula:

$$\text{Seeding rate} = \frac{\text{plant population per acre at harvest}}{(\text{seed germination} \times \text{expected survival})}$$

Seed germination is the percentage germination shown on the seed tag. Most seed corn has a germination rate of 95 percent

or higher. Expected survival is the percentage of plants that you expect to survive to become harvestable plants in the fall. Keep in mind that survival rates for corn are often in the range of 85 to 95 percent but can vary considerably depending on planting conditions and other environmental factors. When early planting is likely to create stressful conditions for corn during emergence (e.g., no-till in early to mid-April), seeding rates 10 to15 percent higher than the desired harvest population are used to ensure that the target population is achieved. Greater plant mortality usually occurs with early spring plantings in temperate regions.

Plant Spacing Configuration

In many grain crops, increased yields can be expected from reducing row width. At a constant plant population, this increase in yield results from more uniform distribution of plants and reduced competition of plants within the row for sunlight, water, and nutrients. Plants in wider rows are more crowded and cannot make as efficient use of sunlight as plants in narrower rows. Narrowing row width may also enhance weed control and moisture utilization. With narrow rows, the crop develops a canopy earlier that restricts light to weeds between the rows. Earlier canopy formation may also shade the soil surface earlier and reduce evaporation of soil moisture, thereby conserving moisture for the crop. In some crops, narrower rows in conjunction with high plant population may create a high humidity microclimate conducive for certain diseases such as *Sclerotinia,* stem rot of soybean. Narrowing row width increases dependence on herbicides because post-emergence mechanical cultivation may injure the crop. In production environments subject to late season droughts, the benefits of narrowing row width may be limited.

Seeding Depth

Crop seed needs to be planted deeply enough to place each seed in contact with seed moisture and prevent rapid desiccation, yet shallow enough to allow rapid emergence of the shoot. Factors affecting seeding depth include seed size, type of emergence (i.e., **hypogeal** or **epigeal**), soil and seedbed conditions, date of planting, and available moisture supply. A rule of thumb for the planting depth of many crops is four to five times the average seed diameter. The seeding depth of small seeded crop species (such as rape) is shallower than larger seeded species (such as corn). Small seeded species have smaller food reserves compared with larger seeded crop species, which may limit shoot elongation and seed vigor. Increasing seed size within a species usually increases final emergence, with the greatest advantage occurring with the deepest planting.

In crops like soybean that are characterized by epigeal emergence and where the cotyledons are brought above the soil surface during emergence, the seeding depth needs to be shallower compared to a crop like corn that exhibits hypogeal emergence and where the cotyledons remain below ground. When soil moisture is limited, it may be necessary to plant deeper than normal to place seed in moisture. Recommended seeding depths generally are greater for the lighter textured (sandy) soils than heavier textured (clay) soils because of lower water-holding capacity and faster drying rate at a given depth. Deeper planting is feasible in these soils because they offer less physical resistance to the emerging seedling than heavier soils. Early spring temperatures in temperate regions decrease rapidly with depth. Therefore, early plantings are usually shallower than later plantings because temperatures are warmer near the soil surface and soil moisture conditions are better early in the season.

WATER MANAGEMENT: IRRIGATION AND DRAINAGE

In areas with insufficient rainfall during the growing season or sandy soils with poor soil moisture retention, irrigation is essential for crop production. Conversely, tile drainage is es-sential throughout much of the US corn belt to allow for timely planting and to minimize saturated soil conditions. Some crops like rice are tolerant of saturated soil conditions.

Most agronomic crops in the United States east of the Mississippi are produced under rainfed conditions. In many areas of the western United States, however, where rainfall is erratic and limited, irrigation is required for successful crop production. Some of the major crops produced under irrigation include corn, sorghum, sunflowers, soybeans, and sugar beets. Most of this irrigated cropland is in the arid regions of the Southwest. Most of these crops are either sprinkle or furrow irrigated.

CROP NUTRITION

Most field crops grow best on well-drained, fertile loam soils well supplied with organic matter. When adequately fertilized, however, field crops perform well in a variety of soils. Most do well on a neutral pH or slightly acid (6.0) soil. Sorghum and cotton are more tolerant of alkaline (high pH) soils and saline conditions than are most field crops. Sugar beets are also more saline-tolerant than other crops, except during the early stages of germination.

In natural ecosystems, most nutrients are cycled within the system. However, for crop production systems to maintain the original fertility level and optimize performance, crops must be fertilized to replace nutrients removed by the grain (Table 16–5) as well as losses that may occur because of erosion, leaching, denitrification, and volatilization. A corn crop that produces 9.4 Mg/ha (150 bu/A) of grain removes 121 kg/ha of nitrogen (N), 49 kg/ha (55 bu/A) of phosphate (P_2O_5), and 36 kg/ha (40 bu/A) of potash (K_2O). Secondary nutrients and micronutrients are also removed and need to be supplied to crops, but in lesser amounts and

TABLE 16–5	Approximate Amounts of Plant Foods Removed in Several Major Crops (kg/ha)*		
Crop	**N**	**P_2O_5**	**K_2O**
Corn at 9.4 Mg/ha (150 bu/a)			
Grain	121 (135)	49 (55)	36 (40)
Stover	210 (235)	22 (25)	143 (160)
Corn silage at 56 Mg/ha (25 T/a)	210 (235)	71 (80)	179 (200)
Oats at 3.6 Mg/ha (100 bu/a)			
Grain	58 (65)	22 (25)	18 (20)
Straw	31 (35)	13 (15)	89 (100)
Sorghum—grain at 8.5 Mg ha (7,600 lbs/a)			
Grain	94 (105)	27 (30)	27 (30)
Stover	71 (80)	45 (50)	205 (230)
Soybean at 3.4 Mg/ha (50 bu/a)	170 (190)	40 (36)	63 (70)
Sugar beets—roots at 56 Mg/ha (25 T/a)	89 (100)	45 (50)	223 (250)
Wheat at 5.0 Mg/ha (75 bu/a)			
Grain	85 (95)	73 (48)	27 (30)
Straw	27 (30)	4 (5)	45 (50)

* Pounds per acre in parentheses.

Source: Adapted from Ohio Agronomy Guide, 13th ed. 1995. Bulletin 472. The Ohio State University Extension.

less often. Nitrogen requirements for grain crops such as corn may be reduced when they are grown in a crop rotation following a legume like soybean that fixes nitrogen. Soybean can convert (fix) adequate atmospheric nitrogen to produce yields of seventy to eighty bushels per acre if well nodulated. Seed of soybean and other grain legumes should be treated with commercial inoculants containing *Bradyrhizobia* bacteria when planting occurs in a field where these crops have not been grown previously or not grown within the past three years.

PEST MANAGEMENT

Weed Control

Weeds are the major pest control problem in many crops. Several factors need to be considered when developing weed-control programs, including soil type, weeds present, crop rotation, and budget. No single control program effectively handles the various weed problems that arise under different environmental conditions. Most weed-control programs use one or more of the following three strategies— prevention (restricting introduction or expansion of a weed problem), eradication (complete elimination of weed plants or seeds from an area), and control.

Several different weed-control methods have been developed. These methods involve cultural practices (rotation, row spacing, and population), mechanical weed control (tillage for seedbed preparation and row cultivation), biological weed control (using natural enemies of weeds), and chemical weed control through the use of herbicides. Most grain producers regard herbicides as the most effective, economical means of controlling weed populations. Several methods of herbicide application are used in producing grain crops, including pre-plant (applying before the crop is planted), preemergence (after the crop is planted but before crop emergence), and post-emergence (after crop emergence). The commercialization of several herbicide resistance crops in recent years (including Roundup Ready® soybeans, corn, and cotton) has increased the popularity of post-emergence herbicide programs.

Insect Control

Despite the number of different pests that feed on crops (Table 16–6), most crops generally exhibit minimal insect problems because the normal activity of most insect populations is below a threshold level; that is, it is too low to inflict significant injury to warrant attention. Economic levels of pest injury do occur, however, and economic losses resulting from sudden outbreaks of pest activity can be prevented if growers are aware of the biology of the various crop pests, monitor fluctuations in pest population activity, and implement timely corrective action. Management of pest populations requires a combination of preventive and timely responsive actions based on the risks associated with various cultural practices and pest activity observations collected through periodic field inspections. Management practices for controlling insect pests include the use of cultural practices (e.g., timing of planting, tillage, crop rotation); seed, soil, and foliar applied insecticides; biological control (reducing pest populations through use of natural predators); and insect resistant crops (varietal/hybrid tolerance to insect pests).

Disease Control

A host of plant diseases limit the yield potential of field crops (Table 16–6). Although some diseases can be controlled by a single practice, such as planting a resistant hybrid, most diseases require a combination of practices to ensure that economic damage is kept to a minimum. Once a disease has been identified, its prevention and/or control depends on understanding its cause(s), the factors that favor the disease, which plant parts are affected, and when the disease organisms are spread. The following is a summary of management practices to prevent yield losses from diseases:

1. Plant high-quality seed, treated with a seed protectant fungicide, in a well-prepared seedbed.
2. Plant cultivars or hybrids with disease resistance or tolerance. Several potentially destructive diseases now cause only minor losses because of the widespread use of cultivars and hybrids with resistance and/or tolerance.
3. Balanced fertility is the key to vigorous, well-developed plants. Use recommended levels of nitrogen, phosphorous, and potassium based on soil tests.
4. Crop rotation and plowdown of infested residues reduce the numbers of disease organisms surviving in the field. To conserve energy and to protect soil from loss through erosion, however, more and more farmers are implementing reduced tillage practices. As a result, these disease-control practices are lost. Other disease practices, such as use of resistant cultivars and hybrids and longer crop rotation, become essential to compensate for this loss.
5. Improve soil drainage in poorly drained soils. This practice reduces water stress and reduces losses from stem, stalk, and root rots.
6. Control insects and weeds in and around fields. Insects such as root worm and stalk borer create wounds that serve as entry points for disease-causing fungi. Some weeds act as reservoirs for overwintering of crop pathogens.
7. Dry shelled grain to a moisture content that restricts fungal growth and mycotoxin production. For corn, this moisture content is 13 to 14 percent. Maintain cool and dry storage conditions to prevent the development of storage molds.

Crop	Diseases	Insect Pests
Agave	Bacterial leaf rot, stem canker	Myriad bugs
Barley	Powdery mildew, stripe mosaic, yellow dwarf, leaf scald, leaf and stem rust, fusarium blight, helminthosporium fungus, smut	Wireworms, cinch bugs, greenbugs, grasshoppers, armyworms, aphids, thrips
Beans	Bacterial brown spot, bacterial blight, halo blight, bean common mosiac virus, anthracnose	Western bean cutworm, grasshoppers, flea beetles, Mexican bean beetle
Castor bean	Fusarium, rhizoctonia, sclerotium, scab, wilt, leaf spot, seedling blight, influorescence rot, pod rot, rust spot, graymold, crown rot, stem canker, leaf blight, bacterial wilt, angular leaf spots	Castor whitefly, red spider mite, pink bollworm, mung moth
Coconut	Lethal yellowing, phytophthora, coconut heart rot, stem bleeding, bud rot, basal stem rot	Coconut caterpillar, coconut leaf hispid, black headed caterpillar, red palm weevil, rhinoceros weevil, slug caterpillar, termites
Corn	Gray leaf spot, northern leaf blight, ear and kernel rots, southern leaf blight, fusarium stalk and ear rot, charcoal rot, maize dwarf mosaic virus	Corn rotworm, European corn borer, cutworm, cinch bugs, slugs, grasshoppers, aphids
Cotton	Damping-off fungi, fusarium, verticillium, bacterial blight, leaf crumple	Boll weevil, pink bollworm, cotton bollworm (corn earworm), thrips, cotton leafworm, fleahopper, tobacco budworm, banded wing whitefly, lygus bug, cotton aphid, spider mite
Flax	Rust, fusarium, pasmo, stem break and browning, seedling blight and root rot, aster yellows, crinkle	Flax bollworm, grasshoppers, cutworms, army cutworm, Bertha cutworm, beet webworm, aphid, aster leafhopper, tarnished leaf bug
Hemp	Gray mold, sclerotinia, fusarium	European corn borer, grasshoppers, Bertha armyworm, flea beetle, hemp borer
Jute	Stem rot, black band, anthracnose, soft rot, powdery mildew, root knot disease, mosaic, leaf spot	Jute hairy caterpillar, jute semilooper
Kenaf	Anthracnose, damping-off	None
Oats	Rust, smut, yellow dwarf, blast, septoria leaf spot	Grasshoppers, cutworms, aphids, thrips, armyworms, greenbugs
Olive	Verticillium wilt, olive knot	Olive fruit fly
Palm oil	Stem rot	Red-striped weevil
Peanut	Leaf spots, southern blight, collar rot, peg rot, black rot	Cutworms, thrips, leafhoppers, corn earworms, armyworms, lesser cornstalk borers, cadelle beetle, carpet beetle, confused flour beetle, flour and grain mites, rice weevil
Rape	Black spot, collar rot, white rot, leaf spot, downy mildew, powdery mildew, clubroot	Cabbage stem flea beetle, turnip flea beetle, rape beetle, pollen beetle, rape stem beetle, cabbage seed weevil, brassica pod midge, cabbage aphid, field slug, cabbage root fly
Rice	Seedling blight, seed rot, brown leaf spot, blast, stem rot	Rice leaf miners, rice water weevils, midges, rice leaf folders, armyworms, leafhoppers, thrips, water scavenger beetles
Rye	Ergot, now mold, leaf and stem rusts, stalk and head rots, blotch, root rots, anthracnose	Grasshoppers, aphids, cutworms, armyworms
Safflower	Rust, phytophtora root rot, verticillium wilt, fusarium wilt, leaf spot, bud rot	Flower thrips, lygus bugs, bean and peach aphids, wireworms, loopers
Sesame	Charcoal rot, alternaria leaf spot, bacterial leaf spot, bacterial blight, phyllody, root wilting	Whitefly, til gallfly
Sorghum	Loose kernel smut, covered kernel smut, head smut, pythium root rot, bacterial spot, crazy top, downy mildew, fusarium stalk rot, leaf blight, milo disease, rhizoctonia stalk rot	Chinch bugs, corn earworm, leaf aphid, corn borer, armyworm, sorghum webworb, southwestern corn borer, sorghum midge, greenbug
Soybean	Phytophtora root rot, pythium rot, rhizoctonia rot, fusarium rot, anthracnose, soybean mosaic, bacterial blight, downy mildew, stem canker, pod and stem blight	Mexican bean beetle, seed corn maggots, seed corn beetles, wireworms, white grubs, thrips, southern corn rootworms, Japanese beetles, grasshoppers, alfalfa hoppers, blister beetles, cabbage loopers, stink bugs, velvet bean caterpillars, spider mites, soybean aphid

TABLE 16-6 Major Pest Problems in Crops Grown for Food, Fiber, Fuel, and Other Industrial Uses

(CONTINUED)

TABLE 16–6 (continued)

Crop	Diseases	Insect Pests
Sugar beets	Beet yellow and western yellow virus, curly top, cercospora leaf spot, powdery mildew, downy mildew, rust, mosiac, rhizoctonia	Aphids, leafhoppers, root maggots, wireworms, white grubs, flea beetles, cutworms, sugar beet crown borers, beet webworms, armyworms, alfalfa loopers, grasshoppers
Sugarcane	Mosiac, gumming disease or gummosos, red rot, smut, ratoon stunting disease	Corn aphid, cane leafhopper, moth borer
Sunflower	Rust, downy mildew, head mold, sclerotinia rot	Salt marsh caterpillar, stalkbores, sunflower moth
Wheat	Stem rust, leaf rust, stripe rust, bunt, loose smut, stinking smut, wheat scab, root rots, crown rots	Hessian fly, wheat stem sawfly, wheat jointworms, strawworms, chinch bugs, aphids, grasshoppers
Wild rice	Bacterial brown spot, bacterial leaf streak, anthracnose, ergot, fungal brown spot, phytophthora, scab, spot blotch, stem rot, stem smut, eye spot, wheat streak mosiac	Wild rice worm, wild rice midge, rice stalk borer, rice water weevil, rice leafminer, wild rice stem maggot

Source: Adapted from Nolte et al. 1976. Bulletin 605. The Ohio State University Extension.

HARVEST AND STORAGE

Most agronomic grain crops are mechanically harvested with combines, which separate the heads, pods or ears from wheat or soybean stems or corn stalks, and shell grain from the pods or cobs in one operation. In the US corn belt, corn and soybeans are harvested in the fall; wheat, barley, and oats are harvested in the late spring or early summer.

For most crops, there is a preferred grain moisture content at which to harvest. The ideal kernel moisture level to harvest corn for dry grain storage is 25 percent. The yield potential can drop considerably if harvesting is delayed much beyond maturity and if poor stalk quality causes stalk lodging. Stalk and root lodging may slow harvest and contribute to yield losses. The loss of one normal size ear per 30 m (100 ft) of row translates into a loss of more than 67 kg/ha (1 bu/ac). In fact, an average harvest loss of 18 kernels/sq m (2/sq ft) is about 67 kg/ha (1 bu/ac). According to an Ohio State University agricultural engineering study, most harvest losses occur at the gathering unit, with 80 percent of the machine loss caused by corn that never gets into the combine. There is a significant grade and price penalty for grain with a high moisture content. Following harvest, grains need to be dried to a moisture content below 14 percent for long-term storage.

Following the harvesting of small grains, some farmers harvest the straw residue, which has become an important income source, especially near urban areas. Straw, especially oat straw, makes excellent bedding for livestock, and the straw is often harvested for that purpose. Some farmers burn the residue and stubble after harvesting the grain to reduce trash and help control some diseases (e.g., rice). This practice is decreasing, however, because the value of residue in conservation tillage is more widely recognized and air pollution laws banning or limiting such fires are becoming more common.

FIELD CROPS GROWN FOR FOOD, FORAGE, FIBER, AND INDUSTRIAL USES

Food Crops

Worldwide food scarcity in the years ahead is a real possibility unless measures are taken to both control population and increase food production. The need for more efficient crop production and reduction of environmental pollution is clear, and a likely area for success with food crops is increased grain production.

Increased food production must come primarily from crop plants with high caloric output per unit area of land. Generally these crops are high carbohydrate producers, and they include cereal grains, potatoes, cassava, and sugar crops. Cereal grains are a concentrated source of energy, and they are easily processed, stored, and distributed. They directly supply the world's population with about 80 percent of its total food calories. The word *cereal* comes from the name of the goddess Ceres, "giver of grain," who is said to have given wheat to the early civilizations. It soon becomes evident that the true cereals are grasses, members of the family POACEAE. Other families are considered cereals because their seeds are used in ways similar to those of the grasses. The cereals are humans' principal food source because of their adaptability to many climates, soils, and handling methods. They are efficient converters of light energy into food, are hardy, produce many seeds per plant, store well, and are readily processed into many uses, including their conversion into meat through livestock feeding. Table 16–1 identified the world's major field crops and the principle uses of each. Several of these crops have multiple end uses. For example, corn (maize) is used for feed and has many industrial uses, including the production of ethanol. Some of the important field food crops are discussed in alphabetical order.

Barley (*Hordeum vulgare* L.) POACEAE

Barley, a widely adapted small-grain cereal, is used for human food, livestock feed, and malting. Barley and wheat are the most ancient of all cereal grains, both dating back 9,000 years. Barley apparently was grown in Mesopotamia in prehistoric times. It is believed to be native to southwestern Asia, but many wild species grow in Ethiopia and southern Tibet. There are two main species: *H. vulgare,* the common six-rowed barley (six rows of kernels per spike) grown mainly for food; and the two-rowed barley *H. distichon,* whose ancestor is probably the wild barley *H. spontaneum,* used in many parts of the world for malting. The crop is classified according to its growth habit as spring (earlier maturing) or winter barley.

Barley is widely adapted because of its drought resistance, tolerance to alkaline and saline soils, and early maturity. Some cultivars grow in subarctic climates, some only in the temperate zones, and others in the subtropics, although barley is not well adapted to hot, humid conditions. The crop may ripen in as few as sixty to seventy days, but normally ninety to 120 days are required. The principal production areas are the former Soviet Union, United States, Europe (Germany, France, United Kingdom, and Spain), and Canada. In the United States, barley is grown mostly in North Dakota, Idaho, Washington, Minnesota, and Montana.

Beans (*Phaseolus* spp. L., *Vigna* spp. Sari, *Vicia* spp. L.) FABACEAE

The seeds of several FABACEAE species of field beans and related field peas or cowpeas are grown for food and are a very important caloric and protein source. These include members of several genera. Various types of beans of the genus *Phaseolus* include the lima, scarlet runner, snap, shell, white, black, pea, blackeye, kidney, tepary, black gram, and mung. Other important beans are the broad bean *Vicia faba*; chickpea *Cicer arietinum*; the cowpea, also known as the blackeye pea *Vigna unguiculata* or *V. sinensis*; the dried pea *Pisum sativum*; lentil *Lens esculenta*; and pigeon pea *Cajanus indicus*. Many other legumes having local importance are grown for human and animal food. Pulse crops are legumes whose seeds and/or pods are consumed.

The blackeye pea *Vigna unguiculata* is one of the oldest crops. It is native to Central Africa, and it spread to Asia and the Mediterranean basin; it is now grown in many areas. The red kidney bean *Phaseolus vulgaris* originated in Central and South America, and was unknown in Europe before Columbus. The broad bean *Vicia faba,* with its origin probably located in North Africa or the Near East, is an old world species grown by the ancient Egyptians and Greeks. The pea *Pisum sativum* was probably domesticated in central and western Asia.

Beans are warm-season crops and are best grown for dry seeds in areas where temperatures are warm but not excessively high, and where the growing season is long enough to avoid frosts. Such areas are found in many temperate regions of North and South America, Europe, Asia, Africa, and Australia. Major production areas are China and India, where dried peas are important. India is also a large producer of lentils. Latin America, the United States, Japan, Italy, and Turkey are also major bean producers. Italy is a leader in the production of broad beans. The major production areas of the United States are Michigan, Nebraska, California, Colorado, Idaho, and North Dakota.

Corn (*Zea mays* L.) POACEAE

The botanical origin of corn *Zea mays* family POACEAE, better known in the world as maize, is vague and is mostly based on evidence from specimens discovered in caves in the Tehuacan Valley of Mexico. The earliest samples appear to date from 5000 B.C., although maize pollen fossils have been found that are about 80,000 years old. The modern history of corn begins with the first voyage of Columbus, who discovered not only the Americas but also corn. Later explorers found Native Americans growing corn in all parts of the Americas from Canada to Chile. The first Indians to plant and cultivate the crop lived in Mexico and South America.

Corn is a tall annual plant with strong erect stalks, a fibrous root system, long narrow leaves spaced alternately on opposite sides of the stem, and separate male and female flowers on the same plant (monoecious). Now grown worldwide in many environments, corn originated in the Americas as a short-day, long-season crop. The increase in diversity of production area environments is the result of the breeding and selecting of cultivars that mature rapidly and are nearly day neutral. Several botanical varieties of *Zea mays* having broad usage and economic importance are grown. The various types of corn can be differentiated on the basis of endosperm composition—usually the relative proportions of soft versus hard endosperm or starch.

- Dent corn (*Z. mays* var. *indentata*) is the principal commercial feed type grown in the United States. The grain is normally yellow, hard, and horny. The kernel is comprised of hard starch with an overlay of softer starch in the crown. The soft endosperm in the crown shrinks on drying more than the hard innermost endosperm, resulting in a dent forming in the top of the kernel. White kernel corn use is increasing in the United States because of increasing Hispanic populations and the popularity of their foods, which use this kind of corn.
- Flint corn (*Z. mays* var. *indurata*) kernels are not indented when dry and are comprised mostly of hard starch.
- Flour corn (*Z. mays* var. *amylacea*) is also known as soft or squaw corn. It has little hard starch. It was preferred because it is easily hand-ground into flour.

◆ Sweet corn (*Z. mays* var. *saccharata*) kernels have sugary kernels when immature; when dried the seeds are wrinkled and somewhat translucent. This type is a popular and important vegetable crop in some nations.

◆ Popcorn (*Z. mays* var. *praecox*) is a type of flint corn with a very high proportion of hard starch endosperm. The kernels and ears are smaller than those of dent corn.

◆ Pod corn (*Z. mays* var. *tunicata*) kernels are enclosed in husks or pods, and the ears are covered with husks. This curiosity is not grown commercially.

Corn has a multitude of food and industry uses. Corn finds use in the production of flour, starch, sugar and syrups, corn oil, alcohol, adhesives, and coatings for paper and fabrics, and so on. Considerable attention is given to improvement of its nutritional value because of its extensive use as food for humans as well as animals. Much of this is aimed at improving its protein value by increasing the content of the amino acids, lysine, and tryptophan, which are normally low in corn grain.

Corn oil content is typically 4 percent. However, high-oil corn contains up to 7 to 8 percent oil. High-oil corn is attractive as livestock feed because it has greater energy than conventional dent corn and can replace more expensive dietary sources of fat. High-oil corn hybrids have not been widely used because their grain yield potential is usually lower than conventional corn hybrids.

Field corn is primarily grown as a grain crop for livestock feed. Some is cut green and made into silage. Sweet corn is grown mostly for human consumption. Corn grain is also processed into a polyunsaturated oil used for cooking and margarine. For a discussion of the culture of sweet corn, see Chapter 18.

Optimal production of corn requires an ample and continuous supply of available soil moisture. The annual precipitation in the US corn belt varies from 60 to 115 cm (24 to 45 in.), with about one-fourth occurring during the summer months; thus, irrigation is usually not provided. Corn is a warm-season crop, requiring relatively high day and night temperatures. The best daytime high temperatures range from 20°C to 27°C (68°F to 81°F), and with night-time lows not less than 14°C (57°F). Seeds do not germinate well at temperatures lower than 10°C (50°F), and temperatures above 40°C (104°F) harm pollination.

In the United States, corn is well adapted to the midwestern states where the term *corn belt* appropriately includes Iowa, Illinois, Nebraska, Minnesota, Indiana, Ohio, Missouri, and the Dakotas. Considerable quantities are also produced in other parts of the United States. In the corn belt areas, a large proportion of the crop is fed to animals, notably hogs and cattle, for meat production. Most corn is used as livestock feed, but an increasing percentage is used for industrial purposes, including ethanol (as a gasoline substitute) and high fructose corn syrup (sugar for soft drinks). On-farm use accounts for about 35 percent of that produced. About 10 percent of the corn grown is converted into silage and/or feed as fresh fodder to animals.

A corn-drying method on the decrease, because it requires extra handling, is the placement of ear corn in covered cribs with slotted sides that permit ventilation. The kernels are shelled at a later time. A moisture level of 20 to 22 percent will permit safe storage in cribs. However, the bulk of the production is stored in enclosed bins or buildings that require the grain be dried to a moisture content below 14 percent for long-term storage. There is a grade and price penalty for corn with higher moisture. High-moisture corn is subject to heat damage and spoilage.

Oats (*Avena sativa* L.) POACEAE

The origin of oats is unclear, but they are thought to have originated in Asia-Minor or southeastern Europe. Oats grew wild in western Europe during the Bronze and Iron Ages. Following domestication, its culture spread to other temperate-zone regions. Oats are principally grown for animal feed because the crop is easily produced, and the grain and foliage are nutritious. Substantial amounts are also processed for human foods. World production has declined because tractors have replaced animal power.

Oats grow to 60 to 120 cm (2 to 4 ft) tall, and have a fibrous root system. The flowers are borne on panicles, either nearly symmetrical or one-sided. The panicle is made up of numerous (20 to 120) small branches and spikelets composed of two glumes and usually two florets (except hull-less cultivars, which contain more).

Oats, like barley, are well adapted and are produced throughout the temperate zones. Important oat-producing countries are the former Soviet Union, the United States, Germany, Canada, and Poland. Major US producers are South Dakota, Minnesota, North Dakota, Iowa, and Wisconsin. Northwestern Europe, some northern areas of the United States, and southern Canada produce some late-maturing common white oats.

Rice (*Oryza sativa* L.) POACEAE

Knowledge of the ancient origin of rice is unclear, but most likely it was southeastern Asia. Some very early Chinese and Indian writings mention rice cultivation for more than 5,000 years in that part of the world. The relative importance of rice, wheat, and corn as human foods cannot be debated. Wheat, rice, and corn are principal food staples for most of the world's population. Each of these major crops prevails in certain areas. Rice is most important in the tropics and semi-tropic zones and has significant production in some temperate regions. Wheat is generally found in the fertile prairies of the Americas, Europe, Asia, and Australia, while most corn is produced in temperate areas that have warm and moisture-rich growing seasons.

Figure 16–2 Young lowland rice plants growing through the water held by earth levees in a flooded rice paddy. *Source:* U.S. Soil Conservation Service.

The cultivated plant, customarily grown as an annual, is a semiaquatic annual grass that grows erect. It has narrow parallel-veined leaves. Spikelets are borne on a loose panicle and each contains one flower enclosed by the lemma and palea. The flower has six stamens and one ovary.

Rice is classified several ways. Based on the chemical characteristics of the starch and grain aroma, rice can be classified into three groups: (1) waxy or glutinous types (starchy endosperm contains no amylose[1]); (2) common types, more or less translucent nonglutinous (endosperm contains one-fourth amylose and three-fourths amylopectin[2]); and (3) aromatic or scented types, grown in India and southeast Asia. The glutinous types, consisting entirely of amylopectin, are grown primarily for a specialty market in the United States but represent about 10 percent of the total production in China and 8.4 percent in Japan. Worldwide, the first two types comprise over 90 percent of the total rice grown.

Rice cultivars are classified as lowland rice (continuously or pond flooded) or upland rice (nonirrigated, or irrigated but not continuously pond flooded) (Fig. 16–2). Upland and lowland do not refer to elevation. More than 80 percent of the world's rice grown is the lowland type. Rice cultivars are also classified on the basis of kernel characteristics into short-, medium-, or long-grain types. The average length of unhulled kernels is 7.2, 8.4, and 9.9 mm respectively. Most northern Asian people who eat rice prefer the short-grain types (also known as pearl rice) because the kernels are sticky when cooked. Most Americans and Europeans prefer the nonsticking, long-grain types.

Another rice classification is based on maturity. The rate of maturity is genetically controlled and is measured by the number of days required for the plants to reach 50 percent heading.

Rice is also classified on the basis of its cultured adaptation as japonica or indica types. The japonica cultivars are usually short-grain and adapted to a temperate climate, while

Figure 16–3 Working the flooded field is the usual method of preparing land for rice in the humid tropics of southeast Asia. Most Asian farmers using simple equipment drawn by water buffaloes plow and harrow the flooded field until the soil is puddled. This helps control weeds, creates a plow pan that reduces water percolation losses, and makes hand transplanting easier. *Source:* R. L. Haaland.

the indica types are long-grain and tropical. Intercrossing of japonica and indica types has led to a breakdown of this traditional classification, and some long-grain varieties are well adapted to temperate areas.

Successful rice production depends on ample water availability, and thus important production areas lie along the great rivers and deltas of Asia and other regions. Asia produces over 90 percent of the world's supply. Major producers are China, India, Indonesia, Bangladesh, Thailand, Japan, Burma, and Brazil. Five states in the United States—Arkansas, California, Louisiana, Texas, and Mississippi—produce 95 percent of the US crop, this being only a little more than 1 percent of the world's total.

Rice grows well on moist soils. For best yields, the fields are flooded for most of the growing season; thus clay, clay loam, or silty clay loam soils are most frequently used to conserve water by minimizing seepage losses. Organic and light-textured soils are used provided they have an underlying hardpan or claypan that prevents water seepage losses. The soil pH ranges between 5.0 and 7.5 for satisfactory plant nutrient availability. Upon flooding, acid or alkaline soils shift slightly toward neutrality. Rice cultivars vary in their tolerance to salinity, but all are adversely affected.

In most Asian countries, the centuries of simple, nonmechanical methods of flooding, terracing, leveling, and other wetland rice cultural operations have produced soils with common characteristics. The immediate soil surface is oxidative and the subsoil strongly reducing (oxygen deficient). These soil conditions create what is called a rice paddy soil.

In the humid tropics, the soil is usually worked wet or flooded to prepare land for rice production. The wet fields are plowed with implements drawn by water buffalo (Fig. 16–3).

---

[1] Amylose is a component of starch characterized by the tendency of its aqueous solution not to gel.

[2] Amylopectin is a component of starch characterized by the tendency of its aqueous solution to set to a stiff gel at room temperature.

Figure 16–4 Most of the world's rice, except in the United States, is transplanted by hand into flooded puddled fields as shown here. The seeds are started in seedling beds; after thirty to fifty days, the young plants are transplanted. *Source:* R. L. Haaland.

Figure 16–5 Small hand-driven tractors have contributed largely to the reduction of hand labor required to produce rice in southeast Asia. Compare the 2,200 labor-hours without tractors necessary to produce one hectare of rice (900 hr/ac), with 1,790 hr/ha (725 hr/ac) with small hand-driven tractors, to the mechanized methods used in the United States, where 1,000 liters of petroleum fuels plus 12 work-hours can produce one hectare (400 gal + 4.75 hr/ac) of rice. *Source:* R. L. Haaland.

The fields are harrowed crosswise and lengthwise until the soil is well puddled (a soft muddy mass). This helps create a hardpan that limits water percolation and facilitates the hand transplanting operation. Hand transplanting is done in tropical areas and much of Asia to establish the crop before seasonal rains (monsoons) occur, or to meet short growing season conditions, and/or to permit double cropping of the soil (Fig. 16–4).

In most developed countries, seedbed preparation is completely mechanized. Direct seeding is practiced either by conventional drilling on dry soil or by seeding, often using aircraft, into flooded fields. The former is practiced predominantly in California. Using this method, seedbeds are finished with large equipment that leaves the surface grooved, or the surface is left rough to prevent seed drift during flooding.

More and more rice farmers worldwide are recognizing and using the improved, high-yielding rice cultivars developed by the International Rice Research Institute (IRRI) in the Philippines or cultivars developed for local adaptation from IRRI germplasm. The choice of seed is an important consideration, and many rice-growing areas have seed certification programs designed to provide high-quality seed. Such seed is varietally pure, produces at least 80 percent germination, and is free of weed seeds and other impurities.

Rice culture has always been and remains a labor-intensive endeavor in many Asian and African countries. Production requires more than 740 labor-hours per hectare (300 hr/ac) in most of Asia and Africa (Fig. 16–5). In the United States and Australia, only 18.5 hours/ha (7.5 hr/ac) are needed because of mechanization. This reduction in back-breaking hand labor is possible because of the substitution of machines for human labor. However, it may not be feasible or even desirable for all countries to mechanize rice production to the extent it is in the United States because of their social and economic situations. Many rice-growing countries

are overpopulated, the people need jobs, and the farms are too small for efficient use of large machines (Fig. 16–5).

Wild Rice (*Zizania palustris* var. *interior* L.) POACEAE

Wild rice, an aquatic member of the grass family, is not closely related to common rice (*Oryza sativa*). Wild rice has been used in the Great Lakes region of the United States and Canada as a food by native Indians for more than 300 years. The seeds sprout under water from grain that fell into the water the previous year, germinating in the spring, and producing a single root and a thin submerged leaf. During the summer a flower stalk elongates and emerges from the water, bearing female flowers at the top of a spike with many drooping male flowers below them. This arrangement almost guarantees cross-pollination between plants. By late summer, the plants are 90 to 120 cm (3 to 5 ft) tall and very leafy. The top grains ripen first, and those below ripen subsequently over a period of about ten days more. In Minnesota, much of the production occurs on wet and marshy public lands, where mechanical harvesting of wild rice is regulated by law. Therefore, much of the harvesting is done from boats by bending the stalks over and into the boat and beating the kernels off with a flail. This process, and the plant's seed-shattering tendencies, results in considerable amounts of seed, more than half, being lost into the water. The portion lost provides seed for the next year's crop. The yield from this kind of culture and harvesting procedure is much smaller than that obtainable by management similar to that for common rice.

In addition to the marshland type of wild rice production (which is rapidly declining), wild rice is also commercially cultivated in California and Minnesota, with cultural techniques similar to those for common rice. New shatter-resistant and semi-dwarf cultivars have been developed.

Rye (*Secale cereale* L.) POACEAE

Southwestern Asia is assumed to be the area of origin for rye. There is evidence of its early cultivation in western Asia, but domestication was relatively recent because rye is not mentioned in early Egyptian records. Until the early 1900s, rye was principally a European crop grown for making bread. Because of its extreme hardiness it was, and is, widely grown in much of Europe, Asia, and North America. Rye does best in a cool dryish climate and tolerates cold temperatures very well. It also performs well on poor and marginal soils. Therefore, it is often grown where environmental conditions are unfavorable for other cereal crops. Because other cereals are preferred, rye is not an important crop except in certain countries.

Rye is used for human consumption, but most is grown for hay, pasture, and stock feed. It also finds wide use as a cover crop and for the manufacture of beverage alcohol.

There are both spring and winter cultivars, but because rye is very winter hardy, most of the rye grown is fall-seeded. Major world production areas are the former Soviet Union, Poland, Germany, the United States, and Canada.

Ergot is a fungal disease often associated with rye. It produces small black bodies called *sclerotia* in the grain. The sclerotia, if present in quantity, are poisonous to animals and humans but can be used in pharmaceutical products.

Sorghum (*Sorghum bicolor* Moench) [*S. vulgare* Pers] POACEAE

Sorghum is believed to have originated in Africa. Grain sorghums are known by several names, such as durra, Egyptian corn, great millet, or Indian millet. In India, sorghum is known as jowar, cholum, or jonna. In the United States, different types of grain sorghum are known as milo, kafir, hegari, feterita, shallu, and kaoliang.

In both Africa and Asia, sorghum is one of the major cereal grains. This important crop was introduced to the United States in the middle 1800s. Major world producers are the United States, India, China, Argentina, Mexico, and Nigeria. Resistant to heat and drought, sorghums having effective and extensive root systems are best adapted to warm and semiarid regions. The principal US production areas are Kansas, Texas, Nebraska, Missouri, and Arkansas, but significant acreages are grown in other states.

Agronomically, sorghum is a common catch-all name applied to all plants of the genus *Sorghum,* which includes numerous grain and foliage types. However, the plants do have markedly different characteristics and uses, and have often been grouped into four types:

1. The grain sorghums (Caffrorum Group) are the non-saccharine plants, including milo, kafir, feterita, hegari,

Figure 16–6 Because sorghums are grown for grain, silage, pasture, syrup, and straw for brooms, the sorghum breeder has a variety of objectives. These short-stemmed, dwarf grain cultivars have been bred for adaptation to mechanical harvesting, early maturity, and resistance to lodging and shattering. These mature heads of grain are ready for harvest.

and hybrid derivatives, among others. These plants are grown for grain used principally for poultry and livestock feed (Fig. 16–6). The grain is similar in composition to corn, but somewhat higher in protein and lower in fat. Grain is ground into meal and made into bread or porridge. Whole grains are sometimes popped or puffed for cereal. Grain sorghums are also used to make dextrose, starch, paste, and alcoholic beverages. The stalks have a dry pith and are not very juicy, except for milo and kafir, which are semijuicy, dual-purpose types for grain and forage.

2. Sweet or forage sorghums or sorgos (Saccharatum Group) are used mainly for forage and silage and to make molasses. The stalks are juicy and sweet. This type is principally grown in the United States and South Africa.

3. Broom corn (Technicum Group) is a sorghum and a panicle woody plant. It has dry pith, little foliage, and fibrous seed branches 30 to 90 cm (1 to 3 ft) long that are often made into brooms.

4. Grass sorghum (*S. Sudanense*) or Sudangrass is grown for pasture, green chop, silage, or hay. Sudangrass is usually ready to pasture in five to six weeks, but to avoid prussic (hydrocyanic) acid poisoning, it should be at least 45 to 60 cm (18 to 24 in.) tall when grazed. The acid in the form of glucosides occurs in young plants and in new shoots. It is highly toxic because it inhibits cellular oxidative processes.

Plant breeders have developed numerous new grain sorghum hybrids and cultivars with specific adaptation to local regions or conditions (Fig. 16–7). Many cultivars are designated by their rates of maturity. The development of hybrid

Figure 16–7 A young field of sorghum, here planted on beds to facilitate furrow irrigation. *Source:* U.S. Soil Conservation Service.

Figure 16–8 The sugar from sugar beets and sugarcane is chemically the same. This field is almost ready for harvest. Many farmers use the tops, which are almost as nutritious as alfalfa hay, for cattle feed. *Source:* F. J. Hills.

seed and short stalk cultivars has greatly improved yields and harvesting efficiencies. The most important production factor in grain sorghum is selection of the correct cultivar for a given area.

Sugar Beets (*Beta vulgaris* L., J. Helm) Chenopodiaceae

Beets are believed to have their center of origin in the Mediterranean area. Cultivated beets were probably domesticated from the wild beet (*Beta maritima*). Within the one species *B. vulgaris,* there are several useful types. These include the coarse mangel-wurzel (also known as fodder beet—grown for animal feeding), the sugar beet, the table beet, chard, and other foliage-type beets. The last three types are commonly used vegetables (Fig. 16–8).

The sugar beet is a biennial plant grown as an annual plant with simple leaves arranged in a basal rosette alternatively on the stem. Leaves are ovate to oblong-ovate. The flowers are borne in clusters in the axils of the leaves. As the ovaries mature, the perianths fuse, resulting in a seed ball containing several ovaries or seeds.

Sugar beets are by far the most important beet grown commercially. Their average sugar concentration is about 15 percent, with many crops exceeding 20 percent. The roots are typically sharply tapered, white-skinned, and white-fleshed.

Major growing areas are the former Soviet Union, France, Germany, the United States, Poland, and Turkey. Leading US producers are California, Minnesota, Idaho, North Dakota, and Michigan.

Hybrid cultivars have been developed that are resistant to bolting (development and growth of a seed stalk) and to certain viral diseases. They were also developed for improved adaptation to specific areas and conditions. Most of these new hybrids are monogerm instead of multigerm; that is, the seed ball contains one seed instead of many. When multigerm seed was planted, one seed ball containing several seeds would produce several seedlings, making thinning necessary as well as difficult. The introduction of monogerm seed helped reduce thinning efforts and also allowed the use of more effective precision planting equipment.

Beets are best harvested when the sugar content is highest, but often the optimum harvest date does not coincide with when the processing mill can or will accept the crop. Because sugar beets are usually grown on contract with a sugar-processing plant, the crop, instead of having the harvest delayed, may have to be harvested before optimum maturity to meet contract terms. The price, acreage, cultivar, harvest date, and other management arrangements are included in the contract.

Sugarcane (*Saccharum officinarum* L.) Poaceae

Sugarcane supplies much of the world's sugar. In many areas where production and labor costs are low, sugarcane provides sugar at a lower cost than the sugar provided from sugar beets. Sugarcane is said to be one of the most efficient converters of solar energy and carbon dioxide into chemical energy, possibly producing more calories per unit of land area than any other crop.

Sugarcane probably originated in India, where it has been grown since ancient times. Its earliest mention is found in Indian writings from about 1400 to 1000 B.C. From India, sugarcane spread to China, Java, and other tropical Pacific islands. It also spread westward from India to Iran and Egypt. Columbus introduced the crop to the West Indies. Sugarcane was first planted in the United States in 1751 near New Orleans.

Sugarcane is a tropical plant that matures in twelve to eighteen months. It is a tall perennial grass, which often attains a height of 2 to 4 m (Fig. 16–9). Sugarcane has essentially the same structure as other members of the grass family.

Figure 16–9 A field of mature sugarcane growing in the delta area of Louisiana. The canes grow to a height of 2 to 4 m (6.5 to 13 ft).

The former low-yielding, low-sugar cultivars have been replaced by highly productive, high-sugar-containing cultivars. Biotechnological advances utilizing tissue culture cloning, and employing sources of genetic diversity such as that provided by somaclonal variation, are used by modern breeders to further improve crop performance.

Sugarcane is grown in tropical regions around the world. The main production areas are the warm, humid, tropical lowland regions of North, Central, and South America, the southern United States, the Caribbean, Africa, Asia, and Oceania, including Australia. The leading countries producing sugarcane are India, Brazil, Cuba, China, Mexico, Thailand, Australia, the United States, and South Africa.

Sugarcane is propagated vegetatively by planting sections of the stems containing three or four nodes into furrows. The stem sections are sometimes planted by hand but are usually planted mechanically. Plantings are made when soil temperature and moisture are suitable. In many areas, a perennial-like cropping is practiced, with the new crop resprouting from previously cut stems left in the soil. However, new plantings of selected and treated stem cuttings are more productive and reduce disease problems.

Before harvest, randomly selected canes are tested for maturity and the juice is assayed for sugar content with a hand refractometer. Testing generally starts four to six weeks before the proposed harvest date. As with sugar beets, the mill needs a steady supply of canes, which farmers provide by coordinating their harvest. Thus, the rate of harvest is governed by the crushing capacity of the mill. Sugarcane is harvested either by hand or mechanically. In the West Indies, the canes are generally sent to the mills remarkably free from undesirable leaves and stems because they are removed when canes are harvested by hand. The cane is cut free at the bottom, topped, and stripped, leaving only the main stalk ready for milling. In other areas, especially where labor costs have hastened the adoption of mechanization, delivery of trashy canes to the mill has been a serious problem, although the trash and waste (**bagasse**) is used as a fuel by the mill. To reduce trashiness and facilitate mechanical harvesting, firing the cane before cutting is a normal practice. Burning mature cane does not harm it unless the fire is exceptionally hot. The work of cutting and loading is

greatly reduced by preharvest burning. The top of the plant and the attached young leaves then need to be removed because they contain invert sugars,[3] nitrogen compounds, and starch—all of which interfere with the extraction of sucrose sugar.

Wheat (*Triticum aestivum, Triticum* spp. L.) POACEAE

Wheat and barley are recognized as being among the most ancient crops. The Egyptians and Mesopotamians grew wheat as well as barley and oats. Two wild species are still found growing in Syria and Turkey, where wheat probably originated and was domesticated. It is known that present-day species originated from the hybridization of several different species.

Wheat leads all of the cereal grains in total volume. Countries that lead in production are China, the former Soviet Union, the United States, India, France, Canada, and Australia. Wheat is grown in almost every state in the United States, with the leading states being Kansas, North Dakota, Oklahoma, Washington, Texas, South Dakota, Minnesota, and Colorado.

Wheat is classified into market classes by the color and composition of the grain and the plant's growing habits; the latter also determines the production area of each class. The classes are (1) hard red spring, (2) durum, (3) hard red winter, (4) soft red winter, and (5) white.

Wheat has a relatively broad adaptation, is very well adapted to harsh climates, and will grow well where rice and corn cannot. Early growth is favored by cool and moist conditions, with warmer and drier weather toward crop maturity.

Generally the winter climate of a particular area determines whether winter or spring types are grown. If winters are severe, spring-type cultivars are planted in the spring to be harvested in the late summer and the fall. If winters are not extreme, winter cultivars are planted in the fall for spring harvest. If winter temperatures are mild, spring-type cultivars are planted in the fall for spring harvest. Another generalization is the different composition of the grain of the different cultivars. The hard wheats contain more protein (13 to 16 percent) and are usually grown in the drier climates. The soft wheats are more starchy, have a lower protein content (8 to 11 percent), and are usually grown in more humid climates. Each of these wheats has different characteristics and different uses.

In the United States, the hard red spring cultivars are grown in the northern Great Plains, the Dakotas, Montana, and Minnesota, and produce a high-grade wheat used principally for bread flour. Durum cultivars, used to make semolina flour for macaroni and other pasta products, are also grown in these areas and some other areas. Hard red winter wheat, which leads in the total volume of production, is used for bread flour and is grown across a broad area ranging from Utah to Illinois. Soft red wheat production is

---

[3]Glucose and fructose are undesirable invert sugars produced upon the decomposition of sucrose.

concentrated in Ohio, Indiana, and southern Illinois. This wheat is milled into flour for cakes and pastries. Grain fed to livestock is most often that of the soft wheat types.

Wheat grain is processed by milling, a procedure that removes the outer bran layer. In doing so, much of the grain's protein is lost because the bran layer contains the highest concentration of protein. This is the concession made to produce white flour with better baking characteristics. Whole wheat refers to wheat that is partially milled, and it has become a much more popular product because of its greater nutritional contribution.

Because of its importance as a basic food staple to so much of the world's population, extensive breeding programs have been undertaken in practically every wheat-growing country in the world. Breeders continue to introduce new cultivars that are more resistant to diseases, insects, drought, lodging, and shattering. Plant breeders have improved quality (size, texture, weight) and increased yields and winter hardiness.

Wheat is grown on a wide range of soils in temperate climates where annual rainfall ranges between 30 and 90 cm (12 to 36 in.). Such areas constitute most of the grasslands of the world's temperate regions. Many of these soils are deep, well-drained, dark-colored, fertile, and high in organic matter, and they represent some of the world's best soils. The prairie soils of the United States and Canada and the steppes of the former Soviet Union are examples of such soils.

Oil Crops

In addition to the use of animal oils, early humans also used various oils from crop plants as a food source and for a primitive fuel. Some of the earliest written records indicated the use of plant oils for illumination, heat for cooking and warmth, and anointing the skin. Olive oil was widely used for these purposes by early Mediterranean civilizations. Linseed oil from flax was recognized for its usefulness in paints. Castor bean oil was used for lubrication and also in paint and varnish products, cosmetics, and as a cathartic. With technological development, the use of various oil crops expanded. These oils function in many applications, many that are very specialized and often not very obvious. After the discovery of petroleum and the technology for its use in the internal combustion engine, the primary source of oil for energy purposes became the earth's underground supply. Concern about the future supply of petroleum compels a strong interest in the oil crop plants as a supplemental source. These crops represent renewable sources, whose importance will increase even more in the future. The better-known uses of plant oils are for cooking, flavoring, margarine, and salad dressings. A multitude of other uses include the manufacture of plastics, paint, varnishes, lacquers, soaps, detergents, inks, cosmetics, lubricants, medicines, fabric, and paper.

Oil is found in all living plants, even in bacteria and fungi. In the plant, a major role of the oil is the retardation of tissue water loss. Certain plant tissues, usually the seed, are the most abundant in oil content. Although some seeds have very little oil, others have an oil content greater than 50 percent. Relatively few plant species provide the majority of the vegetable oil production.

The oil is usually removed by crushing and pressing; other extraction methods, often combined with pressing, include extraction with steam and various solvents.

Vegetable oils, readily seen as small droplets when expressed from tissues, are mostly a mixture of triglycerides with small amounts of mono- and diglycerides that are characterized by the content of their various fatty acids. They become degraded (rancid), some more rapidly than others, because of the breakdown of glycerol into other compounds.

A characteristic of oil is its degree of saturation (amount of single bonding of hydrogen) of the fatty acid molecules. Unsaturated fatty acids have one or more double bonds in their structure and therefore bond less hydrogen. These generally are liquid, or have a melting point about 20°C. The melting point depends on the degree of saturation and also on the molecule's carbon chain length. An increase in saturation and in the number of carbon atoms increases the melting point.

The addition of hydrogen increases the degree of saturation. The level is influenced by the initial saturation and by the amount of hydrogenation. Hydrogenation is a chemical process that can be selected to increase the degree of saturation. This is usually done to solidify an oil to a fat. Natural fats differ from oils in having fatty acidic constituents that are more or less solid at about 20°C. Plant waxes are fatty acid esters of monohydroxyl alcohols rather than trihydroxyl glycerols. Hydrogenation is performed by introducing hydrogen under pressure into heated oil with the use of a catalyst such as finely powdered nickel or other metallic compounds.

Although their end product uses are not always specific, each oil finds uses as food products and/or for industrial purposes. The various crop plant oils are often roughly grouped by their degree of saturation and also by their ability to absorb oxygen (drying characteristics).

♦ Oils of highly saturated fatty acids (nondrying) are largely composed of glycerides of mostly saturated fatty acids such as palmitic and oleic that are found in palm, coconut, peanut, olive, and castor oil.
♦ Oils of highly unsaturated fatty acids (drying) are largely composed of higher amounts of glycerides of unsaturated types such as linoleic and linolenic, as is found in linseed, safflower, soybean, and tung oil.

Castor Bean (*Ricinus communis* L.) EUPHORBIACEAE (*4, 25, 31, 34, 36, 37*)

The castor bean is grown for the oil in its seed. The petalless flowers are produced on racemes, from which brown spine-covered capsular fruits normally containing three large seeds develop (Fig. 16–10). Seed oil content is about 50 percent.

Figure 16–10 The development of castor bean flowers to seed pods. *Source:* University of California Cooperative Extension.

Although commercial hybrid seed is available, seed shatter and highly indeterminate maturity limit harvest mechanization and expansion of production. The seeds contain the alkaloid ricin, which is poisonous and makes the expressed cake unsuitable for animal feeding. The oil has many commercial uses in paints, resins, plastics, inks, cosmetics, greases, and hydraulic fluids.

Coconut (*Cocos nucifera* L.) ARECACEAE (*6, 10, 26, 31, 33, 28*)

The coconut, in addition to being an important nut crop, is an important oil crop.

Olive (*Olea europaea* L.) OLEACEAE

Olives have produced fine oil for centuries, but they are also grown for their fruits.

Rape (*Brassica* spp.) BRASSICACEAE

Rapeseed (also called canola) contains better than 40 percent oil, and the pressed meal cake makes an excellent feedstuff, containing about 40 percent protein. This crop is an important and significant oil crop in many areas because of its low temperature and broad soil adaptation. Breeding research has greatly improved the quality of rape crop oil by reducing its glucosinolate and erucic acid content.

Sesame (*Sesamum indicum*) PEDALIACEAE

The seed, containing about 50 percent or more oil of excellent quality and stability, is most often used as a salad or cooking oil, and for flavoring. The seed is also used as a garnish for bakery products. The pressed cake is an excellent protein source for livestock feeding.

Palm Oil (*Elaeis guineensis Jacq.*) ARECACEAE

This tree was called the prince of the plant kingdom by Linnaeus because of its majestic appearance. The oil palm is botanically related to the coconut palm. When mature, the oil palm tree may attain a height of nearly 30 m (100 ft) but generally it grows no taller than 10 m (32 ft). The trees are monoecious (male and female flowers on the same tree). The flowers grow on a short spadix that develops into a cluster of more than 1,000 drupes (fruits). A **spadix** is a spike with a succulent axis usually enclosed in a spathe. A spathe is the large sheathing bract (or pair of bracts) enclosing an inflorescence, especially a spadix on the same axis, for example, the Jack-in-the-Pulpit flower. The female flower normally has three ovaries, but generally only one is fertilized. The fruit matures about six months after pollination.

The oil palm, native to tropical West Africa, is an important source of vegetable oil. It is grown most abundantly along the west coast of Africa. Oil palms are also cultivated to some extent in the rain forest regions of the Congo, Kenya, Indonesia, and Malaysia. There are plantings in Central and South American countries.

Breeding objectives are to obtain shorter trees and better flowering characteristics. Shorter trees aid in harvesting fruit, and improved flowering produces a higher per tree yield of oil.

The oil palm grows well on a wide variety of soil types. The trees do best on deep, well-drained soils that are neutral to slightly alkaline in reaction, but they are grown successfully on acid soils.

Trees generally respond to phosphorus applications and sometimes to potash. In South Africa, a bronze spotting of leaves seems to be caused by potassium deficiency. Magnesium has been found to be deficient in West Africa and causes a condition known as the orange frond disease.

Irrigation is not necessary for oil palms because they are grown only in tropical areas where rainfall is abundant and frequent.

Of the diseases reported, a fungus-caused stem rot is the most prevalent. The fungi enter the trunk of the tree when the leaves are cut off, and infected trees often die in two or three years. This fungus (*Marasmium palmivorus*) causes the acid content of the oil to increase in ripening fruit. A red-striped weevil is a serious insect pest.

In many southeastern Asia areas, the fruit is harvested throughout the year, heaviest in late summer to early fall, lightest in late winter to early spring. Both the shell (pericarp) and the kernel contain oil.

Kernel oil is light yellow in color and is used principally in the manufacture of edible products such as margarine, chocolate candies, and pharmaceuticals. Palm oil from the pericarp tissues is deep yellow to red-brown in color, and thick in consistency. It is used for making soap, candles, and lubricating greases. It is also used in processing tin plates and as a coating for iron plates.

Peanut (*Arachis hypogaea* L.) FABACEAE

The peanut is native to South America. It was introduced into Africa, where it contributed a large part to the diet of the peoples of East Central Africa. From Africa, the peanut was taken to India, China, and the United States, presently the production leaders. Peanut (groundnut) is a widely grown crop because of its adaptation to tropical, subtropical, and warmer temperate regions.

The peanut is an annual plant with sturdy, hairy branches with growth habits from nearly prostrate to upright. After pollination, the flower withers and drops off, and in a few hours, the base of the flower stalk elongates into a unique structure called the peg. The tip of the peg contains the fertilized ovules, which the growing peg carries down and pushes into the soil. After the peg penetrates into the darkness of the soil, it turns horizontally and the ovary begins to swell and grow into the mature underground fruit (Fig. 16–11).

There are three distinct peanut-growing regions in the United States: the Virginia-Carolina area, which primarily grows the larger-seeded Virginia cultivars; the Georgia-Florida-Alabama area, which grows the Southeastern Runner and some Virginia and Spanish cultivars; and the Oklahoma-Texas district, which grows small runner and the Spanish cultivars. The three types of cultivars are separated on the basis of clearly identifiable agronomic characteristics. These characteristics include branch form, growth habit, pod size and shape, number of seeds per pod, and color of seeds after storage. Typically the Virginia types have both bunch and runner types and two seeds per pod. The Spanish types have erect bunch growth habits and pods are mostly two-seeded. The runner types have true runners, spreading branch growth habits and two to four seeds per pod.

Figure 16–11   Peanut flowers are fertilized above the soil. Fertilization is followed by the development of a subterranean pod after the gynophore (peg) has elongated and buried. Soil penetration provides the necessary moisture and darkness for fruit enlargement, which rarely occurs above ground. *Source:* USDA.

In the United States, peanuts are produced mainly for grinding into peanut butter, for roasted and salted nuts, and for candy and bakery goods. Some limited amounts are used for livestock feed. In other parts of the world, peanuts are grown mainly for their edible oil and as a valuable protein source. Peanut seed has a content of 40 to 50 percent oil, and from 25 to 30 percent protein. After the nuts are harvested, the stems and leaves are often used for hay.

Safflower (*Carthamus tinctorius* L.) ASTERACEAE

Safflower, a relatively new crop to the United States, is actually one of the world's oldest crops. The plant is thought to be native to the Middle East and southwest Asia, where it has been known for centuries. The flowers were first used as a source of red dye for cloth. Safflower is now grown principally for its seed oil, which yields two types of oil: a polyunsaturated (linoleic) type and a monounsaturated (oleic) type, each having varied uses, such as for margarine, salad oil, and mayonnaise. Some cultivars yield a monounsaturated oil similar to olive oil that is used for cooking.

Safflower is a spiny plant that produces a light-colored oil with a high percentage of polyunsaturated fatty acids. Safflower is an annual plant belonging to the ASTERACEAE family.

Safflower is grown commercially in India, Egypt, Spain, Australia, Israel, Turkey, Mexico, Canada, and the United States. The crop was introduced into the United States experimentally in 1925 but has not achieved important status.

Soybean (*Glycine max* [L.] Merrill) FABACEAE

The soybean, also known as the soja or soya bean, is native to eastern Asia. It was cultivated in China and Japan long before written history. Because of its great importance as a high-protein food source, the soybean became one of the five sacred crops of China, joining rice, wheat, common millet, and glutinous millet. Europeans first learned of the soybean about 1700, but not until 1875 was there any great interest in the plant. The soybean in the United States was first mentioned about 1800 in Pennsylvania, where it was reported to grow well.

The soybean is an annual plant of the legume family. It is erect and bushy with many branches, fewer at normal field populations (Fig. 16–12). It varies in height from about 30 to 150 cm (1 to 5 ft) and its root system extends to 150 cm (5 ft) if the soil is permeable. The leaves are alternate and trifoliate except for a pair of opposite simple leaves at the first node above the cotyledons. Most cultivars are pubescent (hairy). Flowers are self-pollinated and are borne on racemes (clusters) of three to fifteen flowers that are either white or purple, or a blend of these colors.

The annual production of soybeans worldwide has more than tripled since World War II. The United States is the leading producer, growing about 55 percent of the world total of more than 130 million metric tons. Because the differ-

Figure 16–12 Soybeans are erect, bushy, branching plants that in their early growth closely resemble garden beans. Most cultivars have deep root systems, which give them more resistance to drought than other bean cultivars. Nearly all cultivars are pubescent. Here an excellent stand of soybeans about 60 cm (2 ft) high is growing on rich prairie soil in Illinois.

ent cultivars of soybean mature their fruits over a wide range of photoperiods, the soybean is adaptable throughout many temperate areas.

The great increase in production in the United States was due to (1) the development of more productive cultivars that are resistant to lodging, seed shatter, and diseases; (2) the extension of soybean production from the US corn belt to the southeastern states; (3) the utilization research done by the USDA and land grant colleges; and (4) the worldwide increase in demand for both the meal and the oil. Other important growing areas are Brazil, China, and Argentina. In the United States, the primary production states include Illinois, Iowa, Minnesota, Indiana, Ohio, Missouri, and Arkansas. States along the Atlantic Coast are also important areas of production.

Soybean cultivars are grouped according to their response to daylength (photoperiod), and it is important to select cultivars adapted to local conditions to take full advantage of the available season. Cultivars with maturity times corresponding with the latitude have been developed. Those requiring long photoperiods are in the northern latitudes; those requiring short photoperiods are in the southern latitudes. Cultivars are also classified as determinate (terminating vegetative development before flowering) or indeterminate (vegetative development continuing for several weeks after the beginning of flowering).

Though few soybeans are now grown for hay, there are forage-type cultivars with fine stems used for this purpose. The soybean hay would be cut when the bean pods are about half filled. At this time, the leaves are about maximum size and the stems are not excessively woody. The hay is cured in the swath for two to three days before it is raked into windrows. Good-quality soybean hay is about equal in feed value to other legume hays, but it is difficult to cure without loss of leaves, reducing quality.

Figure 16–13 The land area devoted to the production of sunflowers is rapidly increasing. The whole seed contains from 25 to 40 percent oil that is high in polyunsaturated fatty acids. The oil is used in margarine, salad oils, and other foods. The oil cake remaining after the oil is extracted contains about 32 percent high-quality protein and is used mostly for livestock feed.

Sunflower (*Helianthus annuus* L.) ASTERACEAE

The sunflower is a native of North America and may have been domesticated before corn. It is a tall annual plant with rough hirsute stems. The flower heads are a compound inflorescence composed of many individual flowers in a large disc, subtended by large ray flowers (Fig. 16–13). In wild specimens, the flower heads are 8 to 10 cm wide, but larger in commercial cultivars. The flowers are cross-pollinated by insects. Wide use of sunflower as an edible oil crop began about 1830 in Russia, and it has become the main source of edible oil in the former Soviet Union, other eastern European countries, and Argentina. Reintroduction to the United States was for its use initially as silage. Its use as an oil crop has greatly increased its production, with more than 1 million hectares now grown. This increase was assisted by the use of Russian germplasm providing oil content above 40 percent and the ability to produce hybrid seed. The introduction and wide usage of hybrid seed improved crop uniformity, yields, and earlier maturity, as well as disease resistance. Breeding for shorter stems (120 to 150 cm) reduced lodging and mechanical harvesting problems.

The cultivated plant has at least three uses: oil, confectionary products, and fodder. The oil form produces a valuable and desirable oil having polyunsaturated and monounsaturated characteristics similar to corn oil and olive oil, respectively. The confection-type cultivars are usually large-seeded and are roasted like nuts, with or without the hull. The smaller whole seed is used as a feed component for pet and wild birds and for small animals. The fodder-type plant is very tall with large stems and leaves, and the entire plant is chopped green for livestock feed. The cake or meal from the oil-pressing process is used for animal feed.

Fiber Crops

Many kinds of plants are cultivated for their fibers, which are used to make yarn, fabrics, rope, paper, insulation, raw cellulose, and hundreds of other products. Fiber-producing plants can be categorized as:

1. Surface fibers—those in which the fibers are produced on the surface of the plant parts in association with floral structures. The principal examples are cotton and kapok, where the fibers develop as outgrowths of the seed coat epidermal cells.
2. Soft or bast fibers—those in which the fibers are located in the stems or, more precisely, in the outer phloem tissues of the bark. Botanically, these are phloem fibers, which are groups of very long, thick-walled cells just external to the conducting phloem sieve tubes. Plants producing such soft-stem fibers include flax, hemp, jute, kenaf, and ramie.
3. Hard fibers—those that are rougher, more lignified strands of veinlike bundles of phloem and xylem supporting cells primarily from leaves of monocotyledons. These are longitudinal rigid fibers. The agave plant is an important example of this group.

For centuries people obtained many needed materials from these plants for clothing, ropes and sails for ships, and cord for fishing nets. In more recent times, numerous other products have been made from such fibrous materials.

Various types of cotton plants were naturally dispersed throughout many of the warm parts of the world for thousands of years, and their lint was spun and woven into cloth. The Egyptians made linen cloth from flax fibers, and ancient writings show that the hemp plant was being grown for its fiber in China north of the Himalaya mountains as long ago as 2800 B.C.

Ramie was cultivated by early civilizations; the cloth made from it was used to wrap mummies in ancient Egypt. Early Chinese literature also mentions the cultivation of ramie for its fibers.

Plants Producing Surface-Fibers in Association with Floral Parts

Cotton (*Gossypium* spp.) MALVACEAE

Cotton has been one of the world's most important crops since the beginning of civilization. It continues to make immense contributions to people's comfort. In spite of competition from synthetic fibers, cotton is still basic in the world's textile industry. Cotton is grown principally for its lint, although the seed produces a valuable and widely used food oil. Cottonseed has an oil content of 30 to 40 percent. The seed residue, called cottonseed meal, is used as a livestock feed and is processed into a high-protein flour.

The various cotton species originated in several warm regions of the world. Tropical by nature, this heat-loving crop has been extended by humans into the warmer temperature regions. There is evidence that cotton was grown and processed into cloth as long ago as 5000 B.C. in Mexico, 3000 B.C. in Pakistan, 2500 B.C. in Peru, and 2000 B.C. in India.

Only four of the twenty or so cotton species are cultivated for their spinnable fibers. Two are diploid Old World species—*G. arboreum* L. and *G. herbaceum* L.—which are believed to have originated in southern Africa. These have a 2n chromosome number of twenty-six and produce a short—1 to 2 cm (0.4 to 0.8 in.)—coarse lint. The other two are tetraploid New World species—*G. hirsutum* L. and *G. barbadense* L. They have a chromosome number of fifty-two. About two-thirds of the cotton grown in the world and almost all that is grown in the United States is known as American Upland cotton and belongs to the species *G. hirsutum*. This species originated in Central America and southern Mexico as a perennial shrub. Improvements through breeding and selection have altered it to an annual plant. American Upland cotton produces fibers varying from short to long—2.2 to 3.1 cm (0.9 to 1.2 in.)—according to the cultivar.[4]

G. barbadense L., originating in the Andean region of Peru, Ecuador, and Colombia, accounts for no more than 1 percent of the cotton grown in the United States, where it is known as American Pima. It is an extralong staple cotton with fibers up to 3.3 cm (1.3 in.) long that are very good for fabrics.

The leading US cotton-producing states are Texas, California, Mississippi, Arizona, and Louisiana.

In some countries, such as China, Egypt, and Sudan, cotton is a major crop and a very important revenue source, even though considerable hand harvesting is still done. Hand harvesting of cotton is laborious but may not be limited to those countries that have the advantage of an abundant sup-

---

[4]In cotton, a cultivar is not a clone, pure line, or a primary mixture of pure lines. It is usually a progeny row selection, bulked and mass multiplied.

ply of low-cost labor. In the United States, the former Soviet Union, Australia, Israel, and other developed countries, production can also be economical and efficient because of the advantages of highly mechanized practices.

In planting, cotton acid-delinted certified seed treated with a fungicide is spaced 8 to 20 cm (3 to 8 in.) apart in rows 76 to 107 cm (30 to 42 in.) apart. Planting is done with a multirow mechanical drill planter. Some planters also apply fertilizers and pesticides in the row. Planting is done after all danger of frost has passed and soil temperatures have reached at least 16°C (61°F), preferably 20°C (68°F). In the United States, this time frame ranges from early March in southern Texas to early May in the northerly areas of the cotton belt.

Maximum productivity is favored with high temperatures during the growing season, high light intensity, ample soil moisture, and good soil fertility. Cotton cannot tolerate frost and requires at least 200 frost-free days from planting to crop maturity. In some areas, as few as 160 to 170 frost-free days will permit production. Generally, the longer growing season provides a greater yield potential.

Optimum day temperatures during the flowering period range from 30°C to 38°C (86°F to 100°F). Because of breeding and selection programs in the temperate regions, cotton is no longer subjected to major photoperiodic control, but high light intensity does promote maximum production of fiber and seeds. Cotton is classified as a C-3 plant and, as such, is not as efficient a user of light energy as C-4 plants.

Depending on the cultivar and environment, flowering branches in cotton originate at about the seventh node on the main stem. Floral buds, called squares, are visible on those branches about three to four weeks before flower opening, which occurs about two months after the seed is planted. The cotton plant initiates far more flower buds than will develop into flowers, and far more flowers than will mature into profitable bolls—the ovaries or fruits that contain the seeds (ovules) bearing the fibers.

The cotton flower is perfect, containing both male and female organs, and is either self- or cross-pollinated. The majority are self-pollinated.

In most commercial cotton, flower petal color is cream to pale yellow on the day the flowers open. The notable exception is Pima cotton, whose flowers have deep purple spots at the base of the corolla, and bright yellow petals. Coincidental with flower opening is anther dehiscence and the release of pollen grains, which germinate about one half-hour after they are deposited on the stigma surface.

Varying slightly among cultivars and influenced by the environment, Upland cottons initiate development of the important lint fibers as protuberances on epidermal cells of the ovule seed coat. This happens soon after fertilization. Fertilization occurs by the end of the first day after flower opening. According to current thinking, fertilization initiates processes that cause the formation of a plant hormone essential to seed formation—and thus fruit set—and continued fiber elongation (Fig. 16–14).

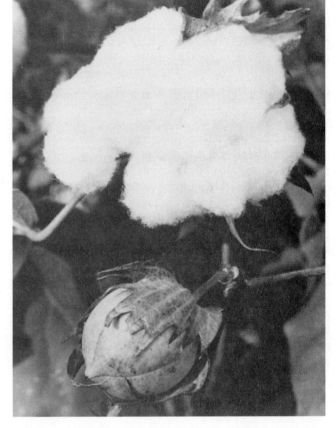

Figure 16–14 Open and unopened cotton fruits (bolls). *Source:* National Cotton Council of America.

All commercial cultivars of Upland cotton produce two types of fibers: lint and fuzz (the latter called linters). As described, lint fibers and fuzz fibers originate as epidermal hairs; lint fibers initiate very soon, but fuzz fibers originate about six to twelve days later. Although variable, fuzz fibers are usually less than one-fifth as long as lint fibers.

Lint fibers elongate and reach their final length in about twenty-one days. Toward the end of this period, secondary wall deposition on the inner surfaces of the cell wall continues to increase the primary constituent, which is cellulose, up to 90 percent. At maturity, the lint dries, and the fibers become twisted and thereby are more suitable for spinning. As bolls dry (Fig. 16–14), they split and open, exposing the lint. Before ginning, lint and seeds are called seed cotton.

At maturity, lint fibers have narrow and delicate shanks at their bases and thus are easily removed during ginning. Fuzz fibers thicken at their bases and are more strongly retained by the seed. After lint removal, seeds for certification and future planting are treated with acid to remove the fuzz fibers, or these fibers are removed by a different set of gin saws. The product is used for mattress stuffing or for cellulose. Removal of fuzz fibers from the seed is essential for the operation of precision planting equipment.

Harvest machinery can be either a spindle-type picker or a stripper. The picker has vertical drums with many revolving steel spindles that engage and twist the cotton and the seeds from the open bolls as it passes along the rows of cotton plants. Any unopened bolls are left for later pickings.

The stripper harvester has roller or mechanical fingers that pull entire bolls, mature and immature, from the plant. This machine can be used effectively only with cultivars whose bolls mature at about the same time. Cotton harvesting can extend over a long time, from midautumn into midwinter if the weather is dry.

Before machine harvesting starts, chemical defoliants (for the spindle-type harvesters) or chemical desiccants (for the stripper-type harvesters) are sometimes applied. The defoliants accelerate leaf drop so that the spindles can remove the bolls easier and lessen contamination (Fig. 16–15). Desiccants rapidly kill the leaves but they stay attached to the plant. Both types of chemicals are applied seven to fourteen days before harvest.

After the cotton has been picked from the plants by the harvesters, it is emptied into large wire-covered trailers for transport to cotton gins. The pressed form is covered with a tarpaulin to protect it from weather. Gins have special equipment to break up the module, and the seed cotton is handled as loose delivered cotton. The cotton (plus seeds) is unloaded using vacuum tubes. Some cotton may require passing through driers to lower its moisture content for easier processing. The cotton is transferred to equipment that removes soil, trash, and foreign material, and then on to the gins. Here the lint is removed from the seed.

The seed-free cotton lint may be further cleaned and is then baled and moved to the textile mills for processing into yarns and fabrics. The seeds left behind are collected and moved to seed-processing houses for fuzz removal. The seed coats are cracked and the kernels removed for oil extraction by pressing or chemical procedures.

Kapok (*Ceiba pentandra* [L.] Gaertn.) BOMBACACEAE

Moisture-resistant fibers called kapok or silk cotton are derived from the seed hairs in the pods of the kapok tree. These trees grow wild in tropical American forests and have been introduced to other tropical regions throughout the world. Almost all commercial kapok production—several thousand tons annually—now comes from the Orient, particularly Thailand, Indonesia, and Kampuchea. Most kapok is sold to the United States, where it is used as insulation in sleeping bags and as filling in life preservers. Synthetic fibers, foam rubber, and plastics are rapidly replacing kapok for such purposes. The hollow kapok fibers are similar to those produced by the cotton plant. The fibers are not twisted and are too smooth and brittle to be spun into cloth but have a remarkable buoyancy and resiliency and are impervious to water. The large kapok trees bear a great many of the football-shaped

Figure 16–15 Mature cotton plants ready for harvest. *Left:* Before defoliation. *Source:* USDA. *Right:* After spraying with harvest-aid defoliant chemical. *Source:* National Cotton Council of America.

pods, up to 15 cm (6 in.) long, which are filled with the fiber-producing seeds. The pods either fall to the ground or are cut from the tree. After drying, the fibers are extracted from the seeds by hand. An edible oil, similar to cottonseed oil, can be pressed from the seeds.

Plants Producing Soft Stem-Fibers

Flax (*Linum usitatissimum* L.) LINACEAE

The flax plant is a slender, herbaceous annual grown in many subtropical and temperate zone countries for its oily seed and for its stem fibers. The fibers are used to make linen cloth, book and cigarette papers, and paper currency.

In the United States, culture of flax for its fibers became unprofitable and ceased in the mid-1950s. Currently, all US flax is produced for seed. Different flax cultivars are used for linseed oil rather than fiber production. Linseed oil is pressed from the seed for paint manufacture, and the residue, called linseed meal, is used as a livestock feed.

The former Soviet Union is by far the leading producer of flax fibers. Poland, France, Czechoslovakia, Rumania, and Belgium are also major producers.

Flax is a cool-season plant. In areas with mild winters, seeds are sown in the fall and the crop harvested early the following summer. In cold-winter regions, seeds are sown in early spring and, if used for fiber production, the stems are harvested eighty to 100 days later, when about half the seeds are mature and the leaves have dropped from the lower two-thirds of the stem. At harvest, weather should be dry to cure the plants properly.

The fibers are obtained from the stems of the flax plant by a process called retting. Commonly, the retting involves simply leaving the stems in the field, where they are exposed to weathering from the dew and to the action of soil-borne bacteria on the straw. This partial rotting for one to three weeks dissolves gums that hold the fibers to the woody xylem tissues and destroys soft tissues around the fibers. A more intensive rapid method is cold water retting by complete immersion of flax stem bundles in cold water for several days. The stems then pass through machines that break up the woody parts but retain the flexible fibers largely intact. Other machines separate the short, woody sections from the fibers, which are then baled and shipped to spinning mills.

Hemp (*Cannabis sativa* L.) CANNABACEAE

The hemp plant is a seed-propagated herbaceous annual adapted to mild temperate zone climates. The unusually stout stem fibers of certain hemp cultivars are used to prepare tough threads, twines, ropes, and textile products. Cloth from hemp can be used in clothing and other products. Most hemp produced for fiber is grown in temperate countries such as the former Soviet Union, Yugoslavia, China, India, Korea, Poland, and Turkey, but various hemp cultivars can be found cultivated or wild in almost all countries in the temperate and tropical zones.

For fiber production, the plants are seeded thickly and grown to a height of 1.5 to 3 m (5 to 10 ft). Hemp seed is planted early in the growing season, and the plants are uprooted or cut off for fiber harvest during the period from start of bloom until full maturity. The stems are hollow except near the base. The fibers are located in the phloem and pericycle tissues of the bark.

The straw is retted like flax to aid in separating the fibers from the other tissues in the stem. Machines are used to separate the fibers further. Manila hemp (*Musa textilis*), also known as abaca, is used in paper manufacturing.

Because certain low-growing types of *Cannabis sativa* are the source of the drugs marijuana and hashish, cultivation of all hemp is prohibited in many countries, including the United States. However, the demand for hemp fiber has changed the laws in some countries such as Canada.

Jute (*Corchorus capsularis* L. and *C. olitorius* L.) TILIACEAE

Plants of these two jute species are herbaceous, seed-propagated annuals cultivated in hot, moist climates with at least 7.5 to 10 cm (3 to 4 in.) of rain per month. The plants grow from 1.8 to 4.5 m (6 to 15 ft) high. The fiber strands are located in the stem just under the bark, some running the full length of the stem, and are embedded in nonfibrous tissue. The fibers are held together by natural plant gums.

As a fiber-producing crop, jute probably ranks next to cotton in importance. It is grown primarily on small farms in Bangladesh and in several districts in India. Jute is chiefly used in the manufacture of bulky, strong, nonstretching twines, ropes, and burlap (Hessian) fabrics, bags, and sacks for packaging many industrial and agricultural commodities. It has a great many other industrial uses—such as backings for carpets and linoleum coverings, webbings for upholstered furniture, packing in electric cables, and interlinings in tailored clothes.

Jute seeds are broadcast over the soil in the spring and the plants are later thinned to about 33/m^2 (3/ft^2). The crop is harvested by cutting off the stems about the time the flowers start to fade. The bundles of stems are left in the fields for a time to shed their leaves.

To release the fibers from other tissues, the stems must be retted. To do this, bundles of the stems are submerged under water in pools or streams for ten to thirty days—long enough for bacterial action to break down the tissues surrounding the fibers, but without damaging the fibers themselves. At the proper time, the fibers are loosened from the wet stems by beating the small bundles with paddles, then breaking the stems to expose the fibers, which are pulled from the remainder of the stems. The fibers are then washed in water and hung on poles or lines to dry. The dried fibers are taken to baling centers, where they are sorted and graded according to strength, cleanliness, color, softness, luster, and uniformity, then pressed into bales. The bales are shipped to local spinning mills or they are exported.

Kenaf (*Hibiscus cannabinus* L.) MALVACEAE

Kenaf has been cultivated for centuries in many places throughout the world between about 45°N and 30°S latitudes. Thailand, India, Brazil, China, and the former Soviet Union are the major kenaf producers. Kenaf competes with jute as a stem fiber crop but has less exacting soil and climatic requirements. Kenaf produces an excellent fiber, tougher and stronger than jute but somewhat coarser and less supple. Kenaf fibers are used for making twines, ropes, and fishing nets and are also suitable for paper manufacture.

The kenaf plant is a herbaceous annual with a strong taproot and a long, unbranched stem reaching to a height of 1.5 to 4.5 m (5 to 15 ft). The plants require a growing season of 100 to 140 days with considerable moisture from either rainfall or irrigation. Harvesting, retting, and drying are done during drier weather. The stems are harvested just as flowering starts by uprooting the plants and tying them into bundles that are placed horizontally in water for retting, as described for jute. Retting may take ten to twenty days. The fibers are then stripped off the stalks and dried.

Ramie (*Boehmeria nivea* [L.] Gaud.-Beaup.) URTICACEAE

Ramie fibers are long strands in the inner bark of the ramie or China grass plant, a many-stemmed perennial shrub with slender shoots about 2.5 cm (1 in.) thick and up to 2.4 m (8 ft) long and heart-shaped leaves along the upper third of the stem. New stems arise from the crown after the older ones are harvested. Three harvests can be obtained annually, and the plant can live for several years before it has to be replaced.

Ramie has been cultivated for thousands of years, and is mentioned in Chinese writings as early as 2200 B.C. It apparently is native to the Chinese area of eastern Asia. It was also cultivated by the ancient Egyptian civilizations. Present-day ramie production centers mostly in warm, humid regions of China, the Philippines, Japan, Indonesia, and Malaysia with fertile soils.

The cells making up ramie fibers are among the longest known—up to 0.3 m (1 ft). The fibers are eight times stronger than cotton and have a fine durable texture and a good color. Ramie fibers are superior in many ways to flax, hemp, and jute. However, the chief problem with ramie has been the difficulty in freeing the fiber bundles from the gummy tissues surrounding them, along with problems in mechanizing the processing of the fibers—the extraordinar-

ily smooth surface of the fibers makes spinning difficult with machinery developed for other fibers.

As with other stem-fiber crops, retting is required to free the desirable fibers from the other stem tissues. This is done by bacterial action, which decomposes the thin-walled surrounding cells and leaves the thick-walled fibers intact. Ramie is more difficult to rett than flax and hemp. Just wetting the stems is insufficient to break down the cementing gums. Pounding and scraping is also required, followed by chemical treatments with acid or lye to remove the tenacious gums and resins to produce completely smooth fibers. After degumming, the fibers are washed, then softened with glycerine, soaps, or waxes. Ramie fibers are generally spun on machinery designed for silk. Ramie fabrics are usually blends with other materials, such as cotton or wool.

Plants Producing Hard Leaf-Fibers

Agave (*Agave sisalana* Perr.) AGAVACEAE

The long hard fibers used in making twines, cords, and ropes are obtained from the 0.6 to 1.2 m (2 to 4 ft) leaves of agave plants. Sisal is produced mainly in East Africa, Brazil, Mexico, and Haiti. A related plant known as henequen (*A. fourcroydes* Lem.) is also grown for its fibers, which are much weaker than sisal. Plants of these two species are similar in appearance to the common century plant (*A. americana* L.).

The agave plant consists of a rosette of stiff, heavy, dark green leaves 10 to 20 cm (4 to 8 in.) wide, 60 to 120 cm (2 to 4 ft) long, and 2.5 to 10 cm (1 to 4 in.) thick, arising from a short trunk. The plant grows very slowly, but after about four years, it reaches full size and harvesting of the lower leaves begins. About 200 leaves can be harvested for fiber extraction before leaf growth ceases and a flower stalk grows upward rapidly to a height of 4.5 to 7.5 m (15 to 25 ft) and bears light yellow flowers. During the next six months, flowering and fruiting take place, followed by the death of the entire plant.

Agaves are vegetatively propagated by suckers arising around the trunk of the plant or by bulbils (small bulbs) that develop in the flower clusters after flowering, thus allowing growers to maintain improved forms.

In harvesting for fiber, the lower leaves are cut off and bundled after the spines are removed. Cleaning machines scrape the leaves, removing the pulp and waste material and leaving the fibers exposed. The fibers are then dried in the sun, or artificially, and are brushed, graded, and baled. The final product is yellow to yellow-white in color, flexible, and strong. Sisal is used mostly in making cords, such as binder twine; it has only limited use in fabric manufacture.

SUMMARY AND REVIEW

Although many different plant species are grown as field crops and each has specific requirements for optimum production, some key concepts are common to all. The first is understanding the types of interactions that occur among all the production factors, the second is to develop management systems that maximize the beneficial aspect of each interaction. Tillage, crop sequencing, variety selection, crop establishment, water and nutritional management, and weed and pest control are the most common factors that have to be considered.

Tillage is the preparation and maintenance of the soil for crop production. The objectives of tillage are to promote seed germination and plant growth, and to control weeds while minimizing soil erosion and fuel costs. Conventional tillage involves extensive preparation of the soil using several different implements to plow under crop residues and weeds and to create a fine-textured soil surface. Since the 1970s, a type of tillage system (no- or low-till) that leaves more crop residues on the surface of the soil and reduces erosion has been developed and is now commonly practiced for many crops.

Cropping sequences can be monoculture (the same crop over and over) or rotational cropping (rotation of crops within a season or annually). Each has advantages and disadvantages.

Variety selection is an area over which a crop producer has considerable control. It is critical to select the variety of the species that is best suited to the area in which the crop will be grown. Varieties are often distinguished by the factor that is most influential in their development. Often this factor is temperature, but it can be sensitivity to daylength, or resistance to pests or other stresses.

Successful crop establishment requires that a grower know when conditions are right for planting and then follow proper planting procedures. These procedures include using the right planting equipment, sowing on the right date, seeding at the proper rate, and planting at the proper depth.

After establishment, water (irrigation) and nutritional management is critical. In much of the United States east of the Mississippi, irrigation of field crops is not necessary, however in the western United States, erratic rainfall usually requires irrigation of most crops. If nutritional elements are naturally available in sufficient amounts, fertilization is required. A grower has to understand what nutrient elements a crop uses and what is removed at harvest from what would be normal nutrient cycling in an undisturbed ecosystem.

Weeds are the major pest in most field cropping systems. No single control program handles all weed problems effectively. Most weed-control problems follow three strategies: prevention, eradication, and control.

Insects are usually not considered to be a serious problem in most field crop commodities because the damage the insects do is below a threshold level of significance. When it does become necessary to control insects, management of pest populations requires a combination of preventive and timely responsive actions based on the risks associated with various cultural practices and pest activity observations collected through periodic field inspections.

Diseases, like weeds, usually require multiple strategies of control. Using high-quality seed is the first step in preventing diseases. Choosing varieties and cultivars resistant or tolerant of diseases that are likely to occur is another strategy. Using crop rotation and proper fertilization and irrigation/drainage practices can reduce disease incidence. Controlling insects can control the diseases spread by those insects. Harvesting grains at the proper state of dryness reduces disease development in storage.

Proper harvesting is as important as the actual growing of the crop. Harvesting must be done when the crop is at the proper stage of development and when weather conditions are appropriate.

FOOD FOR THOUGHT

1. Given the following high and low temperatures (in °F) over a three-day period, use the formula given in the chapter to calculate growing degree day (GDD) accumulation.

	Day 1	Day 2	Day 3
Highs/lows	63/45	74/56	69/88

Why is being able to calculate degree days important to those who grow crops in temperate regions? Why do you think it would be less or even unimportant in tropical regions?

2. Explain why cropping systems that use more than one species in a rotation either annually or within the same year may have fewer insect and disease problems and higher yields than systems that plant the same crop over and over again. What advantage is gained from planting the same crop over and over?

3. Newly planted field soils often develop a hard surface layer (crusting) because of rainwater hitting the finely tilled and exposed soil. Would a small seed or a large seed most likely produce a seedling capable of breaking through the crust? Explain.

SUPPLEMENTARY READING

BEUERLEIN, J., D. ECKERT, D. JEFFERS, J. JOHNSON, P. LIPPS, M. LOUX, E. McCOY, W. SCHMIDT, M. SULC, P. SUTTON, P. R. THOMISON, J. UNDERWOOD, M. WATSON, and H. WILLSON. 1995. *Ohio Agronomy Guide,* 13th ed. OSU Extension Bulletin 472, p. 112.

CORBIN, E. J., and J. E. PRATLEY. 1980. "Cultural practices," pp. 250–292. In *Principles of field crop production.* Sydney, Australia: Sydney University Press.

DRISCOLL, C. J. 1990. *Plant sciences: Production, genetics and breeding.* Ellis Chichester, England: Horwood Ltd., p. 228.

EVANS, L. T. 1993. *Crop evolution, adaptation, and yield.* New York: Cambridge University Press.

HICKS, D. L., and P. R. THOMISON. 2004. Corn management, p. 481–522. In C. W. SMITH, J. BETRAN, and E. RUNGE (eds.) *Corn: Origin, history, technology, and production.* Hoboken, N.J: John Wiley & Sons, Inc.

HOEFT, R. G., E. D. NAFZIGER, R. R. JOHNSON, and S. R. ALDRICH. 2000. *Modern corn and soybean production.* Champaign, Ill.: MCSP Publications.

ILLINOIS COOPERATIVE EXTENSION. *Illinois agronomy handbook.* 1995–1996. Circular 133.

JONES, J. B. 2003. *Agronomic handbook: Management of crops, soils, and their fertility.* Boca Raton, Fla.: CRC Press, p. 450.

METCALFE, D. S., and ELKINS, D. M. 1980. *Crop production: Principles and practices,* 4th ed. New York: MacMillan Publishing Co., Inc., p. 774.

PEARSON, C. J. (ed.). 1992. *Field crop ecosystems.* New York: Elsevier.

VORST, J. 1992. *Crop production,* 2nd ed. Champaign, Ill.: Stipes Publishing Co.

WOLF, B. 1996. *Diagnostic techniques for improving crop production.* New York: Food Products Press.

CHAPTER

17

Forage Crops and Rangelands

Dave Barker and Mark Sulc

KEY LEARNING CONCEPTS

After reading this chapter, you should be able to:

♦ Know the different types of forage and rangeland crops.
♦ Understand principles of hay and silage growing, harvesting, and storage.
♦ Understand the interactions of plants and animals in grazing systems and how animals affect the growing practices of grazing crops.

Forage, browse, or herbage are the edible parts of plants, primarily the leaves and digestible stems that can be fed directly or following storage to livestock. Forage crops, **pasture,** meadows, prairie, **grazing land,** and **range** are a continuum of forage-producing grasslands that vary in many characteristics such as their intensity of production. Generically all are termed grasslands, and although they are usually dominated by grass (**monocotyledonous**) species, they also include **dicotyledonous (legume** and forb) species. Alfalfa is sometimes termed the queen of the forages. Forage crops and rangelands have unique and distinct characteristics compared to other crop systems:

1. Worldwide, more land is devoted to grasslands than to all other crops combined (Fig. 17–1). The proportion of total cropland in grassland varies, depending on geographic features and population pressures. Much of this land is too rocky, hilly, dry, or wet for growing other crops but is suitable for forage production. Grasslands can also have environmental benefits such as stabilizing land against erosion, building soil organic matter, and providing wildlife habitat.

Figure 17–1 Worldwide, more land is devoted to grasslands than to all other crops combined. Grazing of perennial grasslands is productive and economic, and has many environmental and social benefits. *Source:* Photo credit D. J. Barker, Ohio State University.

2. Forage crops and rangelands are predominantly perennial production systems. Often these production systems can support livestock year-round and can support successive livestock crops for many years. For sustainable systems, conservative management to ensure production in the future should be a priority over maximizing production in any single year.

3. Forage crops and rangelands encompass tremendous diversity. These plant systems typically comprise ten to fifty plant species, and worldwide more than 500 plant species are used for forage. These grasslands support a wide diversity of livestock. Forage crops and rangelands occur in every country of the world and on almost all soil types in the world. They encounter the range of climate within and between years and are employed by almost all the world's cultures.

4. Forage crops and rangelands have no immediate value to humans in terms of direct food products. It is only through consumption of the plant by livestock that biological and financial production can be captured from grasslands. In the United States, the value of animal products is more than 25 percent of all farm cash receipts, and the value of the forage crops fed to animals is greater than any other single crop. It is often suggested that it would be more efficient for people to consume plants (e.g., grain) directly rather than having the plants fed to livestock and then consuming the meat and dairy products. However, most types of forage consumed by livestock are high-cellulose vegetative material that can't be digested by humans. Most areas of grasslands can only be converted to human food products by pasturing animals.

UTILIZATION OF FORAGE CROPS

Forages in Confinement Systems

In the United States, livestock are often housed in feedlots (for beef) and milking barns (for dairy) and have all their rations brought to them. These rations usually contain forages as well as other high-energy concentrates and protein, such as by-products from other industries (e.g., brewer's grains, soybean meal) or cereal grains (e.g., corn). A specialized industry has developed around the dedicated production of hay and silage by some farmers, and the blending of these crops into totally mixed rations (TMRs) by feed companies. Although these animals gain weight more rapidly and produce more milk, they could survive totally on forage crops alone. Confinement systems are not found in New Zealand, Australia, South America, and South Africa, where livestock exclusively graze forages. Grazing systems usually have lower animal production than confinement livestock systems, but they are justified because the costs of production are proportionally lower, and proponents argue that they are more profitable. Regardless of the system, forage production is the basis of livestock production.

Grazing

Throughout the world, most forages are utilized by grazing (Fig. 17–2). **Ruminants** (e.g., sheep, goats, and cattle) have four digestive chambers, the first of which (the rumen) has a large population of microorganisms that can digest and break down forage fiber (cellulose, hemicellulose, and pectin). Other groups of livestock, such as camelids (camels and llamas) and monogastrics (horses and pigs), can also thrive totally on forages but employ different digestive mechanisms.

Figure 17–2 Livestock production systems on forages are very flexible. Here, cattle are grazing corn stubble immediately following harvesting. Cattle consume both the crop residue and any grain that escaped harvest. *Source:* Photo credit R. M. Sulc, Ohio State University.

The most important objective in grazing is to match the supply of forage production to the nutritional requirements of livestock. Many management strategies achieve this balance; however, the most important issue is to have the appropriate stocking rate for the rate of forage production. Short-term deficits in forage production (e.g., during drought or over the winter) are usually met by feeding forage that has been conserved and stored (see later in this chapter) during a period of surplus (e.g., spring). Long-term deficits in forage production (i.e., overstocking or overgrazing) result in deterioration of the forage and soil resource.

There are many strategies for forage utilization by grazing; however, these can usually be simplified to variations of either rotational stocking or of continuous stocking. Rotational stocking is the movement of livestock between pastures according to a prescribed strategy. Most rotational stocking employs frequent (daily to weekly) movement of livestock; however, many examples of more and less frequent rotations exist. Once all pastures have been grazed, livestock return to the first pasture grazed, which has had sufficient time to regrow to a harvestable mass (Figs. 17–3 and 17–4), thus completing the rotation. Rotational stocking is usually achieved with temporary electric fences that can be easily moved and replaced to allow livestock access to the correct area of pasture. Continuous stocking involves the allocation of livestock to a fixed area for a prolonged period. In a carefully managed situation, sufficient mass and growth rate support the livestock for the required period.

As a generalization, intensive grazing systems use rotational stocking, while extensive grazing systems use continuous stocking. Many grazers claim higher forage production from rotational stocking; however, it is difficult to identify the precise mechanisms for the response. For exam-

Figure 17–4 The data in Figure 17–3 is redrawn to show the relationship between forage growth rate and forage mass. *Source:* David Barker, The Ohio State University.

ple, in addition to different patterns of forage removal compared to continuous stocking, rotational stocking also includes factors such as more regular livestock inspection, ease of herding livestock, more uniform manure dispersal, and less selective grazing.

Conserved Forage

Forage crops are often conserved or preserved and fed later to livestock. This is the case for confinement systems, where livestock are not on pasture. Conserved forages also provide a useful function by providing forage during periods when grasslands might be dormant or not growing rapidly enough to support livestock needs (e.g., during drought and winter). Forages are conserved in two primary forms: as dry hay or as moist silage.

Hay

Hay is defined as the shoots and leaves—and in some cases, the flowers, fruits, and seeds—of forage plants that are preserved by field drying, harvested, and stored for future feeding to livestock. Hay usually consists of grasses, legumes, or a combination of the two. The harvested material is dried to a moisture level of 15 to 20 percent or less. Hay is the most important type of stored forage and provides considerable flexibility in animal feeding operations. Properly dried and stored, it can be kept for several years with minimal loss of nutritive qualities. In severe-winter regions, it permits feeding livestock a nutritious bulk of material when outside pastures are not growing and are covered with snow or ice. Hay is a cash commodity that can be bought and sold, whereas the value of pasture can be converted only by animal feeding.

In general, the best time to harvest hay for maximum yield and still maintain acceptable quality is just prior to the

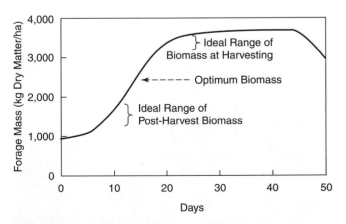

Figure 17–3 Generalized growth curve for forage production. Forages show slow initial growth, a period of high growth rate, and a period of reducing growth rate. After long regrowth periods, forage mass can decrease as leaves low in the canopy wither and fall off. *Source:* David Barker, The Ohio State University.

heading stage. Some forage crops, such as timothy, give only one or two harvests during the growing season, while others, such as alfalfa, give three to ten harvests depending on the region.

Hay quality is determined by plant maturity, leafiness, color, odor, and amount of foreign material. During harvest and transportation, every effort should be made to retain as much leaf as possible because most of the protein is in the leaves. High-quality hay has a fresh green color, a good aroma, and a pliable texture. It is nutritious, digestible, and palatable. Detrimental foreign material includes weeds, poisonous or thorny plants, spiny seeds, and objects such as wire, nails, rocks, and soil.

Hay may be stored loose, chopped, baled, or pressed into cubes, pellets, or wafers. Harvesting is now highly mechanized. Many specialized machines are available for handling hay. The hay is cut, and laid down into **windrows,** dried over two to five days, and baled by mobile machinery (Fig. 17–5). This saves considerable labor over earlier methods of hauling hay to stationary machines. Most hay in the United States is now stored as small square bales (14 × 18 × 38 in., 40 to 60 lbs), large square bales (3 to 4 × 3 to 4 × 7 to 8 ft, 900 to 1,800 lbs), or round bales (4 to 6 ft diameter, 850 to 1,900 lbs). These bales are a commodity that can be bought, sold, or stored for later distribution.

Hay can also be chopped, green or dry, in the field into particles small enough to be blown in an air-stream into trucks. The hay is then transported to feeding areas to be fed green to livestock or to dehydration equipment for processing.

Hay can be cubed, pelleted, or wafered and is more common in regions of the United States where summers have low humidity, such as California, Arizona, New Mexico, and eastern Washington. In these areas, hay can be field-cured down to 12 percent moisture.

Silage

The feeding of silage to livestock is an ancient practice from Europe dating back many hundreds of years, although the first **silo** was not constructed in the United States until 1876. **Silage** is moist forage, preserved by bacterial fermentation under anaerobic conditions. The corn plant is used most often for making silage, but almost any other forage species can be used, alone or in mixtures. Some commonly used species are oats, cereal rye, triticale, sorghum-sudangrass, smooth bromegrass, Italian ryegrass, orchardgrass, timothy, alfalfa, and red clover.

In silage fermentation by the direct cut method, the green chopped forage material—at 60 to 70 percent moisture—is blown into the silo. This is the procedure used for corn silage. A variation of this is haylage or hay crop silage: when the green forage is cut and laid in the field to wilt until it is about 45 to 50 percent moisture. It is then picked up, chopped, and blown into the silo. The practice of field wilting before chopping is required for perennial grasses and legumes, or any standing forage with moisture content above 70 percent (Fig. 17–6).

Silos are airtight structures and are either vertical or horizontal. They may be made of concrete or glass-coated steel plates; they may be large, plastic, baglike containers, or they may be mere trenches in the ground. After the oxygen in the mass of tightly packed chopped material is used up by plant respiration (or aerobic bacteria if spoilage occurs), anaerobic bacteria multiply and act on the carbohydrates in the plant tissue to form lactic acid, which essentially ferments the plant material. The pH drops to 4.2 or below, inhibiting spoilage bacteria and enzyme action, thus preventing deterioration. The fermentation process is completed in two to three weeks. Silage can be kept in good condition for several years if air is kept out, moisture stays high, and the pH remains below 4.2.

Figure 17–5 In spring, forage growth exceeds the requirements of livestock. Surplus forage production can be made into hay and sold or used later during periods of deficit. Round bales suit mechanical manipulation better than the traditional small bales. *Source:* Photo credit R. M. Sulc, Ohio State University.

Figure 17–6 Forages can be conserved by several methods. Here, alfalfa has been wilted to reduce its water content. It is being blown into a wagon and will be transported to a silo to be made into silage. *Source:* Photo credit R. M. Sulc, Ohio State University.

Modern silage preparation is highly mechanized. It starts with cutting the corn plants (or other forage material) by special harvesters that chop and blow the particulate material into trailers for hauling to the silo. At the silo, it is mechanically unloaded and blown into vertical silos, or bags are spread and packed down in bunker silos.

Baleage

With the advent of bale-wrapping technology in the mid-1990s, farmers have the option of replacing silos with individual plastic-wrapped bales. A variation of this technique is wrapping many round bales within a continuous plastic tube in a single row. Wrapped bales have the advantage of being weather resistant (while the wrap isn't broken), thus allowing a shorter drying period in the field compared to hay, and they can be transported. Typically, the fermentation process is similar, although less efficient than in silos because more air (O_2) is usually present in bales than in a well-packed silo. Baleage has become popular on small to medium-size farms in the humid areas of the United States such as the East Coast and southern states, where drying times in excess of three days increase the risk of rainfall damaging traditional hay crops. The additional cost of plastic, disposal issues with the plastic, and the costs of the wrapping machinery make this method less popular in western states of the United States, where field-dried hay is easier to make.

Integrated Systems (Crop Rotations with Cereals, Hay, or Livestock)

The diversity of forage production systems also includes many options for integrating forages into other types of production systems. Forage crops are often included as short-term components within a more complex crop rotation. This technique can work in many situations:

1. Various options for double cropping can be used. When a cereal crop is harvested early (e.g., winter wheat, or silage corn), a short-term annual forage crop might be planted to provide forage for livestock in the period until the main crop is replanted the subsequent year. Forage species that can be used in this situation include warm-season annuals such as sudangrass or cool-season annuals such as cereal rye, oats, annual ryegrass, forage triticale, and brassica species (e.g., turnips, kale, or rape).

2. Forages can be used as part of a rotation with cereal crops. Legume species within grasslands build organic soil nitrogen that can be used by subsequent cereal crops. A possible rotation might include several years in forage crops, followed by several years of grain crops.

3. Hay and grazing systems are often integrated into a single production system. Although examples of grazing-only and hay-only systems are common, these are more generally integrated. Forage growth usually exceeds livestock requirements in the spring, so farmers usually conserve this excess for other periods of the year when livestock demand exceeds supply. These periods include the short (one- to two-month) period of summer drought, and a longer (up to four-month) period of cold during winter when forage growth is slow or zero.

FORAGE QUALITY

Although the single largest influence on the profitability of forage systems is total production (or yield), the next most important factor is forage quality. The quality of a total forage sample is affected primarily by the amount of **fiber** it contains and its digestibility. Fiber in forage can be present as **pectin** and **hemicellulose** (moderately digestible by livestock), **cellulose** (poorly digestible by livestock), and **lignin** (indigestible). Total fiber is measured in the laboratory using neutral detergent fiber (NDF) digestion. Digestibility can be measured directly in animals as the proportion of digested forage compared to total forage consumed (the difference being forage excreted); however, this procedure is time-consuming and expensive. Digestibility can also be predicted following incubation of a ground forage sample in **rumen** fluid in a laboratory and is currently considered the best available laboratory method for estimating forage digestibility. Digestibility of a forage sample is commonly predicted by laboratories from empirical equations based on chemical composition of the forage (such as NDF and lignin).

Protein content, nutrients, and vitamins also affect overall forage quality. In most cases, these are present in sufficient quantities that they have less effect on forage value than does fiber content and digestibility. The main plant factors affecting forage quality are listed below:

- *Forage species.* Forage species vary in their amount of fiber and the chemical composition of that fiber. Generally, grasses have more fiber than legumes, and although grass fiber is generally more digestible (more hemicellulose) than are legumes, it can be insufficient to compensate for the higher amount of fiber. The result is that legumes generally produce better quality forage than grasses. Among grasses is also variation, and species such as perennial ryegrass and timothy are higher quality than orchardgrass and tall fescue.

- *Leaf-to-stem ratio.* Almost without exception, stems have higher fiber (lower quality) than leaves. Vegetative grasses do not have true stems; their pseudostem is actually leaf sheaths wrapped together. Nonetheless, grass pseudostem, the true stems of legumes such as alfalfa, and the reproductive stems of grass all have higher fiber and lower quality than leaves. Good forage management aims to reduce these components in forage and to maximize leaf content.

- *Maturity.* As forages mature, yield increases but quality declines. Stem yield increases dramatically with advancing maturity in most forage species. Flowering initiates developmental changes in all forage species that include higher amounts of fiber. Furthermore, the fiber produced (i.e., lignin) is less digestible. Although one component of this loss in quality is the prevalence of grass seed heads and flowering structures, leaves also have more fiber in flowering forages. Good harvesting strategy is to harvest forages prior to or right as flowering occurs. One field recommendation for obtaining the best balance of yield and quality in a forage crop is to harvest at 10 percent bloom.

♦ *Anti-quality components.* Almost every forage species can contain anti-quality components with varying toxicity to livestock. The list of potential toxic components is lengthy but some important compounds follow:

a. *Leaf saponins.* Most legumes contain saponins that can produce a foam complex that causes **bloat.** Bloat is distortion of the rumen by excess gas production. If left untreated, it will result in death. Livestock should not be allowed pure diets of legume species.

b. *Ergot* **alkaloids** (e.g., ergovaline). These are produced by fungal endophytes of some grass species. Levels vary with heat and drought. In most years, they cause subclinical growth retardation in young livestock, and in extreme cases, can result in animal death.

c. *Nitrates.* Many C4 annual species have the potential to accumulate nitrates that can be toxic to livestock. The greatest risk occurs with the flush of forage regrowth that can follow a temporary drought.

d. *High protein.* Livestock cannot digest protein in excess of 20 percent and forages with levels higher than this can result in impaired livestock growth and even abortion. High-protein forages should be diluted with low protein sources.

e. *Natural estrogens.* Some legume species (e.g., red clover and subterranean clover) can produce natural estrogens that can impair reproductive performance of sheep and cattle. Modern varieties have been selected to avoid this problem.

ESTABLISHMENT

Forage establishment is a necessary component of a forage management and forage improvement program. The many forage establishment methods can be reduced to four main options:

1. *Full cultivation:* The most expensive and most reliable establishment option of forage crops involves destruction of the existing vegetation by mechanical (plowing) or chemical means, a mechanical operation to bury this vegetation (disking or plowing), mechanical treatment of the soil to establish a fine and firm seedbed (by disking and rolling), and placement of the seed at approximately ¼ in. (5 mm) depth.

2. *No-till planting.* As implied in its name, **no-till planting** involves pasture establishment without tillage of the soil. In most cases, existing pasture has desirable physical characteristics (e.g., porosity, infiltration, organic matter), and only the pasture species need to be changed. The existing vegetation is killed with a broad-spectrum herbicide (usually glyphosate, the active ingredient of Roundup®) and seeds are planted directly into the soil. The planter requires strong tynes and furrow openers to ensure that the seeds are placed into the soil, which can make the planter expensive; however, all cultivation steps are eliminated, so the overall operation is cheaper. In general, no-till establishment is less reliable than full cultivation, and higher sowing rates are often used to compensate for the increased risk. No-till planting can also be used to renovate grasslands where the existing vegetation is temporarily suppressed (chemically or by close grazing) and new forage species are sown.

3. *Frost seeding.* Also called **broadcast seeding** and **oversowing, frost seeding** applies seed directly to existing vegetation (or, if applicable, to the soil surface). This method has the lowest success rate but is very inexpensive. Factors that can improve the success rate include: (a) timing—applications in late winter when freeze-thaw cycles occur in some locations can improve the seed-soil contact, (b) sowing rate—typically two to three times higher sowing rates are used, (c) minimizing the existing vegetation with grazing or defoliant herbicides to improve the seed-soil contact, and (d) shortening the post-sowing grazing height to reduce competition from existing vegetation.

4. *Vegetative establishment.* Vegetative forage establishment is a specialized case that is only applicable to a small number of forages. Species such as Bermuda grass and limpograss do not readily produce seed. In addition, these species have prolific **stolon** production (above-ground creeping stems that have the capacity to root and establish new plants). The best option for establishing these species is to harvest stolons, spread them onto a prepared area (cultivated and weed-free), and incorporate them into the soil with a light disc to get partial (but not complete) burial.

PLANT DIVERSITY IN GRASSLANDS

Forages and rangelands typically comprise a mixture of five to fifty species. One important debate is the benefit that this diversity (**biodiversity**) provides to the pattern of forage production. The benefits include better tolerance of environmental stresses, a more uniform forage growth pattern, increased stand persistence, and fewer losses of nutrients to streams and groundwater. Biodiversity can be considered at various levels, and one theory is that biodiversity is maximized by adding species with different and unique functions rather than adding forage species that are essentially similar to those already present. There are many methods or systems for grouping species; however, some common functional groups for grasslands are listed below:

♦ *C3—cool season grasses.* This is the largest and most important group of forages and includes species such as perennial ryegrass, orchardgrass, timothy, Kentucky bluegrass, tall fescue, reed canary grass, crested wheatgrass, and smooth bromegrass (Fig. 17–7). These species can be characterized by typically higher quality for livestock, moderate to poor drought tolerance, good spring and autumn production, and moderate to low tolerance of low fertility. The name for this functional group results from the photosynthetic pathway these species use for fixing CO_2; that is the first step for fixing CO_2 involves 3-carbon molecules and the enzyme **ribulose bisphosphate carboxylase (RUBISCO).**

♦ *Legumes.* Legumes are species of the plant family FABEACEAE and are vital components of most forages and rangelands. Examples of legume species include alfalfa, red clover, white clover, birdsfoot trefoil, and lespedeza. Legumes provide many useful characteristics such as nitrogen fixation by *Rhizobium* bacteria, high digestibility forage, high protein concentration, and often a growth pattern complementary to their companion grass species (Fig. 17–8).

♦ *C4—warm-season grasses.* This group of forages includes species such as Bermuda grass, big bluestem, switchgrass, indiangrass, limpograss, and paspalum. These species can be characterized by excellent tolerance of drought and heat, and high water use efficiency. In addition, many of these species are native prairie species and can have both conservation and productive value. Disadvantages of these species include their difficult establishment and lower forage quality. The name for this functional group results from the photosynthetic pathway these species use for fixing CO_2; that is, the first step for fixing CO_2 involves 4-carbon molecules and the enzyme phosphoenol pyruvate (PEP) carboxylase (see Chapter 11).

♦ *Forbs.* This is a largely maligned group of species often regarded as weeds in pastures. We are realizing that these species might provide useful functions including filling vacant spaces, protecting bare soil, and providing forage (often with high nutrient value) to livestock. Two species relevant to mention are chicory and plantain, which have been selected for

forage production and are being used in forage mixtures. In many extensive pastures, other forb species can provide unique characteristics to livestock products such as unique (niche) flavors to meat and dairy products.

♦ *Annual species.* Annual forages are used at two extremes of forage production.
  a. In intensive systems, annual forages such as cereal rye, wheat, oats, and sorghum-sudangrass can be used for short-term forage production. In these cases, use of these species is often in rotation with traditional forage crops.
  b. In extensive systems, annual forages such as subterranean clover and annual bromegrass can also provide significant short-term forage production. This land often includes stressed environments that encounter low fertility, drought, and cold winters. Annual species can also occur as opportunists within perennial pastures. Perhaps the most common species worldwide is *Poa annua*. Although often characterized as a pasture weed, it does provide forage and useful ground cover when it is growing; however, its propensity to produce seed heads (of low quality) and its failure to produce forage during summer make this species undesirable in most cases.

♦ *Shrubs.* Shrubs can occur in some grasslands such as in tropical grazing systems and rangelands, but they are not common in intensive grasslands in the United States.

SYMBIOSIS WITH MICROORGANISMS

Many forage species form symbiotic relationships with other microorganisms that are important and vital for the success of these plants.

♦ Rhizobia are bacteria that form a symbiosis with legumes, in the form of nodules on the roots. The species of bacteria are unique for each species of legume. The bacteria benefit from a protective host which provides a source of energy and nutrition, while the plant benefits from a source of nitrogen. The nitrogen is fixed from the air and converted by the bacteria into a useful form in the plant. This symbiosis is described in greater detail in Chapter 13.

♦ **Mychorrhiza** comprise a broad class of fungi that form various relationships (mutualistic, symbiotic) with a range of grass, legume, and forb species. In general, the mychorrhiza are more efficient at the uptake of essential nutrients, especially phosphorus, and can aid plant survival in low-fertility soil. In the case of most prairie grasses such as big bluestem and indiangrass, the presence of the appropriate native mychorrhizal species is essential to the persistence of the grass.

Figure 17–7 Some grass species forage crops. *Above* (*left to right*): smooth bromegrass, orchard grass, tall fescue. *Below* (*left to right*): timothy, bluegrass, western wheatgrass.

Figure 17–8 Examples of legume species forage crops. *Above (left to right):* alfalfa, sweet clover, red clover. *Below (left to right):* birdsfoot trefoil, lespedeza, bur clover.

♦ **Endophytes** are a classic seed-borne fungi that form a symbiosis in the intercellular spaces of grass leaves, stems, and seeds. Although there are about fifty fungi species infecting 200 grass species, most interest focuses on the genus *Neotyphodium,* and the grass species tall fescue and perennial ryegrass. When growing in the grass host, fungi produce various alkaloids that are feeding deterrents to both insects and livestock. The symbiosis has the ecological benefit of providing to the host increased production, persistence, insect resistance, and drought/heat tolerance, but it has the practical disadvantage of being toxic to livestock. To avoid the toxic alkaloid effects, the forage seed industry has developed mechanisms for the supply of endophyte-free seed. Since 2001, a new endophyte option has become commercially available. Nontoxic endophyte is a fungal isolate identical in all respects to the toxic endophyte except that it does not produce the toxic alkaloids. Pastures established with nontoxic endophyte have the benefits resulting from the fungus being present, but without the toxic effect of the alkaloid to animals.

GRASSLAND ECOLOGY

Plant Dispersal and Propagation

Although grasslands include many types of vegetation such as turf, hayfields, pasture, rangelands, and prairie, they are all perennial and are expected to last from several to many years. In contrast to this expectation, the individual growth units (**tillers** or **phytomers**) are relatively short-lived and any grassland that seems stable and uniform is actually undergoing a continual cycling of individual plants. Added to this complexity is that most grasslands comprise many different species, each with different patterns of growth. This complex propagation of phytomers is termed population dynamics.

Grassland species can propagate sexually (via seed dispersal) or asexually (via clonal growth and dispersal from tillers, stolons, and **rhizomes**). The functional properties of any grassland for example, the color uniformity of a golf course fairway, stand thickness of a meadow, or survival of a prairie from drought, all depend on an understanding of population dynamics.

Sexual Propagation

One common species in pasture is *Poa annua*—an annual grass species. The life cycle of this plant includes the production of seed in early summer, which germinates during fall, and seedlings grow during spring and produce seed in early summer. Bare areas during summer can result from *Poa annua* dying after seed set, and insufficient water for germination. Many other annual weed species have a similar life cycle, and control of these weeds requires that their seed production be prevented. Other perennial grasses produce seed;

however, this seed does not make a major contribution to the overall plant population.

Tillering

The basic component of a grass is the tiller—and a grass clump is a collection of tillers. These are identical clones of the original parent seedling. Tillering is stimulated by light at the base of the plant, and keeping a grass short (e.g., close grazing) promotes a high tiller density. Allowing it to become tall (e.g., hayfield) results in fewer, larger tillers.

Stolons

Stolons (or runners) are specialized stems of grasses, legumes, and forbs that creep along the ground surface and have the ability to produce roots and initiate new plants. This important plant dispersal mechanism allows plants to explore and exploit areas favorable for growth (e.g., gaps in a grassland). An important species with stolons are white clover and Bermuda grass.

Rhizomes

Rhizomes are similar to stolons, but they exist underground. Rhizomes have the same function as stolons in spread of a species to new areas, but have the advantage of being protected from heat and freezing. Rhizomatous plants are very hard to kill, and the agricultural examples are restricted to a few forage species. The most common rhizomatous plants are the weeds quackgrass and Canada thistle. Reed canarygrass, tall fescue, and Kentucky bluegrass also produce short rhizomes.

Sward Growth Dynamics

With few exceptions, forage crops are plants whose stems and leaves are grown to feed livestock. Strategies for the removal of the herbage have been the subject of extensive research.

The forage mass (or leaf area) present when the crop growth cycle begins determines the initial rate of growth (Figs. 17–3 and 17–4). Overgrazing maintains a grassland with a low mass (<1,000 kg dry matter/ha), slows the growth rate, and reduces productivity. Residual leaf area is required for the plant to continue **photosynthesis.** Some grasslands can tolerate removal of most of the forage (e.g., alfalfa), but the harvests must be spaced far enough apart to allow accumulation of excess energy reserves in the root systems that are used to generate regrowth. Undergrazing has a high pasture mass (>3,500 kg dry matter/ha) and does not result in additional crop growth. In extreme cases, it might even result in yield loss if leaves in the base of the canopy senesce. The ideal range of forage mass ensures continued high growth rate. This ideal range of forage mass also allows maximum harvested yield but does not result in slowed growth rate from shading of leaves lower in the crop canopy.

Expert managers can monitor their crop as near as possible to the optimum forage mass that keeps the crop in a state of rapid vegetative growth. In reality, this balance is more complex than only the pre- and post-harvest forage mass. Other factors such as availability of machinery, labor, and livestock requirements also affect the farmer's decision about when to harvest forage; it is a managerial art to balance the multiple objectives and factors successfully!

Balance Between Species

Grasslands appear to provide uniform green vegetation, but in many cases, they comprise a dynamic balance with different ratios of species in each season. The precise ratios between species depends on which species are present. One of the most important ratios is the balance between grasses and legumes. Legumes have the useful functions of fixing nitrogen from the atmosphere and providing high quality to livestock, but they cause bloat. Most recommendations are to maintain 30 percent legume in a pasture, and pasture managers work continually to manage the balance when it deviates from this ideal. Other important ratios include the proportion of forbs (typically weeds), the balance between desirable and undesirable grasses, and the invasion of woody weeds.

NUTRIENT BALANCES

Grasslands generally have lower fertility than most other cropland; thus, they generally have lower losses of nutrients to the environment than other cropping systems. The low nutrient status occurs because the land is usually of lower quality (shallow soils, sloping hillsides), and grasslands typically have fewer agricultural inputs such as fertilizer. Soil fertility is the most important management tool for control of forage production. Soil fertility also affects the quality of forage for livestock and the balance between desirable and undesirable species. The most significant nutrients in forages are nitrogen (1 to 4 percent by weight), potassium (2 to 4 percent by weight), and phosphorus (0.2 to 0.5 percent by weight).

One of the most important differences between forages and other crops is the contribution of livestock manure (nutrient cycling) to soil fertility. In most cropping systems, nutrients are provided by synthetic fertilizers or organic sources. In forage systems, the inclusion of livestock has a dominating effect on nutrient cycling. While some nutrients are incorporated into the animals' body mass, 98 percent of nutrients pass through the animals and are excreted in urine and dung. The effect that livestock has in removing forage (and the nutrients it contains) from a largely uniform distribution around the pasture and depositing it into piles has a significant effect on the grassland system. Typically, the return of manure by livestock is concentrated in areas where they ruminate, for example, near water sources, and seek shelter, such as near trees and buildings. This redistribution is called nutrient transfer and it can have positive and negative benefits, depending on the manager's objectives.

Important factors affecting hay and silage systems are the nutrients harvested and removed from the field with the crop. Hay and silage crops can yield as much as 10 tons per acre, although yields of 3 to 4 tons per acre are more common. These yields are greater than for any other crop and can represent a huge loss of nutrients from these systems. Sustainable hay and silage production from an area is usually possible only where these nutrients are balanced by inputs of fertilizer.

OTHER USES OF GRASSLANDS

We have focused primarily on the productive capacity of grasslands. As the world's population grows, we are facing increased complexity in our land use, demanding multiple uses from the same unit of land. Our objective as land managers is not only to ensure food production, but also to maximize alternative and often competing land-use options. The dedicated use of grassland for food production only is becoming less and less important as society appreciates the diversity of land-use options that grasslands provide.

♦ *Lifestyle.* Regardless of our cultural background and overriding our detailed biological understanding of plant production systems, the image of contented livestock on pasture is one of the enduring images we all recognize. Stereotypical images of New England, New Zealand, England, and Switzerland are characterized by pastoral landscapes that provide value to society (e.g., economic value through tourism). Even margarine products are marketed with images of cows on pasture! Many small grazing farms list the lifestyle as one of the primary benefits of their business. Values such as working with family members (including children), a belief in doing good for the environment, and making a good profit are based on forage and rangeland plants. Increasingly, products produced in these systems (e.g., free-range product) are marketed as being differentiated from bulk commodity production and thus can gain a market premium.

♦ *Carbon sequestration.* Grasslands have an organic matter content of 2 to 5 percent greater than other cropping systems on identical soil. As we are becoming aware of the environmental impact on the world's rising carbon dioxide (CO_2) levels, there is interest in the use of grasslands to sequester carbon in the soil. Currently, there is no economic incentive to farmers to use this capability; however, if a system of carbon credits were introduced, one option might be to make greater use of forage crops.

♦ *Soil preservation.* Most forages are perennial crops with large root systems. These features provide grasslands with extremely desirable characteristics for year-round soil stability, slowing the surface movement of water and improving water infiltration. Grasses are invaluable in riparian management and are frequently planted in the drainage channels of cropping systems to reduce soil loss during extreme rainfall events.

♦ *Water harvesting.*   As a result of their dense structure and short vegetation, grasslands provide an ideal combination for water harvesting. The yield of water from land and its quality (freedom from soil particulates) are greater than from land producing any other vegetation.

♦ *Wildlife.*   Grasslands have the potential to provide nutritious forage for large livestock such as deer and wild horses, nesting habitat for birds (although well-fertilized introduced species can have excess stand density for some prairie birds), and clean water to streams. Grasslands are an important component of a wildlife management program.

♦ *Conservation.*   Native prairie is threatened vegetation throughout the US plains. Traditionally, prairie supported intermittent grazing by bison and deer; however, the potential for both livestock production and prairie conservation remains underutilized.

♦ *Food quality.*   We are becoming more and more aware of the unique contributions that forages can make to the nutritional value for humans. Quality characteristics such as lean meat, organic produce, conjugated linolenic acid (CLA) levels (in the class of omega-3 fatty acids that have cancer fighting properties in laboratory animals), and vitamin A levels can all be influenced by the amount of forage in an animal's diet.

♦ The raising of forage crops for seed production is a specialized industry that is concentrated in the western states in the United States. Seed crops comprise less than 0.1 percent of total grassland area but are important in providing the seed for newly established grasslands.

SUMMARY AND REVIEW

Forage is the edible parts of plants that can be fed either immediately or after harvest and storage to livestock. Forage crop and rangelands have unique characteristics compared to other crops. They have more land devoted to them than all other crops combined, are predominately perennial production systems, encompass tremendous diversity, and have no immediate value to humans in terms of direct food products.

Animals can consume forages directly when grazing or as preserved food (hay and silage) provided by farmers. Grazing systems generally require little in the way of land preparation and there is no harvesting, but these fields still require considerable attention and management to keep them productive. The production of harvested forage (hay and silage) is much like other field crops and requires land preparation, planting, growing, harvesting, and storing the crop. Hay is the preservation by drying of the edible plant parts. Silage is the moist storage of forage crops in silos or plastic-wrapped bales (baleage) under anaerobic conditions and fermentation. The production of a quality forage crop depends on the grower's understanding of the interactions of the crop and the environment, along with production and grazing or harvest and storage practices, and maximizing the positive benefits of those interactions.

Forage quality is determined by the amount of fiber it contains and the digestibility of that fiber. Fiber can be present as pectin, hemicellulose, cellulose, and lignin. Pectin and hemicellulose are moderately digestible, cellulose is poorly digested, and lignin is indigestible. Forage crop species, plant maturity, and the parts consumed by the animals influence fiber type.

Biodiversity is an important concept in forage and rangeland crop management. C3 and C4 grasses, legumes, and forbs are the primary broad groups of plants that contribute to forage/rangeland biodiversity. The benefits associated with a production system that is biodiverse include better tolerance to environmental stresses, more uniform growth patterns, increased stand persistence, and fewer nutrient losses to runoff or leaching.

Unlike most field crops, forage systems require that plants regenerate from year to year (even hay fields are often perennial in nature). Perennial grassland crops usually propagate naturally from seeds or the spread of tillers, stolons, and rhizomes. Whether animal feeding or harvesting removes the edible parts, forage growers must manage the fields so that the plants have the chance to reproduce adequately for the next season's or year's growth. Limiting the amount of grazing, maintaining the best array of forage species (biodiversity) for the area, and maintaining proper nutrient balance are all essential for forage productivity.

Grasslands have many uses in addition to providing livestock feed. They promote a desirable lifestyle for humans. Their ability to sequester carbon is likely to have an impact in the effort to reduce atmospheric CO_2. They aid in soil and water conservation efforts. Providing wildlife habitat and enabling the conservation of threatened species are additional beneficial uses for grasslands. Improvement in human nutrition can be provided by the use of forages over other types of animal feed.

FOOD FOR THOUGHT

1. One argument in the vegetarian debate is that the efficiency of energy conversion from plants to humans is much less efficient via livestock. What factors make this conversion less efficient than direct consumption of plants by humans? Why might this argument not apply in the case of livestock consuming forages?

2. Grasslands do more than produce food. Can you think of ten benefits that grasslands provide, in addition to livestock production?

3. What role might forages and livestock have in an organic food production system?

4. In contrast to most crop systems, grasslands generally include more than one plant species. What benefits might a mixture of species provide? (*Hint:* What role do legumes have in grasslands?)

5. In crop systems, we often focus on just the plant species; however, grasslands form a complex interaction with other trophic levels. What are some of these interactions? (*Hint:* In what ways do grasslands depend on microorganisms? In what ways do grasslands depend on herbivores?)

6. What are five differences between C3 and C4 forages? How do these characteristics affect the suitability of these species for livestock production?

7. Two main methods of harvesting forage are direct grazing by livestock and conserving forage (e.g., hay, silage) for later use. What are the advantages and disadvantages of each method? How might the two methods be complimentary within the same livestock operation?

8. Compare and contrast nutrient balances within pasture systems and conserved forage (hay, silage) systems. What factors affect sustainability of adequate nutrient balances in each system?

SUPPLEMENTARY READING

BALL, D. M., C. S. HOVELAND, and G. D. LACEFIELD. 2002. *Southern Forages,* third edition. Atlanta, Ga.: Potash and Phosphate Institute (PPI) and Foundation for Agronomic Research (FAR).

BARNES, R. F., C. J. NELSON, M. COLLINS, and K. J. MOORE. 2003. *Forages: An introduction to grassland agriculture,* Vol. I, sixth edition. Ames, Iowa: Iowa State University Press/Blackwell Publishing Company.

HODGSON, J. 1990. *Grazing management—Science into practice.* New York: Longman Scientific and Technical/John Wiley & Sons.

HOPKINS, A. (ed.). 2000. *Grass—Its production and utilization,* third edition. Malden, Mass.: Blackwell Science/Oxford England.

MURPHY, B. 1994. *Greener pastures on your side of the fence,* third edition. Colchester, Vt.: Arriba Publishing.

CHAPTER

18

Vegetable Production

Matt Kleinhenz

KEY LEARNING CONCEPTS

After reading this chapter, you should be able to:

- ◆ Explain how vegetable production benefits society.
- ◆ Understand the differences among field, tunnel, and greenhouse vegetable production.
- ◆ Know the differences and similarities among conventional, organic, and sustainable vegetable production.
- ◆ List the basic steps to successful vegetable production.
- ◆ Name the basic characteristics of the major vegetable crops.

PRINCIPLES OF VEGETABLE PRODUCTION

The Importance of Vegetable Production

Vegetable production is important for many reasons. For example, vegetables are important to peoples' diets. Hippocrates is credited with founding the practice of medicine and offering advice such as "let your medicine be your food and your food be your medicine." This long-ago recognition of the diet-health relationship is supported by a large amount of modern scientific evidence. Some of this evidence points to the specific dietary value of vegetables as derived not only from what they contain—vegetables are a prime, convenient and natural source of minerals, vitamins, fiber, and energy—but also from the apparent ability of a diet rich in vegetables to reduce the risk of the onset of diseases and disorders. Therefore, it is clear why many different health professionals consider vegetable consumption central to nearly all types of healthy eating, including those outlined in the United States Department of Agriculture (USDA) MyPyramid (Fig. 18–1). (For more information about the pyramid, go to **http://www.mypyramid.gov**.)

Figure 18–1 The USDA's MyPyramid is an assessment of personal dietary needs. From left to right, the sections of the pyramid represent grains, vegetables, fruits, milk, and meat and beans. The USDA website at *www.mypyramid.gov* allows you to determine your dietary needs. *Source:* USDA, *www.mypyramid.gov.*

Making vegetables available through production, distribution, and marketing contributes strongly to their consumption.

Vegetable production can contribute to our psychological or emotional well-being as well as the health of our bodies. Kellert and Wilson (1993) provide evidence that our connections with nature are a necessity, not a luxury, a concept asserted earlier by Wilson as biophilia. Certainly, the strength of diverse people-plant connections underscores the biophilia concept (e.g., see US Botanical Garden at **http://www.usbg.gov/** and National Gardening Association at **http://assoc.garden.org/about**). Many people derive pleasure, satisfaction, pride, relaxation, camaraderie, and other benefits from vegetable gardening, whether done individually or in community garden settings (Fig. 18–2). Indeed, in addition to gardening activities done for pleasure or to produce food, gardening activities can add to therapy and rehabilitation programs designed to help maintain or improve emotional stability, self-image, or cognitive function. And vegetable gardening can be an ideal form of individual or group exercise for people of all ages and abilities.

Regardless of location, vegetable gardens and farms can also have positive effects on our environment that, in turn, benefit people. Natural resources can be recycled and maintained on vegetable gardens and farms. Composting and vegetable production are complementary and an example of how outputs of one activity can be used as inputs in another. Soils used in gardening and farming can be maintained for use by future generations in part because soils covered by plants are less likely to move into waterways or the air. While in place, soils used in growing vegetables can help trap and purify water and provide food and shelter to large numbers of organisms. Also, plant photosynthesis captures carbon from the atmosphere. Carbon fluxes within earth's system continue to gather intense research attention because of their apparent link to global climate change (see US Climate Change Science Program at **http://www.climatescience.gov** and US Global Change Research Program at **http://www.usgcrp.gov**). Vegetable gardens and farms help to maintain open, rural, or green spaces—used in various ways, including relaxation and revenue generation—on our landscapes.

The different scales of vegetable farming and the history of vegetable production can give us a better understanding of societies and the natural world. Over time, successes or failures in the production of vegetables and other crops have had major impacts on societies, especially in terms of the health and movement of people. For example, failures of potato crops in the middle nineteenth century, along with other socioeconomic factors, led to the death of many people and emigration of others from Ireland. Today, the likelihood of similar, biologically based crop failures has been reduced through the development of improved production methods. However, some of these methods involve the use of crop protectants (pesticides) that have the potential to harm people or the environment. To protect people and the environment, even newer tactics designed to help reduce the use of pesticides or increase the nutritional value of crops were developed, but some of these tactics involve controversial manipulations of plant genomes. As you can see, advancements in production methods and crop variety development can offer tantalizing benefits but may have other, potentially undesirable consequences. Nevertheless, efficient production of an abundant supply of high-quality food in environmentally sound ways—whether in small gardens or on large farms—is important to all societies. Both amateur and professional growers and citizens of all professions and perspectives will have to contribute to the development and selection of safe methods of growing crops in the future, with everyone learning much in the process (Figs. 18–3 and 18–4). Vegetable plants respond strongly to their growing environment in ways that can be seen and measured, making them an ideal commodity for involving

Figure 18–2 Home vegetable garden. *Source:* Jane Martin, Ohio State University Extension.

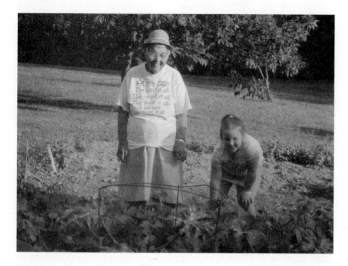

Figure 18–3   Vegetable gardens can be used to educate children about the environment, ecosystems, and nutrition and also to bridge generations. *Source:* Margaret McMahon, The Ohio State University.

Figure 18–4   Professional and amateur vegetable growers can learn about the latest research discoveries and recommended growing practices at field days sponsored by educational institutions. *Source:* Dr. Matt Kleinhenz, State Vegetable Specialist, The Ohio State University.

citizens from all walks of life. Vegetable gardening and farming can be used to create many opportunities to discover the laws of nature and educate people about ways to grow crops safely and efficiently within those laws.

Although home gardening is practiced by many, most of the vegetables we eat are grown on commercial farms and perhaps are processed in plants or factories. Therefore, the fact that vegetable production and processing are large industries is another reason why vegetable farming is important. In fact, economies (family, corporation, community, state, region) throughout the United States and world are based on income generated from the production and sale of vegetables and vegetable products. The heavy financial impact of commercial veg-

etable industries is derived from the sale of products of vegetable farms and processing facilities and the large amount of supplies, equipment, labor, and services they purchase. For example, the total farm-gate value for fresh and processed vegetables grown in the United States in 2004 was estimated at $9.8 and $1.5 billion, respectively. However, these estimates of crop value do not include the total economic impact of the US vegetable industries; for example, additional, significant private and public revenue is generated through vegetable processing.

Overview of Vegetable Farms and Industries

Vegetable farms and industries are (1) widespread but are more prevalent in certain areas; (2) variable in size, crop diversity, and production method; and (3) managed ecosystems. Databases compiled and managed by the USDA at **http://www.usda.gov/nass/**, the United Nations at **http://faostat.fao.org/**, and others underscore that vegetables are produced commercially in all US states and throughout the world. The size of the area used in commercial farming, including for vegetables, is typically measured in acres or hectares. An acre contains 43,560 square feet and a hectare contains 2.47 acres. Acreage devoted to commercial vegetable farming tends to be greatest and most concentrated where the environment (physical, chemical, and biological components) consistently favors crop growth, and natural and other resources required for production are readily available. Physical environments permit crops to be grown in many regions; however, commercial production of many vegetable crops on the largest scale is concentrated in areas where crop marketable yields are consistently high and/or production costs are relatively low.

US farms vary widely in their characteristics, and vegetable farms are no exception. A USDA typology of US farms includes five types of small family farms, two types of larger family farms, and nonfamily farms. Farms, including vegetable, within and among these groups differ in their contribution to overall agricultural production, product specialization, farm program participation, and dependence on farm income. Vegetable farms range in size from less than 1 acre to hundreds of acres, are located in primarily rural or suburban areas, produce hundreds of pounds to thousands of tons of one or more crops, and may or may not be the sole source of income for their owner.

Also, demographic characteristics of the farming population in the United States are changing. Statistics suggest that a rising number of farms are owned and/or managed by women and minorities, that the number of part-time farmers is increasing, and that larger farms tend to have more than one owner-operator.

Characteristics of Vegetable Farming and Gardening Systems

Managed Ecosystems

A defining characteristic of vegetable farms and gardens is that they, like other types of farms, operate as managed ecosystems. The word *ecosystem* is defined in various ways depending on its application to biology, business, or other fields. In biology, ecosystems form through the interactions of abiotic, biotic, and cultural (anthropogenic) components. The community of organisms in an ecosystem has members representing different Kingdoms in the phylogenetic classification system. Ecosystems are functional units and, like all systems, they involve flows of energy and matter in and out of the system.

The word *ecosystem* often raises thoughts of natural spaces that look and behave in ways uninfluenced by humans. Yet few ecosystems are completely uninfluenced by people. Farms are agriculturally based ecosystems (agroecosystems) established and managed to meet specific goals. Vegetable growers establish and manage biological communities within larger natural systems so that a high quantity of desirable vegetables are produced efficiently and profitably, and with as much integration among agro- and other ecosystems as possible.

Agroecosystems combine elements of business (e.g., accounting, finance, marketing) and the natural world (e.g., biology, chemistry, physics). In a business system, relationships among different industries may become mutually beneficial and self-sustaining. Such relationships involving abiotic and biotic participants are also found in agroecosystems. For example, vegetable crop plants interact with nonliving components of soil and air and numerous members of hundreds of micro- to macro-scopic species from different Kingdoms—insects, weeds, pathogenic and nonpathogenic microbes, birds, rodents, and so on.

The genetic potential of vegetable plants for crop yield and quality is often high. However, this genetic potential is rarely realized in commercial production systems or gardens. Plant diseases and pests interrupt or retard crop growth or reduce the marketability of harvested units. Weeds compete with crop plants for light, water, nutrients, and other growth factors. And the levels of light, temperature, nutrients, and/or water may be too low or too high relative to crop need. To capture as much of the genetic potential of vegetable plants as possible, farmers and gardeners supply them with growth factors (e.g., nutrients, water) and protect them from diseases, insects, and weeds. Approaches used by growers to achieve these goals take into account production, market, and personal factors.

Open Field

Vegetable farms and gardens are diverse in overall production scheme. Farming and gardening in open fields is by far the most common approach in the United States and elsewhere. Using this approach, the ability of farmers and gardeners to control the natural environment is limited, and their crops are fully exposed to weather, soil, and pest and disease conditions common to their area. However, these systems usually include an irrigation system (Fig. 18–5), a fertilization program, and utilization of pest- and disease-control strategies to manage the system somewhat.

Climate Controlled

Other farmers and gardeners grow their crops in semi-enclosed and climate-controlled structures (e.g., high tunnels) (Fig. 18–6). Using this approach, vegetable growers limit the exposure of their crops to undesirable natural conditions while, in some cases, creating conditions more supportive of plant growth. For example, placing growing crops within a high tunnel covered by a single layer of clear plastic shields them from wind, rain, and some pests. And it exposes

Figure 18–5 (A) Overhead irrigation systems such as the one pictured here can be used in large-scale vegetable production. (B) Spray-type overhead system. Flood and drip systems are also used in both large and small-scale production facilities. *Sources:* (A) USDA-ARS Image Gallery, *www.ars.usda.gov/is/graphics/photos/;* (B) Margaret McMahon, The Ohio State University.

Figure 18–6 High tunnels are used to extend the vegetable growing season in some areas of the country. *Source:* Dr. Matt Kleinhenz, The Ohio State University.

Figure 18–7 Greenhouse tomato production. *Source:* Margaret McMahon, The Ohio State University.

crops to temperatures higher than those outside the high tunnel. In fact, temperatures in a high tunnel in northern Ohio (USDA Plant Hardiness Zone 5b) in February may closely resemble normal outside temperatures typical of parts of Tennessee (USDA Plant Hardiness Zone 6a). Therefore, vegetables can be grown when conditions outside the tunnel would otherwise not allow it (e.g., early spring, late fall). Semiprotected vegetable production is increasingly common.

Other vegetable growers enclose their crops in fully climate-controlled structures, such as greenhouses (Fig. 18–7). The production of vegetables in greenhouses differs from that in fields or semicontrolled environments in several important aspects. For example, plants grown in commercial greenhouses are often rooted in water only or a manufactured solid but soilless medium that requires a nearly constant supply of water and fertilizer (both are referred to as hydroponic systems). Vegetables grown in fields or semi-enclosed structures are typically rooted in naturally occurring soil. However, plants in greenhouses are often grown in pots or other containers that provide firm, physical boundaries to root growth that is uncommon in field production. The containers may hold a soilless potting mix or a completely inert material that holds little water and no nutrients. In this case, water and nutrients must be supplied constantly. This system is often referred to as hydroponic. Another hydroponic method is to grow the crop (typically lettuce) on floating trays that allow the roots to grow into the nutrient solution that fills the troughs on which the trays float. Greenhouse vegetable production also often relies on artificial lighting of specific wavelengths to supplement sunlight. Commercial greenhouse vegetable production tends to be highly mechanized and involves computerized heating, cooling, lighting, irrigation, spraying, and (increasingly) planting and harvest systems.

Establishing and managing vegetable production systems typically involve numerous interventions by growers. Different strategies and practices are used to meet goals of yield, profitability (in the short- and long-term), and envi-

ronmental stewardship. Collectively, these strategies and practices comprise an overall system or approach to farming and gardening. Current major approaches to farming and gardening, including vegetables, are often described as conventional, organic, and sustainable. Only generalizations are possible regarding the major approaches, which are not formulaic because recipes or blueprints for farming or gardening under all circumstances are unavailable and unrealistic.

Conventional

Overall, in conventional approaches, strategies for protecting crops from diseases, pests, and weeds include the application of synthetic pesticides, which may or may not be based on naturally occurring compounds, as a central component. Pesticides are applied to soils, seeds, plants, and harvested units. Strategies for protecting crops in conventional systems may also include the use of crops genetically modified to resist or tolerate diseases, pests, or herbicides. Some genetic modifications may combine portions of genomes that may not otherwise combine naturally. Crop nutrient management strategies in conventional systems typically involve the application of synthetic fertilizers, often made from the mining and chemical combination of minerals, as a main component. The content of synthetic fertilizers is tailored to meet the needs of specific combinations of vegetable crop, soil, climate, water source, and other factors. Synthetic fertilizers tend to be clearly described and consistent in terms of makeup, are relatively inexpensive, and are widely available.

Organic

Organic agriculture has been defined as "an ecological production management system that promotes and enhances biodiversity, biological cycles, and soil biological activity" by the National Organic Standards Board (NOSB). In the same document, the NOSB outlined the principal goal and general methods of organic production and guidelines for and

Figure 18–8 Certified organic vegetable farm. *Source:* Margaret McMahon, The Ohio State University.

Figure 18–9 Straw and plastic mulch in a tomato field. Mulches help to suppress weeds, moderate soil temperatures, and reduce water loss from the soil. With time, organic mulches such as straw add nutrients and organic material to the soil. *Source:* Dr. Matt Kleinhenz, The Ohio State University.

possible outcomes of this practice. Strict and widely accepted definitions of any farming or gardening system are very difficult to develop. This difficulty arises from the fact that few definitions capture the full, contemporary range of functional production systems, which change over time. With regard to farming, the Organic Foods Production Act of 1990 required the USDA, in conjunction with the NOSB, to develop national standards for organically produced agricultural products, in part to help minimize ambiguity about the organic approach to farming, including standards for food-labeling purposes. Definitions of organic farming have multiple aspects. Because the NOSB program was created through an act of Congress, only certain practices qualify as organic, and labels on food items are used to indicate whether they are certified organic.

In practice, the USDA provides accreditation to local and regional agencies, and they in turn certify whether farmers applying for certification status have met the associated requirements. Approval of an application for certified organic status requires payment and extensive documentation of production practices (Fig. 18–8).

Organic vegetable farming and gardening involves a strong reliance on cultural practices and biological principles for weed, disease, pest, and nutrient management. For example, weed growth is limited through shading (e.g., with a mulch) (Fig. 18–9) and timely mechanical or hand cultivation. Organic growers attempt to limit damage to crops from diseases or pests through the use of natural plant resistance (e.g., variety selection) and fostering increases in the populations of organisms that limit the growth or spread of the diseases or pests through competition, parasitism, feeding, or other natural processes. Vegetable varieties used in commercial organic systems may not be a product of genetic engineering. Crop nutrient requirements in organic systems are met through a combination of practices. Rotation (i.e., planting different crop types in the same soil in alternate seasons),

particularly with legumes such as alfalfa, clover, or vetch, helps provide crop plants with nitrogen. Likewise, cover cropping can help retain soil nutrients in the root zone—nutrients then available to subsequent cash crops. Cover crops may also help maintain or improve soil tilth and habitat diversity and help minimize soil erosion and weed growth. Cover crops are typically not grown to harvest, sell, or consume, but rather for the other benefits they provide. Crop nutrient requirements in organic vegetable systems are also met partially through the application of organic amendments, such as composts and manures, to the soil. To manage diseases, pests, weeds, and nutrients, organic vegetable growers may also attempt to schedule planting and harvest practices so that their crops may avoid periods documented to contain the heaviest pest, disease, or weed pressure or greatest potential nutrient deficiencies. The rate of biological processes (e.g., insect, weed, pathogen growth) is governed by abiotic factors, such as temperature, light, moisture, and nutrient levels. These conditions and the pest and disease pressure they create have been monitored, recorded, and shared with growers in many areas for years and these activities continue today. College and university research extension activities in integrated pest management involving growers have been central to this effort. Equipped with pest and climatic information, growers may have the option of planting or harvesting their crops when the potential damage from diseases, insects, and weeds is lowest. However, having product to sell or consume at nearly any time is important to growers (organic or not), and it precludes them from fully employing alterations in planting and harvest dates as a risk management tactic. Finally, specific chemicals and compounds may be applied in organic production as part of overall pest and disease management strategies. For farms, applied materials must be approved for use by certifying agencies. Examples include copper-based fungicides, *Bt*-based insecticides, and compost teas.

Sustainable

Sustainable approaches to vegetable production integrate components of conventional and organic management. The USDA Sustainable Agriculture Research and Education (SARE) program describes sustainable agriculture as encompassing diverse methods of farming and ranching that are more profitable, environmentally sound, and good for communities (see **http://www.sare.org/coreinfo/consumers.htm**). Sustainable production may involve the selective, targeted use of pesticides (organic-approved, synthetic) to augment biologically and culturally based disease, insect, and weed management, a key component of integrated pest management. Vegetable growers may also release, apply, or attract beneficial organisms that act as competitors, predators, or parasites for organisms that would otherwise damage crops. Likewise, farmers and gardeners employing sustainable approaches often employ a combination of rotation, cover crops, and synthetic fertilizers to meet crop fertility requirements. Sustainable approaches are increasingly common among vegetable growers. Still, it is important to note that sustainability has multiple dimensions (e.g., financial, social, environmental) and may best be considered as a guiding principle for the evaluation of cropping practices within a context larger than production.

Markets for Commercially Grown Vegetables

Once grown and harvested, farm-grown vegetables enter one of two leading markets. In the *fresh* market, vegetables are packaged, at facilities on or off the farm, and distributed to buyers or consumers with relatively minimal pretreatment or preparation. Pretreatment or preparation includes sorting, trimming, washing, cooling, and packaging. Fresh market vegetables may be sold by the farmer directly to consumers

at markets and produce stands, via the Internet, or as shares in a community-sponsored agriculture (CSA) type farm. Fresh market vegetables are also sold to grocers, market managers, restaurateurs, or wholesalers-distributors for resale, possibly on a contractual basis.

Processing markets add value and diversity of use to freshly harvested vegetables. Processing vegetables are packaged and distributed to consumers after being, for example, frozen, precooked, or dehydrated. During processing, vegetables may be used to create another product (e.g., potato chips, french fries). Processed vegetables may be packaged alone or in combination with other vegetables. Processing plants are highly engineered facilities owned by individuals, groups, or multinational corporations employing few to many people. Transactions culminating in the transfer of vegetables from farms to processors often involve contracts set up before harvest.

The distances and time differences between sites of production and consumption of vegetables, particularly fresh market, may be minimal or large. In most cases, consumption occurs many miles from and days after production and harvest, requiring sophisticated harvest and post-harvest procedures to maintain product quality as viewed by consumers. However, a growing percentage of vegetables are being grown and consumed within smaller geographic areas as interest in local food systems increases (see, for example, **http://www.cias.wisc.edu/**; **http://www.sustainable.org/index.html**).

Within a county, state, region, or country, whether a vegetable commodity is grown commercially and at what scale depends on consumer- and production-related factors. Perhaps the most important among consumer-related factors may be the overall demand for the product: greater demand creating a potential need for more growers. Demand for vegetables appears to be driven by their availability, convenience, and diversity, and by consumer perceptions of their eating quality (based on consumer- and product-based traits) and potential health or nutritional value. In general, farmers' interests in the production of various vegetables is based on their potential to generate profit and the farm's ability to meet market demands in terms of quantity, quality, and consistency.

STEPS IN VEGETABLE PRODUCTION

Challenges in vegetable production, particularly outdoors, arise from the fact that production is strongly influenced by the natural environment (e.g., soils, climate, weeds, and plant diseases and pests) and that the natural environment is variable in time and space. Gardeners and farmers deal with these challenges through careful planning and execution in key areas. Successful vegetable farmers must also follow principles integral to the success of other types of businesses, including (1) studying the market and being prepared to meet its requirements, (2) producing a high-quality product efficiently, (3) packaging the product appropriately and delivering on time, and (4) innovating. The key steps in vegetable production are discussed in the following sections.

Site Selection

Factors important in selecting the site for successful vegetable production include climate; soil type; the size and total cost (land, utilities, taxes, etc.) of the site; and the cost and ease of access to water for irrigation, pesticide application, crop washing, and packing. The proximity of the site to suppliers, support industries and personnel (labor), and the market is also important to farmers.

Site Preparation

Once selected, the site's drainage must be checked and improved, if necessary. Installing tile or raised beds or leveling land can improve drainage. Thereafter, in field production, plowing, discing, and other tillage is often needed to prepare the soil for planting operations. Preparing soil usually also involves fertilizer application and steps to minimize weeds and soilborne diseases and pests through, for example, the application of pesticides and fumigants or the use of cultivation, solarization, plastic mulches, or trap crops.

Variety Selection

In summary, the outcome (e.g., marketable yield) of vegetable production at any scale results from the interaction of the crop's genotype with its growing environment (physical, chemical, biological). In farming, fresh and processing markets maintain strict standards for the delivery date, appearance, shape, dimensions, weight, and sensory and chemical properties of crops and products. Crops not meeting standards set by the market typically are unsalable, creating financial and other burdens for farmers. Often, market-based crop standards and desirable outcomes for gardeners can be met only through specific combinations of genotypes and growing environments. Therefore, variety selection is key for all vegetable growers in part because varieties differ in their ability to withstand abiotic (e.g., temperature, moisture, daylength extremes) and biotic (e.g., disease, insect) stress.

Planting

Vegetable propagules include seed and transplants or slips. Crops are established through direct seeding or transplanting, with transplants and slips set by hand or machine. Fertilizers and pesticides may be applied simultaneously. In farming, the population of crop plants per unit area and their arrangement is manipulated to achieve production and marketing goals. Most crops are planted in rows, with the distance between adjacent plants within the same row and adjacent rows specific to the crop, genotype, growing environment, and market.

Management During Crop Growth and Development

Crops are irrigated; fertilized; mechanically or chemically pruned or trained; and protected from damage due to diseases, pests, or weeds. In greenhouses, light and temperature levels are also optimized. The use of remote or on-site scouting or monitoring (soil, crop, weather, pest, disease, weed) tools and services is an established and increasingly common practice in vegetable farming. Information gained through scouting and managing can have significant, variable benefits to individual farmers and regional industries.

Harvest

Crops are visually assessed for harvest readiness. Farmers also may employ instruments (e.g., refractometer, scale) to assist in evaluating market maturity. Careful handling and timeliness of activities are components of harvest procedures that proceed at crop maturity so that crop yield and quality losses are minimized. Keeping crops free of abiotic and biotic contaminants during harvest and post-harvest handling is also important. Crops are harvested by hand and/or machine. In fact, the significant amount of hand labor required to harvest and prepare many vegetable crops for market is an important distinction between vegetables and other crops, including agronomic. The need for production and harvesting equipment specifically designed for use on a single crop is also a characteristic of vegetable production. In contrast, equipment such as planters and combines can be used with only a little modification on several agronomic crops.

Packaging and Post-Harvest Management

Regardless of the market for which they are destined, commercial vegetable crops are placed in containers or trucks before delivery or shipment. They may also be sorted, trimmed, washed, cooled, and packaged in units of market-specified weight or volume.

Record Keeping and Maintenance and Repair of Facilities and Equipment

These activities are needed throughout all aspects of vegetable farming. Vegetable farming requires the tracking, documentation, and analysis of a wide range of production, market, financial, sociopolitical, regulatory, and other information. Farmers use increasingly powerful digital and paper-based tools and resources to assist them in these activities. Vegetable farming also involves the use of specialized buildings and equipment that require ongoing maintenance and repair.

Continuing Education

Vegetable farming is a professional occupation requiring technical and practical understanding of an increasingly complex array of areas (e.g., biology and ecology, chemistry, engineering, marketing and economics, public relations, organizational or association development, and management). Many gardeners also have a strong interest in improving their production systems. Therefore, many growers commit to receiving ongoing education in formal and informal settings ranging from conversations with successful peers to credit courses at universities and other institutions (Fig. 18–4).

VEGETABLE CROPS GROWN FOR FRUITS AND SEEDS

Many of the most popular vegetables we eat, for example, green and dried beans, corn, cucumbers, tomatoes, and many others, are botanically classified as fruits or seeds. Convention, though, has us overlooking the botanical classification in favor of a dietary concept. We generally think of fruits as being eaten for their sweetness and dessertlike quality, while vegetables usually are not very sweet. Some of these types of vegetables, for example, tomatoes, can be eaten raw or cooked, while others such as sweet corn, green beans, and dried beans are usually cooked. The stage of maturity at the time of harvest (harvest maturity) depends on the crop. We want the part we are harvesting to be mature enough to have the qualities we desire (sugar or starch content, flavor, color, etc.). If harvest occurs too early, these qualities may not develop. If harvest occurs too late undesirable traits such as starch formation in sweet corn or softening of the tissues may develop. A product such as tomato can be harvested at a slightly earlier stage, when it is firmer and less easily bruised during harvest and transport. However, the vegetable has to be able to continue to mature after harvest. If maturation stops at the time of harvest, then the vegetable must be harvested when it is at harvest maturity.

Beans, Snap or Green (*Phaseolus vulgaris* L.) and Lima Beans (*Phaseolus limensis* Macf.) FABACEAE

The genus *Phaseolus* includes a number of "vegetable" bean species whose immature fleshy pods and immature seeds, and/or dried seeds are a highly valued food. Dry beans in particular are a nutritionally important protein and carbohydrate food source. The common bean and other *Phaseolus* members are native to Central and South America, and were domesticated and widely used long before the arrival of European explorers.

The better-known common bean types are the snap bean and lima bean, and various types of field beans (shell beans, kidney beans, mung, etc.) grown for their mature dry seeds. Snap beans and limas are grown as annuals and have differing vine types: those usually having indeterminate fruiting and grown on supports (pole), and the bush forms, usually determinate and grown unsupported. Because the immature pods and seed of snap beans are the edible portions, selection was directed toward developing cultivars having thicker, fleshy pods of low fiber and slowly developing seed.

Pod fiber development occurs concurrently with seed maturation and is undesirable. For the dry seed bean types, rapid seed development is the more important criterion, and pod fiber is not of interest. Lima beans, whether grown for fresh market or for processing, are harvested when their seeds have enlarged but before they are fully mature. However, fully mature dried lima bean seeds are also an important product. Beans are self-pollinated and propagated from seed.

Both snap and lima beans are warm-season crops requiring frost-free growing periods, preferably without large temperature fluctuations. Lima beans are very sensitive to both cool and high temperature fluctuations, which interfere with flower fertilization and may cause abortion of blossoms or young pods.

Cucumber (*Cucumis sativus* L.) CUCURBITACEAE

The cucumber is a prostrate, branching vine native to Asia, most probably northern India, and introduced to Africa and Europe before written history. The plant is an annual with hairy leaves and tendrils, commonly with a **monoecious** flowering habit. The first flowers, generally staminate, are followed by pistillate flowers that, when fertilized, develop into fruit. In field production, insect pollinators are necessary. Many newly introduced cultivars have predominantly pistillate flowers, producing many fruit per plant, and therefore are especially desirable for mechanical harvesting of processing-type cucumbers. Cultivars used for glasshouse production are gynoecious as well as parthenocarpic.

Eggplant (*Solanum melongena* L.) SOLANACEAE

The eggplant, also known as aubergine or brinjal, is thought to be native to southeast Asia and has been grown in China for many centuries. Eggplant is grown commercially in many US states, however, with most sold to nearby markets. A great variation of types, varying in skin color and shape, are produced.

Okra (*Abelmoschus esculentus* L. Moench) (*Hibiscus esculentus* L.) MALVACEAE

Okra, also called gumbo and lady's finger, is African in origin, with early introduction into Mediterranean areas and India, where it continues to be very popular and appreciated. A warm-season crop with climatic requirements very similar to those for cotton, okra also requires a long growing season with hot days and warm nights. A significant quantity of okra is grown because of its gummy or thickening characteristics in the preparation of soups and stews.

Peas (*Pisum sativum* L.) FABACEAE

The garden pea, sometimes called the English, green, or common pea, is a viny annual cool-season plant grown for its edible seed, although some cultivars are also grown for their immature edible pods. Neither the wild progenitor nor the early history of the pea is well known. Probable centers of origin are considered to be Ethiopia, the Mediterranean area, and central Asia, with a proposed secondary center of diversity in the Near East. Greek and Roman writers mentioned peas, but not until the seventeenth century were varieties described.

Dried peas, later reconstituted and used in soups and other products, are an important crop worldwide but less so in the United States. In contrast to green peas, which are harvested when immature, dried peas are allowed to mature and dry, which allows for easier storage and transport.

Peppers (*Capsicum annuum* L.) SOLANACEAE

Peppers are an important vegetable commodity highly prized for the flavor, color, vitamin C, and pungency they provide to the human diet. Peppers are a valued spice throughout the world and an important contribution from the Americas to the world's spices. The plants are shrubby perennials, although usually grown as herbaceous annuals in tropic, subtropic, and temperate regions. Peppers are native to the Central and South American tropics. Columbus brought the pepper to Europe. From there, it was introduced to other areas, and into China in the late 1700s.

Capsicum annuum is the most extensively grown species. *C. Frutescens* is represented by the Tabasco type; *C. chinesse, C. pubescens,* and *C. baccatum* are also cultivated. Common black pepper (*Piper nigrum*), also used for seasoning foods, belongs to a different botanical family and should not be confused with the Solanaceous pepper.

The plant is frost-sensitive and grows best at a warmer temperature, preferably within a range of 25°C to 30°C (77°F to 86°F). Excessively high temperatures, above 38°C (100°F), especially at flowering, can result in poor fruit set.

Peppers are interesting in the number of types grown, each having various preferred uses. These types are distinguishable mostly by their characteristic shape and include types commonly known as bell, pimiento, squash or cheese, ancho, anaheim, cayenne, cuban, jalapeno, cherry, wax, and tabasco. Fruit color is also used as a distinguishing characteristic. Peppers are green, yellow, various shades of brown, purple, and red in color. They also differ in pungency; some are extremely pungent, others very mild or sweet. Pungency is contributed by the capsaicinoid content, the principal compound being capsaicin.

Pumpkins and Squashes (*Cucurbita* spp. L.) CUCURBITACEAE

The term *pumpkin* or *squash* does not have a precise botanical meaning. These vegetables are *Cucurbita* species of the *Cucurbitaceae* family, and the terms *squash* and *pumpkin* are used interchangeably. While there is less confusion regarding summer squash (cultivars used for their immature fruit and having soft rinds), there remains confusion between types of winter squash and pumpkins (mature fruit having hard rinds). These commodities are important in the diets of much of the world's population, and they are grown from the tropics and into much of the temperate zones. These commodities, in addition to their multiple uses as a vegetable, are also used for their seed, as ornamentals, and for livestock feeding.

The center of origin of the genus is primarily Central America, ranging into Mexico and the northern parts of South America. The major volume produced is contributed by four species, with *Cucurbita pepo* as the largest contributor, other species being *C. moshata, C. maxima,* and *C. mixta.* Species are distinguishable by their stem, peduncle, seed, and leaf variations. All are monoecious with generally stable sex expression characteristics and are dependent on insect pollinators for fertilization of the flowers. Hybrid seed production is managed through manual removal of unwanted staminate flowers, or suppression of their development through the use of chemical compounds. Some interspecific hybridization is possible and is used to develop new cultivar types and for the incorporation of disease resistance.

Having a tropical origin, the genus is frost-sensitive. Warm temperatures above 25°C (77°F) are more favorable for growth and development.

The winter squashes/pumpkins are generally larger-size fruit. A major distinction is that these are harvested after they have developed hard rinds and well-developed seed. Fruit shape and color vary greatly. These variations along with their irregular surfaces (warts) and their durability make some cultivars popular for use as ornamentals. Some cultivars achieve enormous size, some intentionally grown for exhibition purposes. Single-fruit weights as much as 250 kg (550 lb) have been obtained. Their holding or storage capabilities are highly important. This characteristic is assisted by the development of a hard rind that resists disease and deterioration. Better-known types of winter squash and pumpkins are acorn, butternut, hubbard, banana, marrow, spaghetti, and, of course, the popular pie and Halloween pumpkins.

Summer squashes are harvested at an immature stage when the rind is still soft and the seed are underdeveloped. Delayed harvest results in further development of the fruit with overenlargement, toughness of the rind, and hardening of the seed—all contributing to loss of quality and value. The converse is true with the winter squashes and pumpkins, which are harvested when fruit are fully mature, with hard rinds, firm flesh, and fully developed seed.

Sweet Corn (*Zea mays* L. var. *rugosa,* Bonaf.) POACEAE

Corn, which originated in the Americas, was domesticated at least 5,000 years ago. It was a staple grain crop for the Native Americans and remains a major world grain crop. Only since the middle of the nineteenth century have sweet corn types of corn achieved status as a vegetable commodity. The popularity of sweet corn is at a maximum in the United States, where it is a frequently grown vegetable in many home gardens and where a substantial level of commercial production occurs in many parts of the country. The crop has steadily increased in popularity in other parts of the world, notably in Europe.

The crop thrives in humid environments having hot 30°C (86°F) days and warm nights. Through plant breeding efforts, the temperature range has been broadened so that cultivars are available that perform well at lower temperatures without the penalty of lower crop performance. Extensive use of hybrid cultivars provides producers with greater adaptation, yield, and kernel quality attributes.

The range of characteristics, especially those related to eating quality, available in sweet corn cultivars has steadily increased since the early 1980s. Following the identification of several mutations controlling various kernel traits, plant breeders and others have used an array of techniques to create sweet corn kernels with different combinations of sweetness, aroma, texture, and pericarp thickness. The time through which kernels remain sweet after harvest has also been increased, facilitating the development of long-distance shipping to fresh markets.

Tomato (*Lycopersicon esculentum* Mill.) SOLANACEAE

Tomatoes are one of the most popular, versatile, and widely grown vegetables—known and grown throughout the world and in nearly every home garden. Although not a new crop to the native populations of tropical and subtropical America, the tomato is relatively new to the rest of the world. Columbus returned with tomatoes to Europe, where they were first grown for ornamental purposes. Cultivation for a food crop was soon established along with its dispersion throughout Europe and other areas. The crop began to be cultivated in North America in the early 1700s.

Tomatoes are warm-season perennials grown as annuals. They are frost-sensitive, and many cultivars require about three to four months from seeding to produce mature fruit. Production is generally optimized under conditions that provide warm, above 27°C (86°F), temperatures with clear dry days.

The production of tomatoes in structures such as glass or plastic houses or plastic-covered tunnels is a significant and very important activity. This operation is expensive and highly specialized and requires a high level of technology and management.

VEGETABLE CROPS GROWN FOR FLOWERS, LEAVES, OR STEMS

A diverse and large number of crops are included in the group of vegetables grown for the consumption of their floral, leaf, or stem parts. One similarity is that most crops in this grouping are cool-season vegetables. Some, such as cabbage and lettuce, are produced in significant amounts. However, many others—for example, artichokes, asparagus, or rhubarb—are rather specialized, and their production is relatively limited. These vegetative types of vegetables are not considered to be major caloric providers to human diets. Nevertheless, they do have considerable nutritive value, notably for their contribution of vitamin A, vitamin C, other vitamins and minerals, as well as dietary fiber. Their other attributes are the texture, flavor, color, and variety they provide to meals.

Artichoke (*Cynara scolymus* L.) ASTERACEAE

The globe artichoke, cultivated for over 2000 years, is a thistlelike herbaceous perennial plant native to North Africa and other Mediterranean areas. In early times, the foliage was eaten. However, the preferred edible portion is the enlarged but immature flower bud, comprised of numerous overlaid bracts, and its fleshy receptacle. The globe artichoke is sometimes confused with the Jerusalem artichoke, *Helianthus tuberosus*. Although both are members of the ASTERACEAE family, they are different, the edible portion of the Jerusalem artichoke being the underground fleshy tubers. When eaten, these have a taste that resembles that of the globe artichoke, perhaps a reason for the confusion.

A frost-free climate with cool, foggy days is a requirement for achieving quality production of artichokes. Hot, dry weather causes rapid growth accompanied by rapid development of the internal floral tissues and fiber, both detrimental to tenderness. High temperatures also cause the bracts to separate and spread apart, which is detrimental to market appearance. Freezing temperatures blister the epidermal tissues of the bracts and are a detraction to market appearance. When not seriously frosted, careful and rapid post-harvest handling permits marketing during periods when the crop is in short supply.

Artichokes are propagated vegetatively from stem pieces or from rooted offshoots from the base of older plants. Plantings last for five or more years before reestablishment is necessary. Farmers generally cut back foliage down to the soil surface following final harvest in late spring or early summer. Regrowth during the summer is delayed by restricting moisture for several months. This method allows for scheduling the resumption of production.

Asparagus (*Asparagus officinalis* L.) LILIACEAE

The edible portion of this perennial plant grows from fleshy underground rhizomes called crowns. The plants are dioecious, having either staminate or pistillate flowers (Fig. 18–10). Asparagus is thought to be native to the eastern Mediterranean region.

Better quality and higher yields occur when the crop matures during periods when temperatures are about 16°C to 25°C (60°F to 77°F), especially after a long period of fern growth. In colder climates, the growing season is shortened because of the limited period for fern production. A reason-

Figure 18–10 A female asparagus plant. These plants bear only pistillate flowers. The flowers were fertilized by pollen from a nearby male plant. Note the seed berries. Asparagus is a good example of a dioecious plant (male and female flowers on separate plants).

able period of active fern growth is necessary to supply the crown with the products of photosynthesis that the fern produces following the harvest period. The photosynthate produced is translocated to the crowns to be utilized for the following year's production of spears. In warmer areas, sufficient fern growth occurs and the crowns can sustain a longer harvest period without exhaustion of their food reserves. During warmer periods, spear growth is very rapid, with greater fiber development within the spears and with a tendency for the spear tips to lose some of their compactness characteristics. Plantings are maintained for five to ten or more years.

White asparagus production is a popular choice in markets, although some green asparagus production and consumption exceeds that of white asparagus. Such production depends on blanching the elongating spears, usually with a mounding of soil cover. Harvest is more difficult because it must be done just before the spear tips emerge from the soil. The high labor requirement for such production is the main discouragement for its continuing use.

Broccoli (*Brassica oleracea* L., Italica Group) BRASSICACEAE

The flowers of the cole crops—broccoli, Brussels sprouts, cabbage, and cauliflower—have four petals arranged in the form of a cross. This feature gave the family the name CRUCIFERAE (now BRASSICACEAE), which in Latin means "cross bearers," and from which the English words *crucify* and *crucifix* are derived. The word *broccoli* comes from the Latin word *bracchium*, meaning arm or branch. The tightly grouped fleshy terminal stems, young leaves, and young flower buds and bracts are the plant parts of broccoli that are consumed.

The plant is native to eastern Asia and the Mediterranean areas and was introduced a relatively short time ago into the United States, in about the 1920s. Previously it was not widely cultivated in Europe. However, because of its high nutritive value and ease of preparation, it has become very popular, and its production has expanded to many other parts of the world. The developing inflorescence (head) is hand harvested while it is still immature, compact, and absent of open flowers by cutting from the stem, resulting in a head about 15 to 25 cm (6 to 10 in.) in length, and about 10 to 20 cm (4 to 8 in.) in diameter.

Brussels Sprouts (*Brassica oleraceae* L., Gemmifera Group) BRASSICACEAE

This vegetable, which tastes like cabbage, probably evolved from a primitive nonheading Mediterranean cabbage, after its progression to northwestern Europe. The plant received its name from the city of Brussels, where it was first reported to be grown; it continues to be a popular vegetable throughout that area. The plant develops leafy buds resembling miniature cabbagelike heads about 2.5 to 5 cm (1 to 2 in.) in diameter in the leaf axils along the single tall unbranched stem (Fig. 18–11). The plant grows 60 to 90 cm (2 to 3 ft) tall and requires a cool climate to develop compact, quality buds. The crop is relatively tolerant to cold temperatures and can withstand slight freezing.

Cabbage (*Brassica oleracea* L., Capitata Group) BRASSICACEAE

Cabbage is well adapted for growth in cool climates. It is a biennial plant and tolerant of slight freezing. On the other hand, cabbage is somewhat more tolerant of warmer temperatures than many other cool-season crops. The edible portion is essentially a large vegetative terminal bud, formed by overlapping of numerous leaves developing over the growing point of its shortened stem, and is called the head. Cabbage is believed to have developed from a wild type of a nonheading plant on the east coast of England and on the western European coast. Evidence indicates that the early Egyptians used cabbage for food and medicine, and that the Mediterranean area was its origin.

ACHILLES

Figure 18–11   In the early stages of development, Brussels sprouts closely resemble cabbage seedlings. As the plant matures, however, the main stem elongates 60 to 90 cm (2 to 3 ft), and the axillary buds along the stem develop into small cabbagelike heads about 2.5 to 5 cm (1 to 2 in.) in diameter.

Cauliflower (*Brassica oleracea* L., Botrytis Group)
BRASSICACEAE

The origin of cauliflower is believed to be the eastern Mediterranean area, with subsequent development perhaps from sprouting broccoli in northwestern Europe, where it has been known for centuries. Both annual and biennial forms are cultivated. Among the Brassica crops, cauliflower has the most exacting climatic and cultural requirements. Moderately uniformly cool climates are best for cauliflower, and production sites are frequently near large bodies of water to benefit from their climate-moderating influence. Preferred mean temperatures during growth and crop maturity are 15°C to 20°C (59°F to 68°F). Unlike cabbage, cauliflower is much more intolerant of frost or high temperatures. Cultivars known as tropical types having better warm-temperature

adaptation are grown in large volume in many warmer temperate regions, namely, India and China.

The edible product, called the head or curd, is mistakenly considered to be floral tissue but is instead prefloral, fleshy, apical meristem tissues. The curd consists of the repeatedly branched terminal portion of the main axis of the plant, comprised of a shoot system with short internodes, branch apices, and bracts. High temperatures result in poor head formation and quality, with accompanying market defects such as bract development within the curd, loss of compactness, and ricy curds.

Direct exposure of the curd to sunlight tends to discolor the otherwise white curd surfaces to a creamy yellow, which is a market defect but does not otherwise affect the edible quality. Leaves cover and protect the curd as they form, but as the curd enlarges, the inner leaves are forced apart and the curd is exposed to sunlight. To protect and blanch the enlarging curd, the long outer leaves are gathered together over the top of the curd and tied together to keep out light. Sometimes self-blanching cultivars are used.

Celery (*Apium graveolens* L. var. *dulce* Mill.) APIACEAE

The former family name, *Umbelliferae,* came from the characteristic umbel form of the inflorescence, and the family includes celery, carrots, parsnips, and parsley. Writings indicate that the first use of celery was as a medicine in the fifth century A.D., and as a cultivated food crop in the early 1600s. The plant is thought to be native to the Mediterranean area, although wild types are found widely distributed in other areas. The crop's world importance is relatively small, both as a vegetable and for the use of its seeds and foliage in seasoning foods. Celeriac, A. *Graveolens* var. *rapaceum,* also known as knob or root celery, is closely related and finds similar use, particularly among northern Europeans.

Celery, a biennial plant, has rather exacting climatic requirements. Optimum daytime growing temperatures range between 16°C to 21°C (60°F to 70°F), with night temperatures slightly lower. Exposure of temperatures less than 10°C (50°F) initiates vernalization and seed stalk development and should be avoided.

Chive (*Allium schoenoprasum* L.) AMARYLLIDACEAE

Chives are small, bushy, onionlike perennials grown for their delicate tubelike leaves, which are used in salads, soups, stews, omelets, and various cheese products, and to flavor numerous other food products. It appears that chives are native to the northern hemisphere, and wild forms are found in North America and Eurasia. The crop is of minor economic importance, although it has been cultivated for centuries in Europe and the Orient. Chives are cool-season, cold-tolerant plants with climatic requirements similar to onion and garlic.

Collards and Kale (*Brassica oleracea* L., Acephala Group) BRASSICACEAE

Collards and kale are plants that resemble cabbage but are nonheading. They are closely related to wild cabbage and have been cultivated for many centuries; they are considered to be the least-developed members of the *Oleracea* species. Collards and kales are winter-hardy biennials and tolerate both somewhat lower and somewhat higher temperatures than cabbage. These plants are grown for their abundant foliage, which is usually consumed as a cooked vegetable or as an ingredient in other foods. Some kale cultivars, in addition to having attractive green coloration, also have foliage with color variations ranging from purple, red, and pink to white. They also have very crinkled leaves and find use as ornamental plants.

The distinction between collard and kale is not clear. Collards have a relatively smooth leaf blade, whereas kales, with some exceptions (forage kales), generally have crinkled leaves. Some cultivars have pronounced curly foliage. Kale cultivars also vary in height, with dwarf, medium-tall, and tall types. Some kale cultivars, which include the thicken stem or marrow-stem types, are used for animal forage.

Endive (*Cichorium endivia* L.) and Chicory (*Cichorium intybus* L.) ASTERACEAE

Early records indicate that forms of endive and chicory were grown for many centuries and were well known to the early Egyptians. The probable origin of endive was eastern India, whereas chicory appears to be a native of the Mediterranean region. Both are cool-season crops, somewhat similar in appearance to lettuce. Both endive and chicory initially produce a rosette of leaves that develop from a short stem. When mature, some types form a head, others remain as an open rosette, and some are intermediate between these forms. Additional variation exists with regard to leaf size and shape: some are broad-leaved; others have narrow, divided, and rather crinkled leaves; some have smooth, broad-leaved foliage. Those endives with broad crinkled leaves are commonly known as escarole. Some forms of endive and chicory also develop anthocyanin and have reddish-tinged leaves. Many endive and chicory cultivars are annuals or are grown as annuals, although some important chicories are grown as biennial crops.

Endive is often confused with chicory, *C. intybus,* because many forms of chicory are similar in growth and appearance to endive. One form of chicory, known as witloof and also as Belgium and/or French endive, is a popular crop in northern Europe. Although associated with the name *endive,* it is a chicory. It differs from most endive, mainly in how it is produced. Witloof is the product of the second year's growth of the chicory plant. The purpose of the first year's growth is to produce the thick fleshy storage taproot, used to produce the second season's growth. These roots are harvested at the end of the growing season. The plant's foliage, which strongly resembles cos lettuce, is removed, except that the growing point of the stem is retained and remains attached to the root. The roots are subjected during storage to cold-temperature vernalization at 0°C to 5°C (32°F to 41°F) for several weeks or more and then are placed in soil or peat forcing beds or in containers for **hydroponic** culture. The forcing procedure involves placing the roots in an environment that favors growth of the primary bud of the stem, which, were it allowed to continue growth, would become the seed stalk. The growth of the primary bud is kept compact by controlling forcing conditions and procedures. The market-mature compact bud is called a chicon. The temperature in the area of the roots is maintained at a higher level than the air temperature above the roots. This practice tends to maintain the compaction of the developing bud. Through heating of the soil beds or the hydroponic solutions, along with the use of insulation for regulation of temperature and relative humidity, and, when necessary, the use of cooling equipment, the witloff producer selects and manages the environment needed to initiate bud growth and its development into the compact chicon.

The range of forcing temperatures is 15°C to 20°C (59°F to 68°F) in the root zone, and 12°C to 17°C (54°F to 63°F) in the area of the developing chicons. A compact marketable chicon develops within twenty to thirty days, depending on cultivar and temperatures.

Kohlrabi (*Brassica oleracea* L., Gonglylodes Group) BRASSICACEAE

Kohlrabi is a cool-season biennial plant grown for its turnip-like, above-ground enlarged stem. Compared with other vegetables, it is a relatively recent food crop. Its delicate flavor compares with that of the turnip, although it is milder.

Leek (*Allium ampeloprasum* L., Porrum Group) AMARYLLIDACEAE

Leek, a biennial plant closely related to the onion, is grown for its sheathlike arrangement of elongated, closely overlapped, blanched leaves. The leaves are fleshy and solid, but generally do not form bulbs as onions do. Under long-day conditions, however, a slight thickening near the base of the leaves resembling bulbing may occur. Exposure to low temperatures, if of sufficient duration, may result in bolting. Leek is believed to be a native of the eastern Mediterranean region, where its food use was known as much as 3,000 to 4,000 years ago.

Lettuce (*Lactuca sativa* L.) ASTERACEAE

Lettuce is a plant with an ancient history. Its origin appears to have been in Asia Minor. There is evidence that forms of lettuce were used in Egypt during the period about 4500 B.C. The Romans grew types of lettuce resembling the present romaine cultivars as early as the beginning of the Christian era, and the crop was well known in China about the time of the fifth century.

Cultivated lettuce is an annual plant closely related to the common wild or prickly lettuce (*L. serriola*) weed. The many forms of lettuce that can easily intercross are often grouped into several types. These types include the usually large and dense heading crisphead cultivars that have brittle-textured foliage that is tightly folded. The outer leaves are dark green and the inner foliage pale and mostly absent of chlorophyll. The butterhead cultivars, whose heads are smaller, are relatively soft and less dense, and are comprised of broad, soft, smooth, or slightly crumpled oily textured leaves. The romaine or cos cultivars have elongated coarse-textured leaves that form a loaf-shaped, semi-compact, low-density head. Outer leaves are darker green, coarse textured, with heavy ribs. Inner leaves are smaller and lighter in color, with interior leaves nearly absent of chlorophyll. The leaf lettuce cultivars, which are leafy but loose, are nonheading cultivars of varied color, leaf types, and textures. An additional lettuce type, known by several names, such as celtuce, asparagus, and stem lettuce, is not widely produced but is most popular in the Orient. It is grown for its elongated, thick stems that are peeled and used as a cooked vegetable. The leaf blades are of lesser significance compared to the stem. Lettuce is a cool-season crop and grows best within a temperature range of 12°C to 20°C (55°F to 68°F).

Mustard (*Brassica juncea* L., *B. hirta* Moench., *B. Nigra* Koch, *B. carinata*) BRASSICACEAE

Mustards are cool-season annual plants. Their abundant foliage develops in a rosette pattern on a relatively short stem. Leaf blades are usually broad, with some types having smooth-textured leaves and others extremely curled leaves. The foliage is strongly flavored, although a wide range of pungency levels most often occurs. This flavoring characteristic makes the various mustards popular vegetables for use in salads and as cooked greens. Mustard seeds also vary in flavor and pungency and find use in culinary condiments, and also as a vegetable oil. The origins of the mustards *B. hirta* and *B. nigra* are the eastern Mediterranean and Asia Minor regions, respectively, with *B. juncea* from the central Asia–Himalaya area, and *B. carinata* from Ethiopia. Mustard has been grown in most parts of the world since biblical times.

When grown for its foliage, the plant performs best in cooler climates. For seed production, long, warmer days are preferred.

Parsley (*Petroselinum crispum* Mill.) APIACEAE

Parsley, a biennial plant native to the Mediterranean area, has been known as a cultivated crop for more than 2,000 years. The crop, a close relative of celery, is widely used as an herb or garnish and in flavoring various foods. The plant produces a rosette of divided leaves on a short stem. Distinctive foliage types are produced, such as the plain or single-leaf type used mostly for food flavoring, and the double-leaf type and the

moss or triple-curled leaf type preferred for garnishes. An additional parsley type, known as 'Hamburg Rooted,' is grown for its foliage as well as for its slightly thickened edible root. Parsley is a highly nutritious crop and an excellent source of vitamins A and C.

Rhubarb (*Rheum rhabarbarum* L.) POLYGONACEAE

Rhubarb is a large-leafed plant grown for its thick leaf petioles, which are used as a dessert vegetable in pie fillings or sauces. The attractive deep red coloration of the rhubarb petiole is an important market feature. Rhubarb is forced in the dark to avoid chlorophyll development, which would interfere with the expression of the red pigment. *The leaf blades are poisonous and must not be eaten.*

The plant is a cool-season perennial, and it is one of the first vegetables ready for consumption in the spring. Plant growth is kept inactive by low winter temperatures. By using plant forcing procedures to stimulate rapid growth, a crop can be produced during the late winter–early spring. Temperature control is provided by the use of protective structures or shelters, and if needed, heat can be supplied. In mild climates, warm early spring temperatures encourage early production in field conditions. The crop can also be managed to produce a crop during the fall and winter, in which case the crop is kept dormant through the restriction of moisture during the summer. The plant does not grow well at temperatures above 25°C (77°F), and summer-produced rhubarb is usually of poor quality.

Spinach (*Spinacia oleracea* L.) CHENOPODIACEAE

Spinach is a popular and nutritious vegetable frequently used raw in salads and as a cooked potherb. Spinach is an annual plant, dependent upon wind for pollination of its flowers. The plant is unusual because it exhibits dioecious characteristics, male or female flowers on separate plants, and also is infrequently monoecious, with varying proportions of male and female flowers on the same plant. The variation extends from plants that are considered extreme males, through vegetative males, to female plants. The greater the proportion of male flowers, the greater the tendency is to flower earlier. The male plants have relatively little vegetative growth and die soon after flowering; therefore, they contribute very little to crop production. The preferred types are the female and vegetative male plants, which are much slower to flower. These plant types can produce an abundant amount of foliage, which can usually be harvested before flowering occurs.

Spinach is native to central Asia, most likely in the area of Iran. The absence of records suggests that its domestication is fairly recent. It is a cool-season, long-day plant that produces its best vegetative growth under cool 15°C to 18°C (59°F to 64°F) temperatures and a short daylength. Long days, especially if coupled with higher temperatures above 25°C (77°F), cause the plants to bolt and flower, which is detrimental to production. Spinach plants have some frost tolerance and are grown in many areas as an overwintering crop.

Climate is important in deciding planting dates to ensure that the crop does not mature during periods when high temperatures or long days occur. Low temperatures can partially offset the bolting influence of long daylength. High temperatures intensify the bolting response. While producers try not to exceed the critical daylength, slightly warmer temperatures do accelerate vegetative growth, which is the primary goal.

Types of spinach are distinguished by their foliage characteristics. Some cultivars are smooth-leaved, others are savoy-leaved, and some have intermediate leaf textures. Other important cultivar characteristics are leaf color, leaf blade size, leaf petiole length, growth habit—prostrate or upright, seasonal adaptation, bolting, and disease resistance.

Swiss Chard (*Beta vulgaris* L., Cicla Group)
CHENOPODIACEAE

Swiss chard or chard is a biennial plant that is very similar to the table beet in developing large, crisp, fleshy leaf stalks and large leaves, but it differs in not producing the enlarged tap root. Swiss chard is a widely adapted, cool-season crop that tolerates hot weather better than spinach. It is a well-known crop and one of the easiest to grow.

VEGETABLE CROPS GROWN FOR UNDERGROUND PARTS

Underground plant parts used for food consist of storage roots, bulbs, rhizomes, corms, and tubers. Many plants store part of the excess photosynthate (food) produced during their growth in these plant parts. Rhizomes, corms, and tubers are specialized stems that have buds and the ability to develop roots, leaves, and other stems, and can regenerate fully into new plants. Tubers are thickened stems that exhibit compressed or small internodal elongation. The eyes of the potato tuber are buds that develop into stems, which can also develop into fully functional plants. Bulbs like the onion are comprised of specialized storage leaves attached to a short compressed stem. The base of such leaves enlarge because of the accumulation of food synthesized by the plant and stored in these leaves. This enlargement produces the structure known as a bulb. The sweet potato, carrot, beet, and turnip are examples of vegetables where the enlarged storage root serves as the accumulation point of plant food reserves.

Underground plant parts have worldwide importance as a food source. These crops are staples, extensively used in the food supply of all nations, whether developed or developing. Characteristically, these plants are efficient converters of solar energy into chemical energy, principally as carbohydrates, but also as proteins. For example, the Irish potato plant, using an equivalent land area, produces carbohydrates second to sugarcane and protein second to soybeans. Other plants grown for their edible underground parts are also efficient in their synthesis and storage of food products.

Beets (*Beta vulgaris* L., Crassa Group J. Helm)
CHENOPODIACEAE

Table beets, native to Europe and North Africa, are cool-season biennials grown mainly for their roots. However, the tender young tops are sometimes used as greens. Beets were first used for food about the third century A.D.

Beets are rather hardy and tolerate some freezing, but they also grow well in warm weather. Excessively hot weather, however, causes zoning—the appearance of alternating light and dark red concentric circles in the root. Beet seeds germinate within a broad soil temperature range from 5°C to 25°C (41°F to 77°F).

Carrots (*Daucus carota* L.) APIACEAE

The carrot, a cool-season crop, is grown for its fleshy storage root. It is a native of Europe and parts of Asia. While the carrot is mentioned in some of the ancient Greek writings, it is likely that the crop, as we know it, is relatively new because of improvements by plant breeding. The carrot was introduced in North and South America in the early 1600s. The wild type is an annual, but cultivated types are biennials. The root is slender and varies in length from 5 to 25 cm (2 to 10 in.) when harvested. Carrots are often divided into types according to shape and length of root, with each type represented by several cultivars.

Carrots grow best at temperatures ranging from 15°C to 20°C (59°F to 68°F). In the seedling stage, the plants are sensitive to both high and low temperatures, and mild freezes at harvest cause some leaf damage. Many young plants are injured or killed on hot, sunny days. Long periods of hot weather can cause objectionable strong flavor and coarse root texture. Temperatures below 10°C (50°F) tend to cause longer, more slender, and paler roots. Color develops best at 15°C (59°F).

Celeriac (*Apium graveolens* L., var. *rapaceum* Gaud-Beaup.) APIACEAE

Celeriac and celery are closely related, both belonging to the same species. Celeriac is commonly known as turnip root. The above-ground parts look similar to celery except that the foliage is less compact, more dwarfed, and darker green. In contrast to celery, the leaf petioles are less fleshy and more fibrous. The edible part is the enlarged stem and root axis. Celeriac is mainly used to flavor other foods, such as soups and stews, but it is also used raw or cooked in salads. The crop is not grown extensively in the United States. In Europe, the crop is better-known and more widely used. The climatic requirements for celeriac are similar to those for celery.

Garlic (*Allium sativum* L.) AMARYLLIDACEAE

Garlic is grown especially for its strong and popular flavoring characteristics, which make it a universally important spice. The edible part is a compound bulb consisting of segments called cloves, each surrounded by a thin white or pink sheath. The cloves are formed in the axil of the inner foliage leaves; the outer leaves form the sheath. The leaves are solid and flattened, rather than hollow and round as in onions. Garlic has been grown for centuries. Its origin is believed to be central Asia.

Garlic is a cool-season crop usually planted in the fall or early spring. This time of planting allows the plant to develop sufficient size and to be exposed to low temperatures, both important, prior to experiencing the progressively longer summer photoperiod needed to initiate bulbing. Harvest occurs in the fall or late fall depending on planting dates.

Horseradish (*Armoracia rusticana* P. Gaertn., B. Mey. & Scherb.) BRASSICACEAE

Horseradish is a hardy perennial, native to southeastern Europe. It is a perennial but is commercially grown as an annual to prevent it from becoming a weed pest. The leaves, rhizomes, and roots have been used as food or condiment since the Middle Ages and before that for medicinal purposes. The plant is grown for its pungent compound, allyl isothiocyanate (C_3H_5CNS).

There are two types of horseradish. The common type has broad, crinkled leaves and produces high-quality roots. The Bohemian type has narrow, smooth leaves, is more disease-resistant, but produces lower quality roots.

Onions (*Allium cepa* L.) AMARYLLIDACEAE

Onions are one of the oldest vegetables and are known to have been used before 3000 B.C. Onions are probably a native of the area from southwestern Asia eastward to Pakistan.

Onions are cool-season biennial plants that grow well at temperatures ranging from 13°C to 25°C (55°F to 77°F); they have some frost tolerance. Seeds can germinate from 7°C to 30°C (45°F to 86°F) but do best at about 18°C (64°F). During the early stages of growth (before bulbing) onions grow better at relatively cool temperatures, but during bulbing, harvesting, and curing, higher temperatures and low relative humidity are desirable. Plant growth declines at temperatures greater than 27°C (81°F). Bulbing is initiated primarily by daylength and not by the age of the plant. Once initiated, warmer temperatures hasten bulb growth and enlargement. Bulbing and flowering are not related. Vernalization is required for bolting and flowering.

Green onions are young immature plants, grown at very high populations for fresh market use in salads. Long-day cultivars are grown in short-day environments to inhibit bulbing. These plants are generally hand harvested, several being bunched together, the tops uniformly trimmed, and marketed directly.

Parsnips (*Pastinaca sativa* L.) APIACEAE

Parsnips are considered a native of Europe and Asia and have been used for food and medicinal purposes since the Greek and Roman eras. They were introduced into the United States in Virginia and Massachusetts in the early 1600s. The plants are cool-season biennials but are grown as annuals. Parsnips are slow growing and require a long season (100 to 120 days) to form their white, fleshy edible roots. Parsnips grow best when they are planted and grown during a relatively cool summer and harvested in a cool fall.

Potatoes (*Solanum tuberosum* L.) SOLANACEAE

The potato plant is a bushy, herbaceous annual 60 to 90 cm (2 to 3 ft) in height. The edible tuber is a swollen underground stem. The eyes on the tuber are buds and are more numerous on the apical (distal) end (opposite the end of attachment) than the basal (proximal) end.

As a source for human food, the potato is a leading vegetable crop. Potatoes are South American in origin and were cultivated in Chile and Peru by the Indians long before the arrival of European explorers. Early Spanish explorers transported them to Europe about 1575, and from there they returned to North America with the colonists settling along the Atlantic coast. In those early days, however, potatoes were not a significant food source in North America until the influx of Irish immigration, which began in the early 1700s and reached a maximum later, after the potato famine in Ireland in 1846. Potatoes then became America's favorite carbohydrate food because their cost was low and most people liked them. Today, potatoes are grown on every continent and in every state of the United States. And, like carrots, potatoes are harvested somewhere in the United States every month of the year.

Potatoes are a cool-season crop, slightly tolerant of frost, but easily damaged by freezing weather near maturity. Tubers, called stolons, form on the end of stems and botanically are rhizomes. The tuber is a fleshy stem with buds in the axil of leaf scars. They are often referred to as eyes. Maximum yields of high-quality tubers are produced when the mean temperature is between 15°C and 18°C (59°F and 64°F) during the growing season. Tuberization (tuber formation) is also favored by long days of high light intensity. An optimum temperature for tuber development is about 18°C (68°F). Tuberization is progressively reduced when night temperatures rise above 20°C (68°F) and is totally inhibited at 30°C (86°F).

Flowers are common but fruits rarely set unless precise climate and a long daylength period are met. The fruits are green berries, somewhat similar to small tomatoes, within which seed are produced. True seeds are used in breeding programs but have not, until recently, been planted for commercial production. Recently there has been much interest in using seeds for propagation in developing countries. This interest is with regard to the need and cost of importing disease-free seed potatoes for propagation, where disease-free seed stock cannot be produced.

Radish (*Raphanus sativus* L.) BRASSICACEAE

Of all vegetable crops, the garden radish is one of the easiest and quickest vegetables to grow. A crop can be produced within four to six weeks of seeding. The radish, probably native to the eastern Mediterranean region, was cultivated for centuries. Its use can be documented beyond 2000 B.C.

Radishes are a cool-season crop tolerant of some slight frost. Best growth occurs when the monthly temperature averages about 15°C to 18°C (59°F to 64°F). They are somewhat intolerant of temperatures exceeding 25°C to 27°C (77°F to 81°F). When grown during high temperatures, the interior root tissues tend to be pithy, and the plants may bolt. Overmaturity also causes a softening or pithiness of the normally crisp tissues.

Rutabaga (*Brassica napus* L., Napobrassica Group) BRASSICACEAE

Rutabagas are a cool-season biennial root crop grown mainly in northern Europe, England, and Canada. Limited amounts are grown in the United States for human consumption and some for stock feeding. They are sometimes called Swedes, Swedish turnips, Russian turnips, Canadian turnips, or yellow turnips.

Salsify (*Tragopogon porrifolius* L.) ASTERACEAE

Salsify is a hardy biennial, sometimes called oyster plant or vegetable oyster. It has long, cylindrical roots, used mainly in soups for its delicate oysterlike flavor. The plant grows wild around the Mediterranean Sea, in southern England, and along roadsides in the United States and Canada. During its second season, it bears a purplish flower that looks like an enlarged dandelion head. Salsify is a market vegetable crop of limited use.

Salsify is a cool-season crop tolerant of some freezing weather. Seeds are planted about the time of the average date for the last frost in the spring. This slow-growing plant requires the entire season to develop, and like parsnips, its quality is improved after a heavy frost or cold storage for at least two weeks. The tapered roots are 20 to 23 cm (8 to 9 in.) long; are 2.5 to 4 cm (1 to 1.5 in.) thick; and feature creamy, white flesh.

Sweet Potato (*Ipomoea batatas* Lam.) CONVOLVULACEAE

The moist, soft-textured sweet potato cultivars are often erroneously called yams to distinguish them from the dry-textured sweet potato cultivars. This misnomer is unfortunate because true yams are different plants entirely, not even slightly related to sweet potatoes. True yams belong to the DIOSCOREACEAE family.

Sweet potatoes are native to tropical America, and were transported to the Pacific islands and to Asia early in history. They were cultivated as food in the southern parts of the United States by the Native Americans and were later taken to Europe. There is no evidence that the ancient civilizations of Egypt, China, Persia, or Greece knew of sweet potatoes.

Sweet potatoes are perennial vines that grow prostrate on the ground. They belong to the morning-glory family and are grown for their tuberous roots. They are unique in producing adventitious shoots used to propagate the crop. When grown in temperate and some subtropic regions, the crop is handled as an annual. In the tropics and much of the subtropical regions, the crop is cultured as a perennial.

Sweet potatoes are of particular importance as a food crop throughout subtropical and tropical regions. It is one of the most important carbohydrate sources for many millions of people, particularly those in developing nations.

Sweet potatoes are warm-season plants that do not tolerate frost nor grow well in cool weather. Temperatures below 10°C (50°F) cause chilling injury. They require a long, warm growing season and grow best when mean monthly temperatures are above 20°C (68°F) for at least three months.

Turnip (*Brassica rapa* L., Rapifera Group) BRASSICACEAE

Turnips are one of the easiest vegetables to grow and one of the more widely adapted of the root crops. They are native to northern Asia and are extensively grown in Europe, Asia, and almost everywhere in the United States. They are used as food for both animals and humans. The roots may be eaten raw but are usually cooked, and the leaves make delicious greens. The plant is a cool-season biennial. Seeds are sown either early in the spring (for a fall crop) or in the fall (for a spring crop) for the roots to mature during cool weather. Continued temperatures below 10°C (50°F) are likely to cause bolting. The plants resist frost and mild freezing. In most areas spring and fall crops are grown because only sixty to eighty days are required for the roots to reach maturity.

SUMMARY AND REVIEW

Vegetable production is important for many reasons. Vegetables are major components of a healthy diet. Growing vegetables can contribute to a sense of well-being; gardening and farming vegetables can have positive effects on the environment. The different scales of vegetable farming and the history of vegetable production can give us a better understanding of societies and the natural world. The greatest concentration of area devoted to commercial vegetable production tends to be where the environment consistently favors good crop growth and where the resources needed for production are readily available. Although many large commercial farms are concentrated in specific areas, vegetables are grown in almost all regions of the world, making the growing, processing, and selling of vegetables a major economic factor in local to global economies.

Vegetable farms are managed agroecosystems that are widespread and diverse in many aspects, including size, crops grown, and methods used. Vegetable growers establish and manage biological communities within larger natural systems so that a high quantity of desirable vegetables are produced efficiently and profitably, and with as much integration among other ecosystems as possible. Like other agroecosystems, vegetable-farm ecosystems combine elements of business and the natural world (e.g., biology, chemistry, physics).

Vegetable production can be done outdoors with minimal environment management or in covered structures ranging from simple row covers to sophisticated greenhouses with highly managed environmental systems. Production methods can be traditional, organic, or sustainable. The designations among these systems is not distinct, although to be certified as organic, a farm must meet stringent requirements.

Vegetable farmers must also follow principles integral to the success of other types of businesses, including (1) studying the market and being prepared to meet its requirements, (2) producing a high-quality product efficiently, (3) packaging the product appropriately and delivering on time, and (4) be innovative. Key steps in successful vegetable production are site selection, site preparation, variety selection, planting, management during growth and development, harvest, packaging and post-harvest management, record keeping, and continuing education.

FOOD FOR THOUGHT

1. Soil and water are key resources in vegetable production. Their availability is declining in some areas where vegetable production is a major industry. This decline is due, in part, to greater demands for soil and water for nonfarming uses. What are the pros and cons of shifting natural resources from vegetable production to other industrial or human uses?

2. Farmers often grow more vegetables than they sell. Unsold vegetables are often destroyed because they contain relatively minor imperfections in appearance that do not affect their nutritional value, safety, or usability. However, few consumers overlook these imperfections when making purchasing decisions. The need for vegetables to appear perfect in the marketplace drives grower practices (e.g., use of pesticides). Destroying crops raises production costs. Therefore, should consumers overlook minor imperfections in vegetable appearance? Or should alternative production or post-harvest practices be found that allow consumers to maintain their expectations about product appearance?

3. Production and consumption of organically grown vegetables continue to rise markedly. In your view, what is the basis for the increase in organic vegetable production and consumption? What does a "certified organic" label on a vegetable or vegetable product signify to you and your classmates, friends, and family?

4. Consider the vegetables you eat most during a summer month (e.g., July) and a winter (e.g., February) month. Do the vegetables you eat change with the time of year? Why or why not?

5. Do you purchase the vegetables you eat regularly primarily in a fresh or processed form? Why?

6. Based on your knowledge, rank on a scale of 1 to 10 (1 = none, 10 = extremely high) the technical expertise required to be a successful vegetable farmer. What factors did you consider?

SUPPLEMENTARY READING

AMERICAN HORTICULTURAL THERAPY ASSOCIATION (AHTA). 2005. The history and practice of horticultural therapy, at **http://www.ahta.org/information/**.

DECOTEAU, D. R. 2000. *Vegetable crops.* Upper Saddle River, N.J.: Prentice Hall.

EARLES, R. 2002. *Sustainable agriculture: An introduction.* Appropriate Technology Transfer for Rural America (ATTRA) summary. **http://www.attra.org/attra-pub/PDF/sustagintro.pdf**.

KELLERT, S., and E. O. WILSON. 1993. *The biophilia hypothesis.* Washington, D.C.: Island Press.

KUEPPER, G., and L. GEGNER. 2004. *Organic crop production overview.* Appropriate Technology Transfer for Rural America (ATTRA), fundamentals of sustainable agriculture. *http://www.attra.org/ attra-pub/PDF/organiccrop.pdf*.

LAWRENCE, E. 1995. "Definitions of ecosystems." In *Henderson's dictionary of biological terms,* eleventh edition. New York: John Wiley and Sons. **http://212.187. 155.84/pass_06june/Subdirectories_for_Search/ Glossary&References_Contents/KeywordsContents/ E/Ecosystem.htm**.

MAYNARD, D. N., and G. J. HOCKMUTH. 1997. *Knott's handbook for vegetable growers,* fourth edition. New York: John Wiley and Sons.

NATIONAL ORGANIC STANDARDS BOARD (NOSB). 1995. Definition of "organic." p. 50. **http://www. ams.usda.gov/nosb/archives/ April%2024-28,%201995%20NOSB%20Minutes. pdf**.

ORGANIC FOODS PRODUCTION ACT of 1990. **http:// agriculture.senate.gov/Legislation/Compilations/ AgMisc/OGFP90.pdf**.

POLLACK, S. L. 2001. "Consumer demand for fruits and vegetables: The U.S. example." In *Changing structure of global food consumption and trade.* United States Dept. of Agric. Economic Research Service, WRS-01-1, pp. 49–54. **http://www.ers.usda.gov/ publications/wrs011/wrs011h.pdf**.

POLLAN, MICHAEL, 2001. *Botany of desire.* New York: Random House.

PORTER, R. 1994. *The biographical dictionary of scientists,* second edition. New York: Oxford University Press.

SUSTAINABLE AGRICULTURE NETWORK. 1998. Managing cover crops profitably, second edition. Sustainable Agriculture Network. Handbook Series Book 3. United States Department of Agriculture, Sustainable Agriculture Network, **http://www.sare.org/ publications**.

US DEPARTMENT OF AGRICULTURE. 2003. The 2001–2002 Agriculture Fact Book, pp. 23–35. United States Department of Agriculture. **http://www.usda.gov/ factbook/index.html**.

US DEPARTMENT OF AGRICULTURE. 2005. Vegetables 2004 Summary, January 2005. United States Department of Agriculture, National Agricultural Statistics Service. Vol. 1–2(05), pp. 5, 52. **http://usda.mannlib. cornell.edu/ reports/nassr/fruit/pvg-bban/ vgan0105.pdf**.

VAN DUYN, M. A. S., and E. PIVONKA. 2000. "Overview of the health benefits of fruit and vegetable consumption for the dietetics professional: Selected literature." *Journal of the American Dieticians Association,* 100:1511–1521.

VOLKMER, H. L., D. A. JOLLY, K. S. KELLEY, C. W. ALBERTSON, K. S. ARMSTRONG-CUMMINGS, J. R. BARBER, E. L. BLOUNT, C. D. BOLEN, M. L. BOWLAN, B. F. BURKETT, N. CARRASQUILLO, E. W. COWARD, R. M. DANIELS, R. E. GOMEZ, D. V. GUERRA, G. T. GUNTHORP, J. HARNESS, C. HASSENROOK, D. G. HENDERSON, E. HERNESS, G. B. HOLLAND, F. R. MAGDOFF, J. B. NEELY, J. K. REID, G. E. STITMAN, R. A. STEWART, T. THAO, T. J. TRANTHAM, J. ZIPPERT. 1998. "A time to act: A report of the USDA National Commission on Small Farms." USDA Miscellaneous Publication 1545 (MP-1545).

WILSON, E. O. 1984. *Biophilia.* Cambridge, Mass.: Harvard University Press.

CHAPTER

19

Temperate Fruit and Nut Crops

Joseph Scheerens

KEY LEARNING CONCEPTS

After reading this chapter, you should be able to:

♦ List the factors used in selecting a suitable site.
♦ Understand how to select fruiting cultivars and rootstocks.
♦ Understand how to plant and maintain fruit and nut crops.
♦ Describe important practices in the profitable production of high-quality fruit.
♦ Appreciate the physiological basis for these practices and predict plant response.
♦ Explain the critical role of the environment in fruit crop development.

The first essential step in the successful establishment of a fruit planting is to be certain that the crop to be planted is adapted to the climate and soil characteristics of the region and has an economically viable market. After the suitability of the crop has been ascertained, several horticultural and economic factors should be carefully considered before planting is initiated. Tasks to be accomplished include:

1. Selecting a suitable site.
2. Evaluating market characteristics and developing a market strategy.
3. Projecting, in detail, capital investments, operating costs, and income over time.
4. Thoroughly reviewing and understanding the flowering and fruiting process.
5. Designing the most appropriate cultural management strategy.
6. Choosing cultivars and rootstocks for optimum performance.
7. Preparing the site for planting.

Because the production of most fruit crops is a long-term undertaking, poor initial decisions can be costly or impossible to correct later. Therefore, all available pertinent information should be sought out before final commitments are made.

SITE SELECTION

A thorough study should be made of the following aspects of the proposed site. Any one of them could affect the success or failure of the enterprise.

Climate

It is extremely important to learn as much as possible about the weather patterns of the proposed planting site. Fruit crops currently or historically grown in an area may indicate whether or not the climate is suitable for a particular crop. General information can also be obtained by consulting plant hardiness zone or regional adaptation maps published by the U.S. Department of Agriculture (USDA) or similar documents published by individual states. If recording instruments are located in the vicinity, historical climatic data may be obtained through government weather services. Local cooperative extension or agriculture department agents usually can furnish much valuable information concerning weather patterns in their localities. For some areas, Internet sites may also provide useful local or regional climatic trends. Several weather conditions are of particular importance in fruit growing.

Temperature and the Effect of Site Location

The first determination needed in evaluating a site is the minimum temperature that frequently occurs and if the fruit crop will survive that temperature. For example, if a site frequently experiences a midwinter temperature of $-15°F$ to $-20°F$, peaches or vinifera grapes should not be grown; however, some cultivars of apples would survive these temperatures. In areas subject to severe winter cold, only the most hardy kinds of fruit crops should be attempted.

Second, the length and timing of the growing and dormant seasons are important climatic considerations for two reasons: (1) they may influence the accumulation of heat or chilling units, which affect important physiological processes such as growth, hardening off, dormancy, bud break, and fruit maturation; and (2) they determine the likelihood that maturing fruit or flowers will experience damaging temperatures during the fall and spring transitional periods, respectively.

Climatic extremes experienced by the crop can be greatly influenced by the features of a particular site. Avoid sites in low-lying areas, such as river bottoms or low spots in rolling hills, where cold air settles during frosty nights. This can be particularly dangerous when frosts occur during blooming periods in the spring or for cultivars with late-maturing fruits, which could be damaged by early fall freezes. It is much safer to select an orchard site on the upper portions or slopes of rolling terrain. Orchard heating, wind machines, or irrigation systems can often moderate or prevent low-temperature injury problems in frosty sites. While controlling frost is an additional expense, a more important factor is the cost of crop loss or injury.

In regions with hot summers, site location can influence the temperature; a northern or eastern slope may be a few degrees cooler than southern and western slopes (in the northern hemisphere). For growing crops that cannot tolerate high summer heat, such as sweet cherries and apples, the site location and orientation of tree rows is of considerable importance. (The south side of east-west rows can be excessively hot.)

The location of the proposed planting site in relation to large bodies of water should be considered. A planting site on the leeward side of a large body of water is likely to have a microclimate modified considerably both in summer and winter, with the temperature lower in summer and higher in winter than similar sites at a distance from such bodies of water.

Wind

Avoid sites that have a history of strong winds. Wind can be detrimental from many aspects. Reduced bee activity during windy days in the pollination season can seriously reduce fruit set and yields. Wind can damage young, tender shoots in the spring and can scar and bruise young fruits. It can limit application of pesticides with speeds over 10 miles per hour. Wind can interfere with the uniform application of sprinkler water. Young trees planted in windy areas must be staked and tied. Windbreaks can help reduce this problem, but they can be a costly added expense and can complicate air drainage problems.

Precipitation and Available Water

Fruit plants require adequate soil moisture throughout the growing season to maintain vegetative growth and to produce a full crop of quality fruits. If no supplemental irrigation is possible, attention must be paid to the rainfall history of the proposed site to determine whether the total and summer rainfall is likely to be adequate and consistent. A better situation exists where water supplies for irrigation are available during times of drought. Some of the best fruit-growing areas in the world are located in areas where little or no rainfall normally occurs during the growing season. These regions depend entirely on irrigation.

A pattern of continual rains during the pollination period in the spring could result in poor crops by interfering with bee activity. Continual rains during the fruit harvesting period lead to problems, not only in getting the fruit picked but also in promoting various physical defects (e.g., cracking, epidermal blemishes) or fungal diseases on the maturing fruits. In cold winter areas, accumulated snowfall during the dormant season provides protection against low soil temperatures and potential root injury. Snow can also delay bloom in the spring and reduce the possibility of frost damage to the buds and flowers.

The proposed site for a fruit planting should not be subject to periodic flooding from nearby rivers or streams. Fruit plants will not tolerate water around their roots for any length of time because the water stops air penetration to the roots. Areas with a high water table are usually unsuitable for fruit growing because only the soil mass above the water table is available for root development, and this is usually quite limited.

Hail

Hail can cause serious damage to orchards and other fruit crops from both mechanical injury and disease infections entering open wounds. Small hailstones early in the growing season can reduce crop value, whereas large hailstones can destroy fruit, shred foliage, break spurs and new shoots, and remove bark from branches. Thus, it is very important that the frequency and severity of hailstorms at the proposed site be determined in advance.

Land and Soil Characteristics

Slope of the land is important. Ideally it should be over 2 to 3 percent for good air and water drainage but no greater than 10 to 15 percent for successful operation of equipment without contouring.

Topography should be relatively even, without nondraining depressions and no more than 10 percent grades. A north slope reduces potential for frost losses and sunburn but delays bloom and opportunities for early harvest. West slopes should be avoided except in cold climates or at high elevations where heat units are limiting.

The ideal orchard soil should be a deep (1.8 m or 6 ft) well-drained, nonsaline, fertile silty loam to a fine sandy loam (see Chapter 6). However, a rooting depth of between 2 and 3 feet is usually quite adequate for most fruit crops. The surface should slope gently, allowing good runoff from heavy rains and permitting good infiltration of irrigation water. There should be no impervious hardpans or claypans under the surface. Fine-textured clay soils generally make poor orchard soils and should be avoided.

Planting on unsuitable soil handicaps productivity for the life of the planting and can make the enterprise marginal or unprofitable. If a less than ideal soil type is used, such as a clay loam, it is best to consider planting a fruit species that does relatively well on fine-textured soils, such as plums, pears, or apples, and avoid planting peaches or almonds, which will not tolerate such soils. If the available soil is a loamy sand or sandy loam, plant peaches or almonds (if the climate is suitable). Some species, however, such as grapes and strawberries, do well on a wide range of soil types.

Because soils differ with respect to parent material, ion exchange capacity, organic matter content, and physical properties, the native fertility of the proposed cropping site should be determined as early as possible. Soil samples can be submitted to either state or commercial testing laborato-

ries for physical and chemical characterization. Proper sampling procedures can be obtained through these labs, but recommendations usually include sampling at multiple sites in the plot to a depth of tillage. The soil test information returned often includes a proposed fertilization or liming strategy for rectifying nutrient deficiencies in the soil.

It is also important to consider soil pH (acidity or basicity) because it greatly influences mineral availability and soil solution composition (see Chapter 6). Most fruit crops grow best in mineral soils that are slightly acidic (pH of 5.8–6.5), which provides a balanced window of nutrient availability. In soils that are more acid, the availability of nitrogen (N), phosphorus (P), potassium (K), sulfur (S), calcium (Ca), magnesium (Mg), and molybdenum (Mo) drops off. Conversely, in basic soils, N, P, iron (Fe), manganese (Mn), boron (B), copper (Cu), and zinc (Zn) are less available. Extremes in pH can also cause specific minerals to accumulate in the soil solution at toxic levels. For instance, low soil pH (4.5 or less) results in availability of Mn at superoptimal levels and its uptake at toxic concentrations, one of the interacting factors responsible for internal bark necrosis disorder in apple. On the other hand, certain fruits of the ERICACEAE (heath) family are acid-loving and grow best at a soil pH between 4.0 and 5.5. The most notable temperate zone, ericaceous fruit crops are blueberry and cranberry. Blueberries also prefer a soil organic matter content of 3 percent or greater for adequate root development. Most mineral soils typically contain 2 percent or less decomposed organic matter.

It is essential that all available information be obtained about the soil in the proposed site before planting. A soil map of the area should be consulted at the library of the nearest agricultural college or the office of the cooperative extension or the Natural Resources Conservation Service. The Internet is also a valuable resource for obtaining site-specific information about soil and soil quality. The soil of the area to be planted should be systematically sampled with a backhoe and enough pits opened to inspect the soil profile to a depth of about 1.8 m (6 ft). The sample will divulge any hardpan or rock layers or sandy pockets and will show whether the soil is deep enough to support fruit trees. A shallow soil may support smaller fruit plants such as strawberries or bushberries.

A study of the history of the site, including information of other crops previously grown, can be useful in analyzing soil problems. If cotton, potatoes, or tomatoes have been grown there, expect trouble from verticillium wilt. If an old orchard has been pulled out, the soil could be compacted, infected with *Phytophthora* fungus, or high in pesticide residues. Often it is necessary to fumigate the soil to eliminate fungus and nematode problems before planting a new orchard to obtain good, vigorous tree growth.

In tree fruits and nuts, resistant rootstocks can often reduce problems with soil pests. For example, many sandy loam soils, well suited for growing peaches, are infested with root-knot nematodes, and large-scale fumigation may not be practical. But in some instances, by planting peach trees propagated on a nematode-resistant rootstock, such as 'Nema-guard,' good tree growth and productivity can be obtained in spite of the nematodes in the soil. Often, however, such resistant rootstocks are not winter-hardy in cold climates.

Irrigation Water: Quality and Availability

In areas with low rainfall during the growing season, assurance should be obtained that there is a potential source of ample high-quality irrigation water. The water should not contain high soluble salts, including sodium, chlorides, boron, calcium, and magnesium bicarbonates. Water samples can be analyzed by state or commercial laboratories and unsuitable water sources detected before the planting is made. It is difficult, if not impossible, to correct poor-quality water without high costs.

The quantity of irrigation water to be applied to a fruit planting will vary according to the climatic zone, the crop, and the irrigation system. The frequency of irrigation will depend upon the moisture-holding capacity of the soil. Young trees, coarse-textured soils, and shallow soils require more frequent irrigations but not greater amounts. Most critical are the last thirty days before fruit harvest, when water shortages can significantly reduce fruit size and quality, while increasing various fruit disorders.

For some fruit crops, irrigation systems are also necessary as a means to avoid spring frosts during flowering. As water changes state (from liquid to solid), it releases energy in the form of heat (i.e., the heat of fusion). When the blooming crop is sprinkle-irrigated, this heat energy transfers to open or nearly open flower buds as ice forms on the flower surface. This heat of fusion, along with dissolved sugars and salts in the cell sap, protects the flower tissues from ice damage. For sprinkler irrigation to be effective against frost damage, it must be initiated when temperatures reach 1.1° to 0.6°C (34° to 31°F) and continued throughout the period of freezing temperatures. On windy nights, the effectiveness of irrigation for frost control is greatly reduced.

MARKET AVAILABILITY AND MARKETING STRATEGIES

Utilization of the crop is generally assured for plantings in the home garden—by the family, friends, and neighbors, plus canning, drying, or freezing and, perhaps, sales of surplus fruits and nuts. In a commercial planting, however, it is essential to know that a market will be available for the crop. The potential marketability of a fruit crop is simply a matter of supply and demand. It may be influenced by global, regional, and local production capacity and competition for customers at the time of harvest. In most large supermarkets, fruit is available from countries in the Far East, South America, Europe, and Africa and from all regions of the United States. Locally produced fruit may be in direct competition (e.g., local blueberries versus those from another state) or in-

direct competition (peaches from the United States versus papayas from Brazil) with other fruits if marketed at the same time. Also, trends in fruit crop consumption and use have changed over time. Therefore, it is extremely important to appreciate the worldwide production and marketing trends for any crop being considered for planting (see Tables 19–1 and 19–2 and Figs. 19–1, 19–2, and 19–3).

The marketing situation for most fruit crops is often quite fluid and should be studied thoroughly before heavy planting commitments are made. New plantings of a particular fruit crop in a certain area would be questionable when experienced growers in that area are either not planting or are pulling out existing plantings. Perhaps the influx of an insect or disease problem has added control costs that eliminated the profit margin for that crop. Heavy plantings of a crop in a given area because of enticingly high returns at the moment may lead to market gluts when the plants come into production.

Prospective fruit growers should realize that market demands for their produce can shift over the years, so they should study and determine what are the present, and possibly the future, demands for fruit products—by commodity and by cultivar. Figure 19–3 shows that in the United States from 1979 to 2004, fruits marketed as fresh and as juice increased slightly, while canned products declined. Frozen and dried fruit consumption increased in the early 1980s and then declined in the 1990s. Recent consolidation of large wholesale grocers has dramatically reduced the number of buyers and thus competition. Overall, this has depressed the prices received by fruit producers. Also in the late 1980s, produce sections of supermarkets in the United States and Europe started displaying fresh fruit and vegetables that were unknown a few years earlier, such as kiwifruit, lychees, Asian pears, carambolas, and mameys. This diversity of products coupled with increasing world production and concentration of buyers has created pressure on fruit growers to change cultivars to achieve higher prices. Standard cultivars such as 'Delicious' apples or the 'Hayward' kiwifruit are treated as commodities and receive much lower prices compared with newer cultivars such as Gala or Braeburn apples or yellow-fleshed kiwifruit.

In light of all these factors, marketing strategies for the crop should be considered in advance. Are there marketing cooperatives or private packers and shippers available who will take the crop, or will the grower need to grade, pack, store, and then transport the fruit to city fruit markets? Or, perhaps, in locations with considerable highway traffic, particularly close to large cities, on-the-farm roadside or pick-your-own sales may utilize most or part of the crop. Agrotourism is gaining popularity in many regions of the United States. A new fruit producer considering a pick-your-own marketing strategy might benefit from offering amenities, inducements, or entertainment (e.g., corn mazes, hay rides, etc.), which can greatly increase the pleasure of an on-farm experience for local customers. The type and extent of amenities offered is limited

TABLE 19–1 Characteristics of Common Temperate Tree Fruit and Nut Crops

Species	Origin	Production Centers International	Production Centers US	Plant Habit	Winter Hardiness	Edaphic Requirements
Almond (*Prunus dulcis*) ROSACEAE: edible portion is a seed	S.E. Asia cultivated B.C. in Mediterranean region	United States Spain Turkey Greece Italy	California	Vase-shaped tree (4–6 m)	Requires mild winters	Light well-drained soil preferred
Apple (*Malus X domestica* Borkh.) ROSACEAE: edible portion is a "pome fruit"	W. Asia cultivated in Mediterranean region 1200 B.C.	China United States France Italy Turkey	Washington New York California Michigan Pennsylvania	Central leader tree (height mined by rootstock)	−25° to −29°C	Deep well-drained loam
Apricot (*Prunus armeniaca* L.) ROSACEAE: edible portion is a "stone fruit" or drupe	W. China, cultivated in W. China 2000 B.C.	Turkey Iran Italy France Pakistan	California Washington	Vase-shaped tree (3–8 m)	−29° to −34°C	Deep well-drained, fertile soils
Cherry (sweet) (*Prunus avium* L.) ROSACEAE: edible portion is a "stone fruit" or drupe	S. Central Europe, Asia Minor, cultivated prehistorically in Europe	Iran Turkey United States Italy Spain	Washington Oregon Michigan	Pyramid-shaped tree (4–5 m), moderate control	To −29°C	Deep well-drained loam
Cherry (tart) (*Prunus cerasus* L.) ROSACEAE: edible portion is a "stone fruit" or drupe	S. Central Europe, Asia Minor, cultivated prehistorically in Europe	Russia Poland United States	Michigan New York	Tree more spreading than sweet cherry (<4.5 m)	Greater than sweet cherry	Deep well-drained loam
Hazelnut (*Corylus avellana* L.) BETULACEAE: edible portion is a nut	N. America, Europe, Asia, cultivated in ancient Greece	Turkey Italy United States Spain	Oregon	Multistemmed bush or spreading tree (to 9 m)	−9° to −12°C	Tolerates various soils, well-drained preferred
Peach and Nectarine (*Prunus persica* L.) ROSACEAE: edible portion is a "stone fruit" or drupe	China, cultivated in China 2000 B.C.	China Italy United States France Japan	California S. Carolina Georgia Pennsylvania New Jersey	Vase-shaped tree (to 6 m)	To −23°C	Deep well-drained, fertile soils
Pear (*Pyrus communis* L.) ROSACEAE: edible portion is a "pome fruit" or drupe	W. Asia, cultivated by early Egyptians and Greeks	China Italy United States Japan Spain	Washington California Oregon New York Michigan	Columnar-shaped tree (4–6 m), moderate central leader	To −32°C	Silt loam, somewhat tolerant of heavy soils
Pecan (*Carya illinoensis* Wang) JUGLANDACEAE: edible portion is a nut	S. central United States, cultivated in 1800s	United States	Georgia Texas New Mexico Arizona	Central leader tree (18–24 m)	To −25°C for mature tree, to −5°C for young tree	Deep well-drained loam
Plum (*Prunus* spp.) ROSACEAE: edible portion is a "stone fruit" or drupe	N. Hemisphere, cultivated in Greece 600 B.C.	China United States Romania Germany France	California Washington Oregon Michigan	Vase-shaped tree (6–8 m)	To −30°C highly dependent upon species	Deep well-drained fertile soils
Walnut (*Juglans* spp.) JUGLANDACEAE: edible portion is a nut	Europe, Asia, N. America, S. America, throughout the Roman Empire	China United States Iran Turkey Ukraine	California	Up to 15 m under cultivation, leader	To −42°C highly dependent upon species	Deep well-drained, fertile soils

| Chilling Requirements | Reproductive Behavior | | | | Major Diseases in United States | Major Insect Pests in United States |
	Fruit Producing Wood	Flower Types	Compatibility	Harvest		
200–500 hours	Lateral-bearing on spurs	Complete	Mostly self-incompatible, some cross-incompatible	Mechanized	Blossom and twig blight	Navel orange worm, peach tree borer, various mites
800–1,700 hours	Lateral-bearing on spurs, some terminal-bearing on shoots	Complete	Cultivar-specific compatibility, cross pollination is beneficial	Hand	Scab, fire blight	Coddling moth, plum cucurlio, wooly aphids
400–1,000 hours	Lateral-bearing on spurs	Complete	Self-fruitful except for specific cultivars	Hand	Blossom blast, brown rot	Species of boring insects, leaf rollers, spider mites
1,000–1,500 hours	Lateral-bearing on spurs	Complete	Self-incompatible except for specific cultivars	Hand	Brown rot, Verticillium wilt, canker	Cherry fruit fly, black cherry aphid, plum curculio, spider mites
1,000–1,500 hours	Lateral-bearing on spurs, some terminal-bearing on year-old shoots	Complete	Self-fruitful	Mechanized	Brown rot, Verticillium wilt, canker	Cherry fruit fly, black cherry aphid, plum curculio, spider mites
1,500 hours	Terminal-bearing on year-old shoots	Separate male (catkins) and female flowers on same plant	Self-incompatible, some cross-incompatible	Mechanized	Eastern filbert blight	Filbert aphids, filbertworm
200–1,000 hours	Lateral-bearing on shoots	Complete	Self-fruitful except for specific cultivars	Hand	Peach leaf curl, Verticillium wilt, Armillaria root rot	Plum curculio, borers, scale aphids, Oriental fruit fly
1,200–1,500 hours	Terminal-bearing on spurs, some lateral-bearing on shoots	Complete	Predominantly self-incompatible	Hand	Fire blight, pear decline	Codling moth, mites, pear psylla
500 hours or less	Terminal-bearing on current season shoots	Separate male (catkins) and female flowers on same plant	Requires pollinizers due to protandry or protogyny	Mechanized	Scab, leaf diseases	Pecan weevil, hickory shuck-worm, aphids
700–1,100 hours	Terminal-bearing on short spurs, lateral-bearing on year-old shoots	Complete	Variable and highly dependent upon species and cultivar	Hand (fresh market), mechanized (processing)	Bacterial canker, crown gall, oak root rot, crown rot, plum pox	Scale, peach twig borer, codling moth, mites
400–1,600 hours	Terminal-bearing on current season shoots	Separate male (catkins) and female flowers on same plant	Requires pollinizers due to protandry or protogyny	Mechanized	Crown rot, crown gall, oak root, fungus, walnut blight	Walnut husk fly, navel orange worm, codling moth, aphids, scale

		Production Centers				
Species	**Origin**	*International*	**US**	**Plant Habit**	**Winter Hardiness**	**Edaphic Requirements**
Blackberry (*Rubus* spp.) ROSACEAE: edible portion is an aggregate of drupelets	Europe and N. America, primarily domesticated in N. America in 1800s	United States Canada Chile New Zealand Guatemala	Oregon California	Perennial plant with biennial canes, erect, semi-erect, and trailing types	−18° to −23°C	Deep well-drained, fertile soils
Blueberry (highbush) (*Vaccinium corymbosum* L.) ERICACEAE: edible portion is an epigynous or "false" berry	Eastern N. America, domesticated early in the twentieth century	United States Canada Poland	Michigan New Jersey Oregon Washington	Perennial, deciduous shrub (1–2 m in cultivation)	−21° to −32°C	Sandy, acidic soils high in organic matter
Blueberry (lowbush) (*Vaccinium angustifolium* Alt.) ERICACEAE: edible portion is an epigynous or "false" berry	Eastern N. America, under domestication, harvested from wild stands	United States Canada	Maine	Creeping, rhizomatous shrubs (12–20 cm)	Very hardy	Sandy, acidic soils
Cranberry (*Vaccinium macrocarpon* Ait.) ERICACEAE: edible portion is an epigynous or "false" berry	Eastern N. America, domesticated early in the nineteenth century	United States Canada	Wisconsin Massachusetts New Jersey Oregon Washington	Creeping, rhizomatous evergreen shrubs (12–20 cm)	−20° to −40°C, bogs flooded for winter protection	Humanmade bogs of acidic sand and organic matter
Currant and gooseberry (*Ribes* spp.) SAXIFRAGACEAE: edible portion is an epigynous or "false" berry	Europe, Asia, and N. America, cultivated in Europe in the seventeenth century	Germany Poland Russia United Kingdom Czech Republic	Commercial acreage <100	Perennial, deciduous shrub (1–2 m in cultivation)	−30° to −35°C	Deep well-drained, fertile soils
Elderberry (American) (*Sambucus nigra* s. canadensis L.) CAPRIFOLIACEAE: edible portion is a small "berry-like" drupe	Eastern N. America, domesticated early in the twentieth century	Not applicable	Limited commercial production	Perennial stoloniferous shrub (to 4 m)	Winter injury of roots or year-old canes not reported	Deep well-drained loam
Grape (European) (*Vitis vinifera* L.) VITACEAE: edible portion is a berry	Asia Minor, cultivated prehistorically and spread throughout Europe in B.C.	Italy France United States Spain China	California	Perennial vine, size varies with cultivar and growing system	To −23°C	Variable, dependent on cultivar and use
Grape (Am. Fox and muscadine) (*Vitis labrusca*, *V. rotundfolia*, etc.) VITACEAE: edible portion is a berry	N. America, domesticated by early European settlers	United States	Washington New York Michigan Southeastern states (*V. rotund.*)	Perennial vine, size varies with cultivar and growing system	To −31°C for *V. Labrus*. not less than −12°C for *V. rotund*.	Variable, dependent on cultivar and use
Kiwifruit (*Actinida deliciosa* A. Chev) ACTINIDIACEAE: edible portion is a berry	Mountainous regions of Central China, cultivated in New Zealand in 1930	New Zealand Italy Japan United States France	California	Vigorous, perennial, deciduous, woody vine (18–24 m)	To −15°C	Deep sandy loam, high in organic matter
Raspberry (*Rubus* spp.) ROSACEAE: edible portion is an aggregate of drupelets	N. Hemisphere, First domesticated by Romans in fourth century, gathered from wild B.C.	Russia Serbia/ Montenegro United States Poland	California Washington Oregon	Perennial plant with biennial canes, height varies with cultivar and region	−34° to −40°C	Deep well-drained, loamy, slightly acidic soils
Strawberry (*Fragaria X ananassa* Duch.) ROSACEAE: edible portion is a modified stem or receptacle	Species hybrid created in Europe 1750 from two species native to the Americas	United States Spain Japan S. Korea Poland	California Florida Washington Oregon	Stoloniferous, herblike perennial with short stem (crown), 3–5 cm	−12° to −40°C, highly dependent upon cultivar	Well-drained, sandy or loamy soils

TABLE 19–2 Characteristics of Common Temperate Small Fruit and Vine Crops

Chilling Requirements	Reproductive Behavior				Major Diseases in United States	Major Insect Pests in United States
	Fruit Producing Wood	Flower Types	Compatibility	Harvest		
50–800 hours	Primarily terminal-bearing on current season's lateral growth borne on biennial canes	Complete	Self-fruitful except for specific cultivars, cross pollination beneficial	Hand (fresh market) and mechanized (processed)	*Verticillum* wilt, anthracnose, cane blight orange rust, viruses	Aphids, borers, weevils, Japanese beetle, mites, raspberry fruit worm
150–1,100 hours	Terminal- and lateral-bearing on year-old shoots	Complete	Self-fruitful except for specific cultivars, cross pollination is beneficial	Hand (fresh market) and mechanized (processed)	Mummy berry, stem cankers, *Botrytis* rot, anthracnose, viruses	Blueberry maggot, cranberry fruit worm, plum cucurlio, mites
1,000 hours	Terminal-bearing on year-old shoots	Complete	Self-incompatible	Harvested with hand held rake	Twig and blossom blight, *Botrytis* rot, red leaf disease	Blueberry maggot, beetles, red-striped fireworms, Tussock moths
600–700 hours	Terminal-bearing on year-old shoots	Complete	Self-fruitful	Mechanized (fresh and processed)	Early, end and black fruit rots, leaf spots, twig blight and shoot dieback	Root grubs and weevils, black-headed fireworm, fruit worms
800–1,600 hours	Terminal- and lateral-bearing on year-old shoots, flowers borne on strigs	Complete	Self-incompatible (black currants) or self-fruitful (gooseberries and red currants), cross pollination is beneficial	Hand (fresh market) and mechanized (processed)	Viruses, leaf diseases, oak root rot	Gall mite, clear-winged borer
Not determined	Terminal-bearing on current year's shoots, lateral-bearing on perennial shoots	Complete	Predominantly self-fruitful, cross pollination is beneficial	Hand	Tomato ringspot virus, stem and twig cankers	Elder shoot borer, sap beetles, mites
100–500 hours	Lateral-bearing on current season's growth borne on year-old canes	Complete	Self-fruitful	Hand or mechanized harvest, dependent upon use	Powdery mildew, viruses, Pierce's disease	Borers, beetles, leafhoppers, leaftollers, mites, *Phylloxera*
1,000–1,400 hours—*V. Labrus.*	Lateral-bearing on current season's growth borne on year-old canes	Predominantly complete, some gynoecy	Predominantly self-fruitful	Hand or mechanized harvest, dependent upon use	Anthracnose, black root rot, *Phomopsis* blight, viruses	Borers, beetles, galling insects, moths, grape cucurlio, leafrollers
500 hours or less	Lateral-bearing on current season shoots borne on year-old flowering laterals	Dioecious, separate male and female plants	Cross-pollinated	Hand	Blossom blight, root rots, fruit rots	Leafrollers, scale, root-knot nematodes, passion vine hoppers
800–1,600 hours	Terminal- and lateral-bearing on current season's lateral growth borne on biennial canes	Complete	Self-fruitful, cross pollination is beneficial, requires pollinating insects	Hand (fresh market) and mechanized (processed)	*Phytophthora* root rot, cane blight, viruses, anthracnose, orange rust	Aphids, borers, *Lygus bug*, mites, raspberry fruit worm
200–500 hours	Terminal-bearing from crown or branched crown. Floral buds initiated in previous or current season	Complete	Self-fruitful, pollinating insects are beneficial	Hand	Red stele, anthracnose, *Photophthora* rots, *Verticillum* wilt, viruses	Aphids, mites, thrips, *Lygus* bugs, beetles, strawberry clipper

Source: USDA National Agricultural Statistics Service.

Figure 19–1 Trends in temperate zone (a) tree fruit, (b) tree nut, and (c) small fruit and vine crop production (1989–2004) in the United States. Graphed values represent the percentage increase or decrease in acreage with respect to that reported for 1995 (the reference year). Values in parentheses following legend designations represent the bearing area in hectares × 1,000 reported for 1995. *Source:* USDA National Agricultural Statistics Service.

Figure 19–2 Global trends in temperate zone (a) tree fruit (b), tree nut, (c) and (c) small fruit and vine crop production (1989–2004). Graphed values represent the percentage increase or decrease in acreage with respect to that reported for 1995 (the reference year). Values in parentheses following legend designations represent the bearing area in hectares × 1,000 reported for 1995. *Source:* United Nations FAOSTAT 2005.

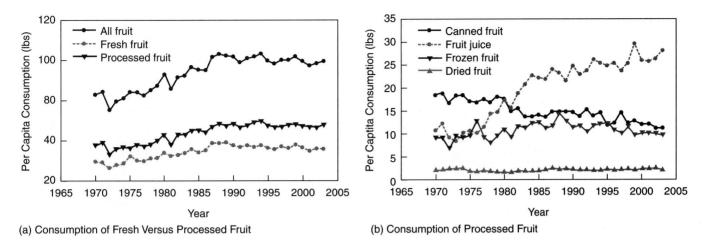

(a) Consumption of Fresh Versus Processed Fruit

(b) Consumption of Processed Fruit

Figure 19–3 U.S. trends in per capita consumption (lbs) of temperate zone fruit and fruit products, 1970–2003. Crops included in the data set are apple, apricot, blackberry, blueberry, cherry, cranberry, grape, nectarine, peach, pear, plum, raspberry, and strawberry. *Source:* USDA Economic Research Service.

only by cost, liability concerns, and the imagination or ingenuity of the producer. Sometimes mail-order or Internet enterprises can be developed with proper advertising.

COSTS IN ESTABLISHING AND MAINTAINING A FRUIT OR NUT PLANTING

There are certain costs to be considered in establishing and maintaining an orchard, vineyard, or berry planting. Unfortunately, many of these costs occur during the planting's establishment. The period from planting to first production can be as short as one season for annually produced strawberries to greater than four or five years for apples, pears, cherries, and many nut crops. Whatever the length of the preproduction phase, fruit producers must develop a business plan that allows for the survival of the enterprise during this period while little or no income is being generated. Once fruit is being produced, yearly expenditures are still necessary to maintain the productivity of the planting and the viability of the business. Costs associated with both pre- and postproduction phases of the planting vary considerably among production regions, localities, crops, and planting systems, and from year to year. These costs are largely the result of:

1. **Capital investment**
   * Land, including water and power.
   * Facilities, including roads, buildings, housing, and irrigation systems, drainage tile, fencing, and alarm systems.
   * Post-harvest facilities, which may include cleaning and packing sheds, refrigerated and/or controlled-atmosphere storage and/or forced air-cooling systems, depending on crop market strategy and size of operation.

   * Equipment, including, tractors; tillage, spray, and harvesting equipment; and shop and farm tools.
   * Plants.
   * Growing system provisions, including staking, trellising, and organic or plastic mulches.
2. **Overhead costs** (during establishment and maintenance periods)
   * Interest, taxes, depreciation, and insurance.
   * Utilities, irrigation water.
   * Management.
3. **Production or direct operating costs**
   * Labor, supplies, equipment, and repairs from preplanting until commercial production.
   * Land preparation.
   * Tree or vine planting.
   * Staking or trellis construction.
   * Pruning and training plants.
   * Pest and weed control.
   * Irrigation, fertilization, and soil management.
   * Harvest supplies, including crates or bins, boxes, cartons, and so on.
   * Post-harvest handling costs.
   * Marketing costs such as advertising, packaging, and transportation to markets.
4. **Labor**
   * Permanent employees for management and farm operations such as weed control, fertilization, irrigation, and so on.
   * Seasonal employees for pruning, harvesting, and packing-shed operations.

In a fruit-growing enterprise, harvest labor is usually the most costly production expense and may determine the profit (or loss) margin. Harvest labor costs and even the availability of harvest labor can fluctuate widely and may remain unknown until harvest is actually under way.

The harvest of some fruit and nut crops such as sour cherries, prunes, almonds, walnuts, pecans, canning peaches, cane fruit and blueberries for processing, and cranberries has been wholly or partly mechanized, and efforts are being made to mechanize the harvest of other fruit and nut crops. Mechanization will greatly stabilize fruit-growing production costs.

Capital investment and production costs can also be influenced by the production system, and in some cases, the cultivars chosen for the new planting. However, before important decisions concerning production strategies or cultivars are made, the producer should ensure that he or she fully comprehends the integrated relationships among plant genotype, horticultural manipulation, and the physiological processes controlling the flowering and fruiting for the new planting's intended crops. A thorough understanding of these relationships is central to the production of commercial quantities of high-quality fruit and ultimately to the success of the enterprise.

UNDERSTANDING THE FLOWERING AND FRUITING PROCESS

There are three initial steps that are critical to the production of large quantities of high-quality fruits. Briefly, these are:

1. The initiation of flower buds in the summer, followed (in most deciduous fruits) by the development of a physiological "resting" condition. This is overcome by chilling winter temperatures, and the flower buds continue development early the following spring.
2. Flower opening and pollination in the spring.
3. Fertilization of the egg in the flower, fruit setting, and the beginning of fruit development.

Flower Initiation

Initiation (also called **differentiation**) involves the change of a vegetative growing point deep inside a bud in the axil of a leaf on a shoot (as in peaches) or on a fruiting spur (as in apples) into miniature flower parts. While some small fruit species produce flower buds in the planting year, initiation does not occur in fruit trees until they have reached a certain size or age (three years or so for some peaches, and up to ten or twelve years for some apples). The transition from the vegetative to the reproductive state and early flower bud development involve a highly integrated series of morphological and physiological events that are highly conditioned by levels of stored carbohydrates (energy), mineral nutrient status, and hormonal balances within the plant during the initiation period. In some fruit plants, this change from a vegetative to a reproductive state is triggered by certain environmental cues. Strawberries, raspberries, and many other fruit crops require short days in the fall (i.e., increasing night length as fall approaches). With most deciduous fruit species, however, no definite environmental factor is known that triggers the change in a bud from a vegetative to a reproductive (flowering) growing point.

Studies many years ago showed that initiation of flower parts in the buds of apples, pears, peaches, cherries,

STRAWBERRIES

The strawberry is an example of a fruit species whose flowering is triggered by a definite, easily defined, environmental factor. Most of the important strawberry cultivars are short-day plants. That is, with the onset of short days (and long nights) in the fall, vegetative growing points in the crown of the plant begin changing to reproductive growing points—or flowers. Such plants then bear a single crop the following spring. When the daylength increases during the summer, flowering stops and the plants become vegetative and start producing runners. However, such cultivars also respond to temperature. These short-day plants grown under long days still produce flowers if the temperature is reduced from 21°C (70°F) to 15°C (60°F). This is the situation in the cool, coastal strawberry districts of central California, where very high yields are obtained because plants of short-day cultivars grown there produce fruit all summer long, even with long days. Certain strawberry cultivars, however, are not responsive to changing daylength. They do not produce runners but continuously form flowers through the long days of summer. These cultivars are termed everbearers.

plums, raspberries, blueberries, and so forth, begins between late spring and late summer during the year preceding bloom. Flower initiation starts after the new shoots have attained a certain diameter and length and a portion of their leaves have matured. Figure 19–4 shows various stages in flower bud initiation and development in the sour cherry.

Horticultural practices that optimize the physiological condition of the fruit plant during the initiation period ensure adequate fruit production for the next fruiting season. In other words, the size of the subsequent fruit crop depends on the number of buds changing from a vegetative to a reproductive state. This, in turn, depends on the general health, hormonal balance, and nutritive condition of the fruit plant.

Trees that were pruned and fertilized heavily the winter before with nitrogen, then copiously irrigated, are likely to become strongly vegetative, producing long, succulent shoots. Buds on such shoots are not likely to form flower buds for the next year. At the other extreme, weakly growing, slender shoots, especially on older trees that are low in vigor, form few flower buds, particularly if the leaf area has been damaged by insect or disease attacks, thereby reducing the tree's photosynthetic capacity.

Severe and prolonged drought during the critical period of flower bud initiation in early summer can create water deficits in the trees that interfere with flower-bud formation. Slight water deficits are not likely to be harmful and can even stimulate flower initiation by reducing shoot elongation and causing carbohydrate accumulation from photosynthesis. Pruning or similar activities, such as the renovation of matted, row-cultured strawberries, should be accomplished in a timely manner to provide an adequate leaf canopy or to avoid drastic disruptions in carbohydrate and hormonal balances at the time of flower initiation. Biotic stresses (insects, disease,

Figure 19–4   Early stages in the development of flower parts in the buds of the sour cherry (*Prunus cerasus*), sampled in Hart, Michigan, as shown by scanning electron micrographs. Sampling dates and magnification are: *Top row* (left to right): June 10 (× 266); June 16 (× 293); July (× 233). *Center row* (left to right): July 30 (× 213); August 15 (× 120); September 29 (× 110). *Bottom row* (left to right): November 4 (× 55); March 27 (× 40). This shows that the cherry flowers that open in the spring have developed slowly in the buds during the preceding summer and fall. *Source:* D. H. Diaz.

weeds, etc.) should also be controlled to maximize photosynthetic capacity. It is evident that the care and management that fruit plants receive strongly influence their productivity.

Rest and Chilling Requirements

Once the flower parts are fully formed in the flower buds of deciduous fruit plants, the buds enter a physiological **rest period** in which they will not open even if the plants are subjected to favorable temperature, moisture, and light conditions. It is important to note that the same is true for vegetative buds (those that did not differentiate into flower buds) on the same tree.

The beginning of this rest period depends on the species and the general vigor of the plant. The rest develops

slowly, reaches a peak, then diminishes; the onset of the rest period can range from midsummer to late fall. In shorter, slow-growing shoots it starts earlier than in the buds of longer, more vigorously growing shoots.

There is some evidence that the physiological resting condition develops because the buds accumulate certain natural growth inhibitors, such as abscisic acid, and lose native growth promoters, such as the gibberellins.

The physiological buildup of such growth blockages in buds of deciduous fruit trees—which also occurs in temperate-zone ornamental deciduous trees and shrubs, as well as in certain bulb species—is an evolutionary development that increases the plants' chances of survival through the winter. Without this self-blocking mechanism for bud growth in the autumn, tender, succulent shoots and tender, newly formed flowers developing from the buds would start to grow during the winter and would be killed by subfreezing temperatures. This physiological internal inhibition of growth (rest) occurs only in the buds—not in the roots.

Chilling of the buds (both vegetative and flower) through the winter is needed to reverse the rest influence. If the chilling is long enough, the influences blocking bud growth disappear. Then, with the beginning of warm spring temperatures, plus adequate soil moisture, both vegetative and flower buds grow rapidly and vigorously. It is believed that the chilling temperatures may act on the buds to lower the levels of growth inhibitors—such as abscisic acid—and increase the amounts of growth promoters, such as the gibberellins.

If the amount of chilling through the winter is marginal, as it can be in regions with normally mild winters, such as the fruit-growing areas of South Africa and the Central Valley of California, bud growth in the spring is erratic. Some flower buds may drop before opening, thus reducing the crop potential, or bud opening in the blooming period may be late and prolonged. Delayed foliation refers to slow development of vegetative buds due to the lack of sufficient winter chilling. A marginal winter chilling may not always be a disadvantage because it can lengthen the blooming period, increasing the chances of good weather during part of the bloom period and giving bees more time to pollinate the flowers. It can also cause the flower buds to open later and avoid being killed by late spring frosts. In areas with long, cold winters, such as the East, Northeast, and Middle West parts of the United States, the buds always receive sufficient chilling and this problem does not exist.

Much study has gone into determining the amount of chilling required by buds of the different fruit species to overcome the rest influence. Tables 19–1 and 19–2 summarize much of this information, which is expressed as the number of hours below 7°C (45°F) but above 0°C (32°F), required by the various species to overcome the rest influence. Subfreezing temperatures are not necessary. High winter daytime temperatures greater than 16°C (61°F) can be detrimental for deciduous fruits, however, because they tend to counteract chilling temperatures. Fog or overcast weather during the winter days can be beneficial in obtaining good bud chilling by keeping direct sun rays from the buds, which would raise bud temperature above the air temperature.

To summarize, the usual sequence of events for temperate-zone plants is for the buds to enter the rest phase in late summer or fall; then, after sufficient winter-chilling, the rest influence is terminated. The buds are then said to be quiescent but will resume growth with the onset of favorable growing conditions in the spring.

However, there are exceptions to the normal pattern of floral initiation, differentiation, rest, and resumption of bud growth in the following spring. Elderberry flowers, for instance, are borne terminally on new canes or arise within a growing season from axillary positions on second-year and third-year canes. Each flower bud produces an inflorescence (cyme) in early summer that measures from 8 to 25 cm in diameter. Each cyme hold hundreds of self-fruitful flowers that develop and mature into very small, dark purple fruit in late summer or early fall.

Moderate to severe biotic or abiotic stresses can trigger precocious flower development. Trees of some species, particularly almonds and plums, will often bloom in late summer or fall if they are partly defoliated by mites or drought and then wetted by rain or irrigation. Some cane and berry fruit cultivars (e.g., strawberries) have been reported to develop a small portion of their flowers (2 to 3 percent) in the fall, presumably from flower buds that have somehow circumvented the dormancy requirement.

In a more dramatic way, strawberry and cane fruit breeders have developed cultivars with the rather unique characteristic of day neutrality. Day-neutral strawberries are

---

TERMS ASSOCIATED WITH FLORAL INITIATION AND DORMANCY

flower initiation   The transition from vegetative bud to floral bud. This transition involves changes in the morphology and physiology of bud meristems and is conditioned by plant age, plant condition, and environmental influences.

short-day plant   A plant that initiates flower buds during periods of long, uninterrupted darkness. In temperate regions, short-day plants have the potential to initiate flower buds in spring or fall.

long-day plant   A plant that initiates flower buds during periods of short nights or when nights are interrupted by short bursts of light.

day-neutral plant   A plant that initiates flower buds regardless of day or night length.

flower bud development   The growth and continued differentiation of floral tissue after flower bud initiation but before anthesis.

dormancy   A general term denoting an inactive state of growth.

rest period   This occurs when buds fail to grow, even under optimal environmental conditions, because of internal physiological blocks. Exposure of the buds to a sufficient amount of above-freezing chilling temperatures can terminate rest.

quiescence   This term describes a condition when nonresting buds fail to grow due to unfavorable environmental conditions such as low temperatures, unavailable water, or an unfavorable photoperiod.

correlative inhibition   In this situation, buds do not grow because of an inhibitory influence of another plant part, for example, failure of lateral buds to grow due to inhibition by the terminal part of the shoot.

chilling requirement   The number of hours a plant (bud) must experience temperatures between 0°–7°C (32°–45°F) to achieve bud-break in the spring. During the accumulation of chilling hours, the physiological (hormonal) conditions affecting dormancy are gradually reversed.

sometimes called everbearers, although the terms may not be synonymous. Unlike their June-bearing counterparts, day-neutral strawberries initiate flower buds in the long, warm days of summer that do not require chilling for continued development. In cooler climates, such as the U.S. Midwest, growers have been able to use day-neutral cultivars to obtain a fruit in the planting year and to extend the harvest season from five weeks to five months. In the establishment year, peak fruit production occurs in late August or September. In seasons thereafter, two fruiting peaks occur in a day-neutral strawberry planting, one in June and the other in August. In warmer production areas such as California, day-neutral varieties have been used to produce fruit earlier in the winter (south coast) and to enhance production in the spring and summer months (central coast).

Cane fruit crops are most often grown as perennials with biennially bearing shoots; as such, most cultivars follow the typical temperate perennial floral development pattern. A newly emerged cane ("a primocane") grows vegetatively during its first season of growth until compound buds containing floral primordia are initiated in the fall in response to day length. The flower buds cease to develop when the cane enters dormancy and then resume growth and flower in the following spring. These two-year-old fruiting canes are called floricanes. However, a small but important percentage of red raspberry and, more recently, blackberry cane fruit cultivars also carry genes for day neutrality. These primocane fruiting (or fall-bearing) cultivars precociously initiate their flower buds in midsummer, and most of them continue to develop without a period of rest or dormancy. Flowering usually proceeds from the apical nodes to the base of the cane. In most primocane

fruiting cultivars, a portion of the flower buds initiated undergo a chilling treatment, then bloom in the following spring.

Location of Fruit Buds and Fruiting Structures

Although most temperate fruit and nut crops are produced from flowers initiated the previous season, the location of flower buds on the plant and the plant structures on which they develop vary widely among fruit-producing species. They can even differ significantly among cultivars within species. Some fruit and nut trees bear their fruit on fruiting spurs rather than laterally or terminally on predominantly vegetative shoots (Fig. 19–5). Spurs are essentially short shoots composed of floral, vegetative, or mixed buds separated by a few shortened internodes. Fruiting spurs can be located throughout the fruiting surface of the tree and can live for several years. Apple spurs terminate in mixed buds that contain leaf, shoot, and floral primordia. As they develop in the spring, these buds produce leaves, at least one new shoot (a bourse shoot), and approximately six to eight flowers, only one or two of which will develop into a mature fruit. In addition to specific apple cultivars, pears and mature pecans and walnuts bear fruit developed from terminal spur buds. Cultivars of almonds, apricots, sweet cherries, and plums also fruit heavily on spurs, but the floral primordia of these species are positioned in lateral buds on the spurs rather than in the terminal bud.

Some fruit and nut trees predominantly set floral buds at lateral or, more rarely, terminal positions along the growing shoot (Fig. 19–6). Predominately shoot-bearing deciduous tree crops include some apple cultivars, peaches and nectarines, sour cherries, and young pecans and walnuts.

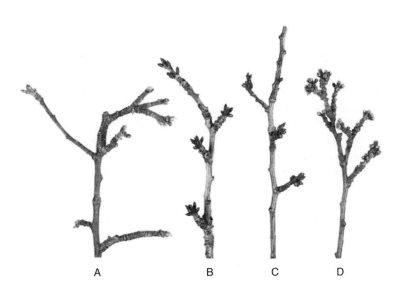

A   B   C   D

Figure 19–5 The development of fruiting buds terminally on (A) fruiting apple spurs and laterally on the fruiting spurs of (B) sweet cherry (C) apricot, and (D) plum.

A   B   C

Figure 19–6 The development of lateral fruit buds on one-year-old shoots of (A) almond, (B) peach, and (C) apricot. Each pair consists of a shoot with vegetative buds (*left*) and fruit buds (*right*). Vegetative buds are small and pointed, whereas fruit buds are large and plump.

Bush fruits commonly set floral buds along new shoots as well. For example, highbush and rabbiteye blueberry shoots differentiate terminal and axillary flower buds midseason, at the end of the last growth flush. These large, spherically shaped buds are visually distinguishable from the smaller, vegetative buds also present in shoot axils.

The number of floral buds developed per shoot appears to be related to shoot thickness, hormonal balance, and cultivar. As a highbush blueberry shoot enters dormancy, it typically possesses a floral bud at its tip and at the four to seven axillary positions immediately below it, although as many as fifteen to twenty total floral buds per shoot have been reported. Each bud will produce from five to ten flowers the following spring after their chilling requirements have been met. The flowering pattern of the lowbush or wild blueberry of New England is similar to that of highbush and rabbiteye blueberries; the shoot terminal becomes senescent after an initial growth flush and flower buds are developed laterally, beginning in the axil closest to the shoot meristem.

In contrast, the extensive decumbent stems or uprights of cranberry initiate floral development in mixed terminal buds in early to midsummer; floral tissues continue to develop in these buds throughout the summer and during the following spring. In spring, the bud breaks and continues to develop vegetatively. Flower stalks bearing solitary, pendulous flowers on stiff pedicels develop at the base of the new shoot growth.

TERMS ASSOCIATED WITH FRUITING AND FRUITING STRUCTURES

apical floral bud   A bud forming at the shoot apex that has undergone a transition to produce floral rather than vegetative tissues.

mixed bud   A bud that contains both vegetative and floral tissues.

compound bud   A bud that contains more than one growing point; usually, one growing point is dominant. The bud may contain floral tissues, vegetative tissues, or both.

shoot-bearing   A term describing a plant that predominantly bears fruit terminally or axially on shoots developed during the current or previous season.

spur-bearing   A term describing a plant that predominantly bears fruit on fruiting spurs.

fruiting spur   A short, thick, fruit-bearing stem terminated by an apical floral, vegetative, or mixed bud at the end of the growing season, depending on the species. Fruiting spurs usually develop flowers, leaves, and new shoot growth.

bourse shoot   Shoot arising from the mixed bud on a fruiting spur of apple. Bourse shoots terminate growth by forming floral or vegetative buds.

upright   A decumbent shoot arising terminally or axially from a cranberry runner that will bear fruit from terminal buds.

strig   A current or gooseberry inflorescence.

primocane   A vegetative cane of raspberries, blackberries, or their hybrids that arises from crown or rhizome buds. Primocanes develop floral buds in response to physiological or environmental cues but typically do not fruit until their chilling requirements have been met.

floricane   A second-year cane (previously a primocane) of raspberries, blackberries, or their hybrids, on which floral buds initiated in the previous season develop flower and fruit.

crown   A strawberry stem with severely shortened internodes that develops a terminal floral bud in response to environmental cues.

branched crown   A branch arising axially from the strawberry stem. Branched crowns also develop terminal floral buds.

scape   A strawberry inflorescence.

Currants and gooseberries also initiate floral buds in midsummer, which develop in the axils of main shoots or at the terminus of secondary shoots. In the following spring, these buds produce short, branchlike fruiting structures called strigs that hold one to three or eight to thirty flowers for gooseberries and currants, respectively.

Grapes produce two types of buds in the leaf axils of the current season's growth: vegetative buds that may give rise to lateral shoots within the same season, and compound buds (eyes) that contain both vegetative and floral primordia for next season. These compound buds are fully developed by late fall and the onset of dormancy. In the spring, after chilling requirements have been satisfied, buds that remain after dormant pruning develop into the new shoots that will bear the grape clusters. The number and position of these clusters on the new shoot is genetically controlled but environmentally and culturally influenced. European grapes and many other grape species often produce only two clusters per shoot, whereas American grapes typically produce two to four clusters borne at nodes two to five of the shoot. French-American hybrid wine grapes are prolific, producing four or more clusters per shoot and additional clusters from buds at the base of the shoot and from latent buds on the cordon (permanent scaffold) of the vine.

As stated above, most cane fruit develop compound buds on first-year canes (primocanes) some time during late summer to fall, or in the case of blackberries, even as late as the following spring. After dormancy is satisfied, these buds develop into fruiting laterals, which can be several nodes long. Flowers are borne on branched inflorescences formed terminally or in leaf axils of the fruiting lateral. Each inflorescence bears from one to many flowers, depending on the species, cultivar, and position on the lateral. In red raspberry, fruiting laterals arise primarily from the leaf axils of the unbranched cane (now a second-year cane, or floricane), with about 70 percent of the fruit-bearing potential concentrated in the central one-half of the cane. In contrast, black raspberries and blackberries and other cane fruits are often cultured to produce vegetative laterals on primocanes because many of the fruiting laterals produced in the following season will arise from compound buds in the leaf axils of these vegetative laterals.

The strawberry crown is essentially a stem with severely shortened internodes resulting in leaves that are arranged in a rosette pattern. Each axillary bud of the strawberry crown may give rise to a branch crown or a runner. However, in response to environmental cues, the terminal growing points of the main crown or branch crowns are transformed into terminal floral buds. In the U.S. Midwest, the floral bud is well-developed by late fall, before entering a dormant period. The strawberry inflorescence or scape (botanically a cyme) develops from this bud in the following spring. Crown growth is re-initiated from a lateral bud, close to the terminal. In continental climates, each crown or branch crown can, at best, produce one scape per year, whereas in warmer regions, several flowering flushes can occur, suggesting several cycles of sympodial crown growth within a season.

Pollination and Pollenizers

When the previously formed flower parts start to develop and the flowers open, a new critical stage begins in the production of the crop. For most fruit species, the flowers must be adequately pollinated before fruit can set and develop. The pollination requirements of the various major fruit species were given in Tables 19–1 and 19–2.

Some fruit species have definite pollination requirements. Successful fruit growers know the needs of their fruit crops and provide for adequate pollination. For example, in both almonds and sweet cherries, the flowers of any given

---

FLOWER AND POLLINATION TERMS

perfect flower A flower that has functional male and female organs.

complete flower A flower that contains all four floral whorls: sepals, petals, stamens, and pistils.

staminate (male) flower A flower that contains functional stamens but no pistils.

pistillate (female) flower A flower that contains functional pistils but no stamens.

monoecious A term describing plants that produce both staminate and pistillate flowers on the same plant.

dioecious A term describing plants that bear staminate and pistillate flowers on separate plants.

pollenizer A plant that is a source of pollen.

pollinator An agent (such as an insect) that pollinates flowers.

pollination Transfer of pollen from an anther to the stigma of a flower.

self-pollination Transfer of pollen from anther to stigma of the same flower or to the stigma of another flower of the same clone.

cross-pollination Transfer of pollen from an anther in the flower of one cultivar to the stigma of a flower in a different cultivar.

parthenocarpy The process by which fruit develops without fertilization of an egg (e.g., some cultivars of seedless grapes). In such fruits, growth of the fleshy parts is stimulated by endogenous hormone levels, most notably auxin (indoleacetic acid) and gibberellins.

---

cultivar must be pollinated by those of a particular different cultivar. This means that pollen must originate in the anther of the flower of the pollenizer cultivar and be carried by bees (or other insects or wind) and deposited on the stigma of the flower of the cultivar that is to produce the main crop. In this example, all almond and sweet cherry cultivars would be self-unfruitful and certain cultivars would be inter-unfruitful.

Some fruit species, such as most apple cultivars, must be cross-pollinated, but the pollen can originate in flowers from any other cultivar of that species. But cultivars selected for cross pollination must have overlapping blooming periods. In contrast, other fruit species, such as sour cherries and apricots, are self-fruitful. The pollen can originate in any flower on the same cultivar, either on another tree, another flower on the same tree, or even from the anther of the same flower. These species are normally self-pollinated, but pollen coming from flowers of different cultivars in the same species may also cause fruit to set.

Before the nursery trees are purchased for a planting of the fruit species, the grower must determine the pollination requirements of his or her proposed cultivars. Where cross pollination is required, the planting should consist of the main fruiting cultivar, interspersed with trees of the pollenizer. Quite often, the latter trees are set in some arrangement, such as every third tree in every third row, or if the main cultivar is one that tends to set fruit heavily, every fourth tree in every fourth row. Where two fruiting cultivars of equal value can be used, harvesting is easier if, for example, blocks of four rows of each of the two cultivars alternate.

Generally, stigmas of fruit tree flowers are receptive to pollination for only about five days after the flowers open. For fruit to set after pollination takes place, the pollen tube must grow through the style into the egg in the embryo sac.

Temperature is an important factor during all the stages of pollination, pollen tube growth, fertilization, and fruit setting. A temperature range of 15.5°C to 26.5°C (60°F to 80°F) is considered optimum for deciduous fruits. Temperatures much above or below this range impair good fruit setting. Temperatures much above 26.5°C (80°F) inhibit pollen germination. Pollen grains themselves are quite stable at low temperatures—when dry, they can be kept viable for years at −18°C (0°F)—but temperatures dropping to −3°C or −2°C (27°F or 28°F) can kill the ovules in the open flowers of most fruit species.

Insects and Pollination

Fruit species with large, showy flowers generally depend on insects to transfer pollen. Bees, particularly honeybees, are the most important type of insect involved in the pollination of commercial crops. Some fruit plants, especially those with nonshowy flowers, such as the walnut, pecan, and filbert, are wind-pollinated. They generally produce large amounts of very light pollen that is carried considerable distances in the wind onto the stigmas of other flowers.

In fruit orchards that require cross pollination, even when the proper mixture of pollenizing cultivars is present, bees must be in the orchard during the blooming period to carry pollen from the flowers of one cultivar to the flowers of the other (Fig. 19–7). For one or two trees in a home garden, enough bees are generally present for successful pollination, but an orchardist with a large number of trees that absolutely require cross pollination (self-unfruitful) will need to have bees brought in. For a rental fee, beekeepers will provide the necessary bees during the blooming season.

Honeybees work the flowers to collect pollen and nectar, which they use as food. Since the bees generally stay within about a 100-yard radius of their hives, hives are placed in an orchard no more than 150 to 180 m (500 to 600 ft) apart. Weather conditions affect bee activity. Below about 13°C (55°F) they are inactive; the optimum temperatures are about 18° to 27°C (65°F to 80°F). Winds much above 15 mi/hr (24 km/hr) keep bees from flying, and continuous rains interfere with their activity, but intermittent showers do not. If bees or other insects are important pollinators, pesticides must be used with extreme caution to protect the pollinators.

CHOOSING A GROWING SYSTEM

Many temperate fruit and nut crops can be cultured in a variety of ways, so producers have the opportunity to tailor a growing system to fit their climate, capabilities, situation, and needs. Growing systems are often chosen to maximize

Figure 19–7 To obtain good crops of many fruit species (e.g., sweet cherries, almonds, plums, and apples) cross pollination between cultivars is necessary. Fruit growers often set hives of bees in their orchards during bloom, as shown here. The bees, working the flowers for nectar and pollen, which they use as food, also transport pollen from tree to tree. *Source:* USDA.

yield, facilitate horticultural practices, or alter fruit harvest to capture market opportunities. Successful use of a growing system is crop- and even cultivar-dependent. For example, both grapes and cane fruits can be grown using a wide variety of trellising techniques that affect the plant's physiology, yield potential, or hardiness, as well as the horticultural practices necessary to effect sustained productivity and planting longevity. Vigorous American or French-American hybrid wine and table grape varieties, such as 'Concord,' are perhaps best cultured using a Geneva double-curtain bilateral cordon or training system (see Training and Pruning later in this chapter), whereas grapes of moderate vigor can be cultured successfully using the popular high bilateral cordon (single-curtain) training system. Both of these systems are adaptable to both hand and machine harvest. In contrast, French vinifera wine grapes are often trained most successfully using the vertical shoot position system because they have a strong tendency toward an upright growth habit.

Strawberry growing systems vary widely among production regions in response to climatic differences (see Fig. 19–8). Additional information is available under Flower Initiation described previously. In much of the Midwest, strawberries are cultured perennially in matted rows established by the setting of runner plants. In southern California and Florida, individual plants are grown as annuals at close spacing on plastic-mulched, raised beds during the winter months, when short daylengths permit multiple flower flushes (cropping cycles) per season. In northern California coastal regions, cool night temperatures also allow for multiple cropping within a single season. The annual strawberry production system is gaining popularity in the southeastern states and some

Figure 19–8 Strawberry production systems: (A) perennial, matted row culture in the Midwest, (B) annual two-row winter production on black plastic in Florida, and (C) annual four-row winter production on clear plastic in southern California. *Source:* Joseph Scheerens, The Ohio State University.

midwestern states as well. In these areas, fall-planted strawberries are overwintered with natural or synthetic mulches, which results in an early spring harvest when market demands are high.

Cane fruit crops are most often grown as perennials with biennially bearing shoots, requiring growers to manage primocanes and floricanes simultaneously. To reduce primocane interference with the fruiting canes, however, a few commercial producers have adopted a two-year cropping system where new primocanes in the second or fruiting year are partially or totally suppressed. All canes are removed after harvest to begin another cycle in the following season. Cycle timing is often staggered among rows or areas within the production field to provide a harvestable crop each year. Pruning activities are simplified and spray costs are reduced using the two-year system, but overall yield is reduced because only half the planting produces fruit in a given year. Reductions in berry size and quality have also been reported using this system.

In recent years, the so-called high-density orchard plantings, particularly with apples and to a lesser extent with pears and stone fruits, have become popular. Trees are planted close together, as hedgerows, or tree walls. Dwarfing rootstocks are used to keep the trees small. The land can be utilized to the maximum by high-density plantings, especially when the trees are young. High-density plantings of the stone fruits—peaches, plums, apricots—have not been as successful mainly because no completely satisfactory dwarfing rootstock is available for these species.

The different categories of planting densities and management systems in use today, particularly for apples, are:

1. *Low density.* Trees are widely spaced (fewer than 250 trees/ha; 100 trees/ac) so that after maturity, each tree has ample space and light contact around it. Pruning is kept to a minimum to allow rapid development of maximum tree size. Dwarfing rootstocks are not used. Maintenance labor is minimal, but the yields and gross returns per unit area are also likely to be minimal, particularly for the first fifteen to twenty years of orchard life, compared to higher-density plantings (of

apples and pears). Fifteen to twenty years may be needed to reach full production. For the stone fruits, tree nuts, and citrus, low-density plantings may be the most profitable, although there has been considerable interest and experimentation in developing high-density management systems with these crops.

2. *Medium density.* Tree spacing (250 to 500 trees/ha; 100 to 200 trees/ac) is at least 1.2 m (4 ft) closer than for low-density plantings, and pruning and training are more intensive. Dwarfing rootstocks, such as 'M.M. 106' or 'M. 7A,' may be used with apple, or the trees may be spur-type cultivars budded on seedling rootstocks. More labor is required in pruning and training the trees, particularly during the early developing years. More care and supervision are required, and the investment per hectare—in nursery stock and perhaps irrigation equipment—is greater than for low-density plantings. However, commercial yields are achieved sooner after planting, which reduces the cost of establishing the planting.

3. *High density.* Trees are planted very close together (500 to 1,235 trees/ha; 200 to 500 trees/ac) and specific training systems are used, such as slender trees, multiple rows, or trellis-trained plantings. Training and pruning are very important. Reduced tree size can be aided by the use of the more dwarfing rootstocks, such as Angers quince with pear and 'M. 9' or 'M. 26' with apple.

Trees grown at high densities or those destined to be harvested mechanically are often intensively trained using a variety of high input (materials and labor) training systems that promote early fruiting and/or the exacting placement of fruiting surfaces. Often these growing systems employ stakes or trellises to control shoot growth carefully. Some examples of high-intensity training systems include the slender spindle, the French axe, the hybrid tree cone, the Heinicke method, the palmette leader, the Lincoln canopy, and the Tatura trellis. Although pruning and training are discussed below, a thorough discussion of these methods, their techniques and applications, and their effects on tree physiology and growth are well beyond the scope of this text. However, the interested reader can explore this topic further using the references listed at the end of this chapter and/or obtaining other appropriate resources.

The yield efficiency of a planting may be described mathematically as marketable yield per unit of production input (cost, time, etc.). Obviously, the choice of a specific growing system will have a marked impact on the yield efficiency. High-density or high-input systems may raise production efficiency, but they are often harder to manage. Because high-input systems usually involve increased establishment and production costs, poor management decisions are more likely to have drastic consequences for the success of the enterprise. High-input system profitability often requires that the crop exhibit an optimum response to horticultural practices to maximize yield. Therefore, an experienced and skilled operator with an in-depth understanding about how a fruit plant's physiology is affected by horticultural manipulation is more likely to succeed with high-input systems than are producers attempting to raise fruits or nuts for the first time.

A discourse on fruit-growing systems would not be complete without a discussion of organic fruit production. The organic food market is thriving, increasing by approximately 20 percent per year in the past decade. Consumer survey data suggest that the choice to purchase organically grown foods over those grown conventionally is complex and involves consumer attitudes and beliefs concerning environmental and agrarian stewardship, family health and safety, and food quality and nutritional value.

Many people equate the terms *organic production* and *pesticide-free,* although this view of organic production systems is perhaps too simple. Organic crop management is based on several overarching principles. The stability of the agricultural ecosystem is ensured by employing inputs and techniques that sustain or increase biodiversity within the cropping site and promote interactivity and ecological balance among plants and other soil-borne, land-based, or airborne organisms. Inputs and techniques that foster sustainability include the following: (1) nutritional and soil resource management through the use of crop rotation, cover crops, and composed mulches and manures rather than synthetic fertilizers, and (2) natural pest management, which involves the use of manual and biological control techniques and pesticides composed of natural products. The integration of farming practices to achieve ecological sustainability and the cycling of resources are also fundamental elements of organic management. The principles of organic production also embrace the integration and diversification of enterprises to deliver products to the consumer that are truly organic in origin. Ideally, the composted materials used in an organic production system would be obtained only from sources that were themselves produced organically. In practice, however governmental regulations only require the use of composts that do not contain specified prohibited compounds. Seed or transplants used to establish production fields should themselves be organically produced. Sources of organically produced nursery stock may not be available for many fruit crops; if conventional nursery stock is planted, the fruit cannot be sold as organic for a period of twelve months following transplanting. The use of genetically modified organisms (GMOs) is not allowed. Finally, organic integrity of the produce must be stringently maintained to prevent contamination with prohibited materials. For instance, organic fields must be isolated from conventionally managed fields at a distance that would prevent accidental spray drift or other sources of contamination. Also, machinery, equipment, and processing and storage facilities must either be designated specifically for organic produce or cleaned extensively after use in conventional systems.

Obviously, suitability for organic production differs among fruit crops, although most can be organically cultured successfully under favorable conditions. In general, the nitrogen requirements of fruit crops are modest; the availability of excess nitrogen actually overstimulates vegetative growth at the expense of fruit production. Therefore, fruit crops are well-suited for culture using natural fertilizers or composts that tend to release nitrogen incrementally over time as they are decomposed by soil organisms. The main challenge for organic fruit growers is perhaps the control of crop pests. Under an organic production system, pests are managed culturally through the use of genetically resistant cultivars, crop rotation, ground covers and cover crops, soil and crop nutrient practices that restrict pest establishment, additional growing practices that promote crop health and limit pest infection or infestation rates (e.g., proper light penetration and air movement), sanitation and physical removal of crop pests and their habitats, the use of biological control organisms and trap crops, and (as a last resort) the application of natural pesticides.

Site characteristics are extremely important indicators of potential economic success for all fruit plantings; this is especially true if a producer is considering an organic growing system for his or her new planting. In general, it is somewhat easier to produce organically grown fruit in arid or semiarid regions because disease pressures are commonly not as severe in these areas. This may be especially true for grapes, for which consumer preference tends toward wine and dessert cultivars that are not highly resistant to insects or diseases that are prevalent in wetter climates. In response to consumer demand, several well-known western U.S. vintners have developed successful, large-scale organic vineyards. Because of the prevalence of fruit-rotting fungal diseases, organic production of bunch grapes is not generally recommended for the U.S. South, except for the indigenous muscadine grape that bears its fruit in small open clusters.

The demand for organically grown strawberries and the premium price they command offset lower yields commonly attained with the organic production system. Regardless of production region or growing system, one of the most serious challenges faced by strawberry growers is the control of weeds. To control weeds, large-scale strawberry production systems often involve the pre-plant application of dangerous and ecologically unsound soil fumigants, such as methyl bromide, and the use of plastic mulches throughout the life of the planting. Although not specifically prohibited in organic production, many in organic agriculture do not consider the use of plastic mulches as weed barriers or the use of other petroleum-based materials such as row, tunnel, hoop, or greenhouse covers used in colder climates to be ecologically friendly or sustainable practices. Under any growing system, weed problems can be diminished by selecting a site where effective weed management has been employed in the past and by using cultural practices that promote plant and soil health. Hand or mechanical removal of weeds is effective, sometimes necessary, and often costly and labor-intensive. Novel approaches to weed control for organically produced strawberries include steam sterilization of the soil prior to planting,

weeder geese when ripe fruit are not present, organic mulches such as straw or shredded newsprint, steam or flame weeding (expensive alternatives), and natural herbicides composed primarily of vinegar with lemon juice or citrus oil additives.

Preliminary studies suggest planting strawberries into living mulches such as Sudan grass as a component of a weed-management strategy. Preliminary evidence also suggests that corn gluten, a by-product of corn milling containing high levels of nitrogen and a natural herbicide, is effective against selected weed pests, especially grasses. Under conditions where the natural herbicide is likely to be leached, however, the increased soil nitrogen supplied by this material may actually promote weed growth. Finally, many organic strawberry growers are opting for shorter life spans for their planting to control the buildup of perennial weeds.

Highbush and rabbiteye blueberries are perhaps the best suited among fruit crops for organic culture. In most production fields, the row middles are sodded and the blueberries are deeply mulched, which discourages weed seed germination and weed growth. Blueberries are more resistant to insects and diseases than most fruit crops, which makes them easier to culture organically. They have a relatively low nitrogen requirement and thrive on organically based fertilizers. Because blueberries are naturally adapted to soils with low pH and high organic matter, the use of composting materials that release nitrogen as NH_4^+ are ideally suited to meet blueberry nutritional and soil requirements. The increased soil organic matter associated with organic-production systems also favors the proliferation of ericoid mycorrhizae, the symbiotic fungi necessary for adequate blueberry nutrient uptake.

Regardless of the crop under consideration, organic fruit farming is a complex and exacting growing system. It is not, as some may believe, farming by benign neglect. The activities and expense of producing fruit organically are over and above the inherent cultural considerations associated with fruit production under any growing system. In the United States, the U.S. Department of Agriculture's National Organic Program (NOP) regulates the labeling, marketing, and record-keeping procedures of all products labeled as organic. Land must undergo a certification process based on exacting requirements before it can be used to grow organic produce. Producers are required to keep detailed records of all materials and activities associated with producing, handling, storing, and marketing the crop. They must also keep abreast of the myriad regulations that govern their activities.

For specific information about the organic culture of fruits, interested growers should contact the National Organic Program under the auspices of the Agricultural Marketing Service of the U.S. Department of Agriculture. A wealth of information concerning organic culture in general and organic fruit production specifically (including production guides) can also be obtained through the ATTRA—The National Sustainable Agriculture Information Service, a public information organization managed by the National Center for Appropriate Technology (NCAT) and funded by the

U.S. Department of Agriculture. Contact information for these organizations is included in the reference list of this chapter.

In summary, the choice of a growing system is an important one. Drastic changes in production systems are often difficult or impossible to effect after a planting is established. Therefore, a grower establishing a fruit or nut planting must remain committed to his or her chosen system throughout the life of the planting and be determined to make it work.

SELECTING FRUITING CULTIVARS AND ROOTSTOCKS

Perhaps one of the more important pre-plant decisions to be made involves cultivar and/or rootstock choice. This decision should not be taken lightly. It should not be left to the local nursery owner, and it should not be based on what surplus nursery stock happens to be on hand for a bargain price at the time. Sometimes this decision must be made and an order placed a year or two in advance so that the propagating wholesale nursery will have time to propagate the desired cultivar on the desired rootstock and to permit obtaining disease-free plant material.

Because cultivar performance affects success at every step of the enterprise, cultivars should be chosen that function well during many facets of the fruit production cycle. Specific characteristics of cultivars to be considered before the choice is made include:

1. *Adaptation response.* Within fruit and nut species, each cultivar has unique physical and physiological traits preconditioned by its genetic profile. The expression of these traits is influenced by the planting environment (e.g., climate, soil characteristics, horticultural factors) to produce a phenotype. The adaptation response of a cultivar can be defined as how well the genetic and environmental components of phenotype complement one another to form a highly productive plant with desirable characteristics. Many cultivars perform well in only specific climatic regions (e.g., warm versus cold, wet versus dry) or in specific types of soils (e.g., well-drained, high organic matter), whereas others are more widely adapted. A widely adapted cultivar, even though highly productive under broad environmental conditions, may vary in some phenotypic traits (e.g., size, vigor) when grown in different regions.

2. *Response to growing system.* A cultivar's adaptation response is also highly dependent on the growing system under which it is cultured. For instance, highly efficient cultivars that partition more of their energy to reproduction than to vegetative growth may be the most suitable for high-density plantings. Some fruit cultivars require extensive support or trellising to perform well, whereas others can be cultured as free standing plants. In some climatic areas, small fruits,

such as strawberries or raspberries, can be grown as annuals. However, successful annual production schemes require the use of cultivars that are uniquely suited to, and sometimes specifically bred for, production under this system.

3. *Growth form (habit), vegetative vigor, and size at maturity.* These three interrelated cultivar characteristics can influence light penetration and air movement within the planting and often direct horticultural factors such as plant spacing, training, trellising and pruning systems, fertilization and irrigation practices, equipment use, and harvest techniques. The importance of considering these characteristics when selecting cultivars is obvious, especially for tree fruit genotypes, which often differ widely in size and vegetative vigor. Apples, cherries, grapes, pears, walnuts, and many other woody fruit and nut species are often propagated by budding or grafting cultivars with desired growth and fruiting characteristics to related rootstocks that exhibit superior climatic and edaphic (soil) adaptation responses. Rootstocks are frequently from a different species within the genus, and they often affect the size, precocity, and reproductive capacity of the cultivar under production. The ability of the East Malling and Malling Merton series of apple rootstocks to produce ultimate tree sizes ranging from dwarf to full stature are perhaps the most well-known examples of this phenomenon (Fig. 19–9). For some fruit species, cultivar choice is conditioned partly by selecting for the most appropriate growth habit. For instance, cane fruit cultivars can be classed as trailing, semitrailing, or erect, and they differ widely with respect to their ability to produce fruit without trellising. Half-high blueberry cultivars (interspecific hybrids of highbush and lowbush species) are short enough in stature that

Figure 19–9   Relative apple tree size produced by selected new and proven rootstocks. *Source: HortScience* 1988, 23(3). Alexandria, VA, American Society for Horticultural Science.

fruit buds are often protected from winter temperature extremes by snow cover. For growers in many northern regions, choosing a half-high cultivar is imperative because cultivars that are standard size are unprofitable to grow.

4. *Flowering and fruiting characteristics.* Beyond a doubt, aspects of flowering and fruiting are extremely important determinants of cultivar performance. For instance, many well-known standard apple cultivars such as 'McIntosh,' 'Delicious,' and 'Golden Delicious' have strains that are spur-bearing, allowing for relatively direct comparisons between the two flowering types. Spur-type strains are generally smaller than their standard counterparts; as such, their within-row spacing is often reduced by comparison. Spur-type strains also exhibit greater fruit-setting ability and higher yields per unit trunk cross-sectional area. Because spur-types often produce a greater yield (i.e., higher number of fruit per tree), however, their fruit, on an individual basis, are generally smaller than those of the standard strains. Fruit from spur-type strains are usually more mature (contain less starch) at harvest, but they develop less red peel color and contain lower levels of sugars and acids. They are firmer after storage but can exhibit a greater amount of green-colored flesh. Other horticultural differences characterizing standard and spur-type strains have been studied. For instance, spur-types are more drought resistant.

Flower initiation, as conditioned by levels of stored carbohydrates (energy) and hormones, is often balanced in the plant with the needs of the developing fruit. The alternate bearing phenomenon in fruit and nut tree crops results from a physiological imbalance usually brought about by the loss of a season's crop to frost or a similar physiological calamity. The lack of a crop during this initial off year shifts the tree's physiology to an extent that an overabundance of flower buds are initiated for the following season (the on year). If not severely thinned the following spring, these flowers will develop excess fruit of poor size and quality. Large numbers of developing fruit during floral initiation will severely limit the number of flower buds produced, thus triggering another off-year. This cyclic phenomenon is difficult to correct once it has been initiated. In particular, apple cultivars differ markedly in their propensity for alternate bearing; 'Golden Delicious' and 'Bradbury' are susceptible, 'Gala' and 'Delicious' are moderately susceptible, and 'Jonathan' and 'Granny Smith' are somewhat resistant to this phenomenon.

Growers must also consider differences in the dormancy or chilling requirements of species and cultivars within species when planning a fruit planting. Plants with relatively low chilling requirements may not be suitable for areas with dramatic climate fluctuations in early spring. These plants often resume growth too early during warm weather spells and then are easily damaged by subsequent frosts. Cultivars with relatively high chilling requirements and that flower later than others are better choices for these areas. Often a flowering delay of seven to ten days is significant.

Conversely, successful fruit growing in southern locations often requires plants with low chilling requirements. Until a few decades ago, blueberry production in southern states of the United States was based entirely on cultivars of rabbiteye blueberry. However, federal and state fruit breeders have developed specific low-chill highbush blueberry cultivars (e.g., 'Sharpblue,' Gulf Coast,' and 'Cape Fear') with improved fruit quality for these regions. These cultivars are steadily gaining popularity among growers.

Self-fruitfulness is also an important cultivar consideration. Most apricots, blackberries, currants and gooseberries, highbush blueberries, grapes, peaches, raspberries, sour cherries, strawberries, and walnuts are predominantly self-fruitful, but some cultivars within this group require cross-pollination for adequate fruit set (Tables 19–1 and 19–2). Even though cultivars may be self-fruitful, fruit production and/or fruit size may be increased when cross pollination occurs. Cultivars of apple, almond, hazelnut, muscadine grape, rabbiteye blueberries, sweet cherries, and other species are considered to be predominantly self-unfruitful, but the degree of unfruitfulness may also vary among cultivars. For instance, the sweet cherry 'Stella' can be produced without additional pollinator cultivars. In contrast, some cultivars of almond must be self-pollinated. Cultivars of European and Japanese plum, pear, and chestnuts are highly variable with respect to their cross pollination requirements. Regardless of self- or cross compatibility, cultivars may vary in the timing or spatial arrangement of pollen shed and stigma receptivity, thus requiring the movement of pollen from one flower to another for adequate fruit set. Species and cultivars may also differ in their requirement for insect vectors to accomplish this task.

5. *Disease and insect tolerance or resistance.* The incorporation of increased insect and disease tolerance or resistance into desirable cultivars and rootstocks has perhaps received more emphasis and effort by both public and private hybridizers than any other fruit crop improvement goal. The diversity and extent of biotic stresses that may affect fruit culture is too large a topic for an adequate discussion of specific examples of tolerant or resistant cultivars or rootstocks in this textbook. However, the major biotic pests of tree and small fruit crops are listed in Tables 19–1 and 19–2, respectively. Climatic and soil conditions affect the incidence and severity of biotic pests. For instance, *Phytophthora* root rots may be common on heavy soils, whereas nematodes are more frequently a problem in sandy soils. Choosing tolerant or resistant cultivars or

rootstocks is always wise, especially if they are available for specific disease or insect problems in the proposed production area. The culture of resistant or tolerant genotypes increases the probability of harvesting a marketable crop of high quality and lowers production inputs for chemical or biological control measures. Environmental concerns surrounding the use of agrichemicals and public awareness of and concern about health and food safety issues emphasize the desirability of tolerant or resistant cultivars and rootstocks for the production of fruit.

6. *Winter hardiness and/or tolerance to abiotic stresses.* **Winter hardiness**—the ability to withstand severe midwinter temperatures—is often the product of a plant's unique physical and physiological characteristics and is conditioned by the success of the hardening-off process as the plant enters dormancy in late fall. Within most temperate fruit species, a broad range in winter hardiness can be found among cultivars and among rootstocks. For example, the crowns of many strawberry cultivars are severely injured or killed if they are exposed directly to temperatures lower than $-12°C$ (15°F), but cultivars like 'Fort Laramie' and 'Ogallala' developed in the United States, in Wyoming, tolerate winter extremes approaching $-40°C$ ($-40°F$). The degree of winter hardiness displayed by a given cultivar or rootstock often limits its range, with only the most hardy cultivars being suitable for planting in colder climates of either hemisphere. For instance, most European wine grape cultivars can be cultured successfully in California and the Pacific Northwest, but only the hardiest among them, such as 'White Riesling,' 'Lemberger,' 'Cabernet Franc,' and perhaps 'Chardonnay,' are recommended for production in northern and central US states. Fruit cultivars and rootstocks also differ in their ability to withstand heat; wind; and adverse soil conditions such as drought, waterlogging, low or high pH, and mineral imbalances.

7. *Ripening period.* Fruit harvest, handling, marketing, and storage are often labor-intensive, time-consuming events, so it is fortunate that cultivars differ substantially with respect to when they ripen their fruit. In most production regions, a selection of cultivars and/or cultivar-rootstock combinations can be planted to extend the harvest season through one to two months, or perhaps longer. For instance, in the US Midwest, early apple varieties (e.g., 'Pristine') are harvested as early as mid-August, whereas late varieties (e.g., 'Fuji,' 'GoldRush') are not harvested until very late October or early November. In the same region, highbush blueberry cultivars such as 'Earliblue' or 'Bluetta' can be harvested mid- to late June, whereas 'Elliott' berries ripen in mid-September. The harvest of primocane fruiting raspberry or blackberry cultivars begins in late summer and can continue until frost. Aside from horticultural advantages (e.g., greatly simplified pruning procedures), primocane fruiting cane fruit cultivars increase the market presence and consumer awareness of these crops. With the range of standard (summer-bearing) and primocane fruiting cultivars available, it is now possible to produce fresh red raspberries in the Pacific Northwest from mid-June to frost.

8. *Quality and desirability.* Fruit quality can be defined in many ways, and the term can be interpreted differently by producers and consumers of fresh or processed fruit products. Producers often consider fruit quality characteristics that affect the harvestability, ease of shelling (in nut crops), storage, shipping, and shelf life of the product, whereas ample evidence suggests that consumers value fruits and nuts for nutritional and health benefits as well as for their flavor and other sensory characteristics. Fruit cultivars often differ considerably in firmness at harvest and susceptibility to bruising during handling, storage, and shipment. Furthermore, differences in post-harvest respiration rates and other physiological characteristics make some cultivars suitable for long-term storage, whereas others cannot be stored for more than a few months, or even days in the case of soft fruits. Rapid cooling and subsequent refrigeration after harvest as well as the employment of controlled atmosphere storage and modified atmosphere packaging can prolong the useful life of most cultivars. Nevertheless, cultivars respond differently to these treatments. For instance, some apple cultivars can be stored successfully for only a few months; others, such as 'Golden Delicious,' can be marketed eight or nine months after harvest if they are stored properly.

Consumers, on the other hand, purchase fruits and nuts based on their appearance first and then on their anticipated nutritional and eating quality. Cultivar name recognition has been a long-standing marketing instrument for crops such as apples, grapes, and pears. Consumers often purchase specific cultivars of these crops based on their intended end use (eating, cooking and baking, canning and sauce or preserve making, etc.). Newer cultivars of apples such as 'Gala' and 'Honeycrisp,' with improved flavor and texture qualities, are demanding a premium price and greater share of the market than older cultivars with good but less desirable characteristics. The use of name recognition as a marketing tool is likely to increase with time. Breeders are releasing many of their new fruit cultivars with trademarked names specifically for this purpose. Many major food outlets have begun identifying the fruit cultivars they market with specific numeric designations in cases where specific names are not available or used. Consideration of a cultivar's quality characteristics, therefore, is extremely important when producers plan new plantings.

This may be especially important if the product is to be marketed directly to consumers via farm markets or pick-your-own operations.

Admittedly, the perfect fruit cultivar does not exist. Therefore, when planning a new production field, a grower must necessarily weigh the importance of various cultivar characteristics to the ultimate success of the enterprise. Often, it is a good idea to plant several cultivars with a balance of characteristics to offset the vulnerability of establishing only one.

Information concerning the various fruiting cultivars can be obtained from several reliable sources. Most extension bulletins produced by state and federal agencies contain cultivar recommendations for the areas they serve. Likewise, reliable cultivar information is often available from organized producer groups and/or individual growers, buyers, and packers in the area. Reputable nurseries also publish accurate catalog descriptions of the fruit crops they offer. Opinions may differ among sources, however, so it is often wise to consult several sources concerning the suitability of various possibilities before commiting to a specific set of cultivars. It is risky to accept new, untried cultivars or to base the decision solely on nursery advertisements.

ACTIVITIES PRIOR TO PLANTING

In addition to choosing an appropriate site, conducting an in-depth market analysis, developing a sound business plan, understanding aspects of flowering, and researching potential cultivar and rootstock characteristics, several important tasks must be completed prior to planting the new production site. Many of these tasks should be completed, or at least begun, one to two years before the anticipated planting date. Because they often require much capital and time to install, the construction of roads and infrastructure (fencing, packing sheds, etc.) necessary for production and handling should be initiated as soon as possible. The site itself should be thoroughly examined for potential problems concerning slope, exposure, drainage and low-lying areas, and other defects in order to correct them in ample time. If the site is on a rolling slope or hillside, terracing and a contour planting system can be employed. This planting arrangement, while subject to problems, permits production from land that otherwise could not be utilized. Considerable care must be taken to stop erosion by heavy rains or by irrigation by diverting the water to run along the tree rows rather than straight downslope. If necessary, windbreaks can be planted.

The soil should be tested by a reputable state or commercially operated soil-testing facility to determine accurately its physical and chemical characteristics. If recommended by test results, the site can be tiled to increase soil drainage, and lime or an acidifying agent can be applied to correct imbalances in soil reaction (pH). If the site is infested with weeds, especially perennial species, control measures should begin at least one year prior to planting. Control measures may include the use of soil fumigants, which may also reduce the presence of disease organisms, nematodes, and insects in the soil.

Fumigation for pests may be especially important if they are expected to be present at potentially harmful levels, based on soil type or cropping history. Harsh fumigants should be employed only when necessary, however, because they often kill beneficial soil biota as well as parasitic or harmful species. Perhaps more important, many soil fumigants have detrimental environment effects, and they are difficult to contain due to their gaseous nature. The planting and maintenance of pre-plant cover crops, especially nitrogen-fixing legumes, is often recommended to improve soil structure and fertility and to help manage weed growth.

Closer to the scheduled planting date, the land should be prepared. For some fruit crops, these operations may include the addition of fertilizers to correct nutrient imbalances, tilling, and/or raised bed construction. Materials needed for plant establishment (e.g., trellis, mulch, etc.) should be purchased and/or constructed.

It is highly advisable to purchase certified virus- and disease-free nursery stock from a reputable nursery rather than obtaining propagules from existing plantings or accepting plants from other sources that may have been previously infected with pathogens. Nursery stock can be purchased within a year of planting, but the delivery date of the stock should be negotiated to coincide as closely as possible with the scheduled planting date. On arrival, the shipment should be checked for clear cultivar and rootstock designations on plants or plant lots and then stored at the recommended temperature and well away from any fruit or other ethylene-generating plant materials in storage. The roots of most, if not all, nursery stock should be continually protected against desiccation by using moist packing material or wood shavings until they are planted.

High-quality, deciduous nursery stock possesses an abundance of well-distributed roots. Tree nursery stock should display a strong, vigorous trunk free of damage from careless handling. A small shallow slice into the trunk should show bright green tissue below the bark with no evidence of brown areas from winter damage or sunburn. A slice into the roots should reveal a moist whitish color. No root tissue should be shriveled or look brown, gray, or black below the bark. In budded or grafted plants the union should be well-healed and strong, with no more than a slight bend at the union. The graft union should be at least 10 cm (4 in.) above the previous soil level so that scion rooting (roots developing above the graft union) is unlikely after planting. Scion rooting can lead to a loss of any desired effects of specific rootstocks such as dwarfing, resistance to nematodes, diseases, and so on.

In deciduous nursery stock, the dormant buds should be plump and well-developed and should look bright green when cut into. Dead buds may indicate low temperature or herbicide injury or lack of water during the growing season.

Cane fruit, bush fruit, and strawberry nursery stock can be purchased as dormant plants or crowns, but they are typically available in actively growing forms such as bare-rooted plants, tissue-cultured transplants, transplant plugs, and containerized plants. Often, this nondormant nursery stock is produced and purchased for use with specific growing systems (e.g., Canadian-produced, leaf-bearing, bare-rooted strawberry crowns used for annual winter production systems in Florida). These materials should arrive with healthy looking leaves that are free from insects or diseases, or evidence of herbicide damage or nutrient deficiency symptoms. Optimal, pre-plant care and storage conditions are specific to each nondormant nursery product available.

This discussion of pre-plant considerations is by no means exhaustive. Specific crops and areas will require additional steps to be accomplished prior to planting. It is best to obtain copies of relevant extension guides or to consult with local producers and producer groups for pertinent pre-plant information concerning the crops to be planted. New producers should also familiarize themselves with all federal, state, and local ordinances governing the production, harvest, and marketing of the prospective crop. Be as ready as you can be!

PLANTING AND CULTURE

Major decisions must be made when the time comes to plant a new orchard, vineyard, or berry planting. Planting distances and patterns must be determined. The distance between plants depends primarily on the choice of growing system and cultivars. However, several factors can modify recommended distances. Greater spacing should be used with conditions of high soil fertility; long growing seasons; vigorous, large-size cultivars: invigorating rootstocks; ample rainfall or irrigation; and heavy use of fertilizers. Spacing should be closer in the opposite situations.

Spacing between rows should also be considered. Between-row spacing should allow for ample sunlight penetration and air movement through the planting and be wide enough to easily accommodate tractors, spray rigs, and harvesting equipment. Some recommended within- and between-row spacings for various fruit and nut crops are listed in Table 19–3. Row orientation should also be considered to maximize sunlight interception and air movement, but protect against potentially damaging winds and or erosion. In planting fruit trees of a species requiring cross pollination to set good commercial crops, it is of the utmost importance that trees of the pollinizing cultivar be appropriately spaced among trees of the principal fruiting cultivar.

Laying Out the Planting

Once the desired plant spacing has been determined and the planting infrastructure has been installed (e.g., raised beds, irrigation systems, etc.), the site can be prepared for planting. Rows and plant positions within the row should be marked just prior to when plants are due to arrive from the nursery. It is important that the rows are lined up properly. This will facilitate

TABLE 19–3	Suggested Planting Distances for Common Temperate Fruit and Nut Crops*
Species	**Planting Distances**
Almond	7.5 × 7.5 to 9 × 9 m (25 × 25 to 30 × 30 ft).
Apple	Highly variable—depends on production system, cultivar, rootstock, soil vitality, and cultural practices.
Apricot	6.6 × 6.6 m (22 × 22 ft) (on plum roots) 7.5 × 7.5 m (25 × 25 ft) (on apricot roots).
Blueberry	1.2 m (4 ft) apart in rows 3 m (10 ft) apart.
Cherry, sour	6 × 6 m (20 × 20 ft).
Cherry, sweet	7.5 × 7.5 to 9 × 9 m (25 × 25 to 30 × 30 ft).
Grape	1.2 to 2.4 m (4 to 8 ft) apart in rows 2.4 to 3 m (8 to 10 ft) apart.
Hazelnut	4.5 × 4.5 m (15 × 15 ft).
Kiwifruit (vines on trellis)	5.4 to 6 m (18 to 20 ft) apart in rows 4.5 m (15 ft) apart.
Peach	3 × 6 to 5.4 × 7.2 m (10 × 20 to 18 × 24 ft).
Pear	6.6 × 6.6 m (22 × 22 ft).
Pecan	9 × 9 to 15 × 15 m (30 × 30 to 50 × 50 ft).
Prune	6 × 6 m (20 × 20 ft).
Raspberry, black	0.6 to 1.2 m (2 to 4 ft) apart in rows 2.1 to 3 m (7 to 10 ft) apart.
Raspberry, red	0.75 m (2.5 ft) apart in rows 1.8 m (6 ft) apart.
Strawberry (matted-row system)	61 to 71 cm (24 to 28 in) apart, permitting a matted row, 38 to 61 cm (15 to 24 in) wide to develop from runners.
Strawberry (double-row bed system)	Beds 96 to 112 cm (38 to 44 in) apart, center to center; two rows in each bed 20 to 30 cm (8 to 12 in) apart. Plants in each row 23 to 36 cm (9 to 14 in) apart.
Strawberry (single-row bed system)	Beds 100 to 107 cm (39 to 42 in) apart, center to center; plants 20 to 25 cm (8 to 10 in) apart in rows.
Walnut (Persian)	6 × 6 m (20 × 20 ft) to 10.5 × 10.5 m (35 × 35 ft) (Paynet type); 10.5 × 10.5 m (35 × 35 ft) to 12 × 12 m (40 × 40 ft) (Hartley type).

* For information on apple spacing, consult regional extension bulletins (e.g., Pennsylvania Tree Fruit Production Guide at http://tfpg.cas.psu.edu/ part1 part13h.htm).

many future operations. A plant out of line can be a target for cultivating disks and other equipment moving down the rows.

Planting the Crop

In temperate climates, the best time to plant the nursery stock is any time during the dormant season, when the ground is not frozen and air temperatures are above freezing. It is important to plant as early as possible so the roots will be well established by the time hot weather arrives and the plants will have a full growing season before they face cold weather.

Alternately, some cultural systems, especially those employed in warmer climates, require planting to be initiated in other seasons. For example, annual production of strawberries in California and Florida are planted from mid- to late fall. In areas with relatively mild winters, such as North Carolina, fall-planted systems are successful when used in conjunction with spun-bonded row covers and other frost-protecting cultural practices. Late summer or fall-planted systems are also used in cooler climates where plants are produced in high tunnels covered with polyethylene or similar materials.

In planting tree, vine, or bare-rooted strawberry or cane crops, it is important that the planting hole be dug to the proper depth. The base of the main supporting roots, which usually have been trimmed back, should rest on solid, undisturbed soil. If the hole has been dug too deep, necessitating some backfilling before planting, then the plant is apt to sink after watering and settling, putting the graft union or plant crown below the soil level and leading to attacks of crown rot fungi, principally of the *Phytophthora* species.

The planting hole should be wide enough to easily accommodate the roots without bending and twisting. Tractor-operated soil augers are often used for digging tree holes. These have the advantage of working fast and saving labor, but if they are not operated properly, the holes can be dug too deep, leading to settling. Also, if the soil is too wet when the auger is used, the sides of the hole become severely glazed, making air and water permeability and root penetration difficult. A shovel should be used to break up the compacted sides of the holes when the trees are planted.

Filling loose soil around the roots once the plant has been set in the hole is an important operation. The soil should be worked around the roots as the hole is filled to ensure good soil-root contact, with the soil pressed firmly about the roots. If there are persistent summer winds from one direction, the plant can be leaned slightly into the wind at planting. Alternately, young plants can be staked to ensure proper growth and development. In systems that employ intensive training regimes, the control of plant form begins at planting or soon thereafter.

Many of the planting techniques for bare-rooted plants are also applicable to the planting of soil-bearing nursery stock such as container plants, tissue-cultured plants, or plant plugs. Typically, the depth of planting should be at or near the depth of the original soil ball. Shallow planting, leaving a significant portion of the soil ball exposed, promotes its dessication; planting at a depth greatly in excess of the original soil ball can expose young tender canes or stems to disease organisms. If the nursery stock was produced or maintained in a controlled environment such as a greenhouse, plants should be hardened off by gradual exposure to full sunlight over a period of a few days or weeks prior to planting in the field. Moisture levels should be monitored carefully during the hardening-off process.

For large-scale plantings, mechanical tree planters are used. The tractor driver must exercise care to ensure that the rows are as straight as possible. In-row spacing is never as accurate with mechanical planting, and it is critical that personnel follow the transplanter to make slight adjustments in graft union or crown height and to firm the soil around the plant.

Some cultural systems require additional operations to be performed at or near the time of planting. For instance, for some raised-bed planting systems, the mechanical planter covers the bed with a synthetic (plastic) much just prior to setting the plants. Synthetic mulches are employed in these systems in part to conserve and regulate soil moisture, moderate soil temperatures, discourage weed growth, and simplify crop harvest. When raised beds are hand planted, mulch can be laid just prior to planting and then slitted by hand or machine to provide openings in which to set the plants. If the crop has specific edaphic requirements, soil amendments may be added at the time of planting. As mentioned above, blueberries thrive best in acidified soil with high levels of organic matter. Therefore, fruit production guides typically recommend adding up to 1 lb of soaking wet peat moss to the planting hole when planting blueberries on mineral soils (soils that typically contain less that 2 percent organic matter). Dry peat moss is difficult to hydrate after planting and could potentially draw water away from the roots during the critical hours after transplanting. Finally, the fibrous root systems of dormant strawberry crowns are often soaked in water for a short period just prior to being set, again in an effort to avoid moisture stress to the root system during the planting process.

Almost without exception, newly set fruit plants require immediate or nearly immediate irrigation. For tree crops, a shallow basin or furrow should be left at the top for filling with water. For those plants with a soil ball around their roots, the top of the soil ball should be slightly exposed at the surface of the basin so that added water can enter directly into and through the root ball. Often the nursery soil found in such root balls is light and porous and if, after planting, they are covered by the heavier clays of native soils, irrigation water will remain in the clays because of their smaller pores and high capillarity, leaving the lighter soil with the roots completely dry. A few days of hot desiccating winds can increase transpiration and severely injure trees planted too deeply, even though the basins have been filled with water.

The soil basins should be filled with water within a few hours of planting to prevent dehydration of the roots and also settle the soil around the roots. Watering by filling the basins should continue until vigorous shoot growth is well underway, when the regular irrigation system or rainfall can be used. In heavy soils with drainage problems, care must be taken to avoid overwatering, which can impede good root aeration and lead to attacks by *Phytophthora* and other soil pathogens.

Bush fruit, cane fruit, and strawberries also require irrigation immediately after planting. However, basins or furrows are not often employed as a means to direct water to the newly planted root systems. Rather, these small fruit plantings are often irrigated with portable or solid-set sprinklers or with drip-irrigation systems. These systems should be operated soon after planting and until the soil is wetted to the depth of the transplant root system.

Newly planted fruit plants often must be shielded from harsh environmental conditions during their period of establishment. The trunks of young trees and grapevines must be protected from sunburn by wrapping them with paper, cardboard, or styrofoam coverings during the summer or painting with whitewash or water-base paint (one part interior white latex paint and one part water). Inspect for bark damage at intervals during the winter by rabbits, mice, or gophers at the soil level and take the necessary preventive measures. Special care should also be given to control weed growth during the establishment of a young fruit planting because competition for moisture; nutrients; and, in extremely weedy conditions, light can prevent adequate crop growth.

It is best not to fertilize or apply herbicides to the trees at planting time. The developing roots could be injured by excessive salts or by herbicides. Generally, the young trees obtain enough mineral nutrients from the soil for their first year's growth. However, recommendations for agrichemical use during the establishment season vary significantly among crops and growing systems. Growers are advised to consult one or more respected state, regional, or federal crop production guides to determine the best management practices specific to their planting as it undergoes establishment.

AN INTRODUCTION TO THE CARE AND MAINTENANCE OF A FRUIT PLANTING

By definition, horticultural crops are intensively cultivated. The care and maintenance of a fruit planting after establishment requires a host of inputs (materials and labor) and horticultural procedures that must be implemented in an integrated fashion by the producer. The ultimate goal of care and maintenance procedures is to optimize the physiological status of the plants, and thus to achieve a sustained production of high-quality fruit throughout the life of the planting.

Common care and maintenance procedures include training and pruning; crop-load management; irrigation; weed-, insect-, and disease-control measures; and fertilization. This list is by no means complete, and most fruit crops require additional inputs that are crop-specific (e.g., the routine organic mulching of blueberries grown on mineral soils). The timing of application for each input and/or procedure is often critical, as is the integration of maintenance strategies, because choices made concerning one often dictate the proper recourse for others. For example, the optimum (balanced) pruning procedure for individual grapevines is determined by last year's cropping history, and the weight of pruned material is often used to determine crop-load adjustments for the following season. Beyond all doubt, the growing system employed by the producer profoundly affects care and maintenance strategies. High-intensity versus moderate-intensity systems, annual versus perennial production systems (strawberries and other small fruit crops), organic versus conventional production systems—all require very

specific sets of management practices. In a perennial fruit planting, management practices are often altered as the plants progress from juvenility, through the peak bearing years, to their eventual decline, to accommodate changes in the crop's physiology during each period.

A complete discussion of all aspects for all crops in all growing systems is well beyond the scope of this introductory text. Inexperienced producers should seek technical information specific to the crops and growing systems they plan to use from state and governmental extension agencies, fruit growers' societies, or qualified consultants before planting. However, even the most well-written and detailed production guides will not substitute for knowledge gained through personal experience. The best advice for a new producer might be to begin with a small planting, grow the crop to the best of his or her capability using recommended practices, be as attentive to detail and as observant of crop behavior in response to care and maintenance procedures as possible, and learn from outcomes that are less than optimal. As the producer experiences the cropping cycle, intuition concerning crop behavior is likely to be gained; once a grower feels a degree of confidence about producing the crop using his or her management strategy, the scale of the planting can be increased.

As mentioned above, crop physiology and management practices are inextricably intertwined. To illustrate this relationship (one of the key learning concepts listed at the beginning of this chapter), two care and maintenance topics will be discussed in moderate detail in subsequent chapter sections. They are the crop-load management strategies of training and pruning.

TRAINING AND PRUNING

Training and pruning of fruit plants is an integral part of the procedures used for sustained and profitable production of high-quality fruits. **Training** can be thought of as a procedure that results in the positioning of limbs, branches, or canes, whereas **pruning** can be considered the act of cutting limbs, branches, or canes. However, because the early training of tree and bush crops almost always involves pruning, these two definitions are somewhat simplistic and confounded. Training procedures are often associated with young (nonbearing) plantings, whereas pruning is practiced on bearing plants to maintain high levels of production. Although this distinction between the two practices accurately describes management practices for tree fruit and nut crops grown by standard production systems, it fails to represent practices for many high-density orchard systems, grape vineyards, or cane fruit plantings where coordinated pruning and training practices continue throughout the life of the planting. Because they are inextricably linked, training and pruning procedures are discussed together herein.

Reasons for Training and Pruning

Fruit and Nut Trees

1. To develop a strong trunk and a balanced scaffold system of branches, well distributed around the tree, which can support heavy loads of fruit without limb breakage. The tendency for limb breakage is partially a function of crotch (branch) angle. In general, crotch angles of 60° to 90° result in strong branches that are vigorous, fruitful, and capable of carrying mature, heavy fruit loads, whereas branches with narrow crotch angles are vegetatively vigorous and inherently weak. Improper pruning can actually increase the frequency of narrow crotch angles. Crotch angles wider than 90° result in branches that are weak and unfruitful and should be removed.

2. To regulate fruit production. Proper pruning encourages development of the type of shoot system that produces the fruit. In older trees with little vegetative growth rejuvenation, pruning can force the development of productive fruiting shoots. Pruning can also be used to limit excess numbers of fruit by removing some fruit-bearing branches, giving a thinning effect that can improve fruit size and quality. Pruning and training also provide important means of balancing vegetative growth to fruiting, which aids in the development of high-quality, high annual production, and increased resistance to disease, heat, and cold. With some fruit tree cultivar/rootstock combinations in high-density training systems, bending or weighting young branches to increased crotch angles alters the hormonal balance, resulting in a significant increase in fruit bud development.

3. To remove dead, broken, or interfering branches.

4. To improve light penetration to the inside and lower parts of the tree.

5. To facilitate insect and disease control by opening the tree, thus increasing penetration of spray materials to the interior branches.

6. To limit tree size and shape to the space allocated to it and to limit tree height to that from which fruit can be conveniently harvested. As extreme examples, intensive tree-training systems in which trees are essentially trellised were developed in part to facilitate the mechanical harvest of fruit. In addition, for some species (e.g., apples) cultured at high densities, tree size can be controlled by yearly mechanical root-pruning treatments, which reduce overall vegetative growth and trunk girth. Some studies have shown root pruning to improve fruit firmness, color, and eating quality; however, reduced fruit size and photosynthetic capacity, and increased susceptibility to drought conditions have also been observed. Root-pruning treatments may be most applicable after the loss of a crop to curtail vigor and reduce alternate bearing tendencies.

Grapevines

1. To establish strong trunk and cordon systems that are maintainable in a form that facilitates vineyard cultural operations.

2. To distribute the fruiting wood to obtain maximum production of high-quality fruit.

3. To maintain vigor and production of fruiting canes and to balance vegetative growth with fruiting for high-quality fruit and to impart greater resistance to heat and winter cold.

4. To aid in control of crop size and increase berry size by reducing the number of fruiting clusters.

5. To increase light penetration throughout the canopy and to increase sunlight exposure for leaves, clusters, and basal buds that will provide next year's crop.

6. To remove old, nonproductive canes.

Bushberries (Raspberries, Blackberries, Blueberries, Currants, and Gooseberries)

1. To remove dead, weak, or diseased canes and old shoots that die following fruiting.

2. To thin out weak canes to give adequate light and space to the remaining canes, resulting in larger, better-quality fruit.

3. To develop lateral fruiting branches by summer topping (of black and purple raspberries and upright blackberries, but not red raspberries). (See the discussion of fruiting habits earlier in this chapter.)

4. To control crop load and increase fruit size.

5. To improve light penetration and photosynthetic capacity. Both training systems and pruning techniques can increase sunlight exposure to actively growing shoots, thus ensuring adequate carbohydrate availability and a physiological balance optimal for the initiation of fruit buds. For example, many trellising systems for cane fruit effect a separation of floricanes from newly emerging primocanes that will produce next year's crop.

6. To control plant shape and the position of fruiting surfaces to ease harvest, especially for crops that are mechanically harvested. For example, blueberries that are to be mechanically harvested are pruned to be more upright with narrower crowns than those cultured for hand harvest (such as for a pick-your-own enterprise). Cane fruit trellising systems have also been designed specifically to accommodate mechanical harvesters and limit damage to mechanically harvested fruit.

Physiological Responses to Pruning

All plants, if not pruned, tend to develop a balance between growth of the shoot and the root systems. Cutting away part of the top, including the plant's photosynthetic apparatus and food storage tissues, together with reducing the number of vegetative growing points and flower buds while leaving the root system intact leads to some interesting physiological reactions. The fruit grower should be aware of these reactions to understand how to prune the plants properly.

1. *Removing shoot terminals causes lateral branching.* Actively growing meristems produce an abundance of the natural growth-promoting substance, auxin (IAA, indole-3-acetic acid), which promotes cell division and elongation and is primarily responsible for the phenomena of phototropism (the bending of a shoot toward light) and apical dominance. In the latter, the downward movement of auxin transported in the phloem inhibits the development and breaking of axillary (lateral) buds and the elongation of side branches. Natural apical dominance, therefore, results in a plant form with fewer branches (and perhaps fewer fruiting surfaces). Heading back a shoot by removing its terminal portion is the most effective pruning cut for promoting the formation of new branches; thinning cuts which remove entire shoots or limbs are generally used to open the canopy to air and sunlight or to remove old, unfruitful, interfering, or diseased wood.

   Newly planted, one-year-old trees of some fruit species—for example, apples, cherries, pecans, pears, walnuts, quince, and European plums—show strong apical dominance. If the terminal bud is not pruned, the tree will grow mostly as a straight, sparsely branched trunk. If the terminal bud is removed (plus several inches of shoot below it for convenience) at planting time, however, the lateral buds remaining below the cut will develop, producing lateral scaffold branches. To cause further branching of these scaffolds the next year, the terminals of the scaffold shoots must again be headed back by several inches.

   Other tree species, such as apricots, peaches, almonds, and nectarines, do not show such strong apical dominance. Lateral buds develop into scaffold branches without heading back of the main shoot terminal growth.

2. *There is a reduction in total vegetative growth.* Removing a portion of the top reduces the total amount of subsequent growth, compared to an unpruned plant. The total number of growing points is reduced, resulting in fewer developing shoots, fewer leaves, reduced photosynthesis, reduced amounts of carbohydrates translocated to the roots, reduced root growth, followed by a reduction in mineral and water absorption, which, in turn, further decreases shoot growth. These effects dwarf the entire plant. Generally, the more severe the pruning, the greater the dwarfing.

   The remaining growing points following pruning, which utilize all the stored foods in the plant, usually show strong shoot growth. The increased vigor of these shoots may lead one to believe that the pruning has caused increased total growth, but numerous experiments have shown that this is not so. (If invigoration of the total plant is necessary, the judicious use of added fertilizers with ample soil moisture, together with only moderate pruning, should be practiced.)

3. *Continual heavy pruning each year of young fruit trees can delay the onset of bearing.* For many fruit species or for specific cultivars, precocity and pruning severity are negatively related, perhaps due to the decreased availability of carbohydrates for flower and fruit development or from reduced development of fruiting structures (e.g., spur development). Although some pruning is required for proper plant training, the extent of tissue removal should be limited to what is necessary. In cultivars with a tendency for prolonged juvenility, flowering, fruiting, and floral bud development for the following season may be increased by judicious and careful use of scoring, ringing, or girdling treatments, which presumably result in a greater retention of carbohydrates and other important metabolites in fruiting and potentially fruiting tissues. This approach is appropriate for some tree species (e.g., apples and pears) but not others (e.g., stone fruits). Flowering and fruiting in most fruit crops can also be stimulated by the application of mineral nutrients, primarily nitrogen, to optimize the plant's physiological condition.

4. *Pruning effects from removal of smaller shoots and branches tend to be localized.* Heading back of shoots during the dormant season, reducing the length of a shoot by 50 percent, for example, results in the development of one or more new shoots from buds just below the pruning cut. The total growth by the end of the summer will not be nearly equal to the total growth of an adjacent shoot that was not dormant pruned. This response is often used in pruning to retard the growth of one of two equal-growing shoots and thus prevent the development of a weak crotch.

5. *Severe pruning in early summer is more likely to weaken the tree and reduce total growth than dormant (winter) pruning, especially with young trees.* Much of the energy for new spring shoot growth comes from stored foods in the roots, trunk, and branches of the tree. These stored foods are not replaced until an appreciable number of new leaves have formed and photosynthesized enough to reverse carbohydrate movement into the larger branches, trunk, and roots. If summer pruning is done, partially removing these new shoots and reducing photosynthesis and the replenishment of foods already utilized, the tree can be intensely debilitated. However, a moderate pinching back of only the tips of the shoots in the spring to direct growth is not likely to be harmful. Tree response to pruning in the late summer (August) is similar to the response from dormant pruning.

6. *Dormant pruning of bearing deciduous fruit trees can be invigorating.* Although vegetative growing points are removed, flower buds are also removed, thus reducing the crop, consequently reducing the demand on the plant's stored foods. The surplus is then utilized for new vegetative growth.

7. *Moderate dormant pruning of bearing deciduous fruit plants can increase production.* Flowers are produced either laterally or terminally on one-year-old shoots and/or laterally or terminally on fruiting spurs (see Tables 19–1 and 19–2, and Figs. 19–5 and 19–6). Such fruit-producing growth (shoots and spurs) must be stimulated to obtain continual crops. Moderate pruning plus fertilizers and ample soil moisture result in such growth stimulation.

Training and Pruning Deciduous Fruit Trees

Over the years, many training techniques have been studied and ultimately employed to raise tree fruit crops. Each method is suitable only for specific crops, cultivar/rootstock, and growing system combinations, and each has unique advantages and disadvantages. Therefore, the compatibility of possible training techniques with other production factors should be carefully considered before a training method is chosen. For instance, as mentioned previously, orchards planted at high densities or those to be mechanically harvested often require high-input, high-maintenance training systems that ensure precocity and proper placement of fruiting surfaces. Growers are also advised to familiarize themselves with all procedures associated with specific methods and to understand fully the physiological ramifications of employing them before training begins. Because training involves the removal of wood, it is difficult, if not impossible, to change methods after training is initiated.

Logically, the choice of training system should consider the natural growth habits of the tree to be trained.

TERMINOLOGY FOR PRUNING AND TRAINING

Pruning and training has its own vocabulary. Here are some important terms and their meanings.

heading back Shoots or limbs are not removed entirely, but the terminal portions are cut off at varying distances from the end. This procedure forces out new shoots from buds below the cut and retards terminal growth of the branch.

thinning Shoots or limbs are completely removed at the point where they attach to the next larger (and older) limb. Thinning out corrects an overly dense area or removes interfering or unneeded branches.

trunk That portion of the tree up to the first main scaffold branch.

primary scaffold branches The main branches arising from the trunk of the tree.

secondary scaffold branches Supporting branches arising from the primary scaffold branches.

central leader training system A training system for fruit trees where one main vertical trunk remains dominant, with all scaffold branches arising from it in an almost horizontal direction.

modified leader training system A central leader continues upward from the trunk, but its identity is lost as it becomes one of the primary or secondary scaffold branches.

open center or vase training system A trunk is developed, then at the top of the trunk (0.6 to 0.9 m; 2 to 3 ft from ground level) several primary scaffolds develop outward and upward at a 30° to 45° angle to form a vase configuration.

weak crotch A situation that can develop when two equal-size branches with a narrow angle between them grow for some years at the same rate. Eventually a considerable amount of bark inclusion develops between the two branches at their junction, with no connecting wood. With heavy loads of fruit—and strong winds—the limbs are almost certain to split apart. This problem can be prevented by completely removing one of the branches when they are young or heading one of them back heavily so the other becomes the main limb and the headed-back branch a lateral.

stub Pruning cuts of the thinning-out type must be properly done. Cutting off the branch leaves a wound that must heal over by callus growth proliferating from the surrounding live tissue or else decay organisms can become established in the wound. If the cut is not made close to the adjacent limb and a stub is left, the wound cannot heal properly.

watersprouts Vigorous shoots usually arising from latent buds in the trunk or older limbs in the lower parts of the tree, growing upright through the center of the tree.

suckers Shoots arising from the underground parts of the plant, usually coming from adventitious buds on roots. In grafted trees, they generally arise from the rootstock below the graft union. They should be removed as soon as they are noticed.

head The aerial terminus of a grapevine trunk; in head-pruning systems, it is the position where canes are selected for fruiting.

cordon A permanent lateral extension of the grapevine trunk.

grape cane A one-year-old woody shoot bearing buds for next year's vegetative and reproductive growth.

grape spur A cane pruned to a length of a few nodes bearing buds for next year's vegetative and reproductive growth.

pre-pruning The initial removal of most one-year-old canes before selection of specific canes or spurs for next year's vegetative and reproductive growth. In the balanced pruning system, the mass of pre-pruned shoots is used to determine the optimum crop load for the upcoming season.

balanced pruning A pruning system that balances vegetative growth and fruiting to maintain optimum vine health and sustained productivity.

hedgerow A cane fruit management system wherein primocanes arise from rhizome buds anywhere within a confined area. The identity of individual plants is not evident.

stool A cane fruit plant managed as an individual, with most primocanes arising at or near the plant crown.

primocane suppression The management system in which early-emerging primocanes are sacrificed to maximize the level of carbohydrate available for developing fruit.

Figure 19–10   Central-leader apple tree after several years' growth. *Left:* Before dormant pruning. *Right:* After pruning. Some lateral branches were removed. In those retained, forked tips were cut to one outward growing branch. Spreader boards have been used with some laterals to develop a wider growth habit. *Source:* Gerdts, M., A. Hewitt, J. Beutel, J. Clark, and F. Cress. 1977. Pruning home deciduous fruit and nut trees. Univ. of Calif. Div. Agr. Sci. Leaflet 21003.

Because of their strong apical dominance and upright growth habits, apples and pears grown in moderate density plantings are most often trained using the central leader or modified central leader training systems (Figs. 19–10 and 19–11). To initiate either of these systems, the top 75 to 90 cm (30 to 36 in.) of newly planted, unbranched nursery stock (whips) are removed to stimulate the development lateral shoots. During the first year of growth, the uppermost lateral is encouraged as the new central leader, and three to four evenly distributed new shoots emerging from the upper half of the trunk, but well below the new leader, are chosen as scaffold branches. Other shoots may be pinched to retard their growth, but pruning should be minimal to avoid stunting the young tree's growth. Shoots that are not chosen to be scaffolds can be removed during the first dormant season of the planting. The secondary scaffold branches are typically developed during the second year of growth, whereas training in the third and fourth seasons is devoted to the development of future fruiting wood on these secondary scaffolds.

Peaches and other stone fruits are more bushlike in growth habit than are apples and pears, and therefore they are often trained using an open-center or vase-training system (Fig. 19–12). If not done so by the nursery, newly planted peach trees should be headed back to force lateral branching that will eventually become the primary scaffolds. A typical tree will possess an open center and about eight scaffold branches distributed evenly to form a vase shape. Traditional recommendations call for heading back the more vigorous lower scaffolds, while upper scaffolds showing less vegetative vigor are not pruned. Many growers have adopted a minimal pruning approach where only weak or inappropriately placed branches are removed for the first three to four years of growth, with a corrective pruning administered only after the tree is nearly full grown. This approach has proven to promote precocity, thus providing an earlier return on investment.

Regardless of the system used, bearing deciduous fruit trees require moderate annual pruning to:

1. Stimulate production of new fruiting wood.
2. Allow light penetration into the tree and prevent the development of dense pockets of vegetative growth. This is especially important for spur-bearing fruit trees because the primary source of carbohydrates for developing fruit are supplied by spur leaves. Fruit set, growth, and ultimately fruit quality are negatively affected if these leaves are in constant shade.
3. Reduce excessive amounts of fruiting wood and lessen the need for expensive hand thinning of fruits.
4. Remove broken, diseased, dead, or interfering branches.
5. Keep the trees from growing so tall that harvesting becomes difficult.
6. Confine the trees to the space available to them.

Figure 19–11 Some fruit trees, such as pears, tend to develop an upright, narrow shape, as shown in the unpruned tree at left. Often, this cannot be corrected by pruning alone. With notched, spreader boards, as shown in the pruned tree at right, the primary scaffold branches can be trained to give a wider tree shape. Inner limbs should be removed and those retained should be pruned to outward facing branches or buds. *Source:* Gerdts, M., A. Hewitt, J. Beutel, J. Clark, and F. Cress. 1977. Pruning home deciduous fruit and nut trees. Univ. of Calif. Div. Agr. Sci. Leaflet 21003.

A light annual pruning of bearing trees is better than a heavy pruning every three or four years because it results in more regular cropping.

Care must be taken not to prune or break off fruiting spurs, although in trees that consistently overbear to the detriment of fruit size and quality, there may be some benefit in partial spur removal. Spurs do not live indefinitely, and some renewal is constantly taking place. Annual pruning tends to stimulate new spur formation.

Fruit species producing fruiting buds laterally or terminally on shoots that developed the previous growing season should be pruned to encourage the growth of such shoots. This entails a moderate annual dormant pruning, together with moderate nitrogen fertilization and ample soil moisture. Excessive pruning plus heavy applications of nitrogen can cause excessive rank vegetative growth, which will not form flowers.

The severity of pruning needed by bearing trees to keep them in a state of optimum fruitfulness can be judged to some extent by the length of annual shoot growth. Table 19–4 gives the amount of growth associated with good fruiting. If the annual growth of the average shoot exceeds these amounts, the

severity of annual pruning should be reduced and nitrogen fertilizer withheld. If shoot growth is much below these amounts, severity of pruning should be increased as well as, perhaps, the amount of fertilizer applied annually.

Training and Pruning Grapevines

In general, grapes are more intensively trained and heavily pruned than fruit and nut trees. Grapes are almost always trellised using a variety of training systems to enhance light penetration and air movement through the canopy and to support the relatively heavy fruit load. On a yearly basis, effective pruning of mature vines requires the removal of most of the previous year's growth to balance this year's vegetative growth against energy devoted to setting and maturing a crop. Grape training and pruning are relatively labor-intensive, exacting processes, but if accomplished properly, they can ensure the productivity and profitability of the vineyard over an extended period of time. For instance, operating 'Concord' vineyards in the US East and Midwest have been cropped continuously for over a century.

Figure 19–12   Young, vigorous peach tree before (*left*) and after (*right*) pruning. Many scaffold limbs have been removed, particularly those that grow low and horizontal. The stronger, moderately upright limbs were selected for the permanent primary scaffolds and were headed back to force out additional branching. *Source:* Gerdts, M., A. Hewitt, J. Beutel, J. Clark, and F. Cress. 1977. Pruning home deciduous fruit and nut trees. Univ. of Calif. Div. Agr. Sci. Leaflet 21003.

TABLE 19–4	Desirable Amounts of Average Shoot Growth for Bearing Trees to Give Maximum Fruit Production			
Species	Young Trees—Under 10 Years of Age		Older Trees—Over 10 Years of Age	
	Centimeters	Inches	Centimeters	Inches
Freestone peaches and nectarines	50 to 100	20 to 40	30 to 76	12 to 30
Clingstone peaches	76 to 100	30 to 40	30 to 76	12 to 30
Apricots	30 to 76	12 to 30	25 to 60	10 to 24
Plums (except prunes) and quinces	25 to 60	10 to 24	23 to 46	9 to 18
Almonds, prunes, and cherries	23 to 46	9 to 18	15 to 25	6 to 10

Proper cultural management (weed and disease management, fertility, etc.) and training are critical during vineyard establishment to achieve optimum vegetative (cane and root) growth necessary to produce strong, straight shoots that will form the trunk and scaffold on which future crops will be borne. Newly planted vines are left intact or are sometimes pruned to six or eight buds, depending on cultivar, vineyard conditions, and the expected growth. If conditions are very favorable, unpruned plants will generally develop more leaf area, thus producing longer, thicker shoots. If less-than-optimal growth is anticipated, plants should be pruned to a few buds, and only the two strongest shoots should be kept. At the beginning of the second year, only the healthiest canes are retained as developing trunks. If the vines are growing from their own roots, there is benefit to choosing canes that arose

from below-ground buds; however, if the cultivar is grafted to a disease or insect-resistant rootstock (e.g., all European or vinifera grapes are grafted to phylloxera-resistant stock), only canes originating from buds above the graft union should be chosen. The number of canes kept depends in part on the training (trellising) system to be used. Often these canes are trained to be straight and perpendicular to the trellis wire using twine or bamboo stakes as temporary plant supports. For all cordon-based training systems, rapidly growing shoots are trained (wound) around the trellis wire to form the horizontal scaffolds or cordons during the second year of growth (Figure 19–13). Some producers remove the apical buds of these shoots when they reach the desired length for the cordon to promote lateral growth. If lateral growth is sufficient and not winter-damaged, it can be pruned to one- or two-node spurs in the

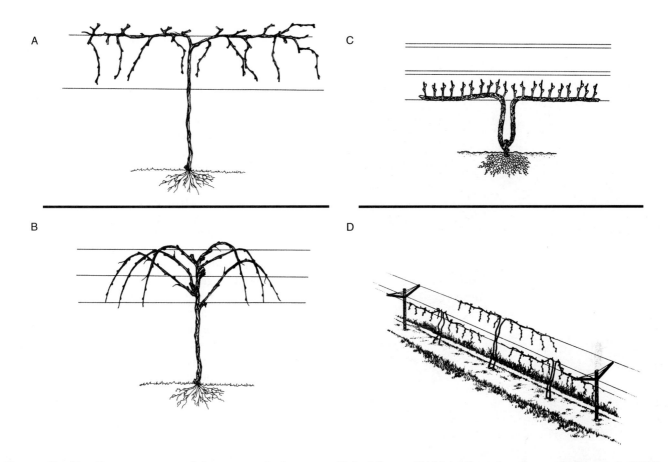

Figure 19–13 Common grape-training systems in the eastern United States: (A) high bilateral cordon or single curtain (HBC—spur-pruned), (B) umbrella Kniffen (UK—cane-pruned), (C) vertical shoot position (VSP—spur-pruned), and (D) Geneva double curtain (GDC—spur-pruned). *Source:* Midwest grape production guide. Bulletin 919. Ohio State University Extension.

third season. The shoots chosen to be cordons can be headed back to half their final length (approximately ten nodes) to promote uniform bud break along the cane. New growth is then wrapped around the trellis wire as it develops.

Common grape-growing systems vary from region to region, but grape growers predominantly choose trellis and training regimes that either capitalize on or compensate for inherent characteristics of specific grape types or cultivars. Vegetative vigor and fruitfulness of the grape to be grown are perhaps the characteristics most often considered when choosing a growing system. Producers must also consider the grapevines' growing habit, cold hardiness, and disease resistance.

Although the positioning of trunks, scaffolds, and fruiting wood differ significantly among training regimes, all systems are based on either cane pruning or spur pruning strategies to affix the correct number of fruitful buds and thus achieve a desired crop load (Figs. 19–13 and 19–14). For spur-pruned systems, one-year-old wood is trimmed back to two to three bud spurs positioned evenly along the length of the cor-

don (Fig. 19–14). Cane-pruned systems achieve the same end, but longer canes arising from near the trunk head are retained, each cane holding from ten to twenty potentially fruitful buds. The number of buds that should be retained for the current fruiting season depends directly on the vigor of last year's vegetative growth, leaf cover, and thus photosynthetic capacity. The weight of pre-pruned (initially removed) one-year-old canes is highly indicative of this vigor. Formulas for balanced pruning procedures based on pre-pruning weights have been developed for many grape cultivars and specific growing regions. For example, for 'Foch,' a popular French-American hybrid wine grape, it is recommended that thirty buds be kept for the first pound of vines removed. Thereafter, an additional ten buds should be retained for each additional pound of vine prunings, up to a maximum of sixty to seventy buds. Balanced pruning ensures that vines will not be cropped beyond their ability to mature the developing grapes adequately and that sufficient vegetative growth will be added to maintain vine health and sustain productivity for the life of the vineyard.

Figure 19–14 Balanced pruning of French-American hybrid grapevines: (A) collection of pre-pruned, year-old wood to be weighed, (B) adjusting spurs to the correct number of buds, and (C) the vine after the completion of the balanced-pruning operation. *Source:* Joseph Scheerens, The Ohio State University.

Among spur-pruned systems, the high bilateral cordon (HBC), or single-curtain, system, is most suitable for grape varieties with moderate vegetative vigor. It is commonly used in eastern and midwestern US production regions, especially for French-American hybrid wine grapes. The advantages of this

system are many: it has low establishment and production costs; it facilitates mechanized pre-pruning and fruit harvest; it places fruiting and renewal zones high in the canopy, thus optimizing sunlight penetration and fruit yield and quality; and it may be less susceptible to frost injury and deer damage. The vertical shoot position (VSP) system is the most commonly used trellis system in the world for European grape cultivars because it is complementary to their natural upright growth habit. In this system, the distance between rows is reduced and overall productivity is improved. It is also easy to manage and mechanize. Although the placement of renewal zones low in the canopy simplifies the renewal of winter-damaged trunks, light exposure to developing grape clusters and their associated leaves is less than it is in HBC, resulting in a lower yield per vine. The fruit is also more susceptible to wildlife predation, and the fruit buds for the following season are more susceptible to frost damage. The Geneva double curtain (GDC) is essentially a double trellis system with four fruiting planes (outer and inner on each trellis wire). It supports approximately 30 percent more vine growth than other systems, and it is commonly used for American grape cultivars like 'Concord' with high vegetative vigor and high potential fruit yields per vine. This system requires wider between- and within-row spacing, with curtains typically 4 ft apart. Fruit quality is enhanced and susceptibility to disease is reduced using this system. In addition to these systems, two head-training systems are used for certain grape cultivars in the western United States. The head-trained, spur-pruned system is used for some table grapes and large-clustered wine grapes. No trellis is required for this system after the trunk is well established. The head-trained, cane-pruned system is frequently used for 'Thompson Seedless' grapes because buds located near the base of the canes are typically not very fruitful for this cultivar. The umbrella Kniffin (UK) and four-arm Kniffin systems were formerly popular, but they are labor-intensive, time-sensitive, and not suitable for mechanical pre-pruning. In addition, tender buds can be damaged easily when fruiting canes are tied into position.

Training and Pruning Cane Fruits

With few exceptions (e.g., erect blackberries, some eastern-grown raspberries), all cane fruit crops are trained to one of various trellising systems. Foremost, the trellis provides support for fruit-laden floricanes and protects all canes from physical damage due to wind, snowloads, and other mechanical stresses. From a physiological standpoint, trellising allows for greater light and air penetration into the canopy, which increases net photosynthesis and flowering and limits disease infection periods and insect population growth. As stated earlier, several trellis designs allow for the separation of primocanes from floricanes, resulting in better primocane growth and ease of harvest. Other horticultural considerations surrounding cane management via trellising include ease of cropping operations, such as spraying and mowing;

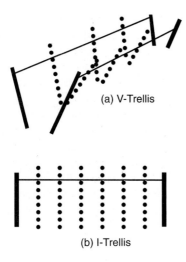

(a) V-Trellis

(b) I-Trellis

Figure 19–15 The I- and V-trellis systems. *Source:* Joseph Scheerens, The Ohio State University.

pruning; and control of row width. Most cane fruit cultivars will tip layer (asexually reproduce) if the cane apex is allowed to be in contact with moist soil.

Trellising systems range from the simple to the complex in design. Perhaps the simplest construction is the I-trellis, composed of periodically placed posts that hold a single horizontal monofilament or metal wire placed approximately 1 to 1.5 m above the soil surface (Fig. 19–15). The I-trellis provides basic support and protects against mechanical damage, but it does little to improve the plant's physiological status or to ease horticultural manipulation of the crop. Trellis system designs such as the V-trellis include two supports that are horizontally separated by approximately 1 m. These trellises allow more light to penetrate through the canopy, allow for the separation in space of primocanes and floricanes, simplify horticultural management and harvesting, and often improve yields. The V-trellis system is versatile because it provides several options for the positioning of floricanes and primocanes. Both the I- and V-trellises can be constructed using T-posts if desired. More complex trellis designs, such as the single-sided shift trellis, the Tatura trellis, and the Lincoln canopy system, have also been examined for cane fruit production under specific circumstances, with some degree of success. For example, scientists at the U.S. Department of Agriculture have engineered a mechanical harvester specifically for eastern-grown thornless blackberry cultivars and concurrently developed a trellis system that is specifically suited to optimize the harvester's performance. The size of the trellis and the strength and extent of trellising materials used depends on the vigor of the crop. Typically, blackberries and blackberry-raspberry hybrids (e.g., boysenberry, loganberry, etc.) require a more substantial trellis than do raspberries. The trellis design and construction should consider other facets of the growing system. For instance, the fruit-laden canes of primocane-fruiting raspberries are often supported by a temporary trellis constructed of T-posts and twine, which can be removed after harvest to facilitate pruning.

Proper training and pruning procedures for cane fruits depend on the growing region, growing system, crop, and cultivars used. Training and pruning are perhaps least complicated for primocane-fruiting cultivars. Most growers of primocane-fruiting berries simply mow off the previous season's growth during midwinter, when plants are deeply dormant. Using this approach, they sacrifice whatever percentage of the crop would have been borne on the floricanes (usually of inferior quality anyway) for ease of management, substantially reduced production costs (i.e., no hand pruning labor costs), winter damage avoidance, stronger primocane growth in the following season, and sustained productivity for the life of the planting. Old canes must be pruned as closely to the ground as possible.

Floricane-fruiting red raspberry primocanes are not generally trained per se, but they can be protected from wind and mechanical damage by the training of floricanes to various two-wire trellises (e.g., the V-trellis described above). Summer pruning of red raspberry primocanes is generally not necessary, except when a cultivar's adaptation response results in excessive, unwieldy vegetative vigor. However, growers may mechanically or chemically remove many of the early-emerging primocanes to limit competition for carbohydrates between vegetative meristems and developing fruit. This practice is called primocane suppression, and it is usually practiced in the Pacific Northwest region, where the growing seasons are long, the climatic and edaphic conditions are generally ideal for cane fruit growth, and midseason primocanes can develop fully prior to dormancy.

Spent floricanes are typically removed sometime after harvest. The timing of their removal is situation-dependent. Under conditions that favor heavy fungal disease or insect pressure, or if the crop is being grown organically, the spent floricanes should be removed as soon as possible after harvest. If left in place until dormant pruning practices are initiated, however, harvested floricanes may contribute to the carbohydrate reserves of the plant and continue to protect primocanes from damage.

Most red raspberry pruning is accomplished in midwinter, while plants are dormant. If still present, spent floricanes are removed first. Then winter-damaged canes or cane portions are removed. Next, the canes that will bear next year's crop are chosen. Ideally, only the most vigorous (thick) and healthy canes are kept, whereas weak and spindly canes are removed. Thinning the number of next year's floricanes is essential; if too many canes are left, the plant may become stressed, resulting in the production of small, poor-quality fruit. If red raspberries are grown in a hedgerow, as is typical in the eastern United States, about three to five canes per foot should be kept (Fig. 19–16). In the western United States, where plants are cultured in hills or stools, about five to eight canes per crown are kept. The chosen canes are then headed back and trained (attached) to a trellis wire simply in a linear fashion or using more complex cane arrangements (see Figs. 19–15 and 19–17).

Steps in the training and pruning of black raspberries, blackberries, and blackberry-raspberry hybrids are similar to those employed for red raspberries, with one important exception. In these crops, primocanes are summer tipped, or

Red raspberry before and after pruning. These canes are planted in a narrow hedgerow.

Figure 19–16 Dormant pruning the red raspberry hedgerow. *Source:* Brambles—production, manage, and marketing. Bulletin 782. Ohio State University Extension.

headed back via the removal of 7–10 cm of growth. Tipping releases the axillary buds from apical dominance and thus stimulates lateral branching. Most fruit buds for the following season will develop on these lateral branches. Black raspberry primocanes are usually tipped when they reach 0.5–0.6 m in length, whereas blackberry primocanes are taller when tipped, approximately 0.9–1.2 m. During the dormant season, lateral branches are trimmed back, and canes are chosen for next year's production (Fig. 19–18).

Training and Pruning Bush Fruits

Most bush fruits are not trained per se but are grown as free-standing bushes. However, their size and shape are often controlled by judicious pruning. As with other fruit crops, bush-fruit pruning is practiced to maintain vigor and productivity, improve growth habit, distribute fruiting wood throughout the plant, increase sunlight to the plant's interior for improved fruit size and quality, and increase air circulation for decreased incidence of disease. Plants are pruned in winter or early spring while dormant. The amount of wood removed depends on the species, the cultivar, plant vigor, and age. Pruning of mature bushes involves the removal of weak, low-growing, or twisted canes or branches; winter-injured tissue; canes that rub together; older canes that are less productive; and branches that receive little light. Pruning cuts should be made as close to the crown or to a main branch as possible to prevent the entrance of disease organisms.

FRUIT THINNING

Many temperate tree fruit crops, especially pome and stone fruits, produce an overabundance of flowers each year, perhaps as an evolutionary consequence of the potential for flower and flower bud losses during spring frosts. If pollinators are active, most of these flowers are pollinated and many are fertilized. After fertilization, tissues surrounding the embryos (which will eventually become the fruit) begin to enlarge in response to

Figure 19–17 (A) Weaving and (B) tepee arrangements of Washington-grown red raspberry floricanes as they are attached to the trellis. *Source:* Joseph Scheerens, The Ohio State University.

hormonal signals from the developing seed. Obviously, the growth and development of both seed and fruit tissues relies on an adequate supply of energy. However, carbohydrates are typically in limited supply, and this supply is not sufficient to grow and mature all the fruit that was initially set. In response to this stress, trees typically experience a period commencing a few weeks after bloom wherein many of the newly developing fruits abscise. This abscission period is called June drop and it can last as long as one month. June drop can be considered a natural thinning process whereby the tree adjusts its crop load to enhance seed and fruit development and thus ensure reproductive success. In some years, when environmental conditions are unfavorable, too many of the fruits drop, and yields are reduced. In some cases, however, the trees retain more fruit than is acceptable from a horticultural standpoint. In the latter case, fruit-thinning procedures are advocated.

Chemical or mechanical fruit-thinning procedures for apples, apricots, grapes, nectarines, peaches, pears, and plums are routinely practiced. The horticultural goal of all fruit-

Figure 19–18 Black raspberry and blackberry pruning: (A) summer tipping of black raspberries, (B) dormant pruning of black raspberries, and (C) dormant pruning of thornless semi-erect blackberries. *Source:* Brambles—production, manage, and marketing. Bulletin 782. Ohio State University Extension.

Figure 19–19 *Left:* Plum fruits on a branch that was properly thinned. Proper spacing has allowed them to develop properly for fruit size, quality, and early maturity. *Right:* Unthinned fruits are small, late maturing, and of poor quality. In addition, such limbs often break from the weight of the fruit.

thinning procedures is to adjust crop load so that it is balanced between overall yield and fruit size and quality (Fig. 19–19). Fruit size is an extremely important commercial issue, with large fruit usually being preferred by consumers. In almost all fruit crops, the size of the fruit harvested is negatively correlated to the number of fruits developed on the plant.

Final fruit size more or less depends on the leaf-fruit ratio on a branch. The more leaves per fruit, the larger the final fruit size. but fruit size is increased at the expense of total yield. Thinning can be highly profitable, however, where a premium is paid for larger-size fruits.

Fruit thinning is the best way to increase fruit size, although flower and bud thinning (by pruning out fruiting shoots) is also effective. Overirrigation or excessive nitrogen fertilization will not increase fruit size; in fact, these approaches can stimulate new shoot growth, which competes with fruits for carbohydrates.

Ripening fruit undergoes a series of chemical and physical changes (e.g., increased pigmentation, increased sugar levels, the development of volatile flavor components, etc.) that determine the eating or processing quality of the product. Typically, if the level of carbohydrate supplied by vegetative plant tissues is inadequate for the number of ripening fruits present, the quality of those fruits will be diminished.

As with training and pruning, maintaining a balance between vegetative and reproductive growth through fruit thinning is an important physiological goal. In addition to its effects on fruit size and quality, removal of excess fruits by thinning results in a greater allocation of carbohydrates for the root and shoot growth that is important for sustaining the productive life of the orchard. Also, if fruit loads are not thinned effectively, floral bud initiation for next year's crop will likely be decreased, inducing the alternate bearing pattern discussed previously in this chapter. Judicious fruit thinning to control excessive fruit numbers during an on year is one of the most effective cultural procedures for an orchardist or a home

REFERENCE DATE

In the stone fruits, fruits that are small early in the season are also small at harvest time. Commercial growers of clingstone peaches and apricots have made use of this relationship to predict much earlier in the season average fruit size at harvest. The young, developing fruits are measured at a specific time, called a reference date. For apricots, this time is determined by cutting through the ends of the young fruits at intervals. The reference date is seven days after the pit begins to harden, when representative fruit samples are measured for size. The reference date for clingstone peaches is the same except that ten days are added to the date that the tip of the pit begins to darken. For clingstone peaches an average suture diameter of 34 to 35 mm (1.36 to 1.40 in) at the reference date is needed to produce an average diameter at harvest of 67 to 68 mm (2.68 to 2.72 in). This is sufficient to make 90 percent of the fruits larger than the minimum harvest size. If the young fruits at reference date are too small, the fruits are thinned to bring the remaining fruits to a satisfactory size by harvest time.

gardener in overcoming alternate bearing. Even if alternate bearing does not occur, however, several consecutive years of overcropping can greatly weaken trees or vines. Last, the inordinate masses associated with an overabundance of fruits may structurally damage or break shoots and even large limbs that cannot withstand the physical stresses the fruits impose.

Thinning Requirements

It may be obvious that a fruit tree is overloaded with young fruits and needs thinning—but how much? It is difficult to judge because several factors are involved, such as the cultivar, time of fruit maturity, availability of water, general vigor and age of the tree, as well as growing conditions. Some general guidelines for fruit thinning can be given, however. Peaches are often thinned to about one fruit every 15 to 20 cm (6 or 8 in.) of shoot. Another rule is twenty-five to forty leaves per fruit for apples and about fifty leaves per fruit for peaches.

Thinning to control yields and fruit size is so important in crops such as peaches and apricots that thinning tables have been developed to determine the optimum number of fruits to leave on the tree. These tables consider the distances between trees, fruit size desired (number per pound), and number of tons of fruit desired per acre. For example, if a 22.4 MT/ha (10 t/ac) harvest is desired for apricots and the trees are set on a 7.2 × 7.2 m (24 × 24 ft) planting distance, no more than about 3,200 fruits should be left per tree to attain a fruit size of at least 200 g (0.4 lb). Fruit counts are made on sample trees throughout the orchard to determine if the fruit set is uniform.

Chemical Thinning

The idea of using chemical sprays applied to trees to remove some of the fruits is definitely appealing, and much research has been given to develop thinning sprays to replace hand thinning. For apples, several natural or synthetic, hormone-based thinning sprays are available and are used commercially. However, the two most effective and widely used thinning chemicals are naphthaleneacetic acid (NAA) and carbaryl (1-naphthyl N-methyl carbamate.)

It is often difficult to obtain consistent results with chemical-thinning sprays, and general recommendations applicable to different situations cannot be made. First, there is often a cultivar-specific response to chemical-thinning agents. For instance, NAA can be a successful thinning agent for late-maturing apples, but its use on summer cultivars results in foliage injury, premature ripening, and fruit cracking. The use of NAA on 'Fuji' and spur-type 'Red Delicious' often results in the formation of pygmy (small, seedless) apples. Carbaryl is a preferred thinning agent for the latter cultivar, but its use with 'Gallia Beauty' and 'Rome' can result in overthinning. 'Honeycrisp' may be overthinned if thinning agents are applied in combination. Environmental conditions also affect the efficacy (potency) of spray compounds. In general, chemical-thinning agents remove more developing fruit if (1) the trees are not vigorous or have been only lightly pruned, (2) the bloom is particularly heavy or the flowers have not been well-pollinated, (3) the humidity is high and the spray's drying time is slow, and (4) the weather is cloudy and cool before or after bloom. There are exceptions to these generalities, however; For example, NAA is a more efficient thinner if it dries quickly. Adding a surfactant (wetting agent) to the tank mix increases foliar absorption rates and minimizes the effects of adverse climatic conditions.

With any chemical-thinning applications, proper timing is critical. Typically chemical-thinning treatments are applied at petal-fall or postbloom (ten to thirty days after full bloom). The critical periods for thinning are both cultivar- and product-specific. In general, it is recommended that late-maturing cultivars be sprayed when the predominant fruit in each cluster is approximately 11 to 13 mm in diameter and that early maturing cultivars be treated at petal-fall. Both NAA and its derivatives and carbaryl have applications as either petal-fall or postbloom sprays, whereas some of the newer thinning agent, such as Accel® [Benzyladenine, N-(phenyl)-1H-purine 6-amine] and Vydate and {oxamyl, 2-(Dimethylamino)-N-[[(methylamino)carbonyl]oxy]-2-oxoethanoimido-thioic acid methyl ester} are labeled for postbloom treatments. With any chemical-thinning applications, proper timing is critical.

Mechanical Thinning

Many other fruit crops are mechanically thinned. Hand thinning is the most precise method of thinning because the decision to remove a fruit is made on a fruit-by-fruit basis. However, it is laborious, time-consuming, and expensive and perhaps is a suitable method for the homeowner with only a few trees to thin.

As an improvement over slow and expensive hand thinning, long poles with sections of rubber hose at the end can be used to hit fruiting branches and knock off some of the fruit.

Hand thinning to remove missed fruiting clusters may follow. Mechanical tree shakers, used in harvesting operations, are sometimes used to shake the trees to remove a portion of the small, immature fruits. Shaking can be satisfactory, but it is not very precise and the desirable larger fruits tend to be removed. A light machine shaking, followed by hand thinning or pole knocking, works well in some cases. Hand thinning gives the best results, however, and should be used whenever practicable.

In addition to tree fruits, some vine crops are also thinned. To obtain better yields of high-quality grape berries, one extra bud on half or more of the fruiting spurs (spur pruning) and an extra cane or two per vine (cane pruning) can be retained. Thinning of grapes can be done as flower thinning—removal of part of the fruiting clusters before bloom—or cluster thinning—removal of clusters after the berries have set. Clusters or fruits should be thinned to reduce the crop to the amount it would have been had the extra fruiting canes not been left. The increased leaf area produced by retaining the extra spurs or canes to support the same number of fruits per vine raises the yield and the quality of the berries.

Because of the production of fruit-bearing shoots from latent or noncount buds, some cultivars of the French-American hybrids (e.g., Seyval blanc, Vidal blanc, and Chambourcin) need annual cluster thinning. This cluster thinning is necessary to maintain sufficient vine growth for subsequent production and to maintain grape quality.

Mature sweet and sour cherries, prunes, almonds, walnuts, pistachio and hazelnuts, cane fruits, bush fruits, and strawberries ordinarily do not require fruit thinning. During plant establishment and juvenility, however, some species benefit greatly from hand removal of flower buds or flowers to conserve carbohydrates for root and shoot development. For instance, it is recommended that flower buds and flowers be hand stripped from newly planted blueberries for the first two years of the planting. Strawberries grown in a matted-row system also benefit from flower removal in the planting year to proliferate runners (stolons) that will be used to establish the solid bed.

FRUIT DEVELOPMENT TERMINOLOGY

maturation The final stage of development while fruits are still attached to the plant. It includes cell enlargement plus the accumulation of carbohydrates and other flavor constituents and a decrease in acids. The flesh may or may not soften.

ripening The changes taking place after full maturation, involving a softening of the flesh, the development of characteristic flavors, and an increase in the juice content. Ripening may occur either before or after the fruit has been picked.

senescence The period following ripening, during which growth ceases and aging processes replace ripening processes. Senescence may occur before or after harvest.

FRUIT MATURATION, RIPENING, AND SENESCENCE

As fruits near their maximum size, significant changes lead to the end product—a fruit with an attractive color, soft enough to be palatable, sweet and juicy, with an accumulation of the other components that give it its own distinctive flavor and aroma. The increase in sugars—or soluble solids—can be measured by a refractometer or hydrometer. Decreasing hardness can be measured by various kinds of pressure testers. Changing color, as the masking chlorophyll disappears, can be measured by color charts or color meters. Changing acids can be determined by chemical means or pH meters. All these procedures are used to determine quantitatively when the fruit is ready for harvest. In some states in the United States, legal maturity standards demand such measurements to determine when fruit can be harvested.

Optimum harvest maturity depends on the intended use of the fruit. For example, strawberries harvested for processing or local markets are picked at or near peak ripeness to maximize their sugar–acid ratio, volatile flavor constituents, and overall eating quality. However, ripe or nearly ripe berries are far too soft for long-distance shipping and prolonged shelf life. Commercial strawberries grown for the larger marketplace are typically picked when they are full size but still very firm and only partially colored.

The home gardener has an advantage over the commercial grower because he or she can allow the fruit to reach optimum maturity right on the tree for eating at the optimum degree of ripeness. For commercial production, however, perhaps involving shipments in containers to great distances, the fruits generally must be picked firm enough to endure shipping stress and arrive at their destination in an acceptable condition.

All fruits on a tree do not mature at the same time. Fruits on the top and outside are usually ready to pick before those on the inside. For the home gardener, this is an advantage because it prolongs the harvest period, but for commercial once-over harvest, some of the fruits are harvested at a less-than-ideal time.

The various kinds of fruits have different ripening patterns and require different storage procedures. An understanding of these is essential to know how to harvest and handle the fruits properly. Most fruits reach maturation and ripen on the tree, vine, or bush, at which time they should be picked. If not picked, they enter a stage of senescence and deterioration. However, other fruits, such as pears, are best harvested when they reach maturity on the tree and are ripened to eating condition off the tree.

It must be emphasized that maturing and ripening fruits are living organisms. Their cells are respiring, and many other complex chemical and physical changes are occurring. Some types of fruits are said to be climacteric and others are nonclimacteric (Fig. 19–20). In both types, the respiration rate of the cells slowly decreases as maturation proceeds. However, in climacteric fruits, the respiration rate rises abruptly as the fruit ripens, reaches a peak (the climacteric point), and then declines. Senescence ensues, leading eventually to the death of

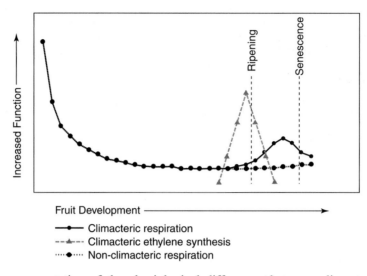

Figure 19–20 Schematic representation of the physiological differences between climacteric and nonclimacteric fruit. *Source:* Joseph Scheerens, The Ohio State University.

all the fruit cells. In nonclimacteric fruits, the respiration rate gradually declines during ripening, with no particular peak.

Apples, pears, peaches, apricots, plums, and blackberries are examples of climacteric fruits. Most climacteric fruits show the same respiration pattern whether ripening on or off the tree. Cherries, strawberries, grapes, and raspberries are examples of nonclimacteric fruits.

Refrigeration prolongs fruit life chiefly because the lowered temperatures reduce respiration rates. Controlled atmosphere (CA) storage—lowering the oxygen levels from 21 percent (the amount in air) to 2 to 3 percent, as well as increasing the carbon dioxide to 1 to 5 percent instead of the usual 0.03 percent in air—can further lower respiration rates and greatly extend the storage life of fruits. The levels of O_2 and CO_2 that can be tolerated vary with species and cultivar. CA storage procedures have come into widespread use in recent years, especially for apples and to a lesser extent for pears. Although harvested mainly in autumn, apples are now available in markets year-round because of CA storage facilities.

Ethylene (C_2H_4), which is produced by all plant tissues, including fruits, can stimulate fruit ripening. An ancient Chinese custom was to ripen fruit in rooms where incense was burned. In the earlier days of this century, kerosene stoves were widely used in California to stimulate color development in lemons. Ethylene, plus other gases, is released by such combustion. Ripe fruits, like apples, give off ethylene and stimulate the climacteric ripening of other fruits confined in the same containers with them. Experiments in the 1960s provide evidence that ethylene is much involved in fruit ripening and is considered a ripening hormone. Applied ethylene is effective in stimulating the ripening of most climacteric and nonclimacteric fruits, although ripening of the strawberry is not affected by ethylene. Gibberellin, a natural plant hormone, is known to counteract ethylene effects and delay fruit ripening.

The biochemistry and physiology of fruit maturation. ripening, and senescence have received considerable study, and various theories have been proposed to explain the ripening mechanisms involved. One theory is that during fruit ripening, new enzymes appear that cause the characteristic changes in the fruit. Energy from respiration provides for the synthesis of the enzyme systems and for their ripening actions. It is believed, too, that in addition to the ripening process itself, hormonal (ethylene, gibberellin, and perhaps others) changes may also make the fruits responsive to ripening signals. The various kinds of fruits develop differently to the point where they are ready for harvesting and ripening. Some examples will show the variability among fruit species and will point out the necessity for understanding these patterns for proper harvesting and fruit ripening.

Table grapes mature and ripen on the vine; the peak of fruit quality is reached and, ideally, the clusters are picked and consumed at that time. Once the clusters are removed from the vine, deterioration sets in. This can be slowed by placing the clusters in cold storage—0°C (32°F) at 95 percent relative humidity. Such stored grapes can remain edible from one to six months, depending on the cultivar. In addition, the European (but not American) grapes require sulfur dioxide fumigation treatments to adequately prevent decay.

Apples also develop optimum maturity and ripeness while on the tree, and senescence begins with harvest. This can be delayed by cooling the fruits to 0°C (32°F) within two to three days after harvest and holding them in storage at a relative humidity of 90 to 95 percent. By retarding the respiration rate, CA storage enables fruits of some apple cultivars to remain in good eating condition for as long as nine months.

Pears ripen differently. On the tree, the fruits accumulate the maximum of stored foods and the firmness of the flesh decreases, and they should be picked. But for ideal eat-

| TABLE 19–5 | Recommended Storage Temperatures and Relative Humidity, Approximate Storage Life, Highest Freezing Point, and Water Content of Fresh Fruits in Commercial Storage |

Commodity	Temperature		Relative Humidity (Percent)	Approximate Storage Life	Highest Freezing Point		Water Content (Percent)
	°C	°F			°C	°F	
Apples	−1–4	30–40	90–95	1–12 months	−1.5	29.3	84.1
Apricots	−0.5–0	31–32	90–95	1–3 weeks	−1.0	30.1	85.4
Berries							
Blackberries	−0.5–0	31–32	90–95	2–3 days	−.7	30.5	84.8
Blueberries	−0.5–0	31–32	90–95	2 weeks	−1.2	29.7	83.2
Cranberries	2–4	36–40	90–95	2–4 months	−.8	30.4	87.4
Currants	−0.5–0	31–32	90–95	1–4 weeks	−1.0	30.2	84.7
Dewberries	−0.5–0	31–32	90–95	2–3 days	−1.2	29.7	84.5
Elderberries	−0.5–0	31–32	90–95	1–2 weeks	—	—	79.8
Gooseberries	−0.5–0	31–32	90–95	3–4 weeks	−1.0	30.0	88.9
Loganberries	−0.5–0	31–32	90–95	2–3 days	−1.2	29.7	83.0
Raspberries	−0.5–0	31–32	90–95	2–3 days	−1.0	30.0	82.5
Strawberries	0	32	90–95	5–7 days	−.7	30.6	89.9
Cherries, sour	0	32	90–95	3–7 days	−1.7	29.0	83.7
Cherries, sweet	−1 to −0.5	30–31	90–95	2–3 weeks	−1.8	28.8	80.4
Grapes, Vinifera	−1 to −0.5	30–31	90–95	1–6 months	−2.1	28.1	81.6
Grapes, American	−0.5–0	31–32	85	2–8 weeks	−1.2	29.7	81.9
Nectarines	−0.5–0	31–32	90–95	2–4 weeks	−.9	30.4	81.8
Peaches	−0.5–0	31–32	90–95	2–4 weeks	−.9	30.3	89.1
Pears	−1.5 to −0.5	29–31	90–95	2–7 months	−1.5	29.2	83.2
Plums and prunes	−0.5–0	31–32	90–95	2–5 weeks	−.8	30.5	86.6

Source: Hardenburg, Watada, and Wang. 1986. The commercial storage of fruits, vegetables, and florist and nursery stock. USDA Agr. Handbook No. 66.

ing characteristics, pears should be harvested before they ripen and then ripened in a cool place (15.5°C to 22°C; 60°F to 72°F) at 80 to 85 percent relative humidity. Or they can be placed directly into cold storage (−0.5°C; 31°F) at 90 percent relative humidity, where they will keep for several months. They can be taken from cold storage to higher temperatures at any time for ripening. Some pear cultivars, like d'Anjou, do not ripen properly unless they are held in cold storage for a time or they are treated with ethylene gas.

When harvested for commercial shipments, plums are too firm for eating. They continue softening and ripening af-

ter harvest to an optimum point for eating, after which deterioration commences. Plums are often stored under refrigeration at 0°C (32°F) just after picking to prolong the life of the fruits for two to four weeks.

Fruits of many species, if stored below certain temperatures but above the freezing point, will develop typical damage symptoms. Such as internal browning, pitting, scald, dull skin color, a soggy breakdown, susceptibility to decay, and failure to ripen properly. Such fruits become worthless. Although influenced by many factors, optimum storage conditions are given in Table 19–5 for most fruit species.

SUMMARY AND REVIEW

The successful production of temperate fruits and nuts requires several considerations, including site selection, marketing and cost analysis, and developing an effective cultural management strategy.

Climate and soil characteristics are factors to consider when selecting a site. Climate determines if temperatures, wind, and precipitation are appropriate for proper growth of the crop. If a crop has a specific temperature requirement, such as chilling, the site must provide that without exposing the plant to deleterious temperatures. Wind can affect pollina-

tion, pesticide applications, and irrigation, and can cause damage to young trees and fruit. Most fruit and nut crops require a deep, fertile soil that drains excess water quickly yet retains enough water to supply the crop. Disease organisms that may be present in the soil from previous vegetation is an important consideration. Precipitation should provide adequate water throughout the growing season if irrigation is not feasible. Precipitation that regularly interferes with proper growth, for example, hail or continual rainfall during pollination, presents a serious problem for fruit and nut production.

The marketability of a crop is influenced by global, regional, and local production capacity and the competition for customers. These factors have to be studied and analyzed before producing a crop at the site under consideration. A crop production cost analysis includes estimating the costs of labor, capital investment, overhead costs, and production or direct operating costs. Many of these costs are incurred before the crop is ready to harvest. In some crops, especially herbaceous or semi-woody plants, there is a relatively short time between investment and harvest. For the tree fruits and nuts, however, several years can elapse between investment and harvest.

Effective cultural management begins with the selection of proper cultivars and rootstocks. They must thrive in the climate conditions and soil of the area. Understanding the growth, flowering, and fruiting physiology of the plants is critical to implementing further cultural management practices. The life cycle (annual, biennial, perennial) of the crop determines many of the cultural practices that are implemented. Pruning practices rely on the pruner understanding when and where the plant develops fruit. A grower has to know if the plants are self-fruitful. If pollenizers are required, they must be planted appropriately.

Collectively, the cultural practices implemented for a fruit or nut crop are called the growing systems. There are major categories of growing systems that a grower can choose from. These are often developed for growing crops in specific regions, or for specific crops or specific purposes. For example, midwestern strawberries are produced as perennials and in Florida, as annuals. A Geneva double curtain bilateral cordon may be best for American and French-American hybrid grapes, while French vinifera wine grapes are often grown most successfully with the vertical shoot position system. Organic and nonorganic are other examples of growing systems.

The appropriate time for harvest depends on the stage of development of the fruits and whether or not the fruit will proceed with ripening after it is harvested. Carbohydrate content, firmness, respiration patterns, and use of the fruit all influence the stage at which a fruit is ready to be harvested.

FOOD FOR THOUGHT

1. Site selection is an important step in producing temperate fruits and nuts. List the criteria to be considered in site selection and explain why each is important.

2. How does temperature influence temperate fruit and nut production? What is chilling, and why is it important for many temperate fruit and nut crops? How does chilling requirement differ from dormancy?

3. Time between significant investment in a fruit or nut production system and income generated by harvest is influenced by crop type (herbaceous, semi-woody, or woody). How does the time between investment and income differ among the types?

4. What is rootstock selection, and why is it important for many cultivars of fruits and nuts?

5. What are the differences between organic and nonorganic growing systems?

6. Describe how the concept of fruit quality can differ between growers and consumers.

SUPPLEMENTARY READING

ATTRA—National Sustainable Agricultural Information Service. **http://attra.Ncat.org**

GALLETA. G. J., and D. G. HIMEBRICK. 1990. *Small fruit crop management.* Englewood Cliffs, N.J.: Prentice Hall.

JACKSON, DAVID I., N. N. LOONEY, and N. E. LOONEY (eds.). 1999. *Temperate and subtropical fruit production.* Cambridge, Mass.: CABI Publishing.

NATIONAL ORGANIC PROGRAM. **http://www.ams.usda.gov/ nop/indexIE.htm**

WESTWOOD, M. N. 1993. *Temperature-zone pomology, physiology and culture,* Third Edition. Portland. Ore.: Timber Press.

CHAPTER

20

Tropical and Subtropical Crops and Crop Production Systems

Richard Pratt

KEY LEARNING CONCEPTS

After reading this chapter, you should be able to:

♦ Recognize the principal components of the tropical environment that differ from those of temperate environments, and how they influence the manner in which crop production is undertaken in the tropics and subtropics.

♦ Understand the rationale for the adoption of diverse techniques and the manner in which these techniques are employed by small-landholder farmers to optimize crop production systems in diverse tropical environments.

♦ Recognize leading tropical fruit and nut species produced on both a large scale for export as well as in small settings near the home.

We often imagine the tropics as lush and full of exotic plants and animals—the tropical paradise evoked by images in film and advertising. In fact, the tropics and subtropics (the region of the earth lying within 30 degrees of the equator—between the Tropic of Cancer and the Tropic of Capricorn) comprise a broad expanse of climatic regions characterized by rich cultural and biological diversity. The majority of the world's population lives in the tropics, and some form of plant agriculture is a primary occupation.

The primary factors determining the feasibility of crop production are land forms, soil conditions, and climate. Climate exerts profound control over the distribution of crops and the manner in which they are produced. Some tropical environments are well suited to crop production, but others present harsh challenges to the producer. In many tropical countries, resource-poor farmers must also cope with poor soils. Crop production is feasible when adapted crops have been selected and/or the limiting factors in the environment can be modified. Many strategies are employed; for example, organic amendments or fertilizers are used to correct soil nutrient deficiencies, multiple crops are grown in the same field, and planting dates and harvest seasons are adjusted to the length of the rainy season.

Tropical plant agriculture assumes countless forms—from complex subsistence-agro ecosystems to large monocultures geared toward global markets. It would be impossible to cover the rich spectrum of crops (Figs. 20–1 and 20–2) and crop systems found in the tropics in one chapter. This chapter will introduce you to the principal tropical crops and to several complex polycropping systems that are of particular importance in the tropical and subtropical regions. This chapter should expand your understanding of how human ingenuity and crop diversity come together in tropical cropping systems.

CLIMATES OF THE TROPICS AND SUBTROPICS

The primary factors giving rise to different climates are the planetary pressure belts and their resulting system of winds and air masses (Strahler, 1975). The most important determinants of suitable climate for crop production are temperature and rainfall. Within the various climatic regions, the duration of arid seasons is important. The humid tropics are typically classified as those with dry periods less than 2.5

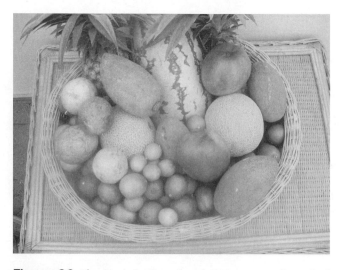

Figure 20–1   Fruit basket showing an array of tropical and subtropical crops. *Source:* Richard Pratt.

Figure 20–2   Market day in Bayaguana, Dominican Republic, with yams (*front*), sweet potatoes (*rear left*), and plantains (*rear right*). *Source:* Richard Pratt.

months per year. In the intermediate tropics, wet-dry seasons (dry 2.5 to 5 months) and tropical savannah (dry 5 to 7.5 months) are characteristic. In the dry tropics, such as the Sahelian region and semideserts, the dry seasons may extend 7.5 to 10 months. True tropical deserts are described by dry periods of ten to twelve months. Alternately wet and dry tropical climates (monsoon climates) occur between 5° to 15° north and south latitude. These intermediate regions have an alternation of a wet season controlled by intertropical convergence zone conditions at a time of high sun and a dry season controlled by continental tropical air masses at a time of low sun. These regions may be found in South America and West and Central Africa. Monsoon climates are wet and dry about six months of each year, depending on prevailing winds.

Wet equatorial climates characterized by deep, moist equatorial air masses, frequent rainfall, constant heat, no marked seasons, and rainforests are found 5°N to 5°S of the equator. Examples include parts of Brazil, Surinam, Guyana, and Venezuela; the Congo River basin; parts of Malaysia, Sumatra, Java, Borneo, the Celebes, New Guinea, and many Pacific islands.

Dry tropical climates are characterized by dry air masses descending and moving outward from the subtropical high-pressure cells over the large land masses; hot, arid climates; and desert areas. They are found between 15° and 30° latitude north and south. Examples include North Africa, the Arabian Peninsula, Iran, northwest India, Australia, and South America.

Variations of the main climatic type (Webster and Wilson, 1980) exist. Areas can have good rainfall (100 to 200 cm/year) distributed in two rainy seasons with only short dry seasons between them (these regions tend to be nearer to the equator, with no large seasonal fluctuations in temperature) or with two shorter rainy seasons and pronounced intervening dry seasons. There may also be considerable variation in annual rainfall (within 60 cm to125 cm/year) distributed as one fairly long rainy season (70 cm to 125 cm/year) and one long dry season, or one short rainy season and a long dry season.

Rainfall and Humidity

Throughout much of the tropics, the rainfall regime is the most important climatic factor influencing agriculture. It frequently has the largest effect on which crops may be grown in particular areas, the farming systems to be employed, and the sequence and timing of farming operations.

Rainfall in the tropics depends mainly on the movements of the air masses that cover the globe. The sun's heat produces a low-pressure zone that circles the earth roughly parallel to the equator and that moves north and south following the sun, with a lag of about four to six weeks. The zone of convergence is often called the inter-tropical convergence zone (ITCZ). When the ITCZ moves toward its northerly or southerly limits, the rate and extent of its movements affects the duration of the rainy and dry seasons. One of the most difficult things farmers have to contend with is the variation in rainfall. Most crop systems in the tropics are closely matched to the local rainy seasons. Shifts in rainfall patterns in the tropics could be devastating (Smith, 1992).

Rainfall generally decreases as distance from the equator increases. This variation is primarily an effect of temperature. At the equator, the atmosphere over one square foot of the earth's surface contains about 22.5 kg (50 lb) of water vapor; at 50°N latitude, it contains about 8 kg (18 lb); and at 70°N latitude, it contains about 2.0 to 2.5 kg (4 lb to 5 lb). The amount of water in the air (water vapor pressure) also falls with increasing elevation. High humidity reduces the evapotranspiration (water loss) demand on crops. High humidity over long periods, combined with high temperatures, favors rapid development and spread of fungal diseases of crops and of molds on stored produce.

Average annual rainfall received in tropical environments can vary widely. Land forms have a distinct effect on rainfall through the effect of altitude and the effect of mountains serving as obstacles to movement of air masses whether moving to (with moisture) or from (without moisture) land masses. In regions of equatorial or wetter monsoon climates—insufficient rainfall is not often a limiting factor to crop production. In parts of the tropics, a large proportion of the rainfall falls in heavy showers and storms of high intensity—much more so than in temperate zones. These heavy and frequent rains become a disadvantage when they increase the leaching of nutrients, carry away soil and applied fertilizers in surface runoff, or interfere with tillage or other operations. Rainfall is also important, especially in areas with wet and dry seasons—how well distributed and how reliable it is will be of paramount importance to crop production.

Light and Temperature

Higher plant population and higher yields are possible in environments with high radiation (assuming water is nonlimiting). Environments with less favorable radiation usually have more favorable soil moisture status.

The surface temperatures of some soils may increase considerably when they are exposed to intensive solar heating called **insolation** (a value expressing the amount of light energy striking one square centimeter of the earth's surface per hour). Soil temperatures may even become hot enough to inhibit or retard the germination and early growth of some crops. The amount of insolation depends primarily on latitude, altitude, season of the year, and a location's proximity to large bodies of land or water. Insolation at the equator varies only slightly from month to month because the angle of the sun's rays does not deviate greatly from vertical with the change in season, as it does at higher latitudes.

If water and nutrients are not limiting, the input of solar energy sets the ultimate limit to dry-matter production by crops. The solar energy input is highest in the tropics, so the potential productivity is highest. In the tropics, the reflection coefficients (loss of light available for photosynthesis due to reflection back to the atmosphere) of crops tend to be lower because the higher elevation of the sun results in light penetrating deeper into the canopy and more being absorbed after reflecting between leaves.

Insolation is less at increased latitudes because of the greater angle at which the radiation waves strike the earth's surface. (There can be some compensation due to longer days at northern latitude.) Because temperature varies with elevation, changes in altitude tend to substitute for changes in latitude. The rate of crop development decreases as altitude increases. For example, in lowland East Java, the time needed for rice to mature (from transplanting to harvest) can range from 90 to 120 days. The same cultivar at 1,600 m (5,250 ft) of altitude could take as long as 220 days because of cooler temperatures at the higher altitude.

Generally, the range of a species is limited by the lowest critical temperature in the most critical part of its life cycle—usually the reproductive stage. These critical temperatures range from chilling (typically a little above freezing) in the tropics to deep-freezing in the boreal regions (Smith, 1992). Low temperature is rarely a factor in tropical crop production except where it is markedly reduced by altitude. Some tropical crops are susceptible to high temperatures, but this susceptibility is usually associated with drought susceptibility. High temperatures combined with long periods of bright sunshine can also cause scorching of certain fruits, notably pineapples.

At the equator, the day length is almost constant at twelve hours throughout the year. Flowering is especially affected by the length of day versus night. Night temperature interacts with the photoperiod in influencing the flowering of many plants. Pineapple, for example, is a quantitative short-day plant in which low temperatures, especially cool nights, may enhance the effects of short days with a small diurnal temperature range (Nakasone and Paull, 1999).

Temperature regimes in the tropic include equatorial, tropical, tierra templada (frostless highlands), tierra fria (nonfrostless highlands), andine, semitropical, marine, temperate, pampean-patagonian, continental, polar, and alpine. An example of an equatorial climate is Madras, India, where the temperature during nine or more months exceeds 33.5°C (92°F). At the other extreme, Quito, Ecuador, just 12 miles from the equator at a height of 2,851 m (9,350 ft) and surrounded by snow-capped peaks, has an alpine climate.

Climatic Regions

Wet climates with tropical windward coasts are characterized by coastlines that are exposed to tradewinds, but with air masses that are shallow due to an inversion layer. Rainfall is variable. Climates of this type occur mainly along the east coasts of the Malagasy Republic (Madagascar), Central America (including the West Indies), and the northern part of South America.

Dry climates of tropical and subtropical west coasts are between 15° to 30° north and south and tend to be arid and cool. Examples include the northwest and southwest African coasts, and the west coast of South America.

The term *savanna* is used to refer to a variety of ecosystems. Tropical savannas include both open grassland with widely spaced shrubs or trees, and more closed woodlands. Rainfall (both amount and distribution), and to some degree the soil water-holding capacity, are the primary determinants of the density of the woody vegetation. Savannas cover many large areas around the world: much of central and southern Africa, western India, northern Australia, large areas of northern Brazil (known as *cerrados*), Columbia and Venezuela (known as *llanos*). Savanna is limited in Malaysia. Sometimes savannas are the result of human degradation of original forest land (e.g. central India).

Savannas typically occur on land with little relief. Soils found in these regions tend to be low in nutrients. This characteristic is partly due to the low value of parent material and also to the process of long weathering. Savanna regions are associated with a warm continental type of climate and are subject to large variation in soil water content. They may become extremely dry, making them prone to fires. The dominant vegetation tends to be fire adapted, and the life cycles of the woody species are usually short (typically not more than several decades). A notable exception is the baobab tree (*Adansonia digitata*).

Tropical rainforests are among the most productive ecosystems in the world and are known for their diversity of plant and animal life. They are restricted mostly to the equatorial climatic zone between latitudes 10°N and 10°S. The rainforests occupy regions of the world where it is warm and

rainfall can be measured by the meter! Annual variation in temperature and rainfall is minimal, but daily variations in heat, rainfall, and humidity may be large. Most tropical rainforests grow below 1,000 m altitude (3,280 ft). There are three main groups of rainforest: Amazonian basin (the largest), Indo-Malaysian area (west coast of India and southeast China through Malaysia and Java to New Guinea), and West Africa (extending to the Congo basin). There are also some small rainforests, for example, on the Hawaiian islands, South Sea islands, and the east coast of Madagascar.

There are many subtypes of rainforests. The most luxuriant is the multilayered lowland rainforest. At higher elevations, the lowland forest grades into mountain forests. Fingers of rainforest, called gallery forests, may follow river courses into savannas. The tropical rainforest can be divided into five general layers, but stratification is usually poorly defined. The large standing biomass is a store of nutrients protected from leaching.

Large areas of the rainforests are now being converted for forestry, ranching, and farmland uses. The relationship between high rainfall and soil erosion is a very important issue when tropical areas are cleared for crop production. The cleared land is exposed and vulnerable to erosion by heavy rains until vegetation can be reestablished. The high intensity of tropical rainfall events is due to the larger size of raindrops and the increased number of drops falling per unit time. When rainfall exceeds the infiltration capacity of the soil and it has high kinetic energy, the likelihood of erosive runoff is high. Charreau (1974, cited in Ker, 1995) determined that, on average, tropical rains have six to ten times more erosive power than temperate rains. Poor soil infiltration capacity can exacerbate the problem.

Considerable areas of tropical land are being increasingly exposed to erosion and nutrient loss as demands for food production increase. As clearing and pressure on the land escalate in response to growing human and livestock populations, more and more land is denuded of its natural vegetation. Cultivation, overgrazing, and the removal of trees all contribute to the dilemma. Eventually, as soils become increasingly less productive, cultivators are forced onto hill slopes and other marginal lands. The increasing proportion of the land that is left bare and exposed to high-intensity rainfall is prone to erosion unless urgent countermeasures are taken (Ker, 1995).

Crops may influence soil erosion potential as well— not only roots and surface organic matter but the canopy too. The greatest potential for agriculture in the tropics is in the 1.5 billion hectares (3,705,000,000 acres) where there is nearly constant temperature and a dry period not longer than four months, and a tropical rainforest (or seasonal tropical forest) as native vegetation (Sanchez, 1976). The major constraint to realization of the potential productivity in these regions is attributable to the poor-quality soils found there.

TROPICAL SOILS[1]

Soil conditions are also somewhat different in the tropics. The major soil groups in the tropics have been derived through a process of weathering or deposition of sediments by streams and floods or through recent volcanic deposits. Most tropical soils differ from those in temperate regions because they are low in organic matter, high in aluminum and manganese, low in cation exchange capacity (CEC), low in base saturation (the percentage of the CEC that is occupied by basic cations such as calcium, magnesium, potassium, and sodium—hence it is a good measure of how much of the CEC is being utilized to store plant nutrients), and low in nutrients (major and minor, especially phosphorus). In the humid tropics, where there is more intensive weathering, there is a more extreme loss of silica and bases.

Also typical of tropical soils is the presence of very little accumulation of organic matter on the surface because the organic matter tends to be rapidly attacked by primary decomposers. Termite mounds are common, especially in Africa. They cycle the organic detritus but may also become pests of crops that are planted following clearing. Trees tend to add some organic matter and nutrients to the soil beneath them.

Many of the parent materials of tropical soils are of granitic origin or of sedimentary rock. They generate soils of poor nutrient status. These parent materials also give soil a low iron content and poor structure. For example, most of Africa is underlain by a vast shield of ancient pre-Cambrian granitic rock. The shield has been lifted up and eroded down repeatedly over time. Kaolinitic clays form in most tropical soils that undergo leaching.

Important Soil Orders in the Tropics

Oxisols (Latosols in Brazilian Classification)

These soils are widespread in the tropics. **Oxisols** are highly weathered and leached, and they lack distinct horizons. They are found most extensively in Latin America and Africa. Brownish-red and reddish soils are common in East Africa. In Malaysia, the soils may be yellow (depends on the degree of hydration of ferric oxide). Most are high in clay content. Oxisols vary from coarse sands to fine clays and have a low CEC, low base saturation, and low pH. Because of leaching, they are usually low in most nutrients. The low nutrient retention capacity of oxic horizons is one of the major weaknesses of oxisols. Some cultural practices can alleviate this problem. CEC can be developed by protecting the soil surface from loss of organic matter through erosion, and by addition of organic matter (high CEC). Oxisols are high phosphate fixers, which means that they do not release phosphorus (P) very easily. In capital-intensive systems, large

quantities of phosphorus (P) fertilizers may be added, with continued applications in subsequent crops. Organic matter breaks down quickly in oxisols and mineralization is rapid. Adding lime is also beneficial as long as it does not unduly accelerate decomposition of the organic matter. Thus, these soils require careful lime and fertilizer management practices. They are free draining and tend not to have good water-holding capacities. Because of the low water-holding capacity, annual crops that grow in deep oxisols are more exposed to drought than in other soils of comparable clay content. Perennial crops (with deeper roots and better access to more stable subsoil moisture) are less exposed to drought stress in oxisols than are annuals.

Ultisols and Alfisols

These soils usually occupy younger geomorphic surfaces than do oxisols. The **alfisols** have considerably higher natural fertility than do the ultisols. The alfisols cover a large area in West Africa and much of East Africa. **Ultisols** are found in relatively more humid areas of Africa such as Liberia, Guinea, Uganda, and the eastern portion of the Democratic Republic of Congo. These are typically red and yellow soils.

Usually within 1 m (3.3 ft) of the soil surface is a zone with a significant increase in clay. The higher clay content of the **argillic** horizon (the most common type of B horizon in soils) can slow down water percolation. The coarse-textured surface may develop a crust during the dry season that may be difficult to till. In general, these soils have more weatherable minerals than do oxisols. Phosphorus fixation can be severe in ultisols because of high amounts of exchangeable aluminum (Al). Potassium is seldom deficient except after several years of intensive cropping. Alfisols can sustain more cropping than ultisols because of higher weatherable minerals, CEC, exchangeable calcium (Ca), and magnesium (Mg).

Vertisols

This soil order is characterized by high clay content and poor physical structure. **Vertisols** typically contain over 30 percent clay. They are known for their propensity to shrink and swell with changes in moisture content. This process creates typical soil heaving and cracking. When they are wet, they become very sticky; and when they are extremely dry, they may crack and break into blocks. These dark gray and black soils have common names including cracking clay, grumusols, black cotton soils, and regurs.

The largest expanses of vertisols are located in Australia, India, Sudan, and parts of Ethiopia—usually on flat or level topography in savanna regions. They occupy approximately 4 percent of the land area of the tropics. Marked dry and rainy seasons are needed for their formation. They tend to be high in Ca, Mg, and potassium (K) but are low in nitrogen (N) and P. Vertisols tend to have high water retention capacity, but because of their high clay content they hold the water so tightly that it is not readily available to plants. Their

---

[1]Wrigley (1982) and Norman et al. (1995) were used as references throughout this section.

cracking and swelling characteristics make them hard to till and those forces may prune and damage crop roots.

Andisols

Andisols are glassy, highly permeable soils derived from volcanic ash. They may occur at any elevation where volcanoes are active. Usually they are very young soils, of low bulk density and variable texture, and with highly stratified layers. Internal drainage is excellent and they possess a high nutrient reserve. These soils are generally the most productive upland soils in the tropics, although N and P deficiencies can be problems. They represent a small fraction of the land, but they tend to be highly farmed because of their suitability for crop production.

Alluvial Soils

Alluvial soils occur in river plains, estuaries, deltas, and low coastal regions. They are the result of relatively recent sedimentary deposits and have little or no horizonation (differentiation of soil profiles). **Alluvial soils** may show considerable stratification and occur in complex patterns because they are deposited by flowing water. Their quality is highly dependent on the source material from which they were derived. They tend to be less desirable than the alluvial soils found in temperate climates. Floodplains may harbor diseases, and flooding is generally uncontrolled, except where rice cultivation is practiced. Major areas of occurrence are along river systems such as the Ganges, Mekong, Congo, Niger, and Amazon.

Inceptisols and Entisols

These are pedologically young soils. The horizons of **entisols** are only slightly developed or underdeveloped. Entisols are widespread in southwest Africa. Some are productive soils, whereas others may require artificial drainage and flood control if they are cropped. **Inceptisols** are present in the Central African Republic, southern Sudan, and parts of India. They are slightly more developed than entisols and not as strongly weathered.

Spodosols (Red and Yellow Podzols in Other Classification)

These soils are found in two unique tropical areas. One is at high altitudes where the areas have often been influenced by volcanic ash. Cool tropical spodosols are typical for mountain slopes that are hit by rain-producing clouds. These soils may also be found in the low tropics (e.g., quartz sand beaches in southeast Asia) and on the lower footslopes of recent valleys in lowland rainforests. **Spodosols** are highly leached, are low in nutrients, and display poor water-retention characteristics. They are not well suited to agricultural endeavors. Erosion is greater than on oxisols because of poor permeability. High elevation spodosols are usually best protected by forest vegetation.

Aridisols (Sirrozems, Desert, and Red Desert Soils)

Aridisols soils are formed in areas where there is low leaching as a result of low rainfall. They are fertile soils, although they lack nitrogen. Calcium carbonate accumulates near the surface.

Tropical soils can present unique challenges to cropping systems. A major problem is their inherent infertility. Nutrient deficiencies are manifested in many regions of the tropics. For example, of the soils in tropical America, 96 percent are phosphorus deficient, and 89 percent are nitrogen deficient.

The oxisols, ultisols, alfisols, and entisols comprise about two-thirds of tropical soils and they are basically infertile, acidic, and with low base saturation. Also, the poor organic matter and poor infiltration characteristics of many tropical soils, and the heightened magnitude of the intensity of tropical storms, makes them especially vulnerable to erosion.

The increasing demands of population pressure and industrial development have resulted in the loss of protective vegetative cover in many regions. The rate of deforestation in the moist tropical regions during the early 1990s was about 8.5 million hectares (20,995,000 acres) per annum. Potentially important consequences of large-scale deforestation, aside from direct erosion losses and loss of biodiversity, are global warming and the resultant potential disruption and increased variability of the wet/dry seasons in tropical areas. If these consequences were to arise, they could have dramatic effects on crop production in the tropics and subtropics. Experts have advised the adoption of sustainable cropping systems and reductions in deforestation. They advise that the consequences of delaying these practices are increased economic and environmental failures in the tropics (National Research Council, 1993).

CROPPING SYSTEMS

Shifting Cultivation

Shifting cultivation is also referred to as slash-and-burn, or swidden, agriculture. This system is highly variable and may be found in the forest, bush, or savanna regions. Traditionally, individuals do not own the land. Instead, the community crops the land together. *Shifting* refers to the fact that people may change the location of cultivated fields around a permanent home. Many populations do not rely exclusively on shifting cultivation and may apply it only on locations more distant from the permanent dwelling.

One essential feature of shifting cultivation is that nutrients held in the vegetative biomass are released by cutting, drying, and burning. Litter fall, comprised primarily of the old leaves of shrubs, woody vines, and trees, is considered the most important because it is nutrient-rich. During the fallow period (when fields are left to return to natural vegetation), nutrients are taken up by the vegetation and then returned to the soil surface as organic litter when the plants die (National Research Council, 1993).

Under this system, in tropical forest regions, woody vegetation is felled and burned during the dry season. The ash is left in place and the crops are interplanted (not in rows) with digging sticks and/or hoes. Livestock are usually not part of the system. Yields usually decline by the third year, and fields are abandoned after three to four years to allow the natural vegetation to regenerate, and the cycle is started again elsewhere. In many areas, the fallow period is no longer of sufficient duration to recharge the nutrient levels to satisfactory levels before they are again brought into cultivation.

In traditional West African forest region systems, lianes (woody vines) and small trees are chopped down, but larger trees are left to provide light shade. Maize is planted into the forest topsoil using a digging stick. Weeds are removed. The maize is intercropped or followed with cassava (*Manihot esculenta*), cocoa (*Theobroma cacao*), yams (*Dioscorea* spp.), tannia (*Xanthosoma* spp.), and bananas (*Musa* spp.). Small areas are planted with a wide range of vegetables.

In a bush system, stumps and large roots are allowed to resume growth and will attain about 6 m (15 ft) height after five years, 10 m (25 ft) after ten years. The resting period may vary from five to twenty years; an eight-year resting period is considered necessary to restore fertility after three years of cropping. The longer the regrowth is allowed, the better is the subsequent level of nutrients available to the crop.

In a tallgrass savanna, the soil needs more vigorous cultivation to deal with the grass roots. Soil is traditionally scraped into mounds using hoes. The mounds are then planted with climbing yams and interplanted with maize (*Zea mays*), beans (*Phaseolus* sp.), and vegetables. The yam mounds are broken down the second year, and sorghum and maize are planted on narrow ridges. Peanuts (*Arachis hypogaea*) are interplanted with millet (*Pennesitum americanum*) in the third year.

The rate of decline of crop yield is related to the natural fertility of the soil, the amount of nutrition contributed by the ash, and how demanding the crops are of nutrients. Usually the cash crops or preferred subsistence crops are grown during the first phase of the cropping cycle, followed by legumes in the later stages as fertility declines, and finally by nutrient scavengers such as cassava toward the last stages of the cropping cycle.

Weed pressure tends to increase during the cycle so taller, robust crops that compete well with weeds may also be chosen toward the latter part of the cycle. Some biennials or perennials may also be grown with the hope that they may persist even after the field has been abandoned. Cassava fits the bill in every regard.

Typically, the most diverse and irregular cropping patterns in shifting cultivation systems are to be found in the more humid, lowland forest areas. Where the soil water regime is less favorable, the crop diversity usually decreases and the planting patterns become somewhat more regimented.

The exploitation of microclimatic conditions in these systems may indeed become "ingenious," as described by Norman et al. (1995). For example, shade-tolerant species are planted on the perimeter of a plot, and moisture-demanding species on the bottom of sloping sites, and fertility-demanding species where the most ash has accumulated. In addition, plants may be planted at various positions on a mound of hoed-up topsoil, vines may be planted on upright tree trunks, and so on. These systems, now considered to be ingenious, were once described as "primitive" in a pejorative way by so-called experts.

Today, there appears to have arisen a greater appreciation for the human-generated precision farming techniques that have been refined for centuries: "The shifting cultivator has always required a detailed knowledge of the environment to survive" (Allan, 1965). The shifting cultivator typically has a wealth of ecological information, such as which types of vegetation indicate physical or nutritional characteristics of the soil, how long a soil may be cropped, and how long it should remain fallow before it can be cropped again.

Shifting cultivation is practiced on about 30 percent of the world's arable land and about 45 percent of the tropical land area. In the past, when population pressure was not high and land tenure was not an issue, these systems were considered sustainable. Shifting cultivation was at one time able to provide sufficient food and products to satisfy the cultivators' needs while maintaining soil fertility for the long term.

The critical elements of the system were that desirable locations were not limited and the regrowth (or fallow) period was long enough to maintain soil fertility. With increasing population pressure came a progression toward increasing cropping intensity, and declining periods for regrowth, reduced species diversity, and increasing soil degradation. Typically, when farmers are confronted with low productivity due to declining soil fertility, they switch to less demanding crops, for example, cassava. Faced with ever worsening productivity however, farmers then begin expanding the practice to newer and usually more marginal lands (e.g., to steeper slopes and drier areas). If that doesn't work, they attempt to increase the nutrient inputs or resort to conservation techniques that slowly build up the fertility of the soil once again. The hesitation to adopt conservation techniques is probably related to the increased labor demands and constraints those practices may place on planting patterns, and so on.

Nye and Greenland (1960) studied three African forest sites and noted that the highest return from litter was of nitrogen (about 180 kg/hectare or 161 lb/acre) and the lowest was of phosphorus (at 6 kg/hectare 5 lb/acre). However, when the vegetation was cut down and burned, much of the nitrogen and sulfur was lost to the atmosphere. After burning, the input of calcium was the highest (at 75 kg/hectare 67 lb/acre), nitrogen was a little less, and phosphorus was still lowest (at 6 kg/hectare 5 lb/acre) (Sanchez, 1976).

Typically, the vast majority of African farmers cultivate 0.2 to 0.4 hectare (0.5 to 1.0 acre) of subsistence crops per person. A family of five to six people would cultivate approximately 1 to 5 hectare (2.5 acre to 12.5 acre) to provide some surplus as well. Problems arise, however, when novices begin the process of slash-and-burn agriculture. Often, in these instances, or when indigenous systems are perturbed (e.g., forced to marginal areas because of increased population or development schemes, etc.) the systems become unsustainable or further degraded.

Making Shifting Systems More Sustainable

Some practices can make these systems more sustainable, for example, low impact land clearing, the use of mulches, plant cover crops, understory crops, and reduced tillage planting techniques. The proportion of legumes as food crops, cover crops, and fields in fallow can also be increased. Improved fallow management techniques may be useful, and breeding crops especially for cropping system, for example, aluminum tolerance, less sensitivity to low pH, and so on, can sometimes have a positive impact. In some areas, farmers practice contour cropping or use integrated pest management (IPM) techniques or even agroforestry. It is also important not to abandon mixed cropping.

The speed and effectiveness of the regeneration of fertility can be improved in response to increasing population pressure. Farmers may sow seeds of, or plant, leguminous trees. The cultivation cycle may be ended with a shade-producing crop such as plantain (*Musa* spp. AAB group) or cassava (*Manihot esculenta*) to aid tree establishment. Also, ending the cultivation cycle with pigeon pea (*Cajanus cajan*) provides nitrogen, and it is a good nutrient accumulator. Farmers can plant selected grass species for the fallow period, and grazing during the resting period can be beneficial.

Fallow System

Where land tenure is held by individuals, shifting cultivation tends to be replaced by a fallow system. The fallow system may be found in the forest, bush, or savanna and is more important in Africa than is shifting cultivation. Grass fallows are common in East African savanna regions. Bush fallows are common in West African and South American savanna regions. Boundaries of fields (2 to 4 hectare or 5 to 10 acre) are usually present near a permanent house compound. Changes in cropping patterns are associated with a fallow system. Intercropping systems are often arranged in a more ordered or geometric pattern than in shifting cultivation. Relay cropping is found in the fallow system (delay planting, overlapping crops) and differs from sequential cropping (one crop following another during the same year).

Brush is burned during the dry season. Rains are then necessary to soften the soil, allowing hoe cultivation. Planting may be delayed, resulting in a missed nitrogen release (often termed a flush) into the system. Cash crops may be grown but the high labor demand for weeding and harvest periods may delay planting of cash crops. Typically cropping systems in the wetter tropical climate areas are based on non-cereal energy corps, which changes to systems based more on cereals in the wet and dry climate areas. The cycle time is

usually four years cropped followed by four years fallow. Animals are usually supplementary, grazing communally on undesirable land.

Permanent Upland Cultivation

Permanent upland cultivation is often practiced in semiarid regions with distinct dry seasons. Planting is delayed until the onset of the rainy season appears to be established. Varieties are chosen as much for reliability as for their yield. In areas with longer rainy seasons, intercropping may be practiced and cattle may be integrated into the overall farm system. Farmers often attempt to irrigate a portion of their fields, but there is typically little investment in agrochemical inputs due to the high-risk factor. Legumes are emphasized in the rotation, as is the use of green manure crops (typically legume crops with vigorous growth that can suppress weeds and provide valuable nutrients for the subsequent crop). Farmers also allow some legume trees or shrubs (e.g., *Acacia*) to remain for fodder and nitrogen fixation. In more humid regions, it is hardly possible to maintain fertility without fertilizers or animal manure. Weeds also tend to grow out of control in permanent systems.

Arable Irrigation Farming—Upland Cropping Systems

This system is typically irrigated by gravity-flow systems with water from a reservoir or stream (sprinkler systems are less common in the tropics), which bring water to fields by a system of canals and ditches. The water is distributed through ridge and furrow systems in fields, and surplus water is distributed through natural drainage systems (e.g., a river). Irrigation allows the production of multiple crops per year and is used for high-intensity production of high-value crops.

EXAMPLES OF TROPICAL CROP PRODUCTION SYSTEMS

Flooded Systems—Tropical Wet Rice

Wet paddy rice systems have supported dense populations in southeast Asia for many centuries. Rice is an ideal crop in many ways—it stores well, needs little fuel for cooking, and is convenient and nutritious. It is also uniquely and highly adapted to flooded conditions. Wet rice systems can be subdivided into various categories: shallow-water rice only; shallow-water rice with upland crops; and long-standing floodwater rice. The degree of water control varies widely. The farmer may grow one or more crops per year.

Rice is the most important food crop in the world. Several species of *Oryza* are cultivated, but *O. sativa* is the most widely grown. The species consists of three ecogeographic races: indica, japonica, and javanica (Vergara, 1992). The indica types have been bred extensively and are now highly productive. They are grown widely in wet-rice systems. The predominant systems are irrigated (wet season) and the rainfed (shallow) system found in Asia.

The average rice farm is rather small and for many years existed as a subsistence system, with little rice sold off the farm (Vergara, 1992). Notable exceptions are the vast wet rice systems of Thailand and Myanmar, where much of the crop is exported. High-yielding varieties were introduced in the 1960s and many of the traditional systems changed. The Green Revolution cultivars were highly responsive to fertilizers and farmers began purchasing off-farm inputs for their systems. The farms' sizes have not changed dramatically, but intensification of the systems has increased dramatically in many areas. The changes have brought with them myriad benefits and costs.

In addition to higher-yielding cultivars, the availability of earlier maturing cultivars has allowed changes to the wet rice cropping systems. Some of the more important one-year systems that have been adopted include (1) one rice crop followed by a season of fallow, (2) one rice crop and one cash crop during the dry season, and (3) two rice crops (where irrigation is available or rainfall is sufficient). A two-year cycle may be planted with up to five continuous crops of rice—although this is a clear invitation to outbreaks of pests or diseases. The main cropping pattern in the rice bowl of southeast Asia and the rain-fed lowland areas of Indonesia and Bangladesh is still one rice crop during the rainy season or two rice crops; one during the rainy season followed by another using irrigation during the dry season.

Land preparation for rice is unique because manipulation of the field to achieve appropriate water management during the cropping cycle is extremely critical. In contrast with farmers in temperate regions who strive to preserve good soil tilth, lowland rice farmers prepare fields for planting by puddling the soil (purposely working the soil while it is wet to break down its structure) to *decrease* water infiltration. Maintenance of optimal water levels in the field is critical to successful rice cultivation, and excess loss through infiltration is to be avoided.

When water has been drained off the puddled soil, rice is then transplanted by hand from nurseries. Direct seeding of dry seedbeds, or casting of pregerminated kernels into shallow water are techniques also practiced. The initial planting operations must be timed to coincide with the commencement of the monsoon season in tropical areas. Puddling of the soil and transplanting are considered to be good practices from a weed-management standpoint. If weeding is not performed during the first forty days after planting, yields can be adversely affected.

Yields during the wet season are actually lower than those during the dry season (assuming irrigation has taken place). The lower yields are ascribed to less than optimal partitioning of energy into the canopy versus the grain during the wet season. Lower night temperatures during the wet season are thought to be related to this phenomenon (Vergara, 1992).

Rice blast is considered to be the most devastating disease of rice. It can kill seedlings and infect the panicles, resulting in appreciable yield losses. Many diseases of rice are incited by viral pathogens, perhaps the most serious of which is rice tungro. Two viruses, vectored by several insects, are causal agents of the disease. Small landholders are still the primary producers, so host resistance is the most appropriate strategy for management of these pathogens.

Harvesting of rice in lowland tropical systems is still primarily by hand. Mechanical threshing and winnowing is gaining in practice, although in many villages, hand or animal threshing is still performed. Sun drying is used to reduce the moisture content to around 14 percent so that the rice may be stored safely.

The limits of expansion are being reached for rice production in Asia, but in Latin America and Africa, additional land could still be brought into production. In Asia, additional population growth will require further intensification of production systems in a manner that could become unsustainable.

Cereal-Based Systems of Semiarid West Africa

The cereal cropping systems of West Africa are based primarily on pearl millet (*Pennesitum americanum*) and sorghum (*Sorghum bicolor*). Maize may be added in the higher rainfall or more fertile environments. Cowpea (*Vigna sinensis*) and peanut (or groundnut; *Arachis hypogaea*) are important grain-legumes traditionally interplanted in the cropping system. These cereal agroecosystems comprise the critical food base for Senegal, Mali, Burkina Faso, Niger, Nigeria, Chad, and Cameroon.

Rainfall is a primary determinant of the cereal of choice, and its distribution shows a steep north-south gradient from 40 to 120 cm (16 to 47 in.) annual precipitation. Pearl millet is grown mainly in the most arid regions receiving only 20 to 80 cm (8 to 31 in.) precipitation, sorghum in those regions receiving between 70 and 140 cm (28 to 55 in.), and maize in those with excess of 100 cm (39 in.) (Fussell, 1992). Rainfall events in this region are typically highly variable and unpredictable due to the randomness of convective storms. Rain frequently occurs in short, intense storms. Droughts of various degrees of magnitude and duration are to be expected.

The high radiation load in the semiarid tropics of West Africa also results in elevated temperatures. The temperatures also increase from north to south in an inverse relationship with annual rainfall. In the Sudano-Sahelian region, mean monthly maximum temperatures can exceed 40°C at both the time of sowing and harvesting. Extremely high temperatures capable of causing seedling death, poor stands, and the need to resow crops sometimes occur. These harsh and uncertain climatic conditions pose a constant threat to stable food production.

Cereals are typically planted in a variety of soils. Aridisols, alfisols, and entisols are found in millet-based systems, and alfisols and small areas of vertisols, entisols, and inceptisols are utilized where sorghum and maize-based systems predominate. Millet is the crop of choice on the sandier soils. A wide variety of soil textures is employed in sorghum production, and maize is the preferred crop on the heavier clay soils, where rainfall is not limiting. Upland soils are generally preferred for millet production, and lowland soils with better organic matter content and water-holding capacity are more desirable for sorghum and maize production. Systematic utilization of the soil variation is a key element of the local farming strategy and is considered a major factor contributing to the stability of crop production (Fussell, 1992).

Cereal-cropping systems in semiarid tropical West Africa experience a seasonal availability of nitrate-nitrogen. The rapid release or flush occurs at the onset of the wet season due to the burst in soil microbial activity. Farmers must work hard to ensure their crops are planted in a timely manner to receive the full benefit of the nitrogen while it is available. Marked deficiencies of phosphorus are common in the alfisols of West Africa. The farmer's strategy must be carefully attuned to the low inherent fertility of sandy alfisol soils where millet is grown. Fussell (1992) summarized studies that have shown phosphorus and nitrogen deficiencies often may be even more limiting than rainfall in the Sahel (the northernmost and driest region of the semiarid cereal zone). In these very poor soils, modest amounts of fertilizer can result in substantially increased yields.

Cultural and economic constraints also limit the degree to which the cereal-based agroecosytems may be altered (Fussell, 1992). The basic social unit in many traditional societies is the compound, or assemblage of living areas organized along familial lines. Compound members may work on both community fields and their own fields. The farms are generally from 2 to 25 hectares (5 to 62 acres) in size. Traditionally, decisions concerning crop production are formed in the context of this complex unit of related families. The size often depends on whether traditional complex family units or simple nuclear family units are operating them. Renting, pledging, leasing, and purchasing of land are becoming more common, but farming remains largely at the subsistence level, with hand labor the primary input. Labor bottlenecks are common during the rainy season, especially for weeding, land preparation, and planting. Little capital is available in most enterprises for either chemical fertilizers or hired labor.

Traditional fallow-rotational systems permit the buildup of organic matter and make nitrogen and phosphorus available. Without a fallow period, reliance on the planting of legumes such as cowpea and peanuts to enhance the system becomes even more critical. Practices such as accumulating removed weeds in mounds or ridges between cereal rows are useful for adding organic matter. Another traditional method is the burning of crop residues and cut brush. Animal wastes are also added to the fields, either by transporting them or through relationships with herdspeople who graze their cattle on the farmers' fields for payment in-kind or for cash. These relationships are more common in areas with lower population and less intensive cultivation (Fussell, 1992).

With the increase of permanent cultivation and decline of shifting cultivation, soil quality has become even more critical. Continuous cropping without fallow periods or added inputs of organic and inorganic fertilizers has lead to the depletion of soil fertility and the lowering of yields (Fussell, 1992).

Environmental conditions are primary determinants of whether or not disease occurs on millet and sorghum. The timing of flowering is critical. The start of the rainy season, and hence the planting date, is always a bit uncertain. Photoperiodicity of local varieties has ensured that flowering commences around the expected end of the rainy season so that grain maturation occurs during dry weather. Disease incidence tends to be higher with increasing rainfall. The primary disease of sorghum and millet is downy mildew (*Sclerospora graminicola*). The primary insect pests of pearl millet are stem borers (*Acigona ignefusalis*) and earhead caterpillar (*Raghuva albipunctella*). An additional stem borer species (*Buseola fusca*) and sorghum midge (*Contarina sorghicola*) attack sorghum (Fussell, 1992).

Mixed Annual/Perennial Systems

In the majority of regions where tropical root crops are cultivated, these crops are typically produced in mixed-cropping systems (Wilson et al., 1992). Among the more important systems are the mixed-root crop systems in wet sub-Saharan Africa (Juo and Ezumah, 1992), the yam-based hill-land system in Jamaica, (Wilson et al., 1992), and the intensive systems of Asia. Asian tuber production systems tend to be intercropped with cereals and legumes and are highly intensive. Planting is usually on fertile alluvial soils, and purchased inputs such as fertilizers and pesticides are frequently utilized (Juo and Ezumah, 1992). The African and Caribbean systems are low-input, extensive systems and are focused less on economic returns than on maximizing benefit from limited resources.

African Mixed Root Crop Systems

Tropical root crops are diverse and ancient crops. Some tropical tubers may well have been the first domesticates (O'Brien, 1972). The major tropical root crop species are cassava, true yams (*Dioscorea* spp.), sweet potato (*Ipomea batata*), the edible aroids dasheen and eddoe (ARACEAE family), tannia (*Xanthosoma* spp.), and white potato (*Solamum tuberosum*). Root crops are traditionally cultivated in mixed systems on small family farms. They are important as an energy source for people, as survival crops, and as industrial crops.

Cassava and both yellow guinea yam (*Dioscorea cayensis*) and white guinea yam (*Dioscorea rotundata*) are of considerable importance in the yam zone (wetter regions) of sub-Saharan West Africa. Globally, cassava production far outstrips that of yams. Nigeria is the biggest producer of both cassava and yams. The next highest world producers of cassava are Brazil, Indonesia, and Thailand. Other world-leading yam producers are Ghana and Côte d'Ivoire (Ivory Coast). Traditional mixed-root cropping systems planted on newly cleared land in the forest zone are generally planted with a mixture of three or more species. Yam, cassava, maize, or upland rice usually predominate in these mixed cropping systems. The choice of yam or cassava is often predicated on the amount of time the field has spent in fallow prior to clearing, and the inherent quality of the soil. Cassava is selected for poorer soils or following shorter fallow periods, whereas yam is planted in the better soils following a longer fallow period. Yam requires more laborious and costly crop husbandry than does cassava (Hahn, 1984). Cassava is easily propagated and more drought-tolerant than yam and this too can affect farmers' selection. Yam is still highly favored in many African societies (Juo and Ezumah, 1992), but cassava cultivation is expanding, whereas yam production is remaining relatively constant.

Wetland yam mound systems are centered around the planting of white yams (*Dioscorea rotundata*) on the top of large mounds averaging 1 meter in height. The associated annual crops and vegetables are then planted at different positions on the sides of the mounds depending on their ability to tolerate waterlogging (Juo and Ezumah, 1992). Wetland rice can be planted between the mounds where flooding is apt to occur during part of the growing season.

Cassava-based mixed systems are planted mainly in the fields most distant to the family dwelling. The system tends to be less diverse than yam-based mixed cropping systems, and frequently only one additional crop is planted to take advantage of the open area prior to ground covering by the cassava crop. Common crops planted with cassava are maize (e.g., in Nigeria and Ghana), upland rice (e.g., Liberia and Sierra Leone), and groundnut (e.g., Democratic Republic of Congo and the Central African Republic.)

Maize and rice are interplanted at a reduced density, and secondary crops such as cocoyams, leafy vegetables, melons, okra, or legumes are used to fill in remaining open areas. The species diversity in this system may sometimes be impressive—with as many as fifty species identified in farmer plots of southeastern Nigeria (Okigbo and Greenland, 1976).

Rotation with bush fallow is the most common practice, but true shifting cultivation is becoming less frequent because of increasing pressure on the fields to produce more food for growing populations (Juo and Ezumah, 1992). Thus, pressure to produce more food decreases shifting cultivation with regenerative fallow periods, resulting in reduced field productivity over time. The precise times of cropping and fallow thus vary considerably depending on soil type and the amount of pressure on the land to produce adequate food supplies.

Caribbean Mixed Root Crop Systems

Major root-crop production in the Caribbean is found in the larger islands of the Greater Antilles (Cuba, Jamaica, and Haiti, and the Dominican Republic). The most important center of yam cultivation in the Caribbean is Jamaica (Wilson et al., 1992). The smaller islands of Dominica and St. Vincent also have intensive production on a per capita basis (Wilson et al, 1992).

Jamaican hill-land culture of yams typically includes multiple varieties representing several species of yam, often intercropped in the same field. Typical planting is on hills or mounds approximately 20 to 90 cm (8 to 35 in.) in height and spaced from 1.5 × 1.5 m to 3.0 to 3.0 m (approximately 5 × 5 ft to 10 × 10 ft) apart. Intercropping is done both on the hills and in the space between hills. Yams are set at the top of the hills, and shorter-season crops such as tomatoes or *Phaseolus* beans are planted at the base of the hills. A variety of crops, including white potatoes, tannia, sweet potatoes, and *Amaranthus* spp., may be planted in irregular patterns between the hills. The land is typically cultivated for three to five years and then left fallow for one or two years.

The system is designed to provide yams year-round by concentrating the planting of rotundata yams from November to March, and greater Asian (*D. alata*), cush-cush (*D. trifica*), and lesser Asian (*D. esculenta*) yams from April to June. Throughout the year cayenensis yams may be planted. Yams are typically staked with poles 1 to 6 m (about 3 to 20 ft) long. Weed control is achieved by mulching and employing hand-weeding as needed. Yams are usually available for harvest approximately six to twelve months after planting. Some of the first yams are used for propagation and then are immediately replanted. A major strength of the system is the inherently different rhythms of the different yams and their adaptation to the bimodal rainfall pattern. Differences in the dormancy and sprouting of tubers, combined with differences in growing seasons, lend themselves to the harmonious rhythm of the system. The major costs include capital for yam planting and staking materials. In this system, the extensive amount of labor required for preparation of the hills limits the extent to which this system can be utilized.

Comparisons Between Systems

The methods of clearing the land are much the same in Jamaica and in Africa, with hand-labor predominating. Seedbed preparation techniques involve the formation of small mounds or low ridges in shallow or infertile soils so that mineral nutrients and organic matter can be aggregated shortly after clearing and burning. In more fertile alluvial soils, large mounds 0.5 to 1 m (3.3 ft) in height are constructed to avoid flooding (Juo and Ezumah, 1992).

Key management considerations of mixed cropping systems hinge on the ability of the farmer to minimize competition for nutrients and light. Where regions are characterized by strongly weathered and infertile soils, the mixed root crop systems offer several advantages over monoculture systems. They include higher food and nutritional production on small areas of land, lower risk of crop failure due to pests and diseases, higher utilization of soil nutrients on land previously left fallow and cleared using traditional methods, and the stability of crop yield.

The major advantage of cassava-based systems is the inherent capacity of cassava to withstand poor environmental conditions and still produce good crops. The major disadvantage is the need to detoxify the tubers. A unique feature of mixed cassava systems is the opportunity to mix several crops of varying life-cycle duration. This technique results in a gain in total yield in time and space. Traditional cassava cultivars are upright and ground coverage is insufficient for adequate weed control. It does permit the opportunity to interplant other crops of shorter maturity. Crops such as maize, melon, cowpea, and beans provide rapid coverage that can suppress weeds, but they complete their life cycle before serious competition with cassava is realized.

Pests and Diseases of Root and Tuber Crops

In Africa, both nematodes and termites are important pests of yams (Juo and Ezumah, 1992). The major disease problem with rotundata yams in the Caribbean is tuber burn, which is induced by nematodes. Colletotrichum leaf spot is the major foliar disease of alata yams in the Caribbean, and both anthracnose and Colletotrichum leaf spot are important diseases in Africa. Most *D. alata* cultivars are susceptible to anthracnose, but there are a few *resistant cultivars*. *D. rotundata* is more susceptible to yam virus than are other yam species. Common diseases and pests of cassava in Africa are cassava mosaic virus (CMV), bacterial blight (*Xanthomonas campestris*), variegated grasshoppers (*Zonocerus* spp.), and silver-leaf white fly (*Bemesia tabaci*) (Juo and Ezumah, 1992). Programs to provide inexpensive disease- and nematode-free planting material can have a tremendous impact on the efficiency of production.

PERENNIAL CROPS

Avocado (*Persea americana Mill.*) LAURACEAE

A majority of the roughly 2,000 or more species in the LAURACEAE family are evergreen trees or shrubs, and many are valued for their ornamental appeal. The genus *Persea* is perhaps the best known due to the evergreen avocado trees that are native to Mexico and Central America. Avocado trees are fast-growing, reaching heights between approximately 9 and 20 m (30 to 66 ft). The tree canopy ranges from low, dense, and symmetrical to upright (some cultivars are columnar). Broad canopies are usually formed, and some varieties may become asymmetrical. Tree limbs are easily broken by strong winds or heavy crop loads. Leaves are variable in shape (elliptic, oval, lanceolate), and they are often pubescent and reddish when young, becoming smooth, leathery, and dark green when mature.

Avocados have been cultivated in tropical America since pre-Columbian times. Their range probably extended from the Rio Grande River in the north to central Peru in the south before the arrival of Europeans. Avocados, also known as *aguacate* (Spanish) or alligator pears, are assigned to three ecological races: Mexican (*Persea americana* var. *drymifolia*), Guatemalan (*P. americana* var. *guatemalensis*), and West Indian (*P. americana* var. *americana*). The races differ in several traits including cold tolerance, salinity tolerance, iron chlorosis tolerance, fruit size, skin thickness, oil content, and flavor. The leaves of Mexican types have a pronounced anise scent when crushed. There are no known sterility barriers among the three races or among any members of the *P. americana* complex.

Mexico is the leading producer of avocados, followed by the United States, Chile, South Africa, Spain, and Israel (FAOSTAT, 2004). Significant avocado industries also exist in many other tropical and subtropical countries of the world. The majority of Mexican avocados are produced in the state of Michoacán. California and Florida are the main producers in the United States. During the last several years, US production has been declining, whereas that of Mexico and Spain has been increasing. Importation has been limited primarily to the United States and Europe. Japan has begun to import large volumes of avocados, but it is the only Asian country to do so thus far. Chile is the principal exporter of avocados to the US market. The chief factors limiting avocado consumption worldwide are lack of familiarity and high cost.

Avocado production is limited by its sensitivity to climatic extremes and its vulnerability to root rot. Trees of the West Indian race produce well in tropical climates, but they freeze at or near 0°C (32°F). The other two races generally fail to flower or set fruit in the tropics. On the other hand, the West Indian race sets little or no fruit in subtropical climates, such as that of southern California. In regions where minimum winter temperatures of −5.5°C to −3.5°C (22°F to 26°F) or below occur, only trees of the Mexican race can be expected to survive. Guatemalan types are native to cool, high-altitude tropics and are hardy to −1°C to −3.5°C (30°F to 26°F). If the proper race and cultivar are chosen, avocados thrive and produce well in climatic conditions from truly tropical to the warmer parts of the temperate zone.

Avocado cultivars must be propagated vegetatively to retain their characteristics. Avocado seeds for rootstocks are commonly planted in large polyethylene tubes filled with sterilized soil, and placed in full sun. The seedlings are not pruned because grafting requires one large stem. Budwood for veneer grafting, side grafting, or cleft grafting is obtained from the terminals of small branches on healthy, vigorous, mature trees of the desired variety. Veneer grafting can be performed anytime that stocks are actively growing and budwood is available. Planting may take place as soon as six months after budding or grafting, but planting is usually delayed for twelve to eighteen months. In California, 'Duke 7' (Mexican) is the leading clonal rootstock. West Indian race seedlings are used predominantly in Florida. A major need in California and some other regions is for a rootstock highly resistant to Phytophthora root rot.

Avocado trees grow well in a wide range of soil types provided drainage is good. Loose sandy loams are ideal. They will not survive in locations with poor drainage, nor will they tolerate drought stress, so supplemental irrigation is important in many drier regions, for example, southern California and Israel. The trees grow well on hillsides and are tolerant of acid or alkaline soil. Avocado tree growth and production is best suited to acidic soils with a pH of 4.5 to 5.5. However, some West Indian race rootstocks allow production on marginal calcareous soils with an alkaline pH, such as those found in Israel and Florida. Fertilization of young trees after one year of growth is usually performed in commercial plantings using a balanced fertilizer and multiple applications. Older trees benefit from feeding with nitrogenous fertilizer applied in late winter and early summer. Adequate, but not excessive, soil nitrogen is conducive to good fruit production. Excess irrigation leads to pronounced root growth. Avocados tolerate some soil salinity, although they show leaf tip burn and stunting of leaves if levels become too high. Deep irrigation leaches salt accumulation. Yellowed leaves (chlorosis) usually indicate iron deficiency, although mature trees may also display zinc deficiency, especially on calcareous soils. Avocado trees of all ages are pruned sparingly, mainly by clipping back upright shoot tips to prevent excessive height. Heavy pruning encourages strong vegetative growth, thus reducing yields.

Lateral inflorescences are borne in a pseudo-terminal position on shoots all over the tree. Several dozen to several hundred flowers are borne on each panicle (inflorescence). An incredible number of flowers per tree is produced, but only one to three fruits per panicle will mature. Avocado flowers are perfect (having both male and female parts), but they exhibit an unusual behavior. Avocado flower behavior can be described as protogynous dichogamy with synchronous daily complementarity. Avocado flowers are either receptive to pollen in the morning and shed pollen the following afternoon (type A), or are receptive to pollen in the afternoon, and shed pollen the following morning (type B). Flowers of a given cultivar tend to behave uniformly as type A or type B. Cross pollination may occur when female and male flowers from type A and type B varieties open simultaneously. New evidence indicates avocado flowers may be both self- and cross-pollinated under Florida conditions. Production is best with cross pollination between types A and B. While isolated trees or large blocks of one cultivar may set well, cross pollination between A and B types can markedly increase fruit set and retention. Self-pollination appears to be caused primarily by wind, whereas cross pollination is secured by large flying insects such as bees, wasps, and hoverflies. Some cultivars bloom and set fruit in alternate years.

The avocado fruit is a berry, consisting of a leathery skin (exocarp), the fleshy mesocarp (which is eaten), and a large seed consisting mostly of two cotyledons. West Indian type avocados produce enormous, smooth, round, glossy green fruits that are low in oil and weigh up to 0.9 kg (2 lb). Guatemalan types produce medium ovoid or pear-shaped, pebbled green fruits that turn blackish-green when ripe. Fruit of Mexican varieties are small, weighing 170 to 280 g (6 to 10 oz), with paper-thin skins that turn glossy green or black when ripe. Early varieties are usually of West Indian and Mexican origin, whereas midseason and late-season varieties are hybrids between the races and have intermediate characters. Varieties vary in the degree of self- or cross pollination necessary for fruit set.

Most commercially grown avocado cultivars originated as selections that traced to chance seedlings. Modern cultivars that are the result of intensive breeding programs are starting to become available. 'Fuerte' was formerly the world's leading avocado cultivar, but it has now been replaced by 'Hass.' 'Fuerte' is a Mexican-Guatemalan hybrid, with green, pear-shaped fruits that mature over a long period of time in California (from November to May). Fruits are high in oil (18 percent) and weigh 227 to 454 g (8 to 16 oz). 'Hass,' a Guatemalan-Mexican hybrid, is the chief summer-maturing cultivar grown in California, and it is the leading cultivar in Mexico as well. At maturity, the skin of the fruit is green but gradually turns black. It weighs 170 to 340 g (6 to 12 oz) and contains a large amount of oil (19 percent) and a fairly small seed. In Florida, the leading cultivars are Guatemalan–West Indian hybrids such as 'Booth 8' and 'Lulu.' The fruits are green and weigh from 280 to 700 g (10 to 25 oz). The Florida avocados are larger than the California types, but they contain less oil (e.g., 12 percent).

Avocado fruits in California are considered mature and ready to harvest only when the oil content of the flesh reaches more than 8 percent and when the seed coats within the fruits change from yellowish white to dark brown. In more tropical areas, fruits are mature with less oil. Cultivars of the three horticultural races differ in oil content: the Mexican race has the highest, the Guatemalan race an intermediate amount, and West Indian the lowest. Mature fruits can be "stored" on the tree for several months. The fruits remain hard as long as they stay on the tree, softening only after harvest. Mature avocado fruits, even though firm, bruise easily and must be handled carefully. Avocados can be held for about a month in cold storage at −7°C (45°F) for 'Fuerte' and 4.5°C (40°F) for 'Lulu' and 'Booth 8.' The fruit softens at room temperature. Mature, soft fruits can be maintained. California harvests and ships avocados throughout the year, although the peak season is usually from March to August. The Florida season is from August through December and into January. Production in Chile (southern hemisphere) compliments the US season. Production in the Mexican state of Michoacán also compliments the US season somewhat. The combination of opportunities for both storage and importation of fruit result in availability of avocados to US consumers year-round.

Avocados are a mainstay in the diets of Mexican and Central American people. Mexican per capita consumption is about 7 kg (15 lb) per person (Plunkett, 1997). Consumption of avocado in salads and as guacamole (avocado sauce or dip) has grown in the United States and other parts of the world as people discover its taste and nutritional benefits. Avocados contain higher quantities of fiber and protein than most other fleshy fruits and are an excellent source of potassium and vitamin A. Avocado fruit do not contain cholesterol. They are high in monosaturated fatty acids (heart-healthy oils), and the oil content of avocados (3 to 30 percent) is, on average, second only to that of olives among fruits.

Pests and Diseases

Dothiorella (*Botryosphaeria*) canker infects the trunk, causing dead patches that spread to maturing fruit, causing darkened, rancid smelling spots in the flesh. Flesh injury begins after harvest and is impossible to detect on the outside. Mexican types are immune to trunk cankers, but the fruit is not. Root rots caused by *Phytophthora, Armillaria,* and *Verticillium* are the major causes of poor tree health and death. These pathogens are easily transported from infected soils. Once a tree is infected (signs include yellowing and dropping leaves), little can be done other than to cut back on water. Some *Phytophthora*-resistant rootstocks are now available, but considerable research is still needed on this problem. Sun blotch is a viral disease that causes yellowed streaking of young stems, mottling and crinkling of new leaves, and occasional deformation of the fruit. The disease also causes rectangular cracking and checking of the trunk, as if it had been sunburned. Sun blotch has no insect vector but is spread by infected scions, contaminated tools, and root grafts between adjacent trees. It is important to use virus-free propagating wood. Cercospora spot on fruits and leaves is the most important avocado disease in Florida. Leaf roller caterpillars (*Tortrix* and *Amorbia* sp.) may destroy branch terminals. Additional pests include greedy scale, thrips, mealybugs, and Lygus bugs. Avocado brown mite can be controlled by dusting with sulfur. The six-spotted mite is very harmful; even a small population can cause massive leaf shedding. Natural predators can help manage the mite population if they are present. Snails can also be a problem in some areas. Where rats and squirrels strip the fruit, trees must be protected with tin trunk wraps.

Banana (*Musa* spp.) MUSACEAE

The genus *Musa* contains about twenty-five species. The edible clones arose from wild relatives whose fruit contain seeds. Only a fraction of banana production enters the world market, yet it is the leading fruit in world trade, with sales totaling more than $2.5 billion annually. To those who have enjoyed only sweet dessert bananas, it may come as a surprise to know that about half the world's production is starchy bananas for cooking. Starchy bananas such as plantains (*platanos* in Spanish) are very popular in South America, the West Indies, and East and West Africa. East African highland bananas are a preferred staple, and Uganda was by far the leading producer of starchy bananas during the last season, with total production of 10 million metric tons

(FAOSTAT, 2004). Per capita consumption of highland bananas (*matooke*), a staple food, reaches an amazing figure of 200 kg/year (441 lb/year) in Uganda. Bananas are now cultivated throughout the tropics and in selected areas of the subtropics, on farms of all sizes. It is an important crop, from the smallest landholders to the largest corporate exporters. The largest producer nations of sweet bananas are India, Brazil, China, Ecuador, and the Philippines.

The taxonomy of bananas is complex, and species names can sometimes be confusing. Groupings within the genus have been established using genomic composition (the genome is one complete set of parental chromosomes). The origin of banana species can be traced back to different ancestral species depending on whether or not it has a genome (compete chromosomal complement) of A, B, or C chromosomes. The cultivated species are polyploid (having more than two [diploid] sets of chromosomes), whereas the wild species are all diploids with only an A, B, or C genome. Cultivated bananas are triploids derived from wild species *M. acuminata,* and *M. balbisiana* belonging to the Eumusa group. The A genome is contributed by *M. acuminata* and *M. balbisiana* contributes the B genome. Various combinations of these parental genomes in the cultivated polyploids give rise to three major triploid groups comprised of AAA, AAB, and ABB genomic combinations. Sweet bananas are AAA and are low in starch and high in sugar when ripe; unripe fruit are sometimes used for cooking. The group also includes red bananas and East African highland bananas. The AAB species, true plantains, are starchy even when ripe and are usually cooked. Starchy ABB species are both cooking and eating bananas and are found in all banana-growing regions bearing local names.

The origin of bananas is in southeast Asia and the southwest Pacific. Early domestication was of *M. acuminata,* crosses with *M. balbisiana* probably followed, then polyploid species developed. *M. balbisiana* probably extended the range to more drought-prone and wet and dry climates. It is likely bananas were taken to the East Coast of Africa about 450 A.D.; and the first introduction in the new world appears to have been in Haiti in 1516. The banana industry grew in the West Indies and Central America in the nineteenth century. It was based primarily on Gros Michel types and other variants (mutants) of the Cavendish type. The other Cavendish types are well flavored but are small and susceptible to bruising.

Growth and Development

Bananas are tall, rhizomatous perennials. Their monocot leaves makes them readily discernible. The Gros Michel group grows about 4 to 9 m (13 to 30 ft) tall, and the Cavendish subgroup grows about 2.5 to 4 m (8 to 13 ft) tall. Dwarf Cavendish cultivars grow about 2 m (7 ft) tall. Short underground rhizomes grow horizontally and rather slowly. Aerial shoots (also known as suckers or followers) arise from lateral buds on the rhizome. Aerial shoots are *pseudostems* built out of overlapping leaf bases rolled tightly around each other. When the pseudostem is about six to ten months old (about forty leaves), the meristem then becomes reproductive. Each shoot is determinate—once a flowering shoot is produced, no further leaves are initiated.

Fruit develop parthenocarpically (without fertilization) from the ovaries. A banana plant produces a bunch every six months in the lowland wet tropics, although fruit development may take as long as 210 days in cooler climates or under overcast conditions (Nakasone and Paull, 1999).

Plants are inherently shallow rooted and are susceptible to drought stress and lodging due to high winds. Rainfall at or above 125 cm (49 in.) per year is needed. The optimum temperature for growth is about 27°C (81°F). Frost causes rapid death of the trees and damages the fruit. Temperatures above 38°C (100°F) stops plant growth and causes leaf burn (Nakasone and Paull, 1999).

Production

Bananas are grown mostly in lowland wet tropics, but they are also grown in wet and dry, and cooler tropical areas. They may be grown successfully in a wide variety of soils, but good soil drainage characteristics are critical. They tolerate a wide range in pH and may even be grown in soils with pH as low as 3.4. (They are not sensitive to aluminum and manganese toxicity, which is common at such low pH.) Bananas also grow in slightly alkaline soils. High organic matter and fertility produce a high yield of bananas, and mulching of bananas is also considered important to maintain constant soil moisture conditions.

Bananas are planted as rhizome pieces or aerial shoots at a density usually at or below 2,000 plants/hectare (8,100 plants/acre). Good crops remove a lot of nutrients from the soil. Balanced fertilizer (NPK) is usually recommended where available. N and K requirements are high, but excess applications result in thin, breakable pseudostems. Drought restricts the root system and diminishes yield; it also predisposes the plant to lodging due to high winds. Waterlogged soils also predispose plants to root lodging.

Root growth ceases and shoots senesce after bunch development. Secondary shoot growth occurs and a second crop, or ratoon, develops. The second crop of followers shift a portion of nutrient accumulation from soil to rhizome and parent plant (as much as 40 percent). Decomposition of parent leaves also adds to the fertility available in the ratoon crop. Thus, there is a lower fertilizer requirement for followers. Plantains generally do not support vigorous ratoon crops in large monocultures. Yields of backyard trees generally do not decline as rapidly perhaps because of the larger inputs of household organic wastes.

Large, monoculture sweet banana plantations are intended primarily for commercial export markets. Production in this regard is usually highly intensive, with both mechanical and hand labor extensively utilized. A single cultivar, 'Grand Nain,' is planted in the majority of export plantations in the Philippines and throughout Central and South America. Small-scale production through subsistence and local marketing systems exist throughout the tropics.

AAA and AAB types are most important in East Africa. In subsistence systems, bananas are popular because land clearance and cultivation inputs are low. Planting preparation usually entails digging a large planting hole and progressively adding organic manure over a period of time. Propagules are placed in the hole and covered with 10 to 20 cm (4 to 8 in.) of soil. Planting material typically consists of a sucker with a piece of corm that contains a growing point. Suckers are young lateral shoots with a rhizome and growing point. The portion of corm is a section of the mother rhizome with a lateral bud for regeneration (Nakasone and Paull, 1999). Peeling and cutting away of old or diseased portions is generally practiced. Dipping in very hot water (52°C for 20 minutes) (126°F) or treatment with pesticides may be practiced in an attempt to eliminate nematodes and borers. In some areas, farmers are fortunate enough to have disease-free clones available to them for planting.

Banana trees are usually intercropped with annuals when first planted. They are also common in adjacent rows or clumps and in mixed culture gardens. Bananas growing around compounds and villages usually continue to receive organic wastes and mulch on an ongoing basis. The benefit of the organic wastes is not immediate, but one sees results in the ratoon crop. The banana crop itself also returns some organic matter to the fields as the large succulent leaves senesce and peels are returned. Fruit are produced within one year and they are typically harvestable for many months. Farmers generally find bananas give a good return per acre in food or cash, although pests or diseases can be serious detractors. Some researchers have proposed that selection for semi-dwarf varieties (and high population density) should be emphasized.

Bananas are harvested when the fruit are fully mature but still hard and solid green. They do not ripen well if left on the plants. The large bunches or, hands, of bananas may be shipped whole to local markets, or they may be cut into clusters of four to sixteen fingers for export in cartons. Upon arrival at distributing centers, the cartons first go to ripening rooms. The fruit is exposed to ethylene for twenty-four hours to induce ripening. The temperature is varied during ripening to achieve the desired combination of color, pulp texture, and flavor, according to market demands. Desert bananas are ready to use when a full yellow color is achieved; however, the best flavor and nutritive value is indicated when flecks of brown appear on the surface of the peels. The banana is a highly nutritious food, low in sodium and fat but high in potassium and easily absorbed carbohydrates.

In East Africa, a common cash crop is the banana-coffee intercrop. Banana trees provide shade for the coffee plants. Banana is also associated with a variety of subsidiary crops. For example, near East African compounds, it is frequently associated with vegetables, and both crops are manured. Farther from the compound, it will likely be planted in association with, for example, beans, maize, groundnuts, or cassava. The banana-coffee balance usually shifts toward coffee at the periphery. Subsidiary crops are highly dependent on regional differences in climate, food preference, and so on.

Utilization

Sweet bananas are enjoyed around the world as desserts, snacks, and breakfast cereal toppings. They are especially important in the diets of infants and young children because of their nutritional value, sweetness, and soft texture. Cooking bananas are important, especially in Columbia and Venezuela. In the West Indies, West Africa, and parts of East Africa, there is an incredible diversity in production and consumption. Cooking typically involves boiling and mashing or frying. Leaves are commonly used as food wrappers in many regions, and they may also be used for brewing beer, fodder, or roof construction.

Pests and Diseases

Diseases have frequently dictated what the commercial industry can and cannot grow. The most serious diseases have been Sigatooka leaf spot (infectious agent *Mycosphaerella fijiensis*), which causes small lesions that coalesce under favorable conditions, resulting in browning or scorching of the leaves. In large plantations, black Sigatooka is managed using frequent applications of fungicides, an unrealistic option for subsistence farmers. Panama disease (illicited by *Fusarium oxysporum*) causes plants to wilt. The Gros Michel group is susceptible to Panama disease. Yellow and black Sigatooka leaf spot can result in almost complete defoliation in susceptible cultivars of banana and plantains. Panama disease and Sigatooka forced prioritization of resistance breeding programs on these pathogens. 'Highgate,' a semi-dwarf Gros Michel type, was used extensively as a Fusarium resistance donor because resistant 'Cavendish' types are seed sterile. Cercospora leaf spot and bunch top virus are also problems in some banana-growing regions. Banana bacterial wilt, or Xanthomonas wilt (causal agent *Xanthomonas campestris* pv. musacearum), has recently become a serious problem in Uganda. Nematodes (e.g., *Radopholus similes*), causal agent of blackhead toppling syndrome, are also a serious problem in some areas, particularly those growing plantain. New cultivars with resistance to pests and diseases are being developed and tested in both the private and public sectors.

Cacao Theobroma (*Cacao L.*) STERCULIACEAE

The center of the genetic diversity of cacao is the Amazon Basin region of South America. The species also seems to have its origin there. Cacao evolved as a perennial evergreen understory tree in the tropical forest. It was naturally dispersed by foraging animals and people, who ate the sweet pulp from the pods and dropped the seeds. It was first cultivated by the Mayan and Aztec civilizations of tropical Central America. Both groups used the ground beans, frequently mixed with chiles, to make a stimulating beverage. Archaeological information suggests that the Mayans cultivated cacao 2,000 to 4,000 years before the arrival of the Spanish explorers. Columbus, on his fourth voyage in 1502, saw cacao beans, but it was Cortés who, a few years later, took the beans back to Spain. From the European perspective, the beverage was improved considerably by the omission of chile and the addition of sugar. Spain kept the secret of this new drink for about 100 years, but it gradually spread throughout Europe and became very popular. This was the origin of our modern hot chocolate. The Dutch invented chocolate candy about 1830. Cacao has mild stimulating properties from its theobromine and caffeine content.

Cacao production has developed into a billion dollar industry of worldwide importance in the manufacture of cocoa and chocolate. Until about 1900, most cacao plantations were in Central and South America and the Caribbean Islands. In 1879, however, cacao was introduced into West Africa, where it started on small farms of 0.8 to 2.0 ha (2 to 5 acres) each. Production in this region increased phenomenally. The Côte d'Ivoire (Ivory Coast) is now the leading producer, followed by Ghana, Indonesia, Nigeria, and Brazil.

Cacao trees are broad-leaved evergreens that grow to a height of 5 to 8 m (16 to 26 ft) under cultivation, but may attain a height of more than 15 m (49 ft) in neotropical forests. They require a truly tropical climate: all production areas are located within 20° of the equator and at altitudes below 300 m (1,000 ft). Cacao trees grow best at a mean annual temperature varying between 21° to 27°C (70° to 80°F). Growth slows at average temperatures of 15°C (59°F) and is minimal at average temperatures of 10°C (50°F). Production is highest in areas with high humidity and rainfall well distributed throughout the year (about 127 to 152 mm [5 to 6 in.] per month) and with little or no dry season. Winds harm cacao trees by increasing water loss from the leaves and causing defoliation; thus, windbreaks are often used. Cacao trees, like some other tropical crops (coffee and tea), are usually grown in the shade of taller trees, planted especially for this purpose or retained when forests are cleared for new plantings. The optimum degree of shade depends on the fertility of the soil. On fertile or fertilized soils, no shade is needed and yields increase.

Cacao trees grow best in deep, slightly acidic soils that are high in organic matter and are well drained but with a high water-holding capacity. Heavy clay soils are unsuitable. Cacao trees respond to nitrogen, phosphorus, and potassium fertilizers when shade is reduced or removed.

Cacao plantings consist largely of seedling trees that develop a wide, branching growth habit and start bearing at two to six years of age. Some high-yielding clones have been developed that are propagated vegetatively by root cuttings. The flowers and fruit are borne directly on the older, leafless parts of the branches and on the trunk. Flowering and fruiting may occur throughout the year, and both are determined largely by the rainfall pattern, but the majority of the fruit harvest occurs from September to March.

Cacao flowers are perfect (with both stamens and pistils) and may be self- or cross-pollinated for fruit setting. Pollination of flowers is by *Forcipomyia* spp. and *Euprojoannisia* spp., small-bodied flying insects in the Ceratopogonidae family. Lack of adequate cross pollination may sometimes limit crop production. A mature cacao tree may produce 50,000 flowers, but less than 5 percent set pods, and many of those do not reach maturity. The podlike oval fruits are 10 to 50 cm (4 to 12 in.) long containing thirty to fifty seeds, each surrounded by sweet, white mucilage. A good yield is 670 kg of dry beans (seeds) per hectacre (600 lb per acre), but yields of over 3,360 kg/ha (3,000 lb/acre) have been obtained from improved cultivars. Pods usually mature in five to six months on cultivated clones and are harvested weekly when they change to a red or yellow color. They are cut open when fully ripe to remove the seeds. The mucilaginous pulp around the seeds is removed by a fermentation process, which takes three to eight days and causes temperatures to rise to about 51°C (125°F). This process kills the embryo in the seeds and develops certain chemical precursors that give the chocolate flavors when the beans are later sun-dried and finally roasted.

Cacao trees are attacked by several fungal pathogens. The pods are susceptible to infection by *Phytophthora palmivora,* which results in a disease called black pod. Another pathogen, *Marasmius* spp., causes witches' broom. Recommended management to reduce the deleterious effects of witches' broom on cacao production includes the use of phytosanitation (removal of diseased plant parts), application of chemical fungicides, and the use of pathogen-resistant varieties.

Citrus (*Citrus* spp.) Rutaceae

Citrus fruits, especially oranges, are among the world's highest ranking fruit crops, along with grapes, bananas, and apples. Commercial citrus species are native to southeast Asia and eastern India. Commercial citrus is produced at low elevations in the world's subtropical climatic zones between 20° and 40° north and south of the equator. Citrus has comparatively little cold resistance and is not grown commercially where minimum temperatures are likely to fall below −6.5°C (20°F).

During the past 100 years, world citrus production has increased from less than 1 million to over 100 million metric tons (MT). The Mediterranean area and North and Central America contain about 55 percent of the commercial plantings. Brazil is the largest producer of citrus fruits—mainly oranges, grapefruit, and lemons—followed by the United States, China, Mexico, and Spain. The United States is the largest producer of grapefruit and pommelos, and Mexico is the largest producer of lemons and limes. About 80 percent of all citrus grown is consumed in the producing countries themselves. Oranges account for the bulk (65 percent) of the citrus produced, with mandarins, including tangerines (17 percent); lemons and limes (9 percent); grapefruit plus pommelo

(5 percent); and other citrus species accounting for the rest. Florida is the leading citrus-producing state in the United States, especially in orange and grapefruit production. The vast majority of the Florida orange crop is used for juice production. California is second in production, and the majority of its orange crop is produced for direct consumption. Citrus production in the United States has declined somewhat during the last several years.

Most present-day orange cultivars have been grown for many years. The famous seedless 'Washington Navel' orange was found in Brazil in the early 1800s, and propagating material was introduced to Washington, D.C., by the US Department of Agriculture (USDA) in 1870. From there, trees were sent to Riverside, California, in 1873 where its important attributes were recognized. It is believed to have originated as a limb sport of the seedy 'Seleta' sweet orange in Brazil. 'Seleta' was introduced to Brazil by the Portuguese via the Iberian Peninsula from settlements in the Orient. Citrus species interbreed easily, and several interspecific and intergeneric hybrids, for example, tangelos, have been developed. New cultivars targeted at specialty markets, for example, "blood" oranges (oranges containing red and/or purple pigments), have recently been introduced in California. The leading cultivars of blood oranges are 'Moro' from Sicily, 'Tarroco' from Spain, and 'Saguinelli' from Italy. The common and scientific names and the important citrus cultivars grown in the United States are provided in Table 20–1.

A deep, well-drained, fertile soil is optimal for growing citrus, but many kinds of soils are used. Soils with low fertility can be made more productive by adding fertilizers. Citrus generally requires enough added nitrogen fertilizer to maintain leaf nitrogen of about 2.5 percent dry weight.

TABLE 20–1 Some Commercially Important Citrus Species in the United States

Common Name	Scientific Name	Important U.S. Cultivars
Grapefruit	*Citrus paradisi* Macf.	'Marsh,' 'Duncan,' 'Redblush,' 'Thompson', 'Henderson and Ray,' 'Rio Ruby,' 'Star Ruby'
Lemon	*C. limon (L.)* Burm. f.	'Eureka,' 'Lisbon,' 'Bearss'
Lime	*C. aura otifolia* (Christm.)	'Tahiti'
Orange (sweet)	*C. sinensis* (L.) Osbeck	'Valencia,' 'Washington Navel,' 'Hamlin,' 'Pineapple,' 'Parson Brown,' 'Marrs'
Mandarin	*C. reticulata* (Blanco)	'Satsuma,' 'Dancy,' 'Temple,' 'Minneola,' 'Orlando,'* 'Nova,'* 'Owari'

* Known hybrids, not pure mandarin.

California soils usually have adequate phosphorus and potassium for citrus, but magnesium is occasionally deficient and the soil may require magnesium nitrate. Zinc and manganese deficiency may appear in California citrus, and it is corrected with zinc sulfate or manganese sulfate sprays. In alkaline or high saline soils, iron deficiency appears and the trees require iron chelate sprays. In Florida, with its less fertile soils, the fertilizer program for citrus includes nitrogen, potassium, magnesium, and possibly phosphorus, as well as micronutrient elements such as boron, iron, copper, manganese, zinc, and molybdenum.

Some areas where citrus grows have long, dry periods that make irrigation mandatory. All types of irrigation are used: furrow, basin, sprinkler, and drip. Because much of the citrus crop is on the trees during the winter months, frost damage is a real hazard. In selecting the site for the grove, low-lying frost pockets should be avoided. Oil-burning heaters or wind machines are often installed in bearing orchards to protect against local radiation frost damage to blossoms, fruits, and trees. Usually heater fires are lit when orchard thermometers drop to a reading of $-2°$ or $-3°C$ ($27°$ or $28°F$) and are turned off at $0°C$ ($32°F$); eighty-seven to 150 heaters per hectare (35 to 60 per acre) may be required. Wind machines are operated at $0°C$ ($32°F$) and below, except when there is no temperature inversion. Where water is available for flood, furrow, or sprinkler irrigation, it may be applied for heat release in cold weather.

Most commercial kinds of citrus set adequate crops without cross pollination. A few of the hybrids do require cross pollination. Bees work among citrus flowers, and hives are placed in citrus groves for the collection of honey.

In subtropical regions with cool winters, most citrus species bloom once a year in the spring. In the tropics and warm-winter areas, the flowering period may be prolonged or it may occur several times during the year. Lemons grown in the mild southern California coastal regions and in southern Florida bloom and fruit year-round.

Navel oranges from California are available in markets from November through June, whereas Valencias are harvested from mid-March to mid-November. Florida Valencias are on the market from March through July, while the Florida early midseason orange cultivars ('Pineapple,' 'Temple,' 'Hamlin,' and 'Parson Brown') are marketed from October to April. Grapefruit harvest periods are: Florida, September through July; Texas, October through April; California, December through October; and Arizona, November through August. Lemons in the United States are produced mostly in California, with much smaller amounts in Arizona and Florida. California lemons are on the market year-round. Arizona lemons are available from October through February. Florida lemons are almost all used in processed forms.

Some seed-propagated citrus trees produce acceptable fruit, but most superior cultivars must be propagated by vegetative methods. In the early days of orange production in Florida (during the late 1800s), seedling trees were grown. In modern times, however, most citrus is propagated by t-budding on selected seedling rootstocks. Citrus cuttings root fairly easily, but this method of propagation is little used commercially.

Seeds of most citrus species produce nucellar embryos by apomixis in addition to the sexual embryo; the nucellar embryos are the same genotype as the female parent and thus maintain the maternal clone. These asexual nucellar seedlings can be separated from sexual embryos in the nursery by their greater vigor and uniformity of character; this permits selection for rootstock uniformity. The type of rootstock used in citrus can strongly influence fruit quality, yield, and tree size of the fruiting cultivar, as well as other horticultural characteristics such as disease resistance and soil adaptation.

Rootstocks generally used are seedlings of rough lemon, sour orange, 'Troyer' citrange, trifoliate orange, 'Cleopatra' mandarin, and 'Rangpur' lime. Because of susceptibility to tristeza (quick decline), sour orange is no longer used as a rootstock for oranges in California, South Africa, or Brazil, but it is still used to some extent in Florida. Sweet orange cultivars on sour orange roots in California, and on 'Rangpur lime' in Brazil, are susceptible to the tristeza virus. This combination can be used in Florida, however, because of an apparently less virulent form of the virus. The use of sweet orange, 'Cleopatra' mandarin, and trifoliate orange roots results in trees that produce high-quality fruits. However, trifoliate and trifoliate hybrid rootstocks are very susceptible to the exocortis virus, and infected trees are dwarfed to some extent, depending on severity of infection. Trees on rough lemon roots tend to produce coarse-textured fruits low in solids and acids, but they are well adapted to light-textured soils.

Citrus trees produce mutations frequently. Because most of these mutations are inferior, propagating material should be taken only from mother trees that have a history of producing high-quality, true-to-type fruits that are free of disease.

Pruning young citrus trees delays bearing and should be done only enough to develop a trunk and strong scaffold system. Bearing orange and grapefruit trees require only light pruning. On the other hand, bearing lemon trees need heavier pruning to facilitate orchard operations and limit their height.

Most citrus fruits store best on the tree, but they can be held for a time under refrigeration. Florida oranges may be kept for about twelve weeks at $0°C$ ($32°F$) and 90 percent relative humidity. California oranges store best slightly above $4.5°C$ ($40°F$). California and Arizona grapefruit keep for six weeks at $15.5°C$ ($60°F$), but Texas and Florida grapefruit require $10°C$ ($50°F$), all with 90 percent relative humidity. Lower storage temperatures cause fruit deterioration in grapefruit. Fully matured and colored lemons can be held for several weeks at $0°$ to $4.5°C$ ($32°$ to $40°F$), but for prolonged storage—four to six months—they should be picked dark green and held at $12°$ to $14.4°C$ ($52°$ to $58°F$). Limes can be stored at $9°$ to $10°C$ ($48°$ to $50°F$) with 85 to 90 percent relative humidity for six to eight weeks. In the latter part of storage, however, the rind color turns from green to yellow.

Most California oranges are marketed fresh, but in Florida about 90 percent are processed, mostly as frozen concentrate juice. Florida and Brazil together produce more than 85 percent of the orange juice consumed in the world. About two-thirds of Florida's grapefruit are marketed in a processed form. Production of Florida frozen concentrate orange juice has been one of the outstanding examples of the development of a new industry by modern technology. Of the total US production of oranges, about 73 percent is sold in processed form. For grapefruit, 34 percent is processed; for lemons, 54 percent is sold processed.

Many insect pests attack citrus. Some important ones are citrus red mite, citrus thrips, citrus mealy bug, red scale, yellow scale, snow scale, purple scale, citrus whitefly, and aphids. In addition, certain species of nematodes attack citrus roots.

Citrus disease problems are more devastating than insect pests, and their management is critical for production of high-quality fruit. Virus diseases include tristeza, exocortis, xyloporosis, vein enation, and yellow vein. All of the viruses are transmitted by budding and grafting with infected wood, and some are distributed by insect vectors. All citrus cultivars and rootstocks are subject to several viral diseases. Oranges in Brazil were especially hard hit by sudden death incited by a strain of tristeza virus. The best control is to use only propagating material or nursery trees produced under conditions free of such viruses. Most citrus-producing countries have set up elaborate certification programs to produce virus-free, true-to-type nursery stock.

A disease caused by a mycoplasmlike organism called stubborn is an important citrus disease in California. The only known control is disease-free propagating material. Fungal diseases attacking citrus are brown rot gummosis due to Phytophthora, brown rot of fruit, Septoria spot, and Armilaria root rot. Bacterial diseases are citrus blast and citrus canker. An outbreak of the latter in Brazil during the 1970s was followed by a devastating epiphytotic in the Florida citrus nursery industry during the mid-1980s, when all infected trees were destroyed. Citrus variegated chlorosis (CVC), incited by a bacterium (*Xylella fastidiosa*) discovered in 1987, is now seriously affecting the Brazilian industry. An insect vector, the sharpshooter, infects new growth with the bacterium, which then moves down the limb toward the main trunk and subsequently throughout the tree. Symptoms are variegated chlorotic spots on leaves and decreased fruit size. Infected trees produce the same number of fruit as healthy trees, but CVC causes the fruit to develop at one-third of normal size, and 75 percent cannot be processed. The result has been a decline in Brazilian orange juice production. Few management options for CVC are available. If 30 percent or more of a grove is infected, the current recommendation is to remove the whole grove or block. The pathogen is confined to South America, and citrus in Argentina, Paraguay, and Uruguay suffers from the disease.

Research determined that field nurseries were instrumental in spreading CVC and Phytophthora. These findings resulted in mandatory certification programs for nurseries that went into effect in 1998. This program required, among other things, that all citrus nursery stock (trees, budwood, and liners) be grown in totally enclosed aphid-proof screen houses. The production of disease-free trees should help to contain the threat to Brazil's industry. A disease of unknown origin called blight is causing serious tree losses in Florida, mostly of trees on rough lemon rootstock.

Coconut (*Cocos nucifera L.*) ARECACEAE

The coconut is by far the most important nut crop in the world. The coconut is a tall, unbranched monocotyledonous tree (palm) grown throughout all the tropical regions. The native home of the coconut is difficult to determine because over the centuries, the nuts (seeds) were easily dispersed by ocean voyagers and by ocean currents from island to island and continent to continent. Some believe it originated in the islands of the Malaysian Archipelago, but others believe it had a Central American origin.

Although the coconut is a tropical plant with commercial production within 15° of the equator, it grows and fruits some distance from the equator, as shown by plantings in southern Florida at latitude of 26° north. The world's leading producers are Indonesia, the Philippines, Brazil, Sri Lanka, and Thailand.

Coconuts withstand some frost. Growth and production are best under high humidity with mean annual temperatures around 26.5°C (80°F) and daily fluctuations of no more than 5°C (9°F). The coconut palm is a light-requiring tree and does not grow well under shade or in very cloudy conditions. Coconuts need ample, well-distributed soil moisture and suffer from long, rainless periods. Annual rainfall of at least 152 to 177 cm (60 to 70 in.) is required, although irrigation can be used to supplement low rainfall. The trees grow in a wide range of soil types—beach sand, coral rock, or rich muck—provided they are at least 1.2 m (4 ft) deep. Coconuts tolerate high salt levels in the soil and salt sprays found along the seashore. Shallow soils with a high water table, which prevents good root development, are unsuitable because the tall palms are likely to blow over in tropical typhoons and cyclones.

The coconut tree has a tall, flexible trunk marked by leaf scars and topped by a crown of leaves. Each leaf is about 6 m (20 ft) long with a central midrib and about 100 opposite leaflets. A growing point in the central crown produces new leaves and flowers. This growing point is the only bud on the palm; hence, if it is killed, death of the entire palm follows. These growing points, called hearts of palm, are a very nutritious and tasty food product, reminiscent of artichoke hearts. Trees of some palm species, including the coconut palm, can be grown specifically for this product, even though extracting it kills the tree.

Nearly all trees are propagated by seed because no part of the tree can be used for vegetative propagation. The coconut palm does not produce offsets as does the date palm.

Tree characteristics of the coconut are reproduced fairly well by seed so that the best strategy is to select seeds for new plantings from trees that produce large crops of high-quality nuts. In some seed-propagated cultivars, such as 'Yellow Dwarf' and 'King,' the seeds are gathered from isolated groves of trees that all have similar characteristics. Self-pollinating dwarfed selections bear early and give high yields, flowering in three to four years from seed. Fully matured nuts (at least twelve months old and still enclosed in the husk) are usually germinated in a seedbed by planting them flat on their sides and covering them with soil. They are kept moist during germination; after eight to ten leaves have formed, the new plant plus the husk is transplanted to its permanent location. Trees are usually set about 7.5 m (25 ft) apart.

Several heterogeneous cultivars are recognized. Tall palms tend to be slow maturing, flowering six to ten years after planting, with a life span of eighty to 100 years. Dwarf coconut palms start flowering in their third year and have a productive life of thirty to forty years. Coconut palms respond to nitrogen and potassium fertilizers and, in some soils, to phosphorus, although in many naturally fertile soils no benefits have been obtained from added fertilizers. Little is known of their need for trace elements.

Both male and female flowers are borne on the same many-branched inflorescence arising from a leaf axil in the top of the palm. Each inflorescence can have up to 8,000 male flowers along the terminal part of the inflorescence and one to thirty globular female flowers near the base. Transfer of pollen from the male to the female flowers by wind, birds, or various insects is required for nuts to develop.

The coconut fruit is classified as a drupe—just as is the fruit of the peach or apricot. The hard shell commonly seen when one buys a coconut in the store is the inner layer of the matured ovary wall of the fruit (the endocarp). Outside this is the husk (the mesocarp plus exocarp), which is usually removed when the nuts are harvested. The husk is now being processed and sold as coir fiber, used in the manufacture of soilless potting mixes. Inside the shell (endocarp) is the true seed with a thin brown seed coat. The white meat of the coconut is part of the endosperm, a food storage tissue. (Much more important is the dried endosperm—known in the trade as copra.) The coconut milk is also endosperm in a liquid form. The tiny embryo is buried under one of the three eyes in the endosperm at the end where the fruit was attached to the plant. The coconut is mature and ready to eat about a year after pollination of the flower.

Depending on the cultivar, the average weight of fruit from tall palms ranges from 1.2 to 2 kg (2.2 to 4.4 lb), with nuts from 0.7 to 1.2 kg (1.5 to 2.6 lb) containing 0.35 to 0.60 kg (0.8 to 1.3 lb) endosperm and yielding 0.20 to 0.29 kg (0.4 to 0.7 lb) of copra. Dwarf palms bear fruits weighing 1.1 kg (2.4 lb), with nuts weighing 0.6 kg (1.3 lb) and yielding 0.20 kg (0.4 lb) of copra. The edible part of the coconut is the white endosperm, usually eaten fresh and raw, plus the liquid endosperm (the milk).

Copra is used principally as a source of coconut oil and of shredded and sweetened products used in confections. It contains 60 to 68 percent oil, of which approximately 64 percent is extractable. To prepare copra, the husk is removed by knives that also split the nut in two. The milk is not saved. The open halves of the nuts are turned upward and stacked in trays to dry. After four to six days, the meat is pried out with special spoons, then further dried in the sun or artificially dried in heated kiln dryers until the moisture content is about 5 percent. The dried product—copra—is then chopped into uniform small pieces, sacked, and delivered to large processing plants for extraction of the oil. Dried copra can be stored under refrigeration below 10°C (50°F) for several months. The copra cake left after oil extraction is ground to a meal that is a high-protein cattle feed. Fresh coconuts can be stored for up to two months at 0° to 1.5°C (32° to 35°F) at a relative humidity of 75 percent or less. They will keep for two weeks at room temperature.

The low content of unsaturated fatty acids makes coconut oil resistant to oxidative rancidity. It is used in the preparation of margarines, shortenings for cooking, frying oils, and imitation dairy products. Coconut stearin is valued as a confectionery fat and as a substitute for cocoa butter. The shredded and dried copra contains 68 to 72 percent oil and less than 2 percent water and is used in confectionery and bakery products. Coconut oil is also used in the manufacture of liquid and solid soaps and detergents, cosmetics, hair oil, and various lubricants.

Pests and Diseases

Coconut trees are highly susceptible to various diseases. Viruses have eliminated entire plantations. Lethal yellowing, possibly caused by a mycoplasmalike organism, is particularly severe in the Caribbean area and has killed thousands of acres of coconut palms. A similar disease has occurred in West Africa. In Florida and some other regions, the common coconut is being replaced with 'Malayan Dwarf' (*Cocos* sp.), which is resistant to lethal yellowing. In the Philippines, a disease called cadang-cadang has wiped out many large coconut plantations. *Phytophthora palmivora* fungus can kill the terminal growing point, leading to the death of the entire palm. Bronze leaf wilt causes death of the older leaves, eventually working its way to the younger leaves. It is a nonpathogenic problem due to unfavorable weather conditions, often appearing after a long drought. Some insect pests attack coconut palms, but only a few are serious and can generally be controlled by sanitary measures.

Coffee (*Coffea* sp.) RUBIACEAE

The genus *Coffea* consists of more than seventy species, but only four species (*Coffea arabica, C. canephora, C. liberica,* and *C. dewevrei*) are involved in world coffee production. *Coffea arabica* (Arabica coffee) and *C. canephora* (Robusta coffee) account for more than 70 percent and 25 percent of the world market, respectively. *C. arabica* is native to the highlands of southwestern Ethiopia and southeastern Sudan. *C. canephora* originated in the African equatorial lowland forests from the Republic of Guinea and Liberia eastward to Uganda, Kenya, and southern Sudan, extending into central Africa. The Arabs were using coffee as a beverage as long ago as 600 A.D. During periods of expansion between the eleventh and sixteenth centuries, coffee appeared in Turkey, the Balkan states, Spain, and North Africa. It found its way into western Europe about 1615 (tea was first used there about 1610 and cocoa about 1528).

Coffee plants were brought to Brazil in 1727. After forty years, it became, and still is, the world's leading coffee grower and exporter, now producing one-quarter of the world's coffee supply. Vietnam ranks second in coffee production, followed by Costa Rica, Indonesia, and Columbia. Latin America produces mostly *C. arabica,* whereas the African continent and southeast Asian coffee production are dominated by *C. canephora.* Although arabica beans represent 70 to 75 percent of the world's production, only about 10 percent qualify as specialty coffees sold by gourmet retailers. Coffee grown in highland areas, where the temperatures are somewhat cooler, produce a milder-type bean from which a better drink can be prepared than that from trees grown at lower and hotter elevations. Robusta beans produce a stronger, less flavorful coffee with a higher caffiene content, but it is now in great demand for use in blending with *C. arabica* coffees and in the manufacture of instant coffees. *C. liberica* is an important species in some areas but produces an inferior coffee. Coffee is the mainstay of the economy of the Central American countries, and it is grown to a limited extent in the West Indies and in Hawaii, where the famous Hawaiian Kona coffee is well known. Young coffee seedlings are occasionally grown outdoors in the warmer parts of the United States and as indoor houseplants, making attractive ornamentals.

The typical *C. arabica* plant is a large bush to small tree with dark green leaves. It is an allotetraploid ($2n = 44$) and differs from other coffee species that are diploid ($2n = 22$). Optimum altitudes for *C. arabica* are 1,000 to 2,000 m and a temperature range of 15° to 24°C (59° to 75°F). *C. arabica* trees withstand temperatures near 0°C (32°F) for only a short period. Frost is one of the principal hazards in growing coffee. Cold-damaged trees that have been exposed to temperatures below freezing have considerable difficulty recovering. Temperatures above 27°C (80°F) tend to reduce flowering and fruiting, while temperatures below 13°C (55°F) cause cessation of growth and tree stunting. *C. arabica* is self-fertile and autogamous. Fruit shape is oval and generally reaches maturity in seven to nine months. It is grown throughout Latin America, in Central and East Africa, in India, and to some extent in Indonesia.

C. canephora prefers altitudes of 0 to 700 m, and optimum temperatures are 24° to 30°C. Thus, disease-resistant robustas are preferred in lowland, wet, humid tropical forests and valleys. It ranges in size from a small shrub to a small tree reaching heights up to 10 m. The rounded fruits mature in ten to eleven months and contain oval-shaped seeds, which are smaller than the seeds of Arabica coffee. *C. canephora* is diploid ($2n = 22$) and self-sterile. It must be cross-pollinated and is highly heterozygous. Robusta coffee is grown in West and Central America, throughout southeast Asia, and to some extent in Brazil, where it is known as 'Conilon.' Because robusta grows at lower elevations, it matures more rapidly than arabica and can thus be cultivated more easily and brought to market less expensively.

Coffee trees need a continual soil moisture supply. Annual rainfall of about 165 cm (65 in.) is required, although supplemental irrigation can be given in drought periods. Optimum precipitation for *C. arabica* is 150 to 200 cm, while *C. canephora* requires 200 to 300 cm annually.

In many countries other than Brazil and Hawaii, coffee trees are planted in the shade of other tall-growing trees. Coffee, apparently, has a low light saturation level for photosynthesis so the trees grow adequately in the shade. However, sun-grown coffee trees give the highest yields provided all other cultural factors, particularly soil fertility, are optimal.

Growth and Development

Although, the coffee plant is technically an evergreen shrub, it is referred to as a tree. One reason is that unless pruned, a coffee plant can grow more than 6 m (20 ft) high. The root system will penetrate down to 4 or 5 m in depth. Coffee trees and their foliage are tender and are quickly injured by strong winds. Wherever such winds blow, rows of windbreak trees must be used.

Most coffee trees now in bearing are seedling trees, grown in a nursery for one year before planting in the field. Coffee trees are planted about 2.4 m (8 ft) apart and are kept low—about 1.8 to 2 m (6 to 7 ft)—by pruning to facilitate harvesting.

The trees grow best in a slightly acid, loamy soil at least 0.9 m (3 ft) deep with good aeration and drainage. To grow and yield, coffee trees require a high level of soil fertility. They readily show symptoms of a lack of any of the essential elements. Plantations managed without added fertilizers show a steady decline in production. As pH moves above 5.1, iron (Fe) deficiency becomes a problem. At pH below, 4.2, calcium (Ca) becomes deficient. Deficiency symptoms of all the essential mineral elements have appeared in coffee plantations.

Coffee plants start bearing full crops on the lateral branches at about five years and reach full production at fifteen years. Clusters of two to twenty perfect flowers (both male and female parts) are produced in the leaf axils. The white flowers are very attractive and fragrant. The blossoms last only about three days, pollination is accomplished by wind or insects, and little berries appear six to nine months later. Coffee yields vary considerably from about 2,240 kg/hectare (2,000 lb/acre) of "clean" coffee (beans removed from the fruit) in Hawaii to only 400 kg/hectare (360 lb/acre) in less productive environments. A coffee tree can produce for up to twenty five years, so it is a must to replace nutrients in the soil.

C. arabica is self-fertile and tends to reproduce fairly true from seed. Thus, several seed-propagated cultivars have been developed such as 'Mondo Novo' and 'Caturra' in Brazil, 'Tico' and 'Hibrido' in Central America, 'Moko' in Saudi Arabia, the dwarf 'San Ramon' and the Jamaican 'Blue Mountain.' *C. canephora* trees are self-sterile and depend on cross pollination; therefore, the seedlings vary considerably.

Botanically, the coffee fruit is a drupe, except that within each fruit is a mucilaginous pulp covering two thin, fibrous, parchmentlike pits (endocarp), each surrounding a single seed. The two greenish seeds (the coffee beans) in each fruit consist mostly of a hard, thick, and folded endosperm food-storage tissue covered with a thin silvery seed coat. The tiny embryo in the seed is embedded in one end of the endosperm. The fruits (called cherries by the growers) first are green, then become dark red when ripe.

Harvesting of coffee begins in Brazil in May at the end of the rainy season. In harvesting the coffee fruits, which are mostly picked by hand, the pickers must remove only the mature red fruits, leaving the green ones for further development. There has been some mechanical harvesting of coffee in Brazil. A coffee tree may have blossoms, green fruit, and ripe fruit all at the same time on the same branch. An average coffee tree yields about 2.25 kg (5 lb) of ripe fruits (converting to about 0.5 kg [1.1 lb] of roasted ground coffee). Following harvesting, the pulp must be removed from the coffee cherries to obtain the seeds (beans) inside. There is both a dry and a wet process for this stage. In the dry process, the fruits are washed, then spread out on concrete slabs in the sun to dry. They are turned several times a day. Dry fruits are then repeatedly run through fanning and hulling machines to free the coffee beans inside. In the wet process, the cherries are run through a depulping machine that breaks them open and squeezes the beans out of the pulpy skin. Then they go into large tanks for twenty-four to forty hours and the jelly like substance surrounding the beans ferments slightly. Then the beans are thoroughly washed, spread out in the sun to dry, and continually turned and mixed. In wet weather, drying machines are used, tumbling the beans in perforated drums through which warm air blows.

Finally, the green coffee beans are polished and graded according to origin, size, quality of preparation, and taste or cup quality. Once quality is determined, and the green beans are inspected to remove any that are defective, they are packed into 60-kg (132-lb) bags for shipment to coffee-importing countries.

Beans on the same tree but picked six months apart may have a perceptibly different flavor. Different growing climates during the year affect the taste of the beans. Bean quality at a given location also varies from one year to the next. Coffee companies must sample beans continuously and buy from different areas to keep the taste of their products the same. Roasting the green coffee beans further develops the characteristic tastes and color and must be skillfully done. Coffee purchases are often based on quality, as determined by highly trained coffee tasters who test-sample lots of roasted and ground coffee.

Millions of bags of unroasted coffee beans are shipped to the consuming countries each year for blending, roasting, grinding, and packing. Marketing of coffee produced by thousands of small farmers must necessarily be done by co-operative organizations. In Colombia, for example, the National Federation of Coffee Growers represents the growers in marketing agreements and conducts research in coffee production and processing.

Many disease and insect pests affect coffee trees. Probably the worst disease is the leaf-rust fungus (*Hemileja vastatrix*). This rust first appeared in Sri Lanka in 1869, destroyed all coffee plantations throughout the island, and spread to other coffee-producing countries in that part of the world. Coffee planters in Sri Lanka dared not risk replanting coffee so they turned to tea, developing Sri Lanka into the world's second-largest tea producer (India is first). Vigorous trees well fertilized with nitrogen seem to resist coffee leaf rust, and sprays with copper compounds aid in controlling the fungus. *Coffea arabica* is susceptible, while *C. canephora* is considered to be resistant.

Another fungus, *Mycena flavida,* is the causal agent of American leaf spot. This disease causes defoliation and actually kills the plant faster than leaf rust. The disease is most prevalent in Mexico, Guatemala, Costa Rica, Colombia, and Brazil, and it can be severe. The spores live indefinitely in the ground. The coffee bean borer (*Stephanoclores*) is a problem in Brazil, where it is referred to as the coffee plague. The coffee leaf miner and the Mediterranean fruit fly commonly attack coffee trees in some countries.

Mango [*Mangifera indica (L.)*] ANACARDIACEAE

The cultivated mango is a member of the ANACARDIACEAE (cashew family). There are roughly 700 species in the genus, and many wild *Mangifera* species can be found in southeast Asia. Mangoes have been cultivated in India for at least 4,000 years. In northeastern India, wild *Mangifera indica* trees are indigenous. Ancestral wild mangoes bear small fruits with only a small amount of fibrous flesh.

The mango has been associated with the economic, cultural, religious, and aesthetic life of the Indian people since prehistoric times. The Buddhist pilgrims Fa-Hien and Sung-Yun mentioned in their travel notes that the Gautama Buddha was presented with a mango grove by Amradarika (500 B.C.) as a quiet place for meditation. The fruit appear to have spread eastward during 400 to 500 B.C. to East Asia and they reached the Philippines in the fifteenth century. Westward spread occurred during the sixteenth century via Portuguese explorers, who carried them to Africa and Brazil. It was introduced into Florida and California in the nineteenth century.

Mango trees are long-lived, erect, and symmetrical perennial trees. They typically reach about 10 to 30 m (33 to 98 ft) in height, but sometimes grow to 40 m (131 ft). Mature leaves are leathery, glossy, and deep green in color. The tree has a deep tap root (to 6 m [20 ft] in good soil) and a wide-spreading feeder root system and several anchor roots. The efficient root system enables the tree to tolerate drought. Mangoes are adapted to a range from 25°N to 25°S, and up to 1,000 m (3,289 ft) elevation. Tropical climates with a wet-dry season (monsoon climate) are preferable because a dry season is best for fruit set.

It may be cultivated in small or large commercial plantings throughout the humid and semiarid lowlands of the tropical world. The tree is nearly evergreen and is a favorite in dooryards and gardens of rich and poor alike. Leading producers are India, China, Thailand, Mexico, and Pakistan (FAOSTAT, 2004).

Although the mango is a tropical plant, mature trees have withstood temperatures as low as −4°C (25°F) for a few hours. Young trees and actively growing shoots are likely to be killed at −1°C (30°F). Flowers and small fruits are damaged if temperatures drop below 4.5°C (40°F) for a few hours. Temperatures of 24° to 27°C (75° to 80°F) are considered optimal for mangoes during the growing season, along with high humidity. They tolerate temperatures as high as 48°C (118°F).

Propagation

Mangoes can be propagated in many ways. Monoembryonic types have only a sexually produced embryo. Mangoes are outcrossing, so these types show considerable variation among seedlings. They must be propagated vegetatively if the cultivar characteristics are to be retained. Polyembryonic types (which come true from seed) can be propagated by seed or by one of several vegetative methods. There is also some apomixis (embryo development without pollination), so vegetative multiplication of the best cultivars is recommended. Seeds are planted in seedbeds, and seedlings are usually transplanted to nursery rows when they are about 15 cm (6 in.) tall. They are typically grafted when about pencil-thick. The various methods employed are: approach grafting (used in India since ancient times), whip grafting, side grafting, veneer grafting, t-budding, chip budding, and patch budding. Mango is difficult to propagate by stem cuttings or layering. In addition, trees obtained from cuttings or by air layering may be weak-rooted.

Mangoes are large trees and should be planted 10.5 to 12 m (35 to 40 ft) apart. Mango is not particular about soil as long as it is well drained. It prefers slightly acid soil and may become chlorotic in alkaline soil. It prefers sun and will not flower in the shade. During tree establishment, phosphorus is important for root development, and nitrogen is applied to sustain the first years of growth. Nitrogen and potassium are needed by bearing trees for good yields. Applications of nitrogen to mature trees should be performed with caution because excess application may prevent blooming. Trees are usually irrigated during the first few years, until roots are established. The amount of rainfall is not as important as when it occurs. The optimal pattern is four months with 750 to 2,500 mm (30 to 99 in.) rainfall followed by a dry season.

The mango inflorescence is an erect, pyramidal, branched terminal panicle up to 0.6 m (2 ft) long, having several hundred to several thousand flowers. A tree may have from 200 to 3,000 panicles so that tremendous numbers of flowers are produced. The flowers are small, may be monoecious or polygamous, and are incompatible. Most of the flowers function as males by providing pollen, but a fraction are hermaphroditic and are able to set fruit. The flowers are visited by fruit bats, and a host of insects including flies, wasps, and bees can effect pollination. Only a few flowers per cluster set fruit. The ratio of male to perfect flowers is strongly influenced by environmental and cultural factors. Flowers during early bloom may be malformed. Rain or heavy dew during bloom are deleterious. High temperature, wind, and low humidity prevent flowering. Flowers have withstood temperatures as low as −4°C (25°F).

Until ten years of age, trees tend to bear fruit each year. After this point, they tend to alternate bear. This tendency may be on a whole tree or branch basis. Some cultivars may flower and fruit irregularly throughout the year. Up to three crops a year are borne by a Thai cultivar. People may try all kinds of tricks (fertilization, irrigation, smudging, pruning, girdling, hacking with a machete, applying salt) to induce flowering. Some of these treatments can have a detrimental effect on vegetative growth. The best approach is to plant cultivars with regular bearing habits like 'Tommy Atkins' (Fig. 20–3), 'Keitt,' 'Kent,' and 'Pope.'

Pruning of well-formed older trees is usually confined to removal of dead branches. Pruning is preferably done after fruiting, before a growth flush occurs. The fruit are variable in color, size, and shape. Mangoes ripen and may be picked when the flesh inside has turned yellow, regardless of exterior color. Fruit may be round, oval, ovoid-oblong, or somewhat kidney shaped. They tend to be somewhat lopsided.

Figure 20–3 'Tommy Atkins' mangoes. *Source:* Richard Pratt.

Some varieties may have a bit of a turpentine odor, and others are very rich and fragrant. Fruit flesh is very juicy and usually light yellow to deep orange in color. The delicious fleshy part, which is eaten, is the mesocarp of the fruit (ovary wall). Chlorophyll, carotenes, anthocyanins, and xanthophylls are all present in the fruit, but chlorophyll disappears with maturity, while anthocyanins and carotenoids increase during ripening. In some cultivars, masses of fibers extend from the stony pit (endocarp) into the flesh, making the fruit difficult to cut. Seed within the stone may be monoembryonic or polyembryonic (usually producing more than one seedling.)

It has been estimated that there are between 500 and 1,000 named varieties in India. Most of them arose through seedling selection; some have arisen through hybridization and selection. Most leading Indian cultivars are yellow skinned. The Indian race, grown in India, Pakistan, and Bangladesh, has highly flavored, brilliantly colored fruits well suited to commercial production. Examples of Indian cultivars include: early season cultivar 'Bombay Yellow'; early to midseason cultivar 'Langra'; midseason cultivar 'Alphonso,' which is known for its high quality; mid- to late-season cultivar 'Rumani,' which is considered to be an off-season bearer; and late-season cultivar 'Fazli.' The Indochina, or Saigon, race produces smaller, less attractive fruits with a yellowish-green color and nonfibrous flesh with a delicate flavor. This type is polyembryonic and generally reproduces true to seed. Cultivars of this type are generally known by the name of the district where they originated, such as 'Cambodiana,' 'Carabao,' and 'Pico.' In Thailand, the leading cultivar is 'Tong dum' (black gold). It has dark green skin with golden flesh. In Brazil, local cultivars and imported ones are cultivated: 'Bourbon' is a yellow-green skin type of nineteenth-century origin, 'Brazil'; 'Extrema' displays large 0.5 kg fruit; 'Haden' from Florida has red skin. 'Ah Ping,' 'Fairchild,' 'Gouveia,' 'Harders,' 'Keitt,' 'Momi K,' 'Pope,' and 'Rapoza' are recommended varieties in Hawaii. There are many varieties of global origin. Active public and private introduction programs are undertaking selection for the following primary traits: succulence, larger fruits, low fiber, low resin, smaller pit (stone).

Utilization

The mango is a highly esteemed tropical fruit. It is generally peeled to be eaten fresh, but it can also be frozen, dried, and canned; used in pies, jams, and jellies; or processed for juices and nectars. Eastern and Asian cultures use unripe mangoes for pickles, chutney and relishes. In India, unripe mangoes are sliced, dried, and made into powder for amchoor, a traditional preparation used for cooking. Seeds can be milled for flour and may also be eaten during periods of food shortages. The fruit are good sources of Vitamins A and C. Vitamin A content (12,500 ug retinol equivalent/kg) is the highest in cultivated crops, even exceeding that of red pepper (6,000 ug equivalent). The timber is used for boats, flooring, furniture, and other applications.

Pests and Diseases

Anthracnose (*Colletotrichum gloeosporioides*) affects fruits, inflorescences, and foliage. Anthracnose may be a problem, especially at fruit-setting time. 'Fairchild' is considered somewhat resistant to anthracnose and is favored for humid areas. Powdery mildew (*Oidium mungiferae*) on inflorescences, common in India, is a problem in Brazil. Trees may also be attacked by mango scab (*Elsinoe* sp.).

Scale is considered the worst pest. Numerous species, including lesser snow scale, Dictyospermum scale, Florida red scale, soft scales (pyriform scale, mango shield scale, acuminate scale, wax scale), and armored scales all attack mangos. Other major insect pests are avocado red mites (*Oligonychus* sp.), tumid mites (*Tetranychus* sp.), and broad mite (*Polyphagotarsonemus*). Mealybugs, stem (or twig) borers, and thrips also can build up to sufficiently high levels on mango trees to cause severe damage and greatly limit fruit production. Mango hopper is a serious pest in India. Mediterranean (*Ceratitis capitata*) and Mexican fruit flies are frequently a problem. To manage fruit flies, many commercial producers have implemented more stringent field sanitation, including immediate plowing under of decaying fruit to destroy breeding areas. Improved lures and bait are now available, and release of sterile flies to mate with wild flies helps to suppress populations. Mango trees are also affected by mango decline, a problem associated with micronutrient deficiency. Internal breakdown of the fruit is an important problem. Its cause has not yet been determined.

Papaya (*Cariaca papaya L.*) CARICACEAE

Common names of cultivated papaya are paw paw (Australia), mamao (Brazil), and tree melon. *C. papaya* plants have never been found in the wild, but a close relative, *C. peltata*, grows wild in southern Mexico and Central America, leading to the supposition that *C. papaya* may have originated in this region. Related species are babaco (*Cariaca pentagona*), mountain papaya (*C. pubescens*), and chamburo (*C. stipulata*). The early Spanish and Portuguese explorers carried the papaya to most tropical countries throughout the world.

Papaya is cultivated primarily for its ripe fruits. It is favored by tropical people as a breakfast fruit, as an ingredient in jellies and preserves, or cooked in various ways. The enzyme papain may also be extracted from green fruit. The world's leading producers are Brazil, Mexico, India, Nigeria, and Indonesia (FAOSTAT, 2004). Most papaya fruits are consumed locally because of the difficulties in long-distance transportation of the tender fruits. The practice of air shipments to the US mainland and to Japan, however, has given impetus to the Hawaiian industry. Most US supplies of papaya come from Hawaii, with smaller amounts from Mexico and the Dominican Republic.

Papaya is reported to tolerate annual precipitation of 650 to 4,300 mm (mean = 1,920 mm, or 76 in.) and annual temperatures of 16.2° to 26.6°C (mean = 24.5°C, or 76°F).

It grows best below 1,500 m in well-drained, rich soil of pH 6 to 6.5. Cultivation is known in soils ranging in pH from 4.3 to 8.0. It is reported to tolerate drought and high pH. It is extensively cultivated—as far north and south as 32° latitude. At extreme latitudes, however, the weather is often too cool in most years for papayas to ripen properly. Papaya becomes almost weedy in some areas of the tropics.

Growth and Development

The papaya is an erect, short-lived, fast-growing, woody tree or shrub. It usually grows to a height of several meters but may reach 8 m to 9 m (26 ft to 30 ft) tall, with a trunk diameter of about 20 cm (8 in.). The hollow green or deep purple trunk is straight and cylindrical with prominent leaf scars. Papaya generally branches only when injured. The leaves emerge directly from the upper part of the stem in a spiral. The leaves remain clustered near the top of the plant. Leaves are large, typically 40 to 60 cm (16 to 24 in.) wide, soft, palmate, and alternate, and are borne on long petioles. Petioles are nearly horizontal, ranging from 30 to 100 cm (12 to 40 in.) long. The leaf blade is deeply divided into five to nine main segments. The leaves have prominent, yellowish ribs and veins. The life of a leaf is generally four to six months. It bears clusters of delicious, mild-flavored large or small fruits varying in shape from spherical to pearlike.

Flowers and fruits appear nearly continuously all year. Papayas are usually dioecious in nature; thus, trees differ from each other because some are either male (staminate) or female (pistillate). Some are perfect (bisexual), having both male and female parts in the same flower. The five-petaled flowers are fleshy, waxy, and slightly fragrant. Male flowers are clustered on panicles 1.5 to 1.8 m (5 to 6 ft) long. At certain seasons, some plants produce short-stalked male flowers, at other times, they produce perfect flowers. This change of sex may occur temporarily during high temperatures in midsummer. Certain varieties have a propensity for producing certain types of flowers. For example, the 'Solo' cultivar has flowers of both sexes 66 percent of the time, so two out of three plants produce fruit, even if planted singly. How pollination takes place in papayas is not known with certainty. Wind is probably the main agent because the pollen is light and abundant, but thrips and moths may assist. Hand pollination is sometimes necessary to get a proper fruit set. For heavy production of high-quality fruits, papayas must be grown in warm, 21° to 26.5°C (70° to 80°F), frost-free climates in full sun. Lower temperatures give poor results.

Fruit are classified as a large yellow to greenish-orange berry, oblong to nearly globose or pyriform (pear-shaped), about 7.5 cm (3 in.) long, and bitter. Wild types grow up to 45 cm (18 in.) long, with flesh 2.5 to 5 cm (1 to 2 in.) thick. Cultivars are sweet, juicy, and of orange color. The seeds, numerous in the central cavity, are rounded, blackish, and about 0.6 cm (0.2 in.) in diameter. Each is enclosed in a gelatinous membrane (aril).

Propagtion

Papayas are normally propagated by seed. Seeds retain their viability for two to three years when kept dry in airtight containers. In areas with summer and winter seasons, seeds may be sown in cold frames or boxes during January or in the open in March. Soil used for seedbeds should be absolutely free from nematodes that cause root knot (*Meloidogyne*). Early planting is much to be desired to make a vigorous plant before the beginning of the following winter. Seeds germinate in two to three weeks. When two or three true leaves have formed, seedlings should be transplanted, spacing them 5 to 7.5 cm (2 to 3 in.) apart in the seedbed. When plants are 10 to 20 cm (4 to 8 in.) tall, they can be set in their permanent places in the field. Seedling papayas do not transplant well.

Trees from seed sown in early spring should fruit the following winter and continue to bear some fruit nearly every month in the year thereafter. Bloom to maturity is five to eight months. The average life of a papaya tree is considered to be two or three years, but trees may live in the wild twenty-five years or more. Yield declines after the first few years. All inferior and wild male trees in a region should be destroyed so that their pollen cannot fertilize blossoms of trees from which seed is to be selected. Seed for new plantings should be saved from perfect-flowered plants whenever possible so that one can be reasonably assured of the source of pollen. Propagation by tissue culture is also possible. Attempts at grafting and rooting shoots have not been successful on a commercial scale.

Production

The usual planting distance is about 3 to 4 m (10 to 13 ft) apart each way, giving about 1,750 trees/hectare (700 trees/acre). The more vigorous growing plants are usually the males, and some may be discarded when selecting plants for field planting. In addition, planting two or three plants on a hill allows another opportunity to eliminate some males when flowers appear. It is necessary to leave about one male plant for each twenty-five females to ensure pollination. Transplants must be watered and shaded. Addition of mulch gives better results than does clean culture. The mulch helps to reduce weeds, preserve moisture, and shade the soil from the hot summer sun. It can also assist in preventing the burning out of humus and nitrates in the top soil layer.

Papayas grow on many soil types, but they do not tolerate salinity in either irrigation water or soil. Good aeration and drainage are necessary. A light, well-drained soil with a pH of 6.5 to 7.0 is optimal. Papaya trees need ample soil moisture at all times but they also can be killed by excess moisture. If rainfall is lacking during parts of the year, irrigation is required to supplement the rain. For good plant growth and production, the various mineral nutrients must be readily available in the soil. The kinds and amount of fertilizers required depend on the soil type. Ideally, the soil should be moist in hot weather and dry in cold weather. A plant that has been injured by frost is particularly susceptible to root rot.

The fast-growing papaya responds to regular applications of nitrogen fertilizers, but exact rates generally are not well established. Phosphorus deficiency causes dark green foliage with a reddish-purple discoloration of leaf veins and stalks. Generous applications of manure or commercial fertilizers are considered beneficial.

Papayas do not need to be pruned, but some growers pinch the seedlings or cut back established plants to encourage multiple trunks. Papaya trees are somewhat delicate and should be protected from strong winds. Recovery of partially lodged trees may require one or two months.

Mature fruits are produced about ten months after planting and then year-round. Individual fruits are harvested when they are at a mature-green stage and show a tinge or more of yellow at the apical end. Fruits can be picked by hand when the trees are young, but as they become taller, their fruits cannot be reached. A tree may bear thirty to 150 fruits per year, each weighing up to 4 kg (8.8 lb) or more. Up to 3.4 MT marketable fruit/ha (1.5 T/acre) is a credible yield, but yields of 100 MT (44.6 T/acre) of fruit are reported. On fertile soils 'Honey Dew' has been reported to yield about 100 MT/ha fruit (44.6 T/acre) with 1,750 bearing trees/ha (709 trees/acre) the first year, eighty-five (40 T/acre) the second, and seventy-five (33 T/acre) the third.

Cultivars

Different types of papaya are grown in the various tropical countries—some large-fruited and some small-fruited. The main commercial cultivar grown in Hawaii is 'Puna Solo,' which was selected from an introduction from Barbados, West Indies, in 1911. It has small fruits of about 0.5 kg (1 lb) each. Fruit are round and shallowly furrowed in female plants, pear-shaped in bisexual plants. Several new papaya cultivars have been developed in Hawaii, originating from controlled crosses. Some of these are 'Waimanolo,' 'Sunrise,' 'Higgins,' and 'Wilder.' 'Sunset' is a Solo type with small to medium-size, pear-shaped fruit, with orange-red skin and very sweet flesh. It is borne on dwarf, high-yielding plants. 'Mexican Yellow' is a very sweet and flavorful, yellow-fleshed papaya. It bears medium to large fruit that can weigh up to 4 kg (8.8 lb). It is generally not as sweet as Hawaiian types. 'Mexican Red' is a rose-fleshed papaya that is lighter in flavor than Mexican Yellow. It bears medium to very large fruit. Some hermaphroditic cultivars grown in Florida are 'Fairchild (no. 745),' 'Graham,' 'Kissimmee,' 'Betty,' and 'Bluestem.' Virus-resistant transgenic cultivars have been recently introduced (see Pests and Diseases section).

Utilization

In the tropics the papaya is used for many purposes. Fruits are eaten fresh and the juice makes a popular beverage; young leaves, shoots, and fruits are also cooked as vegetables. Green papayas should not be eaten raw because they contain latex, although they are frequently boiled and eaten as a vegetable. Many manufactured products are made, especially papaya juice concentrate and dried slices. They are good sources of vitamins A and C and of potassium.

Immature fruits contain papain, a protein-digesting enzyme, which is extracted from the latex of the skin. It is obtained by scoring or injuring the fruit, followed by collecting and drying the latex. Latex drips into a suitable container and is sundried or oven-dried at 55° to 60°C (131° to 140°F). The same fruits are tapped in different cuts at weekly intervals. Tapped fruits are ultimately edible, so that both fruit and latex are harvested. Trees produce about 50 percent of their papain yield in the first year, 30 percent in the second year, and 20 percent in the third year. Yields range seventy to 130 kg/ha (154 to 287 lb/acre). Crude papain represents 5 percent of the fresh latex, and latex yields are close to 1 MT/ha/year, so papaya may yield 20 to 25 kg (44 to 55 lb) dried papain per hectare in the first year; 90 to 100 kg (198 to 228 lb) in the second year, 60 to 90 kg (132 to 198 lb) in the third year, 30 to 40 kg (66 to 88 lb) in the fourth year, and 20 kg (44 lb) or less in the fifth year. Presently Uganda and Tanzania are the principal producers of papain. The United States is the principal importer of papain. It is utilized for meat tenderizers, beer treatment, chewing gum, and in the textile and tanning industries.

Pests and Diseases

The papaya fruit fly (*Toxotrypana*) is a serious pest in some areas, so that bagging fruit while still quite small is the only means of protecting it. Papaya whitefly (*Trialeurodes*), hornworms, and corn earworms (*Heliothis*) may also be pests. The papaya webworm (*Romalopalpia*) produces webs around the fruits and stems therein.

Both fungal and viral diseases attack papayas. Anthracnose (*Colletotrichum*) is a common fungal disease in Hawaii that attacks the fruits. Some local packinghouses use a hot water bath to help minimize anthracnose as a postharvest problem. A *Phytophthora* blight is the most important disease. It attacks above-ground portions of the plant as well as the roots. Powdery mildew (*Oidium*) can sometimes present a problem. Two viral diseases, papaya mosaic and papaya ring spot, caused by papaya mosaic virus (PMV) and papaya ring spot virus (PRSV), respectively occur in Hawaii and have been an ongoing threat to the industry. Initially, papayas were moved from Oahu to Hawaii to escape PRSV infection during the 1960s. By the 1990s, infections were so widespread that growers were replanting trees annually because of the severity of PRSV, which is transmitted by the green peach aphid (*Myrus*). Production had fallen by nearly 40 percent by 1997. A tansgenic cultivar 'Rainbow' (yellow-fleshed) and a red-fleshed transgenic X conventional cultivar ('Sunup') have been made available and the industry has bounced back. The industry in Florida has also been hit hard by viruses and the papaya wasp. Nematodes species isolated from papaya are numerous.

Mites are a serious pest on papayas but can be controlled by sulfur sprays. The melon, Oriental, and Mediterranean fruit flies attack papaya fruits in Hawaii and are the cause of the strict quarantine and fumigation requirements. Before Hawaiian papayas can be shipped to the US mainland, they must be treated to destroy fruit flies. A two-stage hot water dip treatment has been adopted as a quarantine procedure for papayas destined for export. The standard double-dip treatment consists of immersing the fruit in 42°C (108°F) water for thirty to forty minutes, followed by a second immersion in 49°C (120°F) water for twenty minutes. For fruit to qualify for the double-dip method, the blossom end of the fruit should have a Hunter 'b' (color) value of less than 27.5.

Pineapple [*Ananas comosus (L.) Merrill*] BROMELIACEAE

The pineapple is today one of the best-known of all tropical fruits, and it is a leader in both area-cultivated and international commerce. It is a low-growing, herbaceous perennial. Bromeliads are monocotyledonous tropical plants, and pineapple is the only species in the genus producing tasty fruits. The pineapple is thought to have originated in southeastern Brazil, Paraguay, and northern Argentina in central South America, although other researchers argue for a more northern center of origin. Improved cultivars selected and propagated by the indigenous people in this region had become widely distributed throughout tropical America when Columbus encountered the fruit on the island of Guadeloupe in 1493 on his second voyage. It was called anana by the natives who grew it, Pina de Indes by Spanish explorers, and King Pine by European elite who could afford it. The Portuguese and Spanish explorers apparently disseminated the pineapple throughout the world's tropical regions in the 1500s by carrying the fruit crowns, which remained viable throughout the long voyages. Pineapple was grown in Hawaii as early as 1813, but planting on a fairly large scale did not begin until after 1850. The first commercial venture was near Kailua on the island of Hawaii. It was short-lived, however, and it was not until the middle 1880s that a significant industry began to develop. At the turn of the nineteenth century, Florida was the leading producer of pineapples, but the industry was decimated by a presumed disease, which later was found to be mealybugs. Hawaii became the leading producer. Today, the leading producers of pineapple are Thailand, the Philippines, Brazil, China, and India.

Adaptation

Pineapple grows best under uniformly warm temperatures year-round. While plants might survive −2°C (28°F), significant leaf damage would severely weaken the plant. Temperatures higher than 32°C (90°F) can lower fruit quality, and higher daytime temperatures can sunburn exposed fruits. Because pineapple is a xerophytic plant, it uses water very efficiently, so the crop is grown in areas having relatively low rainfall, from 50 to 200 cm annually (20 to 80 in.). The optimum for commercial production is in the range of 100 to 150 cm annually. In the past, irrigation was used only during very dry periods or in very dry areas. It is now recommended whenever less than 5 to 10 cm of rainfall occurs during a one-month period, and particularly as an aid for the establishment of young plantings. Commercial pineapples are generally grown at somewhat higher elevations close to the equator.

Propagation

Pineapple plants are propagated vegetatively using suckers that arise below ground (ratoon suckers) or in the leaf axils, slips growing from the fruit itself or along the stalk below the fruit and crowns. The crown is the vegetative shoot on top of the fruit (Fig. 20–4). New plants from crowns require about twenty-four months to fruit; from slips, about twenty months; and from suckers, seventeen months. Plantings are generally made to coincide with periods of heavy rainfall. Slips and suckers are preferred in commerce. Plants may be dispersed here and there in a compound garden, or they may be planted in long rows on beds under plastic mulch in commercial plantings.

Figure 20–4 Pineapple plant with immature fruit. The uppermost fruit shows the crown, which can be used for vegetative propagation. *Source:* Ken Chamberlain, OSU-OARDC.

Each original plant set out produces one fruit. The pineapple fruit forms in a complex fashion. At the terminal vegetative growing point of the plant, a central axis core develops, bearing an inflorescence of 100 or more closely attached lavender flowers arranged spirally, each subtended by a bract. This central axis core terminates in a crown or rosette of small leaves. All parts of each flower, together with the bract, develop into the fleshy fruitlet tissue. As it develops, the entire infloresence converts to the single pineapple fruit (botanically, a sorosis) by the coalescence of the inner stem tissue and fruitlets (developed ovaries and other parts of the flowers).

Production

Pineapple plants absolutely require soils with good internal drainage. They grow and fruit best in well-drained, slightly to somewhat acidic (pH range 4.5 to 5.0) sandy loams and volcanic soils. Field preparation of large commercial fields is usually done with heavy machinery. Fields are often covered with black plastic and fumigated to manage nematodes. Restrictions and concerns regarding the use of fumigants have encouraged the development of alternative management techniques. Nonvolatile nematocides can be introduced through drip irrigation systems with success. Pineapples need fertile soils and generally require added fertilizers, particularly nitrogen and potassium. Low levels of phosphorus are needed, and excess levels may reduce yields and lower quality. Spiny leaved varieties can make hand weeding difficult. Chemical induction of flowering is frequently practiced in plantations to allow synchronous harvest of fruit.

Cultivars

'Smooth Cayenne' types are grown extensively in the tropics and are the major processing varieties. They are also found fresh in many markets but are the most sensitive to internal browning caused by low temperatures. Internal browning is also known as blackheart or chilling injury. Fruit of 'Smooth Cayenne' weigh 2.3 to 2.7 kg (5 to 6.0 lb) under good culture. The 'Queen' group has superior fresh-market quality, and it is more frequently cultivated in subtropical areas because it is less sensitive to chilling injury. The group is, however, more prone to sun-scalding of the fruit. The 'Natal Queen' cultivar produces fruit for the fresh market weighing 0.9 to 1.4 kg (2 to 3 lb). The 'Spanish' group are generally small to medium; vigorous plants. 'Red Spanish' is a major (white-fleshed) fresh-market pineapple in the Caribbean region, and it is a little hardier than other cultivars. Its fruit weighs 0.9 to 1.8 kg (2 to 4 lb), and its leaves are spiny. 'Singapore Spanish,' 'Singapore Canning,' and 'Nana Merah' are the principal canning pineapples in west Malaysia because of their adaptability to peat soil. They also have brightly colored yellow flesh. There are varieties of pineapple with much better eating quality than 'Smooth Cayenne' or 'Red Spanish'. However, those with better eating quality do not ship very well, so they are not likely to be encountered in export markets. Among the better pineapples for fresh consumption are 'Pernambuco' (pale yellow-flesh) and 'Eleuthera,' weighing 0.9 to 1.8 kg (2 to 4 lb), and 'Abakka,' weighing 1.4 to 2.7 kg (3 to 6 lb). All three of these have spiny leaves, are members of the 'Abacaxi' group, and are leading clones in Brazil. Another cultivar in the Abacaxi group, 'Sugarloaf,' is produced in Mexico. 'Cabezona' is a triploid variety from the Spanish group producing very large fruit up to 6.5 kg (14 lb) that is grown in the state of Tabasco, Mexico, and a small part of Puerto Rico. A clone from Puerto Rico with good shipping qualities, 'PR 1–67,' is thought to have arisen from a cross between 'Red Spanish' and 'Smooth Cayenne.' The 'Maipure' group provides fresh-market fruit in Central and South America and apparently has been underutilized in breeding programs.

Pests and Diseases

The major insect problem is the mealybug (*Dysmicoccus* sp.), which is tended and protected on pineapple plants by big-headed ants (*Pheidole megacephaola*). An effective ant control program makes it possible for the mealybug to be kept under control using baits or by biological means, mostly by predation. Ant-control costs usually are less than those incurred in controlling mealybugs directly. There are no insects that attack the pineapple fruit in Hawaii at the present time. Diseases of pineapple include root rot and heart rot caused by *Phytophthora parasitica* Dastur. and *P. cinnamomi* Rands. Root and heart rot are exacerbated by poor soil drainage and high rainfall. Black rot, caused by *Thelaviopsis paradoxa,* is characterized by a soft, watery rot. The rot is also called butt rot, blister rot, and soft rot. Interfruitlet corking caused by species of *Fusarium* and *Penicillium* fungi, and pink disease caused by a Cetamonas bacteria may also occur periodically. Black heart, and root rots are controlled by a pre-plant dip containing fungicides and by postplant sprays where problems are expected to be severe. Root rot is also controlled to some degree by keeping soil pH low, in the range of 4.5 to 5.5. While some losses due to fruit diseases occur, they are not large enough to justify the costs of control, even though annual losses may reach a few million dollars. Pineapples are also sensitive to attack by nematodes.

SUMMARY AND REVIEW

The tropics and subtropics are characterized by rich cultural and biological diversity. A large portion of the world's population resides in these regions, and agriculture is a primary occupation. Land forms, soil conditions, and climate determine the feasibility of crop production. Temperature and rainfall are the most important determinants for crop production. Humid tropics, intermediate tropics, tropical savannah, dry tropics, and desert describe the range of dry period durations from shortest (less than 2.5 months) to longest (ten to twelve months). The pattern of wet/dry periods is also an important factor in crop productivity. Tropical soils differ from temperate soils in several ways. Tropical soils are low in organic material, high in aluminum and manganese, low in CEC and base saturation, and low in major and minor nutrients. Important soil orders found in the tropics are the oxisols, utisols and alfisols, vertisols, andisols, alluvial soils, inceptisols and entisols, spodozols, and aridisols.

    Several different cropping systems are found in the tropics and subtropics. Shifting cultivation is practiced at the community level. The location of the production fields shifts around a permanent home. During the fallow period, natural vegetation regrows. When put back into production,

nutrients held in the vegetative biomass are released by cutting, drying, and burning. Permanent upland cultivation is practiced in semiarid regions with distinct dry seasons. Planting is delayed until the rainy seasons appears to be established. Arable irrigation farming, as its name implies, relies on irrigation. In this case, water is gravity fed and allows production of multiple high-value crops per year. Flooded systems produce crops of tropical wet rice. The fields may be flooded by rainfall or irrigation. Rice may be the only crop or it may be grown with other crops. One or more crops may be grown per year. Cereal-based systems rely mainly on sorghum and millet, but maize may also be grown where water and nutrients are more plentiful. Cowpeas, beans, and peanuts are legumes that are traditionally interplanted with cereal crops. Mixed annual/perennial systems involve some type of root or tuber crop and another root crop or a cereal or legume.

FOOD FOR THOUGHT

1. Describe how increases in altitude can create subtropical, temperate, and even arctic conditions in tropical areas. What does this difference in altitude mean in terms of the kinds of crops that can be produced in the tropics?

2. What is intercropping? Describe some of the different ways that intercropping is practiced in tropical and subtropical crop production systems.

3. What characteristics generally differentiate tropical and subtropical soils from most temperate soils? Describe the factors that create these differences.

4. What are the different kinds of rainfall patterns in the tropics and subtropics? How do these patterns affect crop production?

SUPPLEMENTARY READING

ALLAN, W. 1965. *The African husbandman.* Edinburgh, U.K.: Oliver and Burgy.

FAOSTAT (Food and Agriculture Organization Statistics). 2004. **http://faostat.fao.org/faostat/**.

FUSSELL, L. K. 1992. "Semi-arid cereal and grazing systems of West Africa," pp. 485–518. In: *Ecosystems of the World, Vol. 18, Field Crop Ecosystems.* C. J. Pearson, ed. The Netherlands: Elsevier Science Publishers.

HAHN, S. K. 1984. "Tropical root crops: Their improvement and utilization." In: Special Publ. No. 2, Ibadan, Nigeria: International Institute of Tropical Agriculture (IITA).

JUO, A. S. R., and H. C. EZUMAH. 1992. "Mixed root-crop systems in wet sub-Saharan Africa," pp. 243–255. In: *Ecosystems of the World, Vol. 18, Field Crop Ecosystems.* C. J. Pearson, Ed. The Netherlands: Elsevier Science Publishers.

KER, A. 1995. *Farming systems of the African savanna.* Ottawa, Canada: International Development Research Centre.

LAMBERTS, M., and J. H. Crane. 1990. "Tropical fruits," pp. 337–355. In: *Advances in new crops.* J. Janick and

J. E. Simon eds. Portland, Ore.: Timber Press. **http://www.crfg.org/pubs/ff/papaya.html**

NAKASONE, H. Y., and R. E. Paull. 1999. *Tropical fruits.* Oson, UK: CABI Publishing, CAB International.

NATIONAL RESEARCH COUNCIL. 1993. Sustainable Agriculture and the Environment in the Humid Tropics. Committee on Sustainable Agriculture and the Environment in the Humid Tropics, Board on Agriculture and Board on Science and Technology for International Development, National Research Council. Washington, D.C.: National Academy Press.

NORMAN, M. J. T., C. J. PEARSON, and P. G. E. SEARLE. 1995. *The ecology of tropical food crops,* second edition. Cambridge University Press.

NYE, P. H., and D. J. GREENLAND. 1960. "The soil under shifting cultivation." Technical Communication 51. Harpenden, U.K.: Commonwealth Bureaux of Soils.

O'BRIEN, P. J. 1972. "The sweet potato, its origin and dispersal." *American Anthropologist* 74:342–365.

OKIGBO, B. N., and D. J. GREENLAND. 1976. "Intercropping systems in tropical Africa," pp. 63–101. In: *Multiple cropping.* Wis.: American Society of Agronomy.

PLUNKETT, D. 1997. "U.S. and Mexican avocado sectors: A comparison." *Agriculture Outlook,* June 1997, pp. 22–23. Economic Research Serivce/ USDA.

ROGER, P. A. 1996. *Biology and management of the floodwater ecosystem in ricefields.* Manila, Philippines: International Rice Research Institute.

SANCHEZ, P. A. 1976. *Properties and management of soils in the tropics.* New York: Wiley Interscience.

SMITH, R. L. 1992. *Elements of ecology,* third edition. New York: Harper-Collins Publishers.

STRAHLER, A. N. 1975. *Physical geography,* fourth edition. New York: John Wiley.

SUPPO, F. R. 1992. *El aguacate.* Planta Alta, Mexico: A.G.T. Editor.

VERGARA, B. S. 1992. "Tropical wet rice systems," pp. 167–182. In: *Ecosystems of the World,* Vol. 18, *Field Crop Ecosystems.* C. J. Pearson, ed. The Netherlands: Elsevier Science Publishers.

WEBSTER, C. C., and P. N. WILSON. 1980. *Agriculture in the tropics,* second edition. London: Longman Group Limited.

WILSON, L. A., L. B. RANKINE, T. U. FERGUSON, N. AHMAD, S. GRIFFITH, and L. ROBERTS-NKRUMAH. 1992. "Mixed root-crop systems in the Caribbean," pp. 205–234. In: *Ecosystems of the World, Vol. 18, Field Crop Ecosystems.* C. J. Pearson, ed. The Netherlands: Elsevier Science Publishers.

WRIGLEY, G. 1982. *Tropical agriculture,* (fourth edition). London: Longman Group Limited.

CHAPTER

Nursery Production

Gary Bachman

KEY LEARNING CONCEPTS

After reading this chapter, you should be able to:

♦ List the factors that go into site and product selection for a nursery.
♦ Explain the principles of field and container (including pot-in-pot) nursery crop production.
♦ Understand the importance of and the methods for testing media fertility for container production.

The nursery industry is an exciting business and provides many opportunities for students interested in plant production. The production of nursery plants for profit has the potential of providing many personal and financial rewards to those who are willing to work hard. On the surface, operating a nursery appears to be fairly simple, but the nursery business is very complex. Operating a nursery requires not only knowledge of and skills in plant basics, but experience in labor-management and marketing of nursery crops.

The nursery industry is diverse. People often think that nurseries produce shrubs and trees, but increasing numbers of nurseries are producing flowering perennials, annual plants, ground covers, and hardscaping materials. **Hardscape** consists of the nonplant components of a landscape, for example, patios, decks, and so on. Always remember that a nursery is a business. Like any other business, success depends on imagination, determination, planning, and good management of resources available to the nursery.

SITE SELECTION AND ANALYSIS

The evaluation of a proposed nursery site is vital to the success of the nursery. Four categories must be considered before establishing a nursery. These categories are ecological, economic, sociological, and biological in nature.

The first ecological consideration that must be examined is water. An adequate supply must be available through the growing season either from municipal systems or from surface or subsurface sources of good quality for irrigation purposes. The second ecological consideration is the physical features of the property. Ideally the land selected should be slightly sloping. Land with the steep slopes and wet areas should be avoided. The land also should have good field drainage and can be checked with a percolation test conducted by a local soil scientist. The quality of the drainage required depends on the type of production system selected. A third ecological consideration is the prevailing weather patterns of the area. Will you need to provide winter protection? What is the average rainfall and flood potential? Is the proposed site in a known hail pocket? Are sources of industrial pollution situated nearby? All these questions must be answered prior to the first plant being put into the ground.

The site analysis includes economic factors, with probably the biggest expense being land cost, depending on the size of the nursery. From the startup situation, a nursery often has to wait three to five years before seeing a return on investment dollars. Labor is another very important consideration when looking at the economics of a nursery, with the availability of labor being the biggest issue. The nursery owner must realize that he or she cannot do everything. Marketing is vital to any nursery business. A retail and/or wholesale nursery requires sufficient space for both growing and sales areas, along with parking for both staff and customers. The landscape nursery provides landscape and landscaping services as well as retail and/or wholesale plant sales. A third marketing decision is the sales area to be serviced: local, regional, and/or national. Bare root plants can be shipped economically nationwide via postal service or commercial carrier due to reduced weight. Balled and burlap plants have a limited shipping range because of the weight of the soil ball containing the root system. This range is usually within 300 to 500 miles of the nursery, and shipping is limited to the plants' dormant season. Container-produced plants have the advantage because they can be marketed all four seasons. Potential niches within the nursery industry must also be explored.

A question that must be answered before any production is begun at a nursery is, which plants should I grow? Or phrased another way, which plants can I sell and to whom do I sell them? There are literally thousands of different species and/or cultivars from which a nursery owner can choose. Not only must the nursery grower decide what species to grow, but he or she must be able to predict the needs of the market anywhere from three to ten years in the future. This timeline should allow ample time for propagation and subsequent growth (in other words, production time) of adequate numbers of the desired plants. A nursery with a good marketing plan can try to create market demand for a particular species of plant, but change in the nursery industry requires time. The rule of thumb for nursery inventory considerations is that 70 percent of production should be industry standard species. These are species for which there is a known market. The remaining 30 percent of the inventory can include plants that are currently being promoted as plants having a bright future in the landscape industry. Plants not to grow include:

♦ Easily produced plants that every other nursery is already producing.
♦ Plants that the nursery grower personally likes.
♦ Slow-growing plants, at least in the initial stages of a nursery.

All plants should pay for themselves regardless of marketing strategy.

Sociological factors must also be taken into consideration. The demographics of an area must be familiar to the nursery. Population numbers and income levels are important if the nursery has targeted local sales. Zoning and local regulations must be adhered to by the nursery. Examples of such regulations include proximity of neighbors, and any restrictions regarding local road weight limitations for streets and bridges.

Biological factors that must be taken into consideration have a more direct production-oriented effect for the nursery. In the initial stages of nursery establishment, it is important to know if a site has a predisposition for certain pests. If the area has a large deer population, you can rest assured that the deer will cause browse damage on your most expensive items.

Once you have decided which crops to grow and have performed a proper site analysis, the next decision is which production system will best suit the potential business: nursery field production; nursery container production; or pot-in-pot, which is a hybrid production system between field and container production strategies. These options will be discussed in the following sections.

NURSERY FIELD PRODUCTION

A field nursery grows nursery crops to marketable size in fields laid out similar to other field crops. Roadways, field borders, fence rows, wooded areas, and areas unsuitable for cultivation can easily account for 30 to 40 percent of a field nursery's land area, leaving only 60 to 70 percent of the space available for actual production. A field is generally divided into blocks, separated by 10- to 12-ft wide grass or gravel-covered roadways. Primary roadways in a field nursery need to be at least 15 to 20 ft wide to accommodate trucks, tractors, and various harvesting machinery used to perform maintenance, planting, and harvesting tasks (see Fig. 21–1).

The design of the production blocks is vital for the success of producing plants through field production. Spacing plants in the field should make efficient use of land, facilitate

Figure 21–1 Field-grown nursery trees in rows with overhead irrigation.

All field stock is harvested during the dormant season. Trees that are transplanted when dormant will establish a new root system capable of supporting the new foliage. Using a tree spade to harvest a tree is a straightforward process. The tree spade is positioned around the trunk of the tree and large blades are hydraulically forced into the soil, cutting all lateral and tap roots. In many cases, 80 to 90 percent of the root system is left behind during this process. The remaining roots and attached soil (rootball) is lifted and placed into a wire basket lined with burlap. Once the tree spade is removed, the burlap is wrapped tightly around the rootball to hold it in place (see Figs. 21–2 and 21–3). A new technology, air spades, has been introduced for digging field-grown stock. The air space removes soil from around the roots during the digging process but does little damage to the roots (Fig. 21–4). After digging, the root mass is wrapped in a moist material such as sphagnum moss for storage and transport.

traffic and maintenance needs, and provide easy access to the field. The most common and most expensive mistake new growers make is to plant liners too close together in very narrow rows in the field. This practice is done with the good intention of digging every other plant early in the production cycle to allow additional growing space for the remaining plants. It is unlikely, however, that every other plant will grow equally with straight, well-branched trunks and canopies. Close spacing also makes harvesting extremely difficult, and digging one tree often damages two or three, making them unsalable.

The distance between the rows and between each plant depends on desired market size, width of maintenance equipment, and the type of harvest (either hand or mechanical). Wider spacing between rows ensures that plants will not be shaded and will have longer and stronger lower side branches, thus achieving a more normal shape. Proper plant spacing allows plants to remain in the field longer, if needed, if sales do not meet projections. Spacing within rows depends on final market size and if intermediate harvesting is to be practiced. A practiced rule of thumb is to plant the trees 3 ft apart for each inch of anticipated trunk diameter. For example, say a grower plants Red Maple liners with 3-ft spacing within the row. When the plants reach an inch in diameter, every other tree is harvested and the remaining trees are grown to a 2-in. diameter because they are now spaced 6 ft apart.

Harvesting field-grown nursery stock requires a mechanized tree spade or laborer experienced in hand digging. Tree spades can be purchased in a wide array of sizes to dig balls from 15 to 96 in. and bigger. The size of the root ball should be in proportion to the diameter of the trunk caliper. Nursery trees are measured 6 in. above the soil. For example, a 1¼- to 1½-inch caliper tree should have an 18-in. ball, a 2½- to 3-inch caliper tree should have a 28-in. ball, and so on.

Figure 21–2 Using a mechanical tree spade to harvest broadleaf evergreen trees.

Figure 21–3 Securing a mechanically harvested tree in a wire basket to hold the root ball together for shipping.

Figure 21–4 Digging plants using an air spade (photo courtesy of Barbara Fair and Ahlum and Arbor Landscaping).

Figure 21–5 View of a large container nursery.

NURSERY CONTAINER PRODUCTION

Almost every species and type of herbaceous and woody ornamental plant produced in a nursery setting can be grown in containers that sit on top of the ground (see Fig. 21–5). Growing plants in containers is vastly different from growing plants in the field and provides production advantages to the nursery grower. The grower can increase the number of plants per unit area when compared to field production: container-grown plants may be placed closer to each other because their root systems are controlled. Also crop rotation issues are eliminated because the plants are not growing in the soil itself. Injury caused by cultivating machinery used in field production is reduced. The grower can produce more uniform plants in containers when compared to plants produced in the field because optimum growing conditions may be maintained more effectively. Container production provides the grower complete control over the media, nutrition, irrigation, light, and spacing.

The root systems of container-grown plants suffer less damage compared to plants that are harvested from the field. A tree that's dug in the field has its root system reduced between 80 and 90 percent. Because of this, field harvesting can be accomplished only when the trees are in a dormant state, typically in late fall through very early spring. Container-grown plants, on the other hand, can be harvested and marketed throughout the year.

Container production does have disadvantages for the nursery grower to consider. The biggest disadvantage is the plant's total dependence on irrigation because the root system is limited to the container size and is not in actual contact with any source of groundwater. Another disadvantage is the limited development of a root system within the container. Normally, the root system grown in the field is approximately two to three times the diameter of the canopy of the plant. Because of this growth potential, plants that remain in containers for too long become root bound.

Extreme heating of the roots during the summer is a problem, especially in southern production facilities. On the other hand, winter protection becomes an issue for the root system, which must be protected from freezing during winter months in cold climates. Various techniques have been used by nursery growers for overwintering containerized nursery stock: clear or milky plastic poly houses or poly tents; consolidation with or without periphery covering; mass consolidation with a plant foam covering; and consolidation with a top covering of poly coated plant foam, white plastic, clear plastic/straw/clear plastic ("the sandwich"), clear plastic/grass, geotextiles, white plastic/microfilm, or microfoam (see Figs. 21–6 and 21–7).

Poly houses are routinely used to provide winter protection for containerized plants. Temperatures within poly houses depend primarily on the amount of sunlight entering the houses. More sunlight means higher temperatures within the structures. Unfortunately poly does a very poor job of holding heat within the houses and tents. On a clear, cold

Figure 21–6 Nursery structure used for winter protection of container-grown herbaceous perennials.

Figure 21–7 Applying opaque white polyethylene across container-grown plants for winter protection.

Figure 21–9 Blowover of container-grown plants. Blowover can be prevented by using pot-in-pot production.

winter day, heat can radiate out of a poly house in late afternoon or early evening as fast as it increases in the morning with the rising sun. Even so, plants in a poly structure generally remain several degrees warmer than the ambient outside temperature. In nearly all situations, except in the most extreme cold temperatures, this winter protection is enough.

POT-IN-POT PRODUCTION

Pot-in-pot production is a recent development in container-grown plant production. In this production system, a planted container is placed into a holder pot that has been permanently placed in the ground. Pot-in-pot production was first started in the southern states to help provide root protection from extreme summer temperatures, but it has really caught on in northern states because of the advantages of root protection during winter (see Fig. 21–8).

The pot-in-pot production system can eliminate many of the difficulties associated with conventional con-

tainer plant production. In conventional container plant production, winter protection for plant roots is costly and time-consuming. Wind tipping or container blowover is another time-consuming, laborious drawback of conventional container production (see Fig. 21–9). It is detrimental to plant quality because top-dressed fertilizers and media are knocked out of the pot. Irrigation applications can be missed or delayed, resulting in drought-stressed plants. Pot-in-pot production can prevent blowover. It also reduces heat stress and root-killing temperatures experienced in conventional container production. The sunken pots require much less irrigation than do conventional containers; with micro-irrigation, each pot has an individual emitter or spray stake, with almost no runoff. Pot-in-pot production requires no overwintering structures because the pots are insulated from freezing temperatures. Pot-in-pot production does require more management than does field production, but less than conventional container production. Pot-in-pot production in a way mimics field production.

MEDIA TESTING FOR CONTAINER-GROWN PLANTS

Certain physical and chemical properties of soilless media are necessary for growing high-quality nursery plants. The physical and chemical characteristics of soils that are important to a field nursery are covered in Chapter 6 of this text.

Each nursery grower should monitor four media properties as a course of normal production management. The four properties are the physical property of the media aeration and the three chemical properties of electrical conductivity (EC), pH, and nitrate nitrogen (NO_3^-) content. Each media property has simple techniques available for on-site nursery testing. The impact of all four on media management, and ultimately on plant quality, is important.

Figure 21–8 Pot-in-pot production field showing socket pot.

The composition of the growing medium is an extremely important consideration for growing plants in containers. Not only is the plant root system restricted to the limited volume of the container, but the perched water table created in the bottom of all containers further restricts the total root growing space. The perched water table is an area where all the pore spaces in the media are filled with water. This saturated area occurs no matter how many drainage holes are in the bottom of the container. It results from equilibrium between gravity and the capillary attraction of the media particles and the water molecules. The deeper the container, the lower the impact of the saturated area at the bottom of container and the greater the overall media aeration. It is recommended that media porosity be measured before any new media formulation is used, and that aeration porosity be checked at least three times, depending on the components of the media mix.

Aeration porosity is a measure of the amount of air space in the media mix at container capacity after a complete irrigation. The optimum aeration porosity for most woody plants is between 20 and 30 percent. Analysis of media porosity is very simple and can be completed by the nursery grower using the following procedure:

1. To measure the container volume, seal the drainage holes in the bottom of the container and fill it with water to the level it would normally be filled with container mix. Record this volume.
2. Empty and dry the container and fill it with growing mix. Slowly add water with a graduated cylinder until the mix is completely saturated. A very thin slick of water will appear on the surface once saturation is reached. Record the total volume of water added as "total pore volume."
3. Place the container over a watertight pan and remove the seals from the drain holes. Allow the free water to drain out of the container. This may take several hours. Measure the amount of drained water and record this amount as "aeration pore volume."
4. The preceding three steps will allow you to calculate total porosity, water-holding porosity, and aeration porosity using the following three equations:

$$\text{Aeration porosity (\%)} = \frac{\text{aeration pore volume}}{\text{container volume}} \times 100$$

$$\text{Total porosity (\%)} = \frac{\text{total pore volume}}{\text{container volume}} \times 100$$

$$\text{Water-holding porosity (\%)} = \text{total porosity} - \text{aeration porosity}$$

ELECTRICAL CONDUCTIVITY (EC) AND pH

Fertilizers and other dissolved salts change the ability of the solution to conduct electricity. This change in electrical conductivity allows growers to measure roughly the fertility of their media. Portable electrical conductivity and/or pH meters can be used by growers to measure the electrical conductivity and pH of the soil solution within their container mixes. These portable meters for the nursery industry range in price from $60 to over $1,000, with the more expensive meters generally being more durable and lasting many years. Growing media should be tested at least every two weeks, preferably every week, for electrical conductivity and pH. Keep a chart of the pH and electrical conductivity values collected. With this information, you can establish trends, for example, whether the electrical conductivity and/or pH is increasing, decreasing, or remaining constant. Nursery growers need to pay close attention to electrical conductivity levels. Growers need to realize, however, that not all electrical conductivity is related to the fertilizers used. It is also important to establish the nonnutritional background content of the irrigation water.

Many nursery growers use controlled-release fertilizers. Frequent monitoring of the media electrical conductivity is the easiest way for a grower to track the nutrient release from controlled-release fertilizers. When the electrical conductivity begins to approach that of the irrigation water, the controlled-release fertilizer is becoming exhausted and supplemental fertilization is required.

The pH scale of zero to 14 is a logarithmic scale, with a pH of 7 being neutral. pH values greater than 7 are considered basic, while pH values less than 7 are considered acidic. Nutrient availability to plants is affected more by pH than by any other factor. Plants in a high-pH soilless media may express micronutrient deficiencies of iron (Fe), boron (B), zinc (Zn), manganese (Mn), and molybdenum (Mo) (Fig. 21–10).

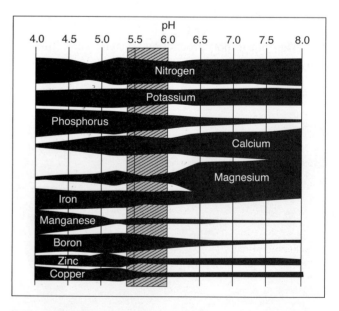

Figure 21–10 pH range of macro- and micronutrient availability in artificial media. The thicker the line, the more available the nutrient.

MEDIA TESTING

Ornamental nursery growers have been slow to conduct regular EC and pH monitoring of container-grown plants. One of the major reasons has been a lack of a uniform and simple media-testing procedure during the crop production cycle. Two media testing methods are now in use: the saturated media extract method and Virginia Tech extraction method.

The saturated media extract (SME) procedure has been used for many years by commercial testing labs and some growers. This procedure is relatively time-consuming, and growers are hesitant to use it. The procedure requires a sample of the media to be removed from the container and placed in a collection vessel of known volume. Water is slowly added to the collection vessel until the media is saturated and there's a slight sheen on top. The water is allowed to equilibrate for about an hour. pH is measured by inserting the pH probe directly into the saturated media. To measure electrical conductivity, the saturated media is placed in a Buchner funnel, and the water is vacuum-extracted, filtered, and measured with a conductivity meter. If you are using controlled-release fertilizers, false readings may occur from damage to the fertilizer prills.

The Virginia Tech extraction method (VTEM), or pour-through method, is gaining acceptance with nursery growers because it does not require any special handling of the container media. Within two hours after irrigation, a minimum of three containers from each block of plants should be selected for testing. The containers are placed on a PVC ring in a collection vessel, and a volume of water with which the crop is normally irrigated is applied. The volume applied depends on the volume of the container; typically about 150 ml of water is added for each gallon size of container, that is, 150 ml/1-gallon container, 300 ml/2-gallon container, and so on. Leachate from the collection vessel is analyzed for pH and electrical conductivity.

For the nursery grower, the Virginia Tech extraction method has five main advantages over other monitoring procedures:

1. The sample extraction time is relatively short, as little as five minutes.
2. pH and electrical conductivity measurements may be made in the field.
3. No media handling is required.
4. Sample extraction requires no special equipment.
5. Controlled-release fertilizer is not used; thus, it does not create false readings.

The Virginia Tech extraction method should be performed every one to two weeks. It is important to keep complete records and to chart the pH and electrical conductivity values measured. Graphically charting the pH and electrical conductivity provides the grower with information about whether the pH and electrical conductivity are changing. The results of a single sample should never be used to make a change in culture and/or management. The strength of the Virginia Tech extraction method is its help in establishing a picture for the crop, thus helping the grower to make informed decisions about watering frequencies, fertilizer needs, and leaching requirements.

Plant growth and ultimately sales have already been lost if the nursery grower analyzes media only when nutritional deficiencies appear. Whichever media testing technique a nursery uses, a media analysis program must be established that is consistent in sample frequency. Table 21–1 serves as a guideline for comparison of different media testing procedures.

TABLE 21–1 Interpretation of Soluble Salt and pH Measurements by Extraction Method*			
Method	Soluble Salt	pH	Electrical Conductivity (dS/M or mMhos/cm)
VTEM	Sensitive crops (liquid feed)		0.50–0.75
	Nursery crops (liquid feed)	5.2–6.2	0.75–1.50
	Nursery crops (controlled-release)		0.20–1.00
Saturated extract method (nursery crops)	Low		0.00–0.74
	Acceptable		0.75–1.49
	Optimum	5.8–6.8	1.50–2.24
	High		2.25–3.49
	Very high		3.50+
Saturated extract method (greenhouse crops)	Low		0.00–0.75
	Acceptable		0.75–2.0
	Optimum	5.6–5.8	2.0–3.5
	High		3.5–5.0
	Very high		5.0+

* The ranges of pH and soluble salt levels should be used as guidelines only. Irrigation water should be < 0.75 dS/M. The soluble salt level of water used in the VTEM procedure should be subtracted from the final leachate value.
Source: The University of Georgia College of Agricultural and Environmental Sciences Cooperative Extension Service.

NITRATE NITROGEN TESTING

Nitrogen is one of the most widely distributed elements in nature, but in container production, nitrogen is the most limiting element related to plant growth. Many plants prefer nitrogen in the form of nitrate (NO_3^-), and many fertilizers supplied this form of nitrogen to plants. Most nitrate ions, like other anions, are easily leached away from the container media by heavy rains and/or irrigation. Because of the importance of nitrate nitrogen to plant growth, the nursery grower should check nitrate nitrogen levels. Measuring electrical conductivity is one way for the nursery grower to get a rough estimate of the nitrate nitrogen in the container media. A correlation exists between electrical conductivity and nitrate nitrogen in the container media when using controlled-release fertilizers. However, this correlation does not give a precise determination of nitrate levels. The nursery grower has two choices for determining nitrate nitrogen levels. The first is to send media samples to a commercial lab and wait for the results. The second choice is to use an ion-specific meter, which is readily available from most horticulture supply companies.

FERTILIZERS

Good nutrition management is required for the production of quality nursery plants, whether plants are produced in the field, containers, or pot-in-pot production systems. This section on fertilization will concentrate on container and pot-in-pot production because the total supply of federal elements available for plant growth is limited by the size of the container. Good container media and nutrition management are basic requirements in the production of quality container grown plants. The decisions involved in providing good nutrition to container stock are complicated by several factors: the multitude of fertilizer products available, the variations in container media, the great number of species involved, and the various cultural practices used.

The impact of fertilizers on the environment and groundwater is an important concern. To minimize environmental impact, nursery growers are using controlled-release fertilizers (CRFs). The two most common methods of using

Figure 21–11   Controlled-release fertilzer top-dressed on a container-grown plant.

controlled-release fertilizers are incorporation and top-dressing. When incorporating controlled-release fertilizers into container media, a previously determined amount is uniformly mixed into the media prior to potting of liner plants. Top-dressing of controlled-release fertilizers is usually performed on plants later in the production cycle, after the original incorporated fertilizer has been exhausted of its nutrients. A previously determined amount is spread evenly across the top of the media in the container. Most controlled-release fertilizer manufacturers supplied growers with scoops or spoons that hold a predetermined amount of the fertilizer to help facilitate application (Fig. 21–11). Growers using CRFs gain several advantages over traditional granular and/or liquid fertilization including: initial fertilizer levels that are lower in the growing media, availability of nutrients during the entire crop cycle, the great reduction in the amount of nutrients last from runoff and leaching, and a localized and efficient supply of nutrients to the nursery crops. Although nitrogen is the primary element of concern in nursery production, growers should be aware that the other required elements need monitoring to make sure no imbalances occur. Commercial laboratories are generally used to test for these elements.

SUMMARY AND REVIEW

Site selection for a nursery includes evaluation of four categories: ecological, economic, sociological, and biological. Ecological factors include water availability and quality, physical features of the property, and weather patterns. Economic factors are land cost, labor, marketing, and choice of products to grow. Sociological factors include demographics of the neighboring populations, the targeted market population, zoning and other development regulations, and restric-

tions that may limit access to the property. The indigenous population of insect and disease organisms and the deer population are biological factors to be considered.

Field production is the growing of nursery stock directly in the ground. Spacing between and within rows is determined by the type of plant being grown and the state of maturity it will reach at harvest time. Access roads into all parts of the field are critical for the movement of trucks, tractors, and maintenance and harvesting machinery. Field stock is harvested during the dormant state. The plants are dug using a mechanical digger, which separates up to 80 to 90 percent of the roots from the plant. The remaining roots and attached soil are wrapped tightly with burlap to hold them in place.

Traditional container production of nursery stock has the containers placed on top of the ground. Spacing is generally more dense than with field production, thus increasing the number of plants that can be produced. Crop rotation is not necessary because the plants are not grown directly in the ground. Another advantage of container production is the reduction in the number of roots lost during harvest. Almost no roots are lost because the plant is removed from its container for transplanting, which means that the plants can be harvested and marketed throughout the year, not just during the dormant season. Disadvantages of container production include the possibility that the plants may become pot-bound, and irrigation systems are an absolute requirement. The likelihood of freeze-damage to the roots occurring

during overwintering is an issue in colder climates, necessitating the use of overwintering structures. Pot-in-pot production was developed to reduce the heating of the container during the summer, but it has become adopted to reduce freeze-damage during the winter. Pot-in-pot production has other advantages, including a reduction in the number of plants blown over by the wind and reduced irrigation frequency.

Management of soilless growing media and its fertility is one of the key elements in successful container production, whether traditional or pot-in-pot. The choice of media is crucial to producing high-quality plants. Media and aeration porosity are important factors to be considered, as are the EC and pH of the media, both at planting and throughout the production cycle. Growers can monitor these factors as well as nitrate/nitrogen levels themselves. Other nutrient elements are usually tested by outside laboratories. Fertilizers can be applied in many different ways, but controlled-release formulations are preferred because they have a minimal impact on the environment.

FOOD FOR THOUGHT

1. What are the four categories to consider when evaluating a potential nursery site?

2. Why are field-grown trees and shrubs harvested during the plants' dormant season?

3. The grower has complete control over what aspects of container-grown plants?

4. What are the disadvantages of growing plants in containers?

5. What are the advantages of growing plants in the pot-in-pot system?

6. Why is it important to monitor container media for EC and pH?

7. What is a media-testing method that is easiest for growers to do on-site?

8. What are the advantages of using controlled-release fertilizers?

9. Why is nitrate-nitrogen considered so important in nursery production?

10. What is an ion-specific probe?

SUPPLEMENTARY READING

AVENT, T. 2003. *So you want to start a nursery.* Portland, Ore.: Timber Press.

DAVIDSON, H., R. MECKLENBURG, and C. PETERSON. 1994. *Nursery management administration and culture.* Englewood Cliffs, N.J.: Prentice Hall Career and Technology.

YEAGER, T., C. GILLIAM, T. BILDERBACK, D. FARE, A. NIEMIERA, and K. TILT. 1997. *Best management practices guide for producing container-grown plants.* Marietta, Ga.: Southern Nursery Association.

CHAPTER
22

Landscape Plants: Evergreen Trees and Shrubs, Deciduous Trees and Shrubs, and Herbaceous Plants

Laura Deeter

KEY LEARNING CONCEPTS

After reading this chapter, you should be able to:

♦ Discuss the proper planting of trees, shrubs, and herbaceous plants.
♦ Explain proper maintenance of trees, shrubs, and herbaceous plants.
♦ Describe how to choose and care for trees, shrubs, and herbaceous plants.
♦ List the environmental characteristics that affect the growth of trees, shrubs, and herbaceous plants.
♦ Know the names and characteristics of some of the common landscape plants.

INTRODUCTION

Close your eyes and picture a tree, a shrub, a flower, and a vine. We all know what these words mean. Or do we? A botanical dictionary tells us that a woody plant has a corky outer surface of bark covering the older stems, which generally survive a dormant period and expand their girth every year. An herbaceous plant lacks wood and bark. The overwintering structures of herbaceous plants are typically underground. An evergreen is a plant that retains its foliage year-round. Evergreens may have thin needles (pines, spruces, junipers) or broad leaves (rhododendron, holly). Evergreens can be 100-ft trees or 1-ft shrubs. Trees typically have one stem (trunk), whereas shrubs typically are multistemmed. We tend to think of trees as tall and shrubs as less than 12 to 15 ft in height. Vines can be woody or herbaceous, depending on the species. Herbaceous plants are often referred to as flowers. Although they can be tall or small, many species are between 1 and 4 ft tall.

Woody plants, whether tree, shrub, or vine, are all perennials. Herbaceous plants have one of three life cycles: perennial, biennial, or annual. Perennials live for more than two years and typically flower each year. They do not die after flowering. Biennials complete their life cycle in two years. The first year they have foliage only. The second year, they flower, set seed, and die. Annuals complete their entire life cycle (germination, foliage, flower, and seed) in one growing season.

Although a botanical dictionary gives us the absolute meanings of these terms (*perennial, biennial,* and *annual*), in reality the boundaries between them can be vague and hard to discern. Some trees have multiple trunks, such as the paper birch (*Betula papyrifera;* Fig. 22–1) or are very small; many cultivars of Japanese maple are under 6 ft tall (*Acer palmatum* var. *dissectum* 'Red Dragon', Fig. 22–2). Carolina silverbells (*Halesia tetraptera*) is a shrub often used as a small tree. Some shrubs get very large. American hazelnut (*Corylus americana*) is a shrub that can reach heights of 20 ft. Some multistemmed shrubs can be pruned or trained to grow as small trees. Crape myrtle (*Lagerstroemia* spp.), Russian olive (*Elaeagnus angustifolia*), and dappled willow (*Salix integra* 'Hakiro Nishiki') are all plants that can grow either as a shrub or a tree. We use some plants as herbaceous plants, but in reality the plants are woody shrubs; examples include roses (*Rosa* spp. (Fig. 22–3) and lavendar bushes (*Lavandula* spp.). Many plants have a perennial life cycle but do not live more than one growing season for environmental reasons. *Pelargonium xhortorum* is not hardy in northern areas; tulips (*Tulipa* spp.) do not survive the summers in the deep South. Careful breeding produces many landscape plants with varying characteristics. Thus, a large shrub may have selections that are dwarf, such as the compact winged euonymus (*Euonymus alatus* 'Compactus'), or even have a vastly different habit from the original. Birdsnest spruce (*Picea abies* 'Nidiformis') is a low-mounding plant rather than a tall conical tree. These cultivated varieties further blur the lines among the categories of tree, shrub, perennial, and annual.

Whether tree, shrub, or herbaceous, all plants have characteristics that vary from season to season. Flowers, fall color, size, growth habit, fruit, ornamental bark, and persistent foliage are all characteristics that provide landscape interest and value. Some trees provide a filtered shade; others provide a dense shade. Some shrubs provide an excellent

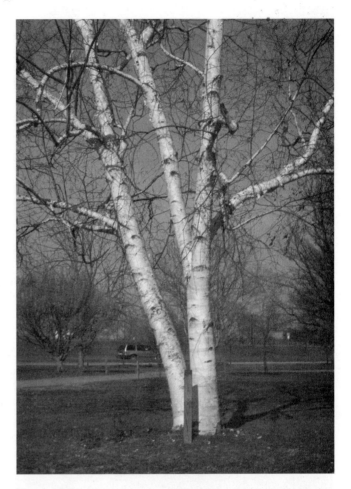

Figure 22–1 *Betula papyrifera,* paper birch, is a tree that can have multiple trunks. *Source:* Laura Deeter, The Ohio State University.

Figure 22–2 *Acer palmatum* 'Red Dragon,' The Red Dragon Japanese maple, is a low-growing cultivar of a small tree. *Source:* Laura Deeter, The Ohio State University.

Figure 22–3 *Rosa* 'Bonica' is a woody shrub that is treated and utilized more often as an herbaceous plant. *Source:* Laura Deeter, The Ohio State University.

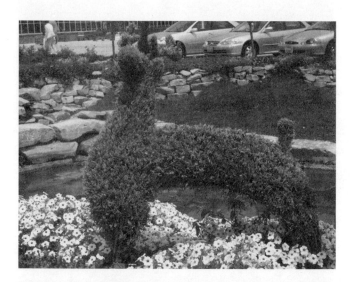

Figure 22–4 Junipers (*Juniperus scopulorum*) can be pruned easily into decorative topiary. *Source:* Laura Deeter, The Ohio State University.

Figure 22–5 Perennials can provide the flashy show in any garden. Many also make great cut-flower arrangements. *Source:* Laura Deeter, The Ohio State University.

habitat for wildlife; others may bear fruit valued by humans. Shrubs can also be utilized as hedges or screens. Many woody plants can be sheared and pruned extensively, thus forcing the plant into topiary or bonsai, for example. Herbaceous plants often provide the flashy show for a garden (see Figs. 22–4 to 22–7). The incredible variety of shapes, sizes, colors, forms, and flowering times of landscape plants provide us with an almost unlimited amount of variety for the yard and garden. Regardless of their size or shape, *all* plants should be selected with care and consideration to ensure their continued survival and success in any landscape.

ENVIRONMENTAL FACTORS AFFECTING PLANT SELECTION

Temperature

Proper plant selection is vitally important to plant success. "Right plant, right place" is a phrase horticulturists use frequently. One part of proper plant selection is plant usage. How the plant will be utilized and what function it will have in the landscape play a vital role in choosing a plant for a site. Trees, shrubs, and herbaceous plants all have different characteristics that can be used to advantage depending on the function the plant needs to fulfill.

Regardless of the type of plant selected or its ultimate use, many environmental factors must be considered before choosing any plant. One of the environmental factors most often mentioned is the US Department of Agriculture (USDA) hardiness zone rating. The USDA hardiness zone is simply a listing of the average minimum winter temperatures for a given region. The USDA hardiness zone (Fig. 22–8) is simply a guideline for selecting plants that are winter-hardy for a given region. However, each broad region contains areas protected by hills, lakes, or even cities that may

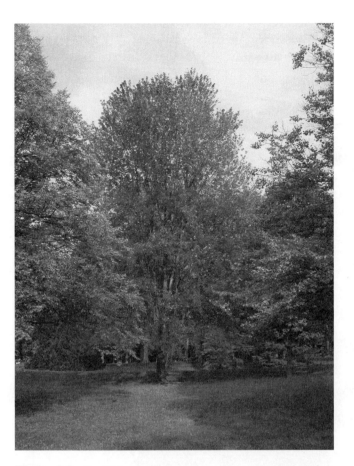

Figure 22–6 This Freeman maple (*Acer xfreemanii*) has spectacular red fall color. *Source:* Laura Deeter, The Ohio State University.

Figure 22–7 Manzanita (*Arctostaphylos manzanita*) has highly ornamental, smooth red bark. *Source:* Laura Deeter, The Ohio State University.

allow the growing of plants otherwise not hardy to that zone. Conversely, exposed sites may allow winter winds to create cold pockets; plants may need to be more cold-tolerant than the USDA hardiness zone indicates. Extremely hot regions may be more concerned with summer temperatures and not as concerned with winter temperatures. We generally think of deserts in terms of the high daytime temperatures. However, the night temperature in the desert can drop to just above freezing. Plants not only must be very heat-tolerant, but cold-tolerant as well.

Some areas of the Midwest often have late frosts, during which many early flowering plants lose their flowers to the cold temperatures. Forsythia (*Forsythia* spp.), star magnolia (*Magnolia xsoulangiana*) and dogwoods (*Cornus* spp.) may have their flowers injured or killed during a late frost. As distressing as it is to lose the flowers of a favorite plant, keep in mind that if the flowers are killed, there will be no fruit later in the season. Apricots, peaches, and almonds are not typically produced in areas with late spring frosts, although the plants are more than winter-hardy. Milder winters in southern regions prevent plants from receiving the cold treatments they need for proper flowering and fruit set. Spring bulbs re-

quire a cold treatment to flower. If the winters are mild, they will not flower properly or at all the following spring.

The American Horticultural Society (AHS) heat zone map was developed in 1997 by Dr. H. Marc Cathey to help describe the impact that heat and drought play in plant survival. While cold causes damage that is immediately apparent, heat is often far more subtle; heat stress can appear in many forms and can be mistaken for other types of stress. Plants suffering from heat stress may not flower properly, flowers may be deformed, foliage may droop or form necrotic lesions, plants may appear stunted or wilted. It is also possible for combinations of these symptoms to occur simultaneously. The continual modification and development of the heat zone map over time will enable gardeners in hot arid regions to predict more accurately which plants might survive the summers.

Light

All plants require light for photosynthesis. Each species is adapted to a certain light regime. Some plants will grow in just about any lighting situation a garden might have. Plants are often listed under one of the following types of light requirement: full sun, partial sun, partial shade, full shade. Most large trees require full sun for proper growth and development. As a general rule, needled evergreens also require full sun to perform properly. Broad-leaf evergreens, shrubs, and perennials are highly variable, and you can find species to fit most garden situations.

By simply determining the direction the garden faces, you can learn much about the amount of light in the area. A south-facing garden is typically a full-sun garden; a north-facing garden is generally considered a partial- to full-shade garden. An eastern exposure receives the milder, less intense morning sun, while a western exposure receives hot, intense afternoon and

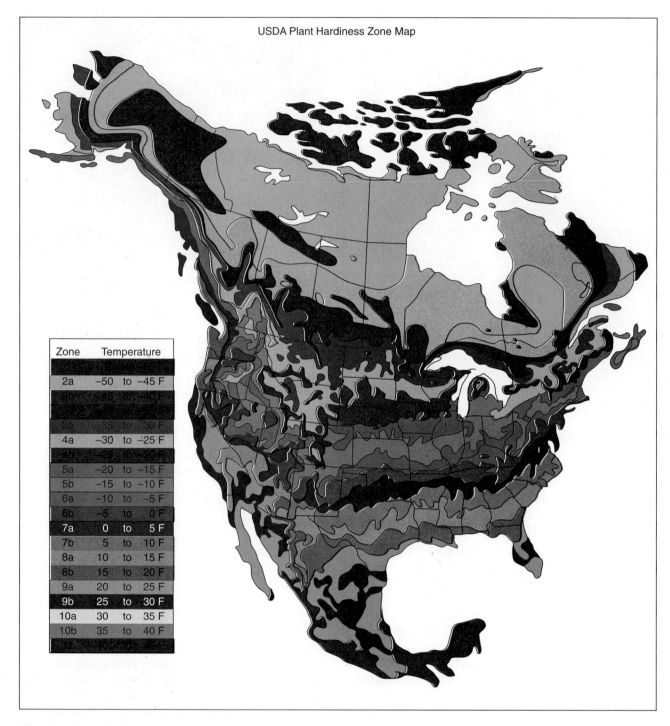

USDA Plant Hardiness Zone Map

Zone	Temperature
2a	−50 to −45 F
2b	−45 to −40 F
3a	−40 to −35 F
3b	−35 to −30 F
4a	−30 to −25 F
4b	−25 to −20 F
5a	−20 to −15 F
5b	−15 to −10 F
6a	−10 to −5 F
6b	−5 to 0 F
7a	0 to 5 F
7b	5 to 10 F
8a	10 to 15 F
8b	15 to 20 F
9a	20 to 25 F
9b	25 to 30 F
10a	30 to 35 F
10b	35 to 40 F
11	40 to 45 F

Figure 22–8   USDA hardiness zone map.

evening sun. It is important to keep in mind that as the season progresses, the amount, intensity, and quality of light will vary greatly. In addition, some areas of the country have many more cloudy days, which also affects the amount of light available.

    Trees, hedges, fences, and buildings all affect the amount of light a garden receives. A large shade tree can easily turn a south-facing garden into a shade garden. As you watch the patterns of the sun over the season, take note of the sunny and

shady areas at various times of the year. Over time, you will develop a talent for "guesstimating" the amount of light that an area will receive and be able to select appropriate plants.

Moisture

Water is absolutely essential to plant life. The amount of rainfall in a region is critical in determining which plants will survive. Many areas of the country receive more than enough moisture over the year to support a large range of plant material; however, the amount that falls during a given month can vary greatly. In some regions of the country, most, if not all, the precipitation comes during the winter months, which means that plants must be tolerant to very dry conditions during the summer months as well as tolerant to large amounts of precipitation during winter months. The Pacific Northwest is an area that receives rain throughout the year. The Midwest and New England generally get rain during spring and summer, with varying amounts of precipitation during the winter. The Southeast is highly variable: some areas get rain throughout the summer; others areas are far more arid. You can find the details of precipitation patterns for your region online using key words such as *precipitation data, climate data,* or *meteorological data,* and your city or region.

Wind

Wind causes plants to utilize more water. In areas of high wind, transpiration increases, which increases the amount of water the plant needs to survive and grow. Windy, exposed sites increase the need for irrigation, especially during hot, dry months. A gentle breeze typically does not increase water usage. A gentle breeze benefits plants by replenishing carbon dioxide levels around the foliage. Strong or sustained winds dramatically increase water usage. In addition, strong winds can damage the structure and even the form of trees and shrubs. We have all seen pictures of wind-swept trees growing on mountain tops. The high strength of wood can be damaged during a storm, causing otherwise healthy trees to break and fall. The natural strength of the wood of different species varies greatly, and this characteristic needs to be considered when choosing plants. Oaks (*Quercus* spp.) typically have strong wood; poplars (*Populus* spp.) generally have weaker wood. This trait can vary within a genus as well. Within the maples (*Acer* spp), silver maple and boxelder maple (*Acer saccharinum* and *Acer negundo,* respectively) are two species with weak wood. Norway maple and red maple (*Acer platanoides* and *Acer rubrum,* respectively) are two species with stronger wood. When working with large trees in an exposed site, it is imperative to consider the inherent strength of the wood (Fig. 22–9).

Though staking new trees is generally not a recommended practice anymore, newly planted trees in exposed, windy areas may require staking to help establishment. High winds not only affect large and newly establishing trees, herbaceous plants are also affected. In highly exposed areas, herbaceous plants may be shorter or go dormant earlier. Staking might be required to keep flowers on tall plants from falling over.

Soil

How many gardeners are aware of the type of soil they have? Successful gardeners are probably keenly aware of many factors in the soil, including texture, pH, and drainage. A

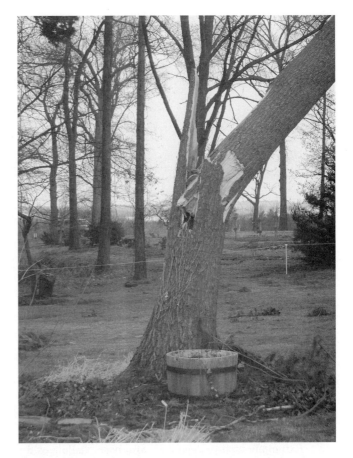

Figure 22–9 It is important to consider the inherent strength of the wood when planting large trees in a windy area; otherwise the tree may break in a storm. *Source:* Laura Deeter, The Ohio State University.

simple soil test tells you a lot about the soil. The pH of a soil greatly affects nutrient availability. Some plants require an acid pH; others prefer the pH to be more alkaline. Most plants perform best at a pH of neutral to slightly acidic (7.0 or slightly below). At this pH level, the most nutrients in the soil are the most available. Despite the marketing of chemicals for altering soil pH, in reality it can be challenging to make long-term modifications. The best solution is to select plants that are adapted to the soil pH in your vicinity. Soil test kits are available at garden centers; soil samples can also be sent to various agencies for testing. Contact your local extension agent for agencies in your area that provide soil testing. It is inexpensive, but it provides a wealth of valuable information about the soil.

Knowing the soil texture (sand, silt, and clay) provides information about how water moves through the soil and gives a basic indication of nutrient availability. Sandy soils are usually very well drained but with limited nutrient availability. Clay soils are often poorly drained but with high nutrient availability. Soils are typically a mixture of all three types in varying percentages. Plants are highly adaptable organisms, and many are capable of tolerating a wide range of

Figure 22–10 *Salix alba* 'Tristis' (weeping willow) grows in standing water. *Source:* Laura Deeter, The Ohio State University.

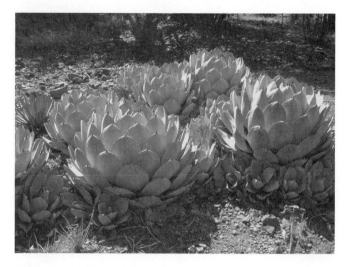

Figure 22–11 *Agave* spp. are extremely drought-tolerant plants for arid regions. *Source:* Laura Deeter, The Ohio State University.

soil types. Plants such as alder (*Alnus* spp) and goat's beard (*Aruncus dioicus*) are tolerant of poorly drained soils. Weeping willow (*Salix alba* 'Tristis,' Fig. 22–10) and yellow flag iris (*Iris pseudacorus*) grow in standing water. Most plants require good drainage, however, and will die if the drainage is poor.

Extremely well-drained soils also pose challenges for landscape plants. Although these plants readily tolerate periods of drought, all plants require water. Newly transplanted plants are more likely to suffer from drought stress and require far more water than older, established plants. Plants with many surface roots, such as honey locust (*Gleditsia triacanthos* var. *inermis*) may suffer from drought more readily than a species that forms a deep tap root, such as black oak (*Quercus velutina*). Plants adapted to very dry situations are more apt to perform better in arid regions than those species adapted to a moister environment (Fig. 22–11).

It is important to remember that all these environmental conditions are highly interactive. The amount of moisture in the soil greatly affects the ability of plants to obtain nutrients. Shady areas often have increased water in the soil; plants not adapted to increased moisture suffer and perform poorly in the landscape. Areas in full sun are often dry; even drought-tolerant plants may require additional moisture to ensure long-term survival. No plant interacts with only one environmental condition at a time. Each condition interacts with and is highly dependent on the others. Keep this fact in mind as you work through the chapter.

TREES AND SHRUBS IN THE LANDSCAPE

Woody plants provide the structure and the backbone of any garden setting. They are generally permanent residents, providing depth, vertical lines, and character as they grow and

Figure 22–12 This mulberry (*Morus alba*) on the grounds of Longwood Gardens, Pennsylvania, was the favorite tree of Pierre DuPont, the main founder of the gardens. The large anchors have allowed the plant to survive much longer than it might have done otherwise. The anchors were installed because of the sentimental value of the tree. *Source:* Laura Deeter, The Ohio State University.

change through the seasons and over the long-term life of the landscape. Each large tree has a unique character, and they affect our lives on an emotional level that is difficult to express. If you have a hard time believing that humans become emotional over trees, take note the next time a large tree is removed from the landscape, either via natural forces or through so-called progress. People will band together and fight a municipality that decides to remove a large tree from public property.

Many will go to great lengths to save a tree with sentimental value, or one that was planted by a famous person (Fig. 22–12).

Trees not only have an emotional or sentimental value, they also have practical value. They provide lumber, shade, and shelter; purify the air; prevent erosion; and provide habitat for wildlife and food for people. Trees can even be utilized to remove harmful heavy metals from the soil. Because of their longevity, it is imperative to choose trees properly from start to finish. This precaution helps the plant perform to its fullest and to the satisfaction of the homeowner. Starting with a healthy plant is the first step.

Woody plants are generally sold in one of three ways: balled and burlaped (B&B), bare root, or container grown. Each has its own unique set of advantages and disadvantages. Balled and burlaped (B&B) plants can be evergreen or deciduous, small or large. They are generally field grown until they reach salable size (Fig. 22–13). The plant is then dug and the root ball is wrapped tightly in burlap to keep the roots together and to keep the roots moist. Plants grown this way can generally be purchased in an array of sizes, ranging from very small evergreens and shrubs to extremely large trees (Figs. 22–14 and 22–15). They are easy for the nursery to care for and can be stored in the B&B form for many seasons provided the root ball is kept moist and winter protection is provided. However, the act of digging and wrapping in burlap can cause undue stress on the plant if proper procedures are not followed. It is also very labor intensive. The root ball cannot be allowed to dry out. In addition, the root ball is unwieldy for landscapers and homeowners to maneuver; a large root ball is extremely heavy, especially when wet.

Bare-root plants are generally small and typically deciduous. Roses, several species of fruits, and some tree seedlings (three years old or less) are sold this way. Bare-root plants are typically field-grown. Once they reach the appropriate size,

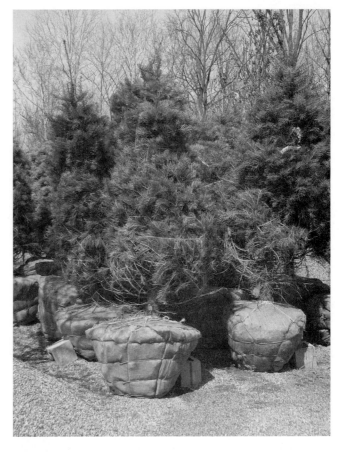

Figure 22–14 Five- to six-foot white pine (*Pinus strobus*) B&B plants. *Source:* Laura Deeter, The Ohio State University.

Figure 22–13 Field production of woody trees. *Source:* Laura Deeter, The Ohio State University.

Figure 22–15 An oak (*Quercus* spp.) with a trunk diameter of 8 ft and a root ball that is over 4 ft tall and 8 ft wide. *Source:* Laura Deeter, The Ohio State University.

Figure 22–16   Purple-leaf birch (*Betula* 'Crimson Frost') in 5-gal containers at a retail nursery. *Source:* Laura Deeter, The Ohio State University.

Figure 22–17   Small perennials in quart containers coming out of winter after overwintering in the container in a greenhouse. *Source:* Laura Deeter, The Ohio State University.

they are dug, and the roots and shoots are pruned if appropriate. The plants are then packaged in a plastic bag filled with moist sawdust because the root system cannot be allowed to dry out. This type of small plant is best suited for late winter or early spring planting. They are generally inexpensive and easy to transport because of their small size. However, it can be difficult for homeowners to discern if the plant is still alive at the time of purchase. The roots should be moist and not moldy at the time of purchase.

Much of the nursery industry is moving away from B&B and bare-root plants to container-grown plants. Plants of all sizes and species are suitable for containers (Figs. 22–16 and 22–17). Plants can be grown directly in the container, eliminating the need for field digging. Containers are easy for the nursery to keep as stock plants for several years, and they are easy for the homeowner to move. Although traditionally this method was restricted to areas with milder winters, the recent development of pot-in-pot systems means plants can be grown in containers in harsh weather with minimal losses. Pot-in-pot systems can be either above-ground or below-ground. Recent studies suggest there is little difference between winter

survival for above-ground pot-in-pot plants compared to below-ground pot-in-pot systems if the container size is 15 gal or greater. This technique eliminates the need for the nursery industry to cover or move containerized plants for winter storage. There are many other advantages to container-grown plants: they can be purchased and planted any time the soil and temperatures are appropriate, a wide range of plant material is available, plants are generally of uniform size and shape, plants are inexpensive to grow and maintain, and specialty containers (reinforced, copper-lined, metal, wood) are available for a wide range of nursery situations. See Chapter 21 for an in-depth discussion on nursery production methods.

Container-grown plants have a few minor disadvantages compared to B&B. The top may become too large for the container; thus, water availability becomes an issue. The root system may girdle the plant if it is left in the container too long. Plants need to have their roots disturbed prior to planting to ensure transplant success, and plants grown in soilless mixes may have a difficult time becoming established in poor soil. Each of these disadvantages has a simple solution. Large plants can have the tops trimmed, or they can be transplanted into a larger container. Copper-lined pots are available in various sizes to help prevent girdling roots. The nursery industry has done a good job informing the public of the necessity of disturbing the roots of container-grown plants. Proper bed preparation and mixing of the container media with the soil around the planting area can also dramatically increase transplant success.

CHOOSING A QUALITY TREE OR SHRUB

When you are standing in the nursery looking at a large number of plants, it may seem as if all the plants are equally healthy. It may feel intimidating to look each plant over carefully to select the best one for you. Remember, these are living creatures that are supposed to last a long time in your landscape. They will provide the backbone and structure in the yard and garden for many years to come. A few minutes spent selecting the best plant available can save time, money, and trouble down the road.

Look for vigorous plants that are free from obvious diseases, pests, or mechanical damage. Trunks should be clean with no nicks; foliage should be a good color throughout the entire plant. The plant should show signs of adequate growth for the season. Poor twig extension (the length the twig grows each year) could be a sign of an underlying problem with the plant. If you have problems assessing plant vigor, ask a nursery or landscape professional for assistance.

Avoid plants with obvious girdling roots. If they can be seen encircling the trunk at the top of the pot or root ball, do not buy the plant. The plant may appear in good condition and in perfect health, but a girdling root system will slowly but surely kill the plant. Once the large roots begin to girdle the trunk, it is very difficult to correct their growth. Root pruning can be done with some species, but others do not tolerate this treatment. In addition, any plant that undergoes root pruning is far more susceptible to drought stress and requires additional water until a new root system has formed. Root pruning is also not a guaranteed method of treating girdling roots. Plants with girdling roots are more prone to fall over in the wind, can take longer to recover from transplanting, and eventually die.

For container-grown plants, be sure to compare the size of the top of the plant to the container. A small root system may not be capable of supporting the number of leaves on the plant. Immediately after planting, the roots are going through several changes, the greatest of which is simply growing more roots. A small number of roots may not be able to supply all the water necessary for a large number of leaves. Plants that are top heavy are also more inclined to blow over while they are still young. Plants that are top heavy but in otherwise excellent condition can have approximately one-third of their foliage and branches removed at planting to help overcome this condition.

Look for plants that are evenly full all around. Plants that are unevenly branched will remain so throughout their life. The overall shape of the plant should be uniform throughout. For trees, be sure the trunk is straight and has a good taper at the base. The trunk should be widest at the soil level and slowly taper toward the top. Plants with little to no taper are less able to withstand wind than those with a good taper to the trunk.

CORRECT PLANTING OF WOODY TREES AND SHRUBS

Trees are best planted in spring when the soil is warming, but the air is still cool and the transpiration rate is low. Under these conditions, the roots will grow into the soil rapidly and provide ample water to the tops. Conifers and many container-grown deciduous and evergreen broad-leaved trees can be successfully transplanted in cold areas (zones 5 to 6) in periods from mid-August to early October. In warm areas (zones 8 to 10), fall planting of many hardy trees is recommended to allow the root system to grow during the mild winters. Bare-root trees are usually planted in midwinter in warm areas and in spring in cold areas.

Care of the Newly Purchased Tree

After purchasing the tree, you should take good care of it before planting. Bare-root trees should be placed in the shade to keep the tops cool, and the roots should always be kept covered and moist. If the trees are not to be planted at once, the roots should be plunged (heeled-in) in moist sawdust, moist soil, or peat moss. Balled and burlapped trees should not be allowed to dry out. A leafy tree continues to transpire in the balled condition, and the soil can dry out quickly, depending on the **evapotranspiration** rate. Keep the tree in a protected location and tie it to keep it from tipping. Extreme care should be taken to avoid breaking the ball when transporting it because the small roots can be broken and lose contact with the soil.

Container-grown plants are perhaps the easiest to maintain temporarily. Still, the plants should be kept in a shady, wind-protected area to reduce evapotranspiration. The containers should be watered well and should be protected from direct sun by shading them with aluminum foil or by plunging them in sawdust. The temperature of the exposed side of a black container can reach 49°C (120°F) in the direct sun. Either method of covering prevents excessive heat buildup on the periphery of the container—which is apt to kill the roots.

The Native Soil

A family usually purchases a home for its proximity to employment and because of the climate and often because it is what they can afford. Soils can range from fertile farm lands to rocky hillsides with a view. Little consideration is given to the garden soil when a family purchases their dream home. Soil problems of all kinds can exist, and it is best to observe the trees that grow best in adjacent neighborhoods with similar soils.

Sometimes soils are compacted by trucks during the building of the house. Such soil should be tilled with a rototiller

(A)

(B)

(C)

Figure 22–18   The ball of the tree should be placed on firm soil in the hole so that the ball will not settle lower than the intended planting depth. If the crown of the tree is below the final soil level, the trunk is subjected to decay organisms. To plant a balled and burlapped tree, (A) place the ball in the hole carefully, taking care that it is oriented properly. Remove ropes and cut the burlap in many places or remove it if it has been treated to prevent decay. (B) If the burlap is decomposable, as most are, cut away the top portion so that no burlap remains above the final soil level. (C) Fill in the hole and tamp the soil firmly until the desired soil level is attained. Water the soil ball and replace soil.

or a spade below the zone of compaction. To this loosened soil, incorporate 10 cm (4 in.) of organic matter into the top 20 to 30 cm (8 to 12 in.) of soil or beyond the limits of compaction. It is then best to irrigate the soil and allow it to settle and dry out to a good **tilth** before planting the tree.

The soil should be probed to 1 m (3 ft) deep to find any natural impervious layer (**hardpan**), which may prevent root penetration that in turn could result in a shallow root system. Similarly, the impervious layer (**subsoil**) may not provide adequate drainage, and the roots will lack adequate oxygen. Layers such as these are often caused by an accumulation of cemented minerals (calcium, iron, or aluminum compounds) and can be penetrated if they are thin. A power soil auger or a posthole digger penetrates layers up to 10 cm (4 in.) thick. These holes can be filled with amended soil or plant residues

so that the roots can enter the open soil layers under the hardpan. Layers thicker than 10 cm (4 in.) are very difficult to penetrate without heavy equipment such as bulldozers. It is possible, however, to haul in a large amount of soil and make a mound at least 1 m deep on which to plant a tree. Often such a scheme can be fit into a landscape plan. In such cases, a **tile drain** should be placed above the impervious layer to drain away the excess water.

Preparing the Hole and Planting

It is to be hoped that the local soil is one that does not pose problems like those discussed above. The hole for B&B (Fig. 22–18) or a container-grown tree should be dug just deep enough to allow the soil ball to set firmly on the native soil and should be

Figure 22–19 Plant a bare-root tree in a hole large enough to accommodate the roots. Prune very long roots for ease in planting. Native soil taken from the hole should be broken up to make backfilling convenient. Firm the soil around the roots by tamping it with a stick as it is replaced. If the soil is properly firmed, the first irrigation will not settle the soil significantly.

about twice the diameter of the ball. When one is digging, it is usually difficult to estimate both dimensions accurately. If the hole is too deep, backfill with soil and tamp it firmly with both feet until the ball can be placed in the hole slightly higher than the original ground line. In heavy clay soils, a glaze from the shovel may appear on the walls of the hole; this glaze should be roughened with a stick or shovel before backfilling to allow good root penetration from the soil ball.

Recent research has shown that it is not necessary to mix in any organic matter into the backfill soil before placing it around the soil ball. The native soil should be broken up with the shovel and gradually placed back and tamped in with a stick or gently with a foot. If the ball is sitting on firm soil, the ball will not settle with subsequent irrigations.

For bare-root trees, make a hole just large enough to accommodate the roots when spread out (Fig. 22–19). Prune broken or very long roots before planting. Gradually fill in the soil and tamp with a stick; finally, only after all the soil is well above the roots, apply a firm foot to press the soil around the tree. Irrigate to settle the soil around the roots. Use the excess soil to make a berm (saucer) around the tree to contain water in a basin. Always plant the **crown** of the tree higher than the surrounding soil to ensure that water

does not accumulate around the trunk. Avoid overwatering newly planted trees before the roots become established in the undisturbed soil. The reduced oxygen in a saturated soil discourages growth of new roots.

When a bare-root tree is planted in a lawn area, enough **sod** should be removed to make room for a basin that should be constructed around the tree. The turfgrass should be kept away from the crown for at least two years to allow the young tree to become established. The bare soil area should be about 75 cm (30 in.) across. Three upright stakes about 30 cm (12 in.) high, spaced 30 cm (12 in.) from the tree trunk, reduces the risk of bark damage from lawn mowers. If the tree is staked, then these short stakes are not required.

Mulching the Soil Surface

All newly planted narrow- and broad-leaved evergreens benefit from organic **mulches** of wood chips, fir bark, pine needles, or sphagnum peat moss. A mulch layer 5 to 10 cm (2 to 4 in.) thick may be placed at the base of the tree, extending out to the edges of the longest branches. Bricks or redwood boards, which resist rotting, can be used to keep the mulch in place. The mulch eliminates most weeds, prevents the soil from caking in the hot sun, and moderates soil temperature and water loss near the surface. Water usually penetrates the soil more easily below a mulch.

Tree Staking and Trunk Protection

The newly planted tree may be staked, but it is not always necessary. The purpose of staking is threefold: to protect the trunk from mechanical damage by mowers or other equipment, to anchor the root system, and to support the tree temporarily in an upright position.

Large leafy B&B trees are usually anchored with guy wires. The top of the tree offers so much wind resistance that the small root ball cannot prevent the tree from toppling. Excess movement keeps destroying any new roots. Three well-placed guy wires (Fig. 22–20) usually maintain stability and also keep machinery from running into the tree. When the roots have grown into the native soil, the supports should be removed, usually after one year's growth.

Many newly planted trees require trunk support because they have been staked too tightly or have been grown too close together in the nursery. Such trees usually have tops that are too heavy for the trunk. The tree supports should be placed as low as possible but still maintain the top in an upright position. This point is found with the method shown in

Figure 22–20   Three well-placed wires or plastic ropes prevent a B&B conifer with dense foliage from toppling in the wind. Wire should be covered to avoid abrasion of the trunk.

Fig. 22–21. The tie is flexible enough to allow some trunk movement. The best material is an elastic webbing that allows movement without abrading the trunk. If the top is leafy and allows much wind resistance, it should be thinned to reduce the wind load.

If the tree lacks small shading branches on the lower trunk, the trunk should be painted with white interior latex paint (diluted with water) or wrapped with strips of burlap. Painting, burlapping, or using paper wrap prevents sun scald of the trunk on bright days. If small branches are still growing along the entire length of the trunk, those up to 15 cm (6 in.) above the soil should be removed, but the ones higher should be left. The small branches shade and supply food to the trunk; however, they should be pinched regularly so they do not develop into large limbs. These lower laterals may be gradually thinned and removed over a two- to three-year period. A wire mesh encircling the tree to about 45 cm (18 in.) above the ground protects the bark against rabbits and mice if needed.

The support system should be checked occasionally. The ties can be removed after one full growing season, but the trunk should again be checked for strength, as shown in Fig. 22–21.

Fertilizing Trees After Planting

There is a chance of applying too much fertilizer at transplanting time, which would burn the roots; therefore, trees are seldom fertilized when planted. However, a *small* handful of nitrate-nitrogen fertilizer, which is readily soluble, can be applied on the soil surface and watered in directly after planting. Fertilizer should never be placed near the crown of the trunk. Trees planted in turf must be fertilized more often than those growing in bare soil (without weeds). Turfgrass strongly competes for mineral nutrients added to the soil, leaving little for the tree roots.

Irrigation After Planting

It is important to irrigate recently planted trees deeply to encourage the roots to grow downward. In dry, hot climates careful irrigation is necessary for the survival of the tree during the first summer. Watering with a soaker hose, which provides low volume over many hours, allows the water to penetrate deeply in many soil types.

Container-grown or B&B trees should be irrigated frequently so that there is always a moisture contact between the soil ball and the backfill soil. This is especially true when the mix in the container is light and porous and the native soil tends to be clay. A bare-root tree, on the other hand, must be watered sparingly, especially during the first winter (warm zones) or spring (cold zones). The soil can become waterlogged with frequent irrigations, which reduces soil oxygen and discourages new root growth or may encourage root diseases, especially near the crown of the tree.

Figure 22–21 *Left:* To determine the point at which a tree should be staked, grip the trunk at various heights and bend the top to determine at which point the trunk springs back. *Right:* Tie the trunk a few cm above the point where the tree trunk returns to an upright position.

Tree Care During the First Year—A Summary

The first year is the most difficult period in the life of the tree. New roots must develop to anchor the tree and feeder roots must develop to supply the tree with water and mineral nutrients. Staking to keep the plant upright is necessary only for trees with weak trunks, trees planted in windy areas, or trees with large top-to-root ratios. Deep irrigation encourages deep rooting if the subsoil is similar to the top soil. Feeding the tree during the first year is usually unnecessary because a small quantity of nutrients can be obtained from the native soil. Overfertilizing young trees the first growing season can damage and seriously hinder root growth. Some **thinning** of the top may be necessary to reduce the wind load on the top. The trunks of some trees need protection from sun scald or rodents, depending on the environment.

PRUNING TREES

Young deciduous or broad-leaved evergreens may have to be pruned after the first year. The purpose of tree pruning is to control the shape or the size, stimulate vigor, or, in the case of fruit trees, to affect the flowering and consequently the fruits that develop later.

The shape of the tree is directed early by selecting or choosing certain branches or buds to grow in a desired direction. The pruner can select one good leader branch and thin out badly placed branches so that a desirable, less-cluttered scaffold system remains (Fig. 22–22). An irregular tree form is sometimes desired, and no one single leader is then chosen. At this time branches with narrow, weak angles (Fig. 22–23) can be eliminated in favor of wide-angle branches. It is possible to prune certain trees so that there are multiple trunks and no single leader (Fig. 22–24). The crape myrtle is an excellent candidate for training in this fashion. A complicated form of pruning that requires much care is the espalier, in which fruit trees are trained against a fence or wall (Fig. 22–25).

The size and foliage density of shade trees may be controlled by pruning at least once a year, but more frequent pruning may be required for vigorous trees. Examples of such pruning control are creating a hedge from closely planted trees to produce a screen or pollarding trees (i.e., cutting them back severely each winter to let in more light in the winter but still to have shade in the summer). Sycamore and fruitless mulberry trees lend themselves to this severe pruning (Fig. 22–26). Because of their vigorous growth, the pruned trees still produce long leafy branches by midsummer. Controlled pruning may be done to shape the plant as animals, clumps, or odd

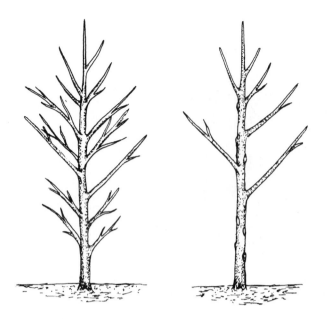

Figure 22–22 *Left:* A young established shade tree before pruning. *Right:* The same tree with those branches retained that will produce the scaffold branches of the tree. It is not a good idea to prune just the leader (topmost) branch to create low branching habits on an ornamental tree.

Figure 22–23 *Left:* Branches with acute angles may split when the branch matures and (*center*) becomes heavy. *Right:* Narrow-angled branches should be removed and wide-angled branches should be encouraged.

forms, as in topiary pruning (Fig. 22–27). The ultimate in controlling tree size is the Japanese treatment of bonsai (Fig. 22–28). These trees are maintained and kept in small containers for many years. Annual pruning of both roots and tops is necessary to maintain the bonsai form.

Even though pruning controls growth, it also invigorates individual shoots by reducing shoot number and opening the center to more light. Thinning operations produce a better top-to-root ratio so that water and nutrients are distributed

Figure 22–24 Crape-myrtle (*Lagerstroemia indica* L.), grown as a single- or multiple-stem tree. If pruned heavily early in its life, this tree can be grown as a shrub.

Figure 22–25 Pruning and maintaining a fruit or an ornamental tree so that it grows in a single plane is called espalier. Shown here are pear trees trained in two different forms. Very often trees are trained on a wall having adequate light. The pruning is done mainly in the winter when the branches are bare, but vigorous troublesome shoots may be removed any time to maintain the shape of the plant. Photographed at Wisley Gardens, Great Britain.

Figure 22–26 Cutting a tree back severely each year during the dormant season (pollarding) is a way of keeping potentially large trees small. This London plane tree is one of the many on Nob Hill in San Francisco that are pollarded to keep their size down. The gnarled branches result from this severe pruning.

Figure 22–27 A type of topiary pruning in the Japanese section of Buchart Gardens, Vancouver Island, Canada. This *Chamaecyparis pisifera* 'Plumosa' often requires pruning during the growing season to remove the small tufts of growth from lateral buds.

Figure 22–28 Normally large trees such as black pine and junipers can be grown in small containers and pruned to reduce their size. These containers average 30 cm (12 in.) long, and the trees are ten to twenty years old. They are known as bonsai trees.

more efficiently. The remaining shoots have less competition for light, water, and nutrients. In hot dry climates, less leaf surface remains after pruning, thus reducing water loss by transpiration. The branch structure is strengthened by eliminating branches that produce narrow, weak angles (Fig. 22–23).

Pruning Methods

The two basic pruning methods of heading back and thinning out are discussed for fruit trees in Chapter 19. Before pruning any tree, one should ask three simple but basic questions:

1. *What do I want to accomplish by pruning this tree?* Shall I head back the plant to keep it small? Shall I head it back to invigorate and strengthen it? Shall I reduce the flowering or fruiting potential?
2. *How will the tree respond to pruning?* Will it greatly alter the growth habit? Will severe pruning markedly reduce or stimulate excessive growth? Will many latent or **adventitious buds** be stimulated to sprout? Based

on the normal habit of growth, is the method I am choosing for this particular tree suitable?
3. *How is it actually done?* Some basic pruning tools are necessary to make pruning easy. Bark splitting and ragged cuts should be avoided to allow for rapid healing.
   a. First remove any dead or dying branches or twigs. They usually have a gray lifeless (shriveled)

Figure 22–29 When small branches are removed, they should be cut with pruning shears to obtain clean-cut surfaces. The cutting blade is held close to the main branch so that very little stub remains.

Figure 22–30 The method of cutting large branches to avoid splitting the trunk below the cut. *Left:* Make a cut on the underside at (A). *Center:* Remove the branch at (B), leaving a stub. *Right:* Remove the stub at (C).

because the new growth will not acclimatize properly and mature before an early frost. Typical methods of cutting twigs or large branches with shears and saw are illustrated in Figures 22–29 and 22–30.

Narrow-leaved evergreens, or conifers, are shaped or kept within certain size limits by top pruning. Pine tops can be reduced in size by nipping the "candles" of the current year's growth to about half size with the thumb and forefinger. The remaining portion develops a group of short needles. The best time to prune the candles is in late spring when they have elongated but the needles are just beginning to grow. A new terminal bud will then form if the pruning is not done too late. Spruce and firs may also be pruned in a similar manner.

Junipers and other conifers without a strong radial symmetry or distant terminal buds may be sheared or irregularly pruned to form any shape or growth habit at most any time of year.

If the leader on a conifer is broken, generally a branch below begins to bend upright. However, one of the adjacent laterals can be encouraged to take over by splinting it upright until it can remain erect without the aid of the splint. One full growing season is ample for this corrective procedure.

appearance. Scratching through thin bark on small branches will show green tissue underneath on live branches.

b. Thin out the cross-over branches or those that are growing toward the center of the crown.
c. Remove narrow angle branches that weaken the tree (Fig. 22–23).
d. Assess what you have done so far, then thin out the crown to give it the desired size and shape. At all times, keep in mind the natural characteristics and shape of the species.
e. If not enough material is pruned this year, you can prune correctively next year after observing the resultant growth of this year's pruning operation.

Ornamental trees and shrubs should be chosen for their growth habits to fulfill the landscape function, assuming that they are known to be hardy for the region. The trees and shrubs chosen to meet the needs of the landscape usually do not require much pruning after the first or second year when the shape is determined. Some pruning may be necessary annually in the case of some flowering shrubs, rose bushes, or conifers.

Pruning can be done almost any time of year except for trees such as the maples and the elms, which should be pruned in the late fall or early winter when the trees are not actively growing and the sap is not flowing. In areas with severe winters, prune after the cold weather has passed. Generally, late summer pruning should be avoided in cold areas

PRUNING SHRUBS

Since shrubs grow differently from trees, they are pruned differently. Shrubs are best pruned just before planting. In the case of bare-root stock, broken roots should be removed and extra long roots should be shortened to facilitate planting. If the top appears too large for the root system, the branches should be cut back to compensate for the root loss.

Container-grown plants often have roots that are circling in the container. Such plants should be rejected but, if used, the roots should be cut in several places to break the circling pattern and to stimulate new growth. The tops of broad-leaved evergreens often need to be pruned to reduce the wind

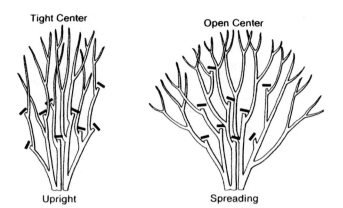

Figure 22–31 Two ways to prune a shrub to create different effects. The cuts are indicated by the black bars. *Source:* Adapted from USDA Home and Garden Bulletin 165.

Figure 22–32 Twigs cut at too sharp an angle (B) may die back to the point that will also cause the bud to die. It is better to cut straight across (or nearly so), well above the bud desired for growth (A). That stub will only die back to a point where it will not interfere with bud growth (shown by the arrow).

Figure 22–33 Rejuvenating old shrubs. *Upper left:* In some cases, it is best to cut back severely (black bars) and develop a new top. *Upper right:* With grafted shrubs, do not cut back below the graft union (knobby growth at bottom). Remove suckers from the rootstock (lines) or thin shoots to open up the plant. *Below:* Cutting or trimming the tops of old shrubs, as indicated by the black lines *(left),* produces an unsightly leggy plant with a bushy top *(right). Source:* Adapted from USDA Home and Garden Bulletin 165.

load and to reduce the transpiration surface. Remove about one-quarter of the leaf area by pruning immediately after planting. The tops of balled and burlapped shrubs should also be pruned for the same reasons, but the roots should not be disturbed other than loosening the burlap and cutting the ropes.

Once the shrubs are established in the landscape, maintenance pruning serves to control their size and shape by removing some stems from the multiple-stemmed crown (Fig. 22–31), improve the shrubs' health, and affect flowering. Some shrub species require more pruning than others because of differences in vigor and flowering habit.

Plant size is reduced by thinning out some of the branches. The tips of the branches may be headed back to increase peripheral branching if a hedge or formal effect is de-

sired. Thinning helps maintain the natural form and shape of the shrub. In thinning shoots, the wood should be cut above a bud that will grow in the desired direction (Fig. 22–32). Selecting the buds to remain allows partial control of growth even in an informal and natural setting of a rural home.

Shrubs may be thinned heavily by cutting some of the old branches back to the crown near ground level. The taller and older branches (i.e., those of greater diameter) should be cut out first, but not more than one-third of the branches should be cut out in any one year unless complete rejuvenation is desired. Old shrubs may be rejuvenated as shown in Figure 22–33 by cutting back the entire plant. After many new shoots appear, weak ones are pruned out to leave only the strongest and most desirable shoots to form the shrub.

Young hedges are pruned gradually by cutting about 6 to 8 in. above the previous cut. Mature hedges that have attained the desired size require frequent trimming to maintain their shape. The shape of the hedge should be slightly tapered, with the base wider than the top. This method of pruning allows light to reach the lower leaves and keep them alive. Keeping a hedge in a formal shape once it has attained the desired size may require as many as four clippings during the growing season. Pruning frequency of a hedge depends on plant vigor, light, temperature, and the water available to sustain vigorous growth.

Diseased wood is pruned out to keep the plant healthy. Diseased branches are cut back to nondiseased wood so that a healthy sprout will grow. This type of pruning should be done when the disease is first noticed, especially in the case of fireblight of pyracantha, cotoneaster, quince, and other members of the rose family.

One of the best reasons for pruning is to affect flowering. Some shrubs initiate their flower buds the summer before flowering (one-year-old wood) and thus should be pruned at a different season than those that flower on current year's wood.

Flowering on One-Year-Old Wood

Many fruit trees and shrubs initiate their flowers in summer, and the flower buds bloom the following spring when the weather becomes favorable. These species are called spring-flowering shrubs. Examples are *Cercis*, dogwood (*Cornus*), *Kalmia* (Fig. 22–34), lilac (*Syringa*), *Pyracantha*, *Rhododendron* (azalea), *Spiraea*, and *Viburnum*. Severely pruning the branches that have the flower buds in winter or early spring cuts away the potential flowers. One should prune such spring-flowering shrubs *soon after* they have flowered (Fig. 22–35). Removing the old flowers and shoots stimulates the plant to branch and produce additional flowering wood that summer

for next year. This flowering behavior is usually identified by observing that the shrub flowers before the leaves appear, such as the flowering quince, flowering peach, forsythia, or *Chimonanthus praecox* (Fig. 22–36). Some species like the lilac produce large flower clusters (inflorescences) instead of single blooms, but these too are borne on the previous year's shoots.

Flowering on Current Year's Wood

The branches of shrubs that flower on the current year's growth (sometimes called summer-flowering shrubs) should be pruned back in late winter or early spring before the buds have sprouted (Fig. 22–37). The resulting shoots have the capability of producing flower buds on the new growth. Some examples of summer-flowering species are *Abelia*, *Clethra*, crape myrtle, *Hibiscus, Nerium* (oleander), roses (floribunda and hybrid tea), and *Vitex*. This type of flowering is identified in some species by shoots that have light green leaves below the flower stalks. Flowers may also appear in several flushes or cycles during the growing season.

Pruning Nonflowering Shrubs

Shrubs not grown for flowers or fruits are pruned almost anytime of year except late summer. Late pruning often stimulates buds to sprout. Such shoots do not have adequate time to mature or "harden" before the onset of winter, leaving them susceptible to winter injury.

Shrubs, especially those grown as specimen plants, should be given ample space to develop properly. Most shrubs usually spread out as they grow taller. It is tempting to plant small shrubs close together to get the desired effect hurriedly. However, it usually becomes necessary to remove alternate plants when they begin to touch. Crowded plants do not develop the grace and beauty of those grown at recommended distances. Shrubs planted next to the house should be set far enough away to keep them from crowding into the building as they mature.

DESCRIPTION OF COMMONLY UTILIZED SPECIES

It is vitally important to choose the right plant for the right place when selecting trees and shrubs. These long-lived organisms are generally thought of as the backbone of the garden or landscape. Thus, their survival is critical to the overall value of the landscape. Unless one enjoys replacing expensive plants on a regular basis, it is critical to select the proper plants for your region. The main decision to make is what function the tree or shrub must fulfill: foundation, specimen, backdrop, screen, and sound barrier are all potential landscape functions. Flowering, fall color, overall appearance, fruiting characteristics, evergreen foliage, and deciduous foliage also need consideration when choosing trees and shrubs. Chinese chestnut (*Castanea mollissima*, Fig. 22–38) is a medium-size tree with edible fruit that will attract squirrels and deer,

Figure 22–34 *Kalmia latifolia* flowers are prized spring blooms. *Source:* Robert A. H. Legro.

Figure 22–35   Spring-flowering shrubs and trees should be pruned after they have flowered and as vegetative growth begins. Unfortunately, many gardeners and homeowners prune these plants in the winter when the twigs are bare, and consequently they cut away the flower buds. Before pruning, study the flowering behavior of the shrub or tree.

Figure 22–36   *Chimonanthus praecox* flowers in the spring before the leaves appear. The flowers were initiated the summer before, and they received their necessary chilling during the winter. *Source:* Robert A. H. Legro.

and has beautiful flowers and a beautiful yellow-bronze fall color. However, the flowers have an extremely strong putrid scent, and the fruit is enclosed in a prickly involucre (covering). Despite the highly ornamental characteristics, this plant may not be the most appropriate for a small yard.

Black walnut (*Juglans nigra*) has edible fruit, is very strong wooded, and provides a nice filtered shade. However, the plant releases toxins into the surrounding soil. Thus, growing other plants under a walnut is a challenge. Stewartia (*Stewartia* spp., Fig. 22–39) are plants with beautiful white flowers, fantastic red fall color, and outstanding exfoliating winter bark. It is a small tree appropriate for small yards. With its four-season interest and small size, it appears to be an ideal choice for just about any landscape situation. However, the plant has exacting cultural requirements: acid pH, extremely well drained soil with high organic content, and protection from winter winds and hot summer sun. If any of these requirements are not met, the plant will suffer and not survive long. It is vitally important, not only for our enjoyment of the landscape but also for the plant's ultimate survival, that we select plants appropriate for the environmental conditions of the given site.

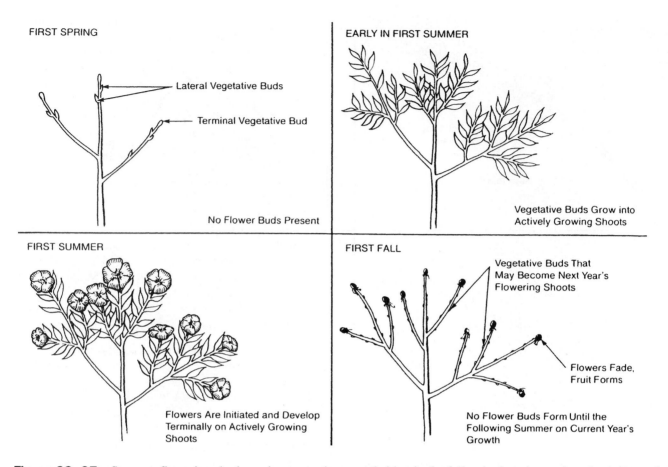

Figure 22–37   Summer-flowering shrubs and trees can be pruned either in the fall or in the winter when the twigs are bare. The flowers form and develop on current second growth that originates from vegetative buds on the previous season's branches.

Figure 22–38   The fruit of *Castanea mollissima,* or Chinese chestnut, is edible, but it is covered in a very sharp involucre (covering). *Source:* Laura Deeter, The Ohio State University.

North America is an extremely large continent with an incredible range of growing conditions. Nursery owners, landscapers, extension agents, and other reputable horticulturists can provide the most accurate information for your region and growing conditions; however, Table 22–1 provides a small sampling of deciduous and evergreen trees and shrubs common for many areas of the country. This table is meant only to illustrate the wide range of sizes, growth habits, and other ornamental characteristics available with woody plants. It is an extremely small sample of what is available in the nursery industry, and you should check with your local experts for a complete list of what grows in your area.

HERBACEOUS PLANTS

Herbaceous plants, especially perennials, have become increasingly popular in recent years. Their popularity is probably due to the fact that they return from year to year; are available in an incredible array of sizes, shapes, flowering time, and color; and can be less expensive than annuals in the

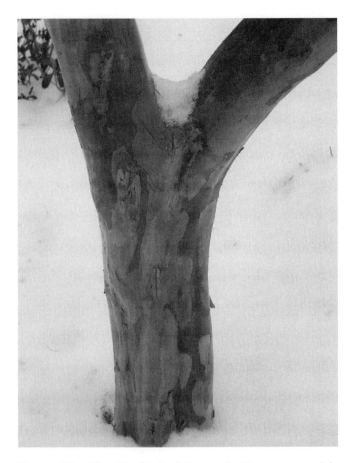

Figure 22–39 The bark of Stewartia (*Stewartia* spp.) is attractive all season, but especially so in winter. *Source:* Laura Deeter, The Ohio State University.

long run. There are *perennials* available for any garden situation: wet, dry, sandy, clay, sunny, shady, large, or small. Perennials allow the gardener to create specialty gardens: butterfly and hummingbird gardens, water gardens, cottage gardens, rock gardens. An almost infinite array of other gardening styles can all be readily accommodated by utilizing perennials (see Figs. 22–40 through 22–44). The possibilities are limitless!

These limitless possibilities, however, are exactly the main source of frustration and perhaps intimidation for many beginning (and advanced) gardeners. How does one choose from a seemingly unlimited number of perennials? Garden magazines offer glimpses of many beautiful gardens, but you might find yourself still wondering where to begin. How did the designer choose the plants? How did they know that placing certain plants next to each other would create the combinations, scenes, and vistas in the photographs? A picture of a cottage garden may seem chaotic and disorganized, with the plants all weaving everywhere. It may be hard to recognize the design and thought that went into the garden. On the other hand, formal gardens may seem to be under such tight control that the amount of work required appears daunting, especially to beginners.

Excellent books have been written on perennial garden design, color theory, and garden maintenance. It is not the intention of this section to supersede the knowledge in these invaluable references. Don't let the sheer numbers of perennials available dissuade you, however, from working with these highly versatile plants. This section should give you a brief overview of the possibilities available to the gardener who incorporates perennials into his or her garden. In addition, the term *perennials* will be utilized throughout this section; however, annuals, bulbs, tender perennials, ornamental grasses, and ferns also follow the same concepts and readily grow with perennials. The term is not meant to be exclusionary; it is used here for convenience.

The concept of "right plant, right place" is still in evidence when utilizing perennials. Just as woody plants do not perform under stressful environmental conditions, neither do perennials. Because the maintenance requirements of these plants are more strict than those for woody plants, however, it is important to analyze your garden more closely. A simple place to start is with honest answers to a few questions:

1. What function (beauty, cut flowers, color, hiding an unsightly feature) will the garden serve in the overall landscape?
2. How big a garden do I really need?
   a. How much work are you able/willing to do?
   b. How much can you afford?
3. What time(s) of the year do I want the garden to look its best?
   a. When am I outside most?
   b. Will the garden be visible from inside the house?
   c. Am I likely to take vacations during certain times?
4. What colors do I like best and which do I dislike: pastels, bright colors, or a smattering of everything?

Once you have determined your needs and desires, you need to look at your growing site. Take a careful inventory of all the environmental conditions. You might think your yard is quite sunny, but careful measurement of the amount of sun may reveal that your yard is shadier than you thought. Will existing trees or shrubs provide shade during certain times of the day or year? How fast will those plants grow and shade out sun-loving perennials? A basic site assessment will save money, time, and trouble in the future.

The same environmental conditions that allow trees and shrubs to grow also affect perennials. Perennials need light, water, soil, and nutrients just as woody plants do. You should be able to answer the following questions about your yard:

1. What is the average date of the first frost in fall or winter?
2. What is the average date of the last frost in spring?
3. Do you have reliable winter snow cover, or is it sporadic?
4. How hot and humid are the summers?
5. How much precipitation do you get and in what months?

TABLE 22–1	A Small Sampling of Common Landscape Trees, Shrubs, and Evergreens			
Scientific Name	**Common Name**	**Habit**	**Height (ft)**	**Ornamental Characteristics**
Acer saccharum	Sugar maple	Tree	60+	Excellent fall color
Araucaria heterophylla	Norfolk Island pink	Tree	50–80+	Evergreen (houseplant in north)
Arctostaphylos manzanita	Manzanita	Shrub/tree	10+	Red bark, pink/white flowers
Betula nigra	River birch	Tree	50–70	Yellow fall color, exfoliating bark
Camellia japonica	Camellia	Shrub	10+	Many colors
Castanea mollissima	Chinese chestnut	Tree	40–60	Edible, yellow-bronze fall color, whitish flowers
Forsythia xintermedia	Forsythia	Shrub	6+	Yellow flowers
Fraxinus pennsylvanica	Green ash	Tree	60–80	Yellow fall color
Hibiscus rosa-sinensis	Rose of Sharon	Shrub	10+	Many colors available
Ilex opaca	American holly	Tree	20–40	Evergreen, red or yellow fruit
Juniperus virginiana	Eastern red cedar	Tree	20–50	Evergreen, can be pruned
Kolkwitzia amabilis	Beautybush	Shrub	10+	Pink flowers, dense plant
Lagerstroemia indica	Crape myrtle	Tree/shrub	10+	Pink or white flowers, exfoliating bark
Laurus nobilis	Laurel	Tree	20–40	Edible
Magnolia grandiflora	Southern magnolia	Tree	15–50+	Evergreen, white flowers
Mahonia aquifolium	Oregon grape-holly	Shrub	6'	Yellow flowers, evergreen, blue fruit
Malus spp.	Crabapple	Tree	15–50	Fruit, pink or white flowers
Picea pungens	Colorado spruce	Tree	60+	Some with blue needles, evergreen
Pinus sylvestris	Scots pine	Tree	50+	Evergreen, orange bark
Potentilla fruticosa	Cinquefoil	Shrub	4	Yellow, white flowers
Pseudotsuga menziesii	Douglas fir	Tree	60+	Evergreen
Pyracantha coccinea	Firethorn	Shrub/vine	6+	White flowers, orange fruit, can be trained to a trellis
Quercus virginiana	Southern live oak	Tree	50–80+	Excellent foliage
Rhododendron catawbiense	Catawba rhododendron	Shrub	10+	Many colors available, evergreen
Sequoia sempervirens	Coast redwood	Tree	100+	Very large tree
Syringa vulgaris	Common lilac	Shrub	10+	Fragrant, many colors available
Taxus xintermedia	Anglojap yew	Shrub/tree	3–50	Evergreen, red fruit, can be pruned

Figure 22–40   These desert plants are perfectly suited to arid conditions. They can also be utilized in a rock garden in more temperate climates. *Source:* Laura Deeter, The Ohio State University.

6. How much sun does the garden get in the various seasons?
7. Will existing plants, buildings, or structures create shade or reflect light?
8. How will you water this garden?
   a. Is the natural rainfall sufficient and spread out enough, or will you need irrigation?
   b. Is there a hose nearby, or will you have to haul water long distances?
9. What are the traffic patterns in the area?
10. Do you have children and/or pets that might affect the garden?

Preparing the Garden

More perennials die from lack of improper bed preparation than any other reason. Proper bed preparation seems to be the obvious way to get your plants to survive over the long run; however, it is not done often. Why? It is a lot of hard work, and it is expensive. It also isn't nearly as much fun as choosing plants! However, hard work in the beginning will pay off many times over in the long run. Bed preparation follows five basic steps:

1. Soil test
2. Elimination of perennial weeds
3. Proper drainage
4. Soil amendments
5. Additional fertilizers

Each step will be discussed in the following subsections.

Soil Test

Companies can perform simple soil tests for gardeners. Soil test kits are also available from local garden centers. Costs

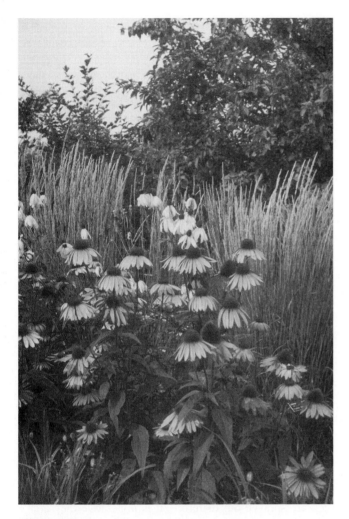

Figure 22–41 These perennials have been utilized to re-create a prairie garden. *Source:* Laura Deeter, The Ohio State University.

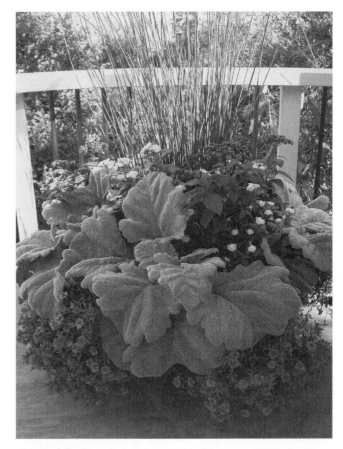

Figure 22–42 Perennials not only mix well with annuals, they can also be grown in decorative containers. *Source:* Laura Deeter, The Ohio State University.

Figure 22–43 A gravel pathway bisects this calming, shady perennial garden. *Source:* Laura Deeter, The Ohio State University.

Figure 22–44 This small container is actually a small water garden filled with specialty perennials. *Source:* Laura Deeter, The Ohio State University.

for professional testing range from $15.00 to $50.00, depending on location and types of testing done. Home kits can be purchased for as little as $5.00 or as much as $50.00, again depending on the complexity of the tests the kit performs. At the very minimum, you should know the soil texture, pH, and drainage of your garden soil. Without going to the expense of professional testing, you can perform some simple tests to determine roughly the texture and drainage. An inexpensive pH kit from the local garden center can provide you with a reasonably accurate measurement of the soil pH. Most perennials prefer a soil pH between 5.5 and 6.5, but in reality they are quite tolerant of anything approximating neutral (7.0). Problems generally occur only when the pH is very alkaline or very acidic. There are exceptions, however. Japanese iris (*Iris ensata*) requires an acid pH to perform well. Baby's breath (*Gypsophila paniculata*) performs much better with a pH above 7.0. In fact, Gypsophila translates as "gypsum (or lime) loving" and refers to the fact that the plant prefers a more alkaline pH.

In the absence of a professional soil test, a homeowner may use two methods to determine soil texture: the soil ball test and the water jar test. For the soil ball test, water the garden thoroughly. Be sure to stop before the soil becomes soggy. Take a small handful of soil and squeeze it. Open your hand and carefully watch what happens to the soil ball. If it stays in a ball, it is predominantly clay. If it disintegrates and feels gritty, it's sandy. If it's somewhere in the middle, it's predominantly silt. The water jar test takes a bit longer, but it is more accurate in determining soil texture. Collect several soil samples (sample the soil, not the organic matter from the top) and remove the debris, pebbles, roots, and other large particles. Put one cup of the soil in a clear quart jar filled with water, seal it and shake vigorously; make sure it's mixed well. After twenty-four hours, examine the soil layers at the bottom of the jar. The heavy sand particles will sink to the bottom, followed by the silt, with the clay on top. Be sure you don't move the jar or the clay particles will float up again. Estimate the percentage of each compared to the whole to determine your soil texture. For example, a soil with one-quarter sand, one-half silt, and one-quarter clay would be 25 percent sand, 50 percent silt, and 25 percent clay.

Knowing the soil texture provides some information about how the garden drains. Again, a very simple test illustrates how well a garden drains. Dig a square hole 12 in. wide and 12 in. deep. Fill it with water and let it drain. Fill the hole a second time and record the length of time it takes to drain the second time. Soil that takes less than one hour to drain is very well drained and may require additional irrigation. If it takes over three hours to drain the second time, the area is extremely poorly drained and may require a drain-tile to move excess water. Between one and three hours is good drainage and would support a wide array of plant species.

Elimination of Perennial Weeds

There is an old gardener's saying: "One year of seeds, seven years of weeds." For a new planting, it is far easier to get rid of the weeds prior to planting rather than dealing with them after the fact. Several methods are available to remove weeds: preemergent and postemergent herbicides, weed barriers, and hand pulling.

Although hand pulling can be effective for a limited number of weeds, it is generally not the most efficient method for large numbers, nor does it generally remove the entire weed. Certain weeds, like Canada thistle, dandelion, and crabgrass, to name a few, are renowned for their ability to return from the tiniest piece of root left in the ground. For an established perennial garden where herbicides may not be an option, hand pulling is often the only choice. A trowel or even a large knife can be very useful in removing as much of the roots as possible, especially for troublesome weeds.

Chemicals are very effective at removing weeds, especially prior to planting. Preemergent herbicides are designed to prevent seed germination and are effective at reducing the amount of seed endemic in the soil. In an established perennial bed, they are often highly useful in keeping weeds under control. Their main downfall is that many species of herbaceous plants self-seed and, indeed, the gardener may need the plants self-seeding to ensure continued presence of the plant in the garden. Johnny Jump-up and columbine (*Viola tricolor* and *Aquilegia* spp) are two perennials that self-seed. A preemergent herbicide will kill those seeds as well. A postemergent herbicide kills the plant after it has emerged and is actively growing. Some of these herbicides kill any plant they are sprayed on; therefore, it is imperative they aren't sprayed on desirable plants. Some postemergent herbicides are very specific, killing only grasses or only broad-leaf plants, and they can be useful for removing certain troublesome pests in the garden. Keep in mind that a grass killer also kills ornamental grasses and a broad-leaf weed killer also kills any of the perennials that it might be sprayed on accidentally. Caution must always be utilized when spraying any chemical; be sure to read and follow all label directions.

Barriers are a good way to kill weeds if you have enough time and you can plan well in advance. In a process called solarization, clear plastic is placed securely over the area and allowed to stand for several weeks during the hot summer months. The light and initial heat encourages plant and seed growth. The lack of water and the rising temperatures eventually kills not only the adult plants, but also the seeds. The only downside to this technique is the time and planning needed. Newspapers piled on the site of the new garden can accomplish a similar effect. Spread 2 in. of newspapers over the new garden bed and pile the soil and amendments on top. The thickness of the papers prevents any weeds from penetrating; the new garden can simply be planted on top of the papers, which will not show once the garden is planted and will disintegrate, allowing the roots of the garden plants to extend deep into the bed. These and similar methods are excellent for eliminating weeds in situations where chemicals are not an option.

Proper Drainage

More perennials are killed from improper drainage than any other reason. Most perennials require well-drained soil to perform well. Thus, the soil drainage test mentioned above is an essential step when planting perennials. Several methods are available for improving the drainage of an area: install a drain tile, raise the bed, or add amendments to the soil.

Except in extreme cases of poor drainage, most gardeners do not have to worry about installing a drain tile. Although highly effective, they are expensive to install and should be done only by a trained professional to ensure that the new piping meets all the sewer and piping codes of your area.

Raising the bed is something that has been practiced since the times of the ancient Greeks and Romans. This technique involves simply adding from 2 to 3 ft up to 12 ft of height to the existing garden bed. A small wall or barrier may be required in cases where 12 ft of height is added to a bed. Adding topsoil or peat moss to an existing bed can raise the level a couple of inches, which will probably not require a barrier of some sort to hold in; the roots of the plants will keep it in place. The last method for improving soil drainage is to add amendments, which is Step 4 in the list for bed preparation and is discussed in the next subsection.

Soil Amendments

Adding amendments to the soil is one of the simplest and easiest solutions to a mild drainage problem. Proper selection of soil amendments can help a soil that drains too quickly just as easily as a soil that drains poorly. Soil amendments can be either organic or inorganic. Inorganic amendments include gypsum, lime, aggregate, or sand. For most garden situations, they are unnecessary additions because organic amendments can provide drainage along with many other benefits.

Organic amendments improve drainage by adding large particles to the soil, thus creating more air spaces. They also improve the water-holding capacity of a quickly draining soil. In dry areas of the country, this water-holding capacity means less irrigation. They create a favorable microhabitat for beneficial soil organisms. In addition, evidence suggests that compost can also dramatically decrease the incidence of several diseases in the landscape. Finally, organic amendments also add small quantities of nutrients to the soil. Regardless of the type of organic amendment utilized, be sure it is well composted and free of weed seed. Several types of organic amendments are available: homemade compost, commercial compost, grass clippings, various manures, peat moss, vermicompost (worm castings), mushroom compost, leaf mold, and so on. These products range in price from extremely inexpensive to highly expensive, depending on the product and the area of the country.

How do you mix the soil amendment into the soil? On an established perennial bed, the easiest method is simply to top-dress around all the plants with approximately 1 in. of amendment. This top-dressing will slowly work its way through the soil over the season. For a new bed, a rototiller is the most efficient method, especially if the area is large. Till the new garden area once prior to adding any amendment. Add half the amount of organic amendment, then till the area again. Add the other half and till a final time. Your bed is ready if you can plant with your hands.

How much amendment should you add? Your soil should be almost 30 percent organic matter. Because most soils have less than 5 percent organic matter naturally, you must add a considerable amount of organic matter. For every 12 in. (1 ft) of soil depth, you must add 4 in. of organic matter. To calculate the amount needed, follow these simple steps:

1. Calculate the square footage of your bed:
   a. Square or rectangle = length × width.
   b. Circle = $\Pi \times r^2$ (Π = 3.14).
   c. Triangle = ½ base × height.
2. Multiply the square footage by .33 (4 in. ÷ 12 in.).
3. Divide this by the unit you are ordering.

Suppose, for example, that your garden bed is 125 ft × 12 ft. How much organic matter does your garden need? Assume you are ordering 4 ft^3 bales of peat moss.

1. Calculate the square footage:
$$125 \text{ ft} \times 12 \text{ ft} = 1{,}500 \text{ ft}^2.$$

2. Multiply by .33 (4 in. ÷ 12 in.):
$$1500 \text{ ft}^2 \times .33 \text{ ft} = 495 \text{ ft}^3$$

3. Divide this by the unit you are ordering:
$$\frac{495 \text{ ft}^3}{4 \text{ ft}^3 \text{ per bale}} = 123.75 \text{ bales of peat moss needed}$$

Additional Fertilizers

This section needs to be prefaced by the fact that perennials don't generally require large quantities of additional fertilizers. If a gardener is top-dressing every year with organic matter, no additional fertilizer is needed. A few exceptions are tulips, hyacinths, roses, and larkspur, which are heavy feeders. However, not all gardeners compost, nor is it always possible to top-dress every year for various reasons. In addition, nutrients may have become depleted and additional fertilizer is needed. Regardless of the reason, perennials don't generally require much fertilizer. But to get the most out of any fertilizer, it is helpful to understand some basics.

All plants require seventeen basic nutrients to survive. They acquire these nutrients from water and air (carbon, hydrogen, and oxygen) and soil. The three needed in the largest quantities from the soil are nitrogen (N), phosphorus (P), and potassium (K), which are the three numbers, such as 10-10-10 or 15-30-15, prominently displayed on a fertilizer bag. The numbers represent the percentage of the chemical contained in a bag of fertilizer. For nitrogen, this number is a direct percentage. Thus, a bag of 8-12-17 complete fertilizer contains 8 percent nitrogen. If you are utilizing a complete fertilizer, there is little need to worry that you will have nutrient problems, except in rare cases. A soil test will tell you if your soil has deficient levels of nutrients. In that case, a special fertilizer may have to be created to overcome that specific situation. This situation is not common, however.

Most commercial fertilizers are sold as either dry granules or as a powder to be mixed with water. Dry granules are often time-release, which means that they will release the various chemicals slowly over a period of weeks or months. Liquids are the fastest way to get nutrients to your plants and are quite handy in situations where your plants are severely nutrient-deficient. It is best to fertilize in the spring, when the plants are actively growing. Perennials generally do not benefit from fall fertilizer applications; it will spur them into growth at a time when they need to be preparing for winter and dormancy.

Unless a soil test reveals a problem with one or more nutrients, most calculations for fertilizer can be based on nitrogen. Nitrogen is highly mobile in the soil and is also needed in large quantities by the plants. The recommended rate for perennials is 1 lb of actual nitrogen per 1,000 ft^2 per year, which is not a large amount. To calculate the correct amount, follow these simple steps:

1. Calculate the square footage of the bed.
2. Divide this number by 1,000 (1 lb N/1,000 ft^2/year)
3. Divide this number by the percentage of nitrogen in the fertilizer you plan to use.

Suppose that your garden bed is 135 ft × 15 ft wide. How much 15-30-15 should be applied to this bed to achieve the recommended rate for the year?

1. 135 ft × 15 ft = 2,025 ft^2.
2. 2,025 ft^2/1,000 = 2.025 lb of actual nitrogen needed.
3. 2.025/0.15 = 13.5 lb of fertilizer needed for the year.

Perennial Maintenance

Once the garden is installed, regular maintenance chores must be performed. Mulching, weeding, checking for diseases and insects, staking, pinching, deadheading, and dividing are common garden chores for various times of the year.

Mulch

A layer of mulch in the garden provides several benefits:

Conserves water

Regulates temperature (soils are cooler in summer and warmer in winter)

Slowly adds nutrients to the soil as they break down

Prevents erosion

Is aesthetically pleasing

Controls weeds (but doesn't prevent them)

Several types of mulch, both organic and inorganic, are available. Gravel, marble chips, volcanic rock, and ground tires are all examples of inorganic mulches. These are good in permanent areas with limited plantings. They are generally *not* recommended for use around perennials. Organic mulches include various types of wood chips, pine straw, hay straw, and living mulches or ground covers. For perennials, organic mulches are the best.

Mulches have a reputation for preventing weeds in the garden. In reality, the actual number of weeds in a mulched and an unmulched garden is almost identical. However, the weeds in a mulched garden are easier to pull because the soil is generally moister. If weeds are blowing in after the mulch is down, the weeds root into the mulch, making them very easy to pull compared to weeds rooted in soil. Most mulches have one disadvantage that is generally not considered by homeowners: they are flammable. Although it's not common for mulches to catch fire, stray cigarettes or matches have been known to set the mulch on fire, occasionally with disastrous consequences. This issue is generally not serious enough to avoid mulching; the benefits greatly outweigh the disadvantages. In certain situations, however, it is a variable that must be considered.

Staking

Staking is nothing more than providing support for those plants that grow too large or whose flowers are too heavy for proper support. There is nothing pleasing about coming home from work in the evening to find that your larkspur or your peonies have flopped over and are flowering beautifully on the ground! Storms are another reason perennials fall over and might require support.

The trick to staking perennials is that the emphasis must always remain on the plant and the flowers, not on the stakes. Another trick with staking is to have the stakes in place before the plant requires support. Placement of the support after the fact invariably looks artificial. A properly staked perennial should not show the support (see Fig. 22–45). Different methods of staking can be used, depending on the growth habit of the plant involved.

Tall flowers are best staked with a thin, single stake (Fig. 22–46). Foxglove and larkspur both have tall spikes of flowers. A tall branch, a thin bamboo stake, or a plastic or metal stake can all serve to hold these plants' flowers upright. Gently tie the flower stalks to the stake, making sure not to damage the succulent growth. As the stem grows, you can add more ties.

Bushy plants, such as aster or garden phlox, require a slightly different method. Circular, metal frames are handy because they last a long time and are sturdy. Small sticks in

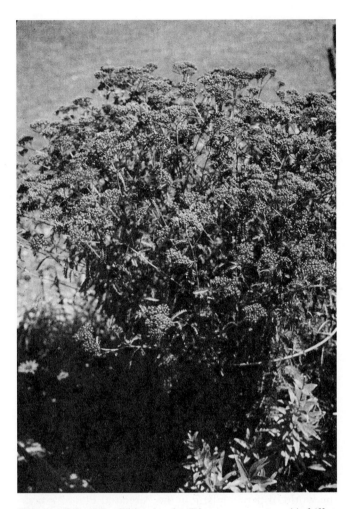

Figure 22–45 This Apple Blossom yarrow (*Achillea millefolium* 'Apple Blossom') is held up by a metal cage. Notice that the cage is hardly visible. *Source:* Laura Deeter, The Ohio State University.

the ground with twine between them can provide the same support, and you won't have to resort to spending money on metal cages. Airy plants, such as baby's breath, are best supported by small sticks set into the ground around the plant. Sticks with a network of string between them are also effective.

Deadheading and Pinching

Deadheading is removing a flower once it has faded. It is not as violent as it sounds! Some perennials drop their old flowers willingly and neatly. Others are not so obliging. Faded flowers and brown or black petals can detract greatly from the overall beauty of the garden (Fig. 22–47). Thus, the prime reason for deadheading is aesthetics; however, the practice yields other benefits as well. Leaving the flowers on the plants only encourages them to make seeds. Unless you are hybridizing your own plants, there is no need for seed production in the garden. Some plants can become aggressive self-seeders, and deadheading prevents this tendency. Others spend energy attempting to reproduce at the expense of root production; thus, the plant may not survive the winter. Deadheading can also prolong the flowering time for many plants.

Pinching is removing new growth prior to flowering. This technique creates more compact, bushy growth (that might not require staking later in the season) and can increase flowering time. Simply pinch the top inch or two from each branch of the plant in mid- to late spring. You can pinch again in a couple of weeks if you want. Be sure not to pinch once you see flower buds or you'll pinch off the flowers.

Pinching can also be done after a plant has flowered to encourage another flush of flowers. Some species, such as

Figure 22–46 Large dahlia (*Dahlia* spp.) are best held up by strong, single stakes. *Source:* Laura Deeter, The Ohio State University.

Figure 22–47 Tickseed (*Coreopsis auriculata* 'Nana') requires continual deadheading. The faded flowers detract from the overall appearance of the plant. *Source:* Laura Deeter, The Ohio State University.

tickseed (*Coreopsis lanceolata*) can be pinched up to three times during the growing season, for three almost full flushes of flowering. Pinching is also done to delay the flowering time. Hardy garden mums are pinched several times during the growing season to promote bushiness and to delay flowering until the fall, when the plants are sold for additional fall color.

Dividing

Dividing perennials may be done for three reasons in the perennial garden. The first is to make more plants. The second is to rejuvenate an old plant. The final reason is to control an aggressive plant that is getting too large.

Plants with multiple stems, such as aster, hosta, garden phlox, and yarrow, are very simple to divide. Plants with a large tap root, such as butterfly weed (*Asclepias tuberosa*) and false indigo (*Baptisia australis*), are not easy plants to divide; it's best to leave them alone. In addition, some plants that we treat and use as herbaceous perennials are actually small woody plants. Lavender (*Lavandula* spp), Russian sage (*Perovskia atriplicifolia*), and zonal geranium (*Pelargonium xhortorum*) are all woody plants and should not be divided. These plants are the exceptions, however, rather than the rule. Most perennials are very easy plants to divide.

To divide a perennial (for any reason), dig the entire clump out of the ground. You might need someone's help: large, old perennials are very heavy. Utilizing a shovel, spade, sharp knife, or even a chainsaw (for large grasses), cut the plant into as many sections as you desire. Be sure to leave a good root system on all sections that are going back into the ground. The smaller you divide the plant, the longer it will take to reach flowering size again. Although some perennials require great care when dividing (sea pinks, or *Armeria maritima,* is one such perennial), most perennials are durable plants. Although a little care is needed, generally you need not worry about causing long-term damage to the plant.

Generally the best time to divide perennials is in the spring (early spring flowering plants are better done late spring or in the fall). This approach gives the plant the most amount of time to recover and put down a good root system again. Regardless of the timing, remember that all plants undergoing division need some extra tender loving care until they become established again. Make sure you water your plants regularly and often until they form a good root system. The following is a checklist to help ensure the future survival and growth of your new divisions:

1. Prepare the site for your new plants prior to dividing them.
2. Use sharp tools. Sharp tools make clean cuts, helping to prevent the spread of disease.
3. Discard any dead stems.

4. For large plants, cut the foliage back by half at least to help conserve water.
5. Replant divisions as soon as you can.
6. Water, water, water.

Perennial gardening is *not* a no-maintenance type of gardening. A small amount of effort and maintenance ensures that your perennial garden will be beautiful and full of flowers for years to come.

The Wonderful World of Perennials

The vast number of perennials available in the nursery industry is staggering and can be intimidating for a new gardener. How do you choose a few plants to start your garden when hundreds are available at your local garden center? In addition, North America is extremely diverse in terms of the plant material that grows across the region. Southern regions have an amazing array of plant material available year-round, and these plants are treated as annuals in northern areas. Plants for full sun in the North may require shade in hot, arid regions. It is impossible to provide a list of perennials that will work in every garden situation across North America. There are perennials for every garden situation: wet, dry, arid, boggy, well drained, clay, sandy, sunny, and shady; small areas, large areas, and hillsides; erosion control; wildlife, butterflies, and hummingbirds; water gardens and herb gardens; to name but a few. A *very* brief listing of several specialty plants can provide you with the range of possibilities available when you include perennials in your landscape. If you are interested in the world of perennials, join the Perennial Plant Association. Visit their website at **www.perennialplant.org** for more information.

The following list was compiled from nurseries across North America as their top ten best-selling plants, and it provides a general listing of plants that grow in as wide a range of growing conditions as possible:

1. *Hemerocallis* 'Stella D'oro'—Stella D'oro Daylily (Fig. 22–48):
   a. Golden-yellow flowers peak in early summer. They rebloom sporadically with deadheading.
   b. 15 to 18 in. tall and 24 in. in diameter.
   c. Grow in full sun or partial shade and well-drained soil.
   d. Foliage remains clean through the season.
2. *Hosta undulata* 'Medio-Variegata'—variegated hosta (Fig. 22–49):
   a. Purplish lilac flowers in midsummer.
   b. Green leaves with a pure white margin.
   c. 15 to 18 in. tall, 24 in. in diameter.
   d. Shade or part shade; moist, well-drained soil.
   e. Makes good cut flowers; hummingbirds are attracted to the flowers.

Figure 22–48 *Hemerocallis. Source:* Laura Deeter, The Ohio State University.

Figure 22–50 *Astilbe. Source:* Laura Deeter, The Ohio State University.

Figure 22–49 *Hosta undulata. Source:* Laura Deeter, The Ohio State University.

Figure 22–51 *Echinacea purpurea. Source:* Laura Deeter, The Ohio State University.

3. *Astilbe* spp.—false spirea, astilbe (Fig. 22–50):
   a. Foliage ranges from dark, shiny green through red and bronze.
   b. Flowers late spring to midsummer depending on type. Flowers are open plume in white, red, pink, or purple.
   c. Plants range in height from 8 in. to 3 ft, with a similar spread depending on type.
   d. Partial shade; moist, well-drained soil.
   e. Great cut flowers; excellent dried seed heads.
4. *Echinacea purpurea* 'Magnus'—purple coneflower (Fig. 22–51):
   a. Pink to purple daisylike flowers in midsummer.
   b. Dark green foliage.
   c. Full sun; drought-tolerant.
   d. 2.5 to 3 ft tall and wide.
   e. Excellent cut flowers; attracts bees.

5. *Rudbeckia fulgida* var. *sullivantii* 'Goldsturm'—black-eye Susan (Fig. 22–52):
   a. Yellow, daisylike flowers with black centers in late summer.
   b. Dark green foliage.
   c. Full sun; drought-tolerant.
   d. 2 to 3 ft tall and 3 to 4 ft in diameter.
   e. Excellent cut flowers.
6. *Coreopsis verticillata* 'Moonbeam'—moonbeam threadleaf tickseed (Fig. 22–53):
   a. Light yellow flowers all summer (requires deadheading).
   b. Full sun to part shade; drought-tolerant.
   c. 18 to 24 in. tall and wide.
   d. Airy foliage is attractive when the plant isn't flowering.

Figure 22–52   *Rudbeckia fulgida. Source:* Laura Deeter, The Ohio State University.

Figure 22–53   *Coreopsis verticillata. Source:* Laura Deeter, The Ohio State University.

Figure 22–54   *Sedum. Source:* Laura Deeter, The Ohio State University.

Figure 22–55   *Heuchera. Source:* Laura Deeter, The Ohio State University.

7. *Sedum* 'Autumn Joy'—Autumn Joy stonecrop (Fig. 22–54):
   a. Succulent, fleshy foliage is bright green.
   b. Flowers start out green in midsummer, change to light pink and finally to dark maroon in early fall.
   c. 2 to 3 ft tall and wide.
   d. Full sun; drought-tolerant.
   e. Great for cut flowers; attracts bees and butterflies; seed heads remain attractive through early winter.
8. *Heuchera* spp.—coral bells (Fig. 22–55):
   a. Foliage colors range from green, purple, maroon, lime, golden, to silver, and to combinations of those colors.
   b. Flowers well above the foliage in pinks, red, and white in late spring, and early summer.
   c. 1 to 2 ft tall and wide.
   d. Partial shade to full sun, depending on the variety.
   e. Excellent for hummingbirds, cut flowers, and foliage color.

9. *Pennisetum alopecuroides* 'Hameln'—dwarf fountain grass (Fig. 22–56):
   a. Fine-textured, short, ornamental grass.
   b. Seed heads are attractive in mid- to late summer.
   c. Excellent winter foliage appearance.
   d. 2 to 4 ft tall and wide (with seed heads); forms clumps.
   e. Full sun; well-drained soil.
   f. Great for winter interest.
10. *Calamagrostis xacutiflora* 'Karl Foerster'—Karl Foerster feather reed grass (Fig. 22–57):
   a. Fine textured, upright ornamental grass.
   b. Tan seeds are attractive from midsummer through midwinter.
   c. 4 to 5 ft tall and 2 to 3 ft wide; forms clumps.
   d. Full sun; well-drained soil; drought-tolerant.
   e. Excellent for winter interest, foliage interest, cut flowers.

Figure 22–56 *Pennisetum alopecuroides. Source:* Laura Deeter, The Ohio State University.

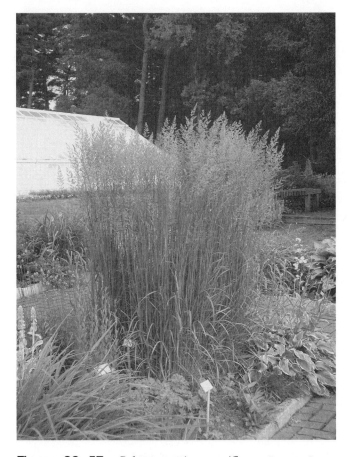

Figure 22–57 *Calamagrostis xacutiflora. Source:* Laura Deeter, The Ohio State University.

Specialty Perennials

In addition to the top-selling perennials, many gardens have specific themes and require plants with specialty characteristics. These garden themes include water, herb, and edible flower.

Figure 22–58 *Nymphaea. Source:* Laura Deeter, The Ohio State University.

Figure 22–59 *Eichornia crassipes. Source:* Laura Deeter, The Ohio State University.

Water Garden Plants

1. *Nymphaea* spp.—water lily (Fig. 22–58):
    a. Floating leaves; some varieties are mottled.
    b. Flowers in a rainbow of colors; some varieties open only at night.
    c. Some are hardy; others are tropical.
    d. Full sun; water gardens.
2. *Eichornia crassipes*—water hyacinth (Fig. 22–59):
    a. Purple flowers in summer.
    b. Not hardy in the northern climates.
    c. Floating plant; does not need a container.
    d. Unique foliage and bladder for floating support.

Herb Gardens

1. *Thymus* spp.—thyme (Fig. 22–60):
    a. White to pink flowers in late spring to midsummer, depending on type.
    b. Upright or ground covers are available.

Figure 22–60   *Thymus. Source:* Laura Deeter, The Ohio State University.

Figure 22–61   *Lavandula. Source:* Laura Deeter, The Ohio State University.

Figure 22–62   *Tropaeolum majus. Source:* Laura Deeter, The Ohio State University.

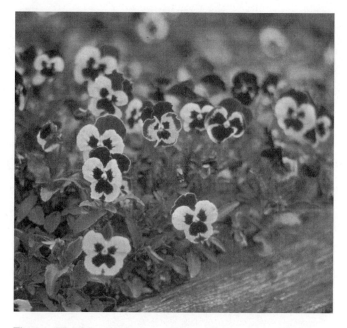

Figure 22–63   *Viola xwittrockiana. Source:* Laura Deeter, The Ohio State University.

c. Edible; highly fragrant with many different scents.
d. Some are tolerant of minimal foot traffic.
e. Full sun; very well-drained soil.

2. *Lavandula* spp.—lavender (Fig. 22–61):
   a. Many species available in various parts of the country.
   b. Purple, blue, red, and white flowers are fragrant and edible; attract bees.
   c. Attractive gray, edible foliage on a small woody shrub.
   d. Can be sheared heavily.
   e. Full sun; extremely well-drained soil.

Edible Flowers

1. *Tropaeolum majus*—nasturtium (Fig. 22–62):
   a. Round foliage can be green or with white variegation.

   b. Flowers all summer in yellow, orange, and red; flowers are edible and have a peppery taste.
   c. Partial shade; well-drained soil.
   d. Annual.

2. *Viola x wittrockiana*—pansy (Fig. 22–63):
   a. Flowers in cool weather (spring and fall in the north) in a rainbow of colors.
   b. Flowers are edible and taste like iceberg lettuce.
   c. Great for adding color to a salad and to Jello, or using as a cake decoration; can be candied.
   d. Partial shade; well-drained soil.
   e. Prefers cool climates or protection from the sun in the south.

SUMMARY AND REVIEW

Ornamental plants provide our landscapes with more than just pleasure. They can serve many purposes and be utilized to solve various landscape problems. Large trees provide shade and shelter, and when used properly, they can even reduce heating and cooling costs. Shrubs provide flowers, fruit, screening, and structure in the landscape. Perennials, annuals, and other herbaceous plants fill in and provide the color and excitement in the garden. They can be utilized in different specialty gardens and to fill the needs of just about any size or shape garden. Properly planted and cared for, ornamental plants can provide years of enjoyment and pleasure in the landscape.

FOOD FOR THOUGHT

1. As you look through the landscapes of your neighborhood, notice the placement of the large trees. Are they appropriate for the sites? Have they become overgrown and in need of professional care?

2. Discuss how improper planting can lead to problems in the future in the growth and survival of trees and shrubs.

3. Soil conditions, water availability, light, and temperature are all interrelated and play a role in plant growth and survival. Discuss how these four environmental conditions are interconnected. What can happen to the other three characteristics if just one, for example, soil pH, were to change dramatically? What might happen to the landscape plants in this instance?

4. Perform the drainage test, soil ball test, and water jar test on a specific area or areas of a landscape. If a soil pH kit is available, measure the pH of the soil. Record the results. Discuss any changes that may be needed to support a garden. If the area is landscaped, discuss current plant performance in light of the soil tests just performed. Are changes needed?

5. Find a garden with perennials. Observe this garden over a period of weeks. Note the changes in flowering, color, and the presence of any wildlife (bees, butterflies, hummingbirds, etc.) in the garden. Record the changes in a garden journal. Note any maintenance that may need to be done in the garden.

6. Walk through a nearby neighborhood. Note the placement of the plants in relation to the houses or other structures. What function(s) is (are) the various plants playing in the landscape? What functions are not fulfilled as well as they could be, if at all?

SUPPLEMENTARY READING

DIRR. MICHAEL A. 1994. *Manual of ornamental woody landscape plants.* Champaigne, Ill.: Stipes Publishing.

DIRR, M. A. 1997. *Dirr's Hardy Trees and Shrubs: An illustrated encyclopedia.* Portland, Ore.: Timber Press.

DISABATO-AUST, TRACY. 1995. *The well tended perennial garden.* Portland, Ore.: Timber Press.

PHILLIPS, ELLEN, and C. COLSTON BURRELL. 1993. *Rodale's illustrated encyclopedia of perennials.* Emmaus, Pa.: Rodale Books.

STILL, STEVEN M. 1994. *Manual of herbaceous ornamental plants.* Champagne, Ill.: Stipes Publishing.

CHAPTER

Floriculture

Jeff Kuehny

KEY LEARNING CONCEPTS

After reading this chapter, you should be able to:

♦ Understand the basic greenhouse structure and components.
♦ Explain how the greenhouse environment is manipulated to regulate plant growth and development.
♦ Understand the principles of growing several greenhouse crops.

THE GREENHOUSE INDUSTRY

Greenhouses (or glasshouses) date from Roman times when wealthy people grew plants in small enclosures, which allowed for spring flowers in the winter and fruit out of season. This desire spread across Europe, helping to create the largest greenhouse industry in the world in the Netherlands. The initial structures were built so that during the day, when the sun shone brightly, the glass could be lifted slightly to allow the trapped hot air to escape. Low glass-covered structures that could be opened were called cold frames by the English. To add heat and make it a hot frame, decaying manure was buried about 0.5 m (18 in.) under the structure (Fig. 23–1). Both the cold and hot frames were constructed to allow a person to reach in and tend to the plants (Fig. 23–2). These units developed into pit houses, which were similar to today's greenhouses and were constructed with a peaked roof, doors at each end, and a center walkway cut about 1 m deep into the ground to accommodate a standing person.

The modern greenhouse with its high sides and glass roof, was developed later (Fig. 23–3). Greenhouses were constructed to allow the rays of the sun to enter the glass roof perpendicularly in the winter but partially reflect them when the sun was high during the summer. Heat was supplied with circulating hot water and, at a much later date, with steam. The sash bars were usually made of thick wood, especially in areas with high snowfalls, to provide struc-

Figure 23–1 A hotbed warmed by decaying manure. Modern hotbeds are warmed by an electrical cable that is thermostatically controlled. The glass sash is lifted during the day to allow excess warm air to escape.

Figure 23–2 Cold frame against the side wall of a greenhouse used to overwinter plants.

Figure 23–3 Detached even-span glass greenhouses constructed in an open area to maximize light intensity and expansion capability.

tural support (Fig. 23–4). Lumber resistant to rot such as cypress and redwood was found to be ideal, but because these woods are not particularly strong, the girth of the sash bars had to be large, creating many shadows and thereby reducing the light intensity.

The even-span greenhouse was the most popular glass greenhouse design for many years (Fig. 23–5). Even-span greenhouses connected along the eave are termed ridge-and-furrow greenhouses (Fig. 23–6). For decades, greenhouses did not vary much in design except that the width of the glass increased as the technology of making glass improved. With improvements in techniques for extruding aluminum, the wide wooden sash bars were replaced with narrow aluminum bars, which were very strong. Narrow sash bars and wide glass allowed more light to enter during the winter, when it was needed most.

As plastics increased in popularity, so did their use as greenhouse glazings. Polyvinyl chloride was used briefly, but it had an unacceptable life expectancy. Fiberglass, with a life expectancy of twenty years, became more popular and is still used if ultraviolet (UV) light-resistant protection is added. Although polyethylene has a short life expectancy (≤ four years) it has become the most popular glazing because of its weight, low heat loss (double layers), and less-expensive price. The quality of polyethylene has increased with added UV inhibitors, infrared inhibitors and absorbers, anticondensate additives, and photoselective inhibitors. Polyethylene can also be purchased in varying thicknesses: 3, 4, 5, or 6 mil. The greater the mil or thickness, the stronger and more durable the film. Single-layer polyethylene is used for cold frames and greenhouses in subtropical or tropical climates. Double-layer polyethylene is used in colder climates;

Figure 23–4 Interior structure of a glass greenhouse consisting of painted cypress sash bars with iron columns providing structural support.

Figure 23–5 Even-span glass greenhouse attached to a head house. The head house is used for storage of planting materials and provides a space for planting and shipping.

Figure 23–6 A ridge-and-furrow greenhouse range with glass covering and polycarbonate end walls, which provide maximum light intensity and good insulation.

Figure 23–7 A squirrel-cage fan pulls outside air through the attached flexible tube and into the air space of the two layers of polyethylene to maintain a constant pressure. *Source:* Jeff Kuehny, Louisiana State University.

Figure 23–8 Quonset-style greenhouse with a single-layer polyethylene covering, exhaust fan for cooling, and unit heater for heating.

it should have a dead-air space between the layers of approximately 4 in. to provide insulating properties. The outer layer of polyethylene is usually 6 mil, while the inner layer is 4 mil. A small fan is used to maintain the air space (Fig. 23–7).

Acrylic and polycarbonate single- and double-layer panels have gained in acceptance. They have many of the same properties as polyethylene. These panels are more rigid, with a life expectancy of up to twenty years.

The use of plastics as greenhouse glazings led to the development of Quonset-style (Fig. 23–8) and gutter-connected greenhouses (two or more Quonset-style greenhouses connected, Fig. 23–9), which are less expensive to

Figure 23–9   Gutter-connected greenhouse range with a double-layer polyethylene covering, polycarbonate end walls, and automatic vents providing natural air circulation and cooling.

build than even-span greenhouses. They remain the most popular greenhouses today, with design changes that continually optimize the growing of plants.

CONTROLLING THE GREENHOUSE ENVIRONMENT

Temperature

Greenhouses are heat traps that are cooled by applying white shading compound to the glass, by applying shade cloth on the outside of the greenhouse, by using automated shade curtains on the interior, by convection ventilation (allowing the hot air to rise through the overhead ventilator and cool air through the side vent), by forcibly removing air with exhaust fans (Fig. 23–10), or by evaporative cooling using a fan-and-

Figure 23–10   Greenhouse range with exhaust fans in the side walls used to pull air through the greenhouse for greater cooling and air circulation.

Figure 23–11   Cross-fluted cellulose cooling pads with water passing over them. As outside air is pulled through the pads, heat is absorbed, thereby cooling the air. *Source:* Jeff Kuehny, Louisiana State University.

pad system. The fan-and-pad cooling system operates by using a pad, usually made of cross-fluted cellulose, with water passing through it along one wall and exhaust fans evenly spaced along the opposite wall (Fig. 23–11). Outside air is pulled through the pads into the greenhouse. The water passing through the pad absorbs the heat from the air, thereby cooling it. The exhaust fans help pull this cool air through the greenhouse and remove the warmer air.

The greatest heat load in the greenhouse occurs when the air is heated by the sun's rays and then trapped by the enclosure of glass or plastic. Short-wave radiation passing through the greenhouse covering strikes the objects in the greenhouse. The reflected radiation in the form of long-wave radiation (heat) does not pass through the covering and is trapped in the greenhouse, causing an increase in the temperature during the day. However, some form of heat must be added at night during the winter season because too little daytime heat is retained for that purpose. In extremely cold climates, supplementary heat is necessary even in daylight hours. A central heating system derives its heat from a boiler circulating hot water or steam (Fig. 23–12). This type of system is primarily used in large greenhouse ranges. Localized heating can be accomplished through the use of unit heaters, forced-air furnaces (Fig. 23–13), infrared radiant heaters (Fig. 23–14), and root-zone bottom heating.

Convection tubes and horizontal airflow fans are also important for heating and air movement in the greenhouse. Convection tubes are made of transparent polyethylene and are attached to a blower on one end and sealed at the other end. These tubes have 5 to 7.6 cm (2 to 3 in.) round holes along the bottom of the tube. The air that is pulled from the blower is forced through the holes at a high velocity, creating a mixing

Figure 23–12   Centralized boilers used to heat water that is piped throughout the greenhouse range for heating. Because centralized heating provides a single source of heat, most greenhouse ranges have at least two boiler units so that if one fails, the other can be used.

Figure 23–13   Single-unit, forced-air furnace used for heating a single greenhouse. Most greenhouses have more than one unit heater to provide even heat distribution. *Source:* Jeff Kuehny, Louisiana State University.

Figure 23–14   These heaters emit infrared radiation that, on striking an object, is converted to heat. Thus, the plants and containers are heated without having to heat the air—saving energy and money.

Figure 23–15   A convection air tube made of transparent polyethylene is attached to a blower on one end and sealed at the other end. Air pulled from the blower is forced through the small holes on the bottom of the tube at a high velocity, thus mixing the surrounding air. *Source:* Jeff Kuehny, Louisiana State University.

or stirring with the surrounding air (Fig. 23–15). The blowers may pull cool air through a gable end vent to cool the air in the greenhouse or pull hot air from unit heaters to warm the air in the greenhouse. This system may also be used simply to circulate the air to prevent stagnation. Horizontal airflow fans (HAF) may also be used to help circulate the air in the greenhouse or help mix warm air from unit heaters with the cooler air. For individual greenhouses, fans should be installed in two rows, one side directing air opposite the other, to help stir the air around the greenhouse. Unit heaters are installed 3 to 4.6 m (10 to 15 ft) behind the first fan at each end of the greenhouse. This positioning helps mix the warm air from the heater with the cooler air of the greenhouse. For a gutter-connected greenhouse, the fans can be installed down the center of each bay and facing the opposite direction for each bay.

Environmental controls for operating the cooling and heating of greenhouses range from very simple, such as a basic thermostat, to more complex computerized systems. The more complex environmental control systems can control cooling, heating, shade curtains, fertilization, irrigation, irradiance, and other environmental systems in the greenhouse.

Night temperatures are considered very important for flowering pot plant production. Green foliage plants and some orchids require night temperatures as high as 21°C (70°F) and day temperatures as high as 29°C (85°F). Temperatures not only affect the growth of plants but also flowering. For example, *Cattleya* orchids grown at a 29/21°C (85/70°F) day/night temperature will flower much more quickly than those grown at a 21/13°C (70/55°F) temperature. Roses, chrysanthemums, and azaleas do best at night temperatures in the 17°C (63°F) range, whereas carnations, snapdragons, cinerarias, and calceolarias grow well at 10°C (50°F). If height control is not a problem, day temperatures are set 0 to 5 degrees above night temperature on cloudy days and 10 to 15 degrees above night temperature on sunny days to maximize photosynthesis and minimize respiration. Maintaining these temperature differences is easy on overcast days but difficult on bright days because of an extreme heat buildup from the sun's radiation. When bright conditions prevail, plants may have to be (a) shaded during the summer months (e.g., May to September), (b) watered more often to compensate for excessive transpiration, or (c) sprayed with an overhead mist system to increase the humidity. All three of these procedures compensate for the high light intensity conditions that may cause sunscald damage to leaves or flowers.

DIF (day temperature − night temperature) can be used to help control the plant height of many greenhouse crops. As the day temperature increases above the night temperature (a positive DIF), internode elongation increases. Inversely, a negative DIF decreases internode elongation.

Light

Three types of light affect plant growth and development: light quality, light quantity, and light duration. Light quality, or the light spectrum, drives photosynthesis and can affect plant shape. Plants use red and blue light most efficiently for photosynthesis, while far-red light affects internode elongation. Plants under high levels of red light have less internode elongation, are shorter, and are usually greener. Plants grown under high levels of far-red light have longer internodes, are taller, and are less branched.

Light quantity or intensity refers to the amount of light that a plant receives and can be measured in foot-candles (fc), photosynthetically active radiation (PAR μmol s^{-1}), or photosynthetic photon flux (PPF μmole s^{-1} m^{-2}). The greater the light intensity, the greater the rate of photosyn-

Figure 23–16 Overhead high-intensity discharge lamps used to supplement the lighting in the greenhouse for optimum crop growth.

thesis until the light saturation point occurs. Supplemental lighting has been used in greenhouse production for many years. The most common form is incandescent lighting, which is used only for photoperiod control because it gives off excessive heat and poor light quality. Because fluorescent lights provide only part of the visible spectrum, they are used primarily for germination chambers, for growth rooms, and in homes for interior lighting of houseplants. High-intensity discharge (HID) lamps are the most common types of supplemental lighting in the greenhouse, providing satisfactory light intensity and light quality (Fig. 23–16).

Light duration, termed **photoperiodism,** is the total light energy that may affect the growth and development of plants. Short-day (SD) plants flower when the dark period is longer than a specified length. Plants such as azaleas, chrysanthemums, poinsettias, and some orchids require SD to initiate flowering. Long-day (LD) plants flower when the light period is longer than a specified length. China asters, Shasta daisies, *Gypsophila,* and *Liatris* require LD. Day-neutral (DN) plants, for example, *Pelargonium* x *hortorum,* require no specific daylength for flower initiation.

Greenhouse growers should remember four dates that will help determine the necessity for photoperiod manipulation: June 21, the longest day of the year; December 21, the shortest day of the year; and September 21 and March 21, the autumnal equinox and vernal equinox, respectively, when the light period is equal to 12 hr. During a long light period, flower induction of SD plants or maintenance of vegetative growth of LD plants can be accomplished by covering the plants with black cloth (excluding all light) from 5 P.M. to 8 P.M. (cooler climates) or 7 P.M. to 7 A.M. (warmer climates). If the light period is short, flower induction of LD plants or vegetative growth of SD plants can be accomplished by night interruption lighting. Incandescent lights can be placed over plants and turned on from 10 P.M. to

Figure 23–17 Poinsettia (*Euphorbia pulcherrima*) grown for the Christmas holidays. The plant on the left was grown without a plant growth retardant (control) and the plant on the right was treated with Bonzi. Bonzi is a common plant growth retardant used to reduce plant size for ease of shipping as well as to provide a more aesthetically pleasing appearance.

2 A.M. (4 hr.) with a light intensity of approximately 2 μ mol (≈10 f.c.).

Control of Plant Growth and Flowering

Control of plant height and growth is important in greenhouse crop production. Plants should be of adequate size for shipping and handling as well as being aesthetically pleasing to the consumer (approximately 1.5 to 2 times the height of the container) (Fig. 23–17). Plant height can be controlled by manipulating the greenhouse environment or production practices. Controlling light quality, quantity, and photoperiod provides a means of controlling plant height and growth. Temperature manipulation using DIF, as discussed earlier, can help control plant growth. Applying a certain amount of water during irrigation and reducing or withholding fertilization can also be used to manipulate plant growth. However, many situations prevent the greenhouse grower from using the aforementioned methods.

Plant growth retardants (PGRs) can be applied to control plant growth, most of which blocks the synthesis of gibberellic acid, inhibiting stem elongation. Plant growth retardants can be applied by spraying the plant or drenching the growing medium (uptake by the root system). Following is a list of popular plant growth retardants:

♦ A-Rest (ancymidol) is labeled for many flowering pot plants and bedding plants, and it has been a popular growth retardant of bulb crops for many years. A-Rest can be applied as a spray or drench.

♦ B-Nine SP (daminozide) is labeled for many flowering pot plants and bedding plants. It can be applied only as a spray to the surface of leaves.
♦ Bonzi or Piccolo (paclobutrazol) is labeled for many greenhouse crops and can be applied as a spray or a drench. Sprays should be applied to the shoots of the plant because the active ingredient is translocated through the stem and reduces internode elongation.
♦ Cycocel (chlormequat) is labeled for most greenhouse crops and can be applied as a drench or a spray. Combining B-Nine and Cycocel in the same tank mix provides often better growth regulation than using either alone.
♦ Florel Brand Pistill (ethephon) is labeled for many floriculture crops and works differently from the previously mentioned PGRs by releasing ethylene to the plant. It has been used for many years to enhance flowering and ripening of fruit. More recently, it has been used extensively to reduce plant height, induce lateral branching and delay flowering.
♦ Sumagic (uniconazole) is labeled for some floriculture crops. It can be applied as a spray or drench and works in a similar manner to Bonzi.
♦ Topflor (flurpimidol) is labeled for most floriculture crops. It can be applied as a spray or drench and works in a similar manner to A-Rest.

Many factors influence the efficacy of these plant growth retardants: timing of application, environmental conditions, dosage, target tissue, pH, uniformity of application, the concentration of the chemical, and the active ingredient.

Graphical tracking has become a popular method of monitoring plant height to help determine the timing of growth retardant application. A standard growth curve is established for a specific crop, and plant height is plotted against the standard curve on a weekly basis. When the plant height begins to exceed the height of the standard curve, a growth retardant application may be necessary to help the plant reach the specified height.

Other plant growth regulators are not growth retardants but they affect other plant processes. Pro-Gibb (gibberellic acid) promotes cell elongation, cell division, and larger flowers. It can be applied as a spray and substitutes for cold in breaking the dormancy of azaleas and hydrangeas or as a bulb dip to promote flowering of calla lily. Fascination (gibberellin$_{4+7}$ and benzyl adenine) can be used as a spray to prevent leaf yellowing of Easter and hybrid lilies. EthylBloc (methylcyclopropene [MCP]) is an antiethylene compound applied as a gas to help prevent flower drop during shipping of flowering plants and cut flowers.

Growing Media

Choosing the proper medium to grow quality greenhouse crops is very important (Fig. 23–18). Although the root sys-

Figure 23–18 Soilless growing media moved by a conveyor for filling containers prior to transplanting.

tems of plants are not visible, the health of the shoot depends on the health of the root. The media must have desirable physical characteristics in terms of bulk density or weight so that the container does not tip over easily. However, if the media weight is too great, the cost of shipping may be prohibitive. The total pore space, water retention, and air-filled porosity also contribute to the health of the root system by providing appropriate amounts of oxygen and moisture for optimum growth. The pH and electrical conductivity (EC) of media are important chemical properties. For most crops, the pH range of media is between 5.5 and 6.5, but some crops must have a more specific pH. The EC should be between 0.75 and 2.0 mmho/cm. If the EC of the media becomes too high (>2.0 mmho/cm), salt accumulation can cause water to be pulled out of the roots, resulting in root burn and death.

Most greenhouse crops are grown in soilless media. The benefits of using soilless media over media that contain soil are reduction in bulk density (weight) and increased uniformity of the media. The most common components of soilless media are sphagnum peat moss, pine bark, perlite, and vermiculite. Sphagnum peat moss is low in salts, does not decompose readily, has high water- and nutrient-holding characteristics, is uniform, and is relatively disease-free as found in nature. Vermiculite is a sterile expanded mica that holds water and nutrients well and does not decompose. It is uniform and reasonably priced. Vermiculite does compress quickly and, therefore, is used most frequently in seedling mixes because of its water-holding capacity. Perlite is a heat-expanded aluminum silicate rock that is porous and lightweight. It does not hold water but provides excellent pore space. Pine bark and hardwood bark make excellent media components. When composted or aged properly, both barks have high water- and nutrient-holding capacities, and can be graded to specific sizes (¼ in. and ⅜ in. are the most popular). Most soilless media are amended with dolomitic limestone

Figure 23–19 Handheld EC meter used to determine the electrical conductivity of irrigation water.

for increasing the media pH, phosphorus, and trace elements so that the nutrient content is adequate for greenhouse crop production. The type of growing medium used for each greenhouse crop can vary significantly from one producer to the next. Therefore, it is extremely important that irrigation and fertilization be managed closely.

Irrigation and Water Quality

Water quality is extremely important to a greenhouse operation due to the use of soilless media, which have much less buffering capacity than that of soil. Irrigation water should be analyzed prior to beginning a greenhouse operation and at least once a year during regular operation. Samples can be submitted to the local extension agent for testing, and they can also be tested by the grower using a small handheld meter (Fig. 23–19). Salinity, the measure of sodium and chloride in irrigation water, can contribute to increased media EC and root damage if in high concentrations. This has been an increasing problem, especially in areas where saltwater intrusion of aquifers occurs. Alkalinity, the

Figure 23–20 Hand watering of greenhouse crops using an extended wand with a water breaker attached to the end. The water breaker helps break up the flow of water to prevent washing the media and/or plants out of their containers.

Figure 23–21 The large tray on the top of this ebb-and-flood bench serves as a reservoir for briefly holding water while it is taken up by capillary action through the bottoms of the containers.

amount of carbonates and bicarbonates in irrigation water, can increase the pH of the growing media. There are methods of controlling salinity and/or alkalinity in irrigation water, but they can be costly if concentrations are high enough.

Irrigation is an art that is not learned quickly. The ability of a grower to look at the plants across an entire greenhouse range and determine the need for irrigation is of primary importance. Hand watering of greenhouse crops can be one of the best methods of irrigation, but it is not economical (Fig. 23–20). However, many growers still water some of their crops by hand. The use of overhead sprinklers for the production of some bedding plants is more economical, but the foliage of most crops must be kept dry to prevent disease. Boom watering can be used effectively on the production of plugs. The use of microtubes has been popular for many years and is an efficient method of irrigating large numbers of plants without getting the foliage wet while minimizing water usage. Mat watering by the use of a capillary mat is another method of irrigating crops when the sizes of containers vary on the same bench. The use of an ebb-and-

flood system and flood floors has become popular in recent years, but it can be cost-prohibitive due to the amount of materials used to build the system (Fig. 23–21). Disease can also spread rapidly in this system; thus, it must be monitored carefully.

Most greenhouse operations use some type of liquid fertilizer in the irrigation system, which is called **fertigation.** This practice adds another factor to irrigation and is discussed in the next section.

Mineral Nutrition

Providing greenhouse crops with all sixteen essential elements is vital for optimizing growth due in part to the use of soilless media and the need to produce a marketable crop as quickly as possible. Most fertilization recommendations are based on the parts per million (ppm) of nitrogen (N) from a complete liquid fertilizer (those fertilizers containing the essential elements) (Fig. 23–22). Therefore, greenhouse fertilizers are formulated to provide the remaining elements in the appropriate amounts.

Some greenhouse crops must have greater amounts of certain elements than other crops and, therefore, fertilizers are specifically formulated for those specific crops. For example, some poinsettia cultivars are inefficient in the uptake of molybdenum (Mo); therefore, most poinsettia fertilizers have increased concentrations of Mo.

Plants have different nutrient needs during different stages of growth, and fertilization must be adjusted to meet those needs. Because most greenhouse crops are flowering plants, the fertilizer concentrations are low at the planting stage of the crop, are increased during rapid crop growth, and are decreased at crop maturity or flowering. Application of fertilizers in too high a concentration without leaching can increase the EC of the media and be deleterious to root growth and thus crop quality. However, leaching can con-

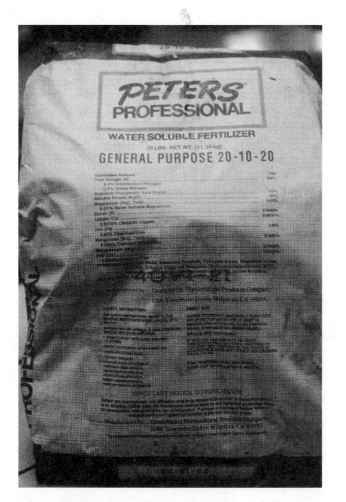

Figure 23–22 Example of the nutrient analysis, 20-10-20 (nitrogen-phosphorus-potassium), on the bag of a general-purpose, liquid-soluble fertilizer.

tribute to runoff of irrigation water and fertilizers, which has become an increasing problem for greenhouse operations. Therefore, growers are trying to match more closely the fertigation schedule to crop growth, thereby reducing waste or runoff and cost.

Two methods are used for monitoring the nutrient status of a plant. The first is testing the substrate by either sending a sample to an extension agent for analysis or taking a pour-through leachate (PTL). A PTL can be conducted easily by pouring a small amount of water over the top of the media in the pot and collecting the leachate from the bottom of the pot. This leachate can also be sent in for nutrient analysis, but these test results are interpreted differently. A foliar analysis is another method for monitoring nutrient status. Recently mature leaves should be taken from the plant. If a plant is showing signs of what could be a nutrient deficiency, then leaves that show the symptoms should be taken separately from those that show no symptoms. Foliar analysis must be interpreted differently according to plant species.

Diseases and Insects

Greenhouses are small microclimates that not only provide an optimal environment for producing quality plants, but also an optimal environment for the growth of diseases and insects. The most common greenhouse insects are aphids, mealy bugs, spider mites, white flies, thrips, fungus gnats, and scale.

Aphids can range in color from transparent to green, to brown. They can be wingless or winged, and they feed on almost all floriculture crops. Aphids feed on young flower buds and newly formed leaves, where most of the soluble sugars are translocated and on which this piercing, sucking insect feeds. Infestation may lead to distorted flowers and leaves.

Mealy bugs are fuzzy white insects with piercing sucking mouth parts. They are primarily a problem on foliage plants. These insects can be difficult to control because of their outer waxy surface.

Spider mites are not insects but belong to the class Arachnida. The two-spotted spider mite seems to be the most prevalent mite and feeds on the foliage of a plant, causing a yellow stippling of the leaves. If the infestation is great enough, they will spin webs around the leaves and flowers of plants. They seem to be most aggressive when environmental conditions are warm and dry.

White flies feed with piercing, sucking mouthparts and are primarily problems on poinsettia, chrysanthemums, lantana, and salvia. The adult form resembles a small white fly. They lay their eggs on the underside of leaves.

Thrips are small winged insects that feed with piercing, sucking mouthparts, but they cause a streaking pattern on leaves and flower petals instead of stippling. They tend to feed most heavily on flower buds and can cause desiccation of the buds. They are also the primary vector for impatiens necrotic spot virus (INSV) and tomato spotted wilt virus (TSWV).

Fungus gnats are gray flies that lay eggs on moist soil surfaces. The adults do not harm the plant. The larva that is hatched from the eggs feed on the roots of plants, causing injury to the roots and thus greater susceptibility to diseases. These insects have become a problem for all floriculture crops and are difficult to get rid of once an infestation has occurred.

Scale insects vary in form and size, but most of them resemble the scales on a fish. Some secrete wax; some have piercing, sucking mouthparts; and some have wings. These insects are most commonly problems on foliage plants and can be difficult to eliminate.

The most common greenhouse diseases are viruses, bacteria, fungi, and nematodes. The most common viruses are INSV and TMSV, mentioned earlier. Because plants do not produce antibodies as do most animals, they cannot recover from a virus or develop immunity. Bacterial diseases are difficult to control, and fortunately there are fewer bacterial than fungal diseases. There are some bactericides but the best method of control is prevention and elimination. Some of the more common bacterial diseases are bacterial blights and *Erwinia* (soft rot). There are many fungal diseases, but they can be controlled through the judicious use of preventive fungicides. Some common fungal diseases are powdery mildew, botrytis blight, pythium, rhizoctonia, and verticillium. The best methods for circumventing these pests are sanitation, integrated pest management (IPM), biological control, preventive fungicides, and spray programs.

Integrated pest management (IPM) is a method of controlling insects and diseases through surveillance, preventive measures, and corrective action. All are integrated for a more efficient use of pesticides. IPM includes understanding of the dynamics of the pest population, biology of the pest and its host, effect of the environment and cultural practices, and use of biological controls. The effectiveness of the program depends greatly on the scout (person in charge of screening plant material for diseases and insects) and the pesticide applicator.

Propagation

Propagation of plant material was commonly done at the same location where the plants were grown to finish for sale. This practice has changed over the years as we increase our understanding of plant growth and development, with advances in technology, and as breeding practices have changed. Most of the rooted or unrooted cuttings and seedlings are now produced by one company to be sold to another. Thus, the floriculture industry has added another facet to production.

For many years, the primary potted plants, such as poinsettias, chrysanthemums, Easter lily, and foliage plants, were the only vegetatively propagated plants. The remaining plants were propagated by seed. Propagation by seed was extremely time-consuming and expensive. The advent of plug production revolutionized floriculture. It involves growing plants in small cells 1.27 to 1.60 cm in diameter and 2.54 cm deep (½ to ⅝ in. in diameter and 1 in. deep) on a plastic sheet that can contain hundreds of cells/plants (Fig. 23–23). Plug production helped the mechanization of growing and transplanting

Figure 23–23   A ninety-eight-cell plug tray. *Source:* Jeff Kuehny, Louisiana State University.

seedlings (reducing labor), shortened production time, increased holding time, and eased the shipping of seedlings.

Growing plugs is a highly specialized area of greenhouse production that requires specialized equipment and facilities and a regimented set of cultural practices that vary from one species to the next. Vegetative propagation of rooted and unrooted cuttings has advanced beyond propagation of the primary potted plants mentioned earlier and is now also important to the bedding plant industry. The increasing number of new bedding plant types and hybridization has facilitated the need for clonal propagation.

A tremendous amount of breeding work has developed new plant forms, flowering colors and shapes, disease and insect resistance, and stress tolerance. This work continues to be important to the industry but is very costly. Plant patents can be obtained on asexually propagated plant material to help protect the breeders' work and recoup some of the costs put into the development of new plant material. Branding of plant material has also become an important tool in advertising and selling plant material from a specific company. Branding can be a form of trademark that can be combined with the plant patent.

GREENHOUSE CROPS

The greenhouse industry in the United States began in the Northeast close to large population centers. Cut flowers—primarily roses, carnations, and chrysanthemums—were the major greenhouse crops produced in the early 1900s. Production of fresh-cut flowers spread slowly across the United States to the Midwest, then to Colorado, and finally into California by the 1960s. During this shift of cut flower production from the East Coast to the West Coast, flowering pot plants began to replace cut flowers. Examples of popular flowering pot plants include poinsettia, chrysanthemums, African violets, azaleas, hydrangeas, Easter lilies, and other flowering bulb crops. Production of flowering pot plants became an even greater part of the greenhouse industry when foreign imports of cut flowers became a major competitor in the late 1960s. During the 1970s and 1980s, green plant or foliage plant production became a considerable part of the greenhouse industry in response to the so-called green revolution. As technology and transportation improved, so did the greenhouse industry. This was most evident in the production of bedding plants, with the introduction of plug technology. Bedding plants now make up a majority of total greenhouse crop production.

This brief history of greenhouse production reflects the dynamics of this field in horticulture. The production guidelines of the following crops are examples of some of the more important greenhouse crops grown.

Cut Flowers

A majority of the cut flower production occurs in California, which produces primarily roses and carnations. *Rosa* L. hybrids flower profusely during the long, warm, sunny days of

Figure 23–24 Roses grown in a greenhouse for cut flower production. The flower stems are a few days from the cutting stage.

Figure 23–25 Cut roses being graded for packing and shipping to a wholesale florist.

California. Greenhouse nighttime temperatures should be maintained at 15°C to 18°C (60°F to 65°F) for optimum growth and flowering. Flowering can be timed by pinching the terminal buds, which produce new flowering shoots in thirty-five to sixty days. The rapidity with which they flower depends on light intensity and temperature. A flower forms on every new shoot after the removal of a flower from the bush (Fig. 23–24). If properly cut and pruned annually, plants may not have to be replanted for five to eight years, depending on the health of the plants and the soil structure. Occasionally, disease hastens the need to replant.

Flowers are cut in the late-bud stage, usually when one or two sepals have turned downward or just as one or two petals have unfurled. They are sorted and graded into uniform stem lengths that also have similar flower head (bud) sizes and quality (Fig. 23–25). Flowers are packaged in bunches of twenty-five and kept in post-harvest coolers at 2°C to 4°C (35°F to 40°F).

More recently, specialty cut flowers (cut flowers that are not grown in large quantities by smaller growers) have increasingly become part of the floral industry. Spe-

cialty cut flowers include crops such as *Achillea, Celosia, Dianthus, Helianthus, Ilex,* and *Solidago.* As the demand for these types of crops increases, so will the need for production of this type of flowering plant in the greenhouse industry.

Flowering Pot Plants

Poinsettias are the number 1 flowering pot plant sold in the United States, even though they are considered a holiday or seasonal crop. *Euphorbia pulcherrima* Wild. Ex Klotzch is a tropical plant native to Mexico belonging to the family EUPHORBIACEA (Fig. 23–26). The poinsettia was introduced into the United States in 1825 by J. R. Poinsett, from whom it gets its common name. Poinsettias have been grown as a Christmas holiday crop mainly due to their photoperiodic response for flower initiation. They are classified as short-day (long-night) plants that initiate and develop flowers as the nights become longer than the days. The inflorescence is characterized by a single female flower, without petals and usually without sepals, surrounded by individual male flowers all enclosed in a cup-shaped structure called a cyathium. The showy red, pink, or white portions of the plant, popularly referred to as the flower, are modified leaves called **bracts.**

Rooted cuttings are planted in well-drained media during August. Plants should be watered immediately after planting, and a fungicide should be used shortly thereafter to prevent disease infection. Poinsettia cultivars are classified according to the amount of time required from the start of short days to flowering, which are called response groups. Response groups range from seven to ten weeks. Therefore, timing of a finished crop can be manipulated by photoperiod control. Some plants, such as 'Freedom Red,' are very sensitive to changes in photoperiod and may need an extended

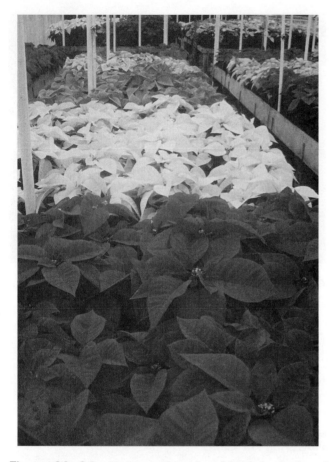

Figure 23–26 Various poinsettia cultivars in shades of red and white grown for the Christmas holidays.

daylength to prevent premature flower initiation. Daytime temperature should be maintained below 25°C (77°F) to maintain leaf unfolding, with nighttime temperature maintained at approximately 20°C (68°F) for flower initiation and development. Reduced night temperatures of 16°C to 17°C (60°F to 62°F) are necessary for bract coloration. Light intensities of up to 6,000 fc should be maintained for good bract coloration. After the roots have grown to the sides of the container (approximately two weeks), the terminal meristem should be pinched to six internodes to ensure lateral branching. Approximately two weeks after pinching (laterals of 2- to 3-in. length), a short daylength should be started for inducing flower initiation. One or two plant growth retardant applications are usually needed to produce a quality finished plant. According to response group, the plants should reach anthesis seven to ten weeks later.

Foliage Plants

Foliage plants or houseplants can be grouped by sensitivity to light levels. Light intensity can be measured easily with a foot-candle or lux meter. The high-light group consists of light levels of 4,000 to 8,000 fc (i.e., *Ixora coccinea, Codiaeum variegatum pictum,* Cactaceae, Crassulaceae), the medium-light group of 1,000 to 3,000 fc (i.e., *Begonia* spp., *Saintpaulia ionantha, Draceana* spp., *Ficus* spp.), and the low-light group of 50 to 500 fc (i.e., *Aglaonema commutatum, Philodendron scandens oxycardium, Spathiphyllum* spp., *Dieffenbachia* spp., *Epipremnum aureum*). Those foliage plants that are moved from an environment with a higher light intensity into one of lower light intensity must adjust to this change. Some plants lose chlorophyll and then drop their leaves soon after being moved to a lower light intensity (i.e., *Ficus benjamina* and *Asparagus densiflorus*). These plants develop new leaves that are usually thinner and broader for more efficient photosynthesis. Many plants, including the aroid, palm, and lily families, do not abscise their leaves. Instead, the leaves become yellow-green and may wither or die but still cling to the stem. Therefore, plants must be acclimatized to the lower light intensities before removing them from environments with higher light intensities. First, reduce the light by 50 percent for several weeks and, finally, reduce the light with shade to 20 percent of the original growing light intensity. The entire acclimatization process should take four to twelve weeks, depending on the species in question. Reducing fertilizer and watering during this period also aids the acclimatization process.

Flowering plants have relatively narrow and specific temperature requirements for flowering. Because foliage plants are not grown for their flowers, they can be grown under temperatures that range from 60°F to 90°F (16°C to 32°C). The lower the temperatures, the slower the growth, and vice versa. The optimum growing temperature, for most houseplants is approximately 70°F to 79°F (21°C to 26°C).

Most foliage plants prefer a relative humidity between 70 and 80 percent. This can be easily achieved in a greenhouse full of plants. However, when foliage plants are moved from the greenhouse to an interior environment, low relative humidity can result in slower growth and insect problems. Misting these plants does not provide enough moisture in the air to compensate for this loss. Keeping plants out of the direct flow of air from heating and/or air-conditioning vents and doorways where large amounts of air exchanges occur, and providing adequate water to the container are the best means of compensating for the loss of relative humidity.

Rapidly growing foliage plants in a greenhouse utilize large amounts of water and fertilizer. However, when moved to an interior environment, growth slows, as does nutrient and water uptake. Thus, houseplants need much lower amounts of fertilizer and water. Overwatering houseplants is one of the most common causes of plant death. Salts may accumulate in the media of plants that are maintained in the same container for long periods of time (greater than one year). Leaching the media once a year or repotting can circumvent this problem. Houseplants can also benefit from rinsing the dust that accumulates on the leaves during repotting.

Figure 23–27   A gutter-connected greenhouse range full of bedding plants on benches in flats (containing six sets of six-packs) and hanging baskets overhead grown for spring sales.

Figure 23–28   A 288-plug tray with *Impatiens wallerana* grown for transplanting in larger containers.

Bedding Plants

The term *bedding plants* was originally used by gardeners who wanted to use annual or perennial plants in a permanent border or in a specially designed flower bed. This term has been expanded to include flowering ornamental plants, vegetables, and herbs (Fig. 23–27). Bedding plants are used in large or small containers, on patios, in window boxes, in hanging baskets, in small gardens, and in beds and borders. The bedding plant industry has responded to the needs of a range of gardeners, from the sophisticated to the novice, by offering a variety of plants in large, medium, or small containers (six-packs are the most common) and a large number of species from which to choose.

Greenhouse production of bedding plants differs from the growing of other greenhouse crops because a larger number of many different species of plants are grown in the same space in a much shorter time. This means quick rotation of crops and rapid return on investment.

Early production of bedding plants was by direct seeding into containers or broadcast seeding or sowing in rows in flats and transplanting into containers. These traditional methods of bedding plant production were improved on and developed into the plug production method, as mentioned earlier. (Fig. 23–28). This technology, along with continuing improvements, has meant continued growth and expansion of bedding plant sales in the greenhouse industry.

Because of the large number of different species of bedding plants, scheduling, fertilization and irrigation, greenhouse temperatures, types and rates of plant growth regulators, and light levels vary significantly. Although bedding plant production can be highly automated and bedding plants are a quick crop to produce, the amount of knowledge that a grower must have to grow these crops is extraordinary.

Bulb Crops

Many species of plants have fleshy underground storage organs capable of carrying the plants through seasonal cold/warm or dry/wet periods. Such structures are popularly called bulbs, but they are defined more accurately as bulbs, tuberous roots, tubers, corms, or rhizomes (Fig. 23–29). They all produce one or more buds for flower production or renewed vegetative growth.

The *true bulb,* such as the lily, hyacinth, *Muscari, Narcissus,* tulip, and onion, has numerous fleshy scales or leaf bases attached to a distinct basal plate (stem) that gives rise to roots and shoots. They may or may not have one or more impervious covering layers (tunic).

Tubers are enlarged fleshy stems with adventitious buds (eyes) near the upper surface, as in the tuberous begonia, Eranthis or Caladium, or in a systematic pattern or arrangement, as in the Irish potato. Some organs are enlarged roots and are classified as *tuberous roots,* such as Agapanthus, Dahlia, and Hemerocallis.

Corms have solid shortened stems with buds systematically arranged under a paper-thin protective covering of the leaf base or scale. Crocus, Gladiolus, and Freesia are examples.

The *rhizome* is a fleshy, horizontal underground stem that grows laterally. Examples are the rhizomatous iris, calla lily, and ginger.

Roots originating from the basal plates or lowest portions on all of these structures are adventitious. The shoots that originate either inside the bulb or on the surface, as with the corm or tuber, give rise to stems that bear the foliage and the flowers. Flowers can be initiated within the bulb during the previous growing season (*Narcissus*) or just after the new shoots begin to emerge from the soil (bulbous iris, lilies, and gladioli). All these types of plants have storage tissues that can produce a flowering stem after going through the seasonal dormant period. The dormancy period ranges from a few months to almost a year under certain environmental conditions.

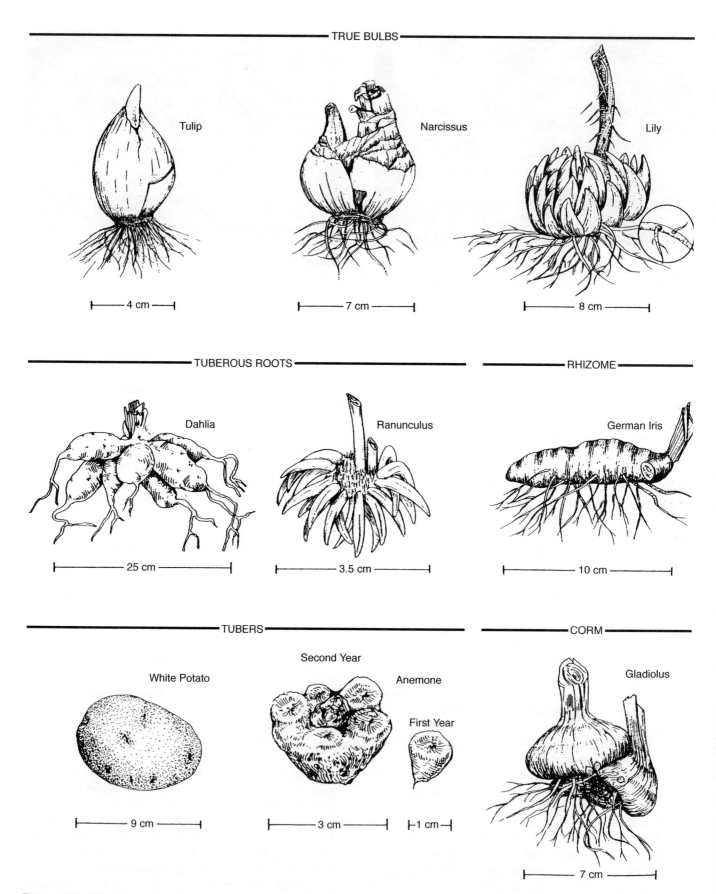

Figure 23–29 Examples of underground food-storage structures found in some herbaceous plants. Sizes given are approximate.

Bulbs can also be categorized according to the pattern of flower formation:

1. Flower buds form during the spring and summer of the previous year (*Narcissus, Galanthus,* and *Leucojum*).
2. Flowers form at the end of the previous growing period or after harvesting and placement in storage (*Hyacinthus, Tulipa,* and *Iris reticulata*).
3. Flowers are initiated after replanting and after a low-temperature treatment (bulbous iris, most lilies, *Triteleia,* and *Ornithagalum*).
4. Flowers are initiated toward the end of storage (*Allium cepa, A. escalonicum, Galtonia,* and some lilies).
5. Flowers are initiated after replanting (*Freesia, Gladiolus, Anemone, Ixia,* and *Ranunculus*).
6. Flowers are initiated more than a year before flowering (*Amaryllis belladonna, Nerine sarniensis,* and *N. bowdenii*).
7. Flower initiation occurs throughout the entire growing season but alternately with leaf formation (*Hippeastrum* and *Zephranthes*).

A method for determining the flowering bud stage is by dissecting the bulb to determine whether flower buds are present and whether they may have been injured by heat or disease.

Because bulbs vary in their stage of flower development, some bulbs such as daffodil already have immature flowers and already may have been precooled and are ready to be planted, rooted, and forced into bloom. Other bulbs must first be given the proper low-temperature treatment (precooling). Then the bulbs must be planted in containers and moved to low-temperature rooms to establish a good root system before they are forced in a greenhouse. Temperatures for precooling and forcing are critical for each species and cultivar.

The Easter lily or *Lilium longiflorum* Thunb. is one of the most important flowering bulb crops in the United States. These true bulbs are native to Japan and grown in California and Oregon. They are dug in September and October and are either shipped to growers for potting or precooled ("case-cooling"). This cooling process or vernalization of the bulbs promotes floral induction and is completed just after Christmas. Bulbs should be planted in well-drained media and drenched with a fungicide. The bulbs are then forced into bloom in the greenhouse. The desired forcing time depends on the date of Easter in that particular year. Growth is measured by counting the number of unfolded leaves during stem elongation. Greenhouse temperatures can be increased or decreased to adjust the rate of leaf unfolding. This allows the grower to adjust the rate of growth so that bloom occurs just prior to Easter. Application of plant growth retardants is necessary to reduce stem height.

Lilium hybrids, both Asiatic and oriental lilies, are becoming popular both as flowering pot plants and as cut flowers. The Asiatic hybrids include flower colors of orange, red, yellow, tan, and white. Their flowers are oriented upward and have little or no fragrance. The oriental hybrids have flower colors of red, pink, and white, and their flowers are oriented horizontally and have a strong fragrance. Both hybrids must be precooled for up to ten weeks prior to forcing. The storage duration and temperature is cultivar-dependent. The bulbs should be planted and forced in a greenhouse, like Easter lilies. Because these lilies are usually not grown for a specific holiday, however, leaf counting is not necessary for finishing. Many of the hybrid lilies do require a growth retardant for producing a quality finished plant. These lilies are most often forced from March to June, and the time to finish depends on the growing conditions. Asiatic hybrids take approximately thirty days to flower from visible bud, and oriental hybrids take approximately fifty days.

Amaryllis or *Hippeastrum* consists of leaf bases and no scales. The genus is indigenous to tropical and subtropical South America. Commercial bulbs 20 to 30 cm (7.8 to 11.8 in.) in circumference are produced by Israel, South Africa, the Netherlands, and Brazil. The bulbs are dug, cured, and shipped at a storage temperature of 9°C to 13°C (48.2°F to 55.4 °F) and then are planted in a container for forcing in the greenhouse or placed in a bulb kit to be forced by the consumer. Depending on the cultivar and its handling, the flower stalk emerges and elongates simultaneously with four leaves. *Hippeastrum* is becoming a very popular Christmas flower in the United States.

SUMMARY AND REVIEW

Greenhouses were initially built to protect plants from cold temperatures while allowing sunlight to reach the plant. Through the years, however, greenhouse structures became much more sophisticated, enabling growers to manipulate the environment and thus control plant growth and development. Greenhouse heating and cooling systems are used to regulate temperatures and consequently plant growth and development throughout the year. Shading systems allow the amount of light reaching a plant to be limited when light intensity is too great. High intensity discharge (HID) lamps can be used to provide supplemental light during low light periods. The timing of water and fertilizer applications and the formulation of fertilizers provide a means to control plant growth and development. Very few modern greenhouse production systems have plants growing in native soil. The plants are most often in a container that can be as large as a bench or as small as a plug tray. The growing media in the container generally contains very little or no soil but is a composition of materials such as sand, peat moss, perlite, vermiculite, and wood products. The media components can be blended to give different characteristics suitable for different needs. When manipulating the greenhouse environment is not sufficient for maintaining desired plant growth, chemical growth regulators can be used. Manipulating the greenhouse environment is also a means to control insect and disease problems, although at times other methods may also be required.

Greenhouse crops can be grouped into four broad categories: cut flowers, potted or container crops, foliage plants, and bulb crops. Cut flowers are plants grown for their flow-

ers, leaves, or stems, which are harvested and mainly used in flower arrangements. Potted or container crops are grown and sold in the container. They include potted crops such as poinsettia, chrysanthemum, and annual and perennial bedding plants. Foliage plants are grown for their foliage rather than their flowers. Many of these plants are tropical and adapted to the warm, humid, and low light conditions found under the canopies of large trees in tropical rainforests. Bulb crops are grown from underground storage sites. They may be used as potted plants or for cut flowers. Each category and the species within that category have specific cultural requirements that must be met in a greenhouse to produce plants of high quality.

FOOD FOR THOUGHT

1. How does a cold or hot frame differ from a traditional greenhouse?

2. What are the different types of glazing material that can be used on a greenhouse and what are the criteria for its selection?

3. What are three methods of cooling and three methods of heating a greenhouse? Which method is the least expensive and most efficient?

4. Why would a greenhouse grower use DIF when growing an Easter lily?

5. What type of photoperiod does a poinsettia need to initiate a flower? What is a bract?

6. How do most of the commercial growth retardants work?

7. What are three components that can be used in a soilless media and what is one attribute of each?

8. What is the difference between salinity and alkalinity?

9. What are two methods of monitoring the nutrition of a bedding plant?

10. What are three types of insects that are problematic for greenhouse growers and how does each affect plant growth?

SUPPLEMENTARY READING

DOLE, J. M., and H. F. WILKINS. 2005. *Floriculture: Principals and species,* second edition. Upper Saddle River, N.J.: Prentice Hall.

JARVIS, W. R. 1992. *Managing diseases in greenhouse crops.* St. Paul, Minn.: APS Press.

NELSON, P. V. 1998. *Greenhouse operation and management,* fifth edition. Upper Saddle River, N.J.: Prentice Hall.

CHAPTER

24

Turfgrasses

David Gardner

KEY LEARNING CONCEPTS

After reading this chapter, you should be able to:

♦ Understand the terms commonly used in turfgrass science.
♦ Explain the principles for establishing and maintaining turfgrasses.
♦ Know the different types of turfgrass and the environmental and cultural requirements of each type.

Growing turf is a multimillion dollar industry. The replacement value of turf in the United States has been estimated to exceed $12 billion, and annual maintenance costs have been estimated at more than $10 billion. Much of the turf is public grass growing in schoolyards, parks, cemeteries, and golf courses; along highways; on military installations; and on other public lands.

Turf is a unique crop because the product is not what is harvested but what remains. The crop is grown densely as an entire population instead of individual plants spaced apart so that each grows vigorously. An appreciation of these differences is basic to being a good turf horticulturist.

There are three principal reasons for growing a lawn: (1) as a carpet, to protect the home from mud and dust, and to soften glare and heat; (2) for recreation; and (3) for beauty and pleasure. Studies have shown that lawns so please us that many of our earliest memories are of grass and trees.

Turf culture differs from other horticultural pursuits in one important way. In nature, plant growth is often limited by competition with many other plants for water, nutrients, and light. Thousands of seeds germinate and die for every one that lives and grows. Successful horticulture comes from spacing plants and eliminating weeds so that each plant has soil, water, and space allotted to it alone. Grass is the exception. In growing a lawn, even more plants are crowded into a given space than would grow there naturally. We strive to grow dense lawns similar to a fine carpet. The more we succeed in growing a dense lawn, the more stress each individual plant gets through competition from its crowding neighbors.

On an infertile dry soil, lack of nutrients and water limits plant growth and results in a poor lawn. We can grow a better, denser lawn by fertilizing and irrigating. But as the number of plants (shoots) increase, there is a point where plants shade each other and compete for sunlight. Beyond that point, individual plants are smaller, have fewer roots, and use nutrients and water less effectively. Thus, cultural practices that result in a dense, tight lawn can also result in individual plants of reduced quality and vigor.

Turf differs from other horticultural plantings because, instead of relieving the stress that results from competition, we increase it. The goal in growing lawns is not to grow individual plants of high quality but a plant population of fine appearance. In addition to suffering stress from competition, most lawn plants also suffer climatic stress at some season of the year. In managing a lawn, we can fertilize, mow, and use other practices to either increase or decrease stress. There is a limit, however, to the stress a plant can withstand before weakening to the point of dying.

The stress concept gives us a tool for evaluating the effects of cultural practices on grass. In a figurative way, there is a stress budget for turfgrass. If plant competition is high and midsummer weather imposes a heat stress, little more allowable stress is left in the stress budget. Cultural practices that reduce stress, such as watering, should then be followed. On the other hand, early fall temperatures are often ideal for the growth of turfgrass. Then certain stressful management practices—for example, power raking to thin out plants—that would have damaged a bluegrass lawn in summer's heat can be used. Throughout this chapter, different management practices are evaluated in terms of the stress they cause. The practices that increase stress are not necessarily bad, but care should be exercised to select and use these practices in appropriate seasons, and not to use several high-stress operations at inappropriate times.

The word *lawn* originally referred to a natural area of grass without trees. Today it refers to any expanse of ground on which grass is growing. Turf originally meant a layer of matted earth formed by soil and thickly growing grass plants. This meaning has gradually changed over the years, so that the word *turf* is used by horticulturists to refer to grass that is mowed and cared for. Turfgrass refers to the grass used in growing a horticultural turf.

Turf culture can be divided into two procedures: (1) establishment, and (2) maintenance for decorative or family use with modest traffic, or for heavy traffic by the public, or for athletic competition.

ESTABLISHING A TURF

A first step in growing a turf is to learn something about the species of plants used and their requirements. The grasses are monocotyledonous flowering plants belonging to the family POACEAE. Within the POACEAE are twelve currently recognized subfamilies. Three of these subfamilies are important in turfgrass management. Subfamily POOIDEAE contains the cool-season turfgrasses adapted to the southern United States are classified into two separate, distinct subfamilies. Grasses of the CHLORIDOIDEAE (such as zoysia grass) are adapted to warm, dry (arid) climates, while grasses of the PANICOIDEAE (such as St. Augustine grass) are adapted to warm, wet (tropical) climates. Differences in the evolutionary adaptation of the grasses used as turf have numerous implications for proper turfgrass management.

Between these two regions is a transition zone where neither cool-season nor subtropical grasses perform at their best. Cool-season grasses grow well during cooler parts of the year, but during hot summer weather, they become dormant or suffer heat injury and disease. Subtropical grasses in the transition zone do well in the hot summer but become dormant and turn brown in winter (Fig. 24–1).

In the western, arid portion of the U.S. plains states, where water is not available for irrigation, native drought-tolerant grasses such as buffalo grass or grama grass are sometimes grown. While their season of growth is limited, they do cover and protect the soil during the dry period.

Commonly used turfgrasses are described in Table 24–1.

TURFGRASS STRUCTURE

Turfgrass anatomy is illustrated in Figure 24–2. Table 24–2 gives a key for identifying common turfgrasses. The key is limited because it distinguishes only among the common cultivated turfgrasses.

Turfgrasses are well adapted to their role as a carpet. As with most monocots, growth of an individual shoot is **determinate** (ends in a flower), and the plants have fibrous roots without tap roots. Each shoot dies after a time but is replaced by new shoots growing from axillary buds. These new shoots produce adventitious roots at the nodes. In this way, turf continually rejuvenates itself. In many grass species, axillary buds produce horizontal creeping stems. If these grow over the surface of the ground, they are called **stolons;** below ground, they are termed **rhizomes.** Both structures enable grass plants to invade open areas and to spread. Other species of turfgrasses grow only new erect shoots or **tillers,** arising from axillary buds. Tillering grasses do spread but slowly.

Turfgrass flowers develop on elongated stems at certain seasons. The rest of the year, the tiller shoots are condensed with the vegetative growing point nestling among the leaves. The growing point produces **intercalary meristems;** that is, regions of cell initiation that lie across a stem or leaf and that, by dividing, interpose—or intercalate—new tissue between existing older tissues. The division and elongation of the intercalary meristems "extrudes" leaves and leaf sheaths. Internodes of the stem of grass plants do not elongate except to produce stolons, rhizomes, or flower stalks.

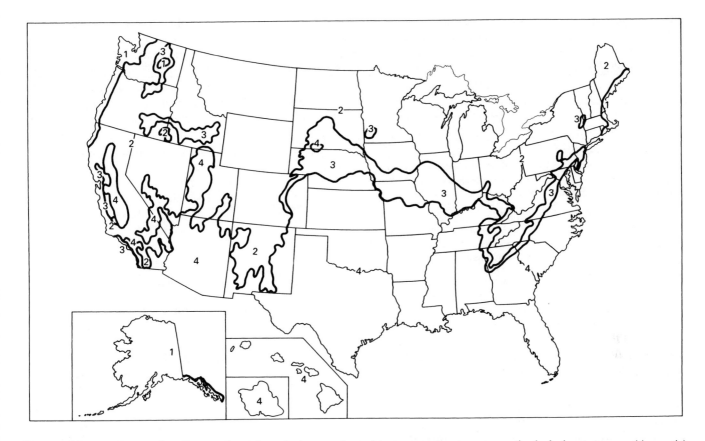

Figure 24–1 Areas of turfgrass adaptation. 1: Areas adapted to temperate grasses, particularly bent grasses (*Agrostis*). 2: Areas of general adaptation of temperate grasses. 3: Transition zone where subtropical grasses suffer winter cold and temperate grasses are stressed by summer's heat. 4: Area of adaptation for subtropical Bermuda and zoysia grasses. The more tender subtropical grasses are limited to hardiness zones 8, 9, and 10 (see Fig. 5–4). This map is based on mean July temperatures, not the low winter temperatures of the hardiness zone map. The boundaries are not sharp and are greatly influenced by local microclimates, especially in the mountain states, where elevation and exposure affect adaptation. Alaska is too cold for all but one or two cultivars of turfgrasses, which survive in milder areas of the state. Hawaii is subtropical to tropical.

CHOOSING TURFGRASSES

Proper selection of species and cultivars is critical to the long-term success of a turfgrass stand. However, selection of adapted species and cultivars does not guarantee long-term success. Proper cultural practices are still required. Local extension information should be consulted when selecting a turfgrass for establishment. Cool-season grasses are commonly established as a blend (two or more cultivars of the same species) or as a mix (two or more turfgrass species). Warm-season grasses are rarely mixed, to avoid problems with segregation. When using a seed blend or mixture, three to five unrelated cultivars should be used to provide adequate diversity. The components of a blend or mix should be similar in appearance, reflect the local environment (heat, shade, etc.), and show resistance to the major diseases and insects (if possible) in a particular region.

For several of the turfgrass species, such as Kentucky bluegrass, perennial ryegrass, tall fescue, fine fescue, creeping bentgrass, and Bermuda grass, there are numerous culti-vated varieties. Each of these varieties was developed because of some unique characteristic, such as darker green color or finer texture. Other grasses, such as centipede grass and Saint Augustine grass, have fewer cultivars. Kentucky bluegrass has the largest number of cultivated varieties. Of the turfgrasses, Kentucky bluegrass is unique in producing a large percentage of seedlings by **apomixis.** In this type of apomixis, the embryo resulting from sexual fusion of an egg and pollen nucleus aborts, but an embryo is produced from somatic cells of the embryo sac (female tissue). In this way it is possible to get seedlings that are genetically identical to each other and to the female parent.

Superior plants are tested for turf characteristics such as low growth, disease resistance, cold tolerance, drought tolerance, and so forth. If they are indeed superior, they can be increased and released as a cultivar. As a result, there are many clonal **cultivars** of bluegrass but few or none of the other turf species.

TABLE 24–1 Characteristics, Regions, Improved Cultivars, and Uses of the Common Turfgrass Species. The First Eight Species Are Northern U.S. Grasses; the Last Six Are Southern.

Turfgrass Species	Turf Characteristics	U.S. Regions Where Used	Desirable Characteristics
Agrostis stolonifera L. (*Agrostis palustris* Huds.) Creeping bentgrass	Able to produce a dense turf of good color and fine texture under extremely close mowing (0.25 in. and less).	North, through the transition zone and, with care, into parts of the South.	Ability to stand low mowing. Withstands salinity.
Agrostis tenius Sibth. Colonial bentgrass	Variable grass adapted to northern coastal climates. Spreads. Color fair. Coarse-textured mowed high; fine-textured mowed low.	Well adapted to north coast, usable throughout the North.	Adaptable. Withstands acid soils of low fertility and wet soils.
Agrostis castellana Boiss, and Reuter Dryland bentgrass	In a mixture, it forms dense patches of blue-green grass of puffy character unless mowed under one inch. Tolerates more heat and less water than other bentgrasses.	Better adapted to transition zone heat and dryness than above.	Performs well in irrigated Mediterranean climate.
Festuca arundinacea L. Schreber. Tall fescue	A coarse-textured bunch grass suited only for growing in pure stands.	Adapted to transition zone. Not fully cold-hardy.	Deep rooted. Tolerates wear, drought, and neglect.
Festuca rubra L. var. *rubra* Creeping red fescue	Fine-leaved, drought-tolerant grass tolerant of some shade. Spreads by rhizomes. Favored by high mowing. Competitive under low fertility.	North. Does poorly in the transition zone.	Fine texture, spreading. Tolerates some drought, shade, infertility.
Festuca rubra L. var. *commutata* Chewing's red fescue	Fine-leaved, drought-tolerant grass. Tolerates some shade. Spreads slowly. Tolerates lower mowing than *F. rubra* L. var. *rubra*.	North.	Tolerates closer mowing than *F. rubra* L. var. *rubra*.
Lolium perenne L. Improved perennial ryegrass	Turf cultivars have good color and fine texture like bluegrass. A patchy hard-to-mow clump grass unless well fertilized and watered.	Transition zone North. Not fully hardy. Winter grass in South.	Good color and texture. Tolerates traffic.
Poa pratensis L. Kentucky bluegrass	A dense, fine-textured, beautifully colored grass spread by rhizomes. Seeds produce uniform apomictic seedlings. The aristocrat of lawn grasses.	Best with cool nights and days under 84°F. Fair in transition zone.	Best color. Sod repairs itself from stolons. Heat and dry dormancy.
Cynodon dactylon (L) Pers. Bermuda grass	A dense vigorous grass creeping by stolons and rhizomes. Withstands neglect but a handsome turf when well cared for. Brown in winter.	South and valleys of Southwest.	Vigorous. Withstands drought, wear, and salinity.
Eremochloa ophiouroides (Munro) Hack Centipede grass	A dense vigorous grass creeping by large stolons. Adapted to low maintenance lawns on acid or infertile soils.	Florida, Gulf states, coastal plains, parts of California.	Withstands low fertility and acidity. Low maintenance.
Lolium multiflorum Lam. Winter grass, annual ryegrass	A coarse short-lived grass used only for overseeding winter dormant grasses for winter color in the South and transition zone.	Winter grass in South.	Grows at low temperatures, near freezing.
Paspalum notatum Fliigge Bahia grass	Forms an open coarse turf.	Gulf states and coastal plains of the South.	Heat tolerant. Low water need.
Stenotaphrum secondatum (Walt.) O. Kuntze Saint Augustine grass	Strong, dense, coarse grass creeping by large stolons. Adapted to shade.	Florida, Gulf states, and parts of California.	Vigorous. Withstands drought, some shade.
Zoysia japonica Steud. Korean grass, *Z. matrella* (L.) Merrill	Extremely dense grower of good color. Spreads by rhizomes and stolons.	*Z. japonica* hardy through zones 6–10, *Z. matrella*, 8–10.	Very dense—crowds out weeds. Good color and texture.

Undesirable Characteristics	Improved Cultivars	Special Care Requirements	Uses
Subject to disease and insect pests. Invasive.	Penncross, Penneagle, A-4, G-2, Crenshaw, Providence, SR1020, L93	Regular day-to-day maintenance, mow short, control pests.	Fine turf for putting and bowling greens
Coarse if neglected. Some disease and insect problems.	Astoria, Exeter, Bardot, Egmont, SW7100	Mow 1 in. or less; use low fertility for low maintenance.	Decorative lawns and utility turf
Does not mix well with other grasses.	Highland	Mow 1 in. or less, dethatch in fall.	Decorative lawns and utility turf
Coarse. Weedy in a mixture.	Arid, Bonanza, Falcon, K31, Mustang, Olympic, Rebel	Mow regularly 1.5 in. plus, use infrequent deep irrigation.	Decorative lawns and utility turf; sports turf
Intolerant of heat, salinity, close mowing.	Boreal, Dawson, Ensylva, Flyer, Fortress, Merlin, Pennlawn, Ruby	Mow 2 in. plus, light fertilizer. Unmowed for erosion control.	Decorative lawns and utility turf; overseeding subtropical grasses for winter color; unmowed for erosion control; sports turf
Intolerant of heat, salinity.	Banner, Dover, Jamestown, Longfellow, Mary, Koket, Wilma	Mow 2 in. or less, fertilize lightly.	Decorative lawns and utility turf; overseeding subtropical grasses for winter color; sports turf
Limited cold tolerance. Difficult to mow clean. Clumpy if unfertile.	Birdie II, Citation II, Manhattan II, Omega II, Palmer II, Prelude, Yorktown III, and others	Mow 1.5 in. plus with sharp mower. Fertilize and irrigate to keep dense.	Decorative lawns and utility turf; sports turf
Disease susceptible.	Too many to name: Adelphi, Baron, Challenger, Eclipse, Flyking, Merion, Midnight, Pennstar, Touchdown, among many good ones	Mow 1.5 in. plus, keep up lime level, main fertilizing in fall.	Decorative lawns and utility turf; sports turf
Invasive. Severe thatch builder. Disease susceptible in humid climate. Brown in winter.	Common, El Toro, Ormond, Santa Ana, Sunturf, Tifway, and others. Only common Bermuda is propagated from seed.	Close frequent mowing, dethatch, overseed for winter color.	Decorative lawns and utility turf; sports turf; fine turf for putting and bowling greens
Coarse. A thatch builder.	Common, Oaklawn	Keep fertilizer low, no lime, dethatch if needed.	Decorative lawns and utility turf
Coarse. Disease susceptible.	Gulf, Tifton 1	Open turf so seed touches soil. Mow as needed.	Overseeding subtropical grasses for winter color
Not cold hardy. Openness favors weed invasion.	Argentine, Pensacola, Tifhi 1, Wilmington	Mow 1.5 in. plus, low fertility. Reduced water need.	Decorative lawns and utility turf; sports turf
Coarse, invasive, disease- and insect-prone. Vegetative planting.	Bitter blue, Floratam, Floratine	Mow closely and dethatch. Control pests.	Decorative lawns and utility turf
Vegetative planting. Slow recovery of injury. Too dense to overseed.	Meyer, Midwest (Z. japonica); Flawn (Z. matrella); Emerald (hybrid)	Low fertility and water need. Mow closely, or neglect for erosion control.	Decorative lawns and utility turf

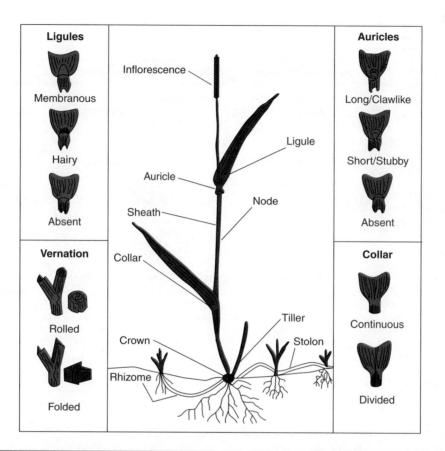

Terms and Definitions	
Phytomere: Basic unit of the grass plant (leaf blade, sheath, internode, axillary bud, adventitious roots, nodel plate).	**Inflorescence:** Terminal reproductive shoot.
Ligule: Appendage at abaxial junction of the blade and sheath.	**Shoots:** Primary—arise from the embryo of the germinating seed.
Auricle: Appendages that extend from the collar and wrap around the leaf.	Lateral—arise from buds on the crown. Tiller—intravaginal lateral shoot (syn.: bunch type).
Collar: Distinct band of tissue at blade/sheath junction.	Rhizome—extravaginal subsurface lateral shoot. Stolon—extravaginal above-ground lateral shoot.
Vernation: Leaf arrangement in the bud shoot.	**Crown:** Area of meristematic tissue between the shoot and the root.
Sheath: Lower part of leaf attached to the node at the crown.	**Roots:** Primary—arise from the embryo of germinating seeds.
Internode: Stem segment between nodes.	Adventitious—arise from nodes on the crown, rhizomes, or stolons.
Node: Joints of stems where leaves are attached.	
Stem appex: Contains the apical meristem.	

Figure 24–2   Key structures, terms, and definitions used in the identification of turfgrasses. Turfgrasses are crown-type plants, and internode elongation on the stem occurs only when an inflorescence is generated, as shown in this diagram. *Source:* David Gardner, The Ohio State University.

Named clones of other grasses can be vegetatively propagated with sod or stolons. Named cultivars are sexually propagated as a genetic mixture of seed from parents carefully selected to a standard of excellence.

Named cultivars of bluegrasses are excellent, but they have the disadvantage of complete genetic uniformity so that a turf is totally subject to any weaknesses. For example, a pure stand of 'Merion' bluegrass appears orange at times be-

cause of spores of a species of stem rust disease to which it is particularly susceptible. Consequently, bluegrass seed is often prepared as a blend. Seeds of four or five cultivars are mixed to take advantage of the good qualities of each while avoiding the extensive damage that could occur when a pure stand of a single cultivar develops a weakness.

With the exception of buffalo grass, none of the species commonly used as turfgrass in the United States are native to

TABLE 24–2	Identification Key to Turfgrass Species

1. Ligule not a fringe of hairs
  2. Folded vernation
    3. Boat shaped leaf tip
      4. Ligule short or absent
        5. Rhizomatous, leaf blade with translucent midrib = Kentucky bluegrass (*Poa pratensis*)
      4. Ligule present
        5. Ligule prominent, sheaths slightly compressed, bunch type = Annual bluegrass (*Poa annua*)
        5. Ligule abrupt, sheath strongly compressed = Canada bluegrass (*Poa compressa*)
        5. Ligule 4–6 mm, stoloniferous, onion-skin sheath = Rough bluegrass (*Poa trivialis*)
    3. Pointed leaf tip
      4. Very fine leaves, bunch type = *Festuca spp.* fine fescues
      4. Very fine leaves, rhizomatous = Creeping Red Fescue (*Festuca rubra*)
      4. Leaves shiny on back, bunch type, red pigment at base of sheath = Perennial Ryegrass (*Lolium perenne*)
  2. Rolled vernation
    3. Auricle present
      4. Long, clawlike auricle, bunch type = Annual Ryegrass (*Lolium multiflorum*)
      4. Narrow auricle, leaf blade 2–5 mm, pubescent upper leaf = Fairway wheatgrass (*Agropyron cristatum*)
    3. Auricle is difficult to detect or is absent
      4. Ligule short or absent
        5. Distinct collar, bunch type, prominent veination = tall fescue (*Festuca arundinacea*)
        5. Ligule evenly truncate, blades rough along edges, bunch type = colonial bentgrass (*Agrostis tenuis*)
      4. Ligule prominent
        5. Stoloniferous, light green color = creeping bentgrass (*Agrostis palustris*)
        5. Rhizomatous, prominent venation = Redtop Bentgrass (*Agrostis alba*)
        5. Swollen base (Haplocorm) = Turf Timothy (*Phleum pratense*)
1. Ligule a membrane with a fringe of hairs
  2. Margins ciliate toward base, thick stolons = Bahia grass (*Paspalum notatum*)
  2. Fine-textured, less aggressive stolons = Seashore Paspalum (*Paspalum vaginatum*)
  1. Ligule a fringe of hairs
  2. Rolled vernation
    3. Light green leaf, surface pubescent on both sides = buffalo grass (*Buchloe dactyloides*)
    3. Distinct collar, evenly spaced internodes = zoysia grass (*Zoysia japonica*)
  2. Folded vernation
    3. Unevenly spaced internodes, vernation may be rolled, thin collar = Bermuda grass (*Cynodon dactylon*)
    3. Evenly spaced internodes
      4. Collar constricted with 90° twist = Saint Augustine grass (*Stenotaphrum secundatum*)
      4. Collar lacks 90° twist
        5. Hairs on lower 1 cm of leaf edge, cottony ligule = centipede grass (*Eremochloa ophiuroides*)
        5. Pubescent sheath, smooth leaf blade = Kikuyu grass (*Pennisetum clandestinum*)

Source: David Gardner, The Ohio State University.

North America. Rather, they were introduced from other parts of the world. For example, Colonial bentgrass (*Agrostis tenuis*) is native to northern Europe. Introduced by colonists to the New World, it quickly took over the coastal regions from Rhode Island to Nova Scotia and from Oregon to British Columbia. It also became widespread along the coasts of New Zealand. Kentucky bluegrass (*Poa pratensis*) is thought to have been introduced at Vincennes, Indiana, by the French about 1720. By the end of the Civil War, it was established throughout the Middle West, and the westward settlers thought it was a native grass. Similarly, Bermuda grass (*Cynodon dactylon*) was introduced into New Mexico in 1750 and within fifty years spread throughout the Southwest.

In any location in the United States, several turfgrasses can be grown and, with good management, all can look good. In a few locations, one certain grass is particularly well adapted to the climate: it will gradually invade and dominate lawns in that area. It might be difficult to choose among several grasses when all grow well in an area or to see any reason why one cultivar or another should be preferred. Among good grasses, differences appear only under conditions of stress such as cold, heat, drought, wet soil, shade, low mowing, or disease; each of these stresses takes a toll at one time or another. The best species or cultivar for a given location is most likely to be determined, not by appearance or growth habit, but by its ability to survive the few days or weeks in the year when growing conditions are unfavorable. Such characteristics are given in Table 24–1.

Heredity endows certain grasses with the ability to grow and survive stress. Culture determines how well a grass achieves its potential.

SOIL PREPARATION

Soil preparation is a first step in planting a turf, and it includes grading and tilling. Grading should result in a convex swell of the soil surface, free from dips, swales, or pockets where surface runoff water can puddle or pond. Tilling breaks up soil to form a seedbed with good porosity and aeration. In time, tilled soil settles back to its original density, but initial root growth of new turf is aided by the loosened soil.

During tillage, fertilizers and chemical or physical soil amendments should be incorporated in the soil. Chemical amendments include materials such as lime or gypsum used to improve the chemical and physical properties of the soil. Physical amendments are mineral or organic and are generally used to improve the physical properties of a heavy soil or to add organic matter to a biologically impoverished one. Because physical amendments must be used at very high rates to have beneficial effects (often 70 to 90 percent amendment), their use is questionable unless the particular soil problem has been thoroughly analyzed.

Organic amendments serve as food for soil organisms and usually improve soil structure. Peat or manure incorporated into the top 1 to 2 cm (0.4 to 0.8 in.) of the seedbed helps seedlings emerge from crusting soils. Incorporation of other organic wastes improves the root zone environment, but some materials can be toxic (e.g., fresh cedar sawdust). Many organic wastes result in severe temporary nitrogen deficiencies in the soil unless the carbon-to-nitrogen ratio in the waste is less than 20:1. Composting organic wastes for a time before application usually corrects such imbalances and toxicities.

When the seedbed is a sterile subsoil, added organic matter and nitrogen fertilizer enhances biological activity. Where the soil is shallow, a few centimeters of additional soil are often added. A sterile soil may be a fill soil, or it may result from leveling or from spreading out soil from a basement excavation. Developers sometimes remove topsoil for sale to boost their earnings from a project.

The construction of a field for a sport such as football may require very sophisticated soil preparation. Synthetic materials are often added to the soil components to add stability and improve drainage. Competing weeds are a major problem in raising a new turf. They shade and suppress the grass, and removing them requires time and effort. It is good practice to irrigate the seedbed to germinate weed seeds before sowing grass seed. Weed seedlings are then killed by cultivation or with herbicides. The seedbed should then not be disturbed to the extent that new weed seeds are brought to the surface, where they could germinate.

Grass seed is available either in mixtures or pure lots. Most mixtures are designed to be sold to home gardeners rather than to professional turf growers. These seed mixtures represent compromises by the seed companies, and the mixtures produce an acceptable lawn regardless of the management given. The label names the species or cultivar of each grass, the percentage of crop and weed seeds, and the percentage of inert matter (Fig. 24–3).

Germination varies from 75 percent for bluegrass seed produced in a poor crop year to over 95 percent for ryegrass seed produced in a good crop year. In general, seed germination should be over 85 percent. Inert matter repre-

sents an inevitable amount of chaff or other material. Noxious weed seeds are of small concern in turf because most such weeds are subsequently destroyed by mowing. Turfgrass seed sometimes includes pasture grasses of no concern in a meadow or pasture, but they form coarse, undesirable, persistent weeds in a fine lawn. To avoid contamination by such grasses, seed of sod quality, which is free of such contaminants, is often purchased at a premium price. When named cultivars of grass are used, certified seed carries a certification statement that the plants were inspected while growing in the seed field and found to be pure and true to type.

In the northern two-thirds of the United States, a typical packaged mixture of high-quality seed is likely to contain a large percentage of several cultivars of Kentucky bluegrass. This blend forms the basic grass. To this is added a blend of red fescue grasses. Red fescue mixes well with bluegrass and grows better in the dry, shady, and less fertile areas. Where winters are not severe, seed of fine-leaved cultivars of perennial ryegrass are also blended with the blue grass to help the mixture resist seasonal disease problems. A low percentage of a bentgrass is usually added. Bentgrass finally predominates if the resulting lawn is mowed too short for survival of the blue, rye, and fescue grasses. In addition, bentgrass often survives better than the others in areas with wet soils.

Such seed mixtures can be diluted, more or less, with seeds of a filler grass to adjust the price of the mixture. A good filler is meadow fescue (*Festuca pratensis*), which looks like bluegrass during its first few months. Later it becomes coarse but tends not to persist in a well-tended turf. Red top (*Agrostis alba*) is often used as a filler but is not desirable because it is coarse and persistent. Some mixtures are blended solely for a low price and often contain large amounts of pasture grasses or the less desirable turfgrasses.

A high-quality seed mixture selected by a knowledgeable horticulturist might consist of a blend of the first three grasses mentioned, but might not include bent or filler grasses. Bentgrass tends to be a weed in bluegrass (and vice versa). Along the coasts of New England, Oregon, Washington, and southern Canada, one might choose a colonial bentgrass alone. Bentgrasses are well adapted in those regions.

In the southern United States a Bermuda grass, centipede grass, Bahia grass, or carpet grass lawn might be started from pure seed. Saint Augustine grass or hybrid Bermuda grass could be started from stolons or sod.

SEEDING

Once the kind of grass is chosen, the rate at which to sow the seed is considered (Table 24–3). Seedlings become crowded and cannot develop properly when seed is sown too heavily; it takes a long time to get a mature usable lawn. Sown too thinly, the plants are far apart, with space left for weeds to start. An initial stand of about twelve to fourteen plants per square inch will develop rapidly into a strong turf. If one

Desirable crop species ⟶

Undesirables not grown as a farm crop ⟶
Anything not seed (soil, stems, etc.) ⟶
Other plants normally grown for profit ⟶
• Often the most difficult weeds to control;
weeds officially declared special problems ⟶

Seed Mixture Analysis	
Fine Textured Grasses	**Germination**
26.24% Glade Kentucky Bluegrass	95%
24.00% Park Kentucky Bluegrass	90%
18.45% Midnight Kentucky Bluegrass	85%
Coarse Grasses	
27.81% Pennfine Perennial Ryegrass	92%
Other Ingredients	
0.45% Weed Seed	Tested 4/05
3.00% Inert Matter	50# net wt.
0.05% Other Crop	
	Seed company
No Noxious Weeds	address

Figure 24–3 Example of a seed tag or label showing the information required by law in the United States. Note that seed quality is influenced by purity and germination. Seeding rate should be based on % pure live seed (PLS) = [(germination% × purity%)/100]. It is also important to note the date on which the seed was tested. *Source:* David Gardner, The Ohio State University.

TABLE 24–3	Characteristics of Seeds of Common Turfgrasses*		
Species	**Number of Seeds per Pound**	**Seeding Rate (Pounds per 1,000 ft²)**	**Germination Time (Days)**
Cool-Season Species			
Creeping bentgrass	6,356,000	0.5–1.0	7–14
Colonial bentgrass	8,172,000	0.5–1.0	7–14
Velvet bentgrass	10,896,000	0.5–1.0	
Turf Timothy	1,135,000	1–2	
Tall fescue	227,000	7–9	5–15
Red fescue	546,000	3.5–4.5	5–12
Chewings fescue	546,000	3.5–4.5	5–12
Hard fescue	546,000	3.5–4.5	5–12
Perennial ryegrass	227,000	7–9	3–10
Annual ryegrass	227,000	7–9	3–8
Kentucky bluegrass	2,179,000	1–1.5	6–28
Rough bluegrass	2,542,000	1–1.5	6–21
Canada bluegrass	2,497,000	1–1.5	6–21
Fairway wheatgrass	318,000	3–5	
Warm-Season Species			
Buffalo grass (burs)	50,000	3–6	10–28
Bermuda grass (hulled)	1,787,000	1–1.5	8–15
Zoysia grass	1,369,000	2–3	10–14
Bahia grass	163,000	6–8	
Centipede grass	409,000	4–6	

* The number of seeds per pound, and therefore the recommended seeding rate, varies considerably among the species. The range of average germination times reflects the amount of time in which careful attention to surface moisture is required and varies depending on the time of year and conditions.
Source: David Gardner, The Ohio State University.

sows off-season (e.g., late fall or early winter) when seed germination is slow, one should increase the rate.

Special machines are available for sowing seed, covering it, firming the soil, and even mulching it. For hand operations, however, either box or cyclone fertilizer spreaders sow seed satisfactorily. The opening is reduced to a size appropriate to the seed. Hand sowing tends to scatter seed un-

evenly because few persons are skilled seed sowers. When hand sowing is necessary, seed should be vigorously thrown forward in a sweeping arc. The falling seed is more apt to drift into a random pattern than when seeds are dribbled out close to the ground.

When both large and small seeds are used, they are best sown separately in two operations. On small areas, rather than calibrate the spreader, the operator may prefer to reduce the seeder opening to a low rate, then cover the area two or three times in different directions to ensure even coverage. Small amounts of seed are often diluted with sand to adapt the volume of seed to the area covered.

Seeding Depth

Another determination is the depth to cover the seed. Turfgrass seeds are small, varying from about 1 million per kilogram for rye- and fescue grasses to over 4 million for bluegrasses and over 10 to 17 million for bentgrasses. Relatively few seedlings emerge from depths over 1 cm (0.4 in), and seedlings' emergence for the smaller seeds is best when they are covered to less than 1 to 5 mm. Uniform coverage is not generally possible. A light raking or dragging of the soil surface after seeds are sown covers seeds at depths from 0 to 5 or 10 mm. Some seedlings on the surface die of desiccation, and some deeper ones fail to grow to the surface, but seeding rates allow for such losses.

Time to Sow

At any given geographical location there is a best week in the year in which to sow grass seed. In the southern parts of the United States, subtropical grasses are best sown in early summer after annual weeds are removed following their principal flush of seed germination. Seeds of temperate zone grasses are best sown in late summer or early fall so that there is time for the seedlings to become well established and to cover the ground before freezing weather arrives. Fine, vigorous grass stands are easiest to obtain with fall-sown seed. Fall weed problems are reduced, and shorter days and cooler temperatures reduce evaporation and the need for frequent irrigation. The next best time to sow grass seed in the temperate zones is as early in the spring as the soil can be prepared. Grass seed germinates and the seedlings grow in cool soil; if a dense stand can be obtained before summer weed seeds germinate, the weed problem is reduced.

After the seed is sown, the seedbed should be rolled, unless you are relying on rainfall to germinate the seed. Rolling firms the soil around the seed and thus encourages capillary movement of available moisture. If there is insufficient moisture for good seed germination, a mulch should be used. A light cover (2 to 5 mm) of clean, weed-free sand or organic waste slows moisture loss from the soil surface. However, light mulching materials such as peat moss tend to wash or blow or to gather in pockets. Such materials are best worked into the top inch (2.5 cm) of soil before seeding.

Seed Germination and Seedling Establishment

Water is the most critical factor during seed germination, whereas nitrogen fertilizer is more critical during seedling establishment. In warm weather, most turf seeds begin to germinate five to seven days after sowing and continue for another week. Bluegrass seed, however, is slower and continues to germinate for a month. Germination is slower in cold weather. During germination, soil moisture is necessary in the surface layer at all times. Differences in available water in the root zone soon appear as differences in color and stand of the grass seedlings. At the same time, excess surface water encourages damping-off pathogens, a complex of *Rhizoctonia, Phytophthora, Pythium, Fusarium,* and other species of fungi. Ideally the top soil layer should remain moist, but if the soil surface is allowed to dry at least once a day, the mycelia of damping-off organisms shrivel and die.

Once germination has occurred, seedling roots begin to explore the soil. The period between irrigations is gradually extended so there is regular drying of the soil surface between irrigations. Irrigations should wet the soil deeper than the roots extend.

As grass seedlings grow, they may deplete the soil of nitrogen. Growth slows, and the seedlings' color becomes pale green to yellowish. Regular feedings of nitrogen fertilizer to provide about 0.25 kg of N per acre (0.5 lb/1,000 ft^2) keep the grass growing vigorously and help it suppress weeds. If seedling growth slows and the blade color is dark green, with some red anthocyanin pigment present, the seedbed probably contains insufficient phosphorus for initial seedling establishment.

As the turf grows and becomes thick and tall enough for the first mowing, the most suitable height for mowing must be decided. The mower should be sharp and set for the correct height. A dull mower pulls up seedling plants or tears leaves instead of shearing them. When the new grass is 2.5 to 5 cm (1 to 2 in.) higher than the desired mowing height, the soil should be allowed to dry for a day or two before the grass is mowed. If the clippings are scattered and the weather is dry, clippings shrivel and fall from sight. But if the clippings form heavy clumps and the weather is moist, they should be removed to prevent smothering and a good environment for diseases.

In the United States, it may take five months in southern states to twelve months in northern states after seed germination for the turf to form a vigorous mature lawn rugged enough for play.

SODDING

Points to Consider When Sodding

Sodding is an alternative to growing turf from seed. Sodding or seeding is a choice that turfgrass managers can make. The advantages and disadvantages of sodding compared to seeding are listed in Table 24–4. The ideal time of establishment from sod is the same as for seeding. However, spring sodding is usually much more successful than spring seeding. Sodding can be done in the summer in climates that do not have extended periods of hot weather (90°+). Sodding can be also done much further into the fall season than can seeding.

Quality sod should be purchased from a reputable dealer. The sod should be free of pest problems (weeds, in-

TABLE 24–4	Advantages and Disadvantages of Establishing a Turfgrass from Sod Versus Seed

Advantages of Sodding	Disadvantages of Sodding
• Sod provides an instant ground cover. • Watering frequency and duration is less at establishment. • Weed invasion is reduced. • Damage from rainfall, animal footprinting, etc., is less likely. • Mulching is not required.	• Sodding is much more expensive. • Sodding is much more labor intensive. • Sod is available with fewer cultivars and mixes. • Improperly laid sod is much more difficult to repair than an improperly seeded area.

Source: David Gardner, The Ohio State University.

sects, diseases). The grass should be in active growth and densely rooted, and the root system should be moist upon delivery. The sod strips should be uniform in width and thickness. The rhizomes should be severed at the bottom of the root mass because new roots form at the ends of severed rhizomes more quickly than fibrous roots can repair themselves.

Sod should be cared for properly after it is delivered. If possible, the sod should be laid the same day it is received to avoid what is called sod heating. Sod heating is an increase in temperature within the sod roll that results from plant and microbial respiration. Heating can result in high temperatures (100°F or higher) that cause death of the grass. Sod may be stored for up to three days during cooler weather if exposed sides and ends are lightly dampened daily, or as needed, and if the sod is lightly covered with a light-colored material to shield it from direct sunlight.

Steps to Follow When Sodding

The soil should be prepared the same way as for seeded areas. The area should be graded to give a slope of more than 1 percent but less than 20 percent and away from buildings. If the soil is compacted, it should be loosened by tilling to at least a 6-in. depth. Debris such as sticks, stones, and weeds should be removed. Starter fertilizer and lime, as indicated by soil test, are then applied and incorporated to a 6-in. depth. The lawn area needs to be leveled to eliminate high and low spots. At the edge of the turf area, the soil should be sloped so that the sod is recessed to the thickness of the sod root system. This practice prevents the edge of the sod from drying out and dying. The final grade should be firm enough to avoid footprinting.

Sod should be planted on lightly moistened soil. The sod should be rolled out by staggering in alternate rows (like the running-bond pattern of a brick wall), with the edges placed tightly together. Overlapping sod pieces should be avoided; instead, the pieces should be cut to fit with a sod knife. Also, one should not pull a strip of sod. Stretched sod will contract to its original size within hours and leave a space between

pieces. Staking may be required on slopes with a grade of 8 percent of more. Three thin stakes (one at each end and one in the center) are placed about 3 in. from the upper edge of the sod strip. The stakes should be placed flush with the top of the sod root system. Laying long, narrow strips (2 to 3 in.) of sod (as may occur at the edges of flower beds) should be avoided. The edges of newly laid sod should not be allowed to dry out.

After the sod is planted, it should be rolled at a 45° angle with a 60- to 75-lb roller to remove air pockets and then thoroughly watered so that water penetrates at least 6 in. into the soil. Sod requires daily afternoon watering for the first seven to ten days. More frequent watering may be required during periods of warm weather. Sod begins to root after ten to fourteen days, and rooting can be tested by lightly pulling up on a sod strip. When the sod is well-rooted, begin deeper and less frequent irrigation to promote deeper rooting. Do not overwater rooting sod. The sod will not root properly if the soil is saturated. Mowing should begin after the grass begins to grow. Fertilization as for a mature turf should begin after the sod has rooted.

Overseeding

A common practice in the southern United States is to overseed dormant subtropical grasses, whose leaves become brown during the winter. To keep the lawn green, seeds of a cool-season species are sown in the dormant turf in the fall. This practice provides winter color until the warming soil and vigorous spring growth of the subtropical grasses crowds and suppresses the winter grass. The traditional grass for such overseeding is annual ryegrass (*Lolium multiflorum*), but red fescue or perennial ryegrasses are also suitable and are finer-textured. Other cool-season grasses are also used.

Timing is important in overseeding. There should be enough warm weather yet to come for germination and seedling growth, but if overseeding is done too early, hot weather is likely to encourage growth of **damping-off** pathogens.

Renovation

An old, thin, or weedy lawn can be renovated by introducing new seed with improved management practices. To germinate and become established, the new seed must contact soil. A seedbed for either overseeding or renovation is prepared in the existing turf by raking vigorously; by power raking; or by using a coring, thatching, or vertical mowing machine to tear out a dense dead thatch and weak plants and to expose soil between the grass plants. Seed is sown, then kept moist during germination by sprinkling. Sown seed should also be lightly top-dressed with compost or sand. As soon as germination begins, a fertilizer should be applied to provide both nitrogen and a small amount of phosphorus. In renovating, an herbicide can be used in advance to kill undesirable grasses in portions of the lawn.

MAINTENANCE

Maintenance begins once the grass is up. Maintenance consists primarily of mowing; fertilizing; irrigating; and controlling weeds, insects, and diseases. These procedures should be programmed to produce a beautiful dense turf and vigorous healthy plants. As noted earlier, these goals are not completely compatible. There must be some compromise area appropriate to the climate, to the equipment available, and to the level of maintenance one is prepared to pursue.

Not only must the chosen program be a compromise between beauty and vigor, but there is the added possibility that a goal can be achieved equally well with different management programs. If the desired result is achieved with reasonable economy of effort and resources, no program is more right or more wrong than another. The lack of positive answers makes turf management comparatively difficult or confusing for some and challenging to others. For this reason, turf culture is best considered in terms of principles rather than applications.

PRINCIPAL TURF MANAGEMENT PRACTICES

There are three principal turf management practices and many secondary practices. The principal ones are mowing, fertilization, and irrigation. Each has a large effect on grass growth, and the manager manipulates them to change grass vigor and appearance.

Mowing

Mowing is a regular chore. There are several choices to make in mowing, for example, the kind of equipment to use, the mowing height, frequency of mowing, and whether to remove clippings or leave them.

Factors That Influence Mowing Frequency

Several interacting factors influence the frequency of mowing required. Shorter cut turf requires more frequent mowing than taller cut turf. The time of year also affects mowing requirements. Cool-season turf requires more frequent mowing in fall and spring than in summer, when conditions are not as favorable for turfgrass growth. Also, when cultural practices such as fertilization and irrigation are optimized, mowing frequency increases. Overapplication of fertilizer or the use of quick-release nitrogen will cause a growth flush. The species used can affect mowing frequency requirements. For example, zoysia grass grows more slowly than do other species.

Recommended Mowing Heights for Turfgrass Species

The recommended mowing height is determined by species and use. For example, tall fescue will appear open and stemmy if mowed shorter than 2 in. In contrast, creeping bentgrass mowed taller than 2 in. will result in puffy and somewhat

course-textured appearance Table 24–5 gives mowing height recommendations for commonly used turfgrasses. Slight variations to these recommendations may occur in different regions, based on climate and other factors. For example, shaded or otherwise stressed turf should be mowed taller. In general, management intensity increases as mowing height decreases.

Mowing Equipment

There are four kinds of modern mowing machines for lawns. Two kinds of power mowers—flail and sickle bar—are used only for large-scale and heavy work.

The flail mower is used on roadsides and in rough park areas. It has a rapidly rotating horizontal axle that swings several vertical knives inside a housing. The mower requires high energy, mows tall grasses and weeds, and reduces them to a mulch.

TABLE 24–5	Recommended Mowing Heights for Commonly Used Turfgrass Species*
Turf Species	**Mowing Height (Inches)**
Creeping bentgrass	0.125–0.5 (high intensity)
(*Agrostis palustris*)	1.0–2.5 (low intensity)
Tall fescue (*Festuca arundinacea*)	1.5–3.0
Fine fescue (*Festuca rubra, F. longifolia*)	1.0–2.5
Perennial ryegrass (*Lolium perenne*)	1.0–2.5
Kentucky bluegrass (*Poa pratensis*)	
Common	1.0–2.5
Improved	0.75–2.5
Buffalo grass (*Buchloe dactyloides*)	1.0–2.0
Bermuda grass (*Cynodon dactylon*)	0.3–2.5
(varies by cultivar)	
Zoysia grass (*Zoysia* spp.)	0.5–1.0
Bahia grass (*Paspalum notatum*)	1.5–3.0
Saint Augustine grass	1.5–3.0
(*Stenotaphrum secondatum*)	
Centipede grass	1.0–2.0
(*Eremochloa ophiuroides*)	

* Mowing height varies depending on species and usage. Mowing height has a tremendous effect on management intensity required.
Source: David Gardner, The Ohio State University.

MOWING STRESS

Mowing places a stress on the grass and reduces its vigor. Root growth is slowed for a few hours to several days depending on the severity of mowing. The total leaf production for the season is reduced by each mowing. Mowing opens the grass canopy to allow more light to enter so more plants can grow in the same area, thus increasing competition. The greatest stress is added to the stress budget when grass is mowed very short. High populations of weak plants result. With exceptionally vigorous and invasive grasses, such as Bermuda grass or kikuyu grass, frequent short mowing can be used to deliberately weaken them.

The sickle bar mower resembles a hair clipper with a 2m (6 ft) horizontal blade. It uses low energy; mows grass, weeds, and woody seedlings; and lays them in a swath with the stems parallel, convenient for raking. This mower is the same type used to mow forage hay.

Two types of lawn mowers are most commonly available: reel mowers and rotary mowers. Reel mowers are used when a high-quality cut or a low cutting height is required, such as on a putting green or fairway. Reel-mower cutting action is similar to that of a pair of scissors. A series of blades forming a cylinder are mounted at a spiral angle to the axis. They spin over a parallel-mounted bedknife. Reel mowers require relatively less power than do rotary mowers.

Rotary mowers are used when a moderate quality of cut at a higher height is acceptable, such as for home lawns, golf course rough, and utility turf. The cutting action is performed by the impact of a horizontal blade rotating at high speed. Rotary mowers can cut taller grass and seed stalks and are easier to maintain than reel mowers. They pose increased safety concerns, however, such as ejection of rocks and debris from the mower chase.

Factors That Influence Mowing Quality

Several factors, such as the species and cultivar utilized, influence the quality of cut when mowing. Certain species, such as perennial ryegrass and zoysia grass, have poor mowing quality because of their leaf structure characteristics. The result is shredded blades and a dull, ragged appearance. Some improved cultivars have a demonstrated increase in mowing quality. Reel mowers provide a much higher quality cut than rotary mowers do in high maintenance situations.

It is also important to mow within the recommended height ranges for a species. Creeping bentgrass and annual bluegrass, when mowed above recommended ranges, produce undesirable thatchy and puffy turf. When all species are mowed below the recommended range, scalping and/or thinning of the turf results.

Properly sharpened blades are very important. Dull blades cause a ragged appearance and result in slower recovery of the turfgrass. When using a reel mower, the relationship between mowing height and the clip of the reel (the distance traveled between clips) is important. The highest quality cut results when the mowing height equals the clip of the reel. If the clip of the reel is greater than the mowing height, then marcelling occurs (the turf takes on a wavy appearance because of the uneven cut). If clip of the reel is less than the mowing height, the efficiency of a reel mower is reduced, resulting in a ragged, nonuniform cut.

Points to Consider When Mowing

To minimize stress to the turf, it should be mowed frequently enough so that no more than one-third of the leaf tissue is removed at one time. In other words, if the desired cutting height is 2 in., the grass should be mowed before the height exceeds 3 in. Mowing should be avoided when the soil is excessively moist because the weight of the mower will cause ruts to form in the turf. One should also raise the mowing height of the mower to avoid delays caused by weather. During excessively wet periods, mowing to remove no more than one-third of the leaf tissue is more important than avoiding ruts in the turf.

It is generally recommended that the mowing height be raised to the maximum for a particular species just before the onset of summer stress and winter. Do not scalp the grass prior to the onset of winter. The mowing height can be lowered to a more desirable height during the favorable growth periods of spring and fall. Lower mowing heights necessitate more frequent mowing. Gradually decrease mowing height (such as would occur in the fall after summer heat and moisture stress) by increasing mowing frequency, not by removing more than one-third of the leaf tissue at once. Alternate the direction or pattern of mowing each time to avoid ruts, uneven cutting of the grass, and clumps of tall grass in the wheel tracks. Select a mowing pattern that minimizes the number of sharp turns required. This practice improves mowing efficiency and decreases injury to the turf.

Collection of clippings is usually not necessary. Clippings contain 85 to 90 percent water and do not contribute to thatch development. Thatch is formed by stem tissue, for example, rhizomes, stolons, and also roots. However, one should collect clippings from diseased turf to help prevent spread of the disease or if uncontrolled weeds are setting seed. Similarly, collect excessively long clippings to prevent shading of the grass.

Cool-Season Turfgrass Growth Regulators

Several growth-regulating compounds are available for use in turfgrass management. These compounds fall into three general categories: Those that are hormones or produce hormones that effect growth, those that inhibit cell division (termed Type 1 materials), and those that inhibit cell elongation (Type 2, or gibberelin synthesis inhibitors).

One should read the label carefully before using growth regulators on turfgrass. Most growth regulators have specific turf species registrations. Certain growth regulators can be utilized on close-cut grass, while others cannot. Some have specialty uses. For example, certain growth regulators are registered for use in an overseeding program.

It is important to determine the economic feasibility of growth regulators before using any of them. Growth regulators are ideal management tools for use on golf course fairways. However, landscape maintenance firms that have a contractual obligation to mow a site on a predetermined (e.g., weekly) schedule may not realize any savings in labor or costs beyond that of reduced clipping hauling and disposal charges. In this situation, the economic benefit of use may be diminished. Growth regulators are also useful in areas that are difficult to mow or along edges of turf.

Growth regulators should not be applied to stressed turf, for example, if there is a disease or insect infestation. In this case, the growth regulator interferes with the turf's ability to recover. Some undesirable yellowing of the turf may occur for a period of time after application. If in doubt, test a small area before making large-scale applications.

Undesirable growth characteristics (elongation of internodes resulting in puffy, thatchy turf) can result from using ethephon on turfgrass grown under shade. Mefluidide is a growth-inhibiting compound (not growth-suppressing, as are other Type I products for landscape use). Mefluidide causes good seedhead suppression.

On the other hand, Type II growth regulators (e.g., trinexepac-ethyl) are not generally effective at reducing seedhead development, though the height of the seedhead may be reduced. These materials suppress vertical growth for longer periods of time compared to Type I materials. In addition, they do not suppress lateral growth (tillering, rhizome) and root growth as much as Type I regulators.

Fertilization

The list of elements required for turfgrass growth varies (depending on the source consulted) from sixteen to eighteen (sodium and nickel are debated). The list can be subdivided into four categories. Carbon, hydrogen, and oxygen are found in highest abundance, yet they are not applied as fertilizer materials. Fertilizer elements are categorized as primary and secondary macronutrients and micronutrients. The primary macronutrients—nitrogen, phosphorus, and potassium—are the elements applied most often by the turf manager. The secondary macronutrients—calcium, magnesium, and sulfur—are generally available in the soil and are not routinely applied as fertilizers. The micronutrients include boron, chlorine, copper, iron, manganese, molybdenum, and zinc. With the exception of iron, these nutrients are generally not applied as fertilizers, because they are often present in sufficient quantities for turf growth in macronutrient fertilizers.

Factors that affect fertility requirements include cultural management (mowing, irrigation frequency), use of the turf, and environmental conditions. It is important to follow the recommended application schedule for the particular species used and the location. These schedules are usually based on pounds of element per 1,000 ft^2. Application when seeding or fertilizing an area for the first time should be based on soil-test results because changes in growth and color of the grass may not be due to a nutrient deficiency. Also, different species have different fertility requirements and application schedules, which may further vary depending on location. Turf managers can consult their state extension literature for specifics. When applying fertilizer, it is important not to apply more than the maximum single application rate, or fertilizer burn will result. It is also a good idea to avoid fertilizing during unfavorable weather conditions such as high heat or drought and to avoid applications to wet foliage because some synthetic fertilizers will cause burn if applied to wet foliage.

TABLE 24–6	Nitrogen Requirements of Common Turfgrass Species*		
Species		**Management**	
		Low	**High**
Cool-Season Species			
Creeping bentgrass (*Agrostis palustris*)		1–3	3–8
Tall Fescue (*Festuca arundinacea*)		1–2	3–5
Fine Fescue (*Festuca rubra, F. longifolia*)		0.5–2	2–4
Perennial ryegrass (*Lolium perenne*)		2–4	4–6
Kentucky bluegrass (*Poa pratensis*)			
Common		1–2	2–4
Improved		1.5–3	3–6
Warm-Season Species			
Buffalo grass (*Buchloe dactyloides*)		0–1	2–3
Bermuda grass (*Cynodon dactylon*)		1–4	3–8
Zoysia grass (*Zoysia* spp.)		2–4	5–7
Bahia grass (*Paspalum notatum*)		0–1	2–4
Saint Augustine grass (*Stenotaphrum secondatum*)		2–4	5–7
Centipede grass (*Eremochloa ophiuroides*)		0–1	2–4

* Higher management intensity increases annual nitrogen requirements.
Source: David Gardner, The Ohio State University.

Nitrogen (N) is the most important element in turfgrass management. Average annual nitrogen requirements vary depending on numerous factors, such as species, use, soil type, and climate. Nitrogen requirements for several turfgrass species are given in Table 24–6.

Points to Consider When Conducting a Soil Test

When conducting a soil test in turf, nitrogen is usually not tested due to rapid fluctuations in the levels of this nutrient. While no accurate test for soil nitrogen exists, many fertility programs on golf courses and other commercially maintained turf areas for other elements are based on the results of soil tests. Soil tests can determine whether mineral nutrients are available in sufficient quantities or if the addition of nutrients with a fertilizer is required. It is important to collect a representative sample for analysis. The usual recommendation is to collect several samples randomly from different parts of the area to be tested. Also, separate tests should be conducted on each soil type and in areas with different characteristics (e.g., separate wet and dry areas, high and low areas, etc.), and the collected cores from each sampling area should be uniformly mixed before being sent to the lab. In most areas, it is necessary only to conduct a complete soil analysis test once every five years. Basic soil tests to monitor conditions and access fertility needs should be conducted more often. Conduct complete tests yearly to monitor progress of pH modification or fertility buildup programs.

There are a couple of different soil test methods: sufficiency level of available nutrients (SLAN) and basic cation saturation ratio (BCSR). SLAN is an older method tradition-

TABLE 24–7	Recommended Phosphorus and Potassium Levels for Turfgrass Soils*			
Nutrient Status of Turfgrass Soils	**Phosphorus**		**Potassium**	
	PPM	**Pounds/Acre**	**PPM**	**Pounds/Acre**
Very low	0–5	0–10	0–40	0–80
Low	6–10	10–20	40–175	80–350
Adequate	10–20	20–40	175–250	350–500
High	>20	>40	>250	>500

*These recommendations vary from many traditional agricultural row crops. It is important to have turfgrass soils tested by a lab familiar with turf management or to interpret the results for P and K using the values listed in this table.

Source: David Gardner, The Ohio State University.

ally used in university labs. It relies on correlation of chemical extracts that approximate a plant's ability to extract nutrients from the soil and calibration of values generated by extraction data. Response curves are generated based on field-response studies. BCSR is used by many private labs. It is based on the concept that the ideal ratio of cations on cation exchange sites produces the best plant response (varies, but usually 65 percent Ca^{++}, 10 percent Mg^{++}, 5 percent K^+, 20 percent H^+). The test results are based on nutrient applications to restore this ratio. The best interpretation of actual fertility needs comes from combining the results from both types of soil test.

It is important for turfgrass managers to recognize that the test recommendations for phosphorus (P) and potassium (K) generated by some laboratories are based on field-crop requirements. The nutrient requirements of mature turfgrass are different than these values suggest. Phosphorus needs are generally overestimated and potassium needs are underestimated. Table 24–7 gives the nutrient status of soils at varying levels of phosphorus and potassium.

The secondary macronutrients—calcium, magnesium, and sulfur—are generally not applied as fertilizers because they are often present in sufficient quantities in the soil. An exception might occur on a golf green with a silica sand root zone. In this case, applications of calcium (in the form of gypsum) may be necessary.

Use of Sulfur to Decrease Soil pH

Sulfur is also sometimes used in areas with high soil pH to effect a temporary reduction in soil pH around the turfroot zone. Soils may also be too alkaline, causing nutrients such as iron and manganese to become unavailable to the plant. It is important to determine if high pH is due to a soil characteristic or application of lime. It is difficult, if not impossible, to lower the pH of naturally alkaline soils.

If the pH is high due to application of lime or other alkaline materials, then acid-forming materials can be used to decrease soil pH. For a given soil, the amount of sulfur re-

quired to decrease the pH one unit is roughly equal to one-third the amount of limestone required to increase the pH one unit. Sulfur may be applied at a maximum of 10 lb per 1,000 ft^2 every eight weeks. However, sulfur oxidizes and mixes with water can form a strong acid that can burn the grassroots. Therefore, it is important not to exceed the maximum application rate nor to apply more frequently than the minimum recommended interval.

Micronutrients

Iron is used by golf course superintendents and landscapers because it causes a rapid greening and darkening of the turf with increasing clipping production. However, too much iron can cause the turf to take on a temporary black appearance. Iron generally is applied as a chelated material to increase availability to the turf. There is some debate about the accuracy of soil tests for other micronutrients, but deficiencies with them are generally uncommon in turf grown on soil with a pH between 5.6 and 7.0.

Points to Consider When Fertilizing

The most important step before applying fertilizer is to read and understand the fertilizer label (Fig. 24–4). A wide variety of fertilizer materials are available, and turfgrass managers can select fertilizer materials according to specific needs or maintenance goals. For example, water-soluble nitrogen sources (quick release) cause rapid green-up and growth, but precise application is more important to avoid fertilizer burn. In contrast, water-insoluble nitrogen results in slower response, long residual activity, lower burn potential, higher cost, low surface runoff and leaching potential, and low frequency of application relative to water-soluble nitrogen.

Certain fertilizers alter the soil pH. For example, if the soil pH is high, it may be beneficial to use potassium sulfate that temporarily acidifies the soil, rather than potassium chloride, which has no effect on pH.

Fertilizers are generally divided into natural and synthetic sources. Most organic fertilizers are slow-release and may yield higher visual quality, but the application rate and cost is higher than that of synthetic fertilizer materials. Also, less nitrogen from natural fertilizers is available in cool weather because of reduced microbial decomposition.

DETERMINING IRRIGATION REQUIREMENTS

All plants require water for growth. Several terms are used to define the overall irrigation requirements of turf. The total irrigation requirement for a turfgrass is the amount of water needed to meet the net irrigation requirement of the grass and also to compensate for losses from evaporation, percolation, and runoff. Irrigation requirements vary according to soil

General Label Information ⟶

Analysis (%N–%P₂O₅– % K₂O) ⟶

Total Product Weight ⟶

Percentage Nitrogen ⟶

%P₂O₅: P₂O₅ × .44 = % Available P ⟶

%K₂O: K₂O × .83 = % Available K ⟶

% Other Nutrients ⟶

Materials Used to Compose the Fertilizer ⟶

Bob's Premium Fairway
24-4-12
50 LB

Total Nitrogen (N) 24.0%
1.6% Ammoniacal Nitrogen
10.8% Water-Insoluble Nitrogen
11.6% Water-Soluble Urea Nitrogen
Available Phosphoric Acid (P₂O₅) 4.0%
Available Potash (K₂O) 12.0%
Sulfur (S) 5.0%

Derived from isobutylidene diurea, ammonium phosphate, and potassium sulfate. Potential acidity 0 lb. calcium carbonate equivalent per ton.

Figure 24–4   Sample fertilizer label. Water-insoluble nitrogen is generally slower-release and provides better results, but it is more expensive. Note that P and K numbers are based on pounds of P₂O₅ and K₂O and require converting to pounds of elemental P and K before determining application rates for these elements. *Source:* David Gardner, The Ohio State University.

type. For example, sandy soils have a high infiltration rate, rapid drainage, and low water-holding capacity. The opposite is true of soils with a high clay content.

The turfgrass species used also influences irrigation requirements. Warm-season grasses, where they can be grown, require much less water than do cool-season grasses. Weather conditions (humidity, temperature, rainfall) and microclimate (amount of shade, slope, etc.) also influence irrigation requirements. Areas that are trafficked more intensely require more water. Cultural practices such as decreasing mowing height and/or increasing fertility also increase irrigation requirements. Finally, irrigation frequency affects irrigation requirements because increased frequency causes increased water use by the turfgrass.

When determining irrigation requirements, adjustments must be made for rainfall received and the efficiency of the irrigation system. It is often better to wait to irrigate until it is required, rather than to irrigate daily with automated irrigation systems. Several signs indicate that irrigation is required. Footprinting is the easiest to use. Moisture-stressed turf is not as turgid and does not spring back to its original position as rapidly, causing footprints to remain for a longer period. Turfgrass should be irrigated just as the plants are beginning to wilt. A slight change in the color (to blue- or grayish green) or the quality of light reflected from the leaf precedes wilting. The appearance of the leaf in cross section can also be used to determine if the turf needs to be watered (Fig. 24–5). Equipment such as a soil probe (core sampler) can aid in determining irrigation requirements. One should irrigate if the soil is dry at a depth of 4 to 6 in. Tensiometers are devices that measure soil moisture potential.

Points to Consider When Irrigating

Turf should be irrigated in the early morning hours if possible. Watering during the day is inefficient because of in-

Just rained   Getting dry   Time to   Drought
water

Figure 24–5   Effect of water status on the appearance of a turfgrass leaf cross section. As the soil dries, the leaves of the grass plant fold and then curl over as a mechanism to conserve water. The difference in how a curled leaf reflects light compared to a flat leaf causes the turf to take on a bluish green cast when dry. *Source:* David Gardner, The Ohio State University.

creased evaporation. Watering in the evening increases the incidence of certain pathogens. On a home lawn, one should commit to regular irrigation or allow the lawn to go dormant. Sporadic watering reduces the carbohydrate reserves of the plant and decreases survival. On lower maintenance surfaces (home lawns), it is traditionally recommended that the turf be irrigated infrequently, that between 1 and 2 in. of water be applied per week. Finally, to avoid water loss due to runoff, the irrigation rate should not exceed the infiltration capacity of the soil.

Methods of Reducing Irrigation Requirements

Use drought-tolerant species or cultivars. Note that the water-use rate does not determine the drought tolerance of a species. For example, Kentucky bluegrass has a low water-use rate but is shallow rooted. It wilts more readily than tall fescue, which has a higher water-use rate but also a deep root system. In areas where warm-season grasses perform adequately (transition zone and south), the water-use rate is about half that of a cool-season grass.

Raise the mowing height to the maximum recommended for the species used. Taller grass has a higher water-use rate, but it develops a deeper root system and will also better shade the soil surface, thus reducing evaporation loss. Mowing frequency should be reduced because significant water loss occurs through mower wounds. Mow with a properly sharpened blade to hasten recovery of the plant.

Apply lower rates of nitrogen and higher rates of potassium during years when precipitation is below normal. Reducing nitrogen reduces growth and thus water requirements. Increasing potassium increases the osmotic potential within the leaves and in turn increases the plant's ability to extract water from the soil, thus improving drought tolerance.

Remove excess thatch to avoid shallow rooting and decreased water penetration to the soil. Cultivate compacted soils to increase water infiltration.

Syringing

Syringing is employed when the turf wilts despite adequate soil moisture. Wilting is caused when the evaporation rate exceeds the plant's ability to absorb moisture. A light amount of water is applied to the shoots of the plant. Not enough water is applied to contribute to soil moisture. Syringing is also employed to remove dew, frost, exudates, and foreign matter from turf leaves.

The use of syringing to alleviate moisture stress is misunderstood. The benefit of syringing is influenced by air temperature, canopy temperature, relative humidity, irridiance, wind, the amount of water applied, timing of application, and water temperature. In areas where cool-season turf is grown, syringing, generally causes about a 1°F to 4°F decrease in canopy temperature for two hours. In southern regions, a return to presyringing temperature is observed within fifteen minutes.

Irrigation Water Quality

Secondary water sources (e.g., effluent, waste, reclaimed or recycled water) are becoming more common irrigation water sources for turfgrass in some parts of the United States. Secondary water is defined as the final liquid product from a sewage treatment plant. Pollutants are removed physically, chemically, or biologically before discharge. Several concerns and sources of increased cost are associated with secondary water use, including additional soil amendments to mitigate high salt and sodium levels, increased fertilizer and pesticide use due to overall poorer turf quality, and specialized irrigation components. In addition, more frequent testing of water quality is necessary.

Several factors affect effluent water quality, such as the amount of salinity, sodium, toxic ions, and bicarbonate. Depending on the impurity found in the water, turf managers can mix gypsum, sulfur, lime, or acidifying agents or blend the effluent water with clean water to ameliorate the impu-

rity. Other concerns to evaluate and monitor include the level of suspended solids, biodegradable organics, pathogens, nutrients, and stable organics.

SECONDARY MANAGEMENT PRACTICES
Pest Control

Problems of pest control are universal. Weeds, diseases, insects, nematodes, and other pests all afflict turfgrasses. Because such organisms tend to occupy the environment to the full extent of its capacity to support them, turf always has insects feeding on grass leaves and roots, fungi consuming dead grass clippings, and weeds filling in any bare spots of soil.

Insects

Insects attacking turf include various caterpillars, beetle grubs, bill bugs, wireworms, flea beetles, chinch bugs, fruit flies, leafhoppers, scales, and aphids. In the spider group, mites attack certain species of grass. Of all these, caterpillars and beetle grubs are likely to affect all lawns.

Caterpillars, the larvae of various species of moths, build silk-lined burrows among the grass crowns and feed nocturnally on the leaves. Caterpillars stay close to their burrows, and each thins out a small area of grass around its burrow, about 2 or 3 cm (0.8 to 1.2 in.) in diameter. Damage is inconsequential until populations build up to several dozens per square meter. At that point, turfgrass is rapidly thinned out as caterpillars mature and their appetites increase. To treat this problem, grass is mowed to reduce leaf area. Leaves are then sprayed with a stomach poison or contact insecticide. Irrigation water is withheld for a day or two so the insecticide is not washed from the leaves.

Grubs that live in the soil and feed on roots are a different problem because it is difficult to get insecticide into the soil at control levels. In their early growth stages, grubs are small and seldom a problem. But in the fall or the spring following their emergence, they are large and hungry and they eat so many roots that patches of grass die. The grubs can be found 2.5 to 5 cm (1 to 2 in.) deep in the soil at the edges of the brown dead patches.

Insecticide chemistry has changed considerably over the last twenty years. Pesticides such as chlordane, which were relatively broad spectrum and long lasting, have been replaced by newer, more selective materials with shorter residuals. Consequently, it is more important than ever to identify the insect pest correctly. In addition, one should have knowledge of the life cycle of the insect and its most vulnerable stage. Insecticides should be targeted for the stage when the insect pest is most vulnerable. For example, one should not try to control adult beetles; rather, the insecticide application should target the grubs. Cultural practices are less effective against insects because they prefer properly maintained turf. However, maintaining a healthy stand of turf by properly mowing, fertilizing, and so on, produces a turf that is more tolerant of insect damage. One should avoid planting tree and shrub species that harbor the adult stage of a turfgrass insect pest.

Weeds

A weedy turf is symptomatic of poor management and often indicates that not enough attention has been given to practices that produce a vigorous turf. Weeds are frequent when turf is undernourished, overwatered, or mowed too short. Soils compacted by heavy traffic tend to grow poor turf, but they do support a large population of certain weed species.

Unwanted perennial grasses are the most difficult weeds to control in turf; broad-leaved annuals are the easiest to control. In some circumstances, a grass that is usually considered useful as a turf can be a serious weed problem. This outcome occurs most often when a turf species is grown out of place or in a situation for which it is not intended. Annual bluegrass is a serious weed in creeping bentgrass putting greens because its lack of tolerance for heat and drought stress causes it to die out during summer stress. Rough bluegrass is sometimes considered a weed because of its poor tolerance of traffic. Tall fescue is a serious weed in Kentucky bluegrass because of its course texture. Creeping bentgrass, when found in Kentucky bluegrass, forms unsightly clumps of thatchy, puffy turf because of its lack of tolerance to higher heights of cut. Annual ryegrass is commonly used in low-maintenance situations or for hillside stabilization during establishment. As an annual, it dies out after one season, resulting in bare patches in the turf. Annual ryegrass is commonly found in lower-cost seed mixes sold to homeowners. Warm-season grasses such as zoysia grass grown north of their adaptive range or when present in a cool-season turf can produce unsightly clumps of straw brown turf during fall, winter, and spring, when temperatures are suboptimal for their growth.

Selective control of weedy turfgrasses is extremely difficult and in many cases impossible. Herbicide selectivity is based on physiological differences between species, and cool-season turfgrasses are all very similar physiologically. Certain herbicides can be used (e.g., Chlorsulfuron for tall fescue control), but success is often limited. In many situations, control with a nonselective herbicide followed by renovation as required.

The first step in weed control is to improve management to encourage the grass to grow vigorously. Then weeds can be dug, pulled, or treated with an herbicide. With the grass growing vigorously, space formerly occupied by a weed fills with grass. Control is then successful. If the space is recolonized by other weeds, management practices should be reexamined as the first step in additional control efforts.

Herbicides

Herbicides used to control turf weeds should be used cautiously and with restraint. They are plant poisons that at recommended rates are more toxic to weeds than to the turf-grasses. Herbicides recommended for weed control on turf also injure and cause stress to the turfgrass plants, though the turf outgrows the injury in time. If herbicides are used repeatedly or at times when the grass is under stress or not growing well, grass growth might be so retarded that weed problems worsen.

Contact herbicides kill plants they touch and are sometimes used as spot sprays to kill a few individual difficult weeds in a turf. The results, however, are unsightly spots that the grass slowly recolonizes.

Preemergence herbicides control annual weeds by soil application before weed seeds germinate. The herbicide kills seedlings as they push through the treated surface soil layer.

Post-emergence selective herbicides are used to remove broad-leaved weeds from grass. Such herbicides are most effective on weeds in the seedling stage. As weeds become older, they become more resistant. Many species of weeds are killed by routine mowing, which continually defoliates them. Weeds that persist tend to be low-growing species that spread out below the mower blades.

If a lawn has only scattered weeds, it is more prudent to dig them out than to mix chemicals and wash spray tanks and thus avoid the risk of spray drift onto garden flowers. The county agricultural agent or local garden center can recommend suitable herbicides to use on lawns.

Annual grasses such as crabgrass should be controlled using preemergence herbicides in late winter or early spring. It is important to apply products uniformly to prevent weed breakthrough. Annual grasses may also be controlled with a post-emergence herbicide. However, the herbicides should be applied while weeds are young (before tillering). Caution must be exercised with post-emergence products because application to stressed turfgrass may result in phytotoxicity. Perennial grasses such as quackgrass must be controlled with nonselective herbicides.

Perennial broad-leaved plants such as dandelion are best controlled with a post-emergence herbicide applied in late fall. Combination herbicide products are used when more than one difficult-to-control weed species is present. With herbicides, fungicides, or insecticides, it is important always to read the label prior to use. One should be aware of reentry period, time until reseeding can occur, which species are controlled with the product, and site and use restrictions.

Diseases

The most important step in disease management is to identify the pathogen correctly. Several characteristics should be used, including plant symptoms and signs of damage, weather conditions, management practices, and so on, when attempting to make the determination in the field. Field identification can be difficult because many diseases have similar appearances. The most accurate diagnosis is achieved when sending a sample to a lab for analysis. However, conclusive results may take weeks to obtain, and sometimes you can formulate an educated guess about the pathogen while you are awaiting lab results.

Three important turf diseases are mildew, rust, and smut. Mildew forms a white dust on grass leaves in the shade and can be controlled only by opening up the area to provide the grass with better light and ventilation. Rust is sometimes prominent in the fall when the orange fruiting bodies discolor grass. Fertilizing with nitrogen usually controls rust. Smut occurs in the late spring or summer as a grass disease that produces a line of black greasy spores along the leaf blade.

Soils contain many saprophytic fungi, which live on dead leaves and other soil organic matter. Some of these are **facultative** parasites that cause turfgrass diseases if predisposing factors are present. These diseases are often most common on the best cared for lawns. For example, a *Pythium* water mold destroys turf when fertility, moisture, and temperature are all at high levels. Excess water soaks the soil, causing stress because of poor soil aeration. Add the stress of high temperatures favorable for the growth of *Pythium,* and the pathogen rapidly kills the nitrogen-rich grass.

Among diseases that commonly appear on the ordinary lawn are *Fusarium* and *Rhizoctonia* brown patch, both of which form rings of dying turf in the summer. These diseases are predisposed by heat stress on the grass. When snow covers turf for a long time, the melting snow reveals patches of dead grass killed by winter-active organisms, such as *Fusarium* and *Typhula* species. Cold and darkness encourage these organisms to cause turf disease. Long periods of wet overcast weather in the spring or fall favor growth of the *Helminthosporium* melting out pathogen. Growth of *Sclerotinia,* or dollar spot, is favored by nitrogen-deficient grass. Organisms causing all of these diseases are generally present but do not infect the grass until a particular factor or combination of factors develops, such as dark overcast days, winter snow cover, high summer temperatures, or excessively wet or compacted soils. The environment must be favorable.

When fungal lawn diseases are a problem, observation shows that the grass areas most affected are those predisposed to fungal attack by factors such as compacted soil, reflected heat onto a lawn by buildings or fences, or excessively short mowing. In the southern states of the United States, hot spots are not a problem for the subtropical grasses; instead, shaded or wet spots lead to diseased turf.

Several cultural management strategies may reduce disease incidence. However, considerable variation exists in the effect of cultural practices on disease severity. For example, increased nitrogen may reduce the severity of dollar spot but increase the severity of another pathogen. Some cultural practices are almost universally beneficial. Attempt to reduce the amount of time that water is on the plant by watering in the early morning hours, not at night. Superintendents may syringe or drag turf to remove dew and guttation fluids. If practical, the tree canopy in densely shaded areas should be pruned to open them up. Varieties of turfgrass with proven resistance to a particular pathogen can be utilized in areas with frequent occurrence of a particular disease. Finally, maintaining a healthy stand of turf by properly mowing, fertilizing, and so on, can make the turf more able to withstand disease.

SECONDARY CULTURAL PRACTICES

Determining Secondary Cultural Practice Requirements

Secondary cultural practices include cultivation, thatch control, and top-dressing. Cultivation and thatch control should not be conducted on a routine basis but only as needed to improve turfgrass stand performance by correcting soil compaction and excess thatch accumulation. The cause of decreased stand performance or stand failure should be determined. Many times coring and aerification are conducted when not warranted or necessary. Cultivation should be conducted if the reduction in turfgrass quality is caused by excess thatch or soil compaction, and if the turfgrass stand is otherwise in reasonable condition, with minimal weed invasion, and no major soil problems are present.

Cultivation

Cultivation is performed to improve root zone conditions, primarily soil compaction, and to reduce excess thatch. Cultivation is also used to promote dense shoot growth. **Compaction** is the pressing together of soil particles into a more dense mass. Complications that arise when turfgrass is grown on compacted soils include reduced water flow, infiltration, and soil aeration. Root growth is restricted due to mechanical resistance. Compaction may be avoided by managing traffic patterns to avoid heavy wear patterns, alternating mowing patterns to avoid ruts, using equipment that is lightweight or with a larger tire width, and restricting traffic when the soil is wet.

Thatch is a layer of undecomposed organic matter located between the soil surface and the zone of green vegetation. A moderate layer of thatch may be advantageous (Table 24–8), but excessive thatch can cause problems. Accumulation of excess thatch occurs when organic matter production exceeds the rate of decomposition: when high amounts of nitrogen are applied, the soil pH is unfavorably low for microbial activity, or if aggressively growing species or cultivars are utilized.

Return of clippings does not contribute to thatch accumulation. Excessive thatch can be prevented by using an appropriate fertilization program, liming if necessary to maintain the soil pH neutral to favor soil microbes, or using species and varieties that are suitable and intended for site and use. Proper mowing practices and avoiding excessive use of pesticides also help control thatch accumulation.

Cultivation can improve growing conditions when turfgrass is grown on subsoil, which occurs when a residential contractor removes and sells the topsoil and uses subsoil excavated from the basement to replace the topsoil. Turf grown on subsoil grows poorly and is shallow rooted, resulting in

TABLE 24–8	Moderate Layer of Thatch Versus an Excessive Layer
Advantages of Maintaining a Moderate Layer of Thatch	**Complications from an Excessive Layer of Thatch**
• More resiliency on sports turfs. • Improved wear tolerance. • Plants that are rooted in the soil may be insulated from temperature extremes. • Thatch may provide a barrier against weeds.	• Thatch does not retain water and nutrients as well as soil. • Thatch is not wetted by upward movement of water from the soil. • Plants rooted in thatch are more subject to drought and nutrient deficiencies as well as extremes of heat and cold. • Scalping from mowers. • Restricted water and pesticide movement into the soil. • Increased potential for disease and insect problems.

Source: David Gardner, The Ohio State University.

frequent drought stress even if the turfgrass is well irrigated. Turf grown on subsoil tends to develop a thick thatch layer. Frequent core cultivation, with return of the cores, may improve the growing medium over time.

Alleviating both compaction and excess thatch is accomplished using core aerification. With core aerification, small soil cores (approximately 0.25 to 0.75 in. diameter) are extracted. The name of the technique implies that aeration is directly improved. This assumption may or may not be accurate because the soil below and between the hole may actually become more compacted. The extracted cores are then broken up by weather or drag mat. The soil that is released incorporates into the thatch layer, causing an indirect improvement in aerification, higher cation exchange capacity, nutrient and water retention, and accelerated decomposition of organic matter (reduction in thatch). The vertical length of the cores varies by machine used and soil strength. Soil moisture is critical when conducting cultivation. Excessively dry soil reduces the depth of penetration of the coring tine. If the soil is excessively wet, soil tilth is destroyed by the cultivation equipment, resulting in sealed soil surfaces that are less permeable to gas exchange.

Several types of machines are available for core aerification. Vertical motion units are slower machines that cause relatively minimal disruption. They are used on putting greens. Circular motion units are faster machines that cause moderate to severe disruption of the surface. They produce shallower cores and are used on athletic fields and home lawns. Drum types are fast and inexpensive, but they are not as effective as the other types. Finally, hydroject types inject water at high pressure. They are used on putting greens to disrupt layers caused by top-dressing. They cause minimal disruption but are actually a type of shatterhole cultivator.

Other methods of cultivation are available, but they do not reduce compaction or excess thatch. Shatterhole aeration involves driving a series of solid tines into the soil to create a shattering effect. Fracturing the soil improves water and air movement but may increase compaction along the sides and bottom of the hole. Slicing is less intense and causes less soil disruption. The turf is penetrated by a series of V-shaped knives mounted on discs. This method cuts stolons and improves tillering. Spiking is not used to aerify but to slice

stolons and promote dense shoot growth. If hollow tines are used, the cores can either be returned or removed. Removal is common on golf course greens where returning cores results in substantial disruption of the playing surface or where sand top-dressing is practiced to keep the sand layer pure. Returning the cores is best, however, because they help to reduce thatch and soil layering.

Cultivation should be conducted during times of active turf growth. This approach allows the plant to respond to increased soil aeration and lack of soil resistance with increased root and shoot growth. Be aware, though, that desiccation damage can result from cultivation in summer because the roots are exposed to high evapotranspiration conditions around the coring holes. Desiccation damage can also result from late-fall cultivation because the coring holes may not close by winter. Desiccation also occurs under windy, low humidity conditions.

Mechanical Thatch Removal

Mechanical thatch removal may be used when a thick layer of thatch exists or if cultivation practices are inadequate. Mechanical thatch removal should not be considered a substitute for core aerification because it does not contribute to thatch decomposition. Current research suggests that mechanical thatch removal is not effective.

Machines used to remove thatch include the verticutter, which has vertically mounted blades that slice into the thatch layer. Solid blades are best, but they damage irrigation lines. However, spring verticutters are not very effective. Power rakes are used instead for areas that may have solid objects buried underground.

When removing thatch, it is best to avoid times favorable for weed seed germination. Dethatching should also be avoided if turfgrass is stressed or not well-established. Excessive layers of thatch and/or shallow-rooted turf should be treated with multiple low-impact dethatching operations, thus allowing full recovery of the turf in between, rather than attempting a single, severe dethatching operation. Thatch removal causes injury to turfgrass; always follow dethatching with proper cultural practices that favor recovery of the grass.

Top-Dressing

Top-dressing refers to the application of a fine layer of soil over an existing turf surface. It is used to smooth and level playing surfaces such as putting greens. Top-dressing helps to control thatch problems by incorporating material into the thatch layer, thus improving the environment for soil microorganisms. Top-dressing can promote recovery of the turf from injury. Top-dressing can be used to alter the turfgrass growing medium (e.g., change a fine-textured, compacted soil to sand-based). Top-dressing can be used during establishment by sod or stolonizing to level uneven areas and make the surface more uniform. It may help protect soil from winter desiccation in regions with mild winters.

Top-dressing may contribute to problems caused when soil layers possess different physical characteristics. Core aerification or shatterhole aeration is usually performed in conjunction with top-dressing to alleviate this problem. Top-dressing with nonsoil materials such as crumb rubber can be effective in softening the soil surface and reducing traffic stress. Once a top-dressing program is initiated, it must be repeated at intervals that match the growth rate of the turf. Selection of the proper top-dressing material is critical to success. The material should be as close in characteristics to the underlying soil as possible. The top-dressing texture should be the same or coarser than the underlying soil.

The top-dressing rate must match the rate of turfgrass growth. If it is applied too slowly, alternating layers of thatch and soil result. If it is applied too quickly, the existing thatch can be buried and cause a barrier to root growth. There is debate about the merits of pure sand top-dressing. It is less expensive and much easier to use than a mix with exacting standards. Research suggests, however, that hydrophobic dry spots, nutrition problems, reduced water-holding capacity, and winter desiccation may result in the long-term use of sand.

Localized Dry Spot

Localized dry spots are small patches of turf where the thatch or soil turn hydrophobic. The most frequent occurrence is on sandy soils where bentgrass is grown, but it can also occur on fine-textured soils with other turfgrasses. The cause of localized dry spot is not entirely understood. One theory is that microorganisms secrete complex polysaccarhides that coat the sand particles, making them hydrophobic. Symptoms of dry spot include areas of localized drought stress that watering does not eliminate. Water tends to runoff rather than infiltrate the affected area. Strategies for dry-spot control include cultivation and the use of commercial wetting agents.

Miscellaneous Management Practices

When frost heaving has raised grass crowns out of the ground in cold-winter areas, they can be pressed back into the soil by rolling.

Dyes are available for coloring brown or off-colored areas of turf. Such cosmetic practices are considered important

when visible turf is part of television presentations, movies, or public spectacles.

Plugs removed from sod can be planted into another turf to introduce a new grass species. Plugs are also used to repair small damaged areas. Devices are available to cut and remove plugs of various sizes and to cut similar holes in the soil to receive the plugs.

Turf growth is often retarded by competition with tree roots. Judicious pruning of the tree's roots often greatly improves a turf.

Salt used to remove snow and ice damages turf bordering sidewalks and roads. The excess salt causes injury to turf and often death. The treatment for this problem is to apply lime on acid soils and gypsum on alkaline soils so that calcium replaces the sodium used in the snow-melting mixture.

Golf, tennis, and bowling greens are precise playing surfaces with grass under severe stress. Special management practices are used to maintain optimum playability of these critical surfaces. These practices are discussed in detail in some of the cited references.

Seasonal Growth

The seasonality of grass growth is illustrated in Figure 24–6. The dotted curve in Figure 24–6 can be applied either to subtropical grasses growing in the southern United States or to temperate grasses growing in northern states or at high elevations. Low temperatures limit winter growth. As the season warms in the spring, growth increases, peaking in the summer months, if drought is not limiting. After late summer, growth declines until it is again stopped by low winter temperatures. Growth of temperate grasses over most of the urban United States is illustrated by the solid line in Figure 24–6. Depending on winter temperatures, growth can either cease or slow

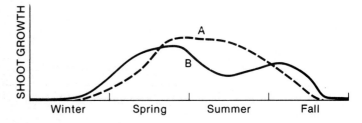

Figure 24–6 Seasonal growth of grasses. The solid line represents a seasonal growth pattern for Kentucky bluegrass, a typical temperate grass. The start of spring growth occurs earlier or later depending on the local climate. Summer dormancy is more or less severe depending on the degree of heat and water stress. The dotted curve represents growth of the subtropical grasses such as Bermuda grass. Most of these make their greatest growth during the hottest weather, provided water is adequate. In the southernmost areas of the United States, growth continues through the winter at a low rate.

down in the transition zone. The solid curve is similar to the dotted curve except that during the summer months, temperature optima are exceeded and growth slows. If high temperatures are accompanied by drought, grass can become semidormant, with little growth in summer. As temperatures cool in late summer, active growth resumes. Fall growth of Kentucky bluegrass differs from spring growth. Fall growth spreads and does not produce as many clippings. Growth comprises mostly new tillers and emerging rhizomes. The spreading form of growth decreases with dropping fall temperatures.

An awareness of the annual growth cycle in local regions can help in planning the best time for various operations. Coring, raking, and thatching are operations that should be done when growth is vigorous so recovery will be rapid. Herbicides are best applied during the rising growth curve in spring, when injury to the grass is low because of cool temperatures, and when rapid growth leads to rapid recovery.

Fertilizing Kentucky bluegrass during the rising growth curve in spring leads to excessive leaf growth. The same fertilizer applied during the rise of the growth curve in late summer results in more tillers and a more spreading growth. New leaves promoted by the fertilizer photosynthesize extra carbohydrate for winter storage.

The falling curve in summer represents a period of stress, and is an appropriate time to apply preventive fungicides—at the time predisposing factors develop. Reduced summer growth results from climatic stress, so the manager should be careful not to add extra stress at this time. One would not reduce mowing height nor increase mowing frequency at this time; instead, mow higher and less often.

It is best not to apply nitrogen fertilizer during summer stress. Heat stress results in part from the depletion of carbohydrates through higher respiration rates accompanying higher temperatures. Mowing also reduces carbohydrate reserves by reducing the leaf surface available for photosynthesis. Nitrogen fertilizer also depletes carbohydrate reserves, diverting them to protein synthesis. Nitrogen fertilizer applied during the summer puts an extra demand on the plant that can weaken its resistance to disease. Thus, during summer, it is best to withhold fertilizer or, if necessary, apply it only in small amounts of 0.125 kg/acre (0.25 lb/1,000 ft^2) or in forms that are slowly available to the plant.

When the growth curve rises in the fall, plants can take extra stress. The low spreading growth pattern and the favorable weather enable the turf to recover from all sorts of abuse.

SUMMARY AND REVIEW

Lawn, turf, and turfgrass in modern use refers to the ground where grass is growing, grass that is mowed and cared for, and grass growing in a horticultural turf, respectively. Determinate, rhizome, stolon, and intercalary meristem refer to the type of growth associated with turfgrasses. Subtropical (warm-season) grasses do well in warm areas of the United States but go dormant in the winter in transition zones. Temperate (cool-season) grasses grow well in the cooler parts of the United States but go dormant in the heat of summer in transition zones. Overseeding refers to the practice of sowing an annual temperate grass on top of a warm-season grass in warm-season transition zones. Sodding is the installation of pregrown mats of turf to areas for a quick covering of grass.

Establishing turf first requires proper preparation of the land. Selection of the appropriate turfgrass species or cultivar(s) is critical. Correct seeding rates and time of seeding determine how well the turf will establish and cover the area.

Maintenance is a demanding part of turf management because both the condition and appearance of the grass must be excellent at all times during use. Mowing must be done with the proper equipment at the appropriate intervals and at the proper height. Irrigation and fertilization programs are determined by the type of grass being grown, the environment, and other cultural practices in use. Weed, insect, and disease control is also a management consideration. Perennial grasses are the most difficult weeds to control in turf. Grubs and caterpillars are the most difficult pests. Fungal diseases are the biggest problem for turf. Wet conditions favor most of the fungal pathogens.

The bentgrasses, fescue grasses, Kentucky bluegrass, and perennial ryegrass are temperate grasses. Bermuda grass, centipede grass, Bahia grass, Saint Augustine grass, and zoysia grass are subtropical grasses. Annual ryegrass is a temperate grass used only for overseeding in the winter.

FOOD FOR THOUGHT

1. Why is zoysia grass not as common in Florida lawns as Saint Augustine grass?

2. What is the transition zone? Approximately where is it located? Why is this area a problem for turfgrass managers? What is a common species used in this area?

3. Use Table 24–4 to choose the method of starting turfgrass that you prefer. Explain why you have this preference.

4. When is the best time of year to establish grass from seed? Give three reasons.

5. True or false: to improve the drainage of a clay soil, it is best to add about 1 in. of sand and till to a depth of 4 in. Briefly explain your answer.

6. Why is mowing a stress to the turf?

7. When is the best time of year to control perennial broad-leaf weeds? Why? List two reasons to explain your answer.

8. Why is it more important to know the identification and life cycle of an insect pest today than it was during the 1960s and 1970s?

SUPPLEMENTARY READING

CHRISTIANS, N. E. 2004. *Fundamentals of turfgrass management,* Second Edition. Hoboken, N.J.: John Wiley and Sons.

DANNEBERGER, T. K. 1993. *Turfgrass ecology and management.* Cleveland, Ohio: Franzak and Foster.

EMMONS, R. D. 2000. *Turfgrass science and management.* Third Edition. Albany, N.Y.: Delmar.

TURGEON, A. J. 2002. *Turfgrass management,* Sixth Edition. Upper Saddle River, N.J.: Prentice Hall.

C H A P T E R

25

Residential and Public Landscapes

Martin Quigley

KEY LEARNING CONCEPTS

After reading this chapter, you should be able to:

- Understand the history and principles of residential and public landscaping.
- Explain the many functions that plants serve in landscapes.

The gardens and landscape ideals of Western culture began with the oases of Babylon, Egypt, and the succeeding civilizations of the Mediterranean region. At the beginning of gardening history, there was no distinction between agriculture and horticulture. Early Middle Eastern gardens were filled only with useful plants, which provided fruit, fiber, medicinal products, and forage for domestic animals. Such gardens were often walled and organized around a well, pool, or fountain. By the Roman era, formal gardens of the very wealthy were deliberately ornamental rather than productive. They were organized around a central water feature with topiary shrubs, groves and bosques planted for shade, and container plantings of colorful shrubs and herbaceous plants as well as herbs and fruiting plants. The spread of Islam in the seventh century brought this kind of garden to southwestern Europe, northern Africa, India, and the Far East. Examples of this kind of oasis garden still exist at the Alhambra in Granada, Spain (Fig. 25–1) and at the Taj Mahal in India. The Spanish brought this garden motif to the New World, and many features of the courtyard and patio gardens of the American Southwest are derived from this tradition.

During the Dark Ages, the Roman gardening style barely survived in the herbal knot gardens and utilitarian vegetable plots of medieval cloisters. Then, at the beginning of the cultural Renaissance in Italy, formal gardening once again blossomed, but plantings were subordinated to terracing, walls, foun-

Figure 25–1   A garden in the Alhambra at Granada, Spain, created before the time of Columbus and still maintained today as a tourist attraction. The fountains are supplied with water from the surrounding mountains to create a feeling of coolness. High walls surround the gardens, which contain small fruit trees, closely clipped shrubs, and plants grown in containers. An Islamic garden such as this is always symmetrical; that is, the left side is a mirror image of the right.

tains, and other architectural features. Geometric landscape designs defined by sheared shrubs and trees soon became the standard of the noble estates, spreading north to Austria, France, and eventually to England. These gardens of the nobility were ostentatious and demonstrated an intense level of maintenance with vast layouts of clipped hedges, gravel walks edged by carpets of bedding plants, and containers of tender trees, such as citrus, that required overwintering in glasshouses or orangeries. The gardens at Versailles are surviving examples of such gardens.

The English nobility soon tired of the rigid, geometric gardens of the rationalist Continentals. By the end of the eighteenth century, under the influence of landscape designers Humphrey Repton and Lancelot "Capability" Brown, the formal parterres and hedges of English estates had been largely replaced with so-called romantic landscapes. These included sweeping meadows and lawns with ponds, streams and grottoes, and naturalistic copses and groves that framed views of ruins and follies built specially for the purpose. Meanwhile, the cottagers of rural England grew vegetables, fruiting trees, shrubs and vines, and annual flowers in great profusion in very small garden areas. It was a hybrid landscape ideal of estate lawns (really pastures for grazing ani-

mals) and open woodlands with very dense, colorful, and productive cottage gardens, which the English settlers brought with them to America and tried to re-create while clearing the native forests of the Atlantic seaboard.

East Asian gardening and landscaping, exemplified in China and Japan, began many centuries before European history and, unlike Western horticulture, was an integral part of the religious and cultural tradition (Fig. 25–2). Completely separate from any production of food or other plant product, Oriental gardens are symbolic miniatures in which forests, rivers, and mountains are represented in formulaic and highly maintained landscapes. In contrast to the public and open nature of Western gardening, the Asian garden tradition has been one of privacy, contemplation, and ceremonial isolation. These elements, too, have been imported to North America, particularly along the West Coast.

Landscaping has many levels of meaning in the United States, from vegetable and flower gardening on the residential scale, to institutional landscapes such as campuses and large public parks in cities, to natural and recreational areas. Landscape architects, horticulturists and garden designers, and private homeowners all contribute to the human-constructed landscape (Figs. 25–3, 25–4, and 25–5).

Figure 25–2   A garden in Kyoto, Japan, using pines and small shrubs that must be pruned often to maintain the same size. Natural rocks and trees play an important part in Japanese gardens. *Source:* Hazel Hartman.

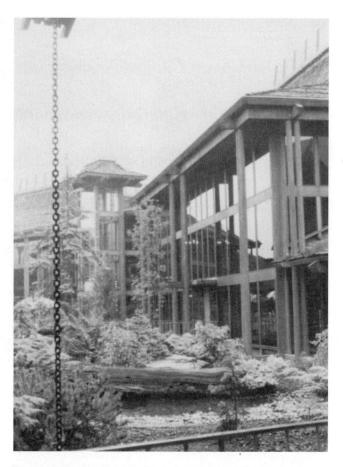

Figure 25–4   A naturalistic landscape is best when composed of both evergreen and deciduous trees and shrubs and ground covers, with central attention to layering of plants in appropriate combinations. *Source:* From Derek Fell, 1998. *Shade gardening.* New York: Friedman Fairfax.

Figure 25–3   It is dangerous to assume that mature trees can be saved when new construction is planned very close to trunks. Any excavation within the drip line, along with compaction and alteration of soil drainage patterns, usually causes the decline or death of existing trees.

THE RESIDENTIAL LANDSCAPE

For better or worse, the lawn is the dominant element of almost every American residential landscape, whether in San Diego, Boston, Grand Forks, or Miami. Although the seeded turf lawn is the least-expensive landscape element to install, it is by far the most expensive and time-consuming to maintain. However, even homeowners who do not set foot outside their house except to mow the lawn generally do not consider any other ground cover suitable. In fact, most municipalities dictate that single-family residences must have lawns maintained to strict specifications of height and weed-free condition. With proper plant selection, it is possible to have seasonal color, changing throughout the year, even in the harshest climate zones. However, given the rather low expectations of most American homeowners—which consist of evergreen foundation shrubs, expansive lawn, and an occasional shade tree—the residential landscapes become maintenance problems or do not often reach their full potential (Fig. 25–6). Ironically, a majority of Americans, when polled, list gardening as their major hobby or pastime.

Edible landscape plants (berry bushes, fruit trees, and annual herbs and vegetables) can be planted in most climates, and organic landscape practices (such as mulching with

Figure 25–6 *Above:* A cluster of recently planted *Thuja orientalis* in a bank parking lot. These are about 2 ft (60 cm) tall but will grow rapidly if given frequent irrigations. *Below:* This, too, is a *Thuja orientalis,* about 30 years after planting, overwhelming the house. Once it starts growing, the home-owner usually does not know how to prune it aesthetically, so it is allowed to take over. Planning is important in knowing what to plant and how to control it.

Figure 25–5 *Top:* A typical, well-maintained highway rest stop providing picnic facilities and pleasant surroundings for automobile travelers. The trees, from left to right, are pines, birches, and purple-leaved plums. *Middle:* One tiny corner of Golden Gate Park in San Francisco, a huge park originally designed in 1877 by John McLaren to keep a three-mile stretch of beach sand from shifting. The park has landscaped facilities for everyone, from playgrounds for children to polo grounds for adults. *Bottom:* The University of California, Davis is a campus of bicycles. Because Davis has a flat terrain, the landscape architect created visual interest by designing two hills and a bikeway with a bridge leading to the library. This effective landscape design was created by first excavating a bike path and then using the soil to make two mounds.

nutrient-rich compost) can benefit the soils and plant growth in all gardens. Flowers can be grown along with food-producing plants, and herbaceous annuals can be set in mixed plantings with trees and shrubs depending on sun and shade requirements. Too often, Americans' gardening practices reflect their agricultural heritage: single-species planting in straight narrow rows maintained by high chemical and mechanical input.

The residential landscape should be considered an extension of the interior living space—an idea exemplified by the work of California landscape architect Thomas Church and well described in his influential book, *Gardens Are for People.* Privacy and definition of property lines, or different areas of activity, can be defined either architecturally with walls, fences, or trellises, or with plant materials—shrubs, trees, or vines (Fig. 25–7). Evergreens can define views and block sound. Many deciduous trees can provide spring flowers, summer shade, fall color, and winter openness so that sunlight reaches windows and living areas only when desired (see Fig. 25–19).

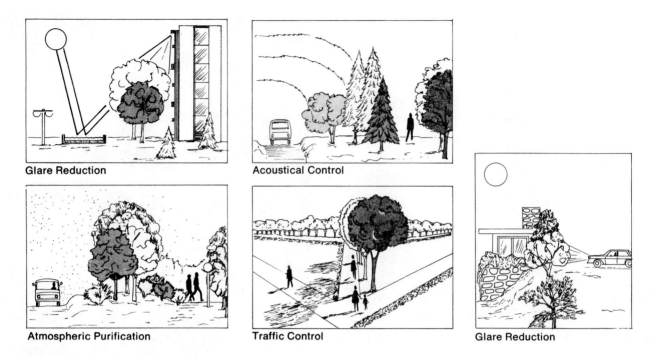

Figure 25–7 Plants can be strategically placed in the landscape to reduce glare, to purify the atmosphere, to control noise, or to direct pedestrian traffic. *Source:* Adapted from Robinette, G. O. 1972. *Plants, people, and environmental quality.* Washington D.C.: U.S. Government Printing Office.

The home landscape should be appropriate to its surroundings. A prairie planting may be out of place in a neighborhood of highly manicured lawns, just as a lawn may be out of context in a naturally wooded area. Most of all, maintenance requirements should be considered in planning: the regular use of pesticides, herbicides, and inorganic fertilizers has potentially negative effects on water quality, wildlife, and human health. The use of native plant materials has become very popular, but when choosing such plants one must remember that they are native to a set of ecological conditions—rainfall pattern, soil type, sun or shade tolerance, and temperature regime—and not a particular state (e.g., Iowa) or other human-drawn boundary. Thus, for example, the popular service berry (*Amelanchier* sp.) is native to forest understory, not to open lawns or parking lot islands, where the delicate flowers are quickly burned by sun and heat. The residential landscape should also reflect the architecture of the house itself: a formal façade looks best with a formal garden or lawn and symmetrical tree plantings, while a cottage-style house calls for a bright, informal mix of perennials and shrubs. Mountain or woodland houses call for a naturalistic garden in which the boundaries are not obvious. Even in the smallest outdoor space, containers can function as a small garden.

Planning the Landscape

In planning the landscape, both the indoors and outdoors should be considered as one unit. This approach provides for greater usefulness of the total space. The private house and garden has four functional areas: (1) the public area or entry,

(2) the public living area, (3) the private living area, and (4) the work or delivery area.

The public entry area includes the front yard or garden area, perhaps a driveway, a walkway to the porch, and the entry. This area may be landscaped to complement the style of the house or to conform to neighborhood landscaping patterns. Shrubs selected for different heights may offer some privacy but are usually used as foundation plantings to allow the house to blend into the landscape.

The living area of a house includes the living, dining, and family rooms and the study. One of these rooms may extend out into the garden by way of a patio or deck. The patio may be used for year-round outdoor living in mild climates or only during the summer months in cold-winter areas. Some shade trees or a patio cover can be provided to make the area more comfortable. A surrounding lawn area reduces both heat and glare. Spring bulbs, flowering shrubs, or flowering annuals can provide the garden, patio, and living room with color.

The work area consists of the workshop, laundry, garage, and storage. Shrubs can hide unsightly storage sheds and work areas.

Functions of Plants in the Landscape

Plants are used in the landscape for many functional purposes: to reduce glare, to aid in air purification, to settle dust, to diminish noise by absorbing sound, to direct pedestrian traffic, and to screen certain areas from public view and provide privacy (Figs. 25–7 and 25–8). Plants must reach their mature effective size before they become fully functional.

Plants can partially moderate climate by reducing radiation frosts (Fig. 25–9) or acting as wind or snow barriers (Fig. 25–10). In hot, dry, clear climates, radiation from the sun makes for hot days and radiation to the clear sky during the night results in cold nights. Plants, particularly shade trees, can be used as radiation barriers to modify such patterns, shading the soil during the day and preventing excess radiation at night (Fig. 25–11). Shade trees lead to a more even soil temperature as compared to the bare soil. The soil thermometers in Fig. 25–11 indicate an equable temperature under the trees. Soil under tree cover has a more even temperature from

Figure 25–8 Plants can solve certain environmental problems such as screening an undesirable view, directing a person's attention to interesting views, or providing privacy. *Source:* Adapted from Robinette, G. O. 1972. *Plants, people, and environmental quality.* Washington D.C.: U.S. Government Printing Office.

Figure 25–9 Plants grown under trees are protected from radiation frosts. Plants are likewise protected from frosts under eaves or patio covers. *Source:* Adapted from Robinette, G. O. 1972. *Plants, people, and environmental quality.* Washington, D.C.: U.S. Government Printing Office.

Figure 25–10 Evergreen conifers act as snow barriers if planted properly in relation to the prevailing winds. *Source:* Adapted from Robinette, G. O. 1972. *Plants, people, and environmental quality.* Washington, D.C.: U.S. Government Printing Office.

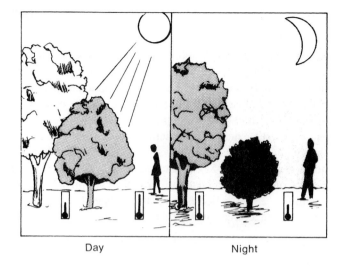

Day   Night

Figure 25–11 Trees and shrubs moderate wide diurnal variation in ground temperatures. They reduce radiation from the sun and prevent excess radiation from the soil to the sky on a clear, cloudless night. *Source:* Adapted from Robinette, G. O. 1972. *Plants, people, and environmental quality.* Washington, D.C.: U.S. Government Printing Office.

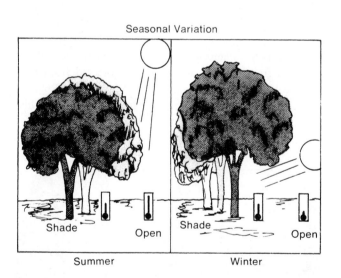

Seasonal Variation

Shade   Open   Shade   Open

Summer   Winter

Figure 25–12 Broad-leaf evergreen trees and shrubs reduce wide seasonal temperature variations by modifying radiation at the soil surface. *Source:* Adapted from Robinette, G. O. 1972. *Plants, people, and environmental quality.* Washington, D.C.: U.S. Government Printing Office.

Figure 25–13 An evergreen tree or shrub planted near a wall facing south or west creates a dead air space that becomes insulation to maintain even summer temperatures. *Source:* Adapted from Robinette, G. O. 1972. *Plants, people, and environmental quality.* Washington, D.C.: U.S. Government Printing Office.

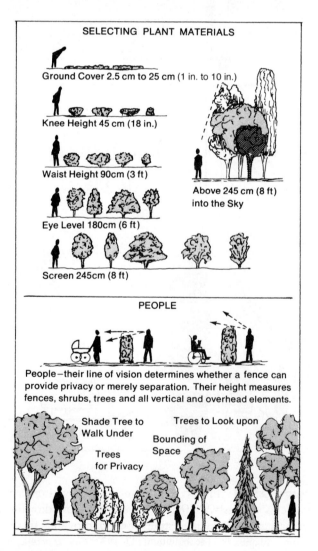

Figure 25–14 Plants of all shapes and sizes can be chosen to create effects or fulfill functions to beautify the environment. *Source:* Adapted from Robinette, G. O. 1972. *Plants, people, and environmental quality.* Washington, D.C.: U.S. Government Printing Office.

season to season than does bare soil exposed to the sun (Fig. 25–12). A concrete wall shaded by ivy or other climbing plants has a much lower temperature than a bare wall open to the full sun. A group of closely planted evergreen trees next to a wall can create a dead air space (Fig. 25–13) much like the dead air space in the walls of the house. This tends to maintain an even temperature between the plants and the wall.

Plant materials that can be used for these various functions are available in many forms, shapes, and sizes. A few are depicted in Figure 25–14. Landscape architects must be aware of the hardiness of plants that can be used functionally in their plans. In addition, they must supply specifications for soil mixes, drainage requirements, and maintenance care. The ultimate size of these plants must be known if they are to be used to their best advantage. Figure 25–15 shows the growth habits of some specimen plants often used in the landscape. These tree and shrub characteristics must be known to choose the most desirable plant for the plan.

*Arrangement and proportioning of openings.

Figure 25–15   Groups of trees or shrubs can produce desired effects if they are chosen wisely. These groups are based on trees of various shapes. *Source:* Adapted from Robinette, G. O. 1972. *Plants, people, and environmental quality.* Washington, D.C.: U.S. Government Printing Office.

Plants are often used to screen undesirable activities or unsightly surroundings (Fig. 25–16). People are becoming more aware of their surroundings, and city, county, and state governments now request certain industries to screen ugly scenes from public view. In the 1960s, a highway beautification act was initiated in the United States to encourage attractive landscaping of federally funded highways.

Areas of high air pollution are usually plagued by unpleasant odors. Plants are useful in ridding the air of such odors (Fig. 25–17) by absorbing them and thus purifying the air. Plants also help remove dust from the air. Dust particles attach to the leaves, and the next rainfall washes the dust away (Fig. 25–18). Carbon dioxide (CO_2) is a waste gas exhaled by humans and animals but ab-

Figure 25–16 Examples of how trees and shrubs can be used to hide undesirable views. *Source:* Adapted from Robinette, G. O. 1972. *Plants, people, and environmental quality.* Washington, D.C.: U.S. Government Printing Office.

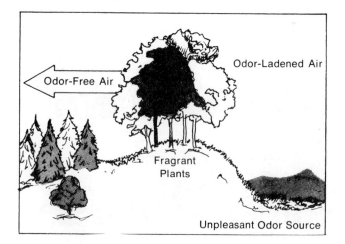

Figure 25–17 Plants can help reduce some unpleasant odors in the atmosphere. *Source:* Adapted from Robinette. G. O. 1972. *Plants, people, and environmental quality.* Washington, D.C.: U.S. Government Printing Office.

Figure 25–18 Plants together with rain can help clean the air. *Source:* Adapted from Robinette, G. O. 1972. *Plants, people, and environmental quality.* Washington D.C.: U.S. Government Printing Office.

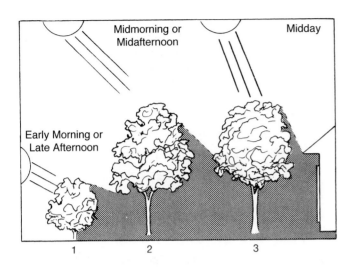

Figure 25–19 Trees and shrubs placed in the proper way can reduce the sun's radiation and glare throughout the day. *Source:* Adapted from Robinette, G. O. 1972. *Plants, people, and environmental quality.* Washington, D.C.: U.S. Government Printing Office.

sorbed by plants. The plants convert the CO_2 to sugars by photosynthesis and give off pure oxygen in the process. This exchange is part of the balancing cycle between plant and animal life.

Glare can be controlled by shading created by trees during different hours in the day (see Fig. 25–19). The mature plant size necessary to be effective can be calculated from the angle of the sun during the four seasons. Shade is

usually desired in summer but not in winter, making deciduous trees and vines most effective (Fig. 25–20). Species of trees vary in the shadow patterns they cast (Fig. 25–21). Density of shade can change with the leaf canopy and species. Plants also vary in the amount of solar radiation they reflect. In general, green plants reflect only about 10 to 25 percent of the radiation (Fig. 25–22), whereas a white building reflects about 90 percent and unpainted concrete about 35 percent.

Glare from street lamps may be blocked by planting trees, as illustrated in Figure 25–23. People who live at the end of a dead-end road or a cul-de-sac find hedges or tall shrubs a necessity to reduce glaring lights from oncoming cars.

Windbreaks of various sorts have been used for centuries to slow or divert strong winds (Fig. 25–24). Tall shrubs and hedges used as windbreaks reduce wind speed by increasing the resistance to the wind flow. Coniferous trees that have thick dense branches clear to the ground (Fig. 25–25) are the best all-year plants for windbreaks in close proximity to buildings.

Plant Selection

The landscape is complete only when the particular plant species have been chosen to fulfill the functions intended in the plan. But a knowledge of plants and their behavior is essential in making landscaping selections. Certain kinds of shrubs can be trained to become formal hedges because they become dense after pruning. Tree selection is based on the trees' ultimate size, shape, shade, and other desirable characteristics. Dense shade trees, like the maples or beeches, discourage or eliminate most plant

Figure 25–20 For maximum comfort, the design of the house and the placement of trees, shrubs, and vines must be considered. The plan shown here gives good protection from the sun in summer and allows penetration through the deciduous trees of the sun's rays in winter. *Source:* University of California Cooperative Extension.

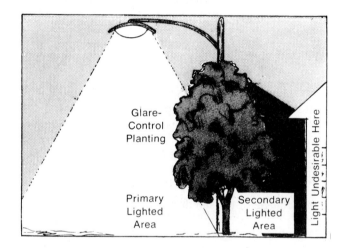

Figure 25–21 Each tree species has a unique shadow pattern. Trees also vary in the density of shade they give. Knowledge of these shade characteristics is valuable in developing a landscape. *Source:* Adapted from Robinette, G. O. 1972. *Plants, people, and environmental quality.* Washington, D.C.: U.S. Government Printing Office.

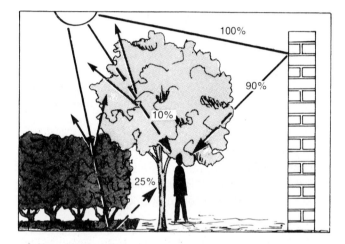

Figure 25–22 Plants can modify solar radiation by diffusion and shading. *Source:* Adapted from Robinette, G. O. 1972. *Plants, people, and environmental quality.* Washington, D.C.: U.S. Government Printing Office.

Figure 25–23 The glare from streetlights can be modified or completely controlled by planting trees in the proper location.

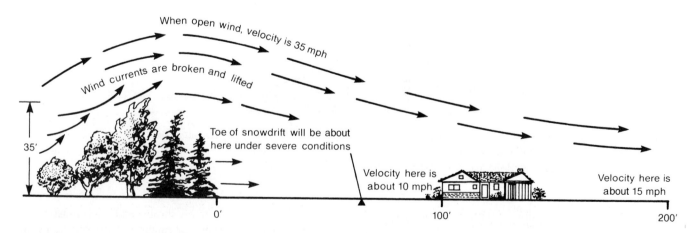

Figure 25–24 Trees and shrubs of varying heights can reduce the velocity of prevailing winds. *Source:* USDA.

534

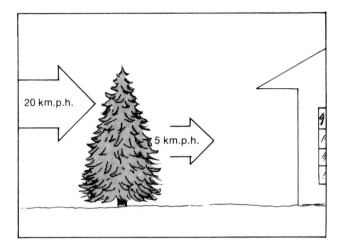

Figure 25–25 Coniferous evergreen trees or shrubs effectively cut wind velocities next to structures when planted a distance from the house that is at least twice the height of the barrier. *Source:* Adapted from Robinette, G. O. 1972. *Plants, people, and environmental quality.* Washington, D.C.: U.S. Government Printing Office.

species under them. However, many kinds of trees produce filtered shade so that certain shade-requiring shrubs or flowering annuals flourish beneath them. Plant selection, of course, must be based on the hardiness zone (Fig. 5–4), but the particular environmental conditions of the planting site must also be considered. Finally, choices need to be made based on the growth habits necessary to enhance the landscape plan. There are, of course, many species and horticultural cultivars that can be used in the various climatic regions. It is very important that a careful, detailed study be made before finally selecting the specific landscape plants to be used for the various aesthetic and functional purposes discussed in this chapter.

SUMMARY AND REVIEW

The history of western landscaping goes back to biblical times and the Hanging Gardens of Babylon. Early landscaping was gardening and farming, with no distinction made among plants grown for food, fiber, and animal feed. Ornamentation was not considered a reason for growing plants. In western culture, formal ornamental gardens began with the Roman Empire, but gardening reverted back to utilitarian purposes during the Dark Ages. With the Renaissance, formal gardening with elaborate ornamental features created with plants was revived and became the standard of European nobility. Meanwhile, in East Asian culture, ornamental gardening had been practiced for centuries as an integral part of religious and social culture. The East Asian garden was designed to promote a peaceful meditative state in those visiting the gardens. In the United States and many other developed countries, many styles of landscape gardens can be found. Landscapers have to understand the history of gardening and landscaping because a client will often request that his or her landscape reflect a historic garden or landscape style.

While aesthetically pleasing, landscape plants can also serve many functions. They can moderate the temperatures of buildings and the ground by providing shade. Annoying street or security light can be blocked by evergreen trees. In windy or noisy areas, they can serve as wind and noise barriers. Privacy can be created with hedges. Landscape plants can also reduce odors by filtering air.

FOOD FOR THOUGHT

1. Develop a list of at least five different uses for landscape plants. Then find an example of each use (intended or unintended) in local residential and public landscapes, including those found on your college or university campus.

2. Sketch a landscape design for a homeowner (real or fictitious). Describe the plants you would use, including their scientific and common names, and why you chose those plants.

3. Suppose you were given the job of designing a landscape that used plants to shade a west-facing window in the summer but allowed light to enter in the winter. Explain what plant or plants you would choose.

4. Describe briefly why it is important to consider the indoors and outdoors as one unit when planning a landscape.

5. Describe the similarities and differences between western and Asian types of gardens.

SUPPLEMENTARY READING

ADAMS, D. W. 2004. *Restoring American gardens.* Portland, Ore: Timber Press.

AUSTIN, R. L. 1984. *Designing the natural landscape.* New York: Van Nostrand Reinhold.

BECKETT, K. A., D. CARR, and D. STEVENS. 1982. *The contained garden.* London: Francis Lincoln.

CREASEY, R. 1982. *The complete book of edible landscaping.* San Francisco: The Sierra Club.

CROWE, S. 1959. *Garden design.* New York: Hearthside Press.

DANNENMAIER, M. 1998. *A child's garden: Enchanting outdoor spaces for children and parents.* New York: Simon and Schuster.

DICKASON, K. S., and C. C. BURRELL. 1996. *Garden styles.* Lincolnwood, Ill.: Publications International.

ECKBO, G. 1950. *Landscaping for living.* New York: F. W. Dodge.

MILLER, R. W. 1996. *Urban forestry: Planning and managing urban greenspaces.* Upper Saddle River, N.J.: Prentice Hall.

MORSE, H. K. 1962. *Gardening in the shade.* Portland, Ore.: Timber Press.

ROBINETTE, G. O. 1972. *Plants, people, and environmental quality.* Washington, D.C.: U.S. Government Printing Office.

ZION, R. L. 1995. *Trees for architecture and landscape.* New York: Van Nostrand Reinhold.

CHAPTER

26

Harvest, Post-Harvest Handling, Storage, Distribution, and Marketing

Michael Knee and Margaret McMahon

KEY LEARNING CONCEPTS

After reading this chapter, you should be able to:

- ♦ Understand the basic principles of harvesting major, field-grown crops.
- ♦ Understand how crop quality changes after harvest.
- ♦ Identify strategies to maintain quality after harvest.
- ♦ Appreciate how the production of crops is linked with consumption through marketing and transport.

People and their domestic animals use plants to obtain energy to live and enjoy life. Every part of the plant serves these needs—roots, stems, leaves, flowers, fruits, and seeds. We have developed efficient methods and machines for harvesting, storing, and preserving the various plant parts.

HARVESTING

The growth and development of agriculture has been greatly stimulated by the mechanization of many operations, from land preparation to processing various commodities. Agricultural practices constantly change to accommodate machines. Plant breeders develop plants that are better adapted to machine harvest operations. Since the beginning of organized agriculture, crops have been

Figure 26–1 Cotton harvester. *Source:* USDA-ARS Image Galley, *http://www.ars.usda.gov/is/graphics/photos/.*

selected for better quality as well as for easier production. A maximum effort in recent years has been directed to the development of labor-saving harvesting machines.

Mechanization of crop harvesting began with the cereal grains. The forerunner of the grain combine was the hand sickle and flail, used as far back as 5,000 years ago. The reaper, threshing machine, and combine followed. Other examples of significant machinery development were the cotton picker (Fig. 26–1) and cotton gin. Harvest mechanization, whether for grain (Fig. 26–2), sugar beets, cotton, peaches, or tomatoes (Fig. 26–3) has greatly reduced the level of human labor previously needed in the harvest of these crops.

Generally, harvest mechanization was developed for crops intended for processing long before those grown for fresh market purposes. For example, essentially all sour cherries and tomatoes for processing are machine-harvested. Fresh market tomatoes and other fresh market products are still harvested by hand to limit physical injury. With regard to processing tomatoes, plant breeders and engineers cooperatively identified and developed plant and machinery characteristics that would make a mechanized system of harvest feasible. This systems approach was applied to other commodities. On the other hand, it is difficult to adapt some crops, such as table grapes, strawberries, or oranges to mechanization, although breeders and engineers are collectively working toward this goal because continuing economic pressures encourage mechanization.

Functions of Harvesting Machinery and Equipment

The functions involved in the harvesting and handling operations of plant products are varied and highly dependent on the particular commodity. These functions include digging, cutting, lifting, separation, grading or sorting, cleaning, conveyance, loading, and so on. Such functions depend on various factors related to the crop, which could include field conditions at the time of harvest, crop maturity, crop use

(fresh or as processed product), plant part or portions involved, and crop worth.

Modern technology incorporates computers and electronic-sensing devices in harvesting equipment to improve harvest efficiency greatly. Global positioning systems (GPS) allow field equipment to be controlled precisely as it moves through a field. Yield sensors monitor the instantaneous yield and slow the harvester down if the yield is high enough to reduce loss due to grain being left in the field (Fig. 26–4). Conversely, if the yield is too low, the harvester speed is increased. Moisture sensors measure and record the moisture content of grain as it is harvested. Harvest can be stopped if the grain is too moist, thus preventing a loss of quality and value of the grain.

Mechanically Harvested Crops

Cotton

All commercial cotton grown in the United States is harvested by machine, and except in some underdeveloped nations, machines are used extensively. The harvester operates by guiding the plant into vertical rollers equipped with rotating spindles (tines) that, by a combining action, remove the fibers from the plant. Special counter rotating drums remove the fibers from the spindles. The fibers are then blown or conveyed into large basketlike trailers for transport to the gin or to storage sites.

A relatively recent practice is to provide for short-term inexpensive storage of harvested cotton until it can be ginned. This is done by compressing the cotton into a compact stack near the harvested field site. The stack is covered with canvas or plastic for moisture protection and then taken to the gin for cleaning.

To facilitate harvest, growers use **growth retardants** to arrest vegetative growth and chemically defoliate cotton plants prior to mechanical harvest to reduce the content of leaf trash that otherwise contaminates and makes cleaning the fibers more difficult. Materials such as ammonia (NH_3), sodium chlorate, ethylene, and various herbicides are used for defoliation.

Grain Crops

Cereal grains—barley, corn, oats, rice, rye, sorghum, milo, and wheat—are mechanically harvested. The principle of harvest machine operation is essentially similar for most grain crops, although slightly different for corn harvested for grain.

The self-propelled combine, cutting a wide swath, enables one operator to rapidly harvest large areas and volumes of grain (Fig. 26–2). The harvester cuts the grain heads from the plant. The heads pass through rotating cylinders that beat and shake the grain kernels from the head. The kernels are separated from other plant parts (hulls, chaff, straw) by sifting through sieves; the straw is conveyed through and out of the machine, and a blower removes the smaller lighter-than-the-grain material. The grain is conveyed to a holding bin

Figure 26–2   (A) Wheat combine. (B) Corn picker. (C) Barley harvester. *Source:* USDA-ARS Image Gallery, *http://www.ars.usda. gov/is/graphics/photos/.*

Figure 26–3   This tomato harvester is equipped with an electronic device that automatically sorts the tomato fruit by color, discarding the immature and green fruit while retaining those of acceptable red color and maturity.

mounted on the machine for later off-loading or is conveyed directly into trailers or trucks that move parallel with the harvester. The grain is then transported to its intended market or to storage. Grains are harvested as mature seed and, when possible, are allowed to dry on the plant to a low moisture content before harvest. A high moisture content can result in poor storage and spoilage. Therefore, the grain must be dried, sometimes artificially, after harvest and before storage.

Hay and Forage Crops

The harvest of hay and forage crops is highly mechanized. For most of the various grasses and legume crops grown for hay, the timing of the harvest is made to optimize both yield and quality. Either a tractor-mounted or pulled mower cuts the hay, depositing it back onto the field or into windrows in one operation. This is done to allow for drying and to facilitate later pickup. Another machine gathers up and compresses the hay into compact bales or rolls and binds them or wraps them in plastic. The bales may later be collected from the field, or equipment may do this operation at the time they are formed.

Silage is a fermented product of the hay and foliage crops. The crop plant harvested for silage is cut but not permitted to dry. Instead, it is placed into structures such as silos,

Figure 26–4 Yield sensors connected to onboard computers automatically adjust the speed of a harvester to increase harvest efficiency. *Source:* USDA-ARS Image Gallery, *http://www.ars.usda.gov/is/graphics/photos/.*

pits, or bunkers, where it is allowed to ferment by **anaerobic** microbial activity. The product, once fermented, is preserved for subsequent animal feeding. It is important that the crop material be well packed to exclude air and to favor microbial activity. In the United States, corn is by far the preferred crop plant for this purpose.

Root and Tuber Crops

To facilitate mechanical harvesting of many root and tuber crops, the above-ground plant parts are often removed prior to harvest of the below-ground portion. The foliage of some root crops (sugar beets) is used for animal feeding, and some (table beets and turnips) are used for human food. It is common for large plantings of potatoes, sugar beets, and carrots to be harvested and handled mechanically (Fig. 26–5). Several root crops, such as radish, table beets, and carrots, are not mechanically harvested when they are marketed with at-

tached tops. Modern root crop harvesters are multiple-row machines, either tractor-drawn or self-propelled. The principal function of the harvester is to pass a broad horizontal blade below the roots or tubers. That blade is angled so that in passing beneath the plant, it lifts and/or allows easy pulling of the crop from the soil and onto conveyors, sorting belts, and so on. On removal from the soil, the product experiences many of the operations common to most root crops, including cleaning; trimming; washing; and being graded, conveyed, cooled, packed, transported, stored, and/or processed (Fig. 26–6). The tops of these crops are generally removed mechanically or chemically before harvesting.

Fruit and Nut Crops

The variable nature of fruit crops presents a challenge to mechanical harvesting. Fruits vary in their characteristics and rate of maturity. Some fruit abscise when mature; others do not. Lack of uniform ripening makes a one-time harvest difficult. The delicate nature of many fruits requires careful handling, and most fruit for fresh market is hand-harvested. Frequently, mechanical aids to harvesting are used. Hydraulic platforms or booms replace ladders and enable workers to be lifted and easily moved about while harvesting. Conveyors are used to transfer the harvested product.

Fruits produced for processing are more likely candidates for mechanical harvesting because the product is processed soon after harvest. If damage occurred during harvesting, it would not be as evident in the processed product, whereas it might make the fresh product unacceptable. The harvest of the tree nut crops—walnuts, almonds, pecans, filberts, and pistachios—is fully mechanized. The nut crop can sustain considerable impact damage of the shell without significant loss of quality or value of the nut meat. These crops are harvested by being shaken from the tree by a machine that attaches to the tree trunk and vigorously shakes the tree. Canvas catching frames are used to catch and collect the nuts, or

Figure 26–5 The harvest of white potatoes is completely mechanized. Tubers are removed from the ground, separated from vines and soil, and then conveyed to trucks alongside the harvester. *Source:* University of California Cooperative Extension.

Figure 26–6 Sorting potato tubers according to size and examination for defects. The removal of defective tubers ensures good quality for packaging and marketing or for storage.

Figure 26–7 Harvesting sour cherries in Michigan for canning. Fruits are shaken from the tree onto catching frames, then conveyed into a bin. *Source:* USDA.

Figure 26–8 Mechanical sweet corn harvester. The ears are pulled from the stalk as the stalk moves the harvester.

they may fall directly on the ground. The nuts are collected by equipment that sweeps and/or vacuums the nuts from the ground into bulk containers. Similar types of shaker machines are used for the harvest of fruit crops such as cling peaches, prunes, and canning cherries (Fig. 26–7). The catching frames are abundantly padded to minimize physical damage. Harvest machines are used for bushberry, raisins, and much of the wine grape crop. The design and development of harvest machinery is a continuing process and it is likely that difficult-to-harvest crops, presently not mechanically harvested, will be harvested with machinery in the future.

Vegetable Crops

The mechanical harvest of tomatoes grown for processing received worldwide attention because it accomplished a seemingly impossible objective. It was admittedly a remarkable achievement, but one that was preceded by the mechanical harvest of other crops, such as spinach, peas, snap and lima beans, sweet corn, celery, cabbage, and onions for processing. For most vegetables, the principal functions of the harvester are to cut and collect the intended crop portion, and to sort, convey, and transport it. Like fruit crops, most of the vegetables for fresh market continue to be hand-harvested. This is continually changing, and harvest machinery for some difficult-to-harvest crops is under development. For example, sweet corn pickers are now available that pull the corn from the stalk (Fig. 26–8).

The tomato harvester is an excellent example of the mutual adaptation of the crop to the equipment. The processing tomato was bred to provide both a determinate and prolific fruiting habit, and thereby maximize yield and uniformity of maturity. The fruit size is small and thick-walled, with a thick skin; thus it is more tolerant to physical handling. The harvester passes a horizontal blade beneath the plant. The entire above-ground plant is conveyed onto chain belts that shake the tomatoes from the vines. The vines are

discharged back onto the ground, and the tomatoes are conveyed past human or electronic sorters before transfer into bulk trailers alongside the harvester (Fig. 26–3). Workers examine the tomatoes, removing those that are immature or otherwise unsuitable. Equipment using photocells rapidly distinguishes color differences and electronically removes immature (green) fruit from the mature (red) fruit.

POST-HARVEST HANDLING AND PRESERVATION

The ability to store food for later consumption was a critical necessity in the development of human societies. This ability enabled people to move beyond the limited area of the earth where food is continuously available. Life in North America and Europe would have been impossible without the ability to store food for the winter. Today, this food may be supplemented with crops from other countries. Transportation is another aspect of post-harvest handling of crops that has gained in importance as people moved from the country, where they produced their own food, to an urban existence, where food comes from the store. The urban lifestyle has led to demand for nonfood crops. Cut flowers, garden plants, and even turf require specific attention to post-harvest handling if they are to reach the consumer in good condition. Although we may often focus on the growing of crops to the point of harvest, the ability to get them from the farm gate to the consumer is equally critical and presents a new set of challenges. Crop losses after harvest remain unacceptably high, and preventing post-harvest deterioration would make a major contribution to improving the world's food supply.

Post-Harvest Deterioration

Crops that are harvested represent all parts of the plant at many different developmental stages (Table 26–1). Crops may be actively growing parts of the plant, such as asparagus

TABLE 26-1	Examples of the Plant Parts That Are Harvested and the Stage of Development at Which the Part Is Harvested for Several Crops

Plant Part	Stage of Development	Examples
Whole plant	Immature	Radish, carrot, bedding plants, turf
Whole plant	Mature	Ornamentals (trees, shrubs, pot plants)
Leaves	Immature	Brussel sprouts, lettuce
Leaves	Mature	Collards, spinach
Stems	Immature	Asparagus
Stems	Mature	Cut foliage ornamentals
Roots	Immature	Radish (includes hypocotyl)
Roots	Mature	Rutabaga, carrot
Flowers	Immature	Broccoli, cauliflower
Flowers	Mature	Cut flowers
Fruits	Immature	Green beans, okra
Fruits	Mature	Tomatoes, squashes, apples, etc.
Seeds	Immature	Green peas, sweet corn*
Seeds	Mature	Pulses, grains*

* Cereals are strictly fruits with a thin, dry ovary wall.
Source: Michael Knee.

stems or spinach leaves; they can be dormant structures with natural longevity, such as seeds, bulbs, or tubers; or they may be terminal structures, such as fruits and flowers that are naturally programmed to senesce and die. Deterioration can occur for many different reasons: physical, chemical, or biological. Biological deterioration may involve physiological processes in the crop itself or pests and pathogens that may be the same as those in the growing crop, but the post-harvest environment can be conducive to a whole new set of enemies.

Physical deterioration often begins at harvest, which for most crops involves separation of the part of the plant that will be marketed and consumed. Damage can occur even when the whole plant is harvested. Vegetables such as radishes and carrots may be uprooted, and ornamental plants may be dug from the field, resulting in breakage and loss of much of the root system. Container-grown plants may suffer the least damage through harvest, but there is still likely to be accidental breakage and loss of plant parts as containers are lifted from the growing area to begin their journey to the consumer. Many fruits and seeds are harvested by taking advantage of a natural separation between them and the parent plant. This natural separation occurs at abscission zones that develop in fruit pedicels, ovary walls, and placentas. The cell walls in these zones weaken as the structure matures, allowing for easy separation and a wound area that heals naturally. Abscission can be promoted with growth regulators to facilitate harvesting of fruits such as apples, tomatoes, and coffee. On the other hand, it may be desirable to retard abscission to avoid loss of crop. Many grain crops were se-

lected long ago for seed retention in the ear so that the whole stem could be harvested and the grain separated by threshing. Although dry seeds and grains are much more robust than most other crops, threshing is of necessity a violent process and causes damage. Other fruits and seeds are harvested in an immature state when the abscission zone is not fully developed and the supporting or enclosing structure must be cut or broken, which results in wound areas. These wounds also occur on ornamental and edible stems, foliage, and flowers that are often harvested by cutting the stem or petiole.

Physical damage at harvest time may be unavoidable, whereas avoidable damage tends to occur during subsequent handling and storage. Depending on the commodity and the handling procedure, damage can include abrasions, cuts, cracks, and bruises. These defects occur through contact between one item of produce and another, and between produce and the edges or surfaces of containers and handling equipment. The kinetic energy of a moving crop item tends to cause damage whenever it experiences sudden deceleration by contact with something else. The damage is directly proportional to momentum (the product of mass and velocity) and inversely proportional to the surface area of contact. Wide contact areas cause less damage than narrow edges or points for a given energy of impact. Of course, the physical properties of the crop itself have a major influence on susceptibility to damage. These physical properties include its size, shape, structure, and the fundamental physical properties of its tissues, including stiffness, hardness, elasticity, and viscoelasticity. Many crops can absorb a certain amount of energy through elastic deformation, which can be greater if the load is applied slowly than when it is sudden. The energy may be given up through a rebound or absorbed over time through deformation of the shape of the product (such as flattening the surface of an orange). After the limit of elastic deformation is reached, the energy causes a bruise, crack, or hole to form in the product.

Much of the avoidable physical damage occurs as crops are being moved along conveyors for packing, sorting, or storage. For delicate crops such as strawberries or cut flowers, this damage can be minimized by harvesting directly into containers in which they remain until the point of retail. Damage can also occur in packaged produce through vibration or sudden changes of direction during transport. Even in a static situation, compression damage can be caused by the mass of crop in a stack. Good design of packaging, handling, and transport systems can minimize mechanical damage, but it remains a major cause of crop loss after harvest. Some of the loss occurs indirectly because cuts and bruises provide entry points for disease organisms.

Post-harvest handling is greatly simplified if the crop is a mature structure that is naturally involved in long-term survival of the plant. Examples include dry grains, onions, and potatoes. However, developmental changes can still decrease the value of the crop. Seeds lose the ability to germinate over time because of subtle changes in cell structure and genetic material. This loss is of less consequence if the seed is to be used for food than if it is required for propagation. Even in a food crop, the deterioration may be associated with the development of off-flavors that render it unpalatable. Another cause of loss in perennating structures such as seeds, bulbs, and tubers is sprouting. Germination of seeds is usually avoided by maintaining dry conditions. Sprouting in bulbs and tubers may be controlled with chemical inhibitors, as with potatoes and onions, or by physical removal of the growing point, as with carrots and rutabagas. If the bulbs or tubers are required for regrowth, natural sprouting may simply be tolerated or manipulated by adjusting the temperature regime during storage.

Immature perennating structures, such as green peas, sweet corn, and salad onions, are among the most perishable crops. Other highly perishable items include immature stems, such as asparagus, and flowers, such as broccoli. All of these products are harvested when they are importing food from the rest of the plant to sustain rapid growth. When this process is interrupted, their metabolism remains rapid but is radically changed so that spoilage tends to be rapid. Spoilage often involves yellowing and premature senescence. Breakdown of chlorophyll and abscission of leaves are common responses of green plant material of all kinds when held in the dark or low-light conditions. These problems apply as much to cabbage and lettuce on the supermarket shelf as to *Ficus* during transit or to evergreen azaleas in winter storage. Immature fruits and flowers are generally highly perishable because they are harvested when they are still dependent on the rest of the plant for food to sustain their development. Mature fruits and flowers are not importing food as rapidly, but they would not naturally persist beyond the time of pollination in flowers or seed maturation in fruits. Many fruits and flowers show a climacteric pattern of development, which involves a burst of respiration and other metabolic changes that are stimulated by ethylene.

Although insects and other field pests may persist and cause damage on fresh plant produce after harvest, they present few special post-harvest problems for these crops. Insects are a major problem for dry seed crops, and several species of beetles are specifically associated with storage of cereals and pulses. Rodents are a problem in primitive storage systems for seed crops and in woody ornamentals held in plastic houses for the winter. Fungi attack crops of all kinds, and the post-harvest environment is often conducive to germination of fungal spores and growth of hyphae. The fungal species involved are often opportunistic pathogens that do not cause major problems in the field. Frequently they are saprophytes that colonize dead plant, persist on equipment and containers, and attack the crop through wounds or other damage occurring during handling and storage. Whereas growing crops tend to be attacked by specific fungal pathogens, after harvest a very wide range of crops can be attacked by a few species of *Botrytis* and *Penicillium*. A few bacterial diseases cause problems, particularly on vegetables, flowers, and herbaceous ornamentals.

Quality

When a crop is growing, attention may be focused on maximizing yield in relation to area and inputs. Maximizing quality may be a secondary concern. After harvest, crop quality generally declines and retarding loss of quality becomes a major objective. Quality can be defined as "fitness for purpose"; clearly, the quality standards required for soybeans as animal feed are different from those required for the manufacture of tofu. For most crops, market quality standards are established by the U.S. Department of Agriculture (USDA) or other bodies such as the American Association of Nurserymen. These organizations set minimum standards for size, shape, color, and freedom from damage. However, consumers have an interest in more subtle aspects of quality that are not easy to identify in a market environment.

Nutritional quality can be a concern for consumers, and federal law requires food packaging to carry nutritional information. Similar information is recommended but not required for fresh produce than it is for manufactured foods. The nutritional content of a particular kind of crop at harvest can be highly variable, and it can change after harvest, particularly for critical components such as vitamins A and C. So when values are presented for nutritional content, they have to be generic and represent minimal expectation for the crop at the end of the marketing chain.

Safety is another rather intangible aspect of crop quality that seems to excite more public interest than nutrition. Samples of crops passing through the markets are analyzed to ensure that pesticide residues fall below accepted levels, and retailers sometimes give assurances of the absence of detectable residues. More commonly, the absence of pesticides is implied by an organic label that is based on production methods rather than analysis. Cases of pesticide poisoning from commercially produced food are almost unknown, but there have been many instances of food poisoning from microbial contamination of plant foods. As with pesticides, samples are routinely screened for contamination, but harmful organisms do occur quite commonly and there is no reason why they should be less common on organic than on conventional produce. The bulk handling and wide distribution methods involved in modern crop marketing are vulnerable to the spread of waterborne organisms such as coliform bacteria. More emphasis on microbial safety is likely in the future.

The commercial success of food manufacturers is often dependent on specific quality attributes of their raw materials, such as the bread-making quality of wheat or the soluble solids content of tomatoes, which is important for ketchup manufacture. It is often up to the grower to meet these quality standards as a condition for being able to sell his or her crop.

All of the quality attributes considered so far are objective: they are capable of being observed or measured, even if this is not commonly done. However, consumer satisfaction also depends on subjective quality attributes. This kind of quality is sometimes called hedonic, which is a word related to *hedonism,* the pursuit of pleasure. Depending on the crop, these attributes can include the aesthetic appreciation of the form and color, the scent or aroma, and eating quality. Eating quality is a complex blend of all the senses, from the visual impression as we are about to bite the item, the texture at the first bite and on subsequent mastication, the taste on the tongue, the flavor perceived as aromatic compounds move from the nose to the mouth, and even the sounds made during eating. A lot of research has tried to measure the individual aspects of eating quality, but it can be assessed only by human subjects. Attempts to improve crop quality often involve panels of experts or general consumers who are asked to report on their subjective assessment of individual samples.

Much post-harvest technology is aimed at preserving crop quality between the point of production and consumption. Sometimes the quality that is important is not readily visible at the time of harvest but will be developed later. These qualities can involve the opening of a flower, the ripening of a fruit, or the conversion of a raw material into a manufactured food.

Post-Harvest Handling

Specific measures must be taken to maintain crop quality from the time of harvest onward. Handling procedures should minimize mechanical injury. Other procedures are commonly applied to prevent physiological deterioration and attack by pests or pathogens. For some crops, a curing period during which they are held at high temperature and low humidity is advantageous to allow wounds incurred in harvesting to heal. Examples of such crops include potatoes and onions. Low moisture content is critical for successful storage of dry seed crops. If it is higher than about 13 percent, hot air is forced through the crop to dry it. Chemical treatments were commonly applied to prevent various kinds of deterioration on a wide variety of crops. Fumigants were used on grain crops to eradicate pests, fungicides were used on many fruits and vegetables, and sprouting inhibitors were applied to vegetables such as potatoes and onions. Although these chemicals were very helpful in minimizing crop losses, public opinion increasingly opposes their use and many have been withdrawn. Irradiation has been investigated and advocated as a replacement for many of the chemical treatments, but it has not been generally approved or widely used. Currently, heat treatment at temperatures above 45°C for a few seconds or minutes is being investigated for insect and fungus control on many fruits and vegetables. Edible oils or waxes are still applied to many fruits and vegetables to improve appearance and reduce water loss. Application is usually preceded by thorough cleaning with detergent and may occur after storage if crops are not to be marketed immediately.

For many fruits, vegetables, and cut flowers, it is critical that the temperature of the crop is reduced quickly after harvest. The need to remove field heat can be reduced if crops are harvested in the early morning when temperatures are naturally low. However, the temperature of bulky crops can increase after harvest because heat is released by crop respiration. Refrigeration is needed to counteract this effect. The importance of this precooling operation is evidenced by the use of mobile coolers that can be taken into the field for immediate temperature reduction in many crops after harvest. Most precoolers rely on convective removal of heat by air or water. Low-bulk items can be cooled by placing them in a refrigerated room (Fig. 26–9), but room cooling is generally too slow for packaged crops. Forced-air cooling is used for delicate, low-bulk crops such as leafy vegetables, berry fruits, and flowers (Fig. 26–10). Packages of product are stacked against a partition in a refrigerated room or container. Powerful fans behind the partition draw cold air through tubes or vents that match perforations in the packages. As well as avoiding mechanical damage, air cooling avoids the spread of pathogens, which is one of the hazards of water used for cooling. Air has a low thermal capacity,

Figure 26–9 Air that has been cooled is drawn into this forced-air cooler through circular vents (upper right of photo) in the wall. After the cold air flows through the produce, it exhausts through holes in the opposite wall. This type of cooler is used extensively for strawberries, but in this case, cauliflower is being cooled. Cut flowers are also cooled by this method. *Source:* Western Grower and Shipper.

Figure 26–10 An example of a large-capacity cold-storage room widely used to store many fruits and vegetables. The cones protruding from the ceiling allow cold air to be introduced. The relative humidity of the air can be regulated. Large fans, such as the one visible on the side wall, circulate the cold air throughout the room, thereby maintaining a low temperature for the stored product. These rooms are not as effective as other methods in the initial cooling of the product, but they are useful for maintaining the temperature and storage of products already cooled.

Figure 26–11 The hydrocooler achieves cooling by allowing the transfer of heat from the product to cold water (contact by immersion or shower of cold water). The water temperature and exposure time influence the level of cooling obtained. *Source:* Robert F. Kasmire.

Figure 26–12 A railroad car, after loading, is having crushed ice applied over and around the packages to maintain low transit temperatures. This technique applies several tons of ice into and around the load. *Source:* Robert F. Kasmire.

however, and air cooling is slow for bulky products. This delay may not matter if the food products are not highly perishable, but water provides much better cooling (hydrocooling) for crops such as sweet corn and asparagus. Typically, packages of crop pass slowly under a falling curtain of refrigerated water (Fig. 26–11). The water is usually recirculated to conserve energy, and it must be treated with chlorine or ozone to prevent the carryover of fungal and bacterial pathogens.

Historically, when crops were transported by rail, it was common practice to top-ice packages (Fig. 26–12). The ice melted during the journey and provided some lasting temperature control. Ice cools more effectively than water because latent heat is absorbed during the phase transition from solid ice to liquid water. However, ice is bulky to transport and meltwater makes top-icing a messy procedure. The use of refrigerated containers obviates the need for additional cooling during transit and the use of top-icing has declined. Nevertheless, the efficacy of ice as a coolant has led to the use of ice slush injection for rapid cooling of boxes of commodities such as broccoli.

Vacuum-cooling is another system that utilizes latent heat to lower temperature (Fig. 26–13). Crops are sealed into a large steel container and the pressure is rapidly reduced by a pump. The temperature of the produce is reduced as water evaporates from its surface. This process can cause wilting of leafy crops, and it is common practice to spray crops with water before vacuum-cooling to minimize this problem.

Figure 26–13   A vacuum cooler can cool lettuce from 30°C to 1°C (86°F to 34°F) in about thirty minutes. While the cooler was originally designed for lettuce, other crops, such as the cauliflower shown here, are now cooled in this manner. *Source:* Western Grower and Shipper.

Preservation by Drying

The drying of vegetables, fruits, and berries by solar radiation was probably the first attempt at food preservation, and its use and popularity remain high. Dates and figs were grown and sundried by early civilizations in the eastern Mediterranean area. Preservation by drying succeeds because decay-causing organisms usually do not grow at moisture contents below 10 to 15 percent.

Some advantages of dehydration over other methods are a reduction in weight and volume, thus reducing handling and storage costs. Storage generally requires minimal or no refrigeration. Solar drying is also less expensive than other methods of dehydration.

A climate having high temperature, clear days, and low humidity is ideal for solar dehydration of various commodities. Crops preserved by solar dehydration include apples, apricots, currants, grapes, peaches, figs, dates, pears, and plums (Fig. 26–14). Sundried tomatoes have become very popular in the United States.

Forced Hot Air Dehydration

The popularity of dehydrated foods, particularly with various fruits, has stimulated the development of efficient methods for rapid dehydration with forced hot air. This method is more appropriate when natural solar drying is not sufficient or in areas of high humidity.

In this process, the product is spread thinly on trays and passed through a tunnel dryer. Sometimes a short-term blanching with steam is used to stop enzyme activity that causes tissue browning. Air heated to 60°C to 70°C (140°F to 158°F) is forced through the tunnel. The time for dehydration varies from a few hours to as much as thirty-six hours, depending on the kind and size of the product.

Figure 26–14   The typically brilliant cloudless summer days in the San Joaquin Valley in California are ideal for drying grapes to make raisins. The grape berries are hand-harvested and laid on heavy brown paper to dry. The use of a continuous roll of paper permits some mechanization of the removal of the raisins after drying and before they are taken to the packing house for processing. A rain during this period can be a catastrophe to raisin producers.

Until recently, vegetables were not dehydrated in any volume, but this situation has changed, and large volumes of onions, potatoes, mushrooms, and other commodities are dehydrated. The use of ready-mixed vegetables for soups has increased the use of dehydrated celery, carrots, peppers, tomatoes, peas, parsley, chives, and other vegetables.

Dehydration by Freezing

Freeze-drying involves removing water by sublimation of ice at temperatures below the freezing point. **Sublimation** is the process by which a solid (ice) passes directly to a vapor without going through the liquid phase. Heat is absorbed in this process. Upon rehydration, the quality of the product is equivalent to that of food preserved by freezing. While freeze-dried products are expensive, they are valuable when weight reduction is necessary. These products are used in special markets where convenience of preparation is desired, such as in backpacking.

Preservation by Processing

The preservation of foods by processing has an interesting early history. In France, about 1795, Napoleon was concerned about the availability and continuing supply of food for his armies. He offered a reward to find a preservation method that would be safe. The prize was won by a Parisian named Appert, who preserved several products in sterilized glass bottles in 1809. Although at that time the process was not fully understood, we now know that its success was due to the use of heat to inactivate the products' enzymes and to destroy microorganisms, followed by providing and maintaining a sanitary or sterile condition. This method of preservation was rapidly adopted and was widely used in many homes; it is now a large and important industry.

Canning

The purpose of canning is to destroy spoilage organisms by heat, although complete sterilization is not mandatory. The product is placed into gastight containers that are sealed. Hermetically sealed containers are in wide use. The major containers are cans made from tin-plated sheet steel or glass bottles. Plastic- or foil-coated paper containers are also used, especially to reduce weight. The interior surface of the cans are coated with acid-resistant lacquers to reduce corrosion and discoloration when high-acid products are canned, or when the products have sulfur compounds that react with the tin surfaces, turning them black if not coated. After sterilization, which is usually performed in a pressure cooker, the cans are quickly cooled, labeled, boxed, and stored in warehouses until they are needed.

Processing with Sugar

Some fruit products are processed with high concentrations of sugar for the production of jams, jellies, marmalades, crystallized fruits, and the like. Sugar increases the osmotic pressure to levels where water is unavailable for microbial activity, thereby reducing spoilage possibilities.

Processing with Salt

The processing of vegetables with brine has changed little over the years. This process preserves cucumbers and gherkins as pickles; cabbage as sauerkraut; and onions, beets, beans, peppers, and olives. Fresh cucumbers or cabbage are allowed to ferment anaerobically in a salt solution concentrated enough to prevent the activity of spoilage organisms but not concentrated enough to destroy the bacteria producing lactic acid. Lactic acid lowers the pH, thus helping to prevent the growth of organisms. Olives in California are usually pretreated before salt storage with a 1 to 2 percent solution of sodium hydroxide (NaOH) to neutralize the bitter glucoside found in the fresh fruit.

STORAGE

In primitive communities, crops were, and still are, stored under a cover of leaves or earth in pits or heaps on the ground. Such simple storage systems have been used for grains, root vegetables, and even fruits such as apples. The system is best suited to root vegetables and can be used for low-cost storage of such crops, particularly when intended for animal feed. Losses of grain crops from pests and sprouting in simple storage can be reduced by raising them off the ground and providing a waterproof cover. This kind of storage can still be seen occasionally on North American farms today, but most grain is stored in purpose-built silos called grain elevators (Fig. 26–15).

Figure 26–15 Extended storage of dried seed grains is commonly made in large "grain elevators," as pictured here. These facilities often have the capability to dry seed further if it is necessary to lower the moisture content. *Source:* James F. Thompson.

Storage of fruit, vegetables, and many ornamental crops nearly always involves refrigeration. Temperature is the most important variable affecting the quality of these crops after harvest. The discovery of the principles of mechanical refrigeration in the nineteenth century was key to the development of our modern systems of food supply. Refrigeration was originally developed for static storage chambers and the production of ice. Now it can be applied to containers used in road, rail, and sea transport. Crop storage, distribution, and marketing can now be achieved through a continuous cold chain from the farm to the point of retail sale.

Refrigeration generally slows down metabolic processes in crops and in the pests and pathogens that effect them, which means that most kinds of deterioration are delayed. As a general rule, the rate of biological processes doubles for every 10°C rise in temperature. A corollary is that the storage life of many crops is doubled on reducing the temperature by 10°C. For example, strawberries might last two days at 20°C, and eight days at 0°C. For most temperate crops, the lower limit of storage temperature is set by their freezing point, which is generally just below 0°C. However, crops of tropical and subtropical origin are injured by temperatures below 10°C. So avocadoes, bananas, tomatoes, and potatoes must be held above a threshold temperature in the range of 5°C to 10°C.

Moisture is another critical storage variable. The importance of dry storage for seed crops has already been mentioned. Fresh produce must be kept close to fully hydrated. Generally a loss of 5 percent of fresh weight results in wilting or shrivel, and the crop becomes unsaleable. Refrigeration tends to cause drying because water condenses from the air onto the surface of the cooler. The crop is slightly warmer than the air around it and tends to lose water even when the humidity is close to 100 percent. Humidifiers can be installed to replenish the water in the atmosphere, but the design of storage units is critical to moisture retention. A poorly designed unit might have a small cooler surface running at a low temperature and little insulation, so that the cooler runs continuously. Condensation on the cooler can be minimized by having a large surface area that can be operated close to the storage temperature and good insulation so that the equipment needs to run for only a fraction of the time. However, high humidity can allow fungal spores to germinate and invade plant tissues. An optimum humidity for a crop balances the risk of infection against the tendency to lose moisture from the crop.

For certain crops, additional control of deterioration can be achieved by raising the carbon dioxide level or lowering the oxygen in the atmosphere in which the crop is stored. This change can be done by allowing the crop to alter the atmosphere through its own respiration in a sealed room or by injecting manufactured gases into the room. Crops need a minimal level of oxygen, usually around 1 to 2 percent, to remain healthy and most cannot tolerate carbon dioxide above 5 percent for long periods. Such controlled atmosphere storage has been applied mainly to apples, pears, and cabbages. It has been tested on a wide variety of edible and ornamental crops but not applied commercially.

The plant hormone ethylene tends to accumulate in the atmosphere during storage, handling, and distribution of crops. Ethylene comes from many sources such as senescent and decomposing crops, gasoline and diesel engines, and even some electrical equipment. It does not escape readily because the crops are generally in enclosed spaces. At levels, from a fraction of a part per million upward, ethylene promotes yellowing and abscission of leaves, distorted growth, flower senescence, and ripening in fruits. Ethylene is used to ripen bananas and tomatoes, but its effects on cucumbers and green beans are not so desirable. Although equipment has been developed to remove ethylene from the atmosphere, the problem can often be minimized by removing the source, whether artificial or natural. Climacteric fruits produce huge amounts of ethylene and they should be kept away from most other kinds of plant material. Many cut flowers are particularly responsive to ethylene, and for many years they have been protected by treatment with silver thiosulfate. A less toxic treatment has recently been discovered. Flowers become insensitive to ethylene if they are treated with another gas: methylcyclopropene. This gas may also find uses on edible crops.

Refrigeration and low-oxygen storage can extend the life of many seed crops, but these technologies have been used only for special purposes such as preservation of crop germplasm. Woody ornamentals have been stored under refrigeration for many years. This practice has its advantages, but it is most useful for bare-root material that has been lifted from field production. The shift of the industry to container production has led to an emphasis on simpler storage technology. Often a plastic cover over the plants is all that is required to protect the root systems in the containers from freezing.

Storage is part of a marketing strategy for crops. Although it may be technically feasible, it makes no sense to put a crop into storage if it can be sold at the same or higher price immediately after harvest. While it would be difficult for a large apple grower to stay in business without the ability to hold a proportion of the crop for up to twelve months, a flower grower may use storage only to hold crop that is harvested for shipment on the following morning.

MARKETING OF AGRICULTURAL PRODUCTS

Primitive societies had little marketing. Individuals or family groups gathered or produced their own food and other necessities. As a barter system and, later, money exchange developed, the variety of products available for exchange widened. Hunters could barter their game with growers of grain, vegetables, and fruits, enabling each to obtain and exchange products. Individuals not directly producing food could buy or trade for food, thus allowing them to participate in their societies in numerous and varied work activities other than agriculture and food production.

In modern agriculture, specialization has progressed beyond the self-reliant community, and greater interaction exists between the farmer and the city dweller. A continuing trend is for fewer farmers to supply more as urban populations increase. More and more farmers tend to concentrate on producing the products best adapted to their particular areas, soils, climate, and markets rather than attempting to supply themselves with the many products they consume directly.

Early in the development of marketing, the producer provided an array of products (eggs, poultry, fruit, vegetables, milk, honey, etc.) directly to the final consumer. To make the exchange of commodities easier, the producers brought their products to a central place (village square or farmers' market), where they and buyers negotiated prices. In much of the world, this arrangement has changed very little. Farmers' markets are a small factor compared to their previous importance, in terms of the total volume of product. For the past fifty years, this type of market has been on a decline. Recently, a growing interest in fresh products has created a resurgence in farmers' markets, roadside outlets, and self-harvest operations. Indications are that the general public will continue to support this form of marketing, even though it will account for only a relatively small portion of the total distribution. For some agricultural commodities, the Internet has become an important means of marketing.

Today most marketing involves a series of intermediates. Intermediates are prevalent in countries with highly developed systems of agricultural marketing that includes imports and exports in addition to the domestic markets. Some intermediates buy the product and further process it, some transport or store it, and others distribute it as wholesalers or retailers. A multitude of intermediate functions oc-

The Internet is increasingly important in the production and marketing of plants and plant products through its business-to-business (B2B) and business-to-consumer (B2C) roles. On the marketing side, these exchanges occur at three levels:

1. **B2B exchange of raw plant commodities throughout the marketing chain.** The Internet's role in exchanging raw plant commodities is the least visible and, perhaps, the most important of the three. Much of this exchange takes place in closed network environments, but it is here that grain prices are set, the availability of most crops is determined by buyers, and exchange deals are developed.

2. **Direct sales of some plants/plant products to end consumers (B2C).** Direct sales range from food crops to flowers and other green products. Indeed florist services were among the early successes in e-commerce. These exchanges offer their own special set of concerns. Sales of many plant products and foods to consumers involve complex signaling of organoleptic traits (smell, color, texture) that do not transfer easily in a digital environment. And yet this form of B2C marketing continues to grow steadily.

3. **Aiding information search for buyers and sellers.** In all types of marketing, the primary role of the Internet is as a provider of information about products and sellers in an attempt to enhance the marketing process. Industry research estimates that for every $1 consumers actually spend, they buy another $6 of product at stores based on their online research. This percentage is likely higher for food and agricultural products given the relatively low adoption rate for B2C at present. This also works well for the rising consumer interest in connecting with food producers, with the Internet used as a method to tell the farmer's story of how food is produced. This story helps to reestablish the connection between agricultural producers and the consumers of their products (see Chapter 1).

As adoption of Internet technologies continues to grow, new marketing applications will evolve, permitting enhanced integration of computer and network technologies with the production side of the industries.

cur in this process. Whereas intermediates are often looked upon as unnecessary, they are essential in the modern orderly distribution of most agricultural commodities.

The Process of Marketing

Marketing plays a key role in the distribution of a product. Neither the modern producer nor the urban consumer can do without it; both depend on the market for their food and other necessities. The process is better understood by discussion of its major components, which are assembly, distribution, transportation, storage, exchange, and financing.

Assembly

Assembly is the concentration of smaller quantities of a commodity into a central place for the necessary handling, processing, or other procedures that must occur before transfer to market. Concentration at convenient locations gives prospective buyers the opportunity to examine products and arrange their purchase. Not all agricultural products need to be assembled and, in fact, a considerable volume is sold directly to processors, wholesalers, or retailers.

Distribution

Systems have been developed to distribute products from places of assembly to places of consumption. The extensiveness of the distribution system is the foundation of marketing in the industrialized countries. The system must be able to adjust the supply of commodities to market demands quickly and easily as either supply or demand changes.

Food sale chains and institutional and fast-food suppliers have captured a large share of food distribution in developed nations. In North America and western Europe, management of the retail distribution of food has become concentrated in relatively few hands. In the United States, large food chains retail food through thousands of stores (Fig. 26–16). Fast-food restaurant corporations prepare a significant proportion of the total number of meals consumed nationally each day in the United States. The same sort of food retailing has occurred to a large degree in Canada, England, and other European countries, and the trend is certain to expand to other areas.

Transportation

An essential part of marketing is rapid, dependable transportation. The products are generally transported from their origin or assembly areas to various markets or processing facilities by the producer or the buyer, or by transport firms contracted to provide such services. Who provides and how the transportation is performed varies considerably.

Storage

Storage must be considered part of the marketing system, and certainly is necessary at various stages in the marketing sequence. Producers often find it necessary to store their products; buyers, whether processors, wholesalers, or retailers, operate storage facilities to control the supply and time of resale of the commodity. Processing plants operate storage facilities to hold raw product stocks to operate the plant when incoming supplies are low as well as to hold the processed product until it is distributed.

Exchange

Before any buying or selling can occur, the buyer and seller must meet or communicate. The exchange process involves two phases: contacting possible buyers and sellers, and negotiating an agreeable price. In many countries, the town square acts as the marketplace, where buyers and sellers

Figure 26–16   Supermarkets have revolutionized the methods of marketing. In the United States, a customer can buy practically any type of fresh fruit or vegetable at any time of the year and at comparatively low prices.

agree on prices and conditions of the transaction. If agreement is reached, a sale is consummated. In some larger markets, sales are negotiated for a fee by intermediates called brokers. They do not take possession of the goods but facilitate the transaction. This setup enables buyers and sellers, who may be separated by great distances and thus unable to physically meet, to arrive at an exchange.

Grade standards developed by the US Department of Agriculture or individual state departments of agriculture are helpful because they allow the buyer to know about the quality of the product without actually seeing it. Specialized market reporting and news carried by the press, radio, and television provide helpful information.

Financing

Financing is essential in all marketing processes. Producers, wholesalers, processors, retailers, and other intermediaries must either have the capital or credit to produce, hold, and handle a commodity until it is sold and payment is received. Market financing entails a certain amount of risk. A major risk is a falling market; another is rapid price fluctuation. Prices for agricultural commodities generally are less stable than those for manufactured goods because supply and demand are subject to greater and less controllable fluctuations. Deterioration of quality entails another financial risk. A buyer must be aware that the quality is the most important attribute of most agricultural commodities

and that the quality of many of these commodities depreciates rapidly.

Commodity Markets and Exchanges

Commodity markets transfer ownership of products from producers to processors, manufacturers, or consumers and determine the price for the exchange. Commodities often handled by this type of market are foodstuffs and nonfood raw materials. Some examples are cereal grains, oil seeds, tea, coffee, rubber, and tobacco, as well as some metals. The commodity market is concerned with trading a given amount of a certain grade of commodity for present delivery. The goods need not be on hand at the time but can be in transit or stored. Terms are arranged, payment is made, and title transferred.

A commodity exchange, or futures market, differs from a commodity market because the exchange involves the purchase or sale of a contract to deliver a certain amount of a given grade of commodity on an agreed date in the future. This arrangement allows each party to hedge against a future market supply and demand situation. Hedging means that one party covers his or her position in the cash market by taking an opposite position in the futures market. In actual practice, the seller of the contract seldom actually delivers the commodity. Usually before the actual transfer date, the contract is dissolved. The futures market can provide contact with a larger number of prospective buyers or sellers than would another type of market, thus giving the parties to the respective contract greater opportunities to either buy or sell the contract before the transfer date. This setup may or may not result in a profit, but more important, it provides a means to avoid excessive losses. The Chicago Board of Trade is an example of a commodity exchange handling mainly agricultural commodities.

Governmental Marketing Services

Most governments participate in the regulation of marketing to some degree. Some go to the extreme of setting prices, and regulating production and distribution of products. For example, the price of bread is regulated in Egypt; that of milk in Denmark. In the United States, California and some other states regulate milk prices, and the production of tobacco and other crops is regulated through acreage allotments. There are many examples of governmental regulations, which often arise during wars or other emergency periods.

Services provided by the federal government are intended for the protection of public health; establishment of standards and grades; and enforcement of weights, measurements, and other regulatory needs. Federal and state governments cooperate to supply production statistics and estimates for various crops, their condition, and other important market characteristics (Fig. 26–17). For some commodities, information on shipments and market prices is provided on a daily basis.

Figure 26–17 A US Department of Agriculture inspector examining potatoes for quality and condition. *Source:* Adel A. Kader.

Grade standards are degrees of quality, each established by definition. The standards define the color, size, and freedom from defects or any other attributes that pertain to quality. These standards are published and maintained by the USDA, although some states also establish grade standards. Regulatory departments of the USDA or state may provide inspection and grade determination services. These official standards permit marketing, even over long distances, without the buyer's actual inspection of the commodity. For example, a buyer in Japan buying rice from a broker in Arkansas specifies the quality grade (USDA No. 1) of rice desired, knowing that the grade ordered will have those quality standards, supported by a certificate of inspection issued at the seller's location. Lettuce from Arizona or Florida oranges are sold to New York buyers who order these commodities by well-defined grade standards, knowing before arrival that the desired quality has been guaranteed for delivery.

Agricultural Cooperatives

Farmers may join together to form businesses known as cooperatives (co-ops), which differ from other businesses because the farmers are owners as well as customers. Co-op members provide the capital, elect officers, and may employ a manager. Farmers use the co-ops to market their products, to procure supplies (fuel, fertilizers, containers, etc.), and to obtain services (insurance, credit, etc.). Co-ops generally do not operate as a profit-making business but function to lower member costs through large volume buying discounts. Through collective marketing and having a large presence in the marketplace, co-ops can often strengthen their selling abilities.

Marketing Boards

Marketing boards were first developed in England and some of the commonwealth nations. Their primary objective was stabilizing producer prices. Examples of marketing boards established in some countries are the National Coffee Board of Ethiopia, the Sri Lanka Tea Propaganda Board, and the Australian Federal Marketing Board. Many other examples exist throughout the world.

In the United States, a similar program, called a marketing order, exists. Used originally for fruits and vegetables in California and for milk in many milk-producing states, marketing orders are agreements among producers to solve a specific problem or to achieve a goal for the common good of the members. Under the authority of a marketing order, either state or federal, an elected administrative board can levy assessments to collect funds to support advisory services, promotional activities, research, or control of supply or sales. Supply control is now largely disallowed for most marketing orders.

TRANSPORTING COMMODITIES

In primitive times, the primary movers of goods were humans. As civilizations developed, movement of goods and people progressed to the use of animals and then to machines. Throughout this progression, various needs were identified and met, among which were those of efficient roadways, rail lines, equipment, and the like. Transportation has not been limited to land surfaces. Rivers, canals, sealanes, and airways are also significant avenues for transportation.

Rail Transport

Rail transport of nonperishable commodities such as grains began during the Civil War era. Perishable commodities began to be shipped in refrigerated rail cars. The earliest of these cars had ice at both ends. Fans powered by the car wheels blew air across the ice and over the products. After World War II, mechanically cooled rail cars replaced those using ice.

The cooling system in the mechanically refrigerated cars is operated by a diesel engine located in one end of the car and operates during the trip. Temperature is regulated by thermostats, with sensors generally located in the return air ducts. Air is distributed throughout the load by fans. If needed, heating can be provided.

Another adaptation is a system of truck-train transportation called the piggyback method, in which a refrigerated truck-trailer is carried on a railroad flatcar. In many aspects, the trailer functions like a refrigerated boxcar. At destination, the trailer is removed from the flatcar and attached to a tractor for further transportation to the final destination. These piggyback units combine the truck's flexibility of delivery with the train's reduced cost of transport.

Truck Transport

Transport by truck and truck-trailer combinations was minimal before World War II, but it expanded rapidly soon after to become the dominant method for transport of goods in the United States. Many shippers, experiencing problems with rail shipment and preferring greater flexibility and direct control, turned to new, large trucks for shipping. Improved highway systems gave the trucks some advantages over the railroads. The truck is as fast as, and in some cases faster than, the railroad, offering dock-to-dock service, and often at less cost. Trucks now haul the major portion of all perishable goods in the United States. Modern trucks for transporting perishables have mechanically operated refrigeration systems similar to those already discussed (Fig. 26–18).

Sea Transport

The history of sea transport and trade is both interesting and colorful. It is impossible to state when early humans first discovered the use of waterways to carry them and their pos-

Figure 26–18 Trucks account for a substantial amount of transportation of all forms of goods and particularly for perishables in the United States. Trucks are equipped to provide refrigeration or heat as needed. *Source:* Yoder Bros., Inc. Barberton, OH.

sessions. The Egyptians in the period about 2000 B.C. were among the earliest to use sea transportation. In modern times, sea transportation remains a vital and considerable factor in the transport of food and other agricultural products.

Starting in the 1940s, there was a shift toward containerized transport. At first, the containers were designed for dry, unrefrigerated cargo, but the apparent advantages of containerization led to the development of refrigerated containers for all types of fresh products. Complementing this development, container ships were designed and constructed specifically to efficiently handle and transport such containerized cargos (Fig. 26–19). Containers for these container-carrying ships are similar to refrigerated trailer vans for trucks. They are transported via flatbed trucks and railcars designed for this purpose to dockside for shipboard loading. Depending on the carrier, the containers are refrigerated by shipboard units or have individual refrigeration units.

Air Transport

Transportation by air, except for small packages and mail, was not important until after World War II. The development in the late 1960s of large-capacity, jet-engine-powered aircraft made such transport somewhat more feasible by reducing the operating cost per unit weight of cargo. Containers especially fitted to the large carriers also contributed to handling efficiencies (Fig. 26–20). Perishable cut flowers, strawberries, pineapples, and various fruits and off-season vegetables are often shipped via this transportation method.

Figure 26–19 Regularly scheduled container ships transport large numbers of sealed containers great distances. The containers can be refrigerated individually if the cargo requires temperature control. *Source:* Matson Navigation Company.

Figure 26–20 Large quantities of cut flowers and some fresh fruits and vegetables are carried throughout the world by air cargo and regularly scheduled passenger aircraft. *Source:* Robert F. Kasmire.

Air transport has opened up previously inaccessible markets. In utilizing this method, other components are required, including precooling, short-term refrigerated storage at departure and arrival destinations, and effective scheduling with unloading and distribution to final destinations. Any delays negate the benefits of the rapid transport provided by aircraft.

In the modern world, transportation is a necessary and important activity compared to earlier times, when distances between producers and consumers were small. Modern methods of transportation allow for a tremendous exchange of products throughout the world and provide access to many new and distant markets.

SUMMARY AND REVIEW

Crop harvesting can be done mechanically or by hand. Mechanical harvesting is done by machines that have been engineered to efficiently collect the part of the plant being harvested and usually separate it from the rest of the plant. At first, harvesters could be used only on uniformly ripening, durable crops that could be harvested in bulk, such as grains, hay, and cotton. Other crops were too fragile, did not ripen uniformly, or were too difficult to harvest in bulk to be har-

vested by machines. Now, with the development of uniformly ripening species and advances in technology such as computers, sophisticated sensors, and robotics, more and more crops are being harvested mechanically. Hand harvesting can be slow and labor intensive, but some crops still require the greater care that it provides. In many parts of the world where labor is inexpensive, it is still the most common method of harvest.

The problems of crop preservation after harvest and the solutions to those problems are influenced by the type of structure that is harvested and its stage of development. Crops are subject to several kinds of deterioration after harvest. Mechanical injury during harvest and subsequent handling is a major cause of loss. Crops also undergo developmental and metabolic changes that may render them unfit for consumption. Insect pests are mainly a problem in the storage of dry crops and some ornamentals. Fungal and, to a lesser extent, bacterial diseases affect many kinds of crop, and the pathogens are often distinct from those occurring in the field. While market standards for crops emphasize visible aspects such as size, color, and shape, consumer quality is determined by other factors such as safety, nutritional quality, and the subjective experience of consumption. Various chemicals have been used to prevent crop deterioration and maintain quality, but their use is declining. Refrigeration is the main technology employed for preservation of fresh produce. Various kinds of equipment are used to remove field heat prior to storage or distribution. The moisture level during storage may need to be high to maintain full hydration in the crop or low to minimize disease. A compromise between these two extremes is often necessary. The storage life of some crops can be extended with atmospheres containing high carbon dioxide or low oxygen concentrations. Ethylene is produced by several sources in the post-harvest environment and can cause damage to a wide variety of crops.

Marketing and distribution of crops involves several steps, including assembly of the product, transportation and distribution, storage, and exchange and financing. Computers and the Internet are becoming an integral part of much of this process. Many commodities have marketing boards that can levy assessments or collect funds to provide advisory services, develop promotional activities, and fund research.

FOOD FOR THOUGHT

1. Many mechanical harvesters are called combines because they combine two or more operations in one machine. What are some of the operations that are combined in these harvesters?

2. Some fruits and vegetables can be harvested mechanically; others require hand picking. What are some characteristics of a crop that determine if mechanical or hand picking is best? What emerging technologies might enable some crops that are currently hand-picked to be harvested mechanically in the future?

3. Research from many American universities and companies led to the development of some cut flowers, such as carnations, roses, and chrysanthemums, that are very durable and can be shipped and stored with little damage. This development in turn allowed the production of these flowers to move from the United States to other countries where production costs are much lower. What do you think American cut-flower growers can do to stay in the cut-flower business in the United States?

4. The first transgenic crop to be marketed commercially was a tomato that was engineered for long shelf life. Can you think of other crops that could be genetically altered to maintain better quality after harvest?

5. Investigate the injury on some of the crops in a local store. When and where do you think the injury occurred and what can be done about it?

6. Is market quality the same as consumer quality? What can be done for specific crops to bring these two standards together?

SUPPLEMENTARY READING

WILLS, R. B. H., B. McGLASSON, D. GRAHAM, and D. JOYCE. 1998. *Postharvest: An introduction to the physiology and handling of fruit, vegetable, and ornamentals,* fourth edition. Sydney, Australia: UNSW Press.

http://agmarketing.extension.psu.edu/Commodity/MrktPln.html
http://www.ams.usda.gov/

Appendix

USEFUL WEBSITES

United Nations Food and Agriculture Organizations (FAO): Global information on all aspects of agriculture.

http://www.fao.org/

United States Department of Agriculture (USDA), National Agricultural Statistics Service (NASS): Gives current and achieved information about agriculture in the United States.

http://www.usda.gov/nass/

USDA Economics and Statistics System: Access to many national and international agricultural information sites.

http://usda.mannlib.cornell.edu/

Locations of all land grant institutions in the United States.

http://www.csrees.usda.gov/qlinks/partners/ state_partners.html

Ohio State University's plant fact sheet database and search engine for universities and departments that teach plant science.

http://plantfacts.osu.edu

NUTRITIVE VALUE OF THE EDIBLE PARTS OF FOODS

Foods, Approximate Measures, Units, and Weight (Edible Part Unless Indicated Otherwise)		Weight (gm)	Water (Percentage)	Food Energy (Calories)	Protein (gm)	Fat (gm)
FRUITS AND FRUIT PRODUCTS						
Apples, raw, unpeeled, without cores:						
3¼-in. diam. (about 2 per lb).	1 apple	212	84	125	Trace	1
Apricots:						
Raw, without pits (about 12 per lb with pits).	3 apricots	107	85	55	1	Trace
Dried, Uncooked (28 large or 37 medium halves per cup).	1 cup	130	25	340	7	1
Avocados, raw, whole, without skins and seeds:						
California, mid- and late-winter (with skin and seed, 3⅛-in. diam.; wt., 10 oz).	1 avocado	216	74	370	5	37
Florida, late summer and fall (with skin and seed, 3⅝-in. diam.; wt., 1 lb).	1 avocado	304	78	390	4	33
Banana without peel (about 2.6 per lb with peel).	1 banana	119	76	100	1	Trace
Blackberries, raw	1 cup	144	85	85	2	1
Blueberries, raw	1 cup	145	83	90	1	1
Cherries, sweet, raw, without pits and stems.	10 cherries	68	80	45	1	Trace
Cranberry sauce, sweetened, canned, strained.	1 cup	277	62	405	Trace	1
Dates:						
Whole, without pits	10 dates	80	23	220	2	Trace
Chopped	1 cup	178	23	490	4	1
Grapefruit:						
Raw, medium, 3¾-in. diam. (about 1 lb 1 oz):						
Pink or red	½ grapefruit with peel	241	89	50	1	Trace
White	½ grapefruit with peel	241	89	45	1	Trace
Grapes, European type (adherent skin), raw:						
Thompson Seedless.	10 grapes	50	81	35	Trace	Trace
Lemon, raw, size 165, without peel and seeds (about 4 per lb with peels and seeds).	1 lemon	74	90	20	1	Trace
Olives, pickled, canned:						
Green	4 medium or 3 extra large or 2 giant	16	78	15	Trace	2
Ripe, Mission	3 small or 2 large	10	73	15	Trace	2
Oranges, all commercial cultivars, raw:						
Whole, 2⅝-in. diam., without peel and seeds (about 2½ per lb with peel and seeds).	1 orange	131	86	65	1	Trace
Papayas, raw, ½-in. cubes	1 cup	140	89	55	1	Trace
Peaches:						
Raw:						
Whole, 2½-in. diam., peeled, pitted (about 4 per lb with peels and pits).	1 peach	100	89	40	1	Trace
Dried:						
Uncooked	1 cup	160	25	420	5	1
Pears:						
Raw, with skin, cored:						
Bartlett, 2½-in. diam. (about 2½ per lb with cores and stems).	1 pear	164	83	100	1	1

Carbohydrate (gm)	Calcium (mg)	Phosphorus (mg)	Iron (mg)	Potassium (mg)	Vitamin A (International Units)	Thiamin (mg)	Riboflavin (mg)	Niacin (mg)	Ascorbic Acid (mg)
31	15	21	.6	233	190	.06	.04	.2	8
14	18	25	.5	301	2,890	.03	.04	.6	11
86	87	140	7.2	1,273	14,170	.01	.21	4.3	16
13	22	91	1.3	1,303	630	.24	.43	3.5	30
27	30	128	1.8	1,836	880	.33	.61	4.9	43
26	10	31	.8	440	230	.06	.07	.8	12
19	46	27	1.3	245	290	.04	.06	.6	30
22	22	19	1.6	117	160	.04	.09	.7	20
12	15	13	.3	129	70	.03	.04	.3	7
104	17	11	.6	83	60	.03	.03	.1	6
58	47	50	2.4	518	40	.07	.08	1.8	0
130	105	112	5.3	1,153	90	.16	.18	3.9	0
13	20	20	.5	166	540	.05	.02	.2	44
12	19	19	.5	159	10	.05	.02	.2	44
9	6	10	.2	87	50	.03	.02	.2	2
6	19	12	.4	102	10	.03	.01	.1	39
Trace	8	2	.2	7	40	—	—	—	—
Trace	9	1	.1	2	10	Trace	Trace	—	—
16	54	26	.5	263	260	.13	.05	.5	66
14	28	22	.4	328	2,450	.06	.06	.4	78
10	9	19	.5	202	1,330	.02	.05	1.0	7
109	77	187	9.6	1,520	6,240	.02	.30	8.5	29
25	13	18	.5	213	30	.03	.07	.2	7

(CONTINUED)

NUTRITIVE VALUE OF THE EDIBLE PARTS OF FOODS (continued)

Foods, Approximate Measures, Units, and Weight (Edible Part Unless Indicated Otherwise)		Weight (gm)	Water (Percentage)	Food Energy (Calories)	Protein (gm)	Fat (gm)
FRUITS AND FRUIT PRODUCTS *(continued)*						
Bosc, 2½-in. diam. (about 3 per lb with cores and stems).	1 pear	141	83	85	1	1
D'Anjou, 3-in. diam. (about 2 per lb with cores and stems).	1 pear	200	83	120	1	1
Pineapple: raw, diced	1 cup	155	85	80	1	Trace
Plums:						
Raw, without pits:						
Japanese and hybrid (2⅛-in. diam., about 6½ per lb with pits).	1 plum	66	87	30	Trace	Trace
Prune-type (1½-in. diam., about 15 per lb with pits).	1 plum	28	79	20	Trace	Trace
Prunes, dried, "softenized," with pits: uncooked	4 extra large or 5 large prunes	49	28	110	1	Trace
Raisins, seedless: cup, not pressed down	1 cup	145	18	420	4	Trace
Raspberries, red: raw, capped, whole	1 cup	123	84	70	1	1
Strawberries: raw, whole berries, capped	1 cup	149	90	55	1	1
Tangerine, raw, 2⅜-in. diam., size 176, without peel	1 tangerine	86	87	40	1	Trace
LEGUMES (DRY), NUTS, SEEDS						
Almonds, shelled:						
Chopped (about 130 almonds).	1 cup	130	5	775	24	70
Beans, dry:						
Common cultivars as Great Northern, navy, and others:						
Cooked, drained:						
Great Northern	1 cup	180	69	210	14	1
Navy	1 cup	190	69	225	15	1
Red kidney	1 cup	255	76	230	15	1
Lima, cooked, drained	1 cup	190	64	260	16	1
Blackeye peas, dry, cooked (with residual cooking liquid).	1 cup	250	80	190	13	1
Brazil nuts, shelled (6–8 large kernels).	1 oz	28	5	185	4	19
Cashew nuts, roasted in oil.	1 cup	140	5	785	24	64
Coconut meat, fresh:						
Piece, about 2 by 2 by ½ in.	1 piece	45	51	155	2	16
Shredded or grated, not pressed down.	1 cup	80	51	275	3	28
Filberts (hazelnuts), chopped (about 80 kernels).	1 cup	115	6	730	14	72
Lentils, whole, cooked.	1 cup	200	72	210	16	Trace
Peanuts, roasted in oil, salted (whole, halves, chopped).	1 cup	144	2	840	37	72
Pecans, chopped or pieces (about 120 large halves).	1 cup	118	3	810	11	84
Walnuts:						
Black:						
Chopped or broken kernels.	1 cup	125	3	785	26	74
Persian or English, chopped (about 60 halves).	1 cup	120	4	780	18	77

Carbohydrate (gm)	Calcium (mg)	Phosphorus (mg)	Iron (mg)	Potassium (mg)	Vitamin A (International Units)	Thiamin (mg)	Riboflavin (mg)	Niacin (mg)	Ascorbic Acid (mg)
22	11	16	.4	83	30	.03	.06	.1	6
31	16	22	.6	260	40	.04	.08	.2	8
21	26	12	.8	226	110	.14	.05	.3	26
8	8	12	.3	112	160	.02	.02	.3	4
6	3	5	.1	48	80	.01	.01	.1	1
29	22	34	1.7	298	690	.04	.07	.7	1
112	90	146	5.1	1,106	30	.16	.12	.7	1
17	27	27	1.1	207	160	.04	.11	1.1	31
13	31	31	1.5	244	90	.04	.10	.9	88
10	34	15	.3	108	360	.05	.02	.1	27
25	304	655	6.1	1,005	0	.31	1.20	4.6	Trace
38	90	266	4.9	749	0	.25	.13	1.3	0
40	95	281	5.1	790	0	.27	.13	1.3	0
42	74	278	4.6	673	10	.13	.10	1.5	—
49	55	293	5.9	1,163	—	.25	.11	1.3	—
35	43	238	3.3	573	30	.40	.10	1.0	—
3	53	196	1.0	203	Trace	.27	.03	.5	—
41	53	522	5.3	650	140	.60	.35	2.5	—
4	6	43	.8	115	0	.02	.01	.2	1
8	10	76	1.4	205	0	.04	.02	.4	2
19	240	388	3.9	810	—	.53	—	1.0	Trace
39	50	238	4.2	498	40	.14	.12	1.2	0
27	107	577	3.0	971	—	.46	.19	24.8	0
17	86	341	2.8	712	150	1.01	.15	1.1	2
19	Trace	713	7.5	575	380	.28	.14	.9	—
19	119	456	3.7	540	40	.40	.16	1.1	2

(CONTINUED)

NUTRITIVE VALUE OF THE EDIBLE PARTS OF FOODS (continued)

Foods, Approximate Measures, Units, and Weight (Edible Part Unless Indicated Otherwise)		Weight (gm)	Water (Percentage)	Food Energy (Calories)	Protein (gm)	Fat (gm)
VEGETABLES						
Asparagus, green:						
Cooked, drained:						
Spears, ½-in. diam. at base:						
From raw.	4 spears	60	94	10	1	Trace
From frozen.	4 spears	60	92	15	2	Trace
Beans:						
Lima, immature seeds, frozen, cooked, drained:						
Thick-seeded types (Fordhooks).	1 cup	170	74	170	10	Trace
Thin-seeded types (baby limas).	1 cup	180	69	210	13	Trace
Snap:						
Green:						
Cooked, drained:						
From raw (cuts and French style).	1 cup	125	92	30	2	Trace
From frozen:						
Cuts	1 cup	135	92	35	2	Trace
French style	1 cup	130	92	35	2	Trace
Yellow or wax:						
Cooked, drained:						
From raw (cuts and French style).	1 cup	125	93	30	2	Trace
From frozen (cuts).	1 cup	135	92	35	2	Trace
Canned, drained solids (cuts).	1 cup	135	92	30	2	Trace
Beet greens, leaves and stems, cooked, drained.	1 cup	145	94	25	2	Trace
Beets:						
Cooked, drained, peeled:						
Whole beets, 2-in. diam.	2 beets	100	91	30	1	Trace
Canned, drained solids:						
Whole beets, small	1 cup	160	89	60	2	Trace
Broccoli, cooked, drained:						
From raw:						
Stalk, medium size.	1 stalk	180	91	45	6	1
From frozen:						
Stalk, 4½ to 5 in. long.	1 stalk	30	91	10	1	Trace
Brussels sprouts, cooked, drained:						
From raw, 7–8 sprouts (1¼- to 1½-in. diam.).	1 cup	155	88	55	7	1
From frozen.	1 cup	155	89	50	5	Trace
Cabbage:						
Common varieties:						
Raw:						
Coarsely shredded or sliced	1 cup	70	92	15	1	Trace
Finely shredded or chopped	1 cup	90	92	20	1	Trace
Cooked, drained:	1 cup	146	94	30	2	Trace
Red, raw, coarsely shredded or sliced.	1 cup	70	90	20	1	Trace
Savoy, raw, coarsely shredded or sliced.	1 cup	70	92	15	2	Trace
Cabbage, celery (also called pe-tsai or wongbok), raw, 1-in. pieces.	1 cup	75	95	10	1	Trace
Cabbage, white mustard (also called bokchoy or pakchoy), cooked, drained	1 cup	170	95	25	2	Trace

Carbohydrate (gm)	Calcium (mg)	Phosphorus (mg)	Iron (mg)	Potassium (mg)	Vitamin A (International Units)	Thiamin (mg)	Riboflavin (mg)	Niacin (mg)	Ascorbic Acid (mg)
2	13	30	.4	110	540	.10	.11	.8	16
2	13	40	.7	143	470	.10	.08	.7	16
32	34	153	2.9	724	390	.12	.09	1.7	29
40	63	227	4.7	709	400	.16	.09	2.2	22
7	63	46	.8	189	680	.09	.11	.6	15
8	54	43	.9	205	780	.09	.12	.5	7
8	49	39	1.2	177	690	.08	.10	.4	9
6	63	46	.8	189	290	.09	.11	.6	16
8	47	42	.9	221	140	.09	.11	.5	8
7	61	34	2.0	128	140	.04	.07	.4	7
5	144	36	2.8	481	7,400	.10	.22	.4	22
7	14	23	.5	208	20	.03	.04	.3	6
14	30	29	1.1	267	30	.02	.05	.2	5
8	158	112	1.4	481	4,500	.16	.36	1.4	162
1	12	17	.2	66	570	.02	.03	.2	22
10	50	112	1.7	423	810	.12	.22	1.2	135
10	33	95	1.2	457	880	.12	.16	.9	126
4	34	20	0.3	163	90	0.04	0.04	0.02	33
5	44	26	.4	210	120	.05	.05	.3	42
6	64	29	.4	236	190	.06	.06	.4	48
5	29	25	.6	188	30	.06	.04	.3	43
3	47	38	.6	188	140	.04	.06	.2	39
2	32	30	.5	190	110	.04	.03	.5	19
4	252	56	1.0	364	5,270	.07	.14	1.2	26

(CONTINUED)

NUTRITIVE VALUE OF THE EDIBLE PARTS OF FOODS (continued)

Foods, Approximate Measures, Units, and Weight (Edible Part Unless Indicated Otherwise)		Weight (gm)	Water (Percentage)	Food Energy (Calories)	Protein (gm)	Fat (gm)
VEGETABLES *(continued)*						
Carrots:						
Raw, without crowns and tips, scraped:						
Whole, 7½ by 1⅛ in., or strips, 2½ to 3 in. long.	1 carrot or 18 strips	72	88	30	1	Trace
Grated.	1 cup	110	88	45	1	Trace
Canned:						
Sliced, drained solids	1 cup	155	91	45	1	Trace
Cauliflower:						
Raw, chopped.	1 cup	115	91	31	3	Trace
Cooked, drained.	1 cup	125	93	30	3	Trace
Celery, Pascal type, raw:						
Stalk, large outer, 8 by 1½ in., at root end.	1 stalk	40	94	5	Trace	Trace
Collards, cooked, drained:						
From raw (leaves without stems).	1 cup	190	90	65	7	1
From frozen (chopped).	1 cup	170	90	50	5	1
Corn, sweet:						
Cooked, drained:						
From raw, ear 5 by 1¾ in.	1 ear	140	74	70	2	1
From frozen:						
Ear, 5 in. long.	1 ear	229	73	120	4	1
Kernels.	1 cup	165	77	130	5	1
Cucumber slices, ⅛ in. thick (large, 2⅛-in. diam.; small, 1¾-in. diam.):						
With peel.	6 large or 8 small slices	28	95	5	Trace	Trace
Endive, curly (including escarole), raw, small pieces.	1 cup	50	93	10	1	Trace
Kale, cooked, drained:						
From raw (leaves without stems and midribs).	1 cup	110	88	45	5	1
From frozen (leaf style).	1 cup	130	91	40	4	1
Lettuce, raw:						
Butterhead, as Boston types:						
Head, 5-in. diam.	1 head	220	95	25	2	Trace
Leaves.	1 outer or 2 inner or 3 heart leaves	15	95	Trace	Trace	Trace
Crisphead, for example, Iceberg:						
Head, 6-in. diam.	1 head	567	96	70	5	1
Pieces, chopped or shredded	1 cup	55	96	5	Trace	Trace
Looseleaf (bunching varieties including romaine or cos), chopped or shredded pieces.	1 cup	55	94	10	1	Trace
Muskmelons, raw, orange-fleshed (with rind and seed cavity, 5-in. diam., 2⅓ lb).	½ melon with rind	477	91	80	2	Trace
Honeydew (with rind and seed cavity, 6½-in. diam., 5¼ lb).	⅒ melon with rind	226	91	50	1	Trace
Mustard greens, without stems and midribs, cooked, drained.	1 cup	140	93	30	3	1
Okra pods, 3 by ⅝ in., cooked.	10 pods	106	91	30	2	Trace

Carbohydrate (gm)	Calcium (mg)	Phosphorus (mg)	Iron (mg)	Potassium (mg)	Vitamin A (International Units)	Thiamin (mg)	Riboflavin (mg)	Niacin (mg)	Ascorbic Acid (mg)
7	27	26	.5	246	7,930	.04	.04	.4	6
11	41	40	.8	375	12,100	.07	.06	.7	9
10	47	34	1.1	186	23,250	.03	.05	.6	3
6	29	64	1.3	339	70	.13	.12	.8	90
5	26	53	.9	258	80	.11	.10	.8	69
2	16	11	.1	136	110	.01	.01	.1	4
10	357	99	1.5	498	14,820	.21	.38	2.3	144
10	299	87	1.7	401	11,560	.10	.24	1.0	56
16	2	69	.5	151	310	.09	.08	1.1	7
27	4	121	1.0	291	440	.18	.10	2.1	9
31	5	120	1.3	304	580	.15	.10	2.5	8
1	7	8	.3	45	70	.01	.01	.1	3
2	41	27	.9	147	1,650	.04	.07	.3	5
7	206	64	1.8	243	9,130	.11	.20	1.8	102
7	157	62	1.3	251	10,660	.08	.20	.9	49
4	57	42	3.3	430	1,580	.10	.10	.5	13
Trace	5	4	.3	40	150	.01	.01	Trace	1
16	108	118	2.7	943	1,780	.32	.32	1.6	32
2	11	12	.3	96	180	.03	.03	.2	3
2	37	14	.8	145	1,050	.03	.04	.2	10
20	38	44	1.1	682	9,240	.11	.08	1.6	90
11	21	24	.6	374	60	.06	.04	.9	34
6	193	45	2.5	308	8,120	.11	.20	.8	67
6	98	43	.5	184	520	.14	.19	1.0	21

METRIC CONVERSION TABLE

INTO METRIC			OUT OF METRIC		
IF YOU KNOW	**MULTIPLY BY**	**TO GET**	**IF YOU KNOW**	**MULTIPLY BY**	**TO GET**
Length					
inches	2.54	centimeters	millimeters	0.04	inches
feet	30	centimeters	centimeters	0.4	inches
feet	0.303	meters	meters	3.3	feet
yards	0.91	meters	kilometers	0.62	miles
miles	1.6	kilometers			
Area					
sq inches	6.5	sq centimeters	sq centimeters	0.16	sq inches
sq feet	0.09	sq meters	sq meters	1.2	sq yards
sq yards	0.8	sq meters	sq kilometers	0.4	sq miles
sq miles	2.6	sq kilometers	hectares	2.47	acres
acres	0.4	hectares			
Mass (Weight)					
ounces	28	grams	grams	0.035	ounces
pounds	0.45	kilograms	kilograms	2.2	pounds
short ton	0.9	metric ton	metric tons	1.1	short tons
Volume					
teaspoons	5	milliliters	milliliters	0.03	fluid ounces
tablespoons	15	milliliters	liters	2.1	pints
fluid ounces	30	milliliters	liters	1.06	quarts
cups	0.24	liters	liters	0.26	gallons
pints	0.47	liters	cubic meters	35	cubic feet
quarts	0.95	liters	cubic meters	1.3	cubic yards
gallons	3.8	liters			
cubic feet	0.03	cubic meters			
cubic yards	0.76	cubic meters			
Pressure					
lb/in.^{2}	0.069	bars	bars	14.5	lb/in.^{2}
atmospheres	1.013	bars	bars	0.987	atmospheres
atmospheres	1.033	kg/cm^2	kg/cm^2	0.968	atmospheres
lb/in.^{2}	0.07	kg/cm^2	kg/cm^2	14.22	lb/in.^{2}
Rates					
lb/acre	1.12	kg/hectare	kg/hectare	0.882	lb/acre
tons/acre	2.24	metric tons/hectare	metric tons/hectare	0.445	tons/acre

TEMPERATURE CONVERSION TABLE*

F	C	F	C	F	C
−26	−32	19	−7	64	18
−24	−31	21	−6	66	19
−22	−30	23	−5	68	20
−20	−29	25	−4	70	21
−18	−28	27	−3	72	22
−17	−27	28	−2	73	23
−15	−26	30	−1	75	24
−13	−25	32	0	77	25
−11	−24	34	1	79	26
−9	−23	36	2	81	27
−8	−22	37	3	82	28
−6	−21	39	4	84	29
−4	−20	41	5	86	30
−2	−19	43	6	88	31
0	−18	45	7	90	32
1	−17	46	8	91	33
3	−16	48	9	93	34
5	−15	50	10	95	35
7	−14	52	11	97	36
9	−13	54	12	99	37
10	−12	55	13	100	38
12	−11	57	14	102	39
14	−10	59	15	104	40
16	−9	61	16	106	41
18	−8	63	17	108	42
				212	100

* Celsius temperatures have been rounded to the nearest whole number.

Fahrenheit to Celsius: Subtract 32 from the Fahrenheit figure, multiply by 5, and divide by 9.

Celsius to Fahrenheit: Multiply the Celsius figure by 9, divide by 5, and add 32.

LIGHT CONVERSION TABLE

Conversion of lux or footcandles (ft-c) from various light sources to watts per sq meter of photosynthetically active radiation (PAR) or to micromols of photosynthetic photon flux (PPF) per sq meter per second.

	To Convert To:			
	Wm^{-2} PAR DIVIDE:		$\mu mol\ m^{-2}s^{-1}$ PPF DIVIDE:	
Light Source	lux	ft-c	lux	ft-c
Daylight[a]	by 247	by 22.9	by 54	by 5.0
Metal halide lamps	326	30.2	71	6.6
Fluorescent, cool-white lamps	340	31.5	74	7.0
Incandescent lamps	250	23.1	50	4.6
Low-pressure sodium lamps	522	48.3	106	9.8
High-pressure mercury lamps[b]	380	35.2	84	7.8

[a] Based on average spectral distribution.
[b] Different types of mercury lamps can vary ± 5 percent.

To convert measurements of light intensity in footcandles (ft-c), or illuminance in lux from various light sources into equivalent watts of photosynthetic radiation per square meter (Wm^{-2} PAR), or micromols of photosynthetic photon flux per square meter per second ($\mu mol\ m^{-2}s^{-1}$), divide the reported measurement by the appropriate constant given for the particular light source.

Example: A light intensity of 2,500 ft-c of daylight recorded in the greenhouse may be converted into equivalent Wm^{-2} PAR or $\mu mol\ m^{-2}s^{-1}$ PPF as follows:

$$\frac{2,500}{22.9} = 109\ Wm^{-2}\ PAR \qquad or \qquad \frac{2,500}{5.0} = 500\ \mu mol\ m^{-2}\ s^{-1}\ PPF$$

Terms Used to Describe Visible and Photosynthetic Light Energy

Lux and footcandles (ft-c) are photometric measurements based on the sensitivity of the eye to various wavelengths of electromagnetic radiation. The *lux* is the illuminance resulting from a light energy (luminous) flux of 1 lumen distributed evenly over one square meter (1 lux = 1 lumen m^{-2}). The *footcandle* is the illuminance resulting from a luminous flux of 1 lumen per square foot (1 ft-c = 1 lumen ft^{-2}).

Watts and joules are *radiometric* measurements that express energy per unit of time per unit area without respect to wavelength. Such measurements made in the PAR waveband are termed W PAR.

In electromagnetic terms, the mol is Avogadro's number of photons. Micro-mol per square meter per second ($\mu mol \cdot m^{-2} \cdot s^{-1}$) is a flux of 6.022×10^{17} photons intercepted by one square meter over one second. Each of the light sources above has its own specific distribution of wavelengths compared with daylight; each, therefore, requires its own conversion factor. Sometimes Einstein (E) is used in place of mol. The terms are equivalent. However, mol is preferred because E is not a *système international* (SI) unit.

Glossary

abaxial   Away from the axis or central line; turned toward the base, dorsal.

abscisic acid   A plant hormone involved in abscission, dormancy, stomatal closure, growth inhibition, and other plant responses.

abscission zone   A layer of thin-walled cells extending across the base of a petiole or peduncle, whose breakdown separates the leaf or fruit from the stem and thus causes the leaf or fruit to drop.

absorption   The taking up of water by assimilation or imbibition. The taking up by capillary, osmotic, chemical, or solvent action such as the taking up of water from air, or taking up of gases by water, or taking up of mineral nutrients by plant roots.

acclimatization   The adaptation of an individual plant to a changed climate, or the adjustment of a species or a population to a changed environment, often over several generations.

achene   A simple, dry, one-seeded indehiscent fruit with the seed attached to the ovary wall at one point only. The so-called strawberry and sunflower "seeds" are examples.

acid delinting   A process used to remove the short fibers (lint) from seed cotton.

acid soil   Soil with a reaction below pH 7; more technically, a soil having a preponderance of hydrogen ions over hydroxyl ions in solution.

action spectrum   A graph indicating the changes in response of an entity to changes in a stimulus.

active transport   Movement of water and solutes into and through a plant by a force that allows a flow through a membrane or against a concentration gradient.

adaptation   The process of change in structure or function of an individual or population caused by environmental changes.

adaxial   Toward the axis or center; turned toward the apex; ventral.

adenosine diphosphate (ADP)   A nucleotide (a nitrogen-based compound) composed of adenine and ribose with two phosphate groups attached. ATP and ADP participate in metabolic reactions (both catabolic and anabolic). These molecules, through the process of being phosphorylated (accepting phosphate groups) or dephosphorylated (losing phosphate groups), transfer energy within the cells to drive metabolic processes.

adenosine triphosphate (ATP)   ATP has three phosphoric groups attached and is the phosphorylated condition of ADP. It conveys energy needed for metabolic reactions, then loses one phosphate group to become ADP (adenosine diphosphate).

adhesion   The molecular attraction between unlike substances such as water and sand particles.

adsorption   The attraction of ions or molecules to the surface of a solid.

adult phase   A growth phase in plants characterized by the ability to flower and reproduce.

adventitious   Refers to structures arising from an unusual place; for example, buds at places other than shoot terminals or leaf axils, or roots growing from stems or leaves.

aeration   The process by which air in the soil is replaced by air from the atmosphere.

aerobic   An environment or condition in which oxygen is not deficient for chemical, physical, or metabolic processes.

agar   A gelatinous substance obtained from certain species of red algae; widely used as a substrate in aseptic cultures.

aggregate   Many primary soil particles held in a single unit as a clod, crumb, block, or prism.

aggregate fruit   Many fruits are attached to a single ovary, for example, blackberry.

agronomy   The art and science of crop production and soil management.

A horizon   The surface layer of varying thickness of a mineral soil having maximum organic matter accumulation, maximum biological activity, and/or eluviation by water of materials such as iron and aluminum oxides and silicate clays.

air-dry   The state of dryness at equilibrium with the moisture content in the surrounding atmosphere.

air layer   An undetached aerial portion of a plant on which roots are caused to develop, commonly as the result of wounding or other stimulation.

air porosity   The proportion of the bulk volume of soil that is filled with air at any given time or under a given condition, such as a specified moisture tension.

albino   A plant or part of a plant lacking chlorophyll. Albinism is usually lethal in higher plants.

albuminous seeds   Seeds that store most of their food in the endosperm, for example, cereal grains.

aleurone   The outer layer of cells surrounding the endosperm of a cereal grain (caryopsis).

alfisol   A fertile tropical soil with a basic argillic horizon.

alkaline soil   Any soil that has a pH greater than 7.

alkali soil   *See* sodic soil.

allele   One of a pair or a series of factors that occur at the same locus on homologous chromosomes; one alternative form of a gene.

allelopathy   The excretion of chemicals by plants of some species that are toxic to plants of other species.

alluvial soil   A recently developed soil from deposited soil material that exhibits essentially no horizon development or modifications. Also, soil comprised of sediment that is deposited by flowing water.

alternate   In taxonomy, a leaf arrangement with leaves placed singly at various heights on the stem; i.e., not opposite or whorled.

alumino-silicates   Compounds containing aluminum, silicon, and oxygen as main constituents. An example is microcline, $KAlSi_3O_8$.

ambient temperature   Air temperature at a given time and place; not radiant temperature.

amendment   Any substance such as lime, sulfur, gypsum, or an organic material like peat moss, sawdust, or bark used to alter the properties of a soil, generally to improve its physical properties.

amino acids   The fundamental building blocks of proteins. There are twenty common amino acids in living organisms, each having the basic formula NH_2–CHR–COOH.

ammonification   The biochemical process whereby ammoniacal nitrogen is released from nitrogen-containing organic compounds.

ammonium fixation   The incorporation of ammonium ions by soil fractions so that they are relatively insoluble in water and nonexchangeable by the principle of cation exchange.

anaerobic   An environment or condition in which molecular oxygen is deficient for chemical, physical, or biological processes.

anchor root   *See* brace root.

andisol   A highly permeable, glassy soil formed from volcanic ash.

angiosperm   One of a large group of seed-bearing plants in which the female gamete is protected within an enclosed ovary. A flowering plant.

angle of incidence   The angle at which an object intersects a surface, relative to that surface.

anion   A negatively charged particle that, during electrolysis, is attracted to positively charged surfaces.

annual ring   The cylinder of secondary xylem added to a woody plant stem by the cambium in any one year.

annuals   Plants living one year or less. During this time, the plant grows, flowers, produces seeds, and dies.

antagonism   Opposing action of different chemicals such that the action of one is impaired or the total effect is less than that of one acting separately.

anther   In a flower, the saclike structure of the stamen in which microspores (pollen grains) are produced; usually borne on a filament.

anthesis   A developmental stage in flowering at which anthers rupture and pollen is shed. A state of full bloom.

anthocyanin   A class of water-soluble pigments that account for many of the red to blue flower, leaf, and fruit colors. Anthocyanins occur in the vacuole of the cell.

antipodal nuclei   The three or more nuclei at the end of the embryo sac opposite the egg nucleus (female gamete). They are produced by mitotic divisions of the megaspore and degenerate following sexual fertilization.

apical dominance   The inhibition of lateral buds on a shoot due to auxins produced by the apical bud.

apical floral bud   A bud forming at the shoot apex that has undergone a transition to produce floral rather than vegetative tissues.

apical meristem   A mass of undifferentiated cells capable of division at the tip of a root or shoot. These cells multiply by division, allowing the plant to grow in depth or height.

apomixis   The asexual (vegetative) production of seedlings in the usual sexual structures of the flower but without the mingling and segregation of chromosomes. Seedling characteristics are the same as those of the maternal parent.

aquifer   Natural underground reservoirs of water.

arable land   Land suitable for the production of crops.

argillic   The horizon in soils characterized by an accumulation of clay particles through leaching.

aridisol   Soil formed in areas with low rainfall and little leaching. It can be fertile but may accumulate salts near the surface.

asexual reproduction   The production of a new plant by any vegetative means not involving meiosis and the union of gametes.

assimilation   The transformation of organic and inorganic materials into protoplasm.

atom   The smallest particle in which an element combines, either with itself or with other elements; the smallest quantity of matter possessing the properties of a particular element.

authority   The abbreviation of the last name of the person who named the organism. The authority follows the genus and species in scientific nomenclature; for example, the scientific name of marigold is *Tagetes erecta* L., and the L. stands for Linneaus.

autogamy   Pollination within the same flower.

autotroph   Organism that can synthesize its own food from inorganic compounds via a process such as photosynthesis. *See also* heterotroph.

auxins   A class of plant hormones that influences plant growth in several ways, including apical dominance, cell growth, and phototropic response.

available nutrient   That portion of an element or compound in the soil that can be readily absorbed, assimilated, and utilized by growing plants. (*Available* should not be confused with *exchangeable*.)

available water (AW)   The portion of water in a soil that can be readily absorbed by plant roots. The soil moisture held in the soil between field capacity and permanent wilting percentage (available water = FC − PWP).

axil   The angle on the upper side of the union of a branch and main stem or of a leaf and a stem.

axillary bud   A bud formed in the axil of a leaf.

backcross   In breeding, a cross of a hybrid with one of its parents or with a genetically equivalent organism. In genetics, a cross of a hybrid with a homozygous recessive.

bagasse   The dry refuse of sugarcane after the juice has been expressed.

bar   A unit of pressure equal to 1 million dynes/cm^2. Equivalent approximately to 1 atm of pressure.

bark   The tissue of a woody stem or root from the cambium outward.

base pair   The nitrogen bases that pair in the DNA molecule—adenine with thymine, and guanine with cytosine.

berry   A simple fleshy fruit formed from a single ovary; the ovary wall is fleshy and includes one or more carpels and seeds. For example, fruits of the tomato and grape are botanically berries.

B horizon   A soil layer of varying thickness (usually beneath the A horizon) that is characterized by an accumulation of silicate clays, iron and aluminum oxides, and humus, alone or in combination and/or a blocky or prismatic structure. Also known as the *zone of accumulation*.

biennial   A plant that completes its life cycle within two seasons. For most biennial plants, the two seasons are separated by an obligate degree of cold temperature sufficient to initiate flowering and fruit formation, after which the plant dies.

binomial In biology, each species is generally indicated by two names; first, the genus to which it belongs, and second, the species name (e.g., *Quercus suber,* cork oak).

biocide A combination of a bactericide and a fungicide used for cut-flower-keeping solutions, or used as a sterilant in horticultural operations.

biodegradable Materials readily decomposed by microorganisms such as bacteria and fungi.

biodiversity The term used to describe the genetic diversity within and among all species.

biology The science that deals with living organisms.

biome Large terrestrial ecosystem characterized by specific communities, usually named after the predominant vegetation found there, e.g., temperate forest.

biotechnology Coined by Karl Erecky in 1919, it refers to any product produced from raw materials with the aid of living organisms. For example, monoclonal antibodies (for immunotherapy), enzymes (lactase and rennin), and human insulin.

biotype A subset of a species that has characteristics distinguishing it from others in the same species.

biuret A compound ($H_2NCONHCONH_2 \cdot H_2O$) formed by the thermal decomposition of urea (H_2NCONH_2) that is phytotoxic to many crops. Therefore, biuret-free urea is used for fertilization of crops.

blade The thin and often flat part of a leaf.

bloat Excessive accumulation of gases in the rumen of some animals.

bolting Rapid production of flower stalks in some herbaceous plants after sufficient chilling or a favorable photoperiod.

botany The science of plants, their characteristics, functions, life cycles, and habits.

bourse shoot Shoot arising from the mixed bud on a fruiting spur of apple. Bourse shoots terminate growth by forming floral or vegetative buds.

brace root Also known as *anchor root.* A type of adventitious root that grows from above-ground parts of the stem and serves to support some plants; for example, corn.

bract A modified leaf, from the axil of which arises a flower or an inflorescence.

branched crown A branch arising axially from the strawberry stem. Branched crowns also develop terminal floral buds.

brassinolides Group of steroids that have been recently identified as a class of plant hormones that appears to regulate cell division and elongation.

breeder seed Seed (or vegetative propagating material; e.g., potato) increased by the originating, or sponsoring, plant breeder or institution and used as the source to increase foundation seed.

broadcast seeding Scattering seed or fertilizers uniformly over the soil surface rather than placing in rows.

brushing The procedure using riverbank brush for constructing a barrier for sheltering crop plants from wind and for modifying the plant's microenvironment to accelerate growth.

bryophytes A division of plants that lack a vascular system. Mosses and liverworts are included in this division.

bud A region of meristematic tissue with the potential for development into leaves, shoots, flowers, or combinations; generally protected by modified scale leaves.

budding A form of grafting in which a single vegetative bud is taken from one plant and inserted into stem tissue of another plant so that the two will grow together. The inserted bud develops into a new shoot.

bud scar A scar left on a shoot when the bud or bud scales drop.

bud sport A mutation arising in a bud and producing a genetically different shoot. Includes change due to gene mutation, somatic reduction, chromosome deletion, or polyploidy.

buffering The ability of some soil components to resist a change in soil pH.

bulb A highly compressed underground stem (basal plate) to which numerous storage scales (modified leaves) are attached. Examples are lily, onion, tulip.

bulk density The mass of a known volume (including air space) of soil. The soil volume is determined in place, then dried in an oven to constant weight at 105°C. Bulk density (D_B) = oven dry weight of soil/volume of soil.

bunch type grass Grass that does not spread by rhizomes or stolons.

burn-down herbicide An herbicide that destroys existing weeds in a field before the field is prepared for planting.

C_3 cycle The Calvin-Benson cycle of photosynthesis, in which the first products after CO_2 fixation are three-carbon molecules.

C_4 cycle The Hatch-Slack cycle of photosynthesis, in which the first products after CO_2 fixation are four-carbon molecules.

calcareous soil Soil containing sufficient calcium and/or magnesium carbonate to effervesce visibly when treated with cold 0.1 normal (0.1N) hydrochloric acid.

calcium pectate An organic calcium compound found in the middle lamella between plant cells and serving as an intercellular cement.

callus Mass of large, thin-walled parenchyma cells, usually developing as the result of a wound.

calorie Also known as *gram calorie.* Unit for measuring energy, defined as the heat necessary to raise the temperature of 1 g of water from 14.5°C to 15.5°C at standard pressure; 1 kilocalorie (kcal) raises the temperature of 1 kg of water 1°C. Thus, 1 kcal = 1,000 cal.

calyx The collective term for the sepals.

cambium A thin layer of longitudinally dividing cells between the xylem and phloem that gives rise to secondary growth.

capillary water The water held in the capillary or small pores of a soil, usually with a tension greater than 60 cm of water.

capsule A simple, dry, dehiscent fruit, with two or more carpels.

carbohydrate Compound of carbon, hydrogen, and oxygen in the ratio of one atom each of carbon and oxygen to two of hydrogen, as in sugar, starch, and cellulose.

carbon dioxide compensation point The level of carbon dioxide available to the leaf for photosynthesis equals the amount of carbon dioxide generated by the plant's respiration. Photosynthesis cannot increase unless the amount of available carbon dioxide is increased by further respiration.

carbon dioxide fixation The addition of H^+ to CO_2 to yield a chemically stable carbohydrate. The H^+ is contributed by NADPH, the reduced (hydrogen-rich) form of $NADP^+$, produced in the noncyclic phase of the light reactions of photosynthesis. The H^+ comes originally from the photolysis of water.

carbon-nitrogen ratio The ratio of the weight of organic carbon to the weight of total nitrogen in a soil or in organic material.

carotene   Yellow plant pigments, precursors of vitamin A. Alpha, beta, and gamma carotenes are converted into vitamin A in the animal body.

carpel   Female reproductive organ of flowering plants. In some plants, one or more carpels unite to form the pistil.

caryopsis   Small, one-seeded, dry fruit with a thin pericarp surrounding and adhering to the seed; the seed (grain) or fruit of grasses.

Casparian strip   A secondary thickening that develops on the radial and end walls of some endodermal cells.

catalyst   Any substance that accelerates a chemical reaction but does not enter into the reaction itself.

cation   A positively charged ion.

cation exchange capacity (CEC)   Also known as *base exchange capacity.* A measure of the total amount of exchangeable cations that a soil can hold; expressed in meq/100 g soil at pH 7.

catkin   A type of inflorescence (a spike) generally bearing either pistillate or staminate flowers. Found on walnuts and willows, for example.

cell   The basic structure and physiological unit of plants and animals.

cell membrane   The membrane that separates the cell wall and the cytoplasm and regulates the flow of material into and out of the cell.

cell plate   The precursor of the cell wall, formed as cytokinesis starts during cell division. It develops in the region of the equatorial plate and arises from membranes in the cytoplasm.

cellulose   A complex carbohydrate composed of long, unbranched beta-glucose molecules, which makes up 40 to 55 percent by weight of the plant cell wall.

cell wall   The outermost, cellulose limit of the plant cell; the barrier that develops between nuclei during mitosis.

Cenozoic   The geologic era extending from about 65 million years ago to the present time.

center of origin   A geographical area in which a species is thought to have evolved through natural selection from its ancestors.

cereal   A member of the POACEAE family grown primarily for its mature, dry seed.

cereal forage   Cereal crop harvested when immature for hay, silage, green chop, or pasturage.

certified seed   The progeny of foundation, registered, or certified seed, produced and handled to maintain satisfactory genetic identity and purity, and approved and certified by an official certifying agency.

character   The expression of a gene in the phenotype.

chelate   Also known as *sequestering agent.* A large organic molecule that attracts and tightly holds specific cations, like a chemical claw, preventing them from taking part in inorganic reactions but at the same time allowing them to be absorbed and used by plants.

chilling injury   Direct and/or indirect injury to plants or plant parts from exposure to low, but above-freezing temperatures (as high as 9°C or 48°F).

chilling requirement   The duration of exposure to cold needed for a plant to overcome dormancy.

chimera   A plant composed of two or more genetically different tissues. Includes *periclinal chimera,* in which one tissue lies over another as a glove fits a hand; *mericlinal chimera,* where the outer tissue does not completely cover the inner tissue; and *sectorial chimera,* in which the tissues lie side by side.

chisel   A tillage implement with one or more cultivator-type shanks to which are attached knifelike units that shatter or break up hard, compact layers, usually in the subsoil.

chlorophyll   A complex organic molecule that traps light energy for conversion through photosynthesis into chemical energy.

chloroplast   Chlorophyll-containing cytoplasmic body, in which important reactions of sugar or starch synthesis take place during photosynthesis.

chlorosis   A condition in which a plant or a part of a plant is light green or greenish yellow because of poor chlorophyll development or the destruction of chlorophyll resulting from a pathogen or a mineral deficiency.

C horizon   A soil layer beneath the B layer that is relatively little affected by biological activity and pedogenesis and is lacking properties diagnostic of an A or B horizon.

chromosome   A specific, highly organized body in the nucleus of the cell that contains DNA.

class   A group of soils having a definite range in a particular property such as acidity, degree of slope, texture, structure, land-use capability, degree of erosion, or drainage. Also, the taxonomic category that falls between phylum (division) and order.

classification   The systematic arrangement into categories on the basis of characteristics. Broad groupings are made on the basis of general characteristics and subdivisions on the basis of more detailed differences in specific properties.

clay   Soil particles less than 0.002 mm in equivalent diameter. Also, soil material containing more than 40 percent clay, less than 45 percent sand, and less than 40 percent silt.

claypan   A compact, slowly permeable layer of varying thickness and depth in the subsoil having a much higher clay content than the overlying material. Claypans are usually hard when dry, and plastic and sticky when wet.

climacteric   The period in the development of some plant parts involving a series of biochemical changes associated with the natural respiratory rise and autocatalytic ethylene production.

climate   Weather conditions such as wind, precipitation, and temperature that prevail in a given region.

climax community   The stable community of plants and animals that results from a complex community replacing simpler ones. *See also* succession.

climax vegetation   Fully developed plant community in equilibrium with its environment.

clod   A compact, coherent mass of soil produced artificially, usually by tillage operations, especially when performed on soils either too wet or too dry.

clone   The aggregate of individual organisms originating from one sexually produced individual (or from a mutation) and maintained exclusively by asexual propagation.

clove   One of a group of small bulbs produced, for example, by garlic and shallot plants.

coenzyme   A substance, usually nonprotein and of low molecular weight, necessary for the action of some enzymes.

cohesion   Holding together; a force holding a solid or liquid together, owing to attraction between like molecules.

colchicine   An alkaloid, derived from the autumn crocus, used specifically to inhibit the spindle mechanism during cell division and thus cause a doubling of chromosome number.

cold frame   An enclosed, unheated covered frame useful for growing and protecting young plants in early spring. The top is covered with glass or plastic, and sunlight provides heat.

coleoptile   A transitory membrane (first leaf) covering the shoot apex in the seedlings of certain monocots. It protects the plumule as it emerges through the soil.

coleorhiza   Sheath that surrounds the radicle of the grass embryo and through which the young developing root emerges.

collenchyma tissue   Elongated, parenchymatous cells with variously thickened walls, commonly at the acute angles of the cell wall.

colloid   Organic and inorganic matter with very small particle size and a correspondingly large surface area per unit of mass.

community   All the plant populations within a given habitat; usually the populations are considered to be interdependent.

compaction   The compression of soil by the passing of heavy machinery or human and animal traffic.

companion cells   Cells associated with the sieve-tubes in the phloem.

companion crop   A crop sown with another crop and harvested separately. Small-grain cereal crops are often sown with forage crops (grasses or legumes) and harvested in the early summer, allowing the forage crop to continue to grow (e.g., oats sown as a companion crop with red clover).

compensation point   The light intensity at which the rates of photosynthesis and respiration are equal.

complete flower   A flower that has pistils, stamens, petals, and sepals, all attached to a receptacle.

compost   A mixture of organic residues and soil that has been piled, moistened, and allowed to decompose biologically. Mineral fertilizers are sometimes added.

compound bud   A bud that contains more than one growing point; usually, one growing point is dominant. The bud may contain floral tissues, vegetative tissues, or both.

compound leaf   A leaf whose blade is divided into a number of distinct leaflets.

cone   The woody, usually elongated seed-bearing organ of a conifer, consisting of a central stem, woody scales and bracts (often not visible), and seeds.

contact herbicide   A chemical that kills plants on contact.

controlled atmosphere (CA) storage   A process of fruit storage wherein the CO_2 level is raised (to 1 to 3 percent) and the O_2 level is lowered (to 2 to 3 percent).

cork cambium   The meristem from which cork develops.

cork tissue   An external, secondary tissue impermeable to water and gases. It is produced by certain kinds of woody plants.

corm   A short, solid, vertical, enlarged underground stem in which food is stored; it contains undeveloped buds (leaf and flower). Examples are crocus, freesia, and gladiolus.

corolla   The collective term for all petals of a flower.

cortex   Primary tissue of a stem or root bounded externally by the epidermis and internally in the stem by the phloem and in the root by the pericycle.

corymb   Inflorescence with outer flowers having longer stems than the inner flowers, resulting in a flat-topped appearance, for example, candytuft.

cotyledons   Leaflike structures at the first node of the seedling stem. In some dicots, cotyledons contain the stored food for the young plant not yet able to photosynthesize its own food. Often referred to as seed leaves.

cover crop   A close-growing crop grown primarily for the purpose of protecting or improving soil between periods of regular crop production or between trees and vines in orchards and vineyards.

critical daylength   The duration of a light period that determines when a photoperiodic plant will flower. *See also* long-day plant; short-day plant.

crop residue   Portion of crop plants remaining after harvest.

crop rotation   Growing crop plants in a different location in a systematic sequence to help control insects and diseases, improve the soil structure and fertility, and decrease erosion.

cross-pollination   The transfer of pollen from a stamen to the stigma of a flower on another plant, except for clones where the plants must be two different clones.

cross section   The surface exposed when a plant stem is cut horizontally and the majority of the cells are cut transversely.

crown   The region at the base of the stem of cereals and forage species from which tillers or branches arise. In woody plants, the root-stem junction. In forestry, the top portions of the tree.

crumb   A soft, porous, irregular, natural unit of soil structure from 1 to 5 mm in diameter.

crust   A surface layer on soils, ranging in thickness from a few millimeters to a few centimeters. It is more compact, hard, and brittle when dry than the soil beneath it.

cryptochrome   Light-receiving molecule that mediates light regulation of plant growth and development.

crystal lattice water   Water that is bound inside certain types of soil particles such as clay and is unavailable to plants.

cubing   Process of forming hay into high-density cubes to facilitate transportation, storage, and feeding.

culm   Stem of grasses and bamboos; usually hollow except at the swollen nodes.

cultivar derived from **"cultivated variety."**   International term denoting certain cultivated plants that are clearly distinguishable from others by any characteristic and that, when reproduced (sexually or asexually), retain their distinguishing characteristics. In the United States, *variety* is often considered synonymous with *cultivar.*

cultivation   The growing or tending of crops.

cure   To prepare crops for storage by drying. Dry onions, sweet potatoes, and hay crops are examples. Dehydration of fruits for storage is not considered curing.

cuticle   An impermeable surface layer on the epidermis of plant organs.

cutin   A clear or transparent waxy material on plant surfaces that tends to make the surface waterproof.

cutting   A detached leaf, stem, or root that is encouraged to form new roots and shoots and develop into a new plant.

cutting propagation   The process in which a piece of tissue regenerates the missing part.

cyme   A type of inflorescence that has a broad, more or less flat-topped determinate flower cluster, with the central flower opening first.

cytochrome   A class of several electron-transport proteins serving as carriers in mitochondrial oxidation and in photosynthetic electron transport.

cytokinesis   Division of cytoplasmic constituents at cell division.

cytokinins   A group of plant growth hormones important in the regulation of nucleic acid and protein metabolism and in cell division, organ initiation, and the delay of senescence.

cytology   The study of cells and their components and of the relationship of cell structure to function.

cytoplasm   The living material of the cell, exclusive of the nucleus, consisting of a complex protein matrix or gel. The part of the cell in which essential membranes and cellular organelles are found.

damping off A pathogenic disorder causing seedlings to die soon after seed germination.

daylength Number of effective hours of daylight in each twenty-four-hour cycle.

day-neutral plant (DNP) Plant capable of flowering under either long or short daylengths.

DDT (dichlorodiphenyltrichloroethane) One of the earliest insecticides of the chlorinated hydrocarbon family. No longer used in some countries because of its persistence in the environment.

deciduous Refers to trees and shrubs that lose their leaves every fall. Distinguished from evergreens, which retain their leaves throughout the year.

decomposition Degradation into simpler compounds; rotting or decaying.

defoliant A chemical or method of treatment that causes only the leaves of a plant to fall off or abscise.

dehiscence The splitting open at maturity of pods or capsules along definite lines or sutures.

dehulled seed Seed from which pods, glumes, or other outer covering have been removed, as sometimes with lespedeza and timothy. Also often ambiguously referred to as hulled seed.

denitrification Biological reduction of nitrate or nitrite to gaseous nitrogen or nitrogen oxides.

deoxyribonucleic acid (DNA) A molecule composed of repeating subunits of ribose (a sugar), phosphate, and the nitrogenous bases adenine, guanine, cytosine, and thymine. Genes, the fundamental units of inheritance on chromosomes, are sequences of DNA molecules.

desalinization Removal of salts from saline soil or water.

desiccant Substance used to accelerate drying of plant tissues.

determinate growth The flowering of plant species uniformly within certain time limits, allowing most of the fruit to ripen about the same time.

diatomaceous earth A geologic deposit of fine, grayish, siliceous material composed chiefly of the remains of diatoms.

dichogamy Maturation of male or female flowers at different times, ensuring cross-pollination. Common in maple and walnuts.

dicot *See* dicotyledons.

dicotyledons The subclass of flowering plants that have two cotyledons. Also known as *dicots*.

DIF Average day temperature minus average night temperature. Used in greenhouses to control stem elongation. Positive DIF (day temperature greater than night temperature) promotes stem elongation. Negative DIF inhibits stem elongation.

differentiation Development from one cell to many cells, together with a modification of the new cells for the performance of particular functions.

diffusion The movement of molecules, and thus a substance, from a region of higher concentration of those molecules to a region of lower concentration.

digestion The breakdown of complex foods to simpler food materials, which are more easily utilized. Digestion requires energy.

dihybrid cross A cross between organisms differing in two characteristics.

diluent An essentially inert or nonreacting gas, liquid, or solid used to reduce the concentration of the active ingredient in a formulation.

dioecious Refers to individual plants having either staminate (male) or pistillate (female) flowers, but not both. Therefore, plants of both sexes must be grown near each other to provide pollen before fruits and seed can be produced. Examples are English holly, asparagus, ginkgo, and date palms.

diploid (2n) Refers to two sets of chromosomes. Germ cells have one set and are haploid; somatic cells have two sets and are diploid (except for polyploid plants).

disease Any change from the state of metabolism necessary for the normal development and functioning of any organism.

disperse To break up compound particles, such as aggregates, into the individual component particles. To distribute or suspend fine particles, such as clay, in or throughout a dispersion medium.

dissolution The process by which constituents of parent material dissolve in water or weak, naturally occurring acids and are leached away.

diurnal Recurring or repeated every day. Going through regular or routine changes daily.

division A propagation method in which underground stems or roots are cut into pieces and replanted.

dolomite $CaCO_3$—$MgCO_3$; a native limestone source having a significant magnesium content.

domain Broadest category of taxonomic classification that separates life forms into three categories: prokaryotes, eukaryotes, and archaea bacteria.

dominant gene The gene (or the expression of the characteristic it influences) that, when present in a hybrid with a contrasting gene, completely dominates in the development of the characteristic. In peaches, for example, white fruit is dominant over yellow.

dormancy A general term denoting a lack of growth of seeds, buds, bulbs, or tubers due to unfavorable environmental conditions (external dormancy or quiescence) or to factors within the organ itself (internal dormancy or rest).

double fertilization The process of sexual fertilization in the angiosperms in which one nucleus from the male gametophyte fertilizes the egg nucleus to form the zygote and a second nucleus from the male gametophyte fertilizes two polar nuclei to form endosperm tissue.

drip irrigation A method of watering plants so that only soil in the plant's immediate vicinity is moistened. Water is supplied from a thin plastic tube at a low rate of flow. Sometimes called trickle irrigation.

drupe A simple, fleshy fruit derived from a single carpel, usually one-seeded, in which the exocarp is thin, the mesocarp fleshy, and the endocarp hard, for example a peach.

dry land farming The practice of crop production without irrigation.

dry matter percentage The percentage of the total fresh plant material left after water is removed. The percentage is determined by weighing a sample of fresh plant material, oven-drying the sample, then reweighing the dried sample. Dry matter percentage equals dry weight divided by fresh weight times 100.

earlywood Also known as *spring wood*. The less dense part of the growth ring. It is made up of cells with thinner walls, a greater radial diameter, and shorter length than those formed later in the year.

ecology The study of life in relation to its environment.

ecosystem A living community and all the factors in its nonliving environment.

ecotype Genetic variant within a species that is adapted to a particular environment yet remains interfertile with all other members of the species.

edaphic Pertaining to the influence of the soil on plant growth.

E horizon The dense layer of soil that forms when ions and organic matter are leached from the A horizon and deposited just above the B horizon.

electrons Negatively charged particles that form a part of all atoms.

element A substance that cannot be divided or reduced by any known chemical means to a simpler substance; ninety-two natural elements are known.

eluviation The removal of soil material in suspension (or in solution) from a layer or layers of a soil. Usually, the loss of materials in solution is described by the term *leaching*.

emasculate To remove the anthers from a bud or flower before pollen is shed. Emasculation is a normal preliminary step in hybridization to prevent self-pollination.

embryo A miniature plant within a seed produced as a result of the union of a male and female gamete and resulting in the development of a zygote.

embryo sac Typically, an eight-nucleate female gametophyte. The embryo sac arises from the megaspore by successive mitotic divisions.

emergy The amount of solar energy represented in a resource or commodity.

endemic Species native to a particular environment or locality.

endocarp Inner layer of the fruit wall (pericarp).

endodermis In roots, a single layer of cells at the inner edge of the cortex. The endodermis separates the cortical cells from cells of the pericycle.

endogenous Produced from within. Opposite of *exogenous*.

endophyte An organism that lives within another organism, often in a symbiotic relationship. Fungal endophytes are associated with many grasses and serve as a deterrent to insects and grazing animals. Endophytes represent a problem in forages because they can be toxic to the animals that graze on the forages.

endoplasmic reticulum (ER) The lamellar or tubular system of the colorless cytoplasm in a cell.

endosperm The 3n tissue of angiospermous seeds that develops from sexual fusion of the two polar nuclei of the embryo sac and a male sperm cell. The endosperm provides nutrition for the developing embryo. A food storage tissue.

energy (kinetic) The capacity to do work. Examples are light, heat, chemical, electrical, or nuclear energy.

entisol A developing soil without distinct horizons.

entomology The scientific study of insects.

enzyme Any of many complex proteins produced in living cells, that, even in very low concentrations, promotes certain chemical reactions but does not enter into the reactions itself.

epicotyl The upper portion of the embryo axis or seedling, above the cotyledons and below the first true leaves.

epidermis The outer layer of cells on all parts of the primary plant body: stems, leaves, roots, flowers, fruits, and seeds. It is absent from the root cap and on apical meristems.

epigeal germination A type of seed germination in dicots in which the cotyledons rise above the soil surface. This occurs in beans, for example.

epigyny Arrangement of flower parts in which the ovary is embedded in the receptacle so that the other parts appear to arise from the top of the ovary.

epinasty A twisted or misshapened stem or leaf. Leaves exhibit downward curling of leaf blade due to more rapid growth on the upper side cell than the lower side.

epiphyte A plant that grows upon another plant yet is not parasitic.

erosion The wearing away of the surface soil by wind, moving water, or other means.

ethephon (2-chloroethyl phosphonic acid) A useful compound in horticultural applications that breaks down to release ethylene gas.

ethylene A gaseous growth hormone (C_2H_4) regulating various aspects of vegetative growth, fruit ripening, abscission of plant parts, and the senescence of flowers.

etiolation A condition involving lack of chlorophyll, increased stem elongation, and poor or absent leaf development. It occurs in plants growing under very low light intensity or complete darkness.

eukaryotic cells Cells that have a nucleus.

evapotranspiration The total loss of water by evaporation from the soil surface and by transpiration from plants, from a given area, and during a specified period of time.

evergreen Trees or shrubs that are never entirely leafless, as in pine or citrus.

evolution The development of a species, genus, or other larger group of plants or animals over a long time period.

exalbuminous seeds Seeds that store little or no food in the endosperm but usually in cotyledons, for example, beans.

exchange capacity The total ionic charge of the adsorption complex active in the adsorption of ions. Also called anion exchange capacity, cation exchange capacity, and base exchange capacity.

exfoliation The peeling of the outer layers of rocks caused by differential rates of contraction and expansion that in turn are caused by temperature changes.

exocarp The outermost layer of the fruit wall (pericarp).

exogenous Produced outside, or originating from or because of external causes. Opposite of *endogenous*.

explant Living tissue removed from its place in a body and placed in an artificial medium for tissue culture.

F_1 First filial generation in a cross between any two parents.

F_2 Second filial generation, obtained by crossing two members of the F_1 generation, or by self-pollinating plants of the F_1 generation.

facultative Referring to an organism having the power to live under a variety of conditions; a facultative parasite is either parasitic or saprophytic.

fallow Cropland left idle for one or more seasons for any number of reasons, such as to accumulate moisture, destroy weeds, and allow the decomposition of crop residue.

family In plant taxonomy, a group of genera.

fascicle A bundle of needle-leaves of gymnosperms such as the pines.

fatty acid Organic compound of carbon, hydrogen, and oxygen that combines with glycerol to make a fat.

fermentation An anaerobic chemical reaction in foods, such as the production of alcohol from sugar by yeasts.

fertigation Practice of adding fertilizer to irrigation water.

fertilization   The union of an egg and a sperm (gametes) to form a zygote. Also, application to the soil of needed plant nutrients, such as nitrogen, phosphoric acid, potash, and others.

fertilizer   Any organic or inorganic material of natural or synthetic origin added to a soil to supply elements essential to the growth of plants.

fibers   Elongated, tapering, thick-walled strengthening cells in various parts of the plant.

fibrous root   Root system characterized by many multidirectional branches.

field capacity   Also known as *field moisture capacity*. The percentage of water remaining in a soil two or three days after having been saturated and after free drainage due to gravity has practically ceased.

filament   Stalk portion of a stamen. *See also* anther; stamen.

fixation field moisture capacity   *See* field capacity. The process in soil by which certain chemical elements essential for plant growth convert from a soluble or exchangeable form to a much less soluble or to a nonexchangeable form.

fleshy fruit   Any fruit formed from an ovary that has fleshy or pulpy (not dried) walls at maturity. Also, those fruits that include fleshy parts of the perianth, floral tube, or the receptacle.

flocculate   To clump together individual, tiny soil particles, especially fine clay, into small granules. Opposite of *disperse*.

flooding   A method of irrigation by which water is released from field ditches and allowed to spread over the land.

flora   A collective term for all the plant types that grow in a region.

floral incompatibility   A genetic condition in which certain normal male gametes are incapable of functioning in certain pistils.

floricane   A second-year cane (previously a primocane) of raspberries, blackberries, or their hybrids, on which floral buds initiated in the previous season develop flower and fruit.

florigen   Term given to a stimulus (suspected of being a protein) that promotes flowering.

flower   Floral leaves grouped together on a stem that, in the angiosperms, are adapted for sexual reproduction.

fodder   Coarse grasses such as corn and sorghum harvested with the seed and leaves and cured for animal feed.

follicle   A simple, dry dehiscent fruit, having one carpel and splitting along one suture. Examples are milkweeds and magnolia.

food chain   The path along which food energy is transferred within a natural plant and animal community (from producers to consumers to decomposers).

foot-candle (ft-c)   A standard measure of light (English system). The illumination of a standard light source of one candle, one foot away, falling at all points on a square foot surface. Photometric unit expressing the illuminance of 1 lumen per sq ft; 1 ftc = 10.76 lux. *See* lux.

forage   Vegetation used as feed for livestock, such as hay, pasture, and silage. The material is fed green or dehydrated.

forcing   A cultural manipulation used to hasten flowering or growing plants outside their natural season.

fossil   Any impression, natural or impregnated remains, or other trace of an animal or plant of past geological eras that has been preserved in the earth's crust.

foundation seed   Seed stocks increased from breeder seed, and handled to closely maintain the genetic identity and purity of a cultivar.

friable   Generally refers to a soil moisture consistency that crumbles when handled.

frost seeding   *See* oversowing.

fruit   A mature ovary; in some plants, other flower parts are commonly included as part of the fruit, e.g., the hypanthium of the apple flower surrounds the ovary.

fruiting spur   A short, thick, fruit-bearing stem terminated by an apical floral, vegetative, or mixed bud at the end of the growing season, depending on the species. Fruiting spurs usually develop flowers, leaves, and new shoot growth.

fumigation   Control of insects, disease-causing organisms, weeds, or nematodes by gases applied in an enclosed area such as a greenhouse or under plastic laid on the soil.

fungicide   A pesticide chemical used to control plant diseases caused by fungi.

fungus   Plural: **fungi**. A thallus plant unable to photosynthesize its own food (exclusive of bacteria).

furrow   Small V-shaped ditch made for planting seed or for irrigating.

furrow irrigation   A method of irrigation by which the water is applied to row crops in ditches.

gamete   A haploid-generation male sperm cell or a female egg cell capable of developing into an embryo after fusion with a germ cell of the opposite sex.

gametophyte   In a seed plant, the few-celled, haploid generations arising from a meiotic division and giving rise through mitosis to the male or female gametes.

gene   A group of base pairs in the DNA molecule in the chromosome that determines or conditions one or more hereditary characteristics.

genetically modified organism (GMO)   Any organism that is the product of genetic recombination, which includes almost all organisms except those produced clonally or parthenogenically. Sometimes different species, such as plant and bacteria, can exchange genetic material, for example, wheat. Includes, but is not limited to, genetically engineered organisms. *See* genetic engineering.

genetic code   The sequence of nitrogen bases in a DNA molecule that codes for an amino acid or protein. In a broader sense, for example, the full sequence of events from the translation of chromosomal DNA to the final stage of the synthesis of an enzyme.

genetic engineering   The process of moving genetic material (DNA) or genes from their normal location and transferring them to another organism or the same organism in a different combination (genetic recombination) or to a different chromosome. The term *engineering* indicates human intervention.

genetics   The science or study of inheritance.

genotype   The genetic makeup of a nucleus or of an individual.

genus   Plural: **genera**. A group of structurally or phylogenetically related species.

geographic information system (GIS)   A computer system that gathers, stores, and analyzes data relative to position on the surface of the earth.

geotropism   Growth curvature in plants induced by gravity.

germination   Sequence of events in a viable seed starting with imbibition of water that leads to growth of the embryo and development of a seedling.

germplasm   The protoplasm of the sexual reproductive cells containing the units of heredity (chromosomes and genes).

gibberellins   A group of natural growth hormones whose most characteristic effect is to increase the elongation of cells.

glacial till   A product of glacial weathering in which rock particles varying in size from clay to boulders are deposited by the glacier on the land surface as it melts and recedes.

global positioning system (GPS)   The use of satellite triangulation and other technology to locate a position on the earth's surface precisely.

glucose   A simple sugar composed of six carbon, twelve hydrogen, and six oxygen atoms ($C_6H_{12}O_6$).

graft   To place a detached branch (scion) in close cambial contact with a rooted stem (rootstock) so that scion and rootstock unite to form a new plant.

graft incompatibility   The inability of a stable graft to form between two plants.

grain (caryopsis)   A simple, dry, indehiscent fruit with ovary walls fused to the seed. The so-called seed of cereal or grain crops such as corn, wheat, barley, and oats is actually a fruit.

grana   Structures in the chloroplasts usually seen as stacks of parallel lamellae with the aid of an electron microscope.

gravitational water   Water that moves into, through, or out of the soil under the influence of gravity.

green chop   Forage that is chopped in the field while succulent and green and fed directly to livestock, made into silage, or dehydrated.

green manure   A crop that is plowed under while still green and growing to improve the soil.

groundwater   Water that fills all the unblocked pores of underlying material below the water table, which is the upper limit of saturation.

group   A taxonomic category of cultivated plants at the subspecies level that have the same botanical binomial but have characteristics different enough to warrant naming them for clarity.

growing degree day (GDD)   The measurement of heat accumulation used to predict crop maturation and other phenological events such as stages in an insect life cycle. A growing degree day is a day that has an average temperature 1° above a baseline. Each degree above that baseline is another growing degree day. For example, a single day that as an average temperature 15° above the baseline is considered to be fifteen growing degree days.

growing medium   Soil or soil substitute prepared by combining materials such as peat moss, vermiculite, sand, or composted sawdust. Used for growing potted plants or germinating seeds.

growth   An irreversible increase in cell size and/or cell number. An increase in dry weight, regardless of cause.

growth regulator   A synthetic or natural compound that in low concentrations controls growth responses in plants.

growth retardant   A chemical that selectively interferes with normal hormonal promotion of plant growth but without appreciable toxic effects.

guard cells   Specialized epidermal cells that contain chloroplasts and surround a stoma.

gully erosion   A process whereby water accumulates in narrow channels and, over relatively short periods, removes the soil from this narrow area to considerable depths.

guttation   Exudation of water in liquid form from plants.

gymnosperm   A seed plant with seeds not enclosed by a megasporophyll or pistil.

gynoecium   The female part of a flower or pistil formed by one or more carpels and composed of the stigma, style, and ovary.

haploid   Having only one complete set of chromosomes; referring to an individual or generation containing such a single set of chromosomes per cell.

hardening off   Adapting plants to outdoor conditions by withholding water, lowering the temperature, or lowering the nutrient supply. This conditions plants for survival when transplanted outdoors.

hardpan   A hardened soil layer, in the lower A horizon or in the upper B horizon, caused by the cementing of soil particles.

hardscape   Nonplant components of a landscape, for example, patios, decks, ponds, and so on.

hardy plants   Plants adapted to cold temperatures or other adverse climatic conditions of an area. *Half-hardy* indicates some plants may be able to take local conditions with a certain amount of protection.

harvest index   The percentage of a plant's biomass that is incorporated into the harvestable part.

hay   Herbage of forage plants, including seed of grasses and legumes, that is harvested and dried for animal feed.

head   A type of inflorescence, typical of the composite family, in which the individual flowers are grouped closely together on a receptacle. Example: sunflower. Also, the aerial terminus of a grapevine trunk; in head-pruning systems, it is the position where canes are selected for fruiting.

heaving   The partial lifting of plants out of the ground, frequently breaking their roots, as a result of freezing and thawing of the surface soil during the winter.

heavy soil   A soil with a high content of the fine separates, particularly clay, or one with a high tractor power requirement and hence difficult to cultivate.

heeling in   Temporary storage of bare-rooted trees and shrubs by placing the roots in a trench and covering with soil or sawdust.

heliotropic movement   Movement of a plant in response to the diurnal movement of the sun.

helix   A spiral form; a term often used in reference to the double spiral of the DNA molecule.

hemicellulose   A type of polysaccharide found in cell walls.

herbaceous   Refers to plants that do not normally develop woody tissues.

herbage   *See* forage.

herbarium   A collection of dried and pressed plant specimens cataloged for easy-reference.

herbicide   Any chemical used to kill plants; an herbicide may work against a narrow or wide range of plant species.

herbivore   Animal that subsists principally or entirely on plants or plant products.

heredity   The transmission of morphological and physiological characteristics from parents to their offspring.

hermaphrodite flower   A flower having both stamens (male) and pistils (female).

heterosis   Also known as *hybrid vigor.* The increased vigor, growth, size, yield, or function of a hybrid progeny over the parents that results from crossing genetically unlike organisms. Also, the increase in vigor or growth of a hybrid progeny in relation to the average of the parents.

heterotroph Organism that depends on the compounds synthesized by autotrophs for nourishment.

heterotrophic Capable of deriving energy for life processes only from the decomposition of organic compounds and incapable of using inorganic compounds as sole sources of energy or for organic synthesis. *Contrast with* autotrophic.

heterozygous Having different genes of a Mendelian pair present in the same cell or organism, for instance, a tall pea plant with genes for both tallness (T) and dwarfness (t).

higher plant parasite Plant that lives on other living plants and takes at least part of its nourishment from the host.

high intensity discharge (HID) A type of lamp that illuminates when an electric current passes through a vapor that contains xenon gas and/or metal halide compounds such as sodium or mercury salts.

hill Raising the soil in a slight mound for planting, or setting plants some distance apart.

hilum The scar on a dicot seed, such as a bean seed, where it was attached to the fruit.

histology The science that deals with the microscopic structure of plant or animal tissues.

homologous chromosomes The two members of a chromosome pair.

homozygous Having similar genes of a Mendelian pair present in the same cell or organism, for instance, a dwarf pea plant with genes for dwarfness (tt) only.

horizon, soil A layer of soil, approximately parallel to the soil surface, with distinct characteristics produced by soil-forming processes.

hormone A chemical substance produced in one part of a plant and used in minute quantities to induce a growth response in another part. For example, auxins are one type of hormone.

horticulture The study of crops that require intense and constant care, from planting through delivery to the consumer.

hotbed A bed of soil enclosed in a low glass or transparent plastic frame and heated with fermenting manure, electric cables, or steam pipes. Used to germinate seeds, root cuttings, and grow other plants for transplanting outside.

hot caps Waxpaper cones, paper sacks, or cardboard boxes with bottoms removed and placed over individual plants in spring for frost and wind protection.

human footprint The impact that human activity has on ecosystems and global resources.

humidity, relative The ratio of the weight of water vapor in a given quantity of air to the total weight of water vapor that quantity of air is capable of holding at a given temperature, expressed as a percentage.

humus The more or less stable fraction of the soil organic matter remaining after the major portion of plant and animal residues have decomposed.

hybrid The offspring of two plants or animals differing in one or more Mendelian characters.

hybridization The crossing of individuals of unlike genetic constitution. Also, a method of breeding new cultivars that uses crossing to obtain genetic recombinations.

hybrid seeds Seeds produced by the crossing of parents whose genetic composition is different from each other.

hydration The addition of molecular water to another compound to form a hydrated material more vulnerable to pulverization.

hydrologic cycle The movement of water from the atmosphere to the earth and its return to the atmosphere.

hydrolysis The chemical reaction in which water participates as a reactant and not as a solvent. Usually, the splitting of a molecule to form smaller molecules that incorporate hydrogen and hydroxyl ions, derived from water, in their structures.

hydroponics Growing plants in aerated water containing all the essential mineral nutrients rather than soil. Also called soilless gardening.

hygroscopic water Water that surrounds and is tightly held by soil particles, making the water unavailable to plants.

hypha Plural: **hyphae**. A filament that forms mycelial growth of a fungus.

hypocotyl Portion of a stem that is located above the root and below the cotyledon.

hypogeal germination In dicots, a type of seed germination in which the cotyledons remain below the soil surface. Peas are one example.

hypothesis A proposition or supposition provisionally adopted to explain certain facts. Once proven by ultimate scientific investigation, it becomes a theory or a law.

igneous rock Rock formed from the cooling and solidification of magma that has not been changed appreciably since its formation.

illuminance The luminous flux (brightness of light) per unit area on an intercepting surface at any given point.

imbibition The absorption of liquids or vapors into the ultramicroscopic spaces in materials like cellulose.

immobilization The conversion of inorganic nitrogen to organic nitrogen.

immune Free from attack by a given pathogen; not subject to the disease.

imperfect flower A flower lacking either stamens or pistils.

impervious Resistant to penetration by fluids or by roots.

inbred line A pure line usually originating by self-pollination and selection.

inceptisol A developing soil that is usually wet or moist and shows alterations in horizons.

incompatibility Failure of a flowering plant to obtain fertilization and seed formation after pollination, usually because of slow pollen tube growth in the stylar tissue. Also, the failure of two graft components (stock and scion) to unite and develop into a successfully growing plant.

incomplete flower A flower that is missing one or more of the following parts: sepals, petals, stamens, or pistils.

indehiscent fruit A fruit that does not split open naturally at maturity.

indeterminate growth The flowers of plants are borne on lateral branches, the central stem continues vegetative growth, with blooming continued for a long period. Examples are alfalfa and fuchsia.

indigenous Produced or living naturally in a specific environment.

indoleacetic acid (IAA) A natural or synthetic plant growth regulator; an auxin.

inferior ovary An ovary imbedded in the receptacle, or an ovary whose base lies below the point of attachment of the perianth.

infiltration rate The maximum velocity at which water can enter the soil under specified conditions, including the presence of an excess of water.

inflorescence   An axis bearing flowers, or a flower cluster (e.g., umbel, spike, panicle).

inheritance   The acquisition of characters or qualities by transmission from parent to offspring.

initiation   *See* differentiation.

inoculate   To induce a disease in a living organism by introducing a pathogen. Also, to treat seeds of leguminous plants with bacteria to induce nitrogen-fixation in the roots.

inorganic compound   A chemical compound that generally is not derived from life processes; compounds that do not contain carbon.

insecticide   Any chemical (organic or inorganic) substance that kills insects.

insolation   The amount of solar energy striking an area of the earth's surface.

integrated pest management (IPM)   Using multiple methods such as cultural, biological, and chemical to control pests in a crop.

integuments   The tissues covering or surrounding the ovule, usually consisting of an inner and outer layer; they subsequently become the seedcoats of the mature ovule.

intercalary growth   A pattern of stem elongation typical of grasses. Elongation proceeds from the lower internodes to the upper internodes through the differentiation of meristematic tissue at the base of each internode.

intercalary meristem   A type of meristem, often found in grasses, that lies across a stem or leaf, resulting in the deposition of new cells between regions of older tissues.

internode   The region of a stem between two successive nodes.

interspecific cross   A cross, natural or intentional, between two species.

in vitro   Latin for "in glass." Living in test tubes; outside the organism or in an artificial environment.

in vivo   Latin for "in living." In the living organism.

ions   Atoms, groups of atoms, or compounds that are electrically charged as a result of the loss of electrons (cations) or the gain of electrons (anions).

IPM   *See* integrated pest management.

irradiation   In genetics and plant breeding, exposing seed, pollen, or other plant parts to X-rays or other short wavelength (gamma) radiations to increase mutation rates.

irrigation   Applying water to the soil, other than by natural rainfall.

isolation   The prevention of crossing among plant populations because of distance or geographic barriers (geographic isolation).

jasmonate   A compound recently identified as a plant hormone involved in plant response to insect attack.

juvenile phase   A growth phase in plants characterized by the inability to flower and reproduce.

lamina   Blade or expanded part of a leaf.

lateral bud   A bud that grows out from the leaf axil on the side of a stem.

latewood   Also known as *summer wood*. The denser part of the growth ring produced late in the season. It is made up of xylem cells with thicker walls and smaller radial diameter, and they are generally longer than those formed earlier in the growing season.

latex   A milky secretion produced by various kinds of plants.

lathhouse   An open structure built of wood lath or plastic screen for protecting plants from excessive sunlight or frost.

layby application   Application of a material with or after the last cultivation of a crop.

layering   A form of vegetative propagation in which an intact branch develops roots as the result of contact with the soil or another rooting medium.

leach   To remove soluble materials from soil or plant tissue with water.

leaf mold   Partially decayed leaves useful for improving soil structure and fertility.

leggy   Weak-stemmed and spindly plants with sparse foliage caused by too much heat, shade, crowding, or overfertilization. *See* etiolation.

legume   Plant member of the family FABACEAE, with the characteristic capability to fix atmospheric nitrogen in nodules on its roots if inoculated with proper bacteria.

lenticel   An opening made up of loosely arranged cells in the periderm that permits passage of gases.

liane   A climbing or twining plant, usually woody.

light compensation point   The level of light intensity where carbon dioxide uptake for daily photosynthesis equals the daily loss of carbon dioxide due to the respiration of a plant.

light flux   The total light energy used by plants in the photosynthetically active region (PAR). The quantity of light (intensity) × duration of light (daylength).

light quality   The composition of light determined by the wavelengths present and their intensity. For example, visible light is composed of wavelengths of the colors of the rainbow at the appropriate intensity for each color.

light reactions   The reactions of photosynthesis in which light energy is required: the photo (light) activation or excitement of electrons in the chlorophyll molecule, transfer of the electrons, photolysis of water, and associated reactions.

light saturation   Light intensity above which photosynthesis activity does not increase as the intensity increases.

light soil   A coarse-textured sandy soil; hence, easy to till.

lignin   An organic substance found in secondary cell walls that gives stems strength and hardness. Wood is composed of lignified xylem cells (about 15 to 30 percent by weight).

lime   An agricultural material containing the carbonates, oxides, and/or hydroxides of calcium and/or magnesium. It is used to increase soil pH and to neutralize soil acidity.

line   Group of individuals from a common ancestry. When propagated by seed, it retains its characteristics. A type of cultivar.

lipid   Any of a group of fats or fatlike compounds insoluble in water but soluble in certain other solvents.

loam   A textural class for soil with prescribed amounts of sand, silt, and clay.

locus   The fixed position of location of a gene on or in a chromosome.

lodging   A condition in which plants are caused to bend for various reasons at or near the soil surface and fall more or less flat on the ground. Most frequently observed in cereals.

long-day plant   A plant that flowers when the light period is greater (dark period is shorter) than a critical duration.

Lucerne   A name for alfalfa used in Europe, Australia, and other regions.

lux (lx)   A standard measure of light (metric system). The illumination impinging upon a surface 1 m^2, each point of which is at a distance 1 m away from a standard light source of one candle. Photometric unit expressing the illuminance of 1 lumen per square meter; 1 lux = 0.093 ft-c. *See* foot-candle.

macronutrient   A chemical element, like nitrogen, phosphorus, and potassium, necessary in large amounts (usually greater than 1 ppm) for the growth of plants.

male sterility   A condition in some plants in which pollen either is not formed or does not function normally, even though the stamens may appear normal.

Malthusian theory   Developed by T. Malthus, a British economist (1766–1834), who speculated that the world's population would grow faster than food productivity and thus world starvation would eventually occur.

Mediterranean climate   A climate characterized by cool, wet winters and warm, dry summers.

meiosis   Two successive nuclear divisions, in the course of which the diploid chromosome number is reduced to the haploid and genetic segregation occurs.

Mendel's laws   A set of three laws formulated by Gregor Mendel; each is generally true but there are numerous exceptions. The laws are (1) characteristics exhibit alternative inheritance, being either dominant or recessive; (2) each gamete receives one member of each pair of factors present in a mature individual; and (3) reproductive cells combine at random.

mericlinal chimera   Plant cell mutates and the mutation is maintained in a portion of a cell layer. The mutation appears only in portions of a leaf or stem.

meristem   Undifferentiated tissue whose cells can divide and differentiate to form specialized tissues, such as xylem or phloem.

meristematic tissue   Region of actively dividing and differentiating cells.

mesocarp   Middle layer of the fruit wall (pericarp).

mesophyll   Parenchyma tissue in leaves found between the two epidermal layers.

Mesozoic   A geologic era beginning 225 million years ago and ending 65 million years ago.

messenger RNA   Ribonucleic acid produced in the nucleus and capable of carrying parts of the message coded in chromosomal DNA. Messenger RNA moves from the nucleus to the ribosomes, where protein is synthesized in the cytoplasm.

metabolism   The overall physiological activities of an organism.

metamorphic rock   A rock that has been greatly altered from its previous condition through the combined action of heat and pressure.

microclimate   Atmospheric environmental conditions in the immediate vicinity of the plant, including interchanges of energy, gases, and water between atmosphere and soil.

micronutrient   A chemical element necessary in extremely small amounts (less than 1 ppm) for the growth of plants. Examples are boron, chlorine, copper, iron, manganese, and zinc.

micropyle   An opening leading from the outer surface of the ovule between the edges of the two integuments inward to the surface of the nucellus.

microspore   One of the four haploid spores that originate from the meiotic division of the microspore mother cell in the anther of the flower and that gives rise to the pollen grain.

microspore mother cell   Diploid cell in the anther that gives rise, through meiosis, to four haploid microspores.

middle lamella   The pectic layer lying between the primary cell walls of adjoining cells.

millimho (mmho)   A measure of electrical conductivity, 1 mmho = 0.001 mho. The mho is the reciprocal of an ohm.

mineralization   The conversion of organic nitrogen to ammonium, nitrate, and nitrite by microorganisms.

mineral soil   A soil whose makeup and physical properties are largely those of mineral matter.

minor element   *See* micronutrient.

mist propagation   Applying water in mist form to leafy cuttings in the rooting stage to reduce transpiration.

mitochondria   Singular: **mitochondrion.** A minute particle in the cytoplasm associated with intracellular respiration.

mitosis   A form of nuclear cell division in which chromosomes duplicate and divide to yield two nuclei that are identical with the original nucleus. Usually mitosis includes cellular division (cytokinesis).

mixed bud   A bud containing both rudimentary flowers and vegetative shoots.

mode of action   The way in which an herbicide or pesticide works to control its target organism.

moisture   A basis for representing moisture content of a product. It is calculated from the net weight of water lost by drying, divided by the dried weight of the material, and the answer multiplied by 100 to obtain the percentage.

mol   Used in plant science when referring to light and when the waveband stated is 400 to 700 nm (PAR); 1 mol = 6.022×10^{23} photons, or 1 micromol = 6.022×10^{17} photons.

mole (M)   Amount of a substance that has a weight in grams numerically equal to the molecular weight of the substance. Also known as *gram-molecular weight*.

molecular biology   A field of biology concerned with the interaction of biochemistry and genetics in the life of an organism.

molecule   A unit of matter; the smallest portion of an element or a compound that retains chemical identity with the substance in mass. A molecule usually consists of the union of two or more atoms, and some organic molecules contain hundreds of atoms.

monocot   *See* monocotyledons.

monocotyledons   The subclass of flowering plants that have only a single cotyledon at the first node of the primary stem. Also known as *monocots.*

monoecious   A plant with separate male and female flowers on the same plant, such as corn and walnuts.

morphology   The study or science of the form, structure, and development of plants.

muck   Highly decomposed organic material in which the original plant parts are not recognizable.

mulch   Any material such as straw, sawdust, leaves, plastic film, and loose soil that is spread on the surface of the soil to protect the soil and plant roots from the effects of rain, soil crusting, freezing, or evaporation.

multiple fruit   A cluster of matured fused ovaries produced by separate flowers, for example, pineapple. *See also* aggregate fruit.

mutation   A sudden, heritable change appearing in an individual as the result of a change in genes or chromosomes.

mycelium   The mass of hyphae forming the body of a fungus.

mycology   A branch of botany dealing with the study of fungi.

mycorrhiza   The association, usually symbiotic, of fungi with the roots of some seed plants.

nanometer (nm) A unit of length equal to one millionth (10^{-6}) of a millimeter or one millimicron; 1 nm equals 10 angstrom units.

naturalized plant A plant introduced from one environment into another in which the plant has become established and more or less adapted to a given region by growing there for many generations.

natural selection Environmental effects in channeling the genetic variation of organisms along certain pathways.

necrosis Death associated with discoloration and dehydration of all or some parts of plant organs.

nectary Flower part that secretes nectar.

nematodes Unsegmented roundworms abundant in many soils; they are important because many species of them attack plants or animals.

net photosynthesis The gain in carbohydrates from photosynthesis after carbohydrate loss through respiration is deducted. *See* total photosynthesis.

neutral soil A soil in which the surface layer is neither acid nor alkaline in reaction.

night-break lighting Low intensity lighting used in the middle of the night to change the photoperiod.

nitrification The conversion of ammonium ions into nitrates through the activities of certain bacteria.

nitrogen assimilation The incorporation of nitrogen into organic cell substances by living organisms.

nitrogen fixation The conversion of atmospheric nitrogen (N_2) into oxidized forms that can be assimilated by plants. Certain blue-green algae and bacteria are capable of biochemically fixing nitrogen.

nitrogenous base A nitrogen-containing compound found in DNA and RNA that, in sequence, specifies precise genetic information. The nitrogenous bases in DNA are adenine, thymine, cytosine, and guanine. In RNA, they are adenine, uracil, cytosine, and guanine.

nodes Enlarged regions of stems that are generally solid where leaves are attached and buds are located. Stems have nodes but roots do not.

no-till planting Also known as *stubble culture*. A cultural system most often used with annual crops in which the new crop is seeded or planted directly in a field on which the preceding crop plants were cut down or destroyed by a nonselective herbicide rather than being removed or incorporated into the soil, as is common in preparing a seedbed.

nucellus A tissue originally making up the major part of the young ovule, in which the embryo sac develops.

nucleic acid An acid found in all nuclei; all known nucleic acids fall into two classes, DNA and RNA.

nucleus A dense body in the cytoplasm essential for cellular development and reproduction.

nut A dry, indehiscent, single-seeded fruit with a hard, woody pericarp (shell), such as the walnut and pecan.

nutrient Element essential to plant growth used in the elaboration of food and tissue.

obligate parasite An organism that must live as a parasite and cannot otherwise survive.

obligate saprophyte An organism obliged to live only on non-living animal or plant tissue.

oedema Intumescence or blister formation on leaves.

opposite An arrangement of leaves or buds on a stem. They occur in pairs on opposite sides of a single node.

order The taxonomic category that falls between class and genus.

organ A part of an animal or plant body adapted by its structure for a particular function.

organelle A specialized region in a cell, such as mitochondria, that is bound by a membrane.

organic In chemistry, the carbon compounds, many of which are associated with living organisms.

organic soil A soil that contains a high percentage (greater than 20 percent) of organic matter.

osmosis The diffusion of fluids through a semipermeable or selectively permeable membrane.

ovary The basal, generally enlarged part of the pistil in which seeds are formed. The ovary, at maturity, is a fruit. It is a characteristic organ of angiospermous plants.

oversowing The practice of sowing seed over an existing crop. Often done on golf courses by sowing a cool-season turfgrass over a warm-season perennial grass to maintain turfgrass growth during the cool season when the warm-season species is dormant.

ovule A rudimentary seed, containing, before fertilization, the embryo sac, including an egg cell, all being enclosed in the nucellus and one or two integuments.

oxidation The reaction of oxygen with soil parent material to form oxides.

oxidation-reduction reaction A chemical reaction in which one substance is oxidized (loses electrons, or loses hydrogen ions and their associated electrons, or combines with oxygen) and a second substance is reduced (gains electrons, or gains hydrogen ions and their associated electron, or loses oxygen).

oxidative respiration The chemical decomposition of foods (glucose, fats, and proteins) requiring oxygen as a terminal electron acceptor and yielding carbon dioxide, water, and energy. The energy is commonly stored in ATP.

oxisol A soil that lacks distinct horizons, contains clay oxides, is highly weathered, and has low fertility.

palatability Term used to describe how agreeable or attractive feed stuff is to animals or how readily they consume it.

Paleozoic A geologic era beginning about 570 million years ago and ending about 225 million years ago.

palisade parenchyma cells The cell layer in leaves immediately below the upper epidermis; it is packed with chloroplasts. Found in dicots but not in monocots.

pallet Rectangular or square platform, usually wooden, designed for ease of mechanical handling and transportation of material (containers) placed on the platform.

palmate Arrangement of leaflets of a compound leaf or of the veins in a leaf. Characterized by subunits arising from a common point much as fingers arise from the palm of the hand.

panicle An inflorescence, common in the grass family, that has a branched central axis. An example is oats.

parasite An organism obtaining its nutrients from the living body of another plant or animal.

parenchyma tissue A tissue composed of thin-walled, loosely packed, unspecialized cells.

parent material Soil component derived from the natural breakdown of rocks by physical and chemical forces. *See also* weathering.

parthenocarpy   Fruit development without sexual fertilization. Such fruits are seedless. Examples are the 'Navel' orange and some fig cultivars.

parthenogenesis   Development of an egg into an embryo without fertilization.

particle size   The effective diameter of a particle measured by sedimentation, sieving, or micrometric methods.

parts per million (ppm)   Weight units of any given substance per 1 million equivalent weight units; the weight units of solute per million weight units of solution (i.e., 1 ppm = 1 mg/l).

pasteurized soil   A soil that has been heated to 82°C (180°F) for at least thirty minutes to kill most of the pathogenic organisms without affecting the saprophytic soil flora.

pasture   Area of domesticated forages, usually improved, on which animals are grazed.

pathogen   An organism that causes disease.

pathology   The study of diseases, their effects on plants or animals, and their treatment.

pearl   The process of grinding off the hull, bran, aleurone, and germ of barley or rice to yield a pellet of endosperm.

peat   Any unconsolidated soil mass of semicarbonized vegetable tissue formed by partial decomposition in water. An example is sphagnum peat moss.

pectin   Polysaccharide from the middle lamella of the plant cell wall; jelly-forming substance found in fruit.

ped   A unit of soil structure such as an aggregate, crumb, prism, block, or granule formed by natural processes.

pedicel   Individual flower stalk of an inflorescence.

pedigree   A record of ancestry.

peduncle   Flower stalk that is borne singly. Also, the main stem of an inflorescence.

percolation   The downward movement of water through soil.

perennial   A plant that grows more or less indefinitely from year to year and usually produces seed each year.

perfect flower   Having both stamens and pistils; a hermaphroditic flower.

perianth   The petals and sepals of a flower, collectively.

pericarp   The fruit wall, which develops from the ovary wall.

periclinal chimera   Mutation in which, mutated plant cells are arranged near the apical dome so that they form a cell layer. The most stable form of chimera.

pericycle   The layer of cells immediately inside the endodermis. Branch roots arise from the pericycle.

periderm   A corky layer formed by the cork cambium at the surface of organs that are undergoing secondary growth.

permanent wilting percentage (PWP)   *See* wilting point.

permanent wilting point (PWP)   The point at which the water potential of the soil is equal to or lower than that of the root, and water cannot move from the soil into the root.

permeable   Referring to a membrane, cell, or cell system through which substances may diffuse.

petal   Part of a flower, often brightly colored.

petiole   The stalk that attaches a leaf blade to a stem.

pH   The negative logarithm of the hydrogen-ion activity of a soil. Also, a measure of acidity or alkalinity, expressed as the negative logarithm (base 10) of the hydrogen-ion concentration. pH 7 is neutral. Values less than this indicate acidity; higher values indicate alkalinity.

phellogen   Cork cambium, a cambium layer giving rise externally to cork and, in some plants, internally to phelloderm.

phenology   The study of the timing of periodic phenomena such as flowering, growth initiation, or growth cessation, especially as related to seasonal changes in temperature or photoperiod.

phenotype   The external physical appearance of an organism.

pheromone   Any of a class of hormonal substances secreted by an individual and stimulating a physiological or behavioral response from an individual of the same species.

phloem   A tissue through which nutritive and other materials are translocated through the plant. The phloem consists of sieve tube cells, companion cells, phloem parenchyma, and fibers.

photomorphogenesis   Plant shape determined by light quality particularly through relative amounts of red, far-red, and blue light.

photomorphogenic response   The reaction of non-shade-adapted plants when they do not receive enough light at the correct red:far-red ratio.

photoperiod   That length of day or period of daily illumination required for the normal growth and sexual reproduction of some plants.

photoperiodism   Response of a plant to the relative lengths of day and night (light and dark), particularly in respect to flower initiation and bulbing.

photophosphorylation   The production of ATP by the addition of a phosphate group to ADP using the energy of light-excited electrons produced in the light reactions of photosynthesis. Photo light, phosphorylation = adding phosphorus.

photosynthesis   The process in green plants of converting water and carbon dioxide into sugar with light energy, accompanied by the production of oxygen.

photosynthetic active radiation (PAR)   Radiation in the 400–700 nanometer (nm) spectral waveband.

photosynthetic photon flux (PPF)   Radiometric irradiance per unit area per unit time at specific wavelengths (400–700 nm waveband for plants). Units micromols $m^{-2}s^{-1}$.

phototropin   A blue-light receptor that senses the direction of light and influences phototropic responses.

phototropism   A change in the manner of growth of a plant in response to nonuniform illumination. Usually, the response, which is auxin-regulated, is a bending toward the strongest light.

Phylum   Plural: **Phyla.** A primary division of the animal and plant kingdom.

physiology   The science of the functions and activities of living plants.

phytochrome   A reversible protein, red/far-red light-sensing pigment occurring in the cytoplasm of green plants. It is associated with the absorption of light that affects growth, development, and differentiation of a plant, independent of photosynthesis (e.g., in the photoperiodic response). *See* photomorphogenesis.

phytomer   *See* tiller.

phytoplasma   Round or ovoid prokaryote organisms that lack rigid cell walls and infect plants.

piggyback   Intermodal system for carrying full trailer loads of product on a railroad flat car. Also known as *trailer on flat car* (TOFC). The trailer is usually moved to and from the rail car by tractor truck.

pigments   Molecules that are colored by the light they absorb. Some plant pigments are water soluble and are found mainly in the cell vacuole.

pinching   The removal of the terminal bud or apical meristematic growth to stimulate branching.

pinnate Leaflets attached to a midrib in a featherlike arrangement.

pistil The seed-bearing organ in the flower, composed of the ovary, the style, and the stigma.

pistillate (female) flower A flower having pistils but no stamens.

pit Channel in plant cell walls that surrounds plasmodesmata.

pith A region in the center of some stems and roots consisting of loosely packed, thin-walled parenchyma cells.

placenta Plural: **placentae.** The tissue within the ovary to which the ovules are attached.

plantae The taxonomic category within the kingdom that includes all plants.

plant growth regulators Natural and synthetic chemicals that influence plant growth and development.

plant pathology The scientific study of plant diseases and their causal organisms.

plasmalemma Membrane that surrounds a plant cell.

plasmodesmata Strand that forms a cytoplasmic connection between two plant cells.

plasmolysis The separation of the cytoplasm from the cell wall because of the removal of water from the protoplast.

plastids The cellular organelles in which carbohydrate metabolism is localized.

plow layer The surface soil layer ordinarily moved in tillage.

plow pan A hard layer of soil formed by continual plowing at the same depth.

plow-plant The practice of plowing and planting a crop in one operation, with no additional seedbed preparation.

plumule The first bud of an embryo or that portion of the young shoot above the cotyledons.

polar nuclei Two centrally located nuclei in the embryo sac that unite with a second sperm cell in a triple fusion. In certain seeds, the product of this fusion develops into the endosperm.

polar transport The directed movement within plants of compounds (usually hormones) mostly in one direction; polar transport overcomes the tendency for diffusion in all directions.

pollen The almost microscopic, yellow bodies that are borne within the anthers of flowers and contain the male generative (sex) cells.

pollenizer A plant that is a source of pollen.

pollen mother cell A 2n cell that divides twice (once by meiosis and once by mitosis) to form a tetrad of four pollen grains.

pollen tube A tubelike structure developed by the tube nucleus in the microspore that helps guide the sperm through the stigma and style of a flower to the embryo sac.

pollination The transfer of pollen from a stamen (or staminate cone) to a stigma (or ovulate cone).

pollinator An agent (such as an insect) that pollinates flowers.

polyembryony The presence of more than one embryo in a developing seed.

polyploidy A condition in which a plant has somatic (nonsexual) cells with more than 2n chromosomes per nucleus.

polysaccharides Long-chain molecules composed of units of a sugar; starch and cellulose are examples.

pome A simple fleshy fruit, the outer portion of which is formed by floral parts that surround the ovary (i.e., apple and pear fruits).

pore size distribution The volume of the various sizes of pores in a soil. It is expressed as a percentage of the bulk volume (soil plus pore space).

pore space The open (nonsolid) portion of soil that holds air or water.

porosity That percentage of the total bulk volume of a soil not occupied by the solid particles.

post-emergence spray A pesticide or herbicide that is applied after the crop plants have emerged from the soil.

potting mixture Combination of various ingredients such as soil, peat, sand, perlite, or vermiculite designed for starting seeds or growing plants in containers.

preemergence spray A pesticide or herbicide that is applied after planting, but before the crop plants emerge from the soil.

pre-plant herbicide An herbicide applied to a field prior to planting.

primary tissue A tissue that has differentiated from a primary meristem.

primocane A vegetative cane of raspberries, blackberries, or their hybrids that arises from crown or rhizome buds. Primocanes develop floral buds in response to physiological or environmental cues but typically do not fruit until their chilling requirements have been met.

primordium An organ in its earliest stage of development, such as leaf primordium.

profile A vertical section of the soil through all its horizons and extending into the parent material.

prokaryotic cells Cells in which DNA is not enclosed in a nucleus, for example, bacteria.

propagation The process of artificially or naturally increasing plant numbers.

propagule Any part of a plant used to generate a new plant.

protein Any of a group of nitrogen-containing compounds that yield amino acids on hydrolysis and have high molecular weights. They are essential parts of living matter and are one of the essential food substances of animals.

Proterozoic The earliest geologic era, beginning about 4.5 to 5 billion years ago and ending 570 million years ago. Also called *Precambrian era.*

protoplasm The essential, complex living substance of cells on which all vital functions of nutrition, secretion, growth, and reproduction depend.

protoplast The organized living unit of a single cell.

provenance The natural origin of a tree or group of trees. In forestry, the term is considered synonymous with geographic origin.

pruning A method of directing or controlling plant growth and form by removing certain portions of the plant.

pure line Plants in which all members have descended by self-fertilization from a single homozygous individual.

raceme An inflorescence in which flowers on pedicels are borne on a single, unbranched main axis.

radial face The wood surface exposed when a stem is cut along a radius from pith to bark, and the cut parallels the long axis of the majority of the cells.

radicle The part of the embryonic axis that becomes the primary root. The first part of the embryo to start growth during seed germination.

range Land and native vegetation that is predominantly grasses, grasslike plants, or shrubs suitable for grazing by animals.

range management Producing maximum sustained use of range forage without detriment to other resources or uses of land.

raphe   Ridge on seeds, formed by the stalk of the ovule, in those seeds in which the funiculus is sharply bent at the base of the ovule.

ray   A narrow group of cells, usually parenchyma, extending radially in the wood and bark.

reaction   *See* pH.

receptacle   The enlarged tip of a stem on which a flower is borne.

recessive gene   A gene that does not express itself in the presence of the contrasting (dominant) gene.

recombination   The mixing of genotypes that results from sexual reproduction.

reduction division   A nuclear division in which the chromosomes are reduced from the diploid to the haploid number.

registered seed   The progeny of foundation or registered seed produced and handled to maintain satisfactory genetic identity and purity, and approved and certified by an official certifying agency. Registered seed is normally grown for the production of certified seed.

regulatory genes   Codes for proteins that regulate structural gene activity. *See also* transcription factors.

relative humidity (RH)   The ratio of the amount of water vapor in the air compared to what the air can hold at that temperature. *See also* saturated vapor pressure (SVP).

replication   In cell physiology, the production of a second molecule of DNA exactly like the first molecule.

reproduction   Sexual reproduction is development of new plants by seeds (except in apomixis). Vegetative reproduction, or vegetative propagation, is reproduction by other than sexually produced seed. Includes grafting, cuttings, layering, and so forth, as well as apomixis.

respiration   The oxidation of food by plants and animals to yield energy for cellular activities.

rest period   An endogenous physiological condition of viable seeds, buds, or bulbs that prevents growth even in the presence of otherwise favorable environmental conditions. This is referred to by some seed physiologists as *dormancy*.

Rhizobium   Genus of bacteria that live symbiotically in the roots of legumes and fix nitrogen that is used by plants.

rhizome   An underground stem, usually horizontal and often elongated; distinguished from a root by the presence of nodes and internodes. Capable of producing new shoots.

ribonucleic acid (RNA)   A single-strand acid, formed on a DNA template, found in the protoplasm, and controlling cellular chemical activities. Whereas DNA transmits genetic information from one cell generation to the next, ribonucleic acid is an intermediate chemical translating genetic information into action.

ribosome   A protoplasmic granule containing ribonucleic acid (RNA) and believed to be the site of protein synthesis.

ribulose bisphosphate carboxylase (RUBISCO)   The enzyme in the non-light-requiring reactions of photosynthesis that fixes atmospheric carbon dioxide into a carbohydrate and in the process converts radiant energy to chemical energy.

ripening   Chemical and physical changes in a fruit that follow maturation.

rock   The material that forms the essential part of the earth's solid crust, including loose incoherent masses such as sand and gravel, as well as solid masses of granite, limestone, and others.

root   The descending axis of a plant, usually below ground, serving to anchor the plant and absorb and conduct water and mineral nutrients.

root cap   A mass of hard cells covering the tip of a root and protecting it from mechanical injury.

root hair   An absorptive unicellular protuberance of the epidermal cells of the root.

rooting media   Materials such as peat, sand, perlite, or vermiculite in which the basal ends of cuttings are placed vertically during the development of roots.

root pressure   Pressure developed in the root as the result of osmosis and causing bleeding in stem wounds of some plants.

rootstock   Also known as *understock* and *stock*. The trunk or root material to which buds or scions are inserted in grafting.

roughage   Plant materials that are relatively high in crude fiber and low in digestible nutrients, such as straw.

rumen   The stomach in ruminate animals in which cellulose is digested.

ruminant   Cud-chewing mammals such as cattle, sheep, goats, and deer, characteristically having a stomach divided into four compartments.

runner   *See* stolon.

rural/urban interface   The influx of nonagricultural residents and nonfarm land uses into rural areas.

salicylic acid   A recently identified plant hormone that plays a role in a plant's response to pathogen attack. Salicylic acid is also the active ingredient in aspirin.

saline soil   A nonsodic soil containing sufficient soluble salts to impair plant growth.

samara   A dry, indehiscent, simple fruit that has winglike appendages on both sides of the ovary. These appendages help carry the wind-borne fruit.

sand   A soil particle between 0.05 and 2.0 mm in diameter.

saprophyte   An organism deriving its nutrients from the dead body or the nonliving products of another plant or animal.

saturated vapor pressure (SVP)   The greatest pressure that water vapor could exert if the air at a given temperature were saturated with water vapor. *See also* relative humidity (RH).

savanna   Grassland having scattered trees, either as individuals or clumps; often a transitional type between true grassland and forest.

scape   A strawberry inflorescence.

scarification   Scratching or abrading the surface of a seed to allow water to penetrate.

scarify   To scratch, chip, or nick the seed coverings of certain species to enhance the passage of water and gases as an aid to seed germination.

scientific method   An approach to a problem that consists of stating the problem, establishing one or more hypotheses as solutions to the problem, testing these hypotheses by experimentation or observation, and accepting or rejecting the hypotheses.

scion   A small shoot that is inserted by grafting into a rootstock.

scion-stock interaction   The effect of a rootstock on a scion (and vice versa) in which a scion on one kind of rootstock performs differently than it would on its own roots or on a different rootstock.

sclereid   A short, thickened cell that provides support or protection in plants. *See also* sclerenchyma tissue.

sclerenchyma tissue   Supporting or protective tissue in which the cells have hard lignified walls.

scutellum   The rudimentary leaflike structure at the first node of the (embryonic) stem (culm) of a grass plant. The single cotyledon of a monocotyledonous seedling.

secondary mineral nutrients   Calcium, magnesium, and sulfur; used in lesser amounts than the macronutrients.

secondary phloem   Phloem cells formed by activity of the vascular cambium. Secondary phloem is found in biennials and perennials, but usually not in annuals.

secondary xylem   Xylem cells formed by activity of the vascular cambium. The development of the secondary xylem accounts for the so-called annual rings seen in most trees.

sectorial chimera   Mutated plant cells are arranged so that they extend through cell layers, thus producing both normal and mutated leaves and stems.

sedimentary rock   Rock formed from material originally deposited as a sediment, then physically or chemically changed by compression and hardening while buried in the earth's crust.

seed   The mature ovule of a flowering plant containing an embryo, an endosperm (sometimes), and a seed coat. Breeder seed (or, sometimes, vegetative propagating material) directly controlled by the originator (or the sponsoring plant breeder or institution) that provides the source for the initial increases of foundation seed. Certified seed is the progeny of foundation, registered, or other certified seed that is handled to maintain satisfactory genetic identity and/or purity and that has been approved and certified by a certifying agency. Foundation seed, or seed stocks, are handled to maintain specific genetic identity and purity, such as may be designated by an agricultural experiment station. Foundation seed is the source of certified seed, either directly or through registered seed. Registered seed is the progeny of foundation or other registered seed handled to maintain satisfactory genetic identity and purity; approved and certified by a certifying agency.

seed bank   The reserve of seeds in the ground.

seedbed   Soil that has been prepared for planting seeds or transplants.

seed coat   *See* testa.

seed potatoes   Pieces of potato tubers or whole tubers that are planted to produce new plants and subsequent commercial crops.

self-fertile   Capable of fertilization and producing viable seed after self-pollination.

self-incompatibility   Inability to produce viable seed following self-pollination. The inability is sometimes due to a pollen-borne gene that prevents pollen tube growth on a stigma with the same gene.

self-pollination   Transfer of pollen from the stamens to the stigma of either the same flower, other flowers on the same plant, or flowers on other plants of the same clone.

self-sterility   Failure to complete fertilization and obtain viable seed after self-pollination.

semiarid   Climate in which evaporation exceeds precipitation, a transition zone between a true desert and a humid climate. Usually annual precipitation is between 250 and 500 mm (10 to 20 in.).

senescence   A physiological aging process in which tissues in an organism deteriorate and finally die.

sepals   The outermost series of floral parts; usually green, leaflike structures at the base of a flower; collectively, they form the calyx.

separation   A type of propagation based on natural breaking apart of plant segments, such as in bulbs or corms.

sequestering agent   *See* chelate.

sessile   Used in reference to flowers, florets, leaves, leaflets, or fruits that are attached directly to a shoot and not borne on any type of a stalk.

sexual reproduction   Development of new plants by the processes of meiosis and fertilization in the flower to produce a viable embryo in a seed.

shoot-bearing   A term describing a plant that predominantly bears fruit terminally or axially on shoots developed during the current or previous season.

shoot meristem   *See* apical meristem.

short-day plants (SDP)   Plant that initiates flowers only under short-day (long-night) conditions.

sidedressing   Applying fertilizer on a soil surface close enough to a plant that cultivating or watering carries the fertilizer to the plant's roots.

sieve tube   Part of phloem tissue formed by the connecting of sieve cells end to end through which nutrients and other soluble substances flow in plants.

silage   Forage that is chemically changed and preserved in a succulent condition by partial fermentation in the preparation of food for livestock.

silique   A dry, one-seeded, dehiscent (infrequently indehiscent) fruit consisting of two carpels that form a bilocular ovary. Common in the BRASSICACEAE family.

silo   A structure for making and storing silage.

silt   A soil textural class consisting of particles between 0.05 and 0.002 mm in diameter.

simple fruit   A fruit derived from a single pistil.

simple leaf   Leaf that is not divided into distinct leaflets. *See also* compound leaf.

sink   The deposition of carbon dioxide in the form of carbohydrates and often carbon compounds in plant tissues and organs.

sink tissue   Tissue that is actively growing or storing carbohydrates and is importing photosynthates. *See also* source leaf.

sod   Top 3 to 7 cm (1 to 3 in.) of soil permeated by and held together with grass roots or grass-legume roots.

sodic soil   A soil that contains so much exchangeable sodium (more than 15 percent) that it interferes with the growth of most crop plants. Also, if the total soluble salts is more than 4 mmho per cm^2, the soil is a saline sodic soil; if it is less than 4 mmho per cm^2, the soil is a nonsaline sodic soil.

soil   The solid portion of the earth's crust in which plants grow. It is composed of mineral material, air, water, and organic matter both living and dead.

soil air   The gaseous phase of the soil; that percentage of the total volume not occupied by solid or liquid.

soil compaction   The reduction of soil pore space and the increase in soil bulk density as a result of pressure from sources such as animals, machinery, or humans moving across the surface of the soil.

soil conservation   A combination of all management and land use methods that safeguard the soil against depletion or deterioration by natural or by human-induced factors.

soil fertilization   *See* fertilization.

soil management   The total tillage operations, cropping practices, fertilizing, liming, and other treatments conducted on or applied to a soil for the production of plants.

soil organic matter   The organic fraction of the soil that includes plant and animal residues at various stages of decomposition, cells and tissues of soil organisms, and substances synthesized by the soil population.

soil pasteurization   Treating the soil with heat [usually steam at 60°C to 70°C (140°F to 160°F)] to destroy most harmful pathogens, nematodes, and weed seeds. Less severe temperature treatment than soil sterilization.

soil profile Distinct layers of soil that develop over time from the weathering of parent material, biological activity, and other factors.

soil salinity The amount of soluble salts in a soil, expressed as parts per million, millimho/cm, or other convenient ratios.

soil series The basic unit of soil classification; a subdivision of a family, comprising soils that are essentially alike in all major profile characteristics.

soil solution The aqueous liquid phase of the soil and its solutes consisting of ions dissociated from the surfaces of the soil particles and of other soluble materials.

soil sterilization Treating soil by gaseous fumigation, chemicals, heat (usually steam) at 100°C (212°F) to destroy all living organisms.

soil structure The arrangement of primary soil particles into secondary particles, units, or peds that act as primary particles. The secondary units are characterized and classified on the basis of size, shape, and degree of distinctness.

soil testing Chemical analysis of soil to determine its nutrient status, pH, electroconductivity (EC), and other compositional factors.

soil texture The relative percentages of sand, silt, and clay in a soil.

soil tilth The physical condition of soil as related to its ease of tillage, fitness as a seedbed, and suitability for plant growth.

soil type The lowest unit in the natural system of soil classification; a subdivision of a soil series.

solarize Cause damage to leaves from overexposure to the sun.

soluble salts Disassociated salts (anions and cations) in the soil that can become toxic to roots when exceeding certain levels.

solute A substance dissolved in a solvent.

solution A homogeneous mixture; the molecules of the dissolved substance (the solute) are dispersed among the molecules of the solvent.

solvent A substance, usually a liquid, that can dissolve other substances (solutes).

somaclonal variation Existing genetic variation that may not be seen until after plant cells have been through aseptic culture, or the culture may force the change.

somatic tissue Nonreproductive, vegetative tissue. Tissue developed through mitosis that will not undergo meiosis.

source leaf Leaf that is producing and exporting more photosynthates than it requires for its own use.

spadix An inflorescence with sessile flowers on a fleshy stalk, for example, *Spathiphyllum*, or peace lily.

species A group of similar organisms capable of interbreeding and more or less distinctly different in geographic range and/or morphological characteristics from other species in the same genus.

specific heat The amount of heat required to raise the temperature of 1 gram of a substance by 1°C.

sperm A male gamete.

spermatophyte A seed-bearing plant.

spike An inflorescence that has a central axis on which sessile flowers are borne. Examples are some grasses, gladioli, and snapdragons.

spiroplasma Helical prokaryote organisms that lack rigid cell walls and infect plants.

spodosol A soil that is highly leached, has low fertility, and retains little water.

spongy mesophyll parenchyma cells The cell layer in a leaf located between the palisade parenchyma and the lower epidermis; these cells have thin cell walls and are loosely packed.

spore A reproductive cell that develops into a plant without union with other cells.

sprigging Vegetative propagation by planting stolons or rhizomes (sprigs).

spur-bearing A term describing a plant that predominantly bears fruit on fruiting spurs.

square An unopened flowerbud in cotton with its accompanying bracts.

stamen The male reproductive structure of a flower. The stamen produces pollen and is composed of a filament on which is borne an anther.

staminate (male) flower A flower having stamens but no pistils.

starch A complex polysaccharide carbohydrate. The form of food commonly stored by plants.

stele The vascular tissue and closely associated tissues in the axes of plants. The central cylinder of the stem.

stem The main body of a plant, usually the ascending axis, whether above or below ground in opposition to the descending axis or root. Stems, but not roots, produce nodes and buds.

stigma In a flower, the portion of the style to which pollen adheres.

stipule Appendage at the base of a leaf where it attaches to the stem.

stock *See* rootstock.

stolon A slender, prostrate above-ground stem. The runners of white clover, strawberry, and Bermuda grass plants are examples of stolons.

stoma (stomate) Plural: **stomata, stomates,** respectively. A small opening, bordered by guard cells, in the epidermis of leaves and stems, through which gases, including water vapor, pass.

stool Sprouts that arise from the base of the plant below ground and become rooted. They are used for vegetative propagation. Also, a cane fruit plant managed as an individual, with most primocanes arising at or near the plant crown.

stool or stooling (Agronomic term). Shoots that arise from below the soil at the base of a plant.

stratification The practice of exposing imbibed seeds to cool 2°C to 10°C (35°F to 50°F) (sometimes warm) temperatures for a period of time prior to germination and thus break dormancy. This is a standard practice in the germination of seeds of many grass and woody species.

stress Plant(s) unable to absorb enough water to replace that lost by transpiration. Results may be wilting, cessation of growth, or death of the plant or plant parts.

strig A currant or gooseberry inflorescence.

strip cropping The practice of growing crops that require different types of tillage, such as row and sod, in alternate strips along contours or across the prevailing direction of wind.

structural genes Code for proteins that regulate the structural component of the plant.

style Slender column of tissue that arises from the top of the ovary in the flower through which the pollen tube grows toward the ovule.

stylet A mouthpart on some insects and nematodes that allows the organisms to pierce the cells of other organisms.

subapical meristem   Region that produces new cells a few micrometers behind an active shoot or apical meristem.

suberin   A waxy, waterproofing substance in cork tissue.

subirrigation   Application of water from below the soil surface, usually from a ditch or a perforated hose or pipe, or by placing a potted plant on a constantly moist surface.

sublimation   The change of phase of a substance from solid to vapor without going through the liquid stage.

subsoil   That part of the soil below the plow layer.

subsoiling   Breaking of compact subsoils, without inverting them, with a special knifelike instrument (chisel) that is pulled through the soil at depths usually of 30 to 60 cm (12 to 24 in.) and at spacings usually of 60 to 150 cm (2 to 5 ft).

succession   The serial replacement of simple communities by more complex communities. *See also* climax community.

sucrose   Table sugar ($C_{12}H_{22}O_{11}$); a carbohydrate formed by chemically joining a molecule of glucose ($C_6H_{12}O_6$) with a molecule of fructose ($C_6H_{10}O_5$).

summer fallow   The tillage of uncropped land during the summer to control weeds and store moisture in the soil for the growth of a later crop.

sunscald   High temperature injury to plant tissue due to the sun's intense rays warming the trunks of trees during winter, cracking and splitting the bark. It can be prevented by shading or whitewashing tree trunks and larger branches. Sunscald also occurs on unshaded vegetable fruits (tomatoes and melons) or to houseplants exposed to direct sunlight.

superior ovary   An ovary situated above the receptacle; all other floral parts develop below the base of the ovary.

surface tension   The physical property of water in the liquid state that is due to the intermolecular attraction (hydrogen bonding) between the water molecules.

sustainability   The ability to be maintained with little input of nonrenewable resources.

symbiosis   An obligate relationship between two organisms of different species living together in close association for their mutual benefit. An example is the mycelium of a fungus with roots of seed plants.

synergids   The two nuclei within the embryo sac at the upper end in the ovule of the flower, which, with the third (the egg), constitute the egg apparatus.

synergism   The mutual interaction of two substances that results in an effect greater than the additional effects of the two when used independently.

systemic   A pesticide material absorbed by plants, making them toxic to feeding insects. Also, pertaining to a disease in which an infection spreads throughout the plant.

systemin   A peptide that has been recently identified as a plant hormone. It is produced in wounded tissue and is transported to remote cells and tissues, where it induces defense genes.

tangential face   The wood surface exposed when a cut is made at right angles to the rays and parallel to the long axis of the majority of cells.

tannin   Broad class of soluble polyphenols with a common property of condensing with protein to form a leatherlike substance that is insoluble in water.

tap root   An elongated, deeply growing primary root.

taxon   Plural: **taxa.** A taxonomic group of plants of any rank such as family, genus, species, and so forth.

taxonomy   The science dealing with describing, naming, and classifying plants and animals.

tendril   A slender coiling modified leaf or stem arising from stems and aiding in their support.

tension   The equivalent negative pressure of water in soil. The attraction with which water is held to soil particles.

terminal bud   A bud at the distal end of a stem.

terrace   A level, usually narrow, plain bordering a river, lake, or the sea. Rivers sometimes are bordered by terraces at different levels. It also refers to a raised, more or less level strip of land usually constructed on a contour and designed to make the land suitable for tillage and to prevent erosion.

testa   Protective tissue derived from the integuments that cover a seed. Also known as *seed coat.*

tetraploid   Having four sets of chromosomes per nucleus.

texture   The relative proportions of the various groups of individual soil grains, of different sizes, in a mass of soil. It refers to the proportions of sand, silt, and clay in a given amount of soil.

thallophytes   A division of plants whose body lacks roots, stems, and leaves (e.g., mushrooms).

thatch   A layer of undecomposed organic material (i.e., roots, leaves, stems) that builds up in turf above the soil and below the green vegetation.

thinning   Removing young plants from a row to provide the remaining plants with more space to develop. Also, the removal of excess numbers of fruits from a tree so the remaining fruits will become larger.

tile drain   Pipe made of burned clay, concrete, or similar material, in short lengths, usually buried at the bottom of a ditch, with open joints to collect and carry excess water from the soil. The newer type polyvinyl tile is made in continuous lengths and is perforated to allow water to enter the pipe for removal from the soil.

tillage   The mechanical manipulation of soil to provide a favorable environment for plant growth.

tiller   A grass stem that arises from an axillary bud.

tilth   *See* soil tilth.

tissue   A group of cells of similar structure that performs a special function.

tolerance   The ability of a plant to be attacked or infested by a pest but show few symptoms.

tonoplast   Membrane that surrounds the vacuole in a plant cell.

top-dressing   Applying materials such as fertilizer or compost to the soil surface while plants are growing.

top-grafting   *See* top-working.

topsoil   The upper layer of soil moved in cultivation.

top-working   Also known as *top-grafting.* To change the cultivar of a tree by grafting the main scaffold branches.

total photosynthesis   The total amount of carbohydrates synthesized from photosynthesis. *See* net photosynthesis.

tracheid   An elongated, tapering xylem cell with lignified pitted walls adapted for conduction and support.

training   The positioning of limbs, branches, and canes.

transcription factors   Proteins that regulate structural gene activity.

transfer RNA   *See* ribonucleic acid.

translocation   The transfer of food materials or products of metabolism throughout the plant.

transpiration   The loss of water vapor through the stomata of leaves.

trifoliate Leaf with three leaflets attached at the same point, for example, clover.

tuber An enlarged, fleshy, underground tip of a stem. The white (Irish) potato produces tubers as food storage organs.

tuberous root An enlarged fleshy, underground root (e.g., sweet potato and dahlia).

turgid Swollen, distended; referring to cells or tissues that are firm because of water uptake.

turgor pressure The pressure within the cell resulting from the absorption of water into the vacuole and the imbibition of water by the protoplasm.

2,4-dichlorophenoxyacetic acid (2,4-D) A selective auxin-type herbicide that kills broad-leaved plants but not grasses.

ultisol A tropical and subtropical soil with acidic, highly weathered layers.

umbel A type of inflorescence in which flowers are borne at the end of stalks that arise like the ribs of an umbrella from one point (e.g., carrot, onion).

unavailable water Water held by the soil so strongly that the root cannot absorb it.

understock *See* rootstock.

understory Those plants that grow under the canopy of other taller plants.

upright A decumbent shoot arising terminally or axially from a cranberry runner that will bear fruit from terminal buds.

vacuole A cavity in the plant's cell bounded by a membrane in which various plant products and by-products are stored.

vapor pressure deficit (VPD) The difference between the amount of moisture actually in the air and the amount the air holds when it is saturated.

variety A subdivision of a species with distinct morphological characters and given a Latin name according to the rules of the International Code of Botanical Nomenclature. A taxonomic variety is known by the first validly published name applied to it so that nomenclature tends to be stable. *See also* cultivar.

vascular bundle A strand of tissue containing primary xylem and primary phloem and frequently enclosed by a bundle sheath of parenchyma or fibers.

vascular cambium A meristem that produces secondary xylem and secondary phloem cells. Vascular cambium is found in biennials and perennials.

vascular plants Plants that have an organized vascular system comprised of xylem and phloem.

vascular system In seed-bearing plants, this system consists of the pericycle, phloem, vascular cambium, xylem, pith rays, and pith.

vector A carrier; for example, an insect that carries pathogenic organisms from plant to plant.

vegetation The plants that cover a region; it is formed of the species that make up the flora of the area.

vegetative growth Referring to asexual (stem, leaf, root) development in plants in contrast to sexual (flower, seed) development.

vein The vascular strand of xylem and phloem in a leaf.

vernalization In reference to flowering, the process by which floral induction in some plants is promoted by exposing the plants to chilling for a certain length of time.

vertisol A soil that has high clay content and poor physical structure. It is prone to shrinking and swelling as moisture content changes.

vessel A series of xylem elements in the stem and root that conduct water and mineral nutrients.

virgin soil A soil that has not been significantly disturbed from its original natural environment.

virus A pathogen consisting of a single strand of nucleic acids encapsuled in a protein coat that can replicate only inside a living cell.

visible light Light with wavelengths of 400 to 700 nm, which is the range visible to the human eye.

vitamins Natural organic substances, necessary in small amounts for the normal metabolism of plants and animals.

volatile Evaporating readily or easily dissipated in the form of a vapor.

water potential The difference between the activity of water molecules in pure distilled water at atmospheric pressure and 3°C (standard conditions) and the activity of water molecules in any other system; the activity of these water molecules may be greater (positive) or less (negative) than the activity of the water molecules under standard conditions.

water table The upper surface of groundwater or that level below which the soil is saturated with water.

watts (W) Radiometric units expressing energy per unit of time and unit area independent of wavelength. Irradiance within the photosynthetic active region (PAR) of 400–700 nm is commonly referred to as W PAR.

wavelength The distance between two corresponding points on any two consecutive waves. For visible light, it is minute and is generally measured in nanometers.

weathering All physical and chemical changes produced in rocks, at or near the earth's surface, by atmospheric agents.

weed A plant not valued for use or beauty. Any plant growing where it is not wanted.

wilting point Also known as *permanent wilting percentage (PWP)*. The moisture content of soil at which plants wilt and fail to recover even when placed in a humid atmosphere.

windbreak A planting of trees or shrubs, usually perpendicular or nearly so to the principal wind direction, to protect soil, crops, homesteads, roads, and so on, against the effects of winds.

windrow Hay, grain, leaves, or other material swept or raked into rows to dry.

winter hardiness The ability of a plant to tolerate severe winter conditions.

wood Secondary nonfunctioning xylem in a perennial shrub or tree.

xanthophyll Yellow carotenoid pigment ($C_{40}H_{56}O_2$) found along with chlorophyll in green plants.

xylem Specialized cells through which water and minerals move upward from the soil through a plant.

zone of accumulation *See* B horizon.

zygote A protoplast resulting from the fusion of gametes (either isogametes or heterogametes). The beginning of a new plant in sexual reproduction.

Index